International Conference on Sustainable Development of Cross-Border Regions 2019

AA007420

IOP Conference Series: Earth and Environmental Science
Volume 395

Barnaul, Russia
19 – 20 April 2019

Editors:

Svetlana V. Lobova
Aleksei V. Bogoviz

ISBN: 978-1-7138-0262-4
ISSN: 1755-1307

Printed from e-media with permission by:

Curran Associates, Inc.
57 Morehouse Lane
Red Hook, NY 12571

Some format issues inherent in the e-media version may also appear in this print version.

This work is licensed under a Creative Commons Attribution 3.0 International Licence.
Licence details: http://creativecommons.org/licenses/by/3.0/.

No changes have been made to the content of these proceedings. There may be changes to pagination and minor adjustments for aesthetics.

Printed with permission by Curran Associates, Inc. (2020)

For permission requests, please contact the Institute of Physics at the address below.

Institute of Physics
Dirac House, Temple Back
Bristol BS1 6BE UK

Phone: 44 1 17 929 7481
Fax: 44 1 17 920 0979

techtracking@iop.org

Additional copies of this publication are available from:

Curran Associates, Inc.
57 Morehouse Lane
Red Hook, NY 12571 USA
Phone: 845-758-0400
Fax: 845-758-2633
Email: curran@proceedings.com
Web: www.proceedings.com

TABLE OF CONTENTS

SUSTAINABLE DEVELOPMENT OF NATURAL AND ANTHROPOGENIC SYSTEMS

IOP Conf. Series: Earth and Environmental Science **395** (2019) 011001 doi:10.1088/1755-1315/395/1/011001

PREFACE

International Conference on Sustainable Development of Cross-Border Regions

SDCBR 2019

April 19-20, 2019, Altai State University, Barnaul, Russia

The emergence of a more integrated world where the national borders are simultaneously opened and closed, neglected and claimed, has given a new impetus to studies aimed at building a better understanding of those challenges the cross-border regions are facing and ways to address them. One of the most widely used concepts in the contemporary scholarship to analyze environmental, economic, social, and other dimensions is the concept of sustainable development, which has a potential to build on both macro- and micro- approaches to design, evaluate, and forecast development of cross-border regions. In turn, sustainability is a complex concept that covers multiple goals of the environment, economy, and society.

Consequently, the main focus of the conference was to discuss how the contemporary scholarship on sustainable development can be used to address the emerging threats and challenges of cross-border regions, whether it would be economic or social issues, environmental degradation and protection, natural resources management, or sustainable governance of large cross-border regions.

Cross-border regions are such establishments where sustainability is most likely to be challenged. More than that, this is where new sustainable solutions are most likely to emerge. The International Conference on Sustainable Development of Cross-Border Regions (SDCBR 2019) was organized to anticipate the voiced interest of the respective research community to review progress in sustainable development, regional studies, cross-border studies, environmental studies, natural sciences and discuss related challenges and innovations in a scientific conference format.

The conference was held in the Altai State University (Barnaul, Russia) and brought over 900 speakers and participants representing various universities, research institutions from Eastern Europe, the Commonwealth of Independent States, and a number of Central Asian nations. We are pleased to present our conference proceedings, containing a selection of high-quality papers from contributed presentations in the conference.

The published materials will be of great interest to scholars, practitioners, and graduate students.

Sincerely,

Svetlana V. Lobova, Editor, Doctor of Economics, Altai State University, Barnaul, Russia

Aleksei V. Bogoviz, Editor, Doctor of Economics, Professor, National Research University Higher School of Economics, Moscow, Russia

Content from this work may be used under the terms of the Creative Commons Attribution 3.0 licence. Any further distribution of this work must maintain attribution to the author(s) and the title of the work, journal citation and DOI.
Published under licence by IOP Publishing Ltd

IOP Conf. Series: Earth and Environmental Science **395** (2019) 011002 doi:10.1088/1755-1315/395/1/011002

Peer review statement

All papers published in this volume of *IOP Conference Series: Earth and Environmental Science* have been peer reviewed through processes administered by the proceedings Editors. Reviews were conducted by expert referees to the professional and scientific standards expected of a proceedings journal published by IOP Publishing.

Content from this work may be used under the terms of the Creative Commons Attribution 3.0 licence. Any further distribution of this work must maintain attribution to the author(s) and the title of the work, journal citation and DOI.
Published under licence by IOP Publishing Ltd

IOP Conf. Series: Earth and Environmental Science **395** (2019) 012001 doi:10.1088/1755-1315/395/1/012001

Designing an additional freshwater source infrastructure to ensure the environmental sustainability of coastal areas

G Degtyarev[1] and O Takhumova[1*]

[1] Kuban State Agrarian University named after I.T. Trubilina, 13 Kalinin str., Krasnodar 350044 Russia

E-mail: takhumova@rambler.ru

Abstract. The work is based on the search for additional sources of supply of coastal areas with fresh water based on resource-saving technologies and the fundamental rationale for the design and application of the proposed structures in real time. The article identifies potential areas for the placement of aboveground and underground reservoirs of freshwater resources in the coastal areas of the Krasnodar region. The article attempts to develop an innovative project for the seasonal freshwater basin, which allows accumulating the necessary volume during the peak consumption season to meet the needs of the population. The emphasis for placing the pool is made in the mountain gorges, taking into account the geographical features of the landscape of the region. The results of the analytical analysis will subsequently help to optimize constructive solutions, with the aim of further practical implementation of reasonable and proposed measures.

1. Introduction

From the 20th century, the issue of reducing the level of fresh water supplies is considered as a global problem of our time and requires special attention to its solution, which is also extensively reflected in the scholarly literature [5, 6, 7]. This is especially true for border resort areas that need to improve the level of environmental safety. According to official statistics, more than 20% of the total population of the planet lives in an area characterized by a shortage of fresh water [8]. Most of them are in developing countries, and there this problem is based on the underdeveloped infrastructure through which water is taken from aquifers and rivers. All this makes it necessary to design new ways to increase freshwater supplies. For many decades, various ways of solving this problem were developing. This was realized through the creation of a system of aqueducts, canals, and dams, which could satisfy the needs of the population of a given territory. Designing engineering systems for the provision of fresh water allowed to get away from the dependence of the created water bodies on the geographical location of water bodies and move to the development of hard-to-reach areas of our planet. But, given the current trends and intensive population growth, the solutions developed can no longer fully cope with the full provision of fresh water.

Regions of the Krasnodar Territory belong to such problem areas, especially the coastal part with resort towns, in which water demand is particularly felt during the rest seasons. This problem can be solved by involving natural water resources based on natural sediments. At the same time, it is necessary to take into account that the mountain gorges should be chosen to accommodate additional infrastructure to provide water sources. Also, the geographical features of the landscape of the Southern Federal District must be considered.

Content from this work may be used under the terms of the Creative Commons Attribution 3.0 licence. Any further distribution of this work must maintain attribution to the author(s) and the title of the work, journal citation and DOI.
Published under licence by IOP Publishing Ltd

IOP Conf. Series: Earth and Environmental Science **395** (2019) 012001 doi:10.1088/1755-1315/395/1/012001

2. Materials and Methods

The human desire to curb nature has become a conductor to the inexhaustible energy of water. 5000 years ago, unknown engineers attempted to pacify the rivers flowing through the lands of Jordan. Dams were built according to the gravitational principle of operation. They allowed to control the spill and fullness of rivers in the flood season and in the dry season.

The main peak of its development is the artificial hydraulic structures received in the period of the emergence of a new structural material (reinforced concrete) and the newly emerged problem of lack of fresh water for the population. The Hoover Dam, located on the Colorado River, is one of the most prominent structures built in the last century. The Six Companies together with the architect Gordon Kaufman acted as the main developers of the dam. The problem of fresh water also sounded in the XVI-XVII centuries. The approach to the solution laid through desalination. The scientific works of F. Bacon and R. Boyle played an undoubted role in technological developments. Their developments were limited to the study of the model of behavior under the pressure of gases in vacuum. To meet the needs of the population in fresh water, the Rosatom Corporation proposes to produce desalination using the hybrid technology, which includes thermal and membrane methods based on the installation of a nuclear power plant. The United Arab Emirates, Saudi Arabia, and other Eastern countries are the main consumers of this technology.

Each of the technologies presented above is applied depending on the qualitative and quantitative factors of the problem being solved. The algorithm for achieving the goal of designing an additional source of press water is developed by the authors of this paper (Fig. 1).

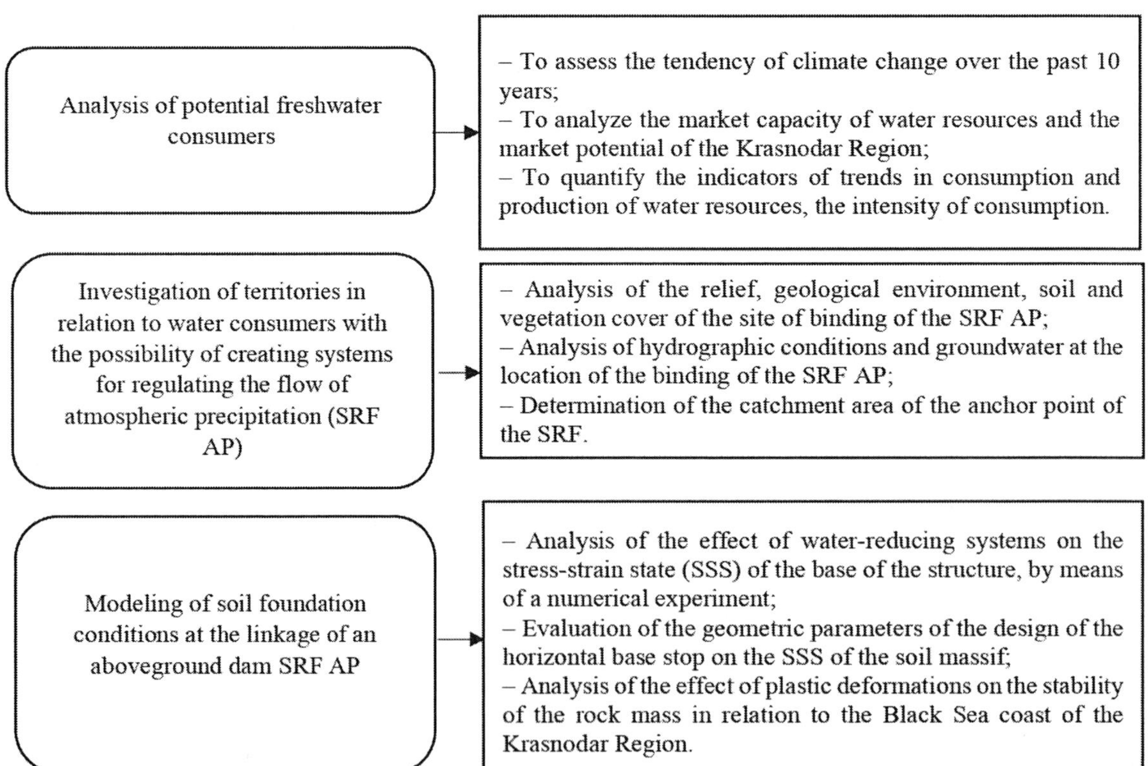

Figure 1. Methodical approaches to designing infrastructure for the supply of fresh water. Source: compiled by authors.

The methods of economic-statistical modeling, mathematical modeling, heuristic evaluation (survey of expert groups) are used for preliminary evaluation.

When designing, we use modern calculation software packages such as Stark ES and ING +, etc.

Traditional approaches to designing SCB often do not exclude filtration of accumulated water. To solve this issue, we use the advanced development and time-tested technical solutions in the field of construction chemistry of such manufacturers as PLASTFOIL, Techno Nicole, Sika, BASF, Carlisle.

3. Results

A distinctive feature of regions with densely populated areas is the excessive consumption of fresh drinking water [8]. This is especially true for the Krasnodar Region, which is part of the South Federal District. According to official data, the district occupies the last place in the overall ranking of the Russian regions in terms of the density of the river network and the lake territory [9].

Over the past few years, the water quality has dropped significantly. For the centralized sources, the inconsistency indicator for microbiological samples exceeded 10%; it reached the level of 40% for the non-centralized ones. The indicator of water resources availability in the Krasnodar Region is 17,039 thousand m3/year, which is lower than the average Russian values by 14,678 thousand m3/year. This confirms the need to find new sources of fresh water. The lowest level of precipitation is observed in September, the average value of this period remains at around 48 mm. December is characterized by significant precipitation (84 mm). As preliminary studies show, this value is sufficient for the accumulation of wastewater in the projected basin to replenish the freshwater reserves in the region.

Figure 2. The study area for the design of an additional source of fresh water (Coast of the Krasnodar Region).

At the initial stage, the zone is determined for the design of the basin (Fig. 2), accumulating atmospheric precipitation (AP). Then, the volume of the proposed basin is calculated (Fig. 3).

Figure 3. Estimated Seasonal Control Basin (SCB) of precipitation.

IOP Conf. Series: Earth and Environmental Science **395** (2019) 012001 doi:10.1088/1755-1315/395/1/012001

During the next stage, a mathematical modeling of the stress-strain state (SSS) of the base-dam system is used to analyze the joint construction of the dam and foundation, taking into account the different filling levels of the seasonal control basin and the effect of seismic load on the structure (Fig. 4).

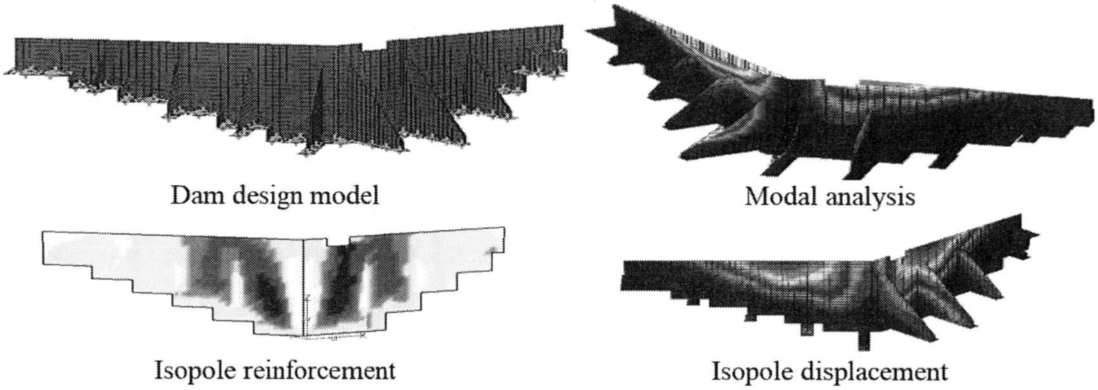

Figure 4. Mathematical modeling of the stress-strain state of the system.

Then, the level of structural and technological regulation is determined (Fig. 5).

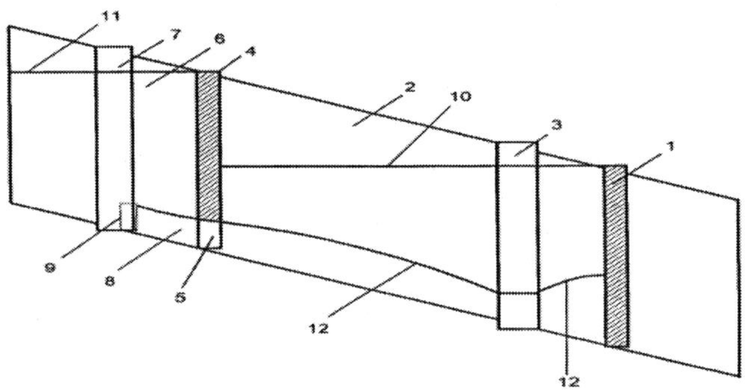

Figure 5. Scheme of constructive and technological level control.

At present, the use of digital technologies allows solving such bold tasks as accumulating precipitation runoff in an integrated way, when the joint work of the underground and above-ground reservoir as well as the ground mass are taken into account. Constructive-technological elements of the system are located on the ground massif, including in the mountainous terrain. As an example, we present the fragments of the element-by-element calculation of the SCB, which is made in the MIDASGTSNX software package. The geotechnical characteristics of the building complex and the topographical survey of the construction site are basic data for the calculation. Based on the data presented, the mathematical model of the soil massif was built in the MIDASGTSNX software package (Fig. 6a). At its base, we can calculate even the main stresses along the axis XX of the structure (Fig. 6b), which is primarily given as an example.

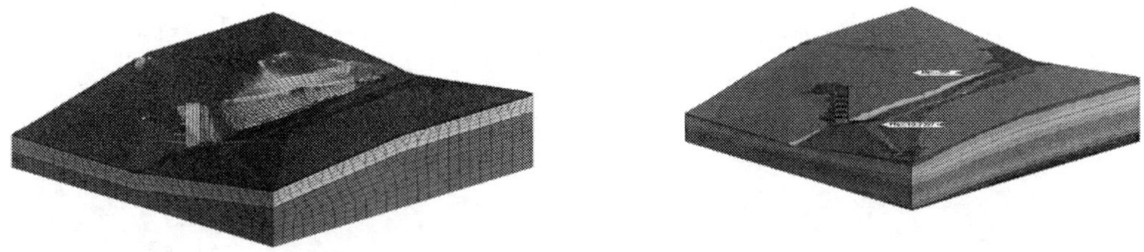

(a) (b)

Figure 6. Mathematical model of the soil massif.

4. Discussion

The device of traditional reservoirs is impossible in the highlands, which represents the Black Sea coast, there is not enough space for the device reservoirs. Also, the device reservoirs in these conditions is very difficult technically and technologically.

Creation of large-scale reservoirs affects the environment, has a strong impact on the landscape, microclimate, flora and fauna. The construction of a reservoir on a large river will lead to changes in its hydrological regime along the course over a hundred thousand meters. The temperature of the water, the ice regime, the flow velocity changes, the height of the wind waves increases. The SCB AP is devoid of all these shortcomings, they are set in mountain gorges. Thus, the creation of a model for the use of natural precipitation sources is of particular interest and requires further development to address the problems of providing the population with water.

5. Conclusion

Analysis of the level of freshwater availability in the region confirms the need to develop its additional sources. Using natural resources by regulating precipitation is one way. Subsequently, the results of the analytical analysis will help optimize design solutions. The achieved purpose was to maximize the economic effect of the introduced resource-saving technologies. This effect was to ensure the environmental sustainability of the territory.

References

[1] Degtyareva O G, Degtyarev G V, Lavrov N L, and Aliev D U 2018 Constructive-technological decisions in regulating the flow of atmospheric precipitation *Magazine of Civil Engineering* **82**(6) pp 32-48 (doi: 10.18720/MCE.82.4.)

[2] Glebova A, Larionova A, and Takhumova O 2019 Organic aquaculture as a promising direction for the production of organic food *Ekoloji* **28**(107) pp 537-543

[3] Gegtyareva O G, Gegtyarev G V 2015 *Patent 2 569 035 "Method of groundwater reserves control"* (Russian Federation, MPK E03B 3/32; E03B 3/06. № 2014134991/13 appl. 26.08.2014; publ. 20.11.2015, Bull. № 32) (Krasnodar, Russia: Kuban State Agrarian University)

[4] Degtyareva O, Degtyarev G, Togo I, Terleev V, Nikonorov A, and Volkova Yu 2016 Analysis of stress-strain state rainfall runoff control system-buttress dam *Procedia Engineering* **165** pp 1619-1628

[5] Kanakoudis V, Papadopoulou A, Tsitsifli S, Curk B C, Karleusa B, Matic B, Altran E, and Banovec P 2017 Policy recommendation for drinking water supply cross-border networking in the Adriatic region *Journal of Water Supply: Research and Technology-Aqua* **66**(7) pp 489-508

[6] Gleick P H 1993 Water and conflict: fresh water resources and international security *International Security* **18**(1) pp 79-112

[7] Kingsford R T 2000 Protecting rivers in arid regions or pumping them dry? *Hydrobiologia* **427**(1) pp 1-11

[8] World Health Organization n.d. *GLAAS 2018/2019 country survey documents* Available at: https://www.who.int/water_sanitation_health/monitoring/investments/glaas-2018-2019-country-survey-documents/en/ (assessed 12 02 2012)

[9] Vörösmarty C J, McIntyre P B, Gessner M O, Dudgeon D, Prusevich A, Green P, Glidden S, Bunn S E, Sullivan C A, Liermann C R, and Davies P M (2010) Global threats to human water security and river biodiversity *Nature* **467** pp 555-556

[10] Rosstat 2019 *Russia in numbers – 2018* (Moscow, Russia: Rosstat)

IOP Conf. Series: Earth and Environmental Science **395** (2019) 012002 doi:10.1088/1755-1315/395/1/012002

Sustainable management of natural pastures in Buryatia

E G Imeskenova[1*], A B Butukhanov[1*] and T MKomendanova[1]

[1] Buryat State Agricultural Academy named after V.R. Pilippov, 8 Pushkin, Ulan-Ude, Russia

E-mail: imesc@mail.ru

Abstract. The article deals with the issues of sustainable pasture resources management on the example of the Tunkinsky district of the Republic of Buryatia. Sustainable pasture management contributes not only to the preservation of forests, but also to the promotion of adaptation to climate change, the preservation of biodiversity, and the improvement of the living standards of the local population. The use of pasture resources provides a long-term balance between livestock numbers, fodder production and pasture productivity, while maintaining the functionality and structure of the entire ecosystem. Pasture ecosystems are dominated by plants from the families *Poaceae, Asteraceae, Fabaceae, Brassicaceae, Chenopodiaceae, Asteraceae.* Lack of grazing leads to the decreasing share of forage species to 47% (on average), as it not used in the ecosystem. For sustainable pasture exploitation, the experimentally established structure of ecological groups is recommended: 70% for forage grass, 15% for medicinal, 7% for ruderal, 5% for poisonous, 1% for others. Unsystematic grazing of animals on pastures leads to a 2.5-3 fold reduction in species of biodiversity (27-30 species), and the family representation becomes 2 times lower. In relation to animal grazing, a total of 5 ecological groups of plants were identified. More than that, their structure, ecosystem, and economic role in stabilizing the ecological situation of pastures were established. In the course of the research conducted in the study area in environmental, economic, financial, and institutional aspects, three scenarios for the management of pasture resources are highlighted: unstable, extensive, and balanced.

1. Introduction

Sustainable pasture resources management is among burning issues throughout the world. Sustainable pasture management contributes not only to the preservation of forests, but also to the promotion of adaptation to climate change, the preservation of biodiversity, and the improvement of the living standards of the local population. As a special ecosystem, pastures are not only a place for cattle grazing, but they also serve as habitat for various wildlife. Therefore, the state of pastures affects the quality of other services provided by these ecosystems. The use of pasture resources provides a long-term balance between livestock numbers, fodder production, and pasture productivity, while maintaining the functionality and structure of the entire ecosystem [11], [13], [14], [16].

Livestock is one of the leading branches of agriculture in Buryatia. About 12% of the area of the region is occupied by pasture land. Recently, as a result of overgrazing and pasture land mismanagement, some pastures were abandoned, while others were degraded, losing vegetation in heavily used areas. In the context of a steady increase in livestock numbers, the use of system approach for livestock grazing, the adoption of a set of measures to improve pasture use, the introduction of forms of environmentally balanced economic activity will generally preserve and improve the state of ecosystems in the Republic of Buryatia [5], [8].

Content from this work may be used under the terms of the Creative Commons Attribution 3.0 licence. Any further distribution of this work must maintain attribution to the author(s) and the title of the work, journal citation and DOI.
Published under licence by IOP Publishing Ltd

On the vast territory of the republic from the mountain peaks to the steppe and semi-desert areas, there is a large variety of landscapes and ecosystems. Therefore, the issues of sustainable management of natural resources should be addressed taking into account this diversity and the characteristics of natural, environmental, and socio-economic conditions. The territories of natural forage lands are not used effectively. As of today, there is a destruction of traditional types of pasture management, all this leads to a decrease in productivity and loss of benefits from year to year, which gradually undermines the economic well-being of the population and leads to the degradation of pasture ecosystems.

2. Materials and Methods

The object of the research is the natural pastures of the Tunkinsky district of the Republic of Buryatia. The research is based on the materials of the field and stationary studies conducted in the district in the period of 2007-2018.

The Tunkinsky district is located in the southwestern part of the Republic of Buryatia, 40 km west of Lake Baikal (Fig. 1). It borders with Mongolia in the south-west, in the north-west it borders in the Oka district of the Republic of Buryatia, in the east there is a border with the Irkutsk region, and the south connects with the Zakamensky district of the Republic of Buryatia. The region is located in the belt of the mountains of Southern Siberia: the Khamar-Daban range, the Munko-Saridak massif, and the Tunkinskie Bold Peaks-Goltsy [3].

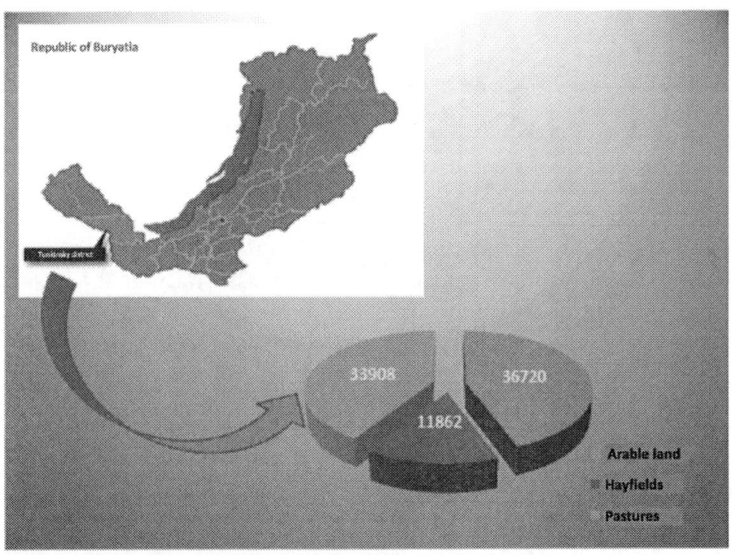

Figure 1. The structure of agricultural land in the Tunkinsky district (as of 2017, ha).

The climate of the region is extreme continental, with large annual and daily fluctuations in temperature, with long, cold winter and little snow in the valley, and large snow cover in the highlands, especially on the Khamar-Daban range. The average temperature in the summer is 24 ° C, the average temperature in the winter period ranges from minus 19 in village Arshan to minus 25.6 in village Kyren. The duration of the period with a temperature below 0 ° C ranges from 174 to 198 days. The average annual rainfall is 300-400 mm. Winter periods are characterized by the predominance of clear days. The soils of the region are alluvial, alluvial-meadow, thin, light loamy, and loamy [2].

Stationary survey sites were established in the Torskaya, Tunkinskaya, and Mondinskaya depressions, where the vegetation cover of 24 pasture associations was studied in detail. The studies were carried out in accordance with the guidelines [1], [4], [10], [15].

3. Results and Discussion

In general, pastures of the studied territory have a discernable seasonal dynamics. The highest feed reserve is formed by the end of the spring and the beginning of the summer period. In the summer it decreases a little and strongly falls in the fall. More than 75% of pastures are of the spring-summer-autumn type. They can be used on average 250 days a year (from March to November). The distant summer pastures make up 20%, they are used only in the summer period (on average 120 days, from May to September) [6], [7].

Monitoring observations of plant biodiversity (Table 1), which were conducted in pastures of varying degrees of degradation with different operating conditions, showed an increase in biodiversity of reclaimed territories by optimizing the composition of life forms of plants and the abiotic environment.

Table 1. Biodiversity of pasture ecosystems.

Indices	Torskaya depression		Mondinskaya depression				Tunkinskaya depression			
	Grazing regime									
	Free		Free		Regulated		Free		Regulated	
Types, units	47		32		63		28		61	
Families, units	14		9		23		10		20	
The most common families (>5%)*	*Po*	21	*Po*	30	*Po*	21	*Po*	32	*Po*	27
	As	29	*Ch*	18	*As*	18	*Ch*	17	*Ch*	17
	Fa	11	*As*	17	*Ch*	12	*As*	12	*Br*	13
	Br	6	*Br, Fa*	10	*Br*	9	*Br*	7	*As*	7
			Bo	6	*Fa*	6	*Fa*	9	*Ca, Fa*	6

* Families: Po – Poaceae, As – Asteraceae, Ch – Chenopodiaceae, Br – Brassicaceae, Fa – Fabaceae, Ca – Caryophyllaceae, Bo – Boraginaceae.

The pasture ecosystems of the Torskoy depression are dominated by plant species from the families *Poaceae, Asteraceae, Fabaceae,* and *Brassicaceae.* The biodiversity of the pasture ecosystems of the Tunkinskaya depression was estimated under free and regulated regimes. The plant cover is dominated by plant species from the families of *Poaceae, Chenopodiaceae, Asteraceae, Brassicaceae,* and *Fabaceae.* Most often in the vegetation cover of pastures in the Mondy depression with free and regulated grazing, there are plants from the following families: *Poaceae, Chenopodiaceae, Asteraceae, Brassicaceae,* and *Fabaceae.*

Table 2. Structure of ecological groups of pasture ecosystems, units in %.

Ecological group	Torskaya depression	Mondinskaya depression		Tunkinskaya depression	
	Grazing regime				
	Free	Free	Regulated	Free	Regulated
Fodder herbs	24-56	28-65	40-57	16-62	53-67
Medicinal herbs	4-8	2-6	13-17	2-8	8-11
Ruderal herbs	13-23	17-23	11-15	5-20	9-13
Poisonous herbs	6-11	1-3	4-5	2-7	6-7
Other	1-2	2-3	5-6	1-3	1-2

Unsystematic grazing of animals on pastures leads to a 2.5-3-fold reduction in species of biodiversity (27-30 species), and the family representation is 2 times lower. It is possible to restore the lost biodiversity in the organization of regulated grazing.

In relation to the grazing animals, 5 ecological groups of plants were identified; and their structure, ecosystem and economic role in stabilizing the ecological situation of pastures were established (Table 2).

A high proportion (17-23%) of ruderal plants is noted with free grazing, as they occupy an ecological niche of medicinal and fodder species that do not withstand the destruction of habitat due to etching and trampling. Lack of grazing leads to a decrease in the share of forage species to 47% (on average), as unclaimed in the ecosystem. For sustainable pasture exploitation, the experimentally established structure of ecological groups is recommended: 70% for feed, 15% for medicinal, 7% for ruderal, 5% for poisonous, and 1% for others.

The pasture productivity depends on the composition of phytocenoses, as well as on topography and wetting conditions, as our research clearly shows.

Table 3. Pasture productivity.

Type of pastures	M+m	lim	σ	M+tm	V, %	Productivity, c/ha (dry matter)
Poaceae - motley grasses	3,0+0,17	2,4-3,3	0,39	2,56+3,36	13,2	3,0
Thread-leaved mixed herbs	5,4+0,26	4,8-6,4	0,25	5,05+6,23	10,2	5,4
Mixed herbs - Poaceae	3,2+0,11	2,8-3,4	0,25	2,96+3,48	7,7	3,2
Steppe mixed hers	5,9+0,41	4,7-6,7	0,91	4,98+6,86	15,4	5,9
Leumusovo - mixed	6,8+21	6,0-7,2	0,48	6,29+7,27	7,0	6,8

The lowest productivity (3.0 centers / dry mass) is characterized by the bluegrass-grass-sedge pasture. It is these pastures that are characterized by the early growth of grasses and good palatability of food throughout the year. Similar types of pastures are widespread in the vicinity of the village Mondy. The most productive pasture in the vicinity was a layland, where a leimus-forb herbage was formed. Our observations showed that the productivity of this grass stand was on average 6.8 c / ha of dry mass.

Based on the research, we can say that the key problems of using natural pastures include: the loss of traditional technology of grazing "rotation / rest"; traditional / usual organization and order of grazing is preserved; the lack of law enforcement by the state in regulating pasture management; an increased number of cattle; an increased degradation of pastures; an increased dependence of livestock grazing on climate change.

In the course of the conducted research in the study area in environmental, economic, financial and institutional aspects, three scenarios of pasture resources management can be distinguished: unsustainable use of pasture, extensive use of pastures, balanced use of pastures.

In the first scenario, the situation leads to the degradation of pastures, to the destruction of the pasture ecosystem, to the decrease in the productivity of pastures and farm animals. This ultimately affects the material well-being of the rural population. This situation is already observed in a number of places (around the villages of Zhemchug, Galbay, Toltoi, Kyren, Khuzhiry), where the load on pastures exceeds the norm by 3-4 times or more. Consequently, this scenario contradicts to the principles of sustainable pasture resources management and can lead to unrecoverable catastrophes in pasture management.

The second scenario does not provide for the development, genetic improvement of animals, strengthening of the forage base and other innovative approaches in the management of pasture resources. It coincides with what is available today. This scenario demonstrates sustainability without growth, in this case the potential of pasture resources will remain unused and stagnation will occur in the economy of the agro-industrial complex.

According to the third scenario, an increase in the livestock population is accompanied by the increased feed reserves. In this case, there is no risk of degradation of grazing lands. It is also necessary to increase the genetic potential of animal productivity, strengthen the forage base and develop pasture farming through the introduction of innovative technologies. In our opinion, this scenario most fully meets the requirements of sustainable pasture resources management and the agricultural sector development.

If we consider the issues of sustainable grazing land management in more details, then in each individual case all aspects (environmental, social, economic, institutional and financial) of sustainable development always appear. Sustainable management requires a balance between its components and only with an integrated approach is it possible to achieve success [15], [16].

4. Conclusions

Given the interrelated character of all aspects of sustainable pasture management, the following conclusions can be drawn:

1. The practiced unsystematic way of using pastures located near settlements is environmentally dangerous and economically unjustified, since it leads to the degradation of pastures and a decrease in the productivity of animals. This type of pasture use is unpromising.
2. In addressing issues of rational management of pasture resources, it is necessary to develop a legislative framework at the level of the national park for their effective functioning.
3. The most optimal is the scheme of seasonal use of pastures as spring-summer-autumn type, with the involvement of distant grazing lands. To this end, it is necessary to take land management decisions on the allocation of distant pasture land for each rural settlement of the district whose residents are pasture users.
4. In addressing the issue of preventing the degradation of pastures and their reclamation, a certain role should be played by the developed rules for grazing farm animals near settlements and in distant pastures.
5. The development of seeded grasslands on fallow and degraded lands by sowing perennial grasses (wheatgrass, alfalfa and others) will allow to improve the forage base.
6. The practical application of the proposed measures will increase the well-being of the local population and reduce the process of pasture degradation near human settlements, thereby contributing to an increase in the environmental and economic effect in grasslands farming.

References

[1] Anenhonov O, Pykhalova T, Osipov K I *et al* 2001 *Key to Buryatia plants* (Ulan-Ude, Russia: Republican Printing House)

[2] Tulokhonova A K Ed. 1995 Comprehensive assessment of the territory of the Tunkinsky Natural National Park: *Series "Land at the Baikal"* (Ulan-Ude, Russia: Publishing house BSC SB RAS)

[3] Ermakova L M 2005 *Traveling through the Tunka valley* (Moscow, Russia).

[4] Geobotanical maps 1984 *All-Union instructions for conducting geobotanical surveys of natural forage lands and drawing up large-scale geobotanical maps* (Moscow, USSR: Kolos)

[5] Imeskenov C S, Korsunova T M, and Imeskenova E G 2016 Natural, social and ethnocultural potential of the Tunkinsky district of Buryatia as an object of agricultural tourism *Bulletin of the Krasnoyarsk State Agrarian University* **4**(115) pp 28-34

[6] Imeskenova E G 2009 *Methods of the use and the quality feed of main types of natural grass stands of Buryatia Thesis for the degree of candidate of agricultural sciences* (Ulan-Ude, Russia: Sib. Research Inst. of feed.)

[7] Imeskenova E G, and Butukhanov A B 2013 Production capacities of pasture ecosystems of the Tunkinsky district *Bulletin of the Buryat State Agricultural Academy V R Filippova* **1**(30) pp 77-81

[8] Imeskenova E G, and Butukhanov A B 2017 Rational use of natural pastures for free-range livestock breeding *Buryat State Agricultural Academy* **4**(49) pp 123-129

[9] Johnson I R, Lodge G M, White R E, The sustainable grazing systems pasture model: description, philosophy and application to the SGS National Experiment *Australian Journal of Experimental Agriculture* **43**(8) pp 711-728

[10] Kutuzova A A, Zotov A A, Trofimov L S *et al* 1996 *Guidelines for conducting scientific research on hayfields and pastures* (Moscow, Russia: VNII of feed)

[11] Larin I V 1990 *et al* Grassland and pasture farming. 2nd ed., amended and add (Leningrad, USSR: Agropromizdat)

[12] Ludi E 2003 Sustainable pasture management in Kyrgyzstan and Tajikistan: development needs and recommendations *Mountain Research and Development* **23**(2) pp 119-123

[13] Robinson S 2015 Results of the First Practical Conference on the Promotion of Sustainable Pasture Management in Central Asia: *Pasture Management in Central Asia* (Bishkek, Kyrgyzstan)

[14] Van Veen S TW, Alimaev I I, and Utkelov B 2005 Kazakstan Rangelands in Transition – the Resource, the Users, and Sustainable Use. Technical Paper. World Bank (Europe and Central Asia Environmentally and Socially Sustainable Development Series, 31348).

[15] VNIIK them V R Williams 1971 *Methods of experiments on hayfields and pastures* (Moscow, USSR: Agropromizdat)

[16] Xinchun Z 2011 Pastoralism in Xinjiang: Implementing the pastoralist settlement programme to promote sustainable development of pastoralism in Xinjiang. In H Kreuzmann, K Abdulalishoev, Lu Zhaohui, J Richter Eds., *Pastoralism and Rangeland Management in Mountain Areas in the Context of Climate and Global Change* (pp 182-188) (Feldafing, Germany: GIZ)

IOP Conf. Series: Earth and Environmental Science **395** (2019) 012003 doi:10.1088/1755-1315/395/1/012003

Growth dynamics of conditionally pathogenic microorganisms *in vitro* under the influence of anthelmintics drugs

S M Alekseeva[1*]

[1] Buryat State Academy of Agriculture, 8 Pushkina str., Ulan-Ude 670024 Russia

E-mail: sayana.a@mail.ru

Abstract. The article presents data on the effect of anthelminthic drugs on the degree of growth in conditionally pathogenic microorganisms. *Belamizole* and *aversect-2* were found to have a stimulating effect on the degree of growth intensity of *Staphilococcus aureus* and *Escherichia coli*. In contrast, *ivertin* has an inhibitory effect on the growth and reproduction of these microorganisms.

1. Introduction

Currently, a close attention is drawn to the fact that some veterinary drugs (anthelmintic) when used can provoke an infectious process. This is due to their side effects on the immune and microbial status of the organism of animals. This is fraught with serious consequences in suppressing the immune system and suppressing the number of beneficial microflora of the gastrointestinal tract of animals, which ultimately may be accompanied by the formation of endogenous spontaneous infections of unknown etiology. As a result, they cause significant damage to livestock breeding associated with the death of animals during anti-invasive measures and the formation of an immunodeficiency state caused by anthelmintics [2]. The clinical picture of outbreaks of individual infectious diseases with economic consequences arises, which requires a timely diagnosis and prevention measures.

Due to the inevitable use of anthelmintics that contribute to the destruction of parasites, the problem of dysbiosis is aggravated. In some animals, a deep dysbiotic change persisted for more than 3 weeks after the use of an antiparasitic drug based on albendazole [1].

In a laboratory experiment, the infection of white mice with Salmonella microbial culture and the simultaneous introduction of anthelmintica Santel caused a large percentage of the death of animals, compared with their infection with Salmonella alone, with a more pronounced pathoanatomical picture [3]. All this necessitates the control of the microbiota of animals during the period of preventive measures (vaccination, anthelmintic treatment) and the development of the means provided for the correction.

In this regard, the need to study the degree of growth intensity of the conditionally pathogenic microorganisms under the influence of anthelminthic drugs in vitro in the prevention of endogenous infections occurs.

Content from this work may be used under the terms of the Creative Commons Attribution 3.0 licence. Any further distribution of this work must maintain attribution to the author(s) and the title of the work, journal citation and DOI.
Published under licence by IOP Publishing Ltd

2. Materials and Methods

The nature of the growth of the studied microbial cultures under the influence of anthelminthic drugs was determined by sowing them on a meat-peptone agar containing 1% solutions of *belamizole*, *aversect-2*, and *ivertin* by nephelometry at KFK - 2. Cultures not treated with the test solutions served as controls. The observations were carried out for 48 hours. The cultivation of the strains with the inclusion of *anthelmintics* in the nutrient medium showed their noticeable influence on the change in the growth pattern of the following microbial cultures: *Staphilococcus aureus* and *Escherichia coli.*

3. Results

When studying the growth of the original cultures incubated with anthelminthic preparations, it was established that *belamizole* and *aversect-2* had a stimulating effect on the growth of the microbial culture of *S. aureus* compared to the control strains (Figure 1, 2).

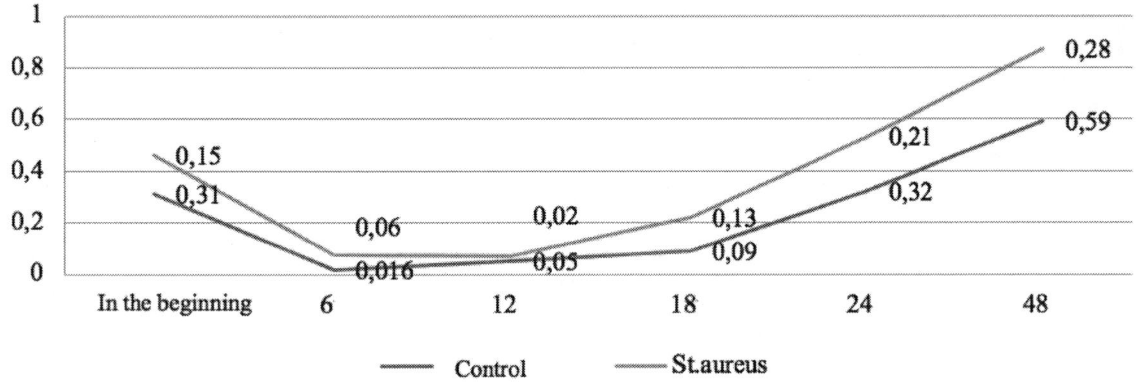

Figure 1. The growth of *S. aureus* under the influence of bezamizole.

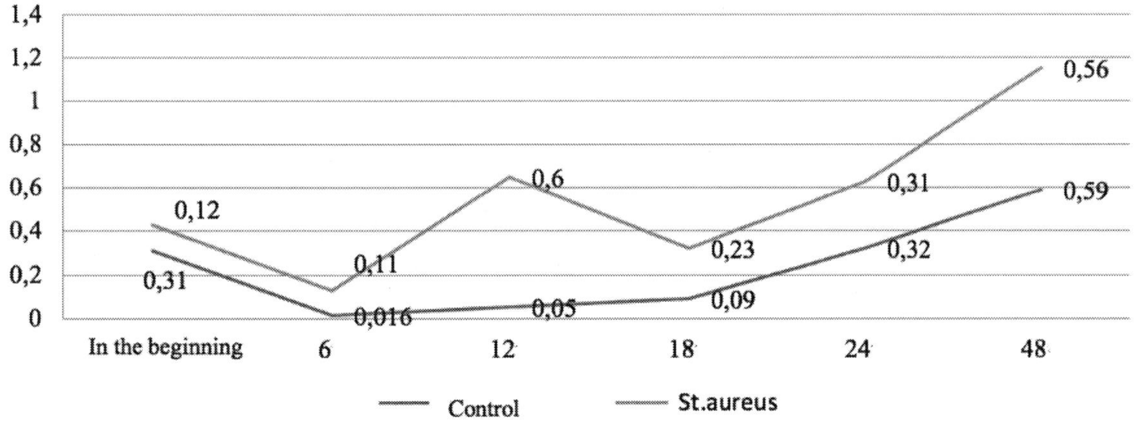

Figure 2. *S. aureus* growth under the influence of *aversect.*

In microbial culture of *S. aureus* under the influence of ivertin, in contrast to the above, the suppression of growth intensity was observed compared with the control strain (without anthelmintic) (Figure 3).

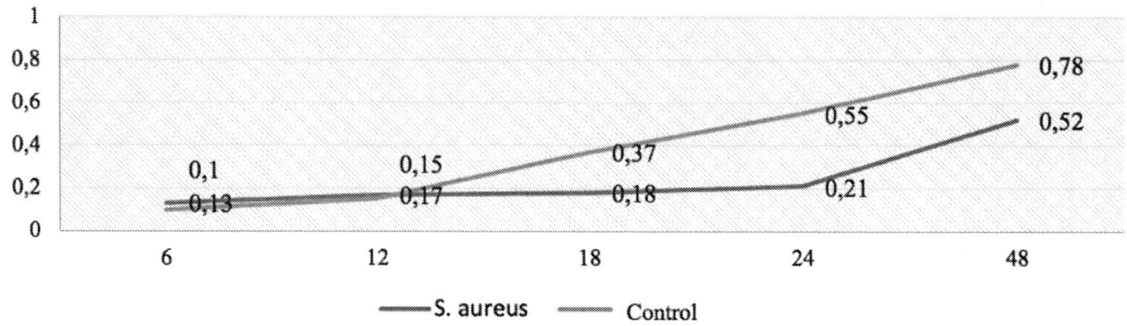

Figure 3. The growth of *S. aureus* under the influence of ivertin.

In microbial culture of *E. coli* under the influence of *belamizole* and *aversect*, an increase in the number of microbial cells was observed compared with the original strain. These drugs had a stimulating effect on the number of microbial bodies throughout the entire experiment for 48 hours. (Figure 4, 5).

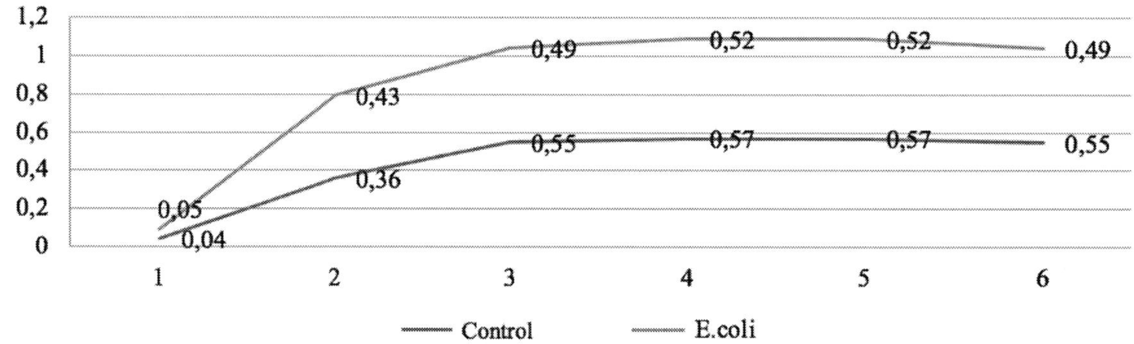

Figure 4. Growth of *E. coli* under the influence of bezamizole.

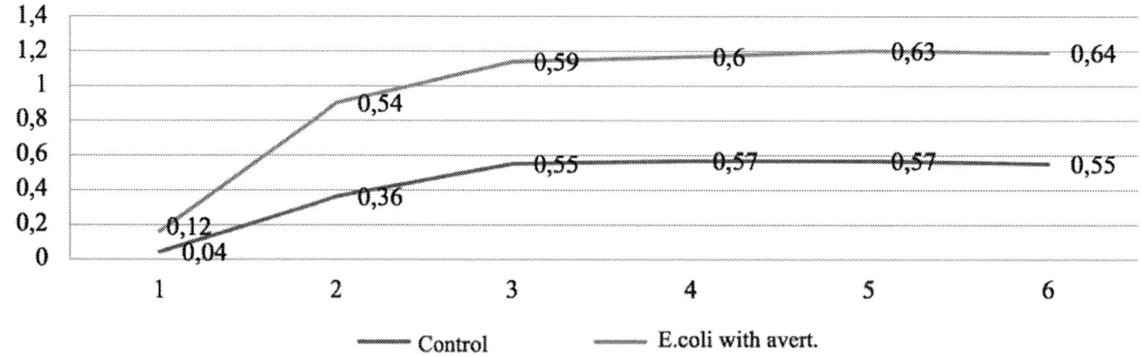

Figure 5. Growth of *E. coli* under the influence of aversect.

Under the influence of *ivertin*, in *E. Coli* more intensive growth of *Escherichia coli* during the logarithmic phase was observed compared with the control strain. In the control, the growth rate indicator also increased (Figure 6).

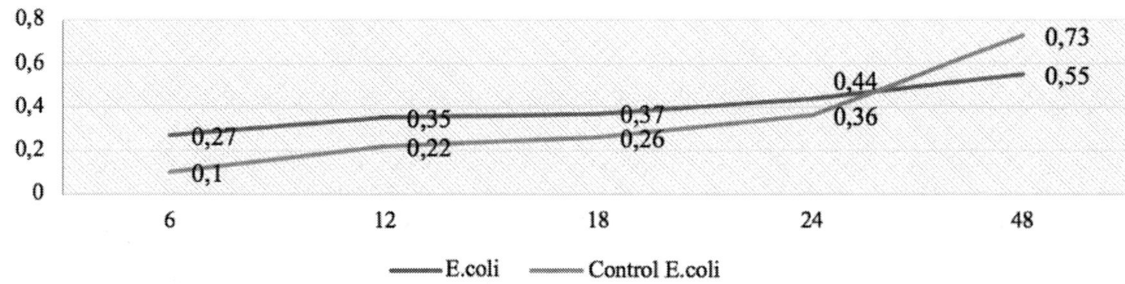

Figure 6. *E. coli* growth under the influence of ivertin.

4. Discussion

Thus, with therapeutic deworming, *belamizole* and *aversect-2* can stimulate the accumulation of conditionally pathogenic microorganisms, for example *S. aureus, E. Coli,* and thus trigger the formation of an infectious process as a potential agent. At the same time, *ivertin* had a negative effect on the growth of microorganisms.

Consequently, *belamizole* and *aversect-2* can provoke endogenous infection with latent bacterial carriers. Reorganization of the enzymatic process of conditionally pathogenic microorganisms is observed under the action of antiparasitic agents. The restructuring is accompanied by an increase in the level of pathogenicity and invasiveness and the appearance of the risk of spontaneous infectious diseases.

5. Conclusion

The results are of practical importance in the study of conditionally pathogenic microorganisms isolated during bacterial carriage in animals subjected to deworming. In this regard, there is a need to develop preventive measures. These preventive measures would exclude the influence of anthelmintics for the prevention of dysbacteriosis caused by deworming, which may result in spontaneous endogenous infections.

References

[1] Danilevskaya N V, and Subbotin V V 2010 *Dysbacterioses in small domestic animals* (Moscow, Russia: Zoomedmet)

[2] Daugaliyeva E Kh, Saveliev A A, and Arinkin A V 2006 The impact of domestic anthelminthic drugs on the immunogenesis of pigs spontaneously infected with nematodoses *Veterinary Pathology* **1** pp 75-79

[3] Tsydypov V Ts, Alekseeva S M, and Ivanova O M 2017 The influence of "Santel" as an inducer on the occurrence of spontaneous endogenous infection *Actual Questions of Veterinary Biology* **4**(36)

Problems of the health status of children and atmospheric air of Gorno-Altaisk under the conditions of increasing transport load

E A Chanchaeva[1*], M G Sukhova[1,2] and S S Sidorov[1]

[1] Gorno-Altaisk State University, 2 Lenkina str., Gorno-Altaysk 49000 Russia
[2] Laboratory of Landscape Water and Ecological Researches and Environmental Management, Institute of Water and Environmental Problems of the Siberian Branch of the Russian Academy of Science, 1 Molodezhnaya str., 656038 Barnaul Russia

E-mail: chan.73@mail.ru

Abstract. The article contains a theoretical study of the problem of the children' health status and atmospheric air under conditions of increasing traffic load. The children's body is highly sensitive to the effects of unfavorable environmental factors, susceptible to the pathogenic effects of even subthreshold concentrations of harmful substances, which makes it possible to consider it as a kind of indicator of the state of the environment. The deterioration of the environmental state of the air of Gorno-Altaysk is associated with an increasing traffic load, solid fuel stationary sources, and transboundary pollution coming from the neighboring regions, as well as the location of the city in a mountain valley with poor natural weathering of the air basin. A comprehensive analysis of the health status of the child population in the region (depending on the degree of air pollution) is also conducted.

1. Introduction

The ecological state of atmospheric air is one of the main problems of environmental health [1]. The state of the air environment depends on many factors, including the transport, industrial, demographic and radiation load, terrain features, etc. [2]. The main sources of pollutant emissions into the air are the solid fuel heating systems and vehicles. If pollution from the first type of sources is seasonal, then emissions from the second type are observed on a year-round basis. In conditions of unfavorable ecological environment, deterioration of health and an increase in deviations in the physical development of children is noted [3-8]. The high sensitivity of the child's body to the effects of adverse environmental factors makes it possible to consider it as a kind of indicator of the state of the environment in the context of increasing environmental threat [8].

The purpose of the theoretical study is to substantiate the relevance of the problem on the ecological state of atmospheric air in Gorno-Altaisk, the need for zoning residential areas of Gorno-Altaisk depending on the degree of air pollution and the health status of the child population.

2. Atmospheric Air Pollution Caused by Road Emissions

The increase in emissions of pollutants occurs due to the use of automotive fuel vehicles that do not meet the requirements of environmental class K_4 and K_5 on the content of pollutants; to the growth in the number of urban vehicles operating in dense urban areas; to the use of coal and firewood as a fuel

Content from this work may be used under the terms of the Creative Commons Attribution 3.0 licence. Any further distribution of this work must maintain attribution to the author(s) and the title of the work, journal citation and DOI.
Published under licence by IOP Publishing Ltd

IOP Conf. Series: Earth and Environmental Science **395** (2019) 012004 doi:10.1088/1755-1315/395/1/012004

by heat supply sources located in the Asian part of Russia (mainly in the Siberian Federal District); to adverse meteorological conditions for dispersion of impurities in the atmosphere [1].

The main components that pollute the air are the fine particulate matter (PM), soot, NO, SO_2, CO, CH, formaldehyde, benzo (a) pyrene, heavy metals. The spectrum of pollutants is very wide, for many of them the norm limits are not defined. For example, in emissions from the road transport, specialists from the Research Institute of Human Ecology and Environmental Hygiene have identified 175 chemicals, 71% of which are not standardized [9]. Despite measures to reduce toxic substances in the exhaust of internal combustion engines due to new fuel standards, the motor transport is one of the leading factors of air pollution [10].

The atmospheric air pollution of cities occurs not only with exhaust gases, but also with wear and tear products of the road and automobile complex (RAC) (tire and roadbed wear). In the formed dust, more than 140 chemical compounds hazardous to health are contained, including polyaromatic hydrocarbons, volatile carcinogens [11]. The diesel engine exhaust is more toxic than gasoline and replacing fuel with ethanol or other types of energy source not only does not give the expected result but aggravates the unfavorable ecological condition of atmospheric air [12]. The use of alternative fuels reduces CO_2, emissions, but it significantly increases PM_{10} and $PM_{2.5}$ emissions [12]. Thus, the motor transport remains one of the leading factors of air pollution, attempts to reduce the content of toxic substances in the exhaust of internal combustion engines do not bring the desired result.

3. Children's Health Exposed to Air Pollutants

According to many authors, the deterioration of children's health and the inconsistency of physical development with age norms are the direct result of the negative impact of environmental factors [3-8]. The children's body is highly sensitive to the effects of adverse environmental factors, is susceptible to the pathogenic effects of even subthreshold concentrations of harmful substances. This allows us to consider the physical development and state of health of children as an indicator of the ecological status of the environment, especially in conditions of increasing environmental threat both in Russia as a whole and in its individual areas [8].

From the period of childhood, the human body is exposed to aerotechnogenic load, reducing the compensatory capabilities. Respiratory organs are most at risk of developing both chronic and acute adverse effects in the population due to suspended substances and NO_2, exposure to formaldehyde [12, 13]. Studies conducted in industrial cities of Siberia have shown that an increase in the incidence of allergic rhinitis and bronchial asthma among children amounts to 28-36%, and more than 30% of those examined have sensitization to this chemical [14]. It has been established that the risk of immunity disorders in children living in industrial cities of the Siberian region is due to the presence of formaldehyde in air [4]. Anthropogenic chemical factors that pollute the air have an ability to reduce the activity of local mechanisms of anti-infective resistance of the upper respiratory tract [3], [5], [6].

Any PM has an ability to adsorb toxic substances that also enter the body's internal environment [11], [12]. A direct relationship is found between the concentration of fine particles (PM_{10} and $PM_{2.5}$) in the air and mortality and morbidity rates. A chronic exposure to PM exacerbates the risk of developing cardiorespiratory diseases [1]. Indicators of medical and demographic losses directly depend on the level of air pollution from PM [2], which trigger pathogenic vascular mechanisms, activate inflammation in the airways and lungs [14]. Studies have shown that the RAC products and exhaust gases determine from 58 to 81% of the diseases of children living in the zone of intensive traffic flow. In particular, the RAC products provoke from 16 to 23% of cases from all diseases, including from 8 to 12% with heavy metal compounds [15].

4. Ecological Condition of the Air in the Conditions of Increasing Transport Load

The Republic of Altai is a recreational and agricultural region, with small industrial enterprises. Air pollution in the region is mainly due to emissions from road transport, boilers, and heating stoves [16-18]. The territory of the city of Gorno-Altaisk is located in the northern part of the Altai mountain region, mainly in the valley-like expansion of the valley of the Maima River. The terrain varies in the range of 250-820 m, with the average height of 400-450 m. In geomorphological terms, the area is located in the foothill and low-mountain zones of Gorny Altai [19].

It is known that the purification of the atmosphere from pollutants is due to meso-and macroscale processes, such as turbulent exchange, a height of the air mixing layer, a wind regime, etc. Repeatability and power of inversions are associated with large-scale atmospheric processes. The highest level of impurity concentration in the atmosphere is noted in slow-moving anticyclones and ridges, on the western periphery of the anticyclone or ridge, with advection of heat in a slow-moving cyclone of small area, in which one and also air mass circulates [2].

The concentration of harmful impurities increases with fogs and haze, which accumulate substances of high toxicity. Fast moving cyclones, strong frontal winds and intense rainfall contribute to the dispersion of harmful impurities. Purification of the atmosphere occurs during the invasion of air masses from the Arctic, carrying clean air.

To characterize the degree of purification of the Gorno-Altaisk air basin from certain pollutants, we used a meteorological indicator of atmospheric self-purification (K_M) [21]:

$$K_M = (P_{SR} + P_T) / (P_O + P_B),$$

where P_{SR} – a wind speed repeatability 0 - m / s; P_T – a repeatability of days with fog, %; P_O – a repeatability of days with precipitation > 0.5 mm,%; P_B – a wind speed repeatability > 6 m / s.

When K_M is less than one, the processes of self-purification prevail over the processes that contribute to the accumulation of impurities. When K_M is more than one, the processes of accumulation of impurities predominate over the processes of self-purification.

It was established that in the valley of the Maima River (where Gorno-Altaisk is located), the meteorological indicator of self-purification of the atmosphere was 1.8 for the winter period, 1.2 for the spring, 1.5 for the summer, and 2.1 for the autumn periods. Consequently, processes that contribute to the accumulation of impurities in the atmosphere prevail throughout the year. This is due to the peculiarities of the relief, the valley-shaped expansion of the valley and slight elevation changes with a significant frequency of anticyclonic weather.

According to the calculation results for 2016, the volume of gross emissions of pollutants from various sources into the atmospheric air was: 33.6 thousand tons/year throughout the republic; 8.3 thousand tons/year on the territory of Gorno-Altaisk, of which 49% are due to road transport and 51% because of fuel and energy enterprises [18]. Despite the city's gasification and reduction of emissions from coal-fired boilers, the total amount of pollutants remains the same. According to the authors of [17], within the city's territory, emissions from solid-fuel heating systems are preserved, as evidenced by the content of pollutants in bioindicators. According to other sources [18], the volume of emissions remains at the same level due to a significant increase in motor transport and unresolved problems in expanding the network of the city's highway, which is currently extremely limited and overloaded with transport. If the task of reducing emissions from coal-fired boilers is solved by switching to gas heating systems, the problems of air pollution due to emissions from motor vehicles remain relevant. During the period from 2013 to 2016, the volume of vehicle emissions increased by 45-50% in Gorno-Altaisk (by 8-10% per year) [18].

According to the literature, the transboundary transfer of air polluting substances is carried out from the territories adjacent to the republic (Altai region, Kazakhstan) [16]. As a result, in the context of low industrial and demographic (63214 people) loads, the unfavorable ecological situation is developing due to the state of atmospheric air. Thus, the deterioration of the ecological state of the air environment of Gorno-Altaisk is associated with an increasing transport load, operating solid stationary sources, and transboundary transport of pollutants from neighboring regions.

Our assessment of the physical development of the 8-years-old children in Gorno-Altaisk shows that 52% of children have lung capacity indicators below the age norm and 48% of children have hand strength values below the age norm. More than that, 22% of children are obese, according to the Quetelet index; 54% of children have an atypical type of reaction of the cardiovascular system to the standard physical activity, a decremental nature of the recovery of heart rate, indicating a low efficiency of the circulatory system [22]. In general, almost half of the children surveyed have physical impairments. According to the authors [3-8], the discrepancy between the physical development of children and age norms often manifests itself in unfavorable environmental conditions.

A comprehensive analysis of the state of health of the population as a whole and of the child population in particular in this region, depending on the degree of air pollution, was not carried out. To actualize the problem, measures are necessary to rank the residential area depending on the degree of air pollution. And also, the analysis of the health status of children, as the most vulnerable part of the population, is necessary, taking into account the environmental situation.

5. Conclusion

Currently, the main sources polluting the atmospheric air in the city of Gorno-Altaisk are motor vehicles and heat power facilities. At the same time, the specificity of atmospheric processes and features of the relief of the territory do not contribute to the active self-purification of the atmosphere. The medico-demographic situation of the child population directly depends on the level of air pollution. Therefore, measures to actualize the problem of air pollution are essential for improving the environmental situation in residential areas and preserving the health of the younger generation.

References

[1] World Health Organization 2018 *Air quality and health: world health urganization newsletter* Available at: http://www.who.int/ru/news-room/fact-sheets/detail/ambient-(outdoor)-air-quality-and-health (Accessed 20 03 2019)

[2] Shcherbatyuk A V 2018 Strategy for optimizing the management of the environmental safety of the air environment of cities in the conditions of intercontinental intermontane depressions *Ecological Safety of Construction and Urban Economy* **1** pp 29-34

[3] Salmanzadeh M, Saeedi M, and Li L Y 2015 Characterization and metals fractionation of street dust samples from Tehran, Iran *International Journal of Environmental Research* **1**(9) pp 213-224

[4] Masnaviyeva L B, Efimova N V, and Kudayeva I V 2018 Effect of chemical admixtures of air on the exposure of toll-like receptors and the level of cytokines in adolescent children *Human Ecology* **3** pp 10-15

[5] Maklakova O A, Ustinova O Yu, and Ivashova Yu A 2017 Features of the cardiorespiratory system in children living in conditions of air pollution with metals (vanadium, manganese) *Bulletin of Perm University (Series: Biology)* **4** pp 427-235

[6] Savchenko O V 2018 Assessment of environmental risks to public health in areas, ranking by degree of technogenic pollution *Human Ecology* **4** pp 10-15

[7] Myachina O V, Zuykova A A, and Pashkov A N 2015 Compensatory abilities of a child's body under conditions of aero-technogenic load *Hygiene and Sanitation* **9**(94) pp 22-25

[8] Yatsyna I V, Sineva E L, Tulakin A V, Zhadan I Yu, and Preobrazhenskaya E A 2015 Child health in an industrialized region *Hygiene and Sanitation* **5**(94) pp 39-44

[9] Rakhmanin Yu A, and Mikhailova R I 2014 Environment and health: priorities for preventive medicine *Hygiene and Sanitation* **5**(93) pp 5-10

[10] Glebov V V, and Kirichuk A A 2014 The possibilities of biomonitoring in the assessment of the ecological status of the ecosystems of the metropolitan metropolis *The World of Science, Culture, Education* **5** pp 339-341

[11] Di Iorio S, Magno A, Mancaruso E, and Dal Bello L 2014 Engine performance and emissions of a small diesel engine fueled with various diesel/RME blends *SAE Technical Paper* 2014-32-0135

[12] Golokhvast K S, Chaplenko T N, Nikiforov P A, Chaika V V, Pamirsky I E, Khristoforova N K, ... Gulkov A N 2013 Granulometric analysis of atmospheric suspensions of the city of Blagoveshchensk *Human Ecology* **7** pp 34-39

[13] Ivanenko A V, Sudakova E V, Skvortsova S A, and Bestuzheva E V 2017 Assessment of the risk to public health from exposure to atmospheric pollution in certain areas of Moscow *Hygiene and Sanitation* **3** pp 206-211

[14] Tabor S M, Shaw C A, Robertson S, Miller M R, Duffin R, Donaldson K … Newby D E 2016 Platelet activation independent of pulmonary in-flammation con-tributes to diesel exhaust particulate-induced promotion of arterial thrombosis. *Particle and Fibre Toxicology* **13**(6) pp 2-14

[15] Rakhmanin Yu A, Rusakov N V, and Samutin N M 2015 Wastes - as an integral environmental-hygienic criterion for a comprehensive impact on the environment and public health *Hygiene and Sanitation* **6**(94) pp 5-11

[16] Robertus Yu V, Udachin V N, Rikhvanov L P, Yusupov D V, Kivatskaya A V, and Lyubimov R V 2016 Indication by the components of the natural environment of the transboundary transport of pollutants to the territory of the Altai Mountains *Izvestiya of Tomsk Polytechnic University. Georesource Engineering* **9** pp 39-48

[17] Robertus Yu V, Rikhvanov L P, and Sitnikova V A 2018 The elemental composition of lichen on slate as a bioindicator of atmospheric pollution in the agglomeration of Gorno-Altaisk *Izvestiya of Tomsk Polytechnic University. Georesource Engineering* **329** pp 70-78

[18] Altai Regional Institute of Ecology 2016 *Report "On the state and environmental protection of the Republic of Altai in 2016"* Available at: http://altai-republic.ru/society/doklad_nature_2016.pdf (Accessed 20 03 2019)

[19] Sukhova M G, and Garms E O 2012 Climatic conditions for the formation of intermundane and mountain-valley landscapes of Altai *The World of Science, Culture, Education* **1**(32) pp 315-318

[20] Sukhova M G 2009 *Bioclimatic conditions of human activity in the Altai-Sayan mountain country* (Tomsk, Russia: Tomsk State University Press)

[21] Selegay T S 1989 Meteorological potential of self-purification of the atmosphere of the Siberian economic district *Proceedings of the West Siberian Regional Research Institute of the State Hydrometeorology Committee* pp 84-89

[22] Sidorov S S, Chanchayeva E A, and Aizman R I 2018 Dependence of the reactivity of the cardiovascular system on physical activity on the level of salivary cortisol in children *Siberian Scientific Medical Journal* **38**(6) pp 130-136

Optimization of acreages as an element of sustainable development of natural landscapes

V Chibis[1,2*] and I Kutyshev[1]

[1] Omsk State Agrarian University named after P.A. Stolypin, Institutskaya square 1, 644008, Omsk, Russia
[2] Omsk Agrarian Research Center, Koralev Avenue 26, 644012, Omsk, Russia

E-mail: vv.chibis@omgau.org

Abstract. The alternation of crops for a long time will remain a fundamental element of the zonal technologies of their cultivation. The use of mathematical modeling methods will allow to reach a new level in solving the issues of effective use of arable land through the use of models that optimize it. Especially, the results of studies based on long-term observations will be valuable. We used data from long-term stationary experiments to study the alternation of crops in crop rotations that took more than five rotations. The stationary experiments are located in the forest-steppe zone of Western Siberia in the fields of the Omsk Agrarian Scientific Center. The purpose of the research was to optimize the use of arable land based on mathematical modeling methods to preserve soil fertility. In the work, a model that takes into account the conditions of the agroecosystem is used (humus balance, labor costs, energy costs, etc.). It allows one to optimize the cultivated area, while maintaining soil fertility, ensuring the production of products at a consistently high level, with increasing economic and energy efficiency. The article shows that in grain-crop crop rotations, the yield is generated due to the natural fertility of the soil, as evidenced by the negative humus balance, which reaches 0.18-0.78 t / ha. During the alternation of the fruit-changing alternation, the crop was formed against the background of the reproduction of the organic matter of the soil by increasing the plant residues from the use of fertilizers. The humus balance was positive (0.15-0.28 t/ha). When using employed steam along with clean steam, rape as siderate, and also straw as an organic fertilizer for grain crops, the amount of 2.01 t/ha of marketable grain was obtained (with a 16.8% share of pure steam, as well as occupied steam (11%) and spring grain (72.2%), of the arable land area). Increasing the area in the crop rotation of Western Siberia using straw, green manure, and occupied vapors as organic fertilizer, the soil fertility reproduction with a certain decrease in the yield of marketable grain up to 2 t/ha can be achieved.

1. Introduction

Mathematical modeling allows one to build an optimization model of the structure of the acreage for the commodity producer, taking into account their needs and the availability of resources. The structure of sown areas should be formed on the basis of a system-energy approach. Research in the mathematical modeling of the production process of plants is being intensively developed, as many published papers in this field show [1-3; 9].

When evaluating agroecosystems, the potential productivity of crop rotations should be determined to identify the structure of anthropogenic energy consumption for growing crops, to analyze in detail, the productivity of agroecosystems to evaluate. Conducting bioenergy analysis will allow to have an

Content from this work may be used under the terms of the Creative Commons Attribution 3.0 licence. Any further distribution of this work must maintain attribution to the author(s) and the title of the work, journal citation and DOI.
Published under licence by IOP Publishing Ltd

objective assessment of the energy potential of the agroecosystem and the feasibility of using anthropogenic energy in the cultivation of crops.

The productivity of crop rotations is determined by the productivity of the crops within their structure, which react differently to rainfall and temperature during certain periods of vegetation. Many scientists in their studies have shown that the rotation of crop rotation and the productivity of crops reliably depend on the natural and climatic conditions and the structure of the sown areas [4, 5].

Thus, the goal of our research was to increase the productivity of field crop rotations based on optimizing the structure of arable land using mathematical modeling to improve the adaptive-landscape farming systems in the forest-steppe of Western Siberia.

2. Materials and Methods

The research was conducted in long-term stationary experiments to study the alternation of crops in crop rotations that took more than five rotations. The stationary is located in the forest-steppe zone of Western Siberia in the fields of the Omsk ASC, former SibNNIISH. Placement of plots in the experiments was randomized; the size of plots was 0.275 ha (110 × 25 m) and 0.138 ha (110 × 12.5 m); it was repeated 4 times. Field crop varieties are zoned for Western Siberia: spring wheat ("Omskaya 36"), spring barley ("Omsky 95"), soybeans ("Golden"), rapeseed ("Yubileynyy"), oats ("Irtysh 21"), corn ("Omka 135"). The soil of the experimental plot is black leached, slightly leached medium-humus, medium heavy, and loamy with a humus content in the arable horizon of 6.5-7%. The system of agrotechnical measures was built taking into account the recommendations of the Omsk Agrarian Scientific Center for the forest-steppe zone of Western Siberia [6]. In the process of building a model of agro-ecosystem performance, linear and non-linear regression analysis was used. The adequacy of the model was tested by the Fisher criterion. The confidence level was assumed to be 95%. To test the hypothesis of the correlation of residues, the Durbin-Watson criterion was calculated. Methods of regression, correlation analysis, and mathematical modeling were used [7].

3. Results

We found that the introduction of grain crops and soybeans into the crop rotation with a share of 15-25% will increase the yield of feed protein units by 0.3-0.7 tons from 1 ha of crop rotational area. Provided that a high level of grain yield in comparison with spring wheat will remain (Table 1).

The yield of grain, KPI, feed units, and digestible protein from one hectare of arable land also refers to the indicators of the efficiency of crop rotation. These indicators are directly dependent on the set of crops in crop rotation and the number of fields. In the conditions of 2010–2018, crop change rotations were effective due to their saturation with legumes and forage crops, mainly.

Table 1. The effectiveness of field crop rotations depending on the length of rotation and saturation of field crops in 2010-2018, tons.

Rotations	**Grain yield**	**Output, 1 hectare of arable land**			
		grains	**feed units**	**digestible protein**	**KPI**
Grain fallow rotations (control)					
Steam - wheat - wheat - oats	2.17	2.01	2.43	0.19	3.21
Occupied steam - wheat - wheat - oats (without chemicals)	1.65	1.31	2.54	0.12	3.91
Occupied steam - wheat - wheat - oats (with chemicals*)	2.32	1.95	3.87	0.31	6.78
Fruiting crop rotations					
Soy - wheat - barley - oats	1.99	2.06	2.55	0.23	4.00
Occupied steam - wheat - barley - oats (without chemicals)	1.87	1.4	2.87	0.28	5.42
Occupied steam - wheat - barley - oats	2.36	2.11	4.18	0.38	9.75

(with chemicals)					
Permanent crops					
Permanent wheat	1.10	1.10	1.92	0.15	2.24
Permanent barley	1.37	1.37	3.36	0.25	5.49

* Fertilizer ($N_{30}P_{30}$) per hectare of crop rotation.

Based on the data on the crop rotations under study and their links, a mathematical model for optimizing the structure of the sown areas has been developed. Data on five crops cultivated in various links of crop rotations are included in the model against the background of $N_{30}P_{30}$ + straw and siderats. The developed model contains 170 variables and 280 constraints.

The areas of individual crops in crop rotations (x_{ik} – the area of the i-th crop in the k-th crop rotation or link), the area of crop rotations (x_k), the total area of arable land, the output of products (Pj – the volume of the j-th type of production), production costs, net income (Yj is the volume of the j-th value indicator; Qj is the volume of the j-th type of resource), energy costs, energy increment, and others are defined as variables.

The maximum production of a j-th product serves as an optimality criterion:

$$Z = \sum_{i \in N_1} \sum_{k \in N_2} b_{jik} * X_{ik} \rightarrow \max \tag{1}$$

where: N_1 – many cultivated crops; N_2 – many crop rotations and their links; b_{jik} – the yield of j-th species from a unit area of the i-th culture of the k-th crop rotation or link.

Obtaining agricultural products is possible under certain conditions described by the system of restrictions. The main restrictions are on the balance of humus, production costs, labor costs, energy costs, energy increment with and without humus. The system also includes restrictions:

- On arable land:

$$\sum_{k \in N_2} X_k = Q_j \qquad j \in M_4 \tag{2}$$

- By crop rotation:

$$\sum_{i \in N_1} X_{ik} <= X_k \quad k \in N_2 \tag{3}$$

- By area of individual crops within crop rotations:

$$\sum_{i \in N_1} X_{ik} = C_{ik} * X_k \quad k \in N_2 \tag{4}$$

- On production resources:

$$\sum \sum a_{jik} * X_{ik} <= Q_j \quad j \in M_4; \quad i \in N_1 \quad k \in N_2 \tag{5}$$

- On production:

$$\sum \sum b_{jik} * X_{ik} => B_j \quad j \in M_5; \quad i \in N_1 \quad k \in N_2 \tag{6}$$

- On production costs:

$$\sum \sum a_{jik} * X_{ik} = Y_j \quad j \in M_6; \quad i \in N_1 \quad k \in N_2 \tag{7}$$

- On gross output:

$$\sum \sum_{i \in N_1 \quad k \in N_2} d_{jik} * X_{ik} = Y_j \quad j \in M_7; \qquad (8)$$

where M_4 – many production resources; c ik – a share of i-th crop in the k-th crop rotation; d ik – a gross output per unit area of the *i*-th crop in the *k*-th crop rotation; a jik – *j*-th resource costs per unit area of *i*-th crop in the *k*-th crop rotation or link; M_5 – many types of products; M_6 – a lot of production costs; M_7 – a lot of gross output; B j – a guaranteed production volume of the *j*-th species.

Currently, the main interest of farmers is to obtain high quality food grains. As a result of solving the problem with the help of the developed model, variants of the structure of sown areas, providing the greatest yield of food grains, were obtained (Table 1, 2).

Table 2. The structure of arable land for obtaining marketable grain at ($N_{30}P_{30}$ and straw) on average for 2010 - 2018.

Rotations Permanent crops	Food grain output, t/ha	Return on grain, %	Energy ratio	Agroecosystem productivity, MJ/dn/GJ	Humus balance, t/ha
Grain fallow rotations					
Steam - wheat - wheat - oats	2.01	87.6	1.84	0.086	-0.78
Steam - wheat - wheat	1.95	94.9	1.72	0.098	- 0.21
Fruiting crop rotations					
Soy - wheat - barley - oats	2.06	91.0	2.01	0.123	+ 0.28
Rape – wheat - barley - soy - oats	2.11	94.9	2.14	0.127	+ 0.15
Permanent sowing					
Permanent wheat	1.10	-12.6	1.34	0.040	- 0.18

Both grain-steam crop rotations and fruit-replaceable cultivation of grain make it possible to obtain from 2.11 to 1.10 t/ha of commercial grain.

4. Discussion

To increase the effectiveness of the use of arable land in the development of animal husbandry, increasing the efficiency of using arable land and introducing into the structure of the area of crops that increase the collection of fodder units and digestible protein is possible (soybeans and grain crops). The introduction of these crops in crop rotations increases the productivity of one hectare of arable land by 20-25% [8].

In general, the introduction of grain crops into crop rotations leads to an increase in grain yield by 0.3-0.5 t/ha, and the KPI increases by 0.5-0.8 tons per hectare of crop rotation area in comparison with spring wheat. The use of intensification tools will increase the productivity of both grain and fruit-replaceable crop rotations. In fruit-replaceable crop rotations, the increase from their use amounted to 0.15-0.22 t/ha of grain, 2.5-3.8 t/ha of digestible protein.

Thus, the proportion of crops should be 67-100%. Using the resulting model will allow one to determine what production is required to ensure the expenditure of material and cash from 1094 to 1636.5 thousand rubles/ha, as well as 23.0-44.6 GJ/ha of total energy and 5.8-10.4 GJ/ha for operating. The profitability of the production of feed units, taking into account by-products, will vary from 17.0 to 43.9%; it is from 87.6 to 94.9% for grain. Only permanently cultivating a culture is unprofitable. The energy efficiency ratio is 1.34-2.14, the agroecosystem productivity indicator is 0.040-0.127 MJ /dn/GJ. The highest return on energy costs (2.14) is noted in the rotation of rape – wheat – barley – soy – oat.

When optimizing the arable land, it is necessary to take into account that the crop formation occurs due to the natural fertility of the soil, as evidenced by the negative balance of humus, which changes by 0.18-0.78 t/ha. Increasing the intensity of land use in alternating-fruit alternation, the crop is formed against the background of the reproduction of soil organic matter by increasing plant residues from fertilizer use, the humus balance was positive, demonstrating 0.15-0.28 t/ha.

5. Conclusion

The introduction of a condition that does not allow a negative balance of humus turns off the possibility of using $N_{30}P_{30}$ + straw, grain-pair crop rotations, where wheat crops are repeated, as they are accompanied by significant losses of humus. Reproduction of soil fertility with a maximum yield of 2.11-2.06 t/ha of food grain is possible with the following structure of sown areas: soybean and rapeseed for seed oil (25-40%) with grain (60-75%). At the same time, the profitability of the production of marketable grain amounted to more than 90%.

Thus, the commodity producer can choose various variants of the structure of sown areas, for example, reducing the share of pure steam to 14.3% and wheat to 43.2%, increasing the share of winter rye and industrial crops (rape, soybean) to 21.3%. This will allow to get 2.15 t/ha of food grain. By increasing the area in crop rotations using straw and colza as a siderate as an organic fertilizer, we can achieve the reproduction of soil fertility with a certain decrease in the yield of marketable grain to 2 t/ha.

6. Acknowledgments

The authors are grateful to the staff of the Agrochemical Service Center at the Omsk Federal State Institution for assisting in the analysis of soil samples for humus content.

References

[1] Blavsky V, Virchenko M, and Shestakova N 2011 Economic-mathematical models in agriculture *Economics* 7 pp 118-123
[2] Penning de Vries F 1982 *Simulation of plant growth and crop production* (Wageningen: Pub Co)
[3] Wit de C T *et al* 1978 Simulation of assimilation and transpiration of crops (Wageningen: Pub Co)
[4] Abramov N and Selyukova G 2001 *Optimization of the structure of acreage on bioenergetic basis* (Ekaterinburg, Russia: Izd-vo USGSHA)
[5] Gilev S, Tsymbalenko I and Surkova Yu 2016 Productivity and economic efficiency of the *korotkorotatsionnyh zernoparovyh* rotations in the central steppe zone of Zauralye *Agriculture* **6** pp 8-11
[6] Yushkevich L and others 2014 *Technological systems of cultivation of grain and leguminous crops: recommendations* (Omsk, Russia: LITERS)
[7] Dospekhov B 1985 A technique of field experiment (with bases of statistical processing of results of researches) (Moscow, Russia: Agropromizdat)
[8] Chibis V V, and Chibis S P 2015 Formation of soil fertility elements during the fertile alternation of field cultures in the forest-steppe zone of Western Siberia *Arable Farming* 7 pp 20-22
[9] Bogoviz A 2019 Transforming the agricultural sector for better sustainable development: Perspectives from Russia as a member state of the Eurasian Economic Union IOP Conference Series: Earth and Environmental Science **274**(1) 012001

IOP Conf. Series: Earth and Environmental Science **395** (2019) 012006 doi:10.1088/1755-1315/395/1/012006

Sustainable development of organic agriculture in the Baikal region, as a cross-border area

O A Altaeva[1*], E G Imeskenova[1] and O M Chibikova[1]

[1] Buryat State Agricultural Academy named after V.R. Pilippov, Ulan-Ude, Russia

E-mail: altaeva_olga@mail.ru

Abstract. The article is devoted to the problem of sustainable development of organic agriculture in the Baikal region, in particular the transboundary territories of the Republic of Buryatia and Mongolia. The article presents a study into the structure of land use in border areas and an analysis of the applied farming systems. The prospects for organic agriculture in the republic are studied and a model of the market for organic products in the Baikal region is suggested.

1. Introduction

The Russian agriculture faces the challenges of not only agrarian and economic problems of increased production, identified in traditional farming systems, but also environmental problems related to environmental protection, environmental management, production of environmentally friendly products safe for health. In this regard, one of the most effective ways to improve the living standard in rural areas, sustainable development of rural areas, and increase the efficiency of production of ecological agricultural products is the development of organic (ecological) agriculture in Russia. Therefore, biologization and ecologization in the Baikal region is recognized as one of the main directions in the development of the zonal farming system and the sustainable development of organic agriculture in the Baikal region as a whole [1, 2].

It implies the adequate provision of the population with ecologically safe food while maximally saving renewable and non-renewable resources, unique natural complexes, and their biological diversity, designing agricultural landscapes to meet environmental requirements, creating an attractive rural environment, greening the processes of obtaining agricultural products that contribute to the successful development of rural areas the main zone of development of agricultural production, using the development of scientifically based methods for the selection of crops and varieties, crop rotation planning, taking into account the effectiveness of organic fertilizers, digital technologies for agro-ecological monitoring of lands, automated technologies in agricultural production [2].

Currently, the problem of sustainable development of organic agriculture in the Baikal region has not been studied enough. The analysis of approaches to solving this problem and the possibilities of implementation in the Baikal region are considered in this article.

2. The Object and Methodology of Research

The Baikal region is located in the southeastern territories of the Russian Federation, bordering Mongolia and China. It covers a total area of 570 thousand km^2. It includes the Lake of Baikal, a water protection zone, a catchment area, specially protected natural areas, as well as the adjacent territory

Content from this work may be used under the terms of the Creative Commons Attribution 3.0 licence. Any further distribution of this work must maintain attribution to the author(s) and the title of the work, journal citation and DOI.
Published under licence by IOP Publishing Ltd

(outside the catchment area), the so-called zone of atmospheric influence. The territory stretches from south-west to north-east along the eastern coast of Lake Baikal. The Republic of Buryatia occupies 73% of this territory. The southern border of Buryatia is the state border of the Russian Federation with Mongolia, in the east it borders on the Trans-Baikal Territory, in the west with the Irkutsk Region, in the extreme south-west with the Republic of Tyva, Russian Federation. The Republic of Buryatia has a common state border with Mongolia in five districts (Okinsky, Tunkinsky, Kyakhtinsky, Dzhidinsky, Zakamensky) [3].

The successful development and formation of cross-border territories is due to the need of establishing mutually beneficial economic ties and cooperation in the social, cultural, and economic spheres between the border areas of neighboring states.

The study of problems in the field of sustainable agriculture in the Baikal region and transboundary territories was carried out using general scientific methods and techniques: a scientific abstraction, analysis and synthesis, comparison, a structural-level approach. The theoretical and methodological basis of the study consisted of the works of national and foreign scientists on the problems of agricultural production of organic products, the development of land relations; scientific research and recommendations of the Russian Academy of Agricultural Sciences, laws of the Russian Federation, Decrees of the President and Decrees of the Government of the Russian Federation, normative legal acts of the subjects of the Russian Federation, Resolutions of the European Union on the development of ecological agriculture, and IFOAM standards.

3. Research Results and Discussion

The development of organic agriculture could be one of the promising areas for Russia as a whole and the cross-border areas of the Baikal region. The analysis shows that the border areas of Russia (Republic of Buryatia) and the adjacent aimaks of Mongolia are similar in the structure of land use, since most of the territory falls on agricultural land. Residential, industrial, and transport land use play a significant role in the Bulgan and Selenge aymaks (the cities of Erdenet, Darkhan and Sukhbaatar) and the adjacent Zakamensk and Kyakhtinsky districts of Buryatia (the cities of Zakamensk and Kyakhta) and constitute a small percentage of economically developed areas [4, 5].

When considering the structure of land use in the border areas, it can be concluded that agriculture dominates with the predominant industry, grazing, where ecological and geographical conditions for the development of traditional environmental management have a great influence.

In the famous research "Border and Transboundary Territories ...", it is stated that "... The structure of agricultural land use in Buryatia is formed by two main types of activity: crop production and animal husbandry. In terms of production, livestock has always been the main branch of agriculture in Buryatia. The development of crop production in Buryatia is primarily subordinated to the interests of animal husbandry. This is especially clearly manifested in the ratio between domestic consumption and government purchases of grain in Buryatia in the gross harvest of grain crops, the crops which in different years occupied from 50 to 90% of the total area of arable land. Obviously, the overwhelming majority of crop production is consumed inside the farms producing these products. The structure of this consumption consists of seeds for crops in subsequent years and the use of crop production as feed for livestock, both public and private. With the almost unchanged area of crops in recent decades, the gross grain harvest varies greatly from year to year. The yield of grain crops in individual years may not exceed an average of 5 centers per hectare, which causes instability of the forage base of animal husbandry in winter. This type of feeding makes livestock farming very risky (lack of food and, as a result worse productive indicators of livestock, its mortality) ..." [5].

The distribution of various types of natural forage lands in the border with Russia of the territory of Mongolia has a rather pronounced geographical pattern. Data on stocks, structure, and dynamics of vegetation indicate a strong decline in the productivity of vegetation in the latitudinal direction [6]. The productivity of natural forage lands in Mongolia depends on weather conditions by years, which is also observed in its regions bordering with Russia.

When analyzing the farming systems that were used in the territory of the republic, as well as in the border areas of Mongolia, it turned out that so far only production systems that mostly depended on the use of mineral fertilizers and pesticides were used in these areas and little or no attention was paid to alternative systems [7].

At the same time, there are prospects for organic agriculture in the territory of Buryatia and they are related to the fact that:

- The territory of the republic is located in the basin of Lake Baikal, which by the decision of UNESCO is classified as a world heritage site and requires a special regime of nature management
- The climate is extreme continental, the soils are low productive, of light grain size distribution, and much of them are subject to erosion processes and degradation;
- Natural forage lands occupy a vast area, and livestock farming is the main branch of agriculture, production of livestock products in the private sector of the republic is increasing [8];
- Perhaps, the wide development of the tourist complex, which will require a significant amount of organic (environmentally friendly) food;
- The use of agrochemicals has decreased dramatically, mainly due to their high cost, as well as the use of organic fertilizers (manure) due to the elimination of livestock farms and the decline in stocks of previous years;
- There is a certain scientific background: various schemes of field and fodder crop rotations have been studied, the effectiveness of applying such legumes as clover and a manure in crop rotations has been considered; siderats, composts, and vermicomposts have been tested in agriculture; a number of studies have been carried out on adaptive-landscape farming systems and nomadic livestock breeding;
- The republic does not have an acceptable strategy of integrating the experience of private, small landowners and large farms in the creation of sustainable livestock production;
- There is a market where organic (organic) food products in demand.

The above indicates that there are certain promises in the republic for the development of organic (ecological) farming systems and the prospects for obtaining organic products. In this regard, we have proposed a model of the market for organic agricultural products in the Baikal region (Fig. 1). Organic farming products in the Baikal region are fundamentally distinguished by a higher level of quality from those produced in the traditional way.

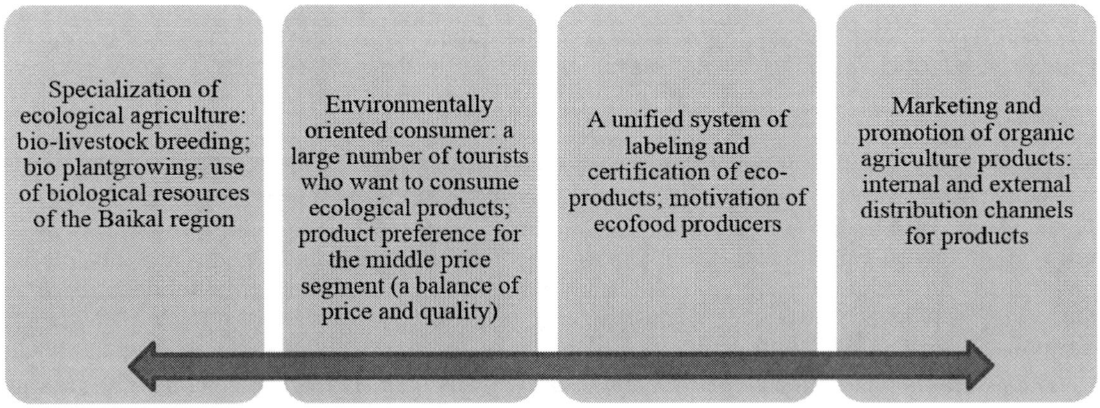

Figure 1. A market model of organic farming in the Baikal region.

Currently, Buryatia is considering ways of transition to organic agriculture. The prospects for organic farming in the Baikal region are being assessed. At the same time, the empty niche of the market for organic (environmentally friendly) products and the considerable land potential for the development of organic farming create all the necessary prerequisites for the formation and development of local agriculture oriented towards the production of organic products.

To ensure the sustainable development of organic agriculture in the territory, it is advisable to use a formalized procedure for analysis, modeling and forecasting, which includes the following steps (Fig. 2).

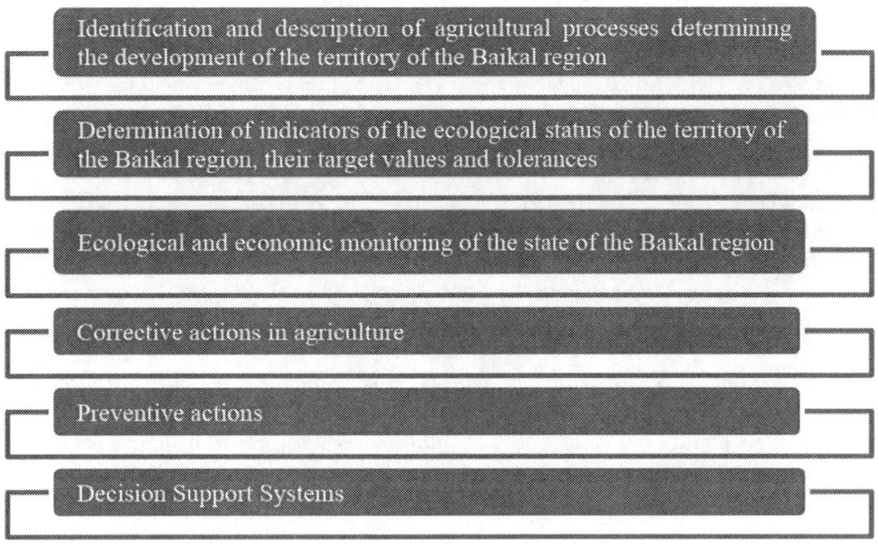

Figure 2. Stages of sustainable development of organic agriculture.

All stages describing the development of organic agriculture in the territory of the Baikal region include the processes of interaction between society and nature and describe the concept of ecological and economic system. This system is a contour formed by the two hierarchical substructures. On the one hand, the economic subsystem affects the environment. On the other hand, the ecological subsystem affects the economic subsystem. The composition of the economic subsystem primarily includes the following elements: agriculture, forestry, water management, environmental protection and quality control, and more.

Thus, sustainable organic farming in the Baikal region is a new stage in the relationship of the producer of agricultural products with the environment. Moreover, the transition to organic farming is a long-term process and the reorientation of agricultural production should be carried out in several directions, among which a special place should be given to personnel training.

4. Conclusions

The agriculture focused on the production of organic products refers to the parallel maintenance of the traditionally established industrial and production system, focused on the production of organic agricultural products, with a gradual increase in the share of the organic sector on the basis of rational, geographically adapted land use with minimum and reasonable use of chemicals in agricultural production. The introduction of methods for sustainable development of the region will allow the use of adapted and environmental technologies for growing high quality crop production in the Baikal region. It will also contribute to the expansion of theoretical knowledge in the field of organic agriculture. It is necessary to create a regional model of organic agriculture, which would promote the use of production technologies in the region built on a local closed cycle (including seeds, processing and consumption). Further, the use and expansion of organic agriculture will lead to the creation of a regional quality mark, the development of convenience stores, the development of social nutrition standards, and the adoption of a regional law on organic products. The creation of this model will favorably affect the development of cross-border areas of the Republic of Buryatia and contribute to

IOP Conf. Series: Earth and Environmental Science **395** (2019) 012006 doi:10.1088/1755-1315/395/1/012006

the establishment of mutually beneficial economic ties and cooperation with the border areas of neighboring states.

References

[1] Diterikh M, and Merzlov AV 2013 *Sustainable development of rural territories* (Moscow, Russia: Ellis Lak)

[2] Imeskenova E, Altaeva O, and Chibikova O 2019 Towards organic agriculture in the Baikal region *Buryat State Agricultural Academy* **1**(54) pp 119-125

[3] Tulohonov A K 1996 *Baikal region: problems of sustainable development* (Novosibirsk, Russia: Nauka)

[4] Gonchikov Ts D *et al.* 2017 Problems of economic regionalization in modern conditions In proceedings of the XII International Scientific Conference *Environment and Sustainable Development of the Mongolian Plateau and Adjacent Territories* (pp. 184-186) (Ulan-Ude, Russia)

[5] A K Tulokhonov, and P Ya Baklanov Eds 2010 *Border and transboundary territories of Asian Russia and neighboring countries: problems and prerequisites for sustainable development* (Novosibirsk: Publishing House of the Siberian Branch of the Russian Academy of Sciences)

[6] Kazantsev T I, and Davaazhamts Ts 1988 *Efficiency of phytocenoses of steppes and deserts of the Mongolian People's Republic: natural conditions, vegetation cover and wildlife of Mongolia* (Pushchino, P.242-256)

[7] Batudaev A P 2018 *Farming system of the Republic of Buryatia: scientific and practical recommendations* (Ulan-Ude, Russia: Izd-vo BGSKHA imeni V.R. Filippova)

[8] Imeskenova E G, and Butuhanov A B 2014 *Rational use of natural grass stands of Buryatia* (Ulan-Udeh, Russia: Izd-vo BGSHA)

Risk analysis of the modern anthropotechnical environment

A Mayakova[1*]

[1] Southwest State University, 94 50 lel Oktyabrya str., Kursk 305040 Russia

E-mail: berryannett@yandex.ru

Abstract. The science and practice offer risk management methodologies that are relevant to modern production, economic and socio-humanitarian realities. Risk management methodologies are successfully developed and implemented in Russia through the creation of scientific and professional communities. Their goal is to solve specific management tasks of identifying, analyzing, controlling, and predicting risks. Russian organizations operating internationally are the first to update foreign risk management methodologies for their managers, and now they are implementing a systematic corporate risk management approach. The world practice also offers standardization in risk management. Such a system successfully operates in the territory of the EU, which is represented by the International Organization for Standardization (ISO). The standards of this organization regulate not only the sphere of risk management, but also the quality management, auditing, production regulation, environmental management, and others. Most of the standards are adapted in the Russian context and are presented in the form of national standards GOST R. The paper considers risk analysis, one of the leading risk management methodologies both in Russia and abroad. We will draw analogies and differences with other modern approaches and risk management methodologies.

1. Introduction

Risk management is a specific management methodology that includes a set of management procedures and methods for solving problems of identification, analysis and evaluation, risk monitoring and information exchange related to risk. The purpose of this methodology is to reduce the costs of an organization and increase its profitability [1]. The risk management objectives cover a whole range of problems of persons and societies: human safety, health protection, environmental safety, certain consequences of financial and economic crises, and others. When considering a risk from the point of view of a particular organization, the task and problem takes the following shape. The problem of risk management is relevant and appropriate, especially in the current conditions of the residual financial and economic crisis and its consequences. Versatile and difficult to formulate risk management tasks involve in their decisions the use of an integrated systems approach, which is contained in the modern methodology of risk analysis [2].

2. Materials and Methods

In the context of this article, it is planned to consider the risk analysis methodology. This methodology is relevant and widely used both in Russia and in the countries of Western Europe. Moreover, the concept of risk analysis is developed precisely in the regulatory documents of the European Union. In this regard, it is advisable to define analogies and differences in approaches and methods of risk management in

Content from this work may be used under the terms of the Creative Commons Attribution 3.0 licence. Any further distribution of this work must maintain attribution to the author(s) and the title of the work, journal citation and DOI.
Published under licence by IOP Publishing Ltd

IOP Conf. Series: Earth and Environmental Science **395** (2019) 012007 doi:10.1088/1755-1315/395/1/012007

Russia and abroad, as well as to argue the relevance and priority of risk analysis and its regulatory documents in comparison with other methods and procedures of risk management.

3. Results

The global practice of applying risk analysis is focused not only on its use in a particular organization, but also on the state level. Thus, the governments of the countries of the European Union have always focused their attention on the safety of citizens, in particular on risks. But lately, risk management has gradually occupied a central position in the activities of the state. As in Russia, in international and European standards, the concept of risk is used to describe a variety of problems and threats. The contemporary society is in constant development, which is accompanied by the constant presence of risk in human life and society. This fact determines the global meaning of the problem of risk management and justifies its relevance: reduction of external risks associated with various kinds of security (health care, physical, and financial security), providing a free access to quality public services [3].

At the same time, the fact that there is a risk at the state level (unforeseen political and economic incidents, the threat of disruption of state programs and projects) will not be disputed. Such uncertainty is not new. Characteristics of risk are subject to change for two main reason. First, the general rapid development of science and technology at the moment is accompanied by constant technological risks [4]. Second, interstate and domestic risks arise from the integration of the global economy, the creation of common communication systems and the generation of common environmental problems. The close interrelationships of the global infrastructure characterize systemic risks: adverse events occurring at various points on the globe affect its inhabitants more than before. At the moment, such risks have a reasonable priority over others. In this regard, the governments of many developed and developing countries are focusing on improving the risk management methods used.

Recently, the situation of mass replication of international, including European standards, has been observed; these standards regulate the procedure for managing risks of man-made factors [1, 5, 6, 7]. In Russia, the document "Risk Management of Organizations. Integrated Model," developed by the Committee of Sponsoring Organizations of the Treadway Commission, received the biggest attention [8]. The regulatory document reflects the conceptual framework and risk management methodology within specific organizations and includes clear recommendations for creating an intra-organizational risk management system. The risk management methodology of a COSO organization consists of eight elements: the definition of the internal environment; setting goals and objectives; risk identification; assessment of each risk; risk "response"; controls; information and communications; monitoring [9].

In this regard, the COSO regulatory document presents a risk management methodology from the perspective of the process approach known in risk management and quality management [10]. Despite the clear methodology of the presented document, Russian managers and specialists have adopted the risk management standard of the Federation of European Risk Management Associations as the basis for risk management methodology in Russia. One of the advantages of this standard over the COSO document is the more acceptable terminology adopted in the documents of the International Organization for Standardization [11]. The concept of "risk" is defined as "a combination of the probability of an event and its consequences" [11]. Risk management is a central part of the organization's strategic management and includes not only a set of measures and procedures for identifying and assessing risks, but also programs for monitoring and minimizing risks.

The risk analysis methodology is such a unified risk management system. The main standard of the Russian Federation regulating the risk analysis methodology is the adapted standard GOST R 51901-2002 "Reliability Management. Risk analysis of technological systems" [1]. Many industrial enterprises have introduced this methodology into their risk management activities; however, many still try to do without it due to the complexity and systematic nature of its implementation. For example, the complexity of implementing risk analysis in the service sector lies in the uncertainty of business processes that must be documented in a risk analysis program.

Let us consider this methodology in more detail. The process of implementing a risk analysis is divided into two major subprocesses. The first is to identify and assess the scale of the risk to be analyzed and managed. This subprocess is called the stage of characterization of risk (nonconformity). The second subprocess involves a detailed risk assessment and the development of a set of measures to minimize and eliminate it. This subprocess is defined as the decision stage. However, the risk analysis process is cyclical and reversible; therefore, a return to the risk characterization stage is possible in the event of a new threat at the decision-making stage.

Risk identification is a meaningful procedure in the decision-making process for risk and strategic planning in general. In most cases, attention is focused on the process of risk analysis itself to the detriment of solving a global problem, an integrated approach to eliminating and minimizing risks on a large scale. Concentration on any one threat leads to the destruction of the entire risk management system. The solution to this problem should be comprehensive, opening the relationship of threats to each other. Only then risk management will be effective. Such a "slogan" underlies the risk analysis methodology.

According to the GOST R 51901-2002 "Reliability Management: Risk analysis of technological systems", risk analysis is a structured process. The purpose of this process is to determine both the likelihood and the size of the adverse effects of the studied action, object, or system [11]. The feature of risk analysis is that this process is structured and contains clear rules for conducting. Moreover, risk analysis necessarily involves the compilation of a risk matrix, in which the interrelations of risks and their assessment are clearly traced. So, risk analysis is divided into the following steps:

1. Determining the scope (process, complex processes, the activities of the organization, region, country, etc.);
2. Risk identification (risk type), which is followed by a potential risk assessment (SWOT, FME (C) A analysis, FTA analysis, and others);
3. Qualitative / quantitative risk assessment (degree of risk influence; probability of risk occurrence);
4. Analysis of the impact of risk on a specific scope (process, complex processes, organization's activities, region, country, etc.);
5. Development of a set of measures to minimize and reduce a risk;
6. Compilation and analysis of the risk matrix.

4. Discussion

According to many experts in the field of risk management (Neumann, G. K., Grace, P., Burns, D., Surridge, M.), "looping" only on critical risks is not the right solution [12]. The situation in all areas of risk must be stabilized through a policy of minimizing risk factors. Naturally, potential threats will remain, especially macroeconomic ones. However, the organization should conduct its activities based primarily on sound strategic planning, one of the components of which is risk analysis. At the same time, risk analysis is a complex procedure that must be carried out periodically. Otherwise, the efficiency and effectiveness of risk management in an organization will be close to zero. The frequency is set individually, depending on the number and characteristics of the risks identified during the first procedure.

5. Conclusion

In conclusion, it should be noted that the systematic integrated risk analysis within the framework of risk management in an organization provides the following opportunities:

1. Improving the organization's strategic planning;
2. Modernization of decision-making policy, which is coupled with a set of specific measures to minimize and eliminate a risk;
3. Improving the organizations' competitiveness in the market;
4. Revaluation of resources, more successful use of positive factors of organization;

5. Effective management of potential risks and inconsistencies;
6. Cost reduction in the production of products and services;
7. Flexibility and mobility in making decisions regarding the organization's activities and the risks associated with it;
8. Development of the organization's innovative activity.

The global practice of risk management is contained in the developed regulatory documents, which contain elements of risk management. The EU Risk Management Recommendations contain a general requirement, which is supported by the Russian Federation, of a detailed expert analysis. This expert analysis is carried out by the employees engaged in risk management activities both within the same organization and at the state and intergovernmental levels. Risk analysis supports this initiative. As can be seen from the risk analysis procedure, the whole methodology is based on the principle of expert analysis. Along with expert analysis, international standards for risk management are similar in the dynamics of this process, its systematicity and consistency, focus on strategic goals, transparency and meaningfulness. Consequently, effective risk management allows one not only to carry out daily activities more efficiently, but also to plan and make strategic decisions on important issues at various levels. Risk analysis within risk management is a documented process that enhances the traceability of decision making.

6. Acknowledgments
Grant of the President MK-240.2019.6 "Transdisciplinary Model of Socio-Cultural Risk Management."

References
[1] Risk management. Risk analysis of technological systems. Official edition (GOST 51901-2002)
[2] Risk management – Principles and guidelines on implementation. Official edition (ISO/DIS 31000:2010)
[3] Knigh K W 2003 *Risk Management: an integral component of corporate governance and good management* (Geneva, Swirtzeland: ISO Bulletin)
[4] Budanov V G n.d. New digital life technoscience – prospects and risks of anthroposphere transformation *Philosophical Sciences* **6** pp 47-55
[5] Interstate Standard Dependability in Technics (27.310-95). Destructive tests on welds in metallic matrials – Etchants for macroscopic and microscopic examination (ISO/TR 16060:2003)
[6] Safety of machinery and equipment. Principles for safety ensuring while designing. Official edition (GOST 54125-2010)
[7] Risk management. Terms and definitions. Official edition (GOST 51897-2011)
[8] Committee of Sponsoring Organization of the Treadway Commission (COSO) 2004 Enterprise Risk Management – Integrated Framework Executive Summary
[9] Committee of Sponsoring Organization of the Treadway Commission (COSO) n.d. Risk management organizations. Integrated model. Summary.
[10] Mayakova A V 2016 Quality of management in the context of modern economic and managerial paradigm. management in reengineering and redesign. *Ekonomicnij casopis-XXI (Economic Annals-XXI)*, **157** (3-4(1)) pp 82-84
[11] Risk management – Vocabulary – Guidelines for use in standards (IDT). Official edition (ISO Guide 73:2009)
[12] Neumann G K, Grace P, Burns D, and Surridge M 2019 Pseudonymization risk analysis in distributed systems *Journal of Internet Services and Applications* **10**(1) pp 32-36

An anthropogenic load of copper and somatic health

I Khlusov[1*], G Slepchenko[2] and E Moiseeva[2]

[1] Siberian State Medical University, 2 Moskovsky tract, Tomsk 634050 Russia
[2] National Research Tomsk Polytechnic University, 30 Lenina avenue, 634050 Tomsk Russia

E-mail: khlusov63@mail.ru

Abstract. The issues of good health and well-being for people have been reflected in the UN Sustainable Development Goal No. 3. The presented paper focuses on the increase in the copper content in the body by 3-4 times within 3-4 months. It is established that such an increase is not accompanied by a change in the biochemical parameters of blood and the structure of somatic pathology characteristic of the population of Tomsk, who are not engaged in industrial production.

1. Introduction

An analysis of a person's hair and nails is considered the most appropriate way for a mass non-invasive assessment of the elemental status and homeostatic capacity of metals [1]. According to A.V. Bgatova et al. [2], any disease can be seen as a consequence of a violation of the internal environment (homeostasis) of the body. At the same time, the official boundaries of the "norm" on the content of chemical elements in the hair of people are not established [3], particularly due to their dependence on the region of human residence [4]. Due to the large scatter of values, the poorly regulated definitions of "recommended levels" and "biologically acceptable levels" [3-5] are proposed. In workers of industrial enterprises, the amount of metals in the body can be increased by 20-50 times according to the methodical recommendation of the USSR Ministry of Health for professional activities (1988). However, there is no data at which excess of the optimal content in the body vital elements become toxic. Dismikroelementozes are distributed in the Russian Federation as a whole and in the Siberian region in particular, having an impact on a low level of population health [1].

The paper presents the results of the analysis of the content of zinc and copper in workers of the industrial enterprise of Tomsk. These elements in certain quantities belong to the biotic group, in the absence of anthropogenic human exposure. Potential linkages of essential (vital) metals with known markers of natural aging of the population have been studied.

2. Materials and Methods

Examination of employees was carried out as part of a preventive medical examination. The examination of healthy volunteers was carried out with the permission of the Ethics Committee (conclusion No. 583 of March 19, 2007) of the Siberian State Medical University (Tomsk). The subjects were informed on the features of the conducted diagnostic procedures in accordance with ethical requirements of the men and women employed in the production of Methanol OJSC (Tomsk, Russia). The hair was cut off at 3-5 points with a thickness of 2-3 mm^2. The total volume of the sample was about 100 mg. Quantitative chemical analysis of hair samples for zinc and copper content is based on a stripping voltammetric

Content from this work may be used under the terms of the Creative Commons Attribution 3.0 licence. Any further distribution of this work must maintain attribution to the author(s) and the title of the work, journal citation and DOI.
Published under licence by IOP Publishing Ltd

IOP Conf. Series: Earth and Environmental Science **395** (2019) 012008 doi:10.1088/1755-1315/395/1/012008

method for determining the mass concentrations of elements in a prepared sample solution [6]. Blood biochemical indices (concentration of glucose, cholesterol, lipoproteins of low (LDL) and high (HDL) density) were detected by standard methods according to the instructions. Semi-quantitative assessment of the state of parenchymal organs was carried out in points according to ultrasound investigation (USI). The absence of structural changes in the parenchyma was taken as 1 point. In the case of detected organic changes in each organ (liver, gallbladder, kidney, pancreas, thyroid, and prostate), an additional score was added.

The results were expressed as mean (X) and standard deviation (SD), median (Me), 1st (Q1) and 3rd (Q3) percentiles. The statistical processing of the results was carried out by the methods of variation statistics using the Mann-Whitney criterion and the Spearman rank correlation coefficient.

3. Results

Analysis of the type of distribution (universal criteria of Lilliefors and Shapiro-Wilk) of the quantitative contents of zinc and copper in the hair of men and in women showed that the results obtained for zinc can be considered close to the normal distribution (Fig. 1a, b).

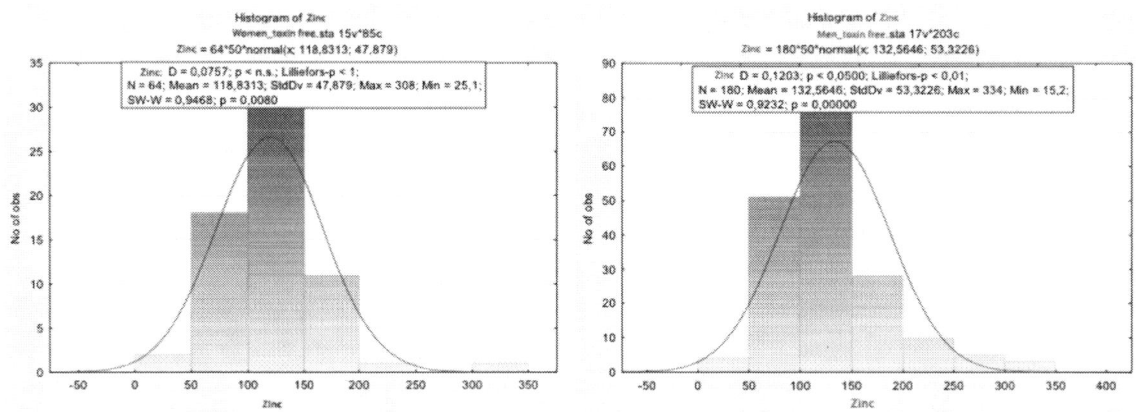

(a) (b)

Figure 1. Histogram of zinc distribution in hair: (a) in women; (b) in men.

When studying the distribution of copper in the hair, a deviation from the normal distribution was noted. The type of distribution obtained indicates a strong impact of external sources of metal in the body (Fig. 2a, b). Considering the types of distribution, the values of elemental load in the hair of the employees of the enterprise were described, both as an average value (X±SD) and as values of centile concentration intervals.

A professional examination of the company's employees (Table 1, 2) showed a statistically significant 3-4-fold excess of the regional indicator for copper concentration in the hair of both men and women. Sexual differences in the accumulation of copper were not detected (Table 1). Therefore, the groups were combined for comparison with the corresponding indicator for the population not engaged in the production of methanol. In the blood, the concentration of cholesterol and its low density fraction (LDL) exceeded the recommended level (Table 2). According to ultrasound data, only 11% of the surveyed employees of the enterprise, mostly young and middle-aged, do not have structural changes in the internal organs.

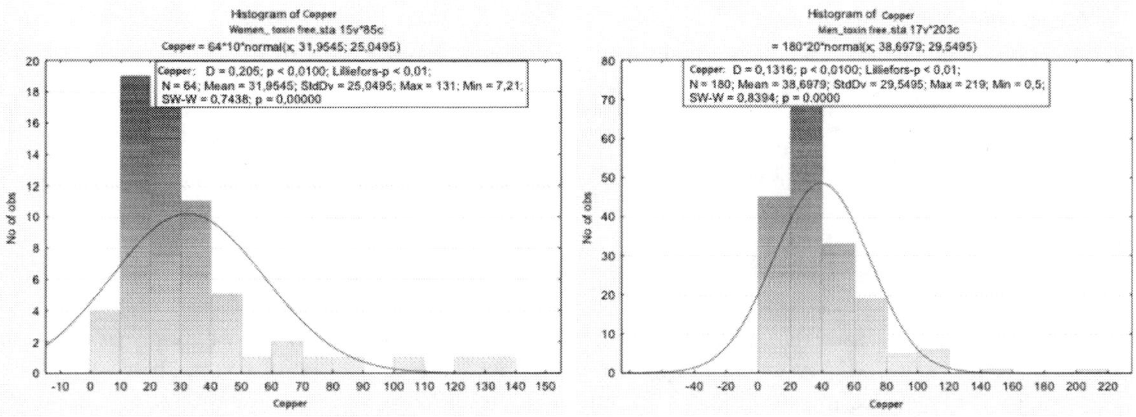

(a) (b)

Figure 2. The histogram of the copper distribution in the hair of (a) women; (b) men.

In other cases, multiorgan (2-3 internal organs) structural changes (Table 2), first of all, the gallbladder (53% of those examined), kidneys (41%), liver (36%), pancreas (6%), were detected.

Table 1. The level of trace elements in the hair of the surveyed, X±SD, Me(Q1-Q3).

Group	Age (years)	Zinc (mg / kg)	Copper (mg / kg)
Recommended levels for men and women [1, 2, 4, 5]	-	140-280 100-250	6.5-15 7.5-80
Male workers, N=180	43.7±11.4	132.6±53.3	38.70±29.52 31.2(20.0-52.4)
Female workers, N=64	42.0±8.5	118.8±47.8	31.95±25.04 24.1(16.1-38.3)
Employees of the enterprise (men and women), N=244	43.3+10.8	129.3±52.2	37.30±28.54* 29.35(19.10-48.90)
Men and women who did not work in the enterprise, N=58	30.6±12.4	113.2±52.9	10.26±10.17 6.05(4.60-12.2)

The predominance of diseases of the excretory systems of the body (biliary excretion, urination) suggests a chronic intake of copper in the body.

Table 2. Biochemical blood parameters (mM) in different categories of patients, X±SD.

Group	Glucose	Cholesterol	LDL	HDL	LDL / HDL	Pathology of organs points
Recommended levels [7]	3.5-6.1	< 5.0	< 3.0	>1-1.2	< 3.0	1.0
Male workers	5.06±1.06 n=174	5.92±1.05 n=176	3.75±0.92 n=173	1.34±0.27 n=176	2.93±0.91 n=173	2.64±1.05* n=173
Female workers	4.92±0.89 n=65	5.96±1.02 n=62	3.64±0.86 n=61	1,62±0,39 n=61	2.43±1.02 n=61	3.27±1.28* n=64

Correlation analysis allowed us to establish a direct link between the age of the company's employees and biochemical markers of aging: in men, with all the blood parameters studied (r=0.17-0.44; p<0.03; n=168-171), except for HDL; also in women (r=0,27-0,42; p<0,04; n=61-62), except for HDL and glucose levels. At the same time, no significant correlations of Zn and Cu content in the hair and blood biochemical parameters were found in the employees of the enterprise (men and women).

4. Discussion

Excess copper in the body can be associated with an increase in its release from the environment or with reduced excretion. An elevated level of the trace element testifies, first of all, in favor of the pathology of the liver (hepatitis, the risk of cirrhosis) and kidneys [5]. However, organ pathology among employees of the enterprise as a whole did not differ from that among the population of the Tomsk region of comparable age. Therefore, the main reason was the occupational hazard associated with the constant use of a copper catalyst in the methanol production process. The direct correlation of copper-zinc ($r = 0.26$; $p < 0.001$; $n = 244$) in the group of surveyed employees of the enterprise indicates a chronic supply of copper from the external environment. Although, these metals in normal conditions (state of rest) are the competitors [4].

It should be emphasized that the relatively healthy volunteers of the city of Tomsk, who participated in the survey, also do not have a correlation of copper and zinc concentrations in the hair with age. An increase in the level of cholesterol and its fractions in the blood is a well-known marker of the natural aging processes.

5. Conclusion

Thus, a 3-4-fold increase in the copper content in the body of workers of an industrial enterprise for at least 3-4 months is not accompanied by changes in the biochemical blood parameters and the structure of somatic pathology characteristic of the population of the region.

6. Acknowledgments

The work was funded by the State Task (science) No. 1.143.2014.

References

[1] Skalnyy A V, and Bykov A T 2003 *Ecological and physiological aspects of the use of macro-and micronutrients in restorative medicine* (Orenburg, Russia: RIK GOU OGU)

[2] Bgatov A V, Bgatov V I, and Novoselova T N 1999 The phenomenon of lithophagy *Natural Minerals in the Service of Human Health: A Collection of Scientific Works* **276** pp 146-154

[3] Skalnyy A V, Gorbachev A L, and Veldanova M V 2004 *Elemental status of children in Northeast Russia* (Orenburg, Russia: RIK GOU OGU)

[4] Skalnyy A V, Bykov A T, Serebryansky Ye P, and Skalnaya M G 2003 *Medico-ecological risk assessment of hypermicroelementoses in the population of a megacity* (Orenburg, Russia: RIK GOU OGU)

[5] Lifshits V M, and Sidelnikova V I 1998 *Biochemical analyzes in the clinic: a Handbook* (Moscow, Russia: MIA)

[6] *Hair: the stripping voltammetric method for measuring mass concentrations of zinc, cadmium, lead, copper, iron, arsenic, manganese, nickel, selenium* (FP.1.31.2006.02273)

[7] Russian recommendations of the Committee of Experts of the All-Russian Scientific Society of Cardiology 2004 *Diagnosis and correction of lipid metabolism disorders for the prevention and treatment of atherosclerosis* (Moscow, Russia)

IOP Conf. Series: Earth and Environmental Science **395** (2019) 012009 doi:10.1088/1755-1315/395/1/012009

Contribution of the forests of the Republic of Buryatia to the development of a "green" economy

S Puntsukova[1*]

[1] Baikal Institute of Nature Management SB RAS, 6 Sakhyanova str., Ulan-Ude 670047 Russia

E-mail: puntsukovas@mail.ru

Abstract. The article presents the results of a carbon balance study in the forests of the Republic of Buryatia based on the materials taking into account the state of the forest fund and key factors (logging, forest fires) that affect carbon absorption and emission. The research clearly demonstrates that the size of the carbon budget in the forests of the republic is determined by the current age structure of forests and the intensity of destructive disruptions because of carbon losses due to logging, forest fires, and the death of plantations. The carbon absorption by forests is characterized by relative stability. Despite the fact that the forests of Buryatia have a great potential for carbon absorption (deposition), the last one decreases in the context of increasing clear-cutting and fire exposure to forests.

1. Introduction

In international documents, a "green" economy is considered in a general way as an economy that improves people's well-being and ensures social justice; at the same time, it significantly reduces risks to the environment and its impoverishment [1]. The forest sector in the implementation of the goals of a "green" economy takes a leading role due to the fulfillment of a complex of functions: economic, social, and environmental. The multifunctional nature of the country's forest complex most fully corresponds to the principles of a "green" economy. Multifunctional character is expressed in the production of not only wood and non-wood products, but also in the provision of vital environmental and social services, such as biodiversity conservation, climate protection, soil, catchment basins, employment. Therefore, there is the task of measuring the contribution of the forest sector to the development of a low-carbon, environmentally friendly, and socially inclusive economy.

In December 2013, at a joint session of the Committee on Forests and Forest Sector of the UN Economic Commission for Europe and the European Forestry Commission of the UN FAO in the Finnish city of Rovaniemi, the Action Plan for the forest sector for the transition to a green economy was developed [2]. Our research objectives are based on the components of this Plan and on the mechanisms for measuring the contribution of the forest sector to the development of a green economy developed by the UNEP, OECD, World Bank, and other organizations collaborating in the framework of the "Green Growth Knowledge Platform" [3]. The work carried out in this area at the global level is consistent with efforts to ensure sustainable forest management, in particular with the criteria and indicators of the Montreal and Helsinki processes for assessing the achievement of sustainability.

According to component B "The low carbon forest sector Action Plan for the forest sector to transition to a green economy," the contribution of forests and the forest sector to climate change

Content from this work may be used under the terms of the Creative Commons Attribution 3.0 licence. Any further distribution of this work must maintain attribution to the author(s) and the title of the work, journal citation and DOI.
Published under licence by IOP Publishing Ltd

IOP Conf. Series: Earth and Environmental Science **395** (2019) 012009 doi:10.1088/1755-1315/395/1/012009

mitigation is manifested through sequestration, carbon storage, as well as substitution and adaptation to it. However, this contribution is not uniform at the regional level. It depends on such factors as the forest area, species, age composition of plantations, the volume of logging, the effectiveness of fighting forest fires, etc.

2. Materials and Methods

The assessment of the contribution of forests and the forest sector of the Republic of Buryatia (RB) to the "green" economy was based on an assessment of the carbon budget. For its assessment, the methodology for regional assessment of the forest carbon budget (ROBUL) developed by the Center for Problems of Ecology and Forest Productivity (CEFP) of the Russian Academy of Sciences and Moscow State University was used [4]. The ROBUL methodology estimates the carbon budget of forests by the flow balance, i.e. by the difference in absorption with increasing carbon pools in forest plantations and losses due to disturbances (logging, fires and other cases of forest death). ROBUL calculations are carried out for 4 main pools of the forest carbon: phytomass, dead wood, litter, and soil organic matter. The choice of this technique is due to the fact that for the calculations of the carbon budget, the materials of the State Forest Register (SFR) are used. This technique has been tested by experts of the UNFCCC ONN and is used in the National Greenhouse Gas Cadastre of the Russian Federation for reporting on the forest sector.

The main source of information on forest resources in the RB is the SFR database as of January 1, 2007 and 2017. The increase in the forest fund area of the republic by 1179.8 thousand hectares in 2017 (compared to 2007) is due to the transfer of former collective and state farm forests to the Rosleskhoz after 2007 (Table 1).

Table 1. Dynamics in the structure of forest land of the Republic of Buryatia.

Land category	2007	%	2017	%
Forested land, thousand hectares	19472.9	75.4	20493.6	75.9
Uncovered land, thousand hectares	852.2	3.3	1019.6	3.8
including woodland	509.2	2.0	569.5	2.1
fires and dead stands	198.9	0.8	290.6	1.1
felling	80.1	0.3	110.5	0.4
Non-forested land, thousand hectares	5505.9	21.3	5497.6	20.4
including arable land, hayfields, and pastures	602.3	2.3	601.4	2.2
swamps	321.4	1.2	315.9	1.2
other lands	4324.9	16.7	4327.3	16.0
All lands, thousand hectares	25831.0	100.0	27010.8	100.0

3. Results

Calculations of the carbon balance in the forests of the RB show that carbon absorption in all pools of the forest ecosystem is characterized with relative constancy. In 2007-2017, it was 9.205 Mt C/year-^{1}. Carbon losses, in contrast to carbon uptake, are characterized by inter-annual variability. Carbon losses from logging have increased from -0.761 Mt C/year^{-1} in 2007 up to -1.078 Mt C/year^{-1} in 2017 as a result of a 1.4-fold increase in cutting area. Losses from forest fires increased since -0.945 Mt C/year^{-1} in 2007 up to -1.417 Mt C/year^{-1} in 2017 as a result of a 1.5-fold increase in the area of fires and dead stands during the period under review (Table 2).

During the study period, the carbon sink in the forests of the RB, which consists of the balance of absorption and losses from logging and forest fires, decreased from 7.344 Mt C/year^{-1} in 2007 to 6.865 Mt C/year^{-1} in 2017.

Table 2. The carbon budget in the forests of the Republic of Buryatia.

Carbon flow	Flow rate, 10^6 t C year $^{-1}$	
	2007	2017
Absorption	9.051	9.360
Losses from clear cutting	-0.761	-1.078
Losses from forest fires	-0.945	-1.417
Budget	7.344	6.865

4. Discussion

First of all, carbon absorption in forests is due to an increase in the area and volume of clear cutting and forest fires. The data characterizing the key factors influencing the carbon balance of forests testify to this (Table 1). During the period from 2007 to 2017, a slight increase in the forest covered area (by 0.5%) was due to an increase in the uncovered area of the forest due to an increase in the area of burning areas and dead plantings by 0.3% (as well as in felling by 0.1%). The increasing uncovered area as a result of logging suggests that the effect obtained during the economic reforms of the early 1990s in Russia, which led to a sharp reduction in timber harvesting, has exhausted itself. So, on the territory of the republic, a total number of 3,757 thousand m^3 of wood was cut in 1990, and the figure dropped to 405 thousand m^3 in 1998 (in 9 times) (Fig. 1). Then, a gradual growth of clear cutting is observed. They harvested 1094 thousand m^3 in 2016, which is 15.7% of the annual allowable volume of timber removal (estimated cutting area).

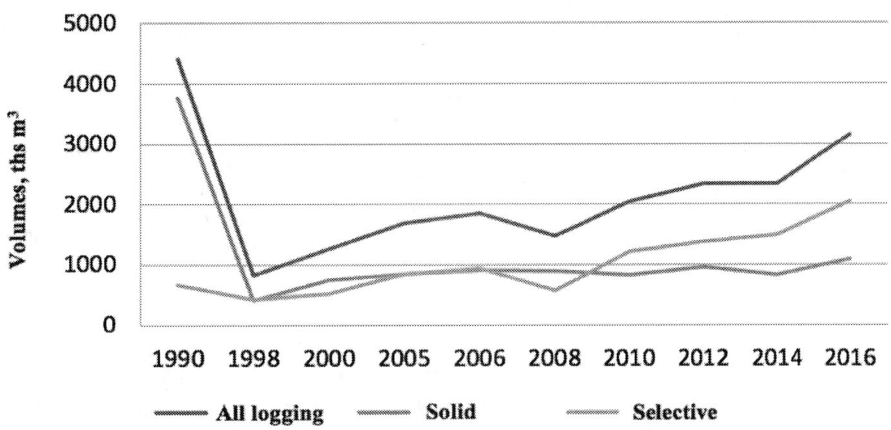

Figure 1. Dynamics of wood harvesting in all types in the Republic of Buryatia, 1990-2016.

The slower growth rate of continuous logging compared with selective logging is due to the introduction of environmental restrictions on the Baikal Natural Territory (BNT) since the end of the last century. The Republic of Buryatia, which is largely part of the BNT, is one of the country's leaders in the environmental regulation of economic activities and the creation of the foundations for environmentally sustainable development. In the region, special standards for permissible impacts on the natural ecosystems of the Lake Baikal and the basins of the main watercourses (Selenga, Upper Angara, etc.) have been developed and approved. Extensive work on the functional ecological zoning of the territories of the central ecological, buffer zones of BNT has been carried out, prohibiting certain activities [5, 6]. Thus, in the region, a number of conditions necessary for the introduction of the principles of a "green" economy into economic activity have already been created.

Environmental restrictions on forest management are expressed in increasing protective forests due to operational ones. In protective forests, depending on the ecological zones of BNT (central or buffer), clear cutting is prohibited or limited, including due to a decrease in the size of cutting areas compared to other regions of the country. Environmental constraints increase requirements for logging operations, the need to introduce environmentally friendly forestry equipment and technology [7].

Thus, the increase in protective forests due to environmental constraints at the BNT inhibits the growth of clear cutting. This contributes to the fact that the removal of wood by clear cutting is going at a slow pace, which leads to an increase in forest areas. Good natural regeneration in logging, characteristic of the Baikal region, also has a positive effect on this process [8].

Forest fires are another key carbon budget impact on forests. Over the past ten years (2007-2016), 10,822 forest fires occurred on an area of 1797.6 thousand hectares, while the average area covered by forest fires for the year was 179.7 thousand hectares [9]. Peaks of large forest fires occurred in 2003 (157.5 thousand hectares), 2009 (242.4 thousand hectares), and 2012 (129.3 thousand hectares). In 2015, the largest fires occurred both in quantity (1,574) and in the area of burnt forests (890 thousand hectares) for the entire observation period. Regarding the catastrophic summer fires in 2003, 2009, and 2015, we can say that the reasons for establishing a high level of forest fires were the imposition of two factors: extreme weather conditions (lack of precipitation and abnormal heat) and lowering of the water level of the Baikal Basin. As a result, the non-accumulation of moisture in the soil, accordingly, accelerated drying of the ground cover and caused an increase in the mass of dry combustible materials in the forest occurs [10].

The deteriorating situation with forest fires is associated both with the problems of organizing and financing the protection of forests and with global climate change, which affects the growth of the number of dangerous hydrometeorological phenomena. Such phenomena include periods of hot and dry weather, creating conditions for catastrophic fires. According to the state service Roshydromet, the number of hydrometeorological hazards has increased by 2 times over the past 15 years [11]. According to one of the climate change scenarios based on the general circulation model of the atmosphere, the duration of the fire hazard season in the middle latitudinal belt of Russia may increase by 50-60 days, i.e. in 30-40% [12].

On the other hand, the prevalence of young and middle-aged stands in the age structure of forests (together they make up 51.2%) contribute to the fact that the value of carbon sequestration by forests has high and relatively stable values. And the presence of a sufficient amount of mature and overmature stands in the forest fund (more than 38%) reduces the amount of carbon deposition. Since, according to experts, ripe and overmature stands of all species are considered the least productive ones, because it gives a minimum increase in wood and phytomass as a carbon accumulator [13].

5. Conclusion

Breed and age composition, structure of forest lands, nature of forest use, organization of forest protection from fires affect the size of the forest carbon budget. Analysis of the carbon budget has shown that the forests of Buryatia have a great potential in the absorption (deposition) of carbon. Despite the gradual reduction of carbon sinks, the forests of Buryatia make a positive contribution to the preservation of the global climate.

In the future, an increase in logging and fire exposure to forests can inevitably lead to an increase in carbon emissions into the atmosphere. Therefore, additional measures to improve the management of both forests and forest management are necessary in order to minimize the negative effects of climate change and to effectively use its possible benefits. The most promising forms of activities to preserve the existing carbon stocks and increase the absorption of carbon dioxide from the atmosphere by forests are the intensification of prevention and control of forest fires, changes in the technologies and regimes of forest use, afforestation of outdated agricultural land, protective forestation in arid areas, etc.

References

[1] UNEP 2011 *Towards a green economy: towards sustainable development and poverty eradication - synthesis report for authorities* Available at: www.unep.org/greeneconomy (Accessed 22 02 2019)
[2] UNECE 2014 *The Rovaniemi Action Plan for the forest sector in developing a green economy* Available at: https://www.unece.org/fileadmin/DAM/timber/publications/SP35R.pdf (Accessed 22 02 2019)
[3] UNECE 2014 *Measuring the contribution of the forest sector to the development of a green economy and providing relevant information* Available at:

http://www.unece.org/fileadmin/DAM/timber/efsos/general/ECE_TIM_2014_5.pdf (Accessed 22 02 2019)

[4] Zamolodchikov D G, Grabovsky V I, and Kraev G N 2011 *Regional forest carbon budget assessment (ROBUL) Version 1.1.* Available at: www.cepl.rssi.ru/regional.htm (Accessed 20 09 2018)

[5] Government of the Russian Federation 2001 *Resolution "On approval of the list of activities prohibited in the central ecological zone of the Baikal natural territory"* (August 30, 2001 No. 643) Available at: http://pravo.gov.ru/ proxy / ips /? docbody = & nd = 102072593 (Accessed 22 02 2019)

[6] State Duma of the Russian Federation 1999 *Federal Law "On the Protection of Lake Baikal"* (January 1, 1999 No. 94-FZ, as amended on June 28, 2014) Available at: http://docs.cntd.ru/document/901732256 (Accessed 22 02 2019)

[7] Puntsukova S D 2012 Formation of an economic mechanism for sustainable forest management in regions with environmental restrictions (Novosibirsk, Russia: Nauka)

[8] Potapov L V, Shagzhiev K Sh, and Varlamov A A Ed 2000 *Buryatia: a conceptual framework for sustainable development strategie* (Moscow, Russia: All year)

[9] Government of the Republic of Buryatia 2008 *Resolution* (October 5, 2017 No. 567) *"On Amendments to the Resolution of the Government of the Republic of Buryatia dated December 31, 2008 No. 608 "On Approval of the Forest Plan of the Republic of Buryatia"* Available at: http://www.alh-rb.ru /documents/section.php?SECTION_ID=90 (Accessed 09 12 2018)

[10] Tulohonov A K, and Puntsukova S D 2016 Forest fires in the Republic of Buryatia in a changing climate *Society: Politics, Economics, Law* **3** pp 72-78

[11] Roshydromet 2010 *The 2009 report on climate features in the territory of the Russian Federation .:* Available at: www.meteorf.ru (Accessed 09 12 2018)

[12] Vorobev Yu L, Akimov V A, and Sokolov Yu I 2004 *Forest fires in Russia: state and problems* (Moscow, Russia: DEKS-PRESS)

[13] Kobak K I, Kukuev Yu A, and Treyfeld R F 2001 The role of forests in the changing carbon content in the atmosphere *Forestry* **2** pp 43-45

[14] Alekseev A N, Lobova S V, Bogoviz A V, and Ragulina, Y V 2019 A critical review of Russia's energy efficiency policies in the construction and housing sector *International Journal of Energy Economics and Policy* **9**(4) pp 166-172

IOP Conf. Series: Earth and Environmental Science **395** (2019) 012010 doi:10.1088/1755-1315/395/1/012010

Experience in studying and assessing hydrological risks in the transboundary conditions of the Selenga River

T A Borisova[1*]

[1] Baikal Institute of Nature Management, Siberian Branch of the Russian Academy of Sciences (BINM SB RAS), 6 Sakhyanovoy str., Ulan-Ude 670034 Russia

E-mail: tabor@binm.ru

Abstract. The article presents the results of a study of hazardous hydrological phenomena on the example of floods in the transboundary basin of the Selenga River. Based on the analysis of long-term observations of water levels, hazard indicators are identified. This allows to calculate a number of risk indicators. In terms of specific indicators, an integrated risk map has been developed. An assessment of the total economic risk from floods is given, and recommendations for their reduction are proposed.

1. Introduction

The transboundary basin of the Selenga River is the most developed and populated territory of both Buryatia (Russia) and Mongolia, where the river plays a very significant role in the socio-economic development of states. On the one hand, it has common natural boundaries with the conditions for the formation of certain processes and phenomena, on the other hand, different priorities of modern economic development, and "vulnerability windows" that are specifically formed on it.

Certain specifics of natural conditions with climatic extremes and the characteristic periodicity of dry and wet years contribute to the occurrence of quite significant hydrological changes, including the occurrence of extreme events. Typically, catastrophic events spanning the entire basin have rather rare occurrences. But looking at the dynamics of emergency statistics according to the main criteria of scale, its frequency of manifestation, damage caused, we can distinguish two main types and their alternation as a result of abnormal climatic phenomena: high water and related floods, as well as low water.

Moreover, in recent decades, the cause of their activation and development of qualitatively new species is the increasing anthropogenic impact. The construction of the Irkutsk Hydroelectric Station and the commissioning of the Angarsk Hydroelectric Station, resulting in the rise of the level of Baikal Lake, led to a modification of hydrological conditions and increased the development of flooding processes in the lower part of the basin. In addition, in recent years, the construction of the following hydropower plants in the Selenga Basin has been seriously considered: Shuren (245 MW), Egiyn-gol (315 MW), Orkhon (100 MW), and "Chargayt" (25 MW), as well as the "Orkhon Water Complex" project, with specific proposals for creating artificial water pools and sending them to the South Gobi belts. In this case, the water level regime in the Selenga River (periods of low water and high water), which is the main river carrying its waters to Lake Baikal, will directly depend on the technical parameters. These phenomena pose a tremendous danger to sections of the middle and lower reaches of the Selenga River, i.e. the Russian part of the basin.

Content from this work may be used under the terms of the Creative Commons Attribution 3.0 licence. Any further distribution of this work must maintain attribution to the author(s) and the title of the work, journal citation and DOI.
Published under licence by IOP Publishing Ltd

Thus, the regulation of relationships between subjects requires a comprehensive and thorough study of the research object, analysis of the occurrence and development of natural and anthropogenic processes in the spatio-temporal dynamics leading to the creation of emergency situations. In this regard, the transboundary basin of the Selenga River is of particular interest for the study of risks in order to minimize them, the possible elimination of international conflicts, the sustainable development of natural and man-made systems.

2. Materials and Methods

Based on the specifics of the natural environment, the features of the territorial organization of the two states' territories, an indispensable condition for researching risks in the transboundary basin is the creation of uniform methodological methods for assessing them to obtain objective quantitative indicators. The results of these studies could serve as an information base for the management system. At the same time, efficiency largely depends on the accuracy and reliability of information about the main sources and recipients of negative impacts, probable future losses, as well as its informatization, which significantly increases the efficiency and literacy of decisions.

Based on the existing experience in studying risks, the accepted theoretical and methodological principles, it is advisable to apply a probabilistic-area approach based on hazard indicators (repeatability, intensity of occurrence of the event and the area of damage), with methodological methods for calculating risk indicators through vulnerability. The calculation of this parameter allows one to get the specific physical risk used in the comparative assessment when mapping and identifying the level of risk. The definition of a physical risk indicator serves as the basis for assessing economic and social risks [1], [2]. An end-to-end process of all work is mapping using GIS technologies and remote methods, which play a key role in the ongoing study.

In the process, the entire volume of primary information and hazard indicators is systematized and compiled into a single database, which allows one to develop scenarios and calculate the total economic risk. In the analysis and calculation of hazard indicators, long-term hydrological data were used for 21 water meter posts in dynamics in Russia since 1931 and for 12 water meter posts in Mongolia. The cartographic base was a vector topobase of scale 1:100000; the high-altitude basis was a digital terrain model, the initial data for the formation of which were the relief contour vector, elevations, and depths of a scale of 1:50000. When projecting a high-resolution satellite image scene onto the DEM, a digital terrain model of high metric accuracy was created. Within the boundaries of settlements, a basis of a scale of 1:25000 is used.

3. Results

The formation of floods in the rivers of the basin depends on climatic factors and the orography of the basin. For all observation points under consideration, analysis of the source data and calculated hazard indicators shows the complex process of floods in the Selenga River basin. The difference in the range of values is very significant. Their repeatability varies from 0.1 to 0.9, middle ones are found with a probability of 0.1-0.3, and extreme ones are found with a probability of 0.05-0.1. According to the studied hydrological posts, the depth of flooding is from 50 to 437 cm. The maximum values are typical for the period of summer floods. Exceptional rises are observed in certain sections of rivers. In conditions of mountainous terrain and large slopes or narrow valleys, they are maximum for rivers; in areas of flattened terrain and floodplain development, they are significantly reduced. At the same time, the size of flooding sharply increases. Frequent recurrence of significant excesses of the water level is most characteristic of the middle-lower flow, starting from the confluence of the Orkhon River, then at the confluence of the Dzhida and Chikoi Rivers. Thus, in the formation of floods in the basin, the decisive influence is exerted directly by the Selenga River and the Orkhon River, located on the territory of Mongolia. And its main tributaries dumping their waters, such as Dzhida, Chikoi and less often Khilok, Uda, significantly enhance the development of floods [1], [3]. The GIS technology served as a modern tool for the implementation of the task of mapping the floodplain and identifying flood zones.

With its help, in the Arc GIS software environment, necessary areas are defined, as well as the settlements and important economic objects that fall into the flood zone [4], [5].

In accordance with the methodology, a total of three forecast scenarios of flood development have been developed. And necessary hazard indicators are defined for each. The generated database made it possible to calculate physical risk. In terms of specific indicators, the territorial ranking is carried out; private maps and a final map of the integrated risk from floods in the transboundary Selenga River basin are developed [6]. The highest risk level was identified in Russia, in the Republic of Buryatia (Fig. 1).

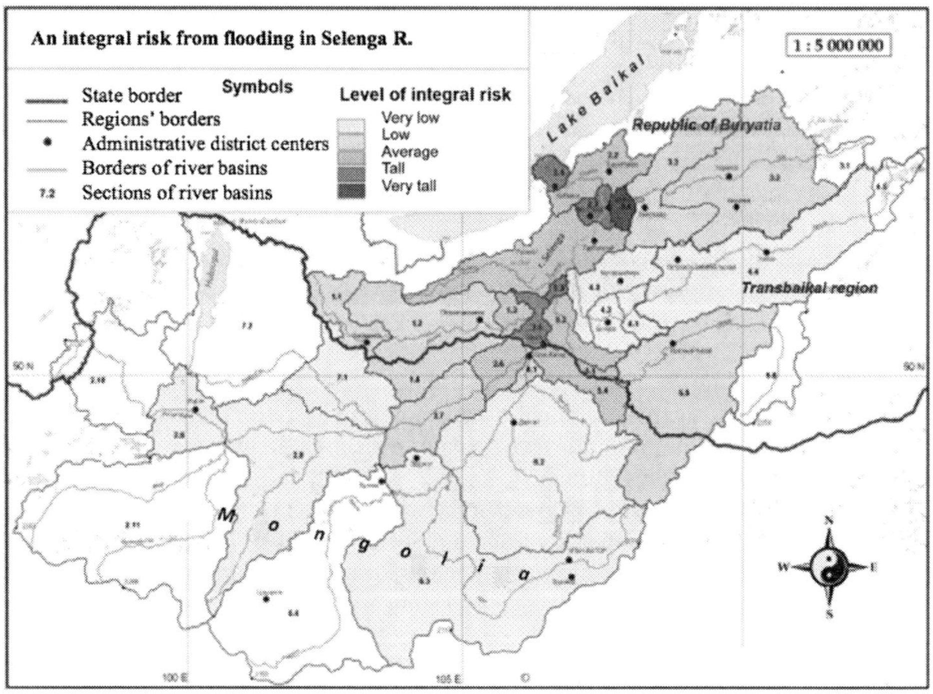

Figure 1. Integral flood risk in the Selenga River basin.

Economic risks are fulfilled for the Russian part of the basin and the territory of the Republic of Buryatia. Also they are identified in the future as possible damage in the absence of protective measures. The calculations were performed using specific indicators of the methodology for assessing probable damage from the harmful effects of water and evaluating the effectiveness of preventive water management measures [7]. To date, indexation of the cost of damage is given using deflator indices and producer price indices by type of economic activity.

The results show that the floods of 480.6 km^2/year affect the basin. High rates are noted on the Selenga river, which is 48%, i.e. almost half of all flooded lands. Its tributaries give the rest. The values are significant for the Chikoi, Uda, and Dzhida rivers basins and relatively small for the Khilok River. The agricultural industry is suffering enormous losses. Annually, 35.9 thousand ha of the agricultural land is temporarily withdrawn from circulation. With the most massive flooding on the Selenga river, direct damage to the republic is estimated at more than 11.2 billion rubles; 0.8 billion rubles for the Chikoi river; 0.7 billion rubles for the Uda river; and 0.3 billion rubles for the Dzhida river. Huge damages within the settlements are noted directly on the Selenga river. Damage to business facilities is more than 1.8 billion rubles per year; damage to the population is 153.7 million rubles, 81% of which is damage to the city of Ulan-Ude, which is practically not protected from floods today.

4. Discussion

In order to reduce damage, an analysis of the possibilities of minimizing risks from the negative impact of water was carried out. A protection system as the most reliable measure is considered. The effectiveness of constructing protective structures is economically justified for large settlements with area affected zones, such as Ulan-Ude, or for those settlements which have protective structures, but they are in disrepair. For most small settlements, carrying out protective measures is not economically feasible. In this case, the option of relocating residents from hazardous areas is preferable, but it also requires serious government participation. It should be noted that the technical flood protection system involves the infusion of significant financial resources of the state, not only provided for construction but also for the trouble-free maintenance of hydraulic structures in the future.

One of the alternative tools to reduce natural risks is the development of an economic mechanism. The introduction and development of a system of compulsory insurance of natural (flood) risks and the formation of an insurance fund, as well as the development of tax incentives, will reduce the financial burden on the state and provide certain financial guarantees to certain sectors of the economy and the population in case of emergency. At the same time, an important component is to take into account the degree of danger of the territory (frequency of floods), the object value, and the lost product.

In modern economic conditions, insurance should become the most effective risk management tool. Improvement of the compensation mechanism is possible subject to the interaction of the state and the insurance company.

5. Conclusion

Thus, the conducted risk assessment from floods in the transboundary Selenga river basin using this methodology can serve as an objective basis for effective management and can be applied to a number of other dangerous processes and phenomena of natural and anthropogenic genesis. The created geographic information system is a technical tool and provides the ability to quickly solve a number of problems at different topological levels and develop evidence-based recommendations.

In conclusion, we can add that the result of the entire study should be the identification of optimal ways to develop an integrated risk management system for the transboundary basin territory and the development of relevant recommendations. For this, the solution of the following most important tasks is necessary:

- Developing interstate programs, intergovernmental agreements on the basis of uniform international legal documents, which should include and take into account all existing problems associated with potential risks;
- Defining the positions of norms and rules on the functional organization of the territory and its further development;
- Creating a unified geographic information space of a transboundary basin;
- Organizing system monitoring based on digital elevation models, space, statistical and operational information;
- Defining a rational version of a single organizational and functional structure as the main governing body in emergency situations;
- Building awareness of actions to eliminate conflicts between states.

6. Acknowledgments

The study was financially supported with a grant from the Russian Science Foundation with the support of the Republic of Buryatia (the project No. 18-45-030020).

References

[1] Borisova T A 2013 *Natural and anthropogenic risks in the basin of Lake Baikal* (Novosibirsk, Russia: “Geo” Publishing House)
[2] Ragozina A L Ed 2003 *Natural hazards of Russia: Assessment and management of natural risks* (Vol 6) (Moscow, Russia: “KRUK” Publishing Company)
[3] Vaskovsky M G Ed 1973 *Surface water resources of the USSR: basin of Lake Baikal (Transbaikalia)* (Vol 16, Issue 3) (Leningrad, USSR: Hydrometeoizdat)
[4] Kauffeldt A, Wetterhall F, Pappenberger F, Salamon P, and Thielen J 2016 Technical review of large-scale

hydrological models for implementation in operational flood forecasting schemes on continental level *Environmental Modelling & Software* **75**(1) pp 68-76

[5] DHI Software n.d. *Modeling the world of water: DHI water & environment* Available at: http://www.dhisoftware.com (Accessed 02 02 2019)

[6] Borisova T A, Kichigina N V, Beshentsev A N, and Davaa G 2014 *Floods: Ecological atlas of the Lake Baikal Basin* (Irkutsk, Russia: Institute of Geography SB RAS)

[7] *Methods for assessing probable damage from the harmful effects of water and assessing the effectiveness of the implementation of preventive water management measures* 2006 (Moscow, Russia: "VIEMS")

IOP Conf. Series: Earth and Environmental Science **395** (2019) 012011 doi:10.1088/1755-1315/395/1/012011

The influence of the cultivation technology elements on the economic performance of the bare barley grain production

N Rendov[1], A Gladkikh[1] and E Nekrasova[1*]

[1] Omsk State Agrarian University named after P. A. Stolypin, 1 Institutskaya square, Omsk 644008 Russia

E-mail: ev.nekrasova@omgau.org

Abstract. At the educational and experimental field of the Omsk State Agrarian University, the evaluation of the influence of elements of the cultivation technology of bare barley of the Omsk golozernyy variety 2 (chemical background, sowing time, and seeding rate) on the indicators of the economic efficiency of grain production was carried out. On the background without the use of chemicals, as well as when using herbicides on the background of applying nitrogen fertilizer, the cost of grain per hectare is higher when sowing the crop in the third decade of May. The use of only herbicides (without fertilizer) provided a greater cost of grain at an earlier sowing. Among the studied seeding rates, the maximum cost indicators were observed when sowing 4.5 million viable seeds per hectare. The chemical background also influenced this indicator. The cost of grain per 1 hectare increased by 3,227 rubles with the use of a tank mixture of herbicides, as well as by 4,513 rubles with their complex use with fertilizers (in comparison with the chemical background). At the same time, the strengthening of the chemical background led to an increase in costs per 1 hectare by 1,559 and 3,543 rubles, respectively. An increase in the seeding rate was also accompanied by an increase in costs. The use of herbicides did not lead to a significant change in the cost of grain, but the additional use of nitrogen fertilizer led to its growth. The level of profitability in the version with the use of a tank mixture of herbicides coincided with the option without the use of chemicals (113.1%). The application of nitrogen fertilizer reduced profitability to 81.5% despite the fact that their use was justified by an increase in grain yield. Thus, when considering specific options for the cultivation of bare barley, sowing on May 14-18 with a rate of 4.5 million viable seeds per hectare while using the Puma Super 7.5 herbicides tank mixture, EMV (0.9 l/ha) and Turbo Pruner, MD (75 ml/ha) should be allocated.

1. Introduction

Grain is of strategic importance and is the basis of the country's food security. Barley can be distinguished among the main grain crops of Russia. In a number of regions, particularly in the Southern Urals [1], Northern Trans-Urals [2], and Novosibirsk region [3], the spread of bare barley is hampered by the possibility of reducing its productivity compared with foamed barley, first of all. With the creation of new varieties of bare forms, the differences in yield are smoothed out [4-6]. At the same time, it is generally accepted that bare barley is superior to filmy barley in terms of the protein, starch, fat, and β-glucanate content [7].

At present, the efficiency of agricultural production largely depends on the choice of its production technology [8]. Thus, at the current level of prices for grain, fertilizers, and pesticides, there is a need

Content from this work may be used under the terms of the Creative Commons Attribution 3.0 licence. Any further distribution of this work must maintain attribution to the author(s) and the title of the work, journal citation and DOI.
Published under licence by IOP Publishing Ltd

to develop a technology for the cultivation of crops, including bare barley, which would ensure high economic efficiency of production.

2. Materials and Methods

In 2011-2014, in the teaching and experimental field of the T&E farm of the Omsk State Agrarian University named after P. A. Stolypin, an assessment of the elements of the cultivation of bare barley varieties of the Omsk Golozornyy 2 was carried out. The teaching and experimental field is located on the soil (meadow-chernozem, medium-power, low-humus, medium loamy) in the southern forest-steppe of Western Siberia.

The crops were located in a crop rotation with the scheme: clean steam – wheat – wheat – barley. The three-factor experience included:

Three chemical backgrounds:

1) Without chemicals (0);
2) Herbicides Puma Super 7.5, EMV - 0.9 l/ha and Turbo Pruner, MD – 75 ml / ha (G);
3) Herbicides + fertilizers at the rate of N_{60} (G + U).

Three sowing times:

1) May 14–18;
2) May 25–28;
3) 4–6 June.

Three seeding rates:

1) 3.5 million germinating grains per hectare;
2) 4.5 million germinating grains per hectare;
3) 5.5 million germinating seeds per hectare.

The flow rate of the working fluid in the processing of crops with herbicides is 200 l/ha.

3. Results

The formed yield level of bare barley in the variants of the experiment allowed to get the results on the following indicators of economic efficiency: cost of grain (Table 1), costs (Table 2), cost of barley grain (Table 3), conditional net income (Table 4), and the profitability of production (Table 5).

Table 1. Cost of grain of bare barley (2011-2014), rub./ha.

Chemical background	Sowing date			Seeding rate			An average chemical background
	I	II	III	3.5	4.5	5.5	
O	12540	13020	11520	11780	12660	12640	12360
G	16440	15160	15160	14780	16120	15860	15587
G+U	16960	17300	16360	16140	17220	17260	16873
The average	15313	15160	14317	14233	15333	15253	14940

Table 2. Costs in the cultivation of bare barley (2011-2014), rub./ha.

Chemical background	Sowing date			Seeding rate			An average chemical background
	I	II	III	3.5	4.5	5.5	
O	5664	5842	5756	5436	5671	6155	5754
G	7392	7289	7289	6933	7344	7664	7313
G+U	9302	9321	9268	8921	9317	9653	9297
The average	7443	7484	7438	7097	7444	7824	7455

Table 3. The cost of grain of bare barley (2011-2014), rub./t.

Chemical background	Sowing date			Seeding rate			An average chemical background
	I	II	III	3.5	4.5	5.5	
O	2783	2691	2998	2775	2764	2932	2882
G	2689	2884	2886	2816	2738	2905	2820
G+U	3294	3233	3400	3317	3247	3362	3309
The average	2922	2936	3095	2969	2916	3066	2984

Table 4. Conditional net income in the cultivation of bare barley (2011-2014), rub./ha.

Chemical background	Sowing date			Seeding rate			An average chemical background
	I	II	III	3.5	4.5	5.5	
O	6725	7178	5763	6343	6839	6485	6556
G	9078	7870	7871	7847	8776	8196	8273
G+U	7658	7979	7092	7219	7903	7607	7576
The average	7820	7676	6909	7136	7839	7429	7468

Table 5. Profitability level in the cultivation of bare barley (2011-2014), %.

Chemical background	Sowing date			Seeding rate			An average chemical background
	I	II	III	3.5	4.5	5.5	
O	115.6	123.2	100.4	116.6	117.4	105.3	113.1
G	123.3	108.0	108.2	113.1	119.4	106.9	113.1
G+U	82.3	85.6	76.7	80.9	84.8	78.8	81.5
The average	107.1	105.6	95.1	103.5	107.2	97.0	102.6

4. Discussion

Indicators of the grain costs varied depending on the factors studied in the experience. So, without the use of chemicals and for all seeding rates on average, the cost of grain in the first term was 12,540 rub./ha (Table 1). When sowing a decade later, the cost of grain increased to 13,020 rubles/ha. A similar trend was observed when using herbicides against the introduction of nitrogen fertilizer. The use of only herbicides was more effective in early sowing (16,440 rub./ha). Among the studied seeding rates for all chemical backgrounds, the maximum indicators of grain cost were observed when using 4.5 million viable seeds per hectare (15,333 rub./ha). Depending on the chemical background, differences in the cost of the obtained grain were also observed. So, when using a tank mixture of herbicides, the cost of grain from 1 hectare increased by 3,227 rubles, and when used in combination with fertilizer – by 4,513 rubles without the use of chemicals).

However, strengthening the chemical background led to an increase in costs per 1 hectare, i.e. by 1,559 and 3,543 rubles, respectively (Table 2). An increase in the seeding rate was also accompanied by an increase in costs. Assessing the cost of cultivating bare barley, one should pay attention to their structure. Without the use of chemicals, the share of technological costs was 72.0%, the share of the remaining costs (seeds) was 28.0%. When using herbicides, they consume 15.6% of all costs. Naturally, the share of expenses for seeds (22.1%) and the share of technological costs (62.3%) decreased. Additional costs for the purchase and application of fertilizers amounted to 18.4%, which reduced the share of technological costs to 52.0%, as well as to 17.4% for seeds and 12.2% for herbicides.

The ratio of the cost of grain and costs allows you to judge the cost of production of grain (Table 3). If the use of herbicides did not lead to a significant change in this indicator, the additional use of nitrogen fertilizer led to an increase in the cost of grain, in all cases, including in relation to the

IOP Conf. Series: Earth and Environmental Science **395** (2019) 012011 doi:10.1088/1755-1315/395/1/012011

background without the use of chemicals. On average, the chemical background in the variants with the joint use of herbicides and fertilizers, the cost of 1 ton of grain amounted to 3309 rubles, which is higher than the background without the use of chemicals and the background with herbicidal treatment by 427 and 489 rubles, respectively. However, due to the higher cost of the obtained grain, the net income exceeded the control results not only against the background of herbicides (by 1,717 rub./ha), but also when they were combined with fertilizers (by 1,020 rub./ha) (Table 4).

In terms of profitability, the use of a tank mixture of herbicides led to the same level as the option without using chemicals – 113.1% (Table 5). The application of nitrogen fertilizer reduced profitability to 81.5% despite the fact that their use was justified by an increase in grain yield. At the same time, taking into account all chemical backgrounds, sowing at a rate of 4.5 million viable seeds per hectare was more profitable. Among the studied sowing dates, the results of sowing in the second decade of May stood out on this indicator. The profitability in that case was 1.5% higher than the sowing in the third decade of May, as well as 12.0% higher than the profitability of the June sowing period.

5. Conclusion

When considering specific options for cultivating bare barley, sowing on May 14–18 with a rate of 4.5 million germinating grains per hectare against the background of the use of the Puma Super 7.5 herbicide tank mixture, EMV (0.9 l/ha) and Turbo Pruner, MD (75 ml/ha) should be highlighted. Here, the maximum indicators of the profitability level (131.0%) and conditional net income (9,700 rub./ha) were obtained, with a minimum cost of 1 ton of grain being at the level of 2,596 rubles.

References

[1] Gryaznov A A, and Lojkova A V 2010 *Bare barley in the Southern Urals* (CHGAA)

[2] Yakubyshina L I, Vydrin V V, and Fajzullina G N 2014 Stability of spring barley yield in various areas of the Tyumen region *Vestnik GAU of Northern Zauralye* **4**(27) pp 30-32

[3] Gomasko S K, Kapinos A I, Stadnik A T, and Stepanenko O Ya 2015 Efficiency of spring barley varieties in the conditions of the forest-steppe of the Novosibirsk Priobye *Vestnik NGAU* **2**(35) pp 25-31

[4] Aniskov N I, and Krolevets S S 2008 Study of naced barley cultivars of the VIR World collection under conditions of the Siberian Irtysh river region *Russian Agricultural Science* **34** pp 293-295

[5] Bogoviz A, Sandu I, Ryzhenkova N 2018 Scientific basis of innovative development of the agro-industrial complex: The mechanism of public-private partnership in venture capital investment *MATEC Web of Conferences* **212** 07006

[6] Bajkalova L P, and Serebrennikov Yu I 2018 *Bare barley and oat in Siberia* (KrasGAU)

[7] Newman R K 2009 *Barley for food and health – scince technology and products* (USA)

[8] Bogoviz A V , Bugai Y A, and Minenko A V 2019 Economic analysis of effectiveness of the existing tools of state support for entrepreneurship in the AIC in the digital economy *Lecture Notes in Networks and Systems* **57** pp 789-794

Cattle leukemia in the territory of the Irkutsk region

A S Batomunkuev[1*], A A Sukhinin[2] and P I Evdokimov[3]

[1] Irkutsk State Agrarian University named after A. A. Ezhevsky, Irkutsk region, Molodezhny township 664038 Russia
[2] Buryat State of Agriculture named after V. R. Philippov, 8 Pushkin str., Ulan-Ude 670024 Russia
[3] St. Peterburg State Academy of Veterinary Medicine, 5 Chernigovskaya str., St. Peterburg 196084 Russia

E-mail: aldar.batomunckuev@yandex.ru

Abstract. The article presents an analysis of the results of diagnostic studies on cattle leukemia in the Irkutsk region for 2004-2017. It has been established that infection with the cattle leukemia virus was 3.7 ± 0.4%; the number of hematological sick animals was 4.2 ± 0.2%. During the analyzed period, the percentage of infected livestock decreased ($p < 0.01$) in both public and individual sectors. The number of hematological animals ranged from 2.3 to 6.1% (with peaks in 2007 and 2015) and was characterized by a slight decrease (by 0.9%).

1. Introduction

Leukemia is an important social and veterinary-biological problem, which is pretty widespread throughout the world [1, 2]. Importation of pedigree cattle in the 40s-60s of the 20th century caused leukemia in the Russian Federation. The level of infection with the cattle leukemia virus (LV cattle) varies considerably in different territories of Russia. Leukemia poses a potential hazard to the gene pool of breeding dairy cattle. The great economic damage caused to animal husbandry by leukemia is related to the culling of animals compromised in relation to the infection, utilization of carcasses, producing young animals for meat, as well as the costs of carrying out recreational activities [3]. This forces many countries to adopt national programs to combat this disease.

2. Materials and Methods

Assessment of the situation of cattle leukemia in the Irkutsk region for the period 2004-2017 was carried out using the reporting forms of the Irkutsk Interregional Veterinary Laboratory (IVL), comprised according to the results of diagnosing leukemia (immunodiffusion reaction in an agar gel (RID) on the basis of the IVL and eight stations for combating animal diseases. Also, assessing the situation on cattle leukemia in the Irkutsk region was carried out according to the results of enzyme-linked immunosorbent assay (ELISA), polymerase chain reaction (PCR), and hematological studies.

During 200402017, cattles from the public and private sectors of the Irkutsk Region, bulls-producers from the State Breeding Association and farms, animals for breeding and procurement, as well as young animals (n = 3481731) were investigated by the serological reaction of the RID. 24206 serum samples for detection of post-infectious antibodies were examined by ELISA; 100 samples of pathological material and 311 blood samples from cattle were examined by PCR. The volume of hematological studies was 173,315, including 15,213 in "leukoformulas." Statistical data processing

Content from this work may be used under the terms of the Creative Commons Attribution 3.0 licence. Any further distribution of this work must maintain attribution to the author(s) and the title of the work, journal citation and DOI.
Published under licence by IOP Publishing Ltd

was carried out in accordance with standard methods [4]. Differences between the compared parameters were considered statistically significant at $p \leq 0.05$. We determined long-term dynamics using the straight-line equation and the regression coefficient b. Significance of these dynamics was determined using the Pearson coefficient r.

3. Results

According to serological studies (RID) for the period 2004-2017, an analysis of the epizootic situation of cattle leukemia in the Irkutsk region showed that the average long-term infection of the cattle leukemia virus (LV) was 3.7±0.4%. Infection of cattle in the public sector (agricultural enterprises) exceeded ($p < 0.01$) the same indicator in the sector of individual animal owners. Among young animals, infection with the cattle LV at the age of 1 to 2 years was higher (3.4 ± 0.5%). The average long-term infection rate of bulls-producers was 1.1 ± 0.1% (Figure 1).

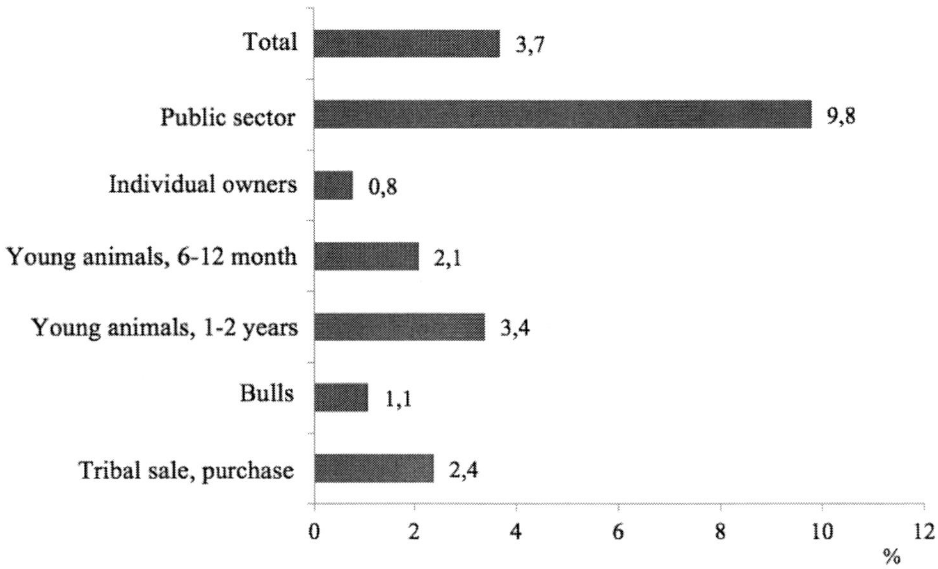

Figure 1. Animals infected with the cattle leukemia virus according to the RID results in the Irkutsk region in 2004-2017 (average long-term indicators), %.

In the long-term perspective, we can note the following: (1) a significant reduction in the infection of cattle in the public (r=-0,806) and individual (r=-0,864) sectors (Figure 2a), (2) young animals of two age groups (r=-0,556 and r=-0,872 respectively) (Figure 2b), as well as animals for breeding and procurement (r=-0,581) identified (Figure 2 c).

In 2004-2005, the highest infection rate of both public and private animals was observed (at the level of 12% and 2%, respectively). In 2017, the lowest infection rate of cattle in the public sector was found (5.4%). Since 2012, the infection of cattle in the private sector did not exceed 0.3-0.4%.

During the analyzed period, the infection of young animals aged from 6 months to 1 year decreased in more than 2 times; and it was in 5 times at the age of 1 to 2 years. The highest percentage of young animals reacting to RIDs (aged from 6 months to 1 year) was recorded in 2004, 2009, and 2013 (over 3%); from 1 year to 2 years – in 2004 (more than 7%), 2005-2007 (at the level of 5%).

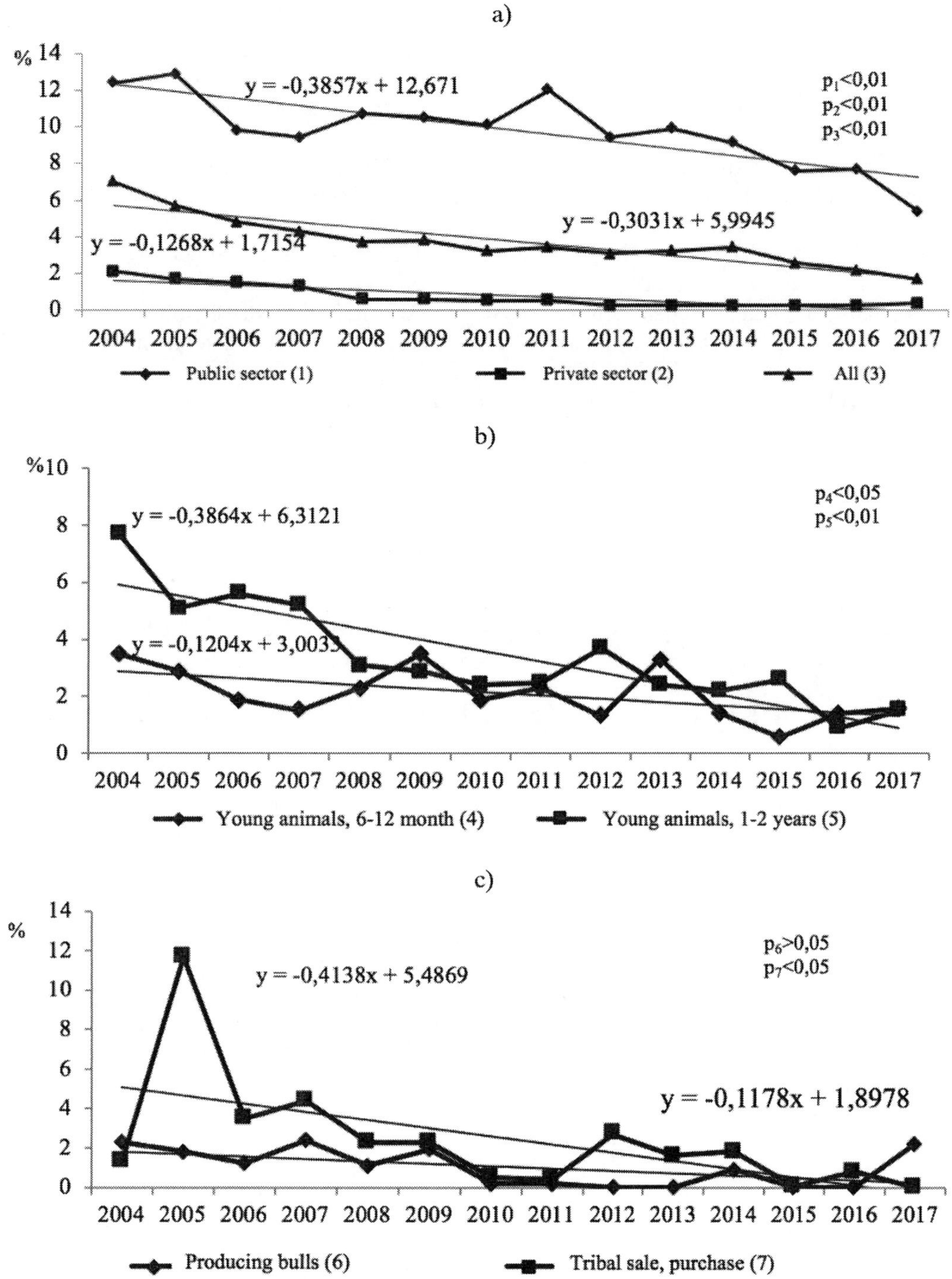

Figure 2 a, b, c. Long-term dynamics of infection with the cattle leukemia virus in the Irkutsk region.

Infection rates among the bulls-producers ranged from 1.1% to 2.4% in 2004-2009. In subsequent years, there was a decrease (at the same time, the bulls-producers being infected with the leukemia virus were not detected in 2012-2013 and 2015-2016), but an increase in this indicator was noted again in 2017.

In 2005, the highest percentage of animals responding to RID for breeding and procurement was identified (11.7%), which was associated with high infection of cattle in Usolsky (27.4%) and Cheremkhovsky (34.4%) districts, mainly because a number of breeding farms in these areas sells animals. In the following years, the cattle infection rate for breeding sales and purchases decreased. Moreover, no infected animals were detected in 2017.

Post-infectious antibodies to the leukemia virus were detected in the serum from cattle by the method of enzyme-linked immunosorbent assay (ELISA) in 42.0% of cases. In the study of samples of the pathological material of cattle by the method of polymerase chain reaction (PCR), all results were negative; in the study of blood samples, positive results amounted to 3.5%, including in 9 cases (3.2%) in 2010 and in 2 cases (7.6%) in 2015.

When analyzing hematological studies, it was established that the number of hematological diseased animals was 4.2±0.2%; suspicious for leukemia – 1.4±0.1%; a total number of sick and suspicious heads for the disease – 5.6±1.2%.

During the observation period, the number of hematological patients varied ($p > 0.05$) from 2.3 to 6.1% (with peaks in 2007 and 2015) and was characterized by a slight decrease by 0.9% in 2017 (if compared with 2004) (Figure 3).

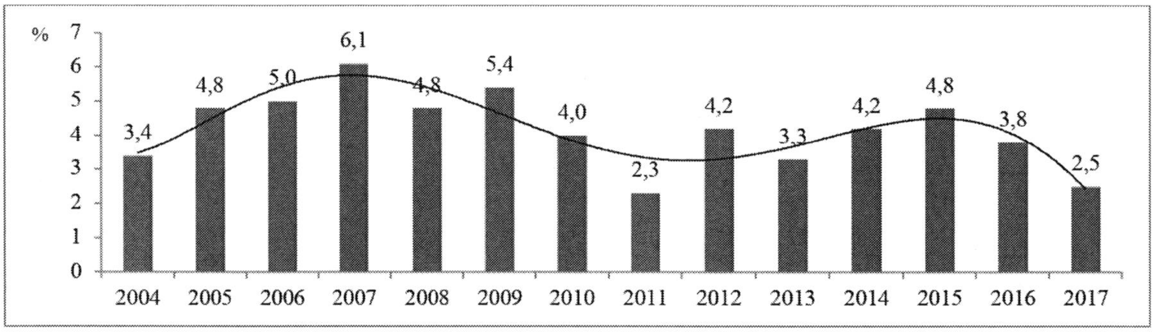

Figure 3. Characteristics of the epizootic situation with cattle leukemia in the Irkutsk region according to hematological studies (percentage of sick animals) for 2004-2017.

The number of animals suspicious of the disease varied ($p > 0.05$) from 0.9% to 2.2% (the highest levels were in 2009 and 2014) and was characterized by a slight increase (by 0.6%).

4. Discussion

Thus, assessing the epizootic situation of cattle leukemia in the Irkutsk region for 2004-2017 showed the following results, which we also would like to discuss. According to the RID results, infection with the cattle leukemia virus was 3.7±0.4% 2.1±0.2% for young animals from 6 months to 1 year old; 3.4±0.5% at the age of 1 to 2 years; 1.1±0.1% for bulls-producers; 2.4±0.7% among the animals for breeding and procurement. In long-term, a significant reduction in the infection of animals in the public and private sectors is observed (in 2.3 and 5.2 times, respectively), including in more than 2 times for young animals aged from 6 months to 1 year, in 5 times for those aged from 1 to 2 years. Also, there is a decrease in the RID percentage among the animals for breeding and procurement. Post-infectious antibodies to leukemia virus by ELISA were detected in 42.0% of cases. In the study of blood samples by the method of polymerase chain reaction, positive results amounted to 3.5%. According to the results of hematological studies, a number of hematological sick animals was 4.2±0.2%, a number of 1.4±0.1% among those being suspicious for leukemia. During the observation period, a number of hematological sick animals experienced a slight decrease (0.9% in 2017 if compared to 2004), and a number of suspicious patients had a slight increase (0.6%).

5. Conclusion

Considering the different levels of infection in different territories of the Russian Federation, our research insights on the epizootic situation of cattle leukemia, its dynamics in the perennial aspect at the regional level, it is highly necessary to identify patterns and characteristics of the spread of this disease, as well as to develop more effective anti-leukemia measures.

References

[1] Bateneva N V, Smirnov P N, Kashkovskiy V G, and Ryabinina V A 2015 Tactical decisions in the study of biology using biochemical test-systems BLV *Veterinary Sciences* **7**(38) pp 86-90
[2] Simonyan G A 2015 Leukemia of cattle *Farm Animals* **3**(10) pp 28-30
[3] Donnik I M, and Dzhailidi G A 2013 Prevention of leukemia of cattle in breeding farms of the Krasnodar region *Veterinary of Kuban* **5** pp 8-10
[4] Ablov A M, Batomunkuev A S, Anganova E V, and Meltzov I V 2014 *The use of statistical methods in the analysis of the epizootic situation of infectious diseases of animals and birds* (Irkutsk, Russia: IrSAA)

The mountain area's economic capacity as the basis for cross-border cooperation

M Yu Shishin[1], O Z Engoyan[1*] and S P Surazakova[2]

[1] Polzunov Altai State Technical University, 46 Lenina avenue Barnaul 656038 Russia
[2] Institute for Water and Environmental Problems of the Siberian Branch of the Russian Academy of Sciences (Gorno-Altaisk Branch of IWEP SB RAS), 8 Mirny lane, Altai Republic, Kyzyl-Ozyok village 649105 Russia

E-mail: shishinm@gmail.com

Abstract. Organization of economic activities in the framework of cross-border cooperation is accompanied by limiting effects. These effects are expressed in periodic or sustained ecosystem digression. In turn, this has a significant impact on dependent communities, including by multiplying these impacts in socio-natural complexes. Achieving a critical level of digression varies for integrable regions (countries). The authors note the need for a systematic approach and innovative solutions in the formation of common cross-border markets in the context of modern turbulence of global economic processes. At the same time, it is necessary to take into account certain requirements of sustainable development, both of individual regions and their industries and enterprises that provide integration.

1. Introduction

Transformation of global socio-economic interrelations requires scientific substantiation of environmental acceptability and economic efficiency in the formation of new economic alliances. As a rule, logic and previous experience underlie such alliances. Cooperation between the regions of the Greater Altai is based precisely on the historically established mutually beneficial socio-economic relationships. At the same time, there are (including observable empirically) restrictions on the organization of economic activities in mountainous areas. For a range of environmental management parameters, mountain ecosystems are vulnerable in a complex of socio-natural relationships. The organization of economic activities is accompanied by limiting effects, expressed in periodic or sustainable digression of ecosystems, which affects dependent communities and has a multiplicative effect. Achieving a critical level of sustainable digression varies for integrable regions (countries).

At the same time, the formation of common markets and participation of enterprises from different countries in the overall technological development requires, on the one hand, the process of synchronization of significant environmental, economic, and technological elements of the integrated countries' systems. On the other hand, the formation of common markets and participation of enterprises from different countries in the overall technological development requires creating conditions for sustainable development of (socio-ecological-economic) integrable countries in the medium and long term.

Content from this work may be used under the terms of the Creative Commons Attribution 3.0 licence. Any further distribution of this work must maintain attribution to the author(s) and the title of the work, journal citation and DOI.
Published under licence by IOP Publishing Ltd

2. Materials and Methods

Cross-border cooperation in mountainous regions (for example, the region of the Great Altai through the Russia-Mongolia line) requires careful consideration of possible consequences in organizing economic activities, since such activities are accompanied by impacts on fragile natural complexes.

When organizing economic activity, the key factor is the means of production. First of all, land resources are such means of production. For the territories under consideration, transhumance is traditional. More than that, this is the most promising and related to strategic areas. This determines the basic nature of land use (pastures and hayfields). Therefore, when doing business in such territories, the non-trivial task of organizing economic cooperation arises in such a way as to ensure sustainability and inexhaustibility of land resources, the means of production.

Analyzing the available data on pasture loading [1; 2], we should emphasize a destructive effect of unregulated exploitation of land resources [3; 4; 5; 6; 7; 8]. Digression of ecosystems leads to a decrease in competitiveness, damage, and other negative effects. Sustainable digression affects the nature of economic activity and certain parameters in the dependent socio-economic system [Mountains of the World], which requires a differentiated approach and, as a result, load regulation on vulnerable elements of socio-natural complexes [9; 10; 11; 12].

At the same time, the accumulated experience shows that innovative approaches can ensure preservation (and in some cases increase) of the territory's capacity. For example, formation of the so-called ecological frameworks [10; 13; 14] allows to reduce wind and water erosion, to increase the productivity of pastures and hayfields [15; 16]. Such a cultivation of landscapes requires a careful study of the long ecological cycles in the territory's development, and the implementation of such projects should be fully justified.

3. Results

Thus, regulating loads on pastures (territory capacity) is the basis for organizing efficient economic activities and for ensuring environmental and socio-economic security (and, consequently, sustainability) in vulnerable cross-border regions. A successful and effective solution of this task is possible only within the framework of a system approach that takes into account integration features of the territories being included and participating in interregional cooperation.

During our study on the territory's capacity for the purpose of producing valuable insights on cross-border cooperation, the following results were obtained.

1. Of course, limiting a region in using a vital resource requires the formation of a system for redistributing funds to compensate for damages associated with the inability to increase competitiveness because of resource depletion. However, current trends indicate the possibility of stopping risks of reduced competitiveness by improving product quality (the so-called "environmental brands") and entering relevant solvent segments of consumer demand [17].
2. Historical experience shows that in the organization of traditional land use, the territorial economic capacity was taken into account (for example, the system of winter and summer pastures in distant pasture). Therefore, the scientific substantiation of the permissible load on specific ecosystems of mountainous territories and the introduction into practice of the principles of traditional environmental management make it possible to stop environmental and socio-economic risks in the organization of economic activities in the framework of cross-border cooperation.
3. An innovative approach allows to increase the territory's capacity through the use of scientific and technical developments (for example, [18]). The introduction of anti-erosion technologies [19; 20], including in combination with the development of horticulture can help preserve (restore and increase) the territory's economic capacity.
4. Of course, investment in innovation implies a return on these investments. This reduces competitiveness (by increasing costs of goods and / or services). The solution of such a task is in preserving the sustainability of a socio-natural complex which is possible under the condition of developing and implementing (1) a system for regulating the level of exploitation and (2) a

system for redistributing resources, ensuring the preservation of the living standard of people employed in relevant industries.

Thus, on the one hand, the economic capacity of the territory is a limiting factor. However, on the other hand, historical experience and modern developments allow us to speak about the fundamental possibility of solving such a problem.

4. Discussion

Land resources are the main means of production for the studied territories (regions of the Great Altai, in particular, Western Mongolia and Gorny Altai). The economic capacity of these territories has objective limitations (primarily, environmental). The excess of such economic capacity is expressed in periodic and / or steady digression of the ecosystem. This leads to a decrease in the productivity of the means of production.

- The significance and role of economic capacity is a poorly developed topic in terms of providing a systematic understanding of the value of territory's capacity in the context of cross-border cooperation.
- Land resources, which are the main means of production for the (distant) livestock industry, need to justify the volume and nature of their operation. This volume and nature of land resources exploitation (in particular, pastures and hayfields) should ensure sustainability of the socio-natural complex in the medium and long term.

When developing program documents (territorial planning schemes, joint ventures, industry holdings, etc.), there is a problem of ensuring environmental safety for sustainable development. Restrictions (primarily, environmental) ensure sustainable development of the socio-natural complex. However, these restrictions entail a decline in competitiveness and a slowdown in economic growth. Consequently, to compensate for such effects, the formation of sustainable development mechanisms should include the introduction of a system for the redistribution of resources that ensure the preservation (maintenance) of the peoples' living standard.

5. Conclusion

In the context of global turbulence of economic processes, configurations of cross-border cooperation vary in different regions. Historical economic relationships remain between different regions. Despite the temporary weakening of such interconnections, their potential is sufficient for the formation of sustainable socio-natural complexes, making possible to check whether environmental safety is ensured. To solve this problem, the territorial economic capacity should be considered when developing cross-border projects (territorial planning schemes, joint ventures, and other program documents).

Acknowledgments

We express our gratitude to the scientists of Khovd State University for their advice and support.

References

[1] Mantai N 2013 Theoretical substantiation of the "Ecological framework" in Western Mongolia *Vestnik of Altai Science* **2-2** pp 44-47

[2] Beket U, Yengoyan O Z , and Shishin M Yu 2015 The problems of desertification of the Great Altai: threats and system solutions *Science and Education of the Big Altai* Aviailable at: http://rectors.altstu.ru/ru/periodical/archiv/Sustainable%20Region%20Development/articles/2_2015/1_1.pdf (Accessed 02 03 2019)

[3] Batomunkuev V S, Darbalaeva D A, Zhamyanov D Ts-D, Mikheeva A S, Osodoev P V, Sanzheev E D, Ariunsanaa B, ... Solongo Ts 2012 The role of anthropogenic factors in the development of desertification processes in Mongolia *Izvestiya of the Irkutsk State University. Series "Earth Sciences"* **5**(1) pp 92-107 Aviailable at: http://isu.ru/journal/downloadArticle?article=_962938e2efbc485ea37591615c1f328d&lang=rus (Accessed 02 03 2019)

[4] Mongolia livestock 2019 *Information for tourists* Aviailable at: http://www.legendtour.ru/rus/mongolia/informations/animal_industries.shtml (Accessed 02 03 2019)
[5] *Mongolia: General information* Aviailable at: http://ru.worldstat.info/Asia/Mongolia/Land (Accessed 02 03 2019)
[6] Urtnasan M 2015 *Pasture digression in the steppes of northern part of central Mongolia* (Dissertation of the Candidate of Biological Sciences) (Kazan, Russia).
[7] Surazakova S P 2010 Contemporary problems of sustainable land use in the mountainous region *Sustainable Development of Mountain Areas* **4**(6) pp 49-51 Aviailable at: http://naukagor.ru/LinkClick.aspx?fileticket=ST1t_8KRR_c%3d&portalid=4&language=ru-RU (Accessed 02 03 2019)
[8] Surazakova S P 2015 Problems of sustainable land use in the mountainous region In *Transition to a Green Economy and Sustainable Development in the Altai Region: Perspectives, Mechanisms, Key Area*s (pp. 121-127) (Barnaul, Russia: Altai State Technical University)
[9] Lu Yu 2016 Agro-pastoralism under climate change: institutions and local climate adaptations in northern China *Land Use Policy* **58** pp 173-182 Aviailable at: http://dx.doi.org/10.1016/j.landusepol.2016.07.022 (Accessed 02 03 2019)
[10] Engler J-O, Abson D J, Feller R, Hanspach J, von Wehrden H 2018 A social-ecological typology of rangelands based on rainfall variability and farming type *Journal of Arid Environments* **148** pp 65-73 Aviailable at: https://doi.org/10.1016/j.jaridenv.2017.09.009 (Accessed 02 03 2019)
[11] Tang J, Davy A J, Jiang D, Musa A, Wu D, Wang Y, Miao C 2016 Effects of excluding grazing on the vegetation and soils of degraded sparse-elm grassland in the Horqin Sandy Land, China *Agriculture, Ecosystems and Environment* **235** pp 340-348 Aviailable at: http://dx.doi.org/10.1016/j.agee.2016.11.005 (Accessed 02 03 2019)
[12] Gomboev B O 2006 *Agricultural use of Inner Asia* (Novosibirsk, Russia: Publishing House of the SB of the RAS)
[13] Kasperson J X, Kasperson R E, and Turner B L 1995 *Regions at risk: comparisons of threatened environments* Aviailable at: http://archive.unu.edu/unupress/unupbooks/uu14re/uu14re00.htm#Contents (Accessed 02 03 2019)
[14] Yengoyan O Z 2014 Economic capacity of a territory and ecosystem reflection *World of Science, Culture, Education* **3**(46) pp 380-382
[15] Ignatovich L I 2004 *The lessons of virgin soil: the experience of combating wind erosion and drought in the Kulunda steppe* (Barnaul, Russia: Az Buka)
[16] Morkovkin G G, Deev N G, and Demin V A 2010 *Agricultural science in Altai: the formation and influence on the social and economic development of the Altai Region* (Barnaul, Russia: AGAU Publishing House)
[17] Shishin M Yu Ed 2013 *Cross-border Altai: ways of international integration and sustainable development* (Moscow, Russia: Institute for Sustainable Development of the Public Chamber of the Russian Federation / Center for Environmental Policy of Russia)
[18] Dryuk V A, Sorochenko S F, and Suvorov S A 2015 *The technology of restoring depleted soils and the unit for its implementation* In *Transition to a Green Economy and Sustainable Development in the Altai Territory: Perspectives, Mechanisms, Key Area*s (pp. 145-149) (Barnaul, Russia: Altai State Technical University)
[19] Krakhmalev V A 2015 *Using the camel thorn to combat soil desertification* In participation *Transition to a Green Economy and Sustainable Development in the Altai Territory: Perspectives, Mechanisms, Key Areas* (pp. 138-141) (Barnaul, Russia: Altai State Technical University)
[20] *Western Mongolia on the way to sustainable development: conceptual design and project proposals 2016* Aviailable at: http://rectors.altstu.ru/ru/periodical/special/ (Accessed 02 03 2019)

Biochemical potential indicators of the transformed sod-podzolic soils of the forest-steppe of the Altai region

S Zavalishin[1] and V Karelina[1*]

[1] Altai State Agricultural University, 98 Krasnoarmeysky prosp., Barnaul 656049 Russia

E-mail: pochva22@mail.ru

Abstract. The paper presents results of a study focused on the biochemical potential of anthropogenically and naturally transformed sod-podzolic soils of the forest-steppe of the Altai Region, depending on the activity of soil enzymes. The research reports that the soils of this territory are affected by forest fires to the greatest and profound changes. Indicators of enzymatic activity serve as an indirect witnesses of the transformation of soil organic matter, the orientation of the soil-forming process. They can be used in the process of biomonitoring the state of the soil cover of forest sod-podzolic soils.

1. Introduction

The territory of the Altai Region is located in the southeast of the West Siberian Lowland in the transboundary limits of the Prealtay and Altai-Sayan soil provinces. A wide range of climatic conditions, in which more than thirty types of soil were formed, is a wide-known characteristic of this area [1]. Their comprehensive study is necessary for understanding the genetic aspects of modern soil formation, and the conditions for the formation of optimal soil properties for the functioning of natural and anthropogenic landscapes.

The study of intrazonal sod-podzolic soils is of particular interest. In addition to the climatic conditions, a number of additional factors affect them, which are actually atypical for zonal chernozem soils of the forest-steppe zone. These factors include: windfalls, forest fires, logging, soil surface mineralization, and a number of other. These processes lead to changes in the morphological structure and physicochemical properties of these soils. Already in the early stages of landscape transformation, the biochemical potential of daytime soil horizons changes significantly, subsequently leading to fundamental changes in the biogeocenosis [2]. Biomonitoring allows one to quickly assess the state of the soil cover, focusing solely on enzymatic activity.

The aim of this research was to determine the biochemical potential of sod-podzolic soils, which are transformed as a result of anthropogenic and natural effects.

2. Materials and Methods

In 2018, field studies were carried out on the territory of Priobsk boron, as well as within the north-eastern part of the Barnaul Ribbon Boron. The object of our research is the sod-podzolic soils that have been subjected to various forms of exposure, such as windfall soils, shifts of the main species from coniferous to deciduous (planting age of 55-60 years), lower forest fires, as well as selective logging. Background, natural sod-podzolic soils served as controls. Full-profile and half-profile soil

Content from this work may be used under the terms of the Creative Commons Attribution 3.0 licence. Any further distribution of this work must maintain attribution to the author(s) and the title of the work, journal citation and DOI.
Published under licence by IOP Publishing Ltd

cuts were used in the studies. In the course of work, the morphological structure of the soil was described in detail, and the sampling from each genetic horizon was carried out. Agrochemical parameters were determined according to generally accepted methods [3].

To assess the soil biochemical potential, we determined the activity of enzymes: (a) the catalase according to A. Sh. Galstyan (gasometric method); (b) the protease by applying the method of decomposing a gelatin layer according to E. N. Mishustin, D. N. Nikitin and I. V. Vostrov; (c) the urease by changing the reaction medium. Also, we determined the activity of polyphenol oxidase and peroxidase for soil according to K. A. Kozlov [4].

More than that, we determined the coefficient of humification relying on the ratio of "polyphenol oxidase activity" to "peroxidase activity."

3. Results

The contemporary scholarship clearly shows that the multifaceted transformation of sod-podzolic soils in the forest-steppe of the Altai Region occurs as a result of various anthropogenic and natural processes. During the field research phase, a change in the morphological structure of the soil profile was noted [5]. Windshields provoke the manifestation of pedoturbation processes [6]. Timber harvesting harms the soil profile because of destructing sod and humus-accumulative horizons; numerous technologies lead to an increase in soil density. When changing the main species from coniferous (Pinus Sylvestris) to broad-leaved (Quércus róbur), we note the structuring of soil aggregates to walnut and granular. A pyrogenic effect on the soil leads to the destruction of forest litter and the development of soil deflation. Also, among the soils prevalent in the watershed areas of the bottom fires, the second humus horizons buried under aeolian sediments were found. Data on the distribution of humus in the soil profile confirm the presence of second humus horizons. The humus content reaches 1.04–2.78% in the horizons lying between the eluvial and illuvial sequences at different depths (from 14 to 36 cm).

In soils of the forest cutting areas, the humus content is reduced to 1.22%, if compared with a background indicator of 2.37%. When changing the main species from coniferous to broad-leaved, with the age of trees at 55-60 years, the content of humus corresponds to the background indicators. In the soils of windfalls, the elluvial horizon A_2 is exposed when a tree falls out. As a result, the humus content in the day horizon does not exceed 0.05%.

A soil solution reaction of the transformed soils under study is close to the background, ranging from acidic to slightly acidic (pH_{KCl} of 5.35-5.98). Weak alkalinization of the soil solution to pH_{KCl}= 6.23 (close to neutral) was observed in soils when changing the main rock.

In humus-accumulative and second humus horizons, the low and very low provision with mobile phosphorus P_2O_5 (close in values to background, natural soils) is noted. The content of exchangeable potassium KCl is also low (1.96-4.48 mg / 100 g of soil).

In transformed soils, changes in the activity of soil enzymes (catalase, protease, urease, polyphenol oxidase, and peroxidase) are observed (Table 1).

Table 1. The activity of enzymes in transformed soils.

Soil horizon, depth	Catalase, O_2 cm³/ g * min	Protease, %	Urease	PPO / PO
Sod-podzolic soil, background				
A_1 (5-15 cm)	4.2	94	1.9	0.18
A_1A_2 (20-30 cm)	1.7	87	0.4	0.14
Sod-podzolic soil, windfall				
A_1 (5-15 cm)	7.2	100	2.0	0.14
A_1A_2 (20-25 cm)	2.4	97	0.5	0.14
Sod-podzolic soil, the change of the main species from coniferous to deciduous				
A_1 (6-16 cm)	5.0	100	2.5	0.21
A_1A_2 (16-23 cm)	1.9	96	2.0	0.25
Sod-podzolic soil, cutting				

A_1 (2-12 cm)	2.8	100.0	2.0	0.15
A_1A_2 (15-25 cm)	1.2	98.2	1.0	0.20
Sod-podzolic soil with a second humus horizon, ashes				
Second horizon (0-10 cm)	0.9	94	0.1	0.00
Second horizon (0-20 cm)	0.9	92	0.0	0.04
Ah (25-35 cm)	0.2	98	0.5	0.28

In the humus-accumulative horizons of windfalls, 30 days after tree falling, the biochemical potential increases due to the increased activity of the detected enzymes in comparison with the background soils (Table 1). In soils after forest felling, catalase activity decreases, but urease and protease activities increase.

Limits in the activity of polyphenol oxidase (PPO) and peroxidase (PO), the coefficient of humification of the PPO / PO ratio in windfall and logging soils are within the activity of these enzymes in soil horizons of the background sod-podzolic soil (Table 1).

A decrease in the biochemical potential of soils was noted in a layer of 0–20 cm, similar to the humus-accumulative horizon of background sod-podzolic soils, in the watershed areas of the 2004 ashes. Soils with a buried second humus horizon have a feature related to the structure of the biochemical profile, which consists in increasing the activity of PPO and PO down the soil profile, having with a peak in the horizon Ah and an increase in the humification coefficient of PPO/PO. Also, an increase in the coefficient of humification is observed in soils in the territory with changes in the main breed.

4. Discussion

The greatest changes in the biochemical potential were observed in soils with the second humus horizon in the watershed areas of the 2004 ashes. The sod horizon is poorly developed in the territory affected by pyrogenic influence (anthropogenic origin). As a result, deflationary processes are activated; also, inspired soils are formed with a low content of organic matter necessary for the functioning of soil biota and microorganisms. At the same time, the activity of soil enzymes largely depends on the qualitative and quantitative microbiological composition. We assume that preserving the second humus horizon in the elluvial profile is caused by a small amount of coniferous plant litter on the surface of the inspired horizons. The second reason is the ratio of the activity of polyphenol oxidase to peroxidase, higher than in humus-accumulative horizons of background soils. Probably, an increase in the activity of polyphenol oxidase is associated with a qualitative change in the microbiological composition and concentration of oxygen acceptors.

An increase in enzyme activity in sod-podzolic soils with an impaired morphological structure and a structure of genetic horizons (windfall soil) is associated with an increase in aeration and water permeability as a result of the turbulence in genetic horizons during mechanical impacts on the soil.

An increase in the biological potential and an increase in the PPO/PO ratio in the humus-accumulative soil horizons (which are common in the area with the changing main species from coniferous to broadleaf) indicates a change in the direction of transformation of organic matter.

The enzymatic activity of protease and urease in cutting soils is close in meaning to background soils. Catalase activity is an indirect indicator of the quantitative composition of soil microflora. The decrease in the activity of this enzyme in our soil cuttings indicates the suppression of microorganisms' activities. A number of authors [7, 8] also speak about a decrease in the microbial and enzymatic potential of cutting soils. A simultaneous increase in the percentage of gelatinous film layer decomposition is a consequence of an increase in the percentage of fungal microflora. The set of indicators of enzymatic activity reflects changes in the total biological activity of these soils. They indicate a shift in soil-forming processes towards podzol formation.

5. Conclusion

In the course our research on the activity of soil enzymes of sod-podzolic soils in the areas subjected to anthropogenic and natural transformations, it was possible to establish the following:

1. Pyrogenesis, or the effects of forest fires, have the greatest impact on the biological potential determined by soil enzymatic activities.
2. In the first year, a mechanical effect on the soil, as a result of windfalls, leads to an increase in enzymatic activity.
3. When changing the main species from coniferous to broad-leaved, progradation of enzymatic activity is noted, as well as changes in the direction of transforming soil organic matter.
4. A decrease in catalase and an increase in protease activity in the soils of selective forest felling indirectly indicate a change in the quantitative and qualitative composition of the soil microflora.

Research on the soil biological potential of enzymatic activity reflect the nature of changes not only in the general biological activity, but also in the direction of soil-forming processes. Along with indicators of microbiological analyzes, these studies can be used as the basis for biological monitoring of the transformed forest sod-podzolic soils.

6. Acknowledgments
This work was supported by RFBR grants №18-34-00421 мол_а, №18-04-00866 А.

References
[1] Burlakova L M, and Tatarintsev L M 1988 *Soils of the Altai region* (Barnaul, Russia: Altai Agricultural Institute)
[2] Tarasov P A, Mikhno A S, and Sizina A F 2011 Evaluation of the pyrogenic effect on the soils of Altai belt pines *Vestnik of Krasnoyarsk State Agrarian University* **1** pp 26-31
[3] Arinushkina E V 1970 *Soil chemical analysis guide* (Moscow, Russia: Moscow State University Publishing House)
[4] Zvyagintsev D G 1991 *Methods of soil microbiology and biochemistry* (Moscow, Russia: Moscow State University Publishing House)
[5] Zavalishin S I, Sokolova L V, Chernyshkov V N, and Karelina V S 2017 Influence of soil morphology on the number of Pinus Sylvestris L. seedlings under the conditions of the right coast of the Ob river in the Altai region *Vestnik of the Altai State Agrarian University* **9**(155) pp 24-27
[6] Zavalishin S I, and Patrushev V Yu 2014 Changing the morphology of sod-podzolic soils of strip forests of the Altai Region as a result of windfall *Forestry Bulletin* **1** pp 161-164
[7] Caldwell D 2005 Enzyme activity as component of soil biodiversity *Pedobiologia* **49** pp 637-644
[8] Vinogradova Yu A, Lapteva E M, and Perminova E M 2014 Microbial communities of podzolic soils in clearings of middle taiga spruce forests *Izvestiya of the Samara Scientific Center of the Russian Academy of Sciences* **16**(5) pp 74-80

IOP Conf. Series: Earth and Environmental Science **395** (2019) 012015 doi:10.1088/1755-1315/395/1/012015

An algorithm of adaptive stimulation of ecological and socio-economic development of a cross-border region

L Kirhmeer[1*]

[1] Orenburg State University, 13 Pobedi prospect, Orenburg 460018 Russia

E-mail: lkirhmeer @mail.ru

Abstract. Ecological, social, and economic development of a cross-border region is determined by regulated and unregulated factors, some of which are formed outside the region and even countries. Environmental pollution cannot be limited to territorial boundaries, and this requires taking into account this fact when developing programs and activities aimed at the regional ecological and socio-economic development.

1. Introduction

The Orenburg region is one of the subjects of the Russian Federation. It is located on the border with the Republic of Kazakhstan, having the longest part of the Russian-Kazakhstan border. The Orenburg region belongs to the mining regions, with a rather complex environmental situation and a high level of mortality and morbidity from environmentally related diseases. There are problems associated with the development of natural resource potential, despite the existing advantages of cross-border cooperation with Kazakhstan [1]. Ecological and socio-economic development of the Russian region and territories of Kazakhstan adjacent to the Orenburg region is influenced by a variety of internal and external, exogenous (controllable) and endogenous (uncontrollable, formed within the region as a system) factors. For this reason, environmental and socio-economic development should be stimulated, taking into account not only internal factors and changes, but also looking into the situation in neighboring regions and countries. To make effective management decisions, we need to adapt not only to the existing state of ecological, social, and economic development, but it is necessary to also take into account its prospects [12; 13]. In this connection, the development of an algorithm to adaptively select a number of tools to stimulate the ecological and socio-economic development of a cross-border region is relevant.

2. Materials and Methods

S. L. Vasenev, Kh. B. Badarchi, D. S. Gordeeva, D. F. Dabiev, V. V. Borodkina, and others devoted their research to the assessment of the ecological-social and economic development of regions [2-5] These researchers use on a system of indicators that characterize the ecological and socio-economic state of regions in modern conditions. The pressure-state-response model was taken as the basis in our study. The model is based on the thesis that human activity exerts "pressure" on the environment, forcing the government to react, and changing social agents' behavioral patterns [6]. We have developed a methodology for monitoring and forecasting (in short-term) the regional ecological and socio-economic development. The methodology outcome is a set of strictly limited indicators of environmental and socio-economic forecasting, taking values from 0 to 1 [7]. The results are used in our proposed roadmap [8]. The roadmap structure for environmental and socio-economic development includes the time axis

Content from this work may be used under the terms of the Creative Commons Attribution 3.0 licence. Any further distribution of this work must maintain attribution to the author(s) and the title of the work, journal citation and DOI.
Published under licence by IOP Publishing Ltd

IOP Conf. Series: Earth and Environmental Science **395** (2019) 012015 doi:10.1088/1755-1315/395/1/012015

(short, long, and super long-term perspectives) and the road map layers (global, in-country, and intra-regional challenges). Choosing incentive instruments and their priorities adaptively changes, depending on the levels achieved and the scenarios implemented. The transition from one scenario to another occurs due to correctly selecting and implementating incentive tools.

3. Results

Let's consider a number of scenarios describing the ecological and socio-economic development of the Orenburg region:

1. "A mining resource-dependent depressed region." Such a scenario is negative. In the long term (15 years or more), the dominant type of economic activity ("mining and quarrying," having a high proportion in GRP) is preserved and strengthened. At the same time, the contribution of this type of activity to the "Pressure" indicator block remains unchanged or increases. Indicators of the block "State" and "Reaction" are in a negative trend. Interaction with Kazakhstan within the framework of changes in "Pressure" and "Reaction" indicators is minimal or absent.
2. "A multi-industry moderately developed region (economic diversification)." In the long run, the role of other types of economic activity increases, without any predominance. Mineral mining continues, but with a tendency to reduce and completely eliminate this type of activity. Improving indicators for all "Pressure-state-response" blocks are observed. A really working mechanism for reducing pressure on the environment is being established with Kazakhstan.
3. "An agrarian region with a green economy (moderate growth)." In this region, the role of agriculture with an increasing share of the population engaged in this type of economic activity increases. Innovative methods are being introduced; they are effective for those areas with risky agriculture. Transition to such a scenario in WTO conditions is possible, as the region belongs to the zone of risky farming (thus, it is subjected to drought). Accordingly, requirements for reducing and canceling state support do not apply to the region. At the same time, transition to the agrarian specialization of a competitive level is possible only with significant government support [9]. Mining continues, but it has a tendency to reduce and completely eliminate this type of activity. Food security is ensured. Positive changes are observed in all "Pressure-state-reaction" blocks. A really working mechanism for reducing pressure on the environment is being established with Kazakhstan.
4. "Innovative development (optimistic variant)." The region is reorienting to new high-yielding economic activities that are not typical for the region's economy at present, developing existing economic activities and ensuring food and economic security. The impact of both internal and external pressure factors on the environment can be leveled out or reduced to a minimum in the region, by achieving the same results in neighboring regions and in Kazakhstan.

A management scheme of the region's ecological and socio-economic development is demonstrated in Figure 1. It begins with assessing the region's state. As shown in Figure 1, the presented scheme suggests an iterative solution to the issue of managing the region's ecological and socio-economic development. Also, it is based on the scenarios and methods for monitoring and forecasting the ecological and socio-economic development within the framework of our roadmap.

Without describing the tools used in detail (which can be found in the works of Yashalova [10] and IDegtyareva [11]), we move on to our proposed adaptive algorithm for stimulating the region's ecological, social, and economic development (Figure 1). Ranking tools in order of importance is required in advance, and this is done on an expert basis. At the same time, in the conditions of a cross-border region, experts from the Orenburg region and Kazakhstan should be involved. Each degree of importance of these tools is conditionally denoted by the following letters: A, B, C, D, etc. The degree of different instruments' importance may coincide, and their list may vary depending on the scenario being implemented. When switching to another scenario, the corresponding importance degree is reviewed. The effect is estimated on the basis of monitoring and forecasting the region's ecological and socio-economic development.

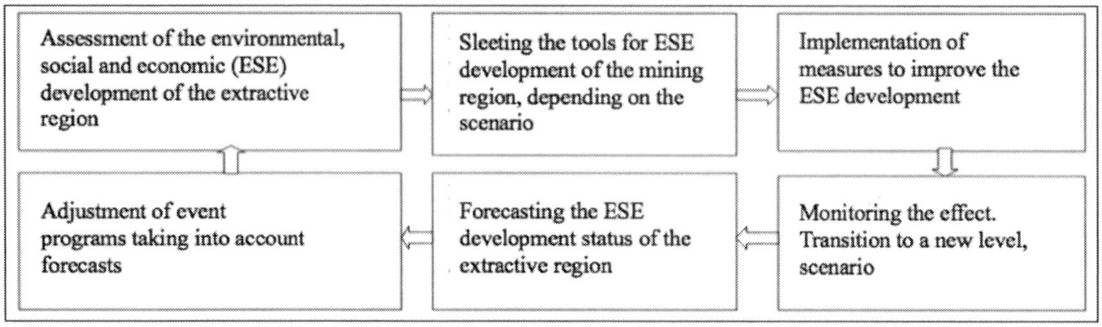

Figure 1. A management scheme of the region's ecological and socio-economic development. Source: Developed by authors.

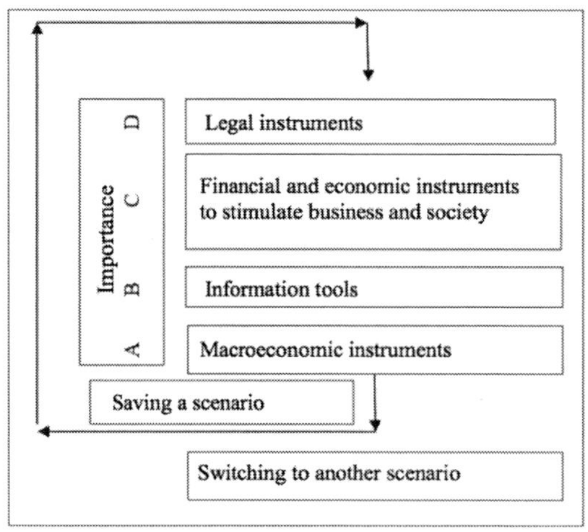

Figure 2. An algorithm for choosing a tool to stimulate the region's ecological and socio-economic development. Source: Developed by authors.

Adaptation of the algorithm implies recalculating the importance of tools, as well as their composition depending on the implementation of a particular scenario.

4. Discussion

The adaptive nature of the choosing instruments for stimulating the ecological and socio-economic development of a cross-border region is also determined by the fact that shifts in the weight of control levers to the regional level are observed [1]. Close cooperation in the field of environmental protection between Kazakhstan and the Orenburg region is initiated at the regional level. The solution of such major tasks as the transition to a "green" economy is possible only in the long term with an integrated approach that takes into account many factors of a controlled and uncontrolled nature.

The proposed approach requires clarification of incentive tools. The selected incentive tools will be projected through programs and activities to improve environmental and socio-economic development. In the future, the development of projects of specific programs is planned with definiting the program life cycle, costs of implementing program's activities, as well as effectiveness of its implementation.

5. Conclusion

Thus, the cross-border region's ecological and socio-economic development depends not only on internal factors, but it is also determined by the external "pressure." To improve the whole situation, the Orenburg region should adopt measures aimed at addressing the causes of environmental pollution. In our opinion, it is also necessary to work closer with Kazakhstan on such issues, including reaching bilateral treaties and establishing common monitoring mechanisms.

Within this research, we present a roadmap for ecological and socio-economic development, which is based on a system of indicators within the framework of the "Pressure-state-reaction" model. More than that, an adaptive incentive algorithm is described in the paper, which would allow the region to move to a new level of development that meets the concept of sustainable development.

References

[1] Semenov E A, Grigorevsky D V, and Chibilev A A 2017 Modern integration processes in the territory of the Orenburg-Kazakhstan cross-border region *Izvestiya of the Orenburg Branch of the Russian Geographical Society* **9**(42) pp 17-24

[2] Badarchi Kh B, and Dabiev D F 2013 Factor analysis of the impact of mining development on the ecological situation in the regions *Economic analysis: theory and practice* **46**(349) pp 13-26

[3] Borodkina V V 2015 Indicators for assessing the sustainable environmental and socio-economic development of the region *Basic Research* **5** pp 384-389

[4] Vasenev S L 2012 Ecological, social and economic monitoring of the Russian regions *Engineering Vestnik of the Don* **20**(2) pp 527-536

[5] Gordeeva D S, Pluzhnikova I I, Demtsura S S, and Izyumnikova S A 2017 Problems of assessing the ecological and economic development of a region in modern socio-economic conditions *The Azimuth of Scientific Research: Economics and Management* **6**(3(20)) pp 105-109

[6] Ponomareva M A 2011 Adaptation of foreign methods to the development of a system of indicators for sustainable development of Russian regions *Izvestiya of the Ural State University of Economics* **4**(36) pp 102-107

[7] Kirhmeer L V 2019 Monitoring ecological and socio-economic development of the mining region *The Azimuth of Scientific Research: Economics and Management* **8** (1(26)) pp 173-175

[8] Kirhmeer L V 2019 Roadmap as a tool for managing the ecological and socio-economic development of the mining region *Financial Economy* **1**(1) pp 67-70

[9] Lobova S V, and Churina E V 2010 State support of agriculture in the Altai region in modern conditions *Regional Economics: Theory and Practice* **1** pp 16-20

[10] Yashalova N N 2011 Organization of investment promotion of environmental activities in the region *Regional Economics: Theory and Practice* **37** pp 61-67

[11] Degtyareva I Ya 2015 Approaches and tools to stimulate the economic development of regions in the face of increasing external influences and uncertainty *Notes of the Institute of Public Administration* **3** pp 1-48

[12] Bogoviz AV, Lobova S V, Ragulina J V, Vypryazhkina I B, Boldyreva I N 2018 Educational incentives for creating high-performance jobs *Quality - Access to Success* **19**(S2) pp 57-61

[13] Bogoviz A V, Litvinova T N, Lobova S V, Ragulina Y V 2018 Infrastructural provision of international entrepreneurship: theory and practice *International Journal of Trade and Global Markets* **11**(3) pp 190-199

Agroecological substantiation of the sustainable development of arable soil fertility of the Altai dry steppe

A Kudryavcev[1*], G Stetsov[2] and E Toropova[3]

[1] Altai State Agrarian University, 98 Krasnoarmeysky avenue, Barnaul 656049 Russia
[2] Altai State University, 61 Lenin avenue, Barnaul 656099 Russia
[3] Novosibirsk State Agrarian University, 160 Dobrolyubova str., Novosibirsk 630039 Russia

E-mail: kae5959@mail.ru

Abstract. The soil resources of the Altai dry steppe (and their fertility in particular) degrade in their evolutionary anthropogenic development. Transformation of fertility is occurring due to a combination of phenomena, such as natural conditions, intensive use, farming system practiced in the absence of science-based regimes. It is established that the humus content and humus horizon thickness are statistically significantly reduced due to the activation of deflation and dehumification processes. More than that, areas of alkaline soils are increasing. The authors also point out to the degrading granulometric composition. Also, the amount of physical clay, silt fractions, cation exchange capacity, exchangeable calcium, potassium, and mobile phosphorus decreases. Statistical processing of fertility parameters of allowed us to develop a scale of process dynamism and establish the degree of dynamism in the considered soil properties, as well as to assess the rate of increasing adverse changes or their absence. Intensity of degradation processes and indicators determining it contributed to the development of levels of agro-ecological state, such as "Norma", "Risk", "Krizis" and "Bedstviye" (based on GIS).

1. Introduction

Preservation and reproduction of arable soil fertility is the key to sustainable development of the agro-industrial complex, in general, and the Altai dry steppe in particular. It has been established by practice that the intensive use of arid territories leads to global catastrophes, such as the tragedy that occurred in the Stone Steppe in the late 19th century. Decades have been spent to eliminate its effects. A series of dust storms in the prairies of the United States and Canada of the 1930s (where the fertile layer was completely lost) are to be remembered for a long time [1]. Today, those soil resources being once subjected to the processes of deflation and practically turned into a desert, are restored in their original state. However, huge costs are required for this restoration.

Quite a few works on the causes of deflationary catastrophes and recovery mechanisms have been published. In these works, the authors analyzing events come to the conclusion that it is necessary to create conditions for preservation and (under favorable circumstances) restoration of fertility The fundamental factor in the suspension of this process is the organization of territories on an agro-ecological basis and the use of farming systems that would allow to hinder deflationary processes when introducing elements of innovation, such as binary crops and cover crops [2]. According to foreign scientists, about 28% of the land resources of the world are subject to the process of deflation, it progresses over time. Today, about 2 billion hectares of productive soil resources are irretrievably lost

Content from this work may be used under the terms of the Creative Commons Attribution 3.0 licence. Any further distribution of this work must maintain attribution to the author(s) and the title of the work, journal citation and DOI.
Published under licence by IOP Publishing Ltd

[3, 4]. Arid territories of Altai used in agriculture should not comprehend such negative phenomena. To prevent such a phenomenon and maintain a stable productivity of agrocenoses, modern science-based agro-ecological approaches are necessary. The results of implementing these approaches are presented in our study.

2. Materials and Methods

Experimental studies are carried out in the arid area of the Kulunda steppe, the soil cover of which develops on the parent rocks of the sandy-sandy sediments with lake-alluvial origin, on steppe landscapes of tipchak-wormwood-feather grass vegetation. Its development is accompanied by the continental climate, which is characterized by harsh winter and hot summer, with a sum of active temperatures up to 2400°C and with the annual amount of precipitation reaching 240-300 mm, more than 70% of them occur during the growing season [5]. A soil cover of this area consists in 42% of chestnut and dark chestnut soils. Light chestnut and light chestnut solonets soils occupy 14%. 8% fall to the share of chestnut and dark chestnut solonets, they are used in arable land [6]. Intensive use of soil resources indicates that significant changes in fertility have occurred over time [7].

Analyzing archival materials of 1899-1913, 1980-1990 and using results of our own research for 2016-2018, we established the rate of degradation of fertility processes of agrogenic soils [8; 9; 10]. Using MapInfo Professional 10.0, we linked to archival materials a number of digitizing field soil maps and set the coordinates of cuts, making it possible to open them according to the “cut in the cut” principle. In field conditions, 35 cuts opened and showed the structure of soil cover, allowing to estimate the load intensity on soil resources. Selected soil samples were analyzed in the laboratory “Agrogenesis and Soil Fertility” of the Altai State Agrarian University (Barnaul, Russia). The results obtained allowed us to use the comparative chronological method, providing us with estimates of changes in fertility parameters over time, such as a humus horizon thickness, a humus content, a medium response, an amount of absorbed bases, as well as compositions of absorbed sodium, calcium, magnesium. To confirm the reliability of evolutionary changes in fertility over time, we used a mathematical method for statistical processing (Statistica 6.0). As a research outcome, we established dynamics and indicators of fertility parameters, as well as a degree of degradation. Using geographic information systems (GIS), a number of levels in the agro-ecological state and modes of their use have been developed and analyzed.

3. Results

The contemporary scholarship clearly shows that the intensive use of soil resources provokes agro-ecological tensions. It is established that in the modern era, soil formation proceeds along the steppe type with a significantly significant participation of the anthropogenic factor. Together with natural factors (climate, topography, vegetation, hydrography, and hydrology), this factor causes the development of such degradation processes as deflation, salinization, and alkalinization, for which the suspension of tools that allows one to quickly bring the agricultural soil to a progressive level of evolutionary development is necessary.

Assessing the deflation process intensity, it was found that over a century-long period of time, the humus horizon thickness in arable soil had changed significantly. On average, soil resources of the Kulunda plain have lost about 9 cm of the fertile layer. In modern conditions, the humus content was statistically significantly reduced by over 2%; while at the beginning of the last century, its content was more than 3%. The growing anthropogenic impact on the soil resource significantly stimulates the process of alkalinization. It has been established that the areas of alkaline spots are multiplied in size due to the activation of evaporation processes when plowing. The general trend of degradation of such indicators of fertility as the granulometric composition, cation exchange capacity, absorbed calcium, exchangeable potassium, mobile phosphorus under the existing anthropogenic load can be clearly seen. At the same time, the statistical processing of empirical data made it possible to establish that these changes are unreliable. It should be noted that the dark chestnut subtypes of agrosoils are less susceptible to this phenomenon than chestnut and light chestnut. This is due to the lower anti-deflationary stability of the latter, acquired in the evolutionary development.

We believe that the considered soil properties can serve as indicators that determine the intensity in the development of degradation processes or their absence. The analysis of these indicators would allow developing a scale of dynamism, evaluating the evolution of elementary soil-forming processes, establishing the intensity of degradation processes, and determining the dynamism of the considered fertility indicators. Those cases when the soil properties under anthropogenic impact do not undergo significant transformation in time should be attributed to the natural process of soil formation and the first degree of dynamism. These are weakly dynamic deviations, which correspond to changes not higher than 0.2% per year. Moderately dynamic deviations correspond to the second degree of dynamism and range from 0.2 to 0.3% deviation per year. The degradation of fertility parameters of arable soil from 0.3 to 0.5% per year is attributed to the average dynamic, this is the third degree of dynamism, and above 0.5% per year is the fourth degree of dynamism, which is called "highly dynamic deviations."

The proposed assessment of the dynamism of the negative evolution of agricultural soils fertility is acceptable for spatial indicators characterizing anthropogenic disturbance in general. For a more objective assessment of the evolution of fertility, we suggest using changes in the considered soil property within the boundaries of the lower and upper values of the confidence interval. The developed dynamism scale states that fertility parameters undergo negative changes over time. The obvious manifestation of degradation processes can be traced by the example of alkalinity, humus content, humus horizon thickness, partly physical clay, silt fraction, and mobile batteries.

The conducted research is a scientifically based foundation for organizing a territory on an agro-ecological basis, taking into account degradation processes and characterizing the current state of the Altai agralandschaft dry steppe as a whole. For agroecological assessment of agrosoils fertility, indicators of fertility parameters have been developed. It allows one to characterize the degree and intensity of degradation processes in elementary soil ranges (ESR). In essence, this is an individual approach to ESR, in which not all soil properties and profile have degraded. Those properties that have significant deviations from the natural soil-forming process inside the ESR are called "indicators." The boundaries of indicator values signal the maximum permissible development of the degradation process, which determines the state of fertility and indicates critical changes that are occurring, which makes it possible to draw a boundary that determines levels of the agro-ecological state. In turn, the levels of agroecological condition of arable land is a set of ESR, they characterize a structure of the soil cover within the limits of permissible temporary deviations of fertility indicators, causing the degree of degradation and taking into account peculiarities of the territory. In fact, this is an on-farm land management on an agro-ecological basis, which allows to increase the efficiency of arable soil use and halt the development of degradation processes caused by human activities [11].

For the sustainable development of the dry steppe of the Altai, such levels of agro-ecological state as "Norma", "Risk", "Krizis", "Bedstviye" have been developed and tested [12]. The agroecological state level "Norma" corresponds to the natural soil-forming process, in which the deviation of soil properties in arable horizons does not exceed 10%. Changes in the parameters of fertility at the maximum deviation from the "Norma" by 20-40% characterize the agro-ecological state level "Risk". Arable soils related to the agroecological state level "Krizis" are characterized by a change in the state of fertility up to 80% of the "Norma." Those ESRs located on low relief elements and near water bodies are related to the agro-ecological state level "Bedstviye." Because of intensive use, evaporation processes causing degradation processes of salinization and alkalinization are intensified.

For preserving and reproducing fertility of crisis areas, progressive methods and approaches in the use of arable soil are necessary. At the agroecological state level "Norma," the use of the Strip-Till farming system is desirable, while it is necessary to limit the use of the traditional farming system based on flat-cutter. We believe that in crop rotation, the steam field should be an exception, since it is the cause of intensive development of deflationary processes. At the selected agroecological state level "Risk," the No-Till farming system is sufficient to apply. It will allow to stop the degradation processes due to the constant presence of "covering material" in the form of stubble. The "Krizis" provides for the mandatory introduction of cover crops into the No-Till farming system, which will be a vegetative "covering material" during the spring and autumn periods. This will ensure the prevention of deflation processes, improve the processes of humification, and create positive changes in the microbiological and phytosanitary background of arable soil. For all levels of the agro-ecological state, we recommend using binary (poly-species) crops. Selecting binary crops should be based on the compatibility of plants, such as rapeseed and legumes, for example. This will increase the productivity of agrocenoses, create favorable conditions for developing a microbiological component of soils, improving their phytosanitary condition [13].

4. Discussion

We believe that the solution of the problem of preserving and reproducing the agrosoil fertility of the Altai dry steppe should be carried out on an agro-ecological basis, highlighting the levels of the agro-ecological state. And using modes should be developed for each level individually. In these modes, all elements of the farming system must be combined. Time is needed for introducing new farming systems. Of course, every farming system has its positive and negative sides, which cannot be strengthened without scientific support. Each proposed farming system requires practical implementation. An integrated plant protection system, a fertilizer system, breeding and seed production, etc. should be developed. Of course, Strip-Till and No-Till are the farming systems of the future involving biologization, preservation, and reproduction of fertility. At the present stage, the practice of testing innovative farming systems has shown that in the conditions of the Altai dry steppe, the grain yield of 3 t/ha and more can be obtained. Until recently, such a yield was considered an unattainable result, it can be considered random or implemented as a pattern.

5. Conclusion

Thus, the conducted studies allow us to conclude that, at the present stage, the arable soil of the Altai dry steppe undergo significant changes in fertility. It has been established that the agroecological stress of agrosoils is caused by anthropogenic load in combination with natural conditions, contributing to the development of the deflation process, salinization, and alkalinization. Of course, in order for the area under study not to suffer the fate of the “dust storm syndrome,” the proposed scientific developments with outlining the agro-ecological state levels and the modes of their use must be implemented on an agro-ecological basis. This will allow for a differentiated use of the soil fertility potential in the dry steppe of the Altai.

6. Acknowledgments

This work was supported by the Russian Foundation for Basic Research (project 18-416-220007p_a).

References

[1] Lagunina I 2009 *Soil scientist Hugh Bennett and the dusty cauldron, or the history of soil conservation in America* Available at: https://www.svoboda.org/a/1492493.html (Accessed 02 03 2019)

[2] Drydiger V K 2016 *Practical recommendations on the development of the technology of cultivation of crops without tillage in the arid zone of the Stavropol region* (Saratov, Russia: Amirit)

[3] Buschiazzo D E, and Zobeck T M 2008 Validation of WEQ: RWEQ and WEPS wind erosion for different arable land management systems in the Argentinean Pampas. *Earth Surf. Processes Landf.* **33** pp 1839-1850

[4] Federal Register 2010 Notice of implementation of the Wind Erosion Prediction System for soil erodibility system calculations for the Natural Resources Conservation Service *Federal Register* **75**, No. 234 (7 December 2010)

[5] *Weather in Altai Region* (2019) Available at: https://rp5.ru/ (Accessed 02 03 2019)

[6] AltaiNIIGiprozem (n.d.) *Soils of the Altai region* Available at: http://altgipro.ru/ (Accessed 02 03 2019)

[7] Burlakova L M, Yashutin N V, Gribov S I, and Bivalkevich V I 2000 *The concept of the rational use of agricultural land in the Altai region in modern conditions* (Barnaul, Russia: ASAU Publishing house)

[8] Vydrin I P, and Rostovskiy Z I 1899 *Materials on the study of the soils of the Altai Okrug* (Barnaul, Russian Empire: Tipo-Lithography under the Head Office of the Altai Okrug)

[9] K D Glinka (Ed) *Soil-Botanical expeditions to investigate the colonization areas of Asian Russia conducted by A I Khainsky in 1912-13* (St. Petersburg, Russian Empire)

[10] Archival materials of "AltaiNIIGiprozem" OJSC 1980-1990 (n.d.) *Soil essays of districts of the Altai Region* Available at: http://altgipro.ru/ (Accessed 02 03 2019)
[11] Kudryavtsev A E 2013 *Agroecological state of fertility of arable soils of the Altai Priobye and intermontane depressions of the Altai* (dissertation of the doctor of biological sciences) (Barnaul, Russia)
[12] Kudryavtsev A E, Belyaev V I, and Volnov V V 2014 *Agroecological aspects of arable soil fertility of the dry steppe of the Altai*. In *Ecological and Economic Strategies for Sustainable Land Use in the Steppes of Eurasia in the Context of Global Climate Change* (Barnaul, Russia: ASU Publishing)
[13] Toropova E Yu, and Sokolov M S, *et al* 2016 Induction of soil suppressiveness is the most important factor in limiting the severity of root infections *Agrochemistry* **8** pp 46-55

IOP Conf. Series: Earth and Environmental Science **395** (2019) 012017 doi:10.1088/1755-1315/395/1/012017

Using mathematical modeling in predicting the economic efficiency of lake-commercial fish farming in cross-border areas of northern Kazakhstan

G Baryshnikov[1*], I Fomin[2] and T Nasarova[3]

[1] Altai State University, 61 Lenina Ave., Barnaul 656049 Russia
[2] Tyumensky State University, 6 Volodarsky str., Tyumen 625003 Russia
[3] L. N. Gumilyov Eurasian National University, 2 Satbaeva str., Almaty, Nur-Sultan, 010000 Kazakhstan

E-mail: bgj@geo.asu.ru

Abstract. On the basis of the developed mathematical forecasting model, the prospects for developing the lake commercial fish farming. Also, its effectiveness in the cross-border regions of northern Kazakhstan is considered in details.

1. Introduction

In the Republic of Kazakhstan, the Program for the Development of the Agro-Industrial Complex in the Republic of Kazakhstan for 2013-2020, Agribusiness 2020 was adopted for the development of the agro-industrial complex, including the commercial fish industry [1]. This Program envisaged the creation of lake-commercial fish farms whose goal is to achieve sustainable economic and social development. One of the main ways of rational, highly efficient use of biological resources of inland waters is to organize commercial fish farming on their basis. This method of farming is caused by the violation of the natural conditions for the reproduction of fish stocks, which include the destruction of spawning grounds due to shallowing of water bodies, the overlapping of rivers by hydraulic structures, etc. To determine the annual increase in fish stocks, we developed a mathematical model to determine the ability of the reservoir to provide such an increase.

2. Material and Methods

To determine the possible production of fish products from water bodies, we rely on the data given in [2], which uses a set of formulas to calculate the annual production of herbivorous fish, planktophages, benthophages, and crustaceans [3]: $P_{rast}=0{,}006P_1$; $P_{planktofagov}=0{,}10P_{zoopl}$; $P_{bentofagov}=0{,}2P_{bentos}$; $P_{rakoobr}=0{,}04P_{rakooobr}$, where P is the annual fish production, P_1 is the production of aquatic vegetation, P_{zoopl} is the production of zooplankton, P_{bentos} is the benthos products. To determine the annual production of predatory ichthyophages, the formula was proposed [4]:

$$X_{isch} = (P - P_1) / 3, \quad (1)$$

where X_{isch} is the products of ichthyophage predators, kg / ha, P – a calculated production of non-predatory fish, kg / ha, P_1 – desirable commercial production of non-predatory fish, kg / ha, 3 – a feed coefficient for ichthyophages. Based on the data, we have developed a mathematical model.

Content from this work may be used under the terms of the Creative Commons Attribution 3.0 licence. Any further distribution of this work must maintain attribution to the author(s) and the title of the work, journal citation and DOI.
Published under licence by IOP Publishing Ltd

3. Results

For practical use of the model, we used the data obtained in the study of the Lake of Bolshoy Tarangul. Knowing the main morphometric characteristics of the reservoir, as well as the results of hydrobiological analyzes, determining an annual increase in ichthyomass is possible. Consequently, according to the state of the predicted data, not only those species provided with food in the reservoir can be identified, but also the maximum number of fish necessary for stocking can be determined.

The Lake of Bolshoy Tarangul is located in the North-Kazakhstan region of the Republic of Kazakhstan, 100 meters north-west of the village of Korneevka. The pond has an oval shape and stretches from northwest to southeast. The length of the lake is 9.4 km, the width is 4.6 km, the maximum depth is 5 m, with an average depth of 3 m. The volume of water is 102 million m^3. The water surface area is 3400 ha.

The temperature of summer water fluctuates in open and coastal parts within 22-24°C. Transparency was 0.7-1.0 m, which according to the complex classification, corresponds to the indicator "a quite clean water." The water color on the Forel scale belongs to the XII category. The reaction is medium/slightly alkaline, pH = 8.4. Electrical conductivity corresponded to $PEEP_B$, and it is 1.73-1.74 mCm / cm. The concentration of dissolved oxygen satisfied $PEEP_B$ and was 9.6 mg/dm^3. BOD_5 was lower than the $PEEP_B$ and amounted to 1.6 mg O_2/dm^3, which corresponds to the category of "sufficiently clean" waters [5].

According to the magnitude of mineralization (2148 mg/dm^3), the lake waters are the β-mesogalin brackish. Water salinity exceeds $PEEP_B$. According to the ionic composition in the classification of O. A. Alekin, this water belongs to sodium chloride type II. Chlorides prevail among anions (318 mg/dm^3). Concentrations of these ions do not exceed $PEEP_B$ (350 mg/dm^3 according to GOST 4245-72). The water hardness is average, exceeded $PEEP_B$ and amounted to 32 mol/dm^3. In 2012, the first hydrobiological studies of the lake were carried out by the "Ecosphere" Ltd. In 2018, data was updated as a result of our summer field research [6].

According to the quantitative development of zooplankton, the Lake of Bolshoy Tarangul is medium fodder, and it is characterized as β-mesosaprobic. Also, according to the quantitative development of zoobenthos, the studied water body belongs to mesotrophic. All calculated indicators relating to the annual increase in ichthyomass are calculated by the method of V. N. Abrosov [4]. The results of the calculation are shown in Table 1, calculations are made in Excel.

Based on these calculations, it is found that in terms of a feed base, the Lake of Bolshoy Tarangul is able to provide an annual increase in ichthyomass of non-predatory fish up to 60 tons (or 17.8 kg/ha). In this regard, in order not to undermine the natural forage base, we recommend catching no more than 60.5 tons of fish products annually. However, it should be borne in mind that this value is not the total biomass of non-predatory fish, but only their annual increase. According to the adopted legislation, the generally accepted optimal amount of withdrawal does not exceed 30%. Consequently, the total biomass of non-predatory fish is 200 tons, and the assessment of predator biomass growth could be calculated. Hence, the annual production of predator ichthyophages in the Lake of Bolshoy Tarangul can be 47.1 tons. The annual increase in fish stocks in polyculture can be 107.6 tons.

However, in addition to the annual increase in fish stocks, the proposed model calculates a number of other indicators that are important for commercial fish farming. Namely, according to hydrobiological data and on the basis of the above recommendations [7, 8], biological standards for stocking fish of different fish species are also calculated according to the nature of nutrition.

Table 1. Mathematical model of the Lake of Bolshoy Tarangul.

Reservoir	**Bolshoy Tarangul**		
Square, ha	3400		
Average depth, m	3		
Degree of growth, %	15		
Feed base of the investigated reservoir			
Sampling point	Phytoplankton Biomass, mg / m^3	Zooplankton Biomass, mg / m^3	Zoobenthos Biomass, mg / m^3
Coastal area	9.8	1.79	2.82
Open zone	6.9	1.72	0.71
Average	8.85	1.76	1.77
This reservoir is able to provide an annual increase in ichthyomass, kg			
Herbivorous fish	30605		
Planktophage	17901		
Benthophagous	12002		
Predatory fish	47062		
The total increase in ichthyofauna can be, t			
107.6			

4. Discussion

Thus, as a kind of water resources, the lakes play an important role in economic development of a region. In order to make rational use of these resources and obtain maximum economic profit, it is advisable to use models that make it possible to predict the effectiveness of lake-commercial fish farming.

The main direction of increasing fish productivity in inland freshwater reservoirs is the transition from the simple exploitation of fish stocks to the creation of highly efficient regulated fishery in lakes and reservoirs. By their potential, small and medium-sized lakes can produce significantly more fish and better quality if there is the creation of intensive-type managed lake farms.

5. Conclusion

Transition to the forms of lake-commercial fish farming for the northern transboundary areas of the Republic of Kazakhstan will contribute to the sustainable economic and social development of the agro-industrial complex, and hence the economy as a whole.

The proposed model will allow to calculate the economic effect of the fisheries reservoir, taking into account various conditions and factors, during its transition to the forms of lake-commercial fishery. The Lke of Bolshoy Tarangul is one of the promising water bodies of Northern Kazakhstan, and its fish stocks are confirmed by means of a mathematical model created by us.

References

[1] Government of the Republic of Kazakhstan 2013 *Resolution "On Approval of the Program for the Development of the Agro-Industrial Complex in the Republic of Kazakhstan for 2013-2020"* (February 18, 2013 No. 151) Available at: http://tengrinews.kz/zakon (Accessed 29 10 2015)
[2] Bessonov N M, and Privyazentsev Yu A 1987 *Fisheries hydrochemistry* (Moscow, USSR)
[3] Salazkin A A, and Ogorodnikova V A 1984 *Methodological recommendations on the collection and processing of materials during hydrobiological studies: Tasks and methods of studying and using the fodder base of fish* (Leningrad, USSR)
[4] Abrosov V N 1972 Determination of ichthyomass in lakes and its annual increase In *Elements of aquatic ecosystems* (pp 225-237) (Moscow, USSR)
[5] Kirillov V V, Zarubina E Yu, Beletskaya N P, Vilkov V S, and Lipchanskaya M A 2011 *Aquatic ecosystems of Northern Kazakhstan* (Petropavlovsk, Kazakhstan: NKSU named after M. Kozybayev)
[6] Dmitriev P S, and Fomin I A 2012 *Assessing the state of fish resources in fisheries reservoirs and features of the genesis of hydrobiocenoses of the North Kazakhstan region* (Petropavlovsk, Kazakhstan: NKSU named after M. Kozybayev)

[7] Kolomin Yu M 2006 *Fishes of Northern Kazakhstan* (Petropavlovsk, Kazakhstan)
[8] Shashulovsky V A, Makarov S N *et al* 2008 *Recommendations for the production of commercial fish* (Saratov, Russia)

IOP Conf. Series: Earth and Environmental Science **395** (2019) 012018 doi:10.1088/1755-1315/395/1/012018

Microbial preparations in the cultivation of grain crops in the Priobskaya zone of the Altai region

V Kursacova[1] and L Stupina[1*]

[1] Altai State Agricultural University, 98 Krasnoarmeysky, Barnaul 656049 Russia

E-mail: stupina-liliya@mail.ru

Abstract. The article examines the action of biologics of associative nitrogen-fixing bacteria for non-leguminous crops in spring soft and durum wheat and spring barley crops in the steppe zone of the Altai Priobye. The positive effect of biological products in all years of research is established. In different years, the increase in yield of different varieties of wheat from inoculation with monopreparations amounted to 8.0-53.5%, with a minimum tillage of 68.0-89.7%. In drought conditions and in the context of yield reduction, the effect of drugs was higher than in a moister period. The drugs effectiveness increased with mineral fertilizers (a dose of $N_{30}P_{60}K_{60}$) and mycorrhiza.

1. Introduction

According to I. A. Tikhonovich, and A. A. Zavalin, "To preserve soil fertility and prevent the most dangerous forms of its degradation, an environmentally safe direction of land use is envisaged, based on the need for the maximum possible approach of agriculture to natural analogs..." [1]. In recent decades, microbiological preparations based on rhizosphere bacteria have been developed in Russia. They are distinguished by high physiological activity and have positively proven themselves both in Russia and abroad [2, 7].

Long-term trials of drugs show their high efficiency in a variety of cultures in different soil and climatic zones. Productivity increases on average by 20-40% or more. Being inoculated with preparations, it is comparable or often even exceeds the yield with a full mineral fertilizer NPK with a dose of active substances 60 kg/ha [3, 4, 5].

In Siberia and Altai Region, grain crops occupy more than half of the arable land. However, their potential is not fully realized. Therefore, increasing crop yields (such as wheat and barley) is an important public task. This problem should be solved by applying contemporary achievements of science based on the widespread introduction of qualitatively new technologies using environmentally friendly biological preparations [3, 4, 5, 6, 7].

The purpose of this study is to evaluate the effectiveness of microbial preparations on cereals in the Priobskaya zone of the Altai Region within the framework of the Geographic Test Network.

2. Materials and Methods

Studies have been carried out in small-scale experiments on the training and experimental field of the Altai State Agrarian University since 2000. And also, research was carried out in industrial crops in farms of the region. Years of research were characterized as insufficiently moist (2001, 2007, 2008, 2009, 2012, 2014, 2015), wetted (2016, 2017), and dry (2000, 2010, 2011, 2013).

Content from this work may be used under the terms of the Creative Commons Attribution 3.0 licence. Any further distribution of this work must maintain attribution to the author(s) and the title of the work, journal citation and DOI.
Published under licence by IOP Publishing Ltd

IOP Conf. Series: Earth and Environmental Science **395** (2019) 012018 doi:10.1088/1755-1315/395/1/012018

The predominant soil of the zone is leached chernozem, which is characterized by low humus content of about 5%, pH close to neutral, with a sufficient content of mobile compounds of phosphorus and potassium and low nitrogen.

Over the years, we studied the effect of associative nitrogen-fixing bacteria: Bioplant-K (contains a culture of bacteria *Klebsiella planticola*), Mobilin (*Klebsiella mobile*), Mizorin (*Arthrobacter mysorens*, strain 7), Rizoagrin (*Agrobacterium radiobacter*, strain 204), Flavobacterin (*Flavobacterium sp.*, strain 30), Azorizin (*Azospirillum lipoferum*), strain Ya-2 (*Corinebacterium freneyi*), as well as the mycorrhiza preparation (contains fungus *Glomus intraradices*, strain 8). In the experiments, we used varieties of spring soft and durum wheat and spring barley, as highly recommended for the zone. Studies were performed according to generally accepted methods in crop production [2, 8, 9].

3. Results

Inoculation of seeds of cereal crops with preparations of rhizosphere diazotrophs helps to stimulate seed germination, increase root biomass, increase the supply of nutrients to the root system, increase growth and, consequently, plant productivity [3, 4, 5, 10].

The results of the 2000-2001 experiment are presented in Table 1. On average, the yield of the spring soft wheat variety Tselinnaya 60 was significantly increased from the use of preparations of diazotrophic bacteria. A less significant effect was obtained from seed inoculation with Rizoagrin (19.2%). And the greatest responsiveness was noted for Agrofil and Mizorin preparations – 32.2 and 34.4%, respectively.

The results of studying the effect of biopreparations on the yield of spring soft wheat varieties Altayskaya 530 and Altayskaya 325 in 2007–2008 showed that inoculation also increased the yield of wheat of both varieties in both years of research. On average, over two years, Rizoagrin was more effective (Table 1).

Table 1. Productivity of spring wheat in the years of research.

Option	Tselinnaya 60 variety, average 2000-2001			Altayskaya 530 variety, average 2007-2008			Altayskaya 325 variety, average 2007-2008		
	t/ha	increase		t/ha	increase		t/ha	increase	
		t/ha	%		t/ha	%		t/ha	%
Control	0.84	-	-	2.42	-	-	2.45	-	-
Agrofil	1.11	0.27	32.2	-	-	-	-	-	-
Mizorin	1.13	0.29	34.4	-	-	-	-	-	-
Rizoagrin	1.00	0.16	19.2	2.80	0.38	15.7	3.00	0.55	22.4
Flavobacterin	1.05	0.21	24.8	2.74	0.32	13.2	2.95	0.50	20.4
Azorizin	-	-	-	2.75	0.33	13.6	2.85	0.40	16.3
HCP_{05}		0.15			0.19			0.11	

The Altayskaya 325 variety responded more actively to the processing. In this variety, the reaction to inoculation was higher by 2.7-7.2% than in the Altayskaya variety 530.

In 2009, 2010, and 2012, the test of Bioplant-K biopreparation on spring soft wheat varieties Omskaya 36, Altayskaya 100, and Altayskaya 105 was performed (Table 2). Bacterization of wheat seeds Omskaya 36 with Bioplant-K has contributed to a significant increase in its yield throughout the years of research. Despite the low level of yield in 2010 and 2012 dry years, the increase in inoculation was higher than in humid 2009 and amounted to 73-83%. A slightly smaller gain from the inoculation with Bioplant-K was obtained on wheat varieties Altayskaya 100 and Altayskaya 105 (Table 2). In 2009, it was 10.9-12.3%, but already 53.5-21.6% in 2010. Higher increases were obtained with the use of mineral fertilizers $N_{30}P_{60}K_{60}$. On average, they amounted to 61.4-24.5% over 2 years. At the same time, they were higher than on the background.

Table 2. Productivity of spring wheat in the years of research.

Option	Altayskaya 100 variety, average 2009-2010			Altayskaya 105 variety, average 2009-2010			Omskaya 36 variety, average 2009, 2010 and 2012		
	t/ha	increase		t/ha	increase		t/ha	increase	
		t/ha	%		t/ha	%		t/ha	%
Control	2.64	-	-	3.83	-	-	2.00	-	-
Bioplant-K	3.36	0.72	27.3	4.44	0.61	15.9	2.93	0.93	46.5
$N_{30}P_{60}K_{60}$-background	3.18	0.54	20.5	4.30	0.47	12.3	-	-	-
Background + Bioplant	4.26	1.62	61.4	4.77	0.94	24.5	-	-	-
HCP_{05}	-	0.34	-	0.39	-	-	0.66	-	-

In 2011-2013, we tested the effect of the preparations Mizorin, Rizoagrin, and Flavobakterin on two varieties of durum wheat, namely Altaysky yantar and Aleyskaya (Table 3). Bacterization with biological preparations contributed to an increase in the yield of both varieties. Inoculation provided an increase in the Aleiyskaya variety yield on average by 8.8-30.9 over 3 years, while inoculation provided an increase in yield by 28.4-43.9% in the Altaysky yantar variety. The maximum effect was obtained from the drug Flavobakterin and a triple combination of drugs.

Table 3. Productivity of spring durum wheat, on average in 2011-2013.

Option	Aleyskaya variety			Altaysky yantar variety		
	t/ha	increase		t/ha	increase	
		t/ha	%		t/ha	%
Control	1.81	-	-	1.48	-	-
Mizorin	2.23	0.42	23.2	1.90	0.42	28.4
Rizoagrin	1.97	0.16	8.8	2.04	0.56	37.8
Flavobacterin	2.37	0.56	3.9	2.11	0.63	42.6
Mizorin + rizoagrin + flavobacterin	2.27	0.46	25.4	2.13	0.65	43.9
HCP_{05}	-	0.21	-	-	0.18	-

In the experiment conducted in 2015-2017 on spring wheat varieties Stepnaya volna and spring barley varieties Signal, the highest yields were obtained on binary mixtures of Mobilin and strain Ya-2 with mycorrhiza (Table 4). The yield of wheat on the mixture of Mobilin and Mycorrhiza preparations exceeded control by an average of 58.1%; it was by 63.3% after combining strain Ya-2 with Mycorrhiza, amounting to 3.32 and 3.43 t / ha, respectively. On spring barley, a greater effect in the binary combination turned out to be on a mixture of Mobilin and Mycorrhiza (33.2%). When mono-using on wheat, the drug Mobilin was more effective, and on barley was strain Ya-2.

Table 4. Productivity of spring wheat and spring barley, on average in 2015-2017.

Option	Wheat Stepnaya volna variety			Barley Signal variety		
	yield, t/ha	increase		yield, t/ha	increase	
		t/ha	%		t/ha	%
Control	2.10	-	-	2.17	-	-
Mobilin	2.66	0.56	26.8	2.36	0.19	8.8
Strain Я-2	2.47	0.37	17.6	2.84	0.67	30.9
Mycorrhiza	2.24	0.14	6.7	2.27	0.10	4.6
Mobilin + mycorrhiza	3.32	1.22	58.1	2.89	0.72	33.2
Strain Ya-2 + mycorrhiza	3.43	1.33	63.3	2.80	0.63	29.0
HCP_{05}	-	0.12	-	-	0.26	-

In the experiment conducted under production conditions in the Aleysky region in the "Mashchensky" farm, a significant increase in the yield of Altayskaya wheat 530 from inoculation with biological preparations is also shown (Table 5).

Table 5. Wheat yield, Altayskaya 530 variety with minimum tillage at the "Mashchensky" farm in Aleysky district, 2016-2017.

Option	Yield, t/ha		Yield, average 2016-2017, t/ha	Increase	
	2016	2017		t/ha	%
Control	0.88	1.06	0.97	-	-
Rizoagrin	1.78	1.69	1.73	0.76	78.3
Flavobacterin	1.53	1.73	1.63	0.66	68.0
Mizorin	1.83	1.84	1.84	0.87	89.7
HCP_{05}	0.16	0.19	-	0.17	-

On average, the yield without using drugs was 0.97 t/ha over 2 years and 1.63-1.84 t/ha for biological products, providing an increase of 68.0-89.7%. The drug Mizorin proved to be a more effective drug in these conditions.

Calculating economic efficiency shows that the use of drugs of associative bacteria on grain crops in the Priobskaya zone of the Altai Region is quite profitable. On average, compared to the control, the cost of production of spring soft wheat is reduced by 71.4-1482.2 rubles/ha, and by 104.4-445.4 rubles/ha in spring barley. More than that, the level of profitability of wheat increases by 2.8-86.1% and of barley by 19-84%, respectively.

4. Discussion

In our experiments, the response of wheat plants to inoculation with biological products in different years was slightly different, but the yield on inoculated variants exceeded the control level. The effectiveness of binary combinations of associative rhizosphere bacteria with the Mycorrhiza mushroom preparation is far superior to the effects of monodrugs. Such a productive symbiosis is due to their different mechanism of action on plants, which in sum better provides plants with nutrients, growth stimulants, and fungicidal metabolites, as was already reported in the works of I. A. Tikhonovich and A. A. Zavalin [1, 10], as well as by other authors [7].

The level of grain yield increase depends on agrotechnical methods and soil and climatic conditions; however, biologics help plants to cope with stressful conditions (drought, waterlogging, sudden temperature drops), developing directly in the root zone. High gains with minimal tillage are obtained through the activity of microorganisms in the composition of drugs. In the untreated soils, microbiological activity is strongly suppressed, plants lack many nutrients and nitrogen, especially which reduces the yield. Diazotrophic microorganisms not only provide plants with nitrogen, but they also stimulate the activity of beneficial soil microflora, which mineralizes organic matter and improves plant nutrition.

5. Conclusion

The research conducted allows to make the following conclusions:

1. Studying the effect of biopreparations of root diazotrophs in spring soft and durum wheat crops in different weather conditions in the steppe zone of the Altai Territory showed their rather high efficiency. The increase in wheat yield from inoculation in different years was 8.0-53.5%, and 68.0-89.7% with minimal processing.

2. The plant response to inoculation was very dependent on the variety and weather conditions. In drought conditions, the effect of the action of the drugs was higher than in a moister period.

3. With the mineral fertilizers and Mycorrhiza, the drug effectiveness increases. The effect of drugs on yield is slightly higher than the effect of mineral fertilizers in a dose of $N_{30}P_{60}K_{60}$.

4. The economic effect from using biological preparations of root diazotrophs is on average from 100 to 800 rubles/ha.

6. Acknowledgments

We thank Candidate of Biological Sciences A. P. Kozhemyakov, the head of the Laboratory of Ecology of Microorganisms at the All-Russian Research Institute of Agricultural Microbiology, for providing us with drugs. We are grateful to V. V. Mashchensky, the head of the "Mashchensky" farm, for providing us with space and assistance in conducting experiments.

References

[1] Tikhonovich I A, and Zavalin A A 2016 Prospects for using nitrogen-fixing and phytostimulating microorganisms to increase the efficiency of the agro-industrial complex and improve the agro-ecological situation in the Russian Federation *Fertility* **5** pp 28-36

[2] Tikhonovich I A 2005 *Biological products in agriculture: methodology and practice of using microorganisms in crop production and feed production* (Moscow, Russia: Russian Agricultural Academy)

[3] Zavalin A A 2005 *Biopreparations, fertilizers and harvest* (Moscow, Russia: VNIIA)

[4] Kozhemyakov A P, Schott P R, and Litvintseva T A 2008 Testing new strains of nitrogen-fixing bacteria on brewing barley crops *Agrarian Science - Agriculture* **1** pp 85-88

[5] Shott P R, Litvintsev P A, Litvintseva T A, and Kozhemyakov A P 2010 The use of preparations of root diazotrophs in the cultivation of grain in the Altai *Achievements of Science and Technology in the AIC* **6** pp 29-31

[6] Zavalina A A 2000 *Evaluation of the effectiveness of microbial drugs in agriculture* (Moscow, Russia: Russian Agricultural Academy)

[7] Azevedo J, Maccheroni J, Pereira O, and Ara W B 2000 Endophytic microorganisms: a review on insect control and recant advances on tropical plants *Electronic Journal Biotechnologies* **3** pp 40-65

[8] Dospekhov B A 1985 *Methods of field experience* (Moscow, Russia: Agropromizdat)

[9] State Commission for the Testing of Agricultural Crops at the Ministry of Agriculture of the USSR 1971 *Methods of state variety testing of agricultural crops* (Moscow, Russia: Kolos)

[10] Tikhonovich I A, Zavalin A A, Blagoveshchenskaya G G, and Kozhemyakov A P 2011 Biopreparations use is an additional source of plant nutrients *Fertility* **3** pp 9-13

Prospects for using post-mortal genetic materials on the example of sable to ensure the biodiversity in natural systems

I Silkin[1*], A Popov[2] and D Dashko[1]

[1] Irkutsk State Agricultural University named after A. A. Ezhevsky, Molodezhny village, Irkutsk region, Irkutsk district 664038 Russia
[2] V. R. Philippov Buryat State Agriculture Academy, 8 Pushkina str., Ulan-Ude 670024 Russia

E-mail: ivsi@list.ru

Abstract. The article presents a technique of taking a genetic material from a postmortem material of appendage and testes of sables, conducted research to determine the mobility of sperms, morphological assessment and survival of sperms outside the animal body. When introducing artificial insemination in the sable industry, the use of post-mortem genetic material obtained from male sables is a promising direction in the reproduction of sable as a particularly valuable commercial species, as well as for its introduction into the zoo culture in order to ensure the biodiversity of natural systems where these animals live. It can also be used in the development of programs for the reproduction and conservation of populations of other particularly valuable, rare, and endangered species of mammals.

1. Introduction

In accordance with Article 2 of the International Convention on Biological Diversity [4], biological diversity means the variability of living organisms from all sources, including, among others, terrestrial, marine, and other aquatic ecosystems and ecological complexes, of which they are part; this concept includes diversity within a species, between species, and a variety of systems.

Biological diversity is the result of almost four billion years of development. Many forms of life and ecological processes ensure the continuation of biological evolution, which is a prerequisite for human well-being. In the framework of the Strategy and Action Plan for the Conservation of Biological Diversity of the Russian Federation [12], the ecosystem service for the production of hunting products is of regional and local importance, primarily. Also, the recreational component of this service is great. Besides, it is important for preserving the traditional lifestyle of the indigenous peoples of the North, Siberia, and the Russian Far East.

According to the fifth national report, *Biodiversity Conservation in the Russian Federation*, prepared by the Ministry of Natural Resources and Ecology in accordance with the identified important problems and unresolved obstacles that have persisted in recent years in connection with Russia's fulfillment of its obligations under the International Convention on Biological Diversity, one of the priority tasks is to ensure the sustainable use of biodiversity resources in natural resource industries - agriculture, hunting, fisheries, and forestry [7].

Currently, the development of a bioresource strategy is becoming increasingly important for animal reproduction biotechnology. This strategy could provide a high yield of useful products and, at the same time, the preservation of natural diversity and a balanced abundance of zoocomponents of

Content from this work may be used under the terms of the Creative Commons Attribution 3.0 licence. Any further distribution of this work must maintain attribution to the author(s) and the title of the work, journal citation and DOI.
Published under licence by IOP Publishing Ltd

natural ecosystems. This global problem can be solved by developing technologies of zooculture for those species not yet being involved in it, as well as improving the existing ones. In principle, the zoo culture technology should be developed for all living animal species [1].

Sable is one of the most valuable in the fur production of animals. Sable pelts are widely known and in constant demand at various international fur auctions. Commercial sable pelts are more in demand [5]. Very often, wild sable pelts are exported from Russia illegally, suggesting an oversupply of demand before offering [6].

As an independent branch of fur farming, the history of sable breeding in Russia has more than 75 years. Despite this, significant problems are noted in the biotechnology of sable breeding. For example, in animal farms, sables mature at 2-3 years, and in their natural habitat at 1 year. Among scientists, there is no consensus on this issue, some scientists believe that sable males become sexually mature at the age of one year [3], others point to the puberty of sable males at a later perios [2], [9], [10]. There are the most numerous populations of subspecies (ridges) of sable, such as the Barguzin, Yenisei, Sayan, Sakhalin, Tobolsk, Kamchatka, Yakut, Far Eastern and Tungus [11].

Based on the above, we believe that the results of our research presented in this article in the future can solve the problem of reproduction of the highly valuable commercial animal. And globally, it will help to more effectively address the problems of ensuring the biodiversity of other species in various natural systems.

2. Materials and Methods

Seed plants with appendages of annual males of sables of the Yenisei and Barguzin ridges obtained by shooting from 5 animals in February 2019 in the Irkutsk district of the Irkutsk region (Russia) served as materials for our research. The experimental material taken was placed in special containers with ice and delivered to the laboratory of the Department of Special Veterinary Disciplines of the Irkutsk State Agricultural University named after A. A. Ezhevsky within 10-12 hours. After that, the post-mortal genetic material was placed in a thermostat at 37°C for 2 hours, in order to prevent a "cold" shock in sperm.

The technique of sperm aspiration from the testis and its appendage is as follows: a 2 ml disposable syringe needle is inserted through the testis head for its entire length along the axis of the testis network. This needle position allows aspiration from the tubule network (rete testis), which is a complex system of tubules connecting the appendage with the testis, and which is functionally justified because it contains the largest number of sperms. After insertion of the needle, the latter slowly retracts back, at this moment, a drop of unclear liquid appears in the needle cannula. A 2 ml syringe containing 0.3 ml of a solution of neutralized 2.9% sodium citrate is attached to the needle, after which, at the same time, the aspiration of the contents of the testis network is carried out and the needle is removed. Additionally, the genetic material was taken from the anterior portion of the testicle. For this, the puncture of the anterior portion of the testis was carried out using a butterfly needle, the tube of which was connected to a 2 ml disposable syringe, creating a vacuum in it using a piston, additionally pressing fingers on the testicle, changing the position of the needle tip at 15 different points inside the testis tissue.

For the evaluation of the obtained seminal fluid from the postmortal material of the testis and the appendage of sable males, the method for evaluating the freshly obtained sperm from farm animals was used [8], since in the literature available to us, we have not found methods for assessing sperm in sables and other commercial animals, as well as methods for evaluating sperm obtained from the post-mortal material.

3. Results

Evaluation of the results was carried out according to three criteria: determination of sperm motility, morphological assessment of sperm and its survival outside the body [8].

First, determination of sperm motility was conducted. The visual microscopic evaluation was performed in a darkened field of view with an increase in the obtained material 140 times in dilution with 2.9% sodium citrate (1:3). According to the results of the analysis carried out on all five samples with a generally recognized assessment of the sample, the results obtained from the epididymis averaged 4.3±0.3 points, where each point corresponds to 10% of sperm with translational motion. According to the results of the analysis carried out on five samples taken from the testis tissues, the results averaged 2.7±0.3 points. Here, a large number of immature sperms at the stage of round spermatids marked.

Second, we conduct a sperm morphology assessment. For morphological evaluation of sperm, a drop of the obtained material in dilution with 2.9% sodium citrate (1: 3) was placed on the edge of a defatted slide, a thin smear, prepared, dried, then fixed in 96% ethanol for 3 min and stained with eosin. Evaluation of sperm in a stained smear was performed under a microscope at a magnification of 400 times. On average, the percentage of viable sperm taken from the epididymis was 47.5 ± 3.3% and 24.3 ± 3.0%, respectively.

Third, the survival of sperm outside the body is one of the objective indicators of the fertilizing ability of sperm in the biotechnology of animal reproduction. According to the results of our research, in samples taken from the epididymis of the testis, it averaged 6.9 ± 0.6 hours and 4.1 ± 0.4 hours, respectively.

4. Discussion

Our studies of postmortem genetic material taken from sable males and the original methods tested by us for collecting seminal fluid from the tissues of the epididymis and testis allowed us to obtain the following results.

According to the results of our studies on the generally accepted methods for assessing sperm in animals with respect to determining the ability of sperm to progressive movement on a 10-point scale, we can say that the sperm of male sables obtained from the post-mortal material is quite suitable for artificial insemination. Its goal is to ensure the unique biodiversity of sable in various natural systems and to preserve its gene pool as a highly valuable commercial species. We believe that the results obtained (4.3 ± 0.3 and 2.7 ± 0.3) on a 10-point scale, taking into account the high level of modern biotechnologies, are quite acceptable results, considering the fact that the material was post-mortal, and 12-14 hours had passed since it was delivered to the laboratory (on average), taken, and placed in an optimal biological environment.

In our opinion, while conducting a morphological assessment of sperm, the percentage of viable sperm is relatively high, given the fact that the material was obtained after the death of animals. The survival rate of sperm outside the body was at a low level, but one should take into account the fact that we used the technique adopted to evaluate the freshly obtained sperm from live animals. When performing artificial insemination in order to preserve the biodiversity of valuable and endangered species, we consider that this assessment methodology is unacceptable for the above purposes.

5. Conclusion

Based on our research and the results obtained, we can say that when artificial insemination is introduced into sable breeding, the use of post-mortal genetic material obtained from sable males is a promising direction in the reproduction of sable as a particularly valuable commercial species. Also, the use of post-mortal genetic material is a promising direction for its introduction into the zooculture in order to ensure the biodiversity of the natural systems where this animal lives. Research in this direction should be continued, improving the methods of obtaining post-mortal material from male sables, increasing the methods of preservation, survival and fertilized ability of sperm cells using advanced technologies in biotechnology and animal reproduction, as the most optimal methods of rational use and management of biological resources. The results of our research can be used in the development of programs for the reproduction and preservation of a population of other especially valuable, rare, and endangered mammalian species.

IOP Conf. Series: Earth and Environmental Science **395** (2019) 012019 doi:10.1088/1755-1315/395/1/012019

References

[1] Blokhin G I 2005 Zooculture in solving the problem of conservation of biological resources *Collection of Materials of Scientific and Practical Conference* (pp 9)
[2] Erstenyuk T A, and Silkin I I 2018 Puberty sable at the artificial breeding *Scientific Research and Development to Introduction in Agriculture* pp 193-196
[3] Zakharova S A 1988 *Postnatal ontogenesis of sable testes* (Dissertation of the Candidate of Biological Sciences)
[4] *Convention on Biological Diversity (adopted June 5, 1992)* 1992 Available at: https://www.cbd.int/convention/text/default.shtml (Accessed 08 04 2019)
[5] Luzina S N 2004 Winter auction season of 2003-2004 *Breeding and Farming* **2** pp 16-17
[6] Lyapustin S N, and Fomenko P V 2003 *Fight against smuggling of fauna and flora objects in the far East of Russia* (Orange Pub Co)
[7] Ministry of Natural Resources and Ecology of the Russian Federation 2015 *Fifth National Report "Biodiversity Conservation in the Russian Federation"* Available at: https://wwf.ru/upload/iblock/6d3/www_natdoklad_final.pdf (Accessed 08 04 2019)
[8] Polyantsev N S 2015 *Veterinary obstetrics, gynecology and bioengineering* (Moscow, Russia: Lan' Pub Co)
[9] Ruchkina Z S 2005 *Reproductive potential of annual sables in the conditions of cellular content* (Dissertation of the Candidate of Biological Sciences)
[10] Snytko E G 1973 Reproductive ability of male sables and their influence on fertility of daughters. In *Proceedings of the conference of young scientists on animal husbandry and rabbit breeding* pp 46-50
[11] Sokolov G A 1998 The Problem of rational use of sable in Russia In *Materials of scientific and practical conference* (pp 145-154)
[12] Ministry of Natural Resources and Ecology of the Russian Federation 2014 *Strategy and Action Plan for the Conservation of Biological Diversity of the Russian Federation* Available at: https://www.cbd.int/doc/world/ru/ru-nbsap-v2-ru.pdf (Accessed 08 04 2019)

IOP Conf. Series: Earth and Environmental Science **395** (2019) 012020 doi:10.1088/1755-1315/395/1/012020

Features of seasonal migrations and wintering of epy elks *(Alces alces)* in the Kuznetsk-Salair mountain region

N Skalon[1*], P Stepanov[1] and A Prosekov[1]

[1] Kemerovo State University, 6 Krasnaya str., Kemerovo 650000 Russia

E-mail: nskalon@kemsu.ru

Abstract. The article presents an analysis of elksэ seasonal migrations in the Kuznetsk-Salair mountain region, where a largely independent population of the species has developed. Its core is located in the administrative boundaries of the Kemerovo region. In snowless and little snow periods, elks are diffusely located in forest areas of the region. In the Kuznetsk-Salair mountain region, seasonal migrations of moose are caused by the need to move from the snowy slopes and foothills of the Kuznetsky Alatau, the Gornaya Shoria, and the Salairsky ridge to the little snowy slopes of the eastern exposition, to the steppe and forest-steppe parts of the Kuznetsk Basin and the Mariinsky-Tisulsky forest-steppe areas. In some places, these migrations are cross-border in nature, affecting the administrative territories of the Altai and Krasnoyarsk Regions, the Republics of the Gorny Altai and Khakassia, and the Novosibirsk and Tomsk Regions. After the depression of the beginning of the last century, up to 24 winter camps were formed when the elk reappeared on the territory of the Kemerovo region (Kuzbass). Local animals from the Kuznetsk Alatau and Gornaya Shoria and animals from the adjacent areas of the Altai, Novosibirsk, and Tomsk regions migrated regularly to these sites. At the same time, a part of elks from the Kuzbass southern regions migrated to the territory of Altai and Khakassia. In the 1960s and 1980s, the elk winter sites were clearly localized, up to 300-500 animals were concentrated in the largest of them. In the Kemerovo region, a system of regional specially protected natural areas (SPNA) was created. In the 1990s, in the period of the decline in the number of elks, about 15 winter sites remained. And since the 2000s, many wintering sites have become diffuse, seasonal migrations have changed. Currently, with the exception of 4 large sites, in the Kuznetsk Basin, most elks do overwinter in small herds from 2-3 to 20-30 head. They are dispersed in forest areas with low snow cover, well protected or difficult to access for snowmobile transport.

1. Introduction

The Kuznetsk-Salair mountain region occupies the north-western part of the mountains of Southern Siberia, and it is clearly divided into four physiographic provinces [1]. The Kuznetsk forest-steppe basin is located in the center. It is framed by the mountains from three sides: the Salair ridge from the west, the mountain ridge system of Shoria from the south, and the Kuznetsk Alatau from the east. About 80% of the geographical Kuznetsk-Salair mountainous region is within the boundaries of the administrative Kemerovo region (Kuzbass). The western slope of the Salair ridge is divided between the Novosibirsk region and the Altai region, the southeastern slope of the Kuznetsk Alatau, and the Abakan Range is located within the Republic of Khakassia.

In the period of depression in the first half of the twentieth century in Western Siberia, the formerly complete elk habitat got fragmented and narrowed to the southern taiga subzone. At the same time, in

Content from this work may be used under the terms of the Creative Commons Attribution 3.0 licence. Any further distribution of this work must maintain attribution to the author(s) and the title of the work, journal citation and DOI.
Published under licence by IOP Publishing Ltd

the basin of the upper course of the Ob, an area gap of more than 500 km wide from west to east has been formed. The whole territory of the Kuznetsk-Salair mountain region entered this gap [2]. The elks did not reside permanently in this territory [3, 4]. Only isolated visits from the northeast were recorded [5].

In the late 1940s, the reintroduction of elks into the territory of the Kemerovo region and their subsequent penetration into the Kuznetsk-Salair mountain region began. In 1947-1949, the elks were noted in the northern and eastern regions of Kuzbass, which indicates their introduction from the territory of the Krasnoyarsk and Tomsk Regions. The number of elks grew rapidly. Already in 1952, it was proposed to open a hunt for them in the north-eastern Tyazhinsky and then in the Tisulsky districts. However, hunting for elks in the Kemerovo region was opened only in 1960, when their number reached 2 thousand. The number of elks exceeded 7 thousand individuals by the end of the 1970s. They settled all suitable habitats and formed about 24 large winter sites in low-mountain and low-snow areas in the territory of Kuzbass. In the 1980s, the decline in elks' numbers began as a result of overfishing. By 1994, the elk population had declined to 1,850 individuals with increasing poaching and an increase in the number of wolves from 200 to 400 individuals. In the 2000s, the nature of wintering and the migration routes of elks changed significantly. Since 2005, a stable increase in the number of elks begins in the conditions of improving the work of hunting users, state hunting authorities and the almost complete extermination of wolves. In 2017, a number of elks exceeded 5 thousand individuals and continues to grow.

2. Materials and Methods

From 2014 to 2019, the authors conducted field studies to study the migration of elk in the Kuznetsk-Salair mountain region, analyzed the results of departmental winter route surveys (WRS). In 2019, the UAV (Supercam, S250 model) was used to observe the migration of elks in the territory of the Salair reserve (zakaznik). In addition, a survey of district inspectors, hunting specialist, huntsmans, SPNA staff, local hunters, and local historians was conducted. Scientific publications on the topic of research and departmental reports on the registration of the number of elks, on their official prey and other reasons for their death in the Kemerovo region from 1950 to 2018 were also analyzed [6].

3. Results

In the snowless summer-autumn period, the elks are diffusely distributed throughout the forest areas of the region, both in the plains and in the mountains. In winter, the elks are forced to concentrate on limited areas with snow cover not exceeding 70-80 cm. Accordingly, the seasonal migrations of moose in the Kuznetsk-Salair mountain region are caused by the need to move from the snowy slopes and foothills of the Kuznetsk Alatau, Gornaya Shoria and the Salair ridge, to the relatively little snowy slopes of the eastern exposure, and to the low snowy steppe and forest-steppe areas of the Kuznetsk basin and the Mariinsky-Tisulsky forest steppe areas. In some places, these migrations are cross-border in nature, affecting the administrative territories of the Altai and Krasnoyarsk regions, the Republics of the Gorny Altai and Khakassia, and the Novosibirsk and Tomsk regions (Fig. 1).

In the second half of the twentieth century, the elks formed up to 24 winter camps with the most favorable wintering conditions in the Kemerovo region. Animals from the western macroslope of the Kuznetsk Alatau, Gornaya Shoriam and the foothills of the Salairsky, including from adjacent areas of the Altai and Novosibirsk Regions, began to migrate there regularly. At the same time, part of the elks from the southern regions of Gornaya Shoria migrated to the territory of Altai and Khakassia. In the 1960-1980s, the elk winter sites were clearly localized. In a limited area, up to 300-500 animals were concentrated in the largest of them.

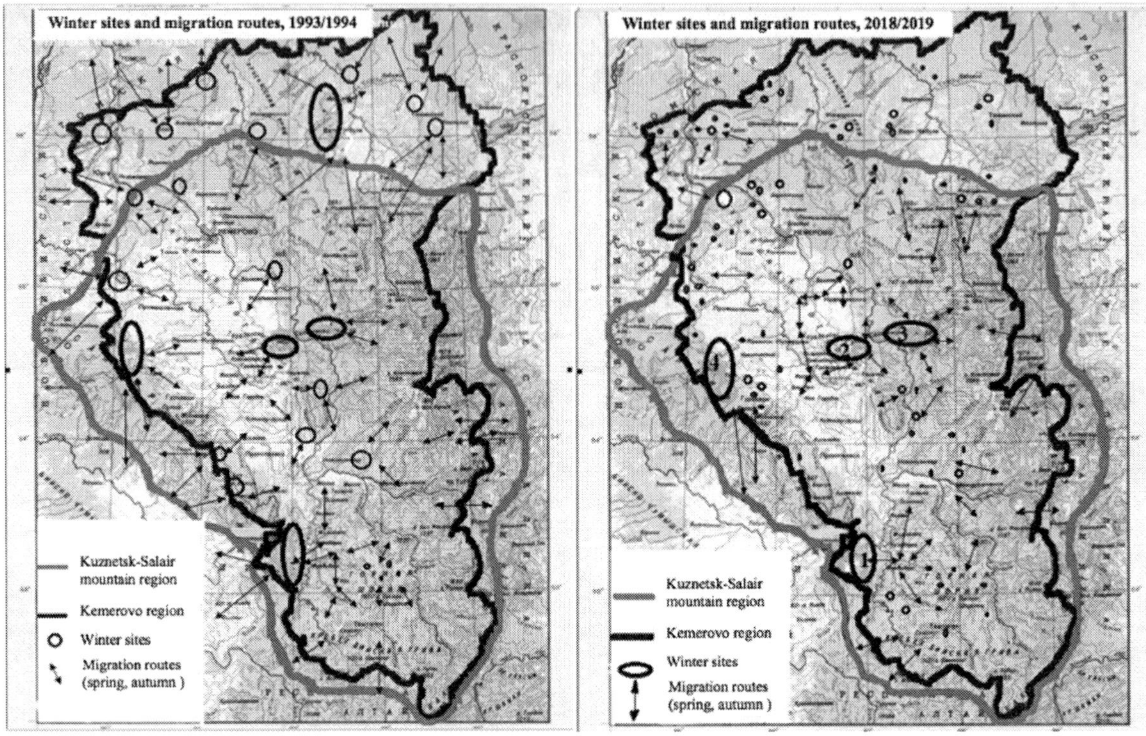

Figure 1. Migration routes and winter camps (sites) in 1993/1994 and in 2018/2019 (Winter camps: 1 – Sary-Chumyshskaya, 2 – Bungarapsko-Azhendarovskaya, 3 – Saltymakovskaya, 4 – Salairskaya).

In the 2000s, only 4 large and long-standing winter sites remained in the Kemerovo region. The largest of them is located on the border with the Altai Region in the basin of the Sary-Chumysh River. The accumulation of more than 300 elks was noted here in the winter season 2018 - 2019.

The second site is located on the Azhendarovsky Range, jutting into the central part of the Kuznetsk Basin, within the boundaries of the Bungarapsko-Azhendarovsky reserve. In the season 2018-2019, the 80-90 elks are counted here. The Zhendarovsky range is separated by the Tom River from Saltymakovsky, which branches off from the Kuznetsk Alatau in the western direction. The Saltymakovsky reserve is located here. In the season 2018-2019, about 70-80 elks hibernated in it. The fourth winter site is located in the northeastern part of the Salair ridge along the border with the Novosibirsk and Altai regions, where about 60 elk are counted. The territory of the winter sites is guarded within the boundaries of the Salair reserve.

In recent years, most elks winters in the Kuznetsk Basin and in the northern lowland areas of the Kemerovo region in small groups of 2-3 to 7-10 individuals, sometimes up to 20-30 individuals. At the same time, animals keep in relatively small low snow forests with good fodder and protective properties. In accordance with changes in wintering patterns, the elk migration routes have also changed significantly.

Currently, a number of single long-distance migration routes with a length of more than 100 km are well defined. On these routes, the elks gather at large sites in the Sary Chumysh river basin and on the Salair ridge. The trajectory of the routes through the Kuznetsk Basin, which crossed major highways, including the highway under construction between Kemerovo and Leninsk-Kuznetsky, has changed significantly. The number of elks migrating along this route has decreased.

4. Discussion

In southeastern Western Siberia, the migration of elks was largely cross-border in nature, affecting the administrative territories of the Altai and Krasnoyarsk Regions, the Republics of Altai and Khakassia,

IOP Conf. Series: Earth and Environmental Science **395** (2019) 012020 doi:10.1088/1755-1315/395/1/012020

and the Novosibirsk and Tomsk regions. The main part of the elk population living in the Kemerovo region during the period of seasonal migrations did not practically go beyond the limits of the Kuznetsk-Salair mountain region. So, the elks from the western slope, administratively related to the Novosibirsk and Altai regions, came to the winter stands located on the eastern slope of the Salair ridge. In the past, in the Sary-Chumysh river basin dividing the Salair ridge and Gornaya Shoriya, the elks migrated from the Altai region. At the same time, wolves came after them. At the same time, part of the elks from the eastern regions of the Gornaya Shoria migrate to Khakassia, from the southern regions to the Republic of Altai and from the western areas of the Gornaya Shoria to the Altai Region [6], [7].

Since the 2000s, the cross-border migrations of elks have decreased significantly. In summer time, the elks inhabiting the Kuznetsk Alatau, Kuznetsk Basin, Mariinsky Taiga, practically do not go beyond the administrative borders of the Kemerovo Region. In recent years, the moose migrating to the Sary-Chumysh valley from the Altai Region ceased to be noted, and the elks migration from the Krasnoyarsk and Tomsk regions to the northern Kuzbass regions ceased.

5. Conclusion

Over the past 20 years, the nature of migrations and winter sites has undergone significant changes associated with anthropogenic factors: the impact of hunting, anxiety from snowmobile transport, construction of high-speed motorways, and an increase or decrease in wolves. Since the late 1990s, large sites have almost gone. Most elks in the Kuznetsk Basin hibernate in small herds, usually in 3-10 individuals of heads, dispersed in forest areas that are favorable for wintering and difficult to access for snowmobile vehicles and with low snow cover. In addition, a number of co-wintering elks depends on the area and feeding area.

We believe that the population of elks living in the territory of the Kuznetsk-Salair mountain region and in the northern regions of the Kemerovo region has adapted to the new conditions that have developed in the past 15 years. They are characterized by the absence of wolves and a changed factor of human anxiety. With the widespread use of snowmobiles and their illegal hunting, large concentrations of elk in the winter camps have become more vulnerable. Especially if these sites are poorly guarded. It is easier for small herds and groups to find remote areas for the winter camp. A small group of elks can feed themselves on a small area without making long-distance crossings and remaining inconspicuous for hunters and poachers.

References

[1] Mikhailov N I 1968 *Mountains of Southern Siberia: Physical-geographical zoning of the USSR* (Moscow, USSR: Moscow State University Press)

[2] Heptner V G, Nasimovich A A, and Bannikov A G *1961 Mammals of the Soviet Union: Even-toed ungulates and Non-even-toed ungulates (vol 1)* (Moscow, USSR: High School)

[3] Baranov P V, and Zaitseva A E 2002 Status and prospects of using the elk population in the Kemerovo region *InEka Eco-Bulletin* **6**(77) pp 16-17

[4] Yanushevich A 1931 *Inspection of the hunting industry of Gornaya Shoria: Materials on the study of Siberia (vol 3)* (Tomsk, USSR)

[5] Laptev I P 1958 *Mammals of the taiga zone of Western Siberia* (Tomsk, USSR: TSU Publishing House)

[6] Website Archive fund of the Kemerovo region 2019 *Reports on the work of the hunting management department at the Kemerovo Regional Executive Committee and reports of the Department for the Protection of Objects of Fauna of the Kemerovo Region (1951-1998, 2008-2018)* Available at: http://afond.kuzbassarchives.ru/index.php?act=fund&fund=181 (Accessed 02 03 2019)

[7] Bondarev A Ya, Zhuravlyov V B, and Petrov V Yu 2013 On the dynamics of the number of wolves and elks in Western Siberia *Vestnik of the Altai Agricultural University* **7**(105) pp 56-62

IOP Conf. Series: Earth and Environmental Science **395** (2019) 012021 doi:10.1088/1755-1315/395/1/012021

Optimization of processing branches, needles, and leaves ("woody greens") of dark-coniferous breeds of thin tree stands using a modified plant

E Petrenko[1*], V Parshikova[1] and E Zaychenko[1]

[1] Siberian Federal University, 2 L. Prushinskaya str., Krasnoyarsk 660075 Russia

E-mail: evp.2011@yandex.ru

Abstract. The waste-free technological scheme is proposed in the paper. It is based on an installation which allows to optimize the integrated processing of woody greens (branches, needles, and leaves) from fir and spruce. This scheme takes into account their contribution to the mixtures made, as well as the age of plantations used to produce an essential oil and pine extract. An installation is designed with an internal and external condensation circuit of the vapor stream serving as the basis, the use of which increases yields and significantly reduces the time needed to release the water-soluble products being processed into coniferous extract.

1. Introduction

When felling in Siberian forests, at least half of the tree biomass remains in the cutting area. In addition to the fact that logging waste poses a serious fire and biological hazard, the cost of their harvesting places a burden on the cost of wood products, reducing its competitiveness in the global market [1-3]. To date, only a small part of the green fir wood from their huge mass is processed to produce essential oils with a useful use of raw materials at 1-3% [4]. Developments, testifying to the possibility of using coniferous extracts in the production of furniture plates [5], [6], open up the prospect of large-scale development of this wood chemical direction. Its implementation provides for the involvement in the processing of green wood both fir and spruce, with the development and an increased yield of 15-17% of the second marketable product of coniferous extract. Along with the utilization of the generated logging waste, which increases the profitability of the wood processing industries, the development of this direction creates conditions for the functioning of small businesses in forest depleted areas with the emergence of new jobs. This seriously improves the social status of rural settlements.

The aim of the research is the development of the apparatus, technology, and processing lines for the wood greenery of fir and spruce in the forest areas covered by logging with the production of essential oil and coniferous extracts.

2. Materials and Methods

In Siberia, only green fir wood is processed on a semi-industrial scale. In rare cases, the processing is accompanied by the production of fir extract in small quantities consumed in feed and recreational purposes [1]. Occasionally, CO_2 extracts of coniferous shoots are obtained at pilot plants in a small amount. Approximately in the same volumes, St. Petersburg State Forestry Academy's employees produce a large range of wood-chemical products by means of water-gasoline extraction and

Content from this work may be used under the terms of the Creative Commons Attribution 3.0 licence. Any further distribution of this work must maintain attribution to the author(s) and the title of the work, journal citation and DOI.
Published under licence by IOP Publishing Ltd

IOP Conf. Series: Earth and Environmental Science **395** (2019) 012021 doi:10.1088/1755-1315/395/1/012021

fractionation from pine-spruce wood green. This range does not yet find consumption due to lack of demand [7]. In this connection, and also because of the comparative complexity of the technology, its implementation is hardly justified on the forest territory in Siberia. At the same time, the modification of coniferous extracts of adhesive masses, which improve the ecological consumer properties of furniture, indicates the possibility of the appearance of large-scale demand for them and, consequently, the urgency of developing a technology for producing such products.

The specificity of the raw material, which consists in reducing the content of essential oil with the almost unchanged contribution of water-soluble substances [4], indicates the feasibility of improving the apparatus for their selection.

3. Results

The basis of the developed apparatus, the laboratory setup for the extraction of plant materials was taken [9]. A diagram of the developed apparatus for separate distillation of the essential oil and leaching of extractive substances with water or ethanol (and another solvent) is shown in Figure 1. The extraction chamber and the refrigerator cover with flanges form the basis of the apparatus, as in the prototype [9]. For better sealing and convenience, the gasket between the flanges is replaced with a rubber cord placed in their grooves. Raw materials are located on interchangeable grilles of the chamber. Water (or other extractant) is poured into its bottom part, where the internal heater is placed. Filling is carried out either before loading the raw material or after it through a pipe with a plug in the bell of the lid of the refrigerator, mounted when an installation is improved. The bottom solution is discharged through the pipe at the bottom of the bottom part of the chamber. To complete the drain, its bottom is raised to 10°C on the opposite side of the pipe. Due to the tilting of the bottom, the camera body is mounted on a stand for stability. An additional heater, which allows to force heating, is placed in the resulting space. For process control and sampling, including in the vapor phase, two nozzles with plugs are in the chamber through which the corresponding equipment is introduced.

The main difference of the developed device from the prototype [9] is its addition of the florentine with a collection of essential oil to the external circuit consisting of a refrigerator, as well as a system of pipelines connecting florentin to the camera. Its connection to the main working device is carried out by replacing the regulating hole at the top of the lid-cooler of the prototype with a branch pipe with a valve that is connected to the entrance of an external refrigerator. During operation of this circuit, the steam oil flow is condensed in an external refrigerator with the internal refrigerator-cap disconnected. The florentine condensate is separated into florentine water and essential oil, which is collected in a tank. Removing oil from the condensate allows one to select both of the main commercial products of production in a pure form. This mode is especially important in the processing of raw materials rich in essential oil. Data on the yield of essential oil from fir-spruce wood greens with different ratios in the mixture contribute to the choice of the mode of processing of raw materials. The deoiled florentin water returns to the chamber through a piping system through a nozzle in the conical part of the fridge lid, irrigating the treated green wood and closing the water cycle.

The final stage of processing the oil-depleted raw material is carried out in the internal circuit of the installation during the condensation of the main mass of the stream in the tubes of the cooler lid. In addition, some of it is condensed in the above-pipe part of the cap. For a better discharge of the formed liquid, the upper part of the tubes is made flush with the plate tightening them. The condensate, flowing down the refrigerator's tubes, flushes out the products it carries from the flow that rises along them and enters the cubic part of the chamber, having passed a layer of green wood. Thus, during the operation of both the external and internal circuits of the installation, multiple extraction of raw materials takes place with almost "pure" reagent.

At the end of the process, the solution in the bottom part of the process (bottom condensate) is sent to the tank and then to the evaporator. Its concentration when bringing to the extract in connection with the viscosity of the product is possible only with external heating of the evaporator. In order to avoid burning and the resulting quality deterioration, this is done with constant mixing and control. An evaporating extractant (water) is condensed in the refrigerator. Then, the liquid is collected in the tank and through a pipe embedded in the pipeline with the florentine water, is sent to the chamber through a pipe in the conical part of the cooler lid.

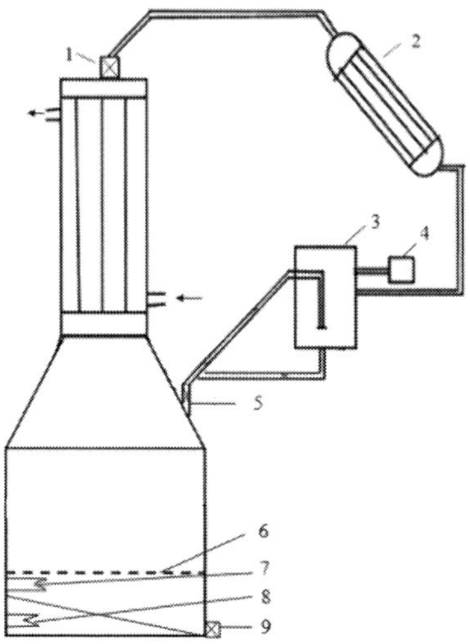

Figure 1. A scheme of the installation of the extraction processing of wood greens.

The technology of complex processing of fir-spruce wood greens with obtaining essential oil and pine extract from raw material harvesting to utilizing its spent solid residue is presented in Figure 2. Before felling, the species composition and age of plantations are additionally estimated for making a decision on the choice of the mode of processing raw materials (1). With this in mind, at its completion, the green wood of felled trees is crushed to a predetermined size and accumulates in the bunker of the self-propelled unit. Its delivery to the warehouse of raw materials is carried out by the installation itself (2) or by specialized transport. Loading of raw materials into the working apparatus (5) is carried out by a fan or conveyor (4), using a belt conveyor (3). Wood greens entering the cyclone of the installation are loaded onto the grids with the help of a flexible hose, due to which the mechanized unloading of the spent residue is ensured with the help of a hoist. Upon completion of loading, the chamber is closed with a cooler lid, sealed, and prepared for launching according to the chosen option.

When raw material enriched with essential oils is loaded, the steam-oil flow, which is formed when hot steam passes through wood greens, is processed in the external circuit (6) of the installation. Steam is generated by heating the water in the bottom part of the chamber. In this case, the use of an autonomous steam generator is unrealistic, since the cyclical nature of the treatment of green wood with an agent is violated. Through the uncooled cooling lid and nozzle, the steam-oil flow enters an external cooler, and there it condenses. The resulting condensate in a florentine is divided into oil flowing into the tank and florentine water. The water with traces of oil is returned to the chamber through the nozzle in the conical part of the lid. Irrigating raw materials on the grid, it flows into the bottom part and a new cycle begins.

In the case of the using the depleted essential oil of wood greens after the distillation of its main quantity, the treatment of the steam-oil flow in the external circuit is replaced with the internal one. Such a translation consists in connecting the lid-cooler to the cooling system, which provides condensation in it of almost the entire mass of the steam-oil flow. The condensate is poured into the chamber onto the wood green through the tubes of the refrigerator, due to which the efficiency of leaching of extractive substances increases. The non-condensed part of the flow can be processed in the external sector, which allows high-boiling volatile products to be isolated.

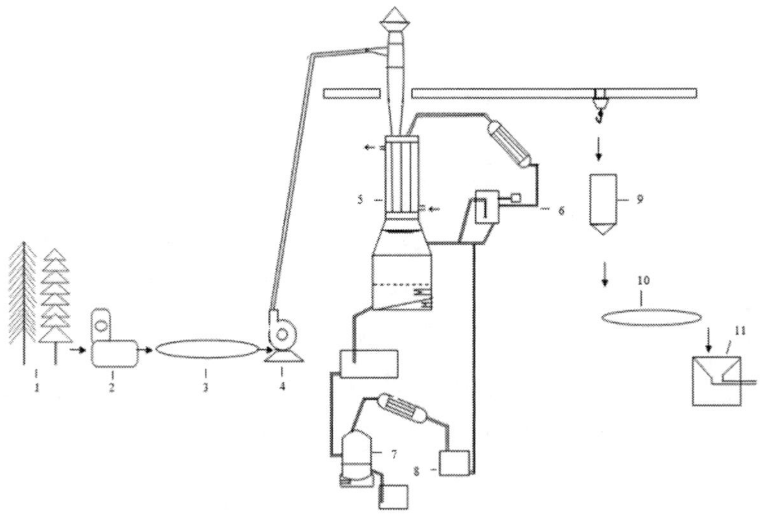

Figure 2. Technological scheme of the line for the integrated processing of fir-spruce wood green: (1) a stand, (2) a self-propelled installation, (3) a 10-belt conveyor, (4) a high-pressure fan, (5) a working unit, (6) an external circuit, (7) an evaporator, (8) a capacity, (9) a bin of spent raw materials, (11) a receiving device.

Processing the bottom condensates consists in concentrating them to the required density and returning the evaporated extractant to the chamber. Technologically, this process consists in accumulating the bottom solution in the collection and pumping it into the bottom part of the evaporator (7). The vapors formed here are condensed in the refrigerator, from where they enter the intermediate tank (8), and then periodically through the pipeline into the chamber. Extracts merge into a tank, are conditioned like an essential oil into a marketable product.

The spent green wood is transferred to the bunker (9) by means of hoists from the gratings, from where, when being used for economic purposes, it is loaded onto transport or during further processing by a conveyor (10), moving to the feed meal production workshop (11).

4. Discussion

A significant drawback of the installation, taken as a prototype, is the simultaneous isolation of the extract and essential oil, which, being constantly in the high-temperature zone, undergo negative changes. To avoid this, the oil should be separated from the rising steam-oil flow, which is advisable to produce outside the device, in the external circuit (which was implemented in this case). Thanks to the collection of the bottom condensate, in the case of using waste materials, wasteless processing of green wood is achieved.

Connecting the external circuit is also advisable when processing the depleted essential oil of wood greens. However, in this case, its functioning is justified during the initial period, when the main amount of terpenoids distills. The study of the dynamics of oil release indicates that 85-90% of its volume accumulates in the first third of the process duration.

The possibility of large-scale application in the production of furniture panels of coniferous extract and the development of equipment for its rational production allow, along with green fir, to have an economically profitable to recycle and seed shoots of other conifers, especially spruce. This increases a proportion of the useful use of this raw material from 1-3 to 15-17%.

The waste mass can be used as feed for animals and poultry, as a basis for fertilizing, structuring and refining the soil, etc. The softening of wood greens during hydrothermal processing and the removal of resin, phenolic and terpenoid compounds from its composition, which impede microbiological decomposition, contributes to this. The absence of volatile terpenoids favors the processing of waste raw materials into feed meal, and this fact reduces the fire hazard and the shredding of green wood.

The development of this direction opens up the prospect of organizing a wide network of small businesses in the wood chemical industry.

5. Conclusion

The high importance of coniferous extract as a multi-tonnage modifier of adhesive compositions of furniture panels indicates the feasibility of joint processing of fir-spruce green trees. With this in mind, waste-free technology, optimizing the choice of the mode of processing of raw materials, depending on the ratio in the mixture of wood green fir and spruce and the age of the stand, was developed on the basis of the modernized installation.

6. Acknowledgments

The work submitted for publication was carried out under the direction of Stepen Robert Alexandrovich (1936-2017), Doctor of Biological Sciences, Professor of the Siberian State Technical University, and the work is dedicated to his memory with gratitude for many years of effective cooperation.

References

[1] Medvedev S O, Soboleva S V, and Stepen R A 2010 *Possibilities for the rational use of wood waste in the Lesosibirsk timber industry complex* (Krasnoyarsk, Russia: SibSTU)
[2] Can Baser K H, Buchbauer G 2010 *Handbook of essential oils: science, technology, and applications* (Boca Raton, FL: CRC Press)
[3] Galyavetdinov N R, Safin R R, Voronin A E, and Shaikhutdinova A R 2015 *The mathematical modeling of the wood greenery extraction processes* In *International Conference on Mechanical Engineering, Automation and Control Systems (MEACS)* (IEEE)
[4] Stepen R A, Nevzorov V N, and Nevzorova T V 2010 *Organization of fir oil production* (Krasnoyarsk, Russia: KrasSAU)
[5] Petrenko E V, Parshikova V N, and Stepen R A 2012 Hydrothermal processing of spruce-fir wood green *New Advances in Chemistry and Chemical Technology of Vegetable Raw Materials* pp 520-521
[6] Petrenko E V, Parshikova V N, and Stepen R A 2010 Use of an alcoholic extract of green wood as an adhesive material *Actual Problems of Contemporary Science and Ways to Solve Them* pp 187-190
[7] Yagodin V I 2000 Fundamentals of wasteless technology of wood greens *Problems of Chemical Processing of Raw Wood* pp 50-58
[8] Chernyaeva G M, Dolgodvorova S Ya, and Stepen R A 1987 *Disposal of woody biomass* (Institute of Forest and Wood SB AS USSR)
[9] Parshikova V N, Stepen R A, and Demina L N 2008 Patent for invention (No. 2281135) Laboratory installation for the extraction of vegetable raw materials

An anthropogenic impact on the atmosphere and measures taken to protect it in 1988-2017: data from the industrial Kemerovo region (a historical aspect)

A Kovalevskiy[1*], Ye Luchnikova[1] and V Ilyashenko[1]

[1] Kemerovo State University, 6 Krasnaya str., Kemerovo 650000 Russia

E-mail: passer125@yandex.ru

Abstract. The development of industry in the Kemerovo region was accompanied by a significant increase in anthropogenic pressure, including in the form of air pollution, which ultimately leads to a decrease in the quality of life of the local population. Due to environmental legislation adopted at the federal level, a favorable trend in reducing emissions of pollutants into the atmosphere can be traced. However, in practice, the measures taken did not lead to a noticeable improvement in the quality of atmospheric air in large cities of the region. On the positive side, we should note that the environmental legislation and the centralized system for monitoring and controlling the quality of the environment were laid towards the end of the USSR existence. In the Russian period of history, financial incentives for the transition to cleaner production technologies began to be applied. The right to a favorable environment is enshrined in the Constitution of the Russian Federation.

1. Introduction

In accordance with Article 42 of the Constitution of the Russian Federation, "Everyone has the right to a favorable environment, reliable information about its condition, and to compensation for the damage caused to its health or property by an environmental offense" [1]. Consequently, atmospheric air protection is one of the pressing problems facing public authorities, especially in industrialized regions. Due to their peculiarities, various substances emitted into the atmosphere can be spread over many hundreds and thousands of kilometers, leading to negative consequences in territories that are far beyond the borders of an industrial region. At the same time, in residential areas adjacent to industrial facilities, concentration of various pollutants reaches such values that living here becomes unsafe. Constant stay in conditions of high levels of air pollution leads to increasing incidences of respiratory organs, blood systems, cancer pathology. Ultimately, it increases mortality rates and contributes to the decreasing life expectancy, constantly increasing the burden on medical health authorities [2, 3].

In connection with the states above, the research purpose is to study the measures taken with a purpose to improve an atmosphere in an industrialized region on the example of the Kemerovo region.

2. Materials and Methods

The study is based on a retrospective and chronological analysis of statistical, scientific, and legislative sources related to the Kemerovo region. Also, the paper presents unique own data on assessing the state of the environment based on the level of asymmetry of a birch by *Betula pendula*

Content from this work may be used under the terms of the Creative Commons Attribution 3.0 licence. Any further distribution of this work must maintain attribution to the author(s) and the title of the work, journal citation and DOI.
Published under licence by IOP Publishing Ltd

IOP Conf. Series: Earth and Environmental Science **395** (2019) 012022 doi:10.1088/1755-1315/395/1/012022

Roth. The quality of the environment is its condition, which is necessary to ensure the health of humans and other species [4]. The chronological scope of the study is limited to 1988, since it was precisely this year when the Decree of the Central Committee of the CPSU and the Council of Ministers of the USSR No. 32 "On the Fundamental Restructuring of Nature Conservation in the Country" was published on January 7 [1]. The research is also limited by the year of 2017 for which statistical data on the state of the environment has been published [3]. Geographically, the study is limited to the administrative boundaries of the Kemerovo region.

3. Results

Stability of development as an ability to develop normally (without disturbances and errors) is a sensitive indicator of the state of natural populations and allows to estimate the total value of anthropogenic load. Our studies in the city of Kemerovo in 2013-2017 indicate the critical state of the environment within the entire city. To assess the quality of the environment, we used a birch tree, according to the methodological guidelines [4].

When considering dynamics of the total volume of polluting emissions into the atmosphere of the Kemerovo region, the general trend of their gradual increase is traced from 1,475 kilotons (kt) in 1990 to 1,719 kt in 2017 (Fig. 1) [2, 3].

In the dynamics of the analyzed indicators, we find that that in the 90s of the twentieth century, there was a decrease in the volume of coal production, industrial and chemical production by 60-80% if compared to 1990 [5, 6]. Consequently, the amount of pollutant emissions in the atmosphere dropped by about 20%. The technical deterioration of equipment, coupled with the failure to implement the planned measures for the construction of wastewater treatment plants against the background of an acute social and economic crisis, did not lead to the expected proportional reduction in emissions.

Analyzing the structure of emissions of pollutants by sources of pollution, we can note the general trend of reducing emissions from mobile sources from 320.0 kt in 1990 to 227.8 kt in 2016. In the Kemerovo region, the main mobile source of pollution is motor transport.

In relation to stationary sources of air pollution, the general trend of increasing emissions is traced. However, over the past three decades, in the structure of the volumes of pollutants, the share of hydrocarbons, which are mainly represented by methane, has increased to 55% of the total emissions of pollutants (Fig. 2). The main supplier of methane is coal mining enterprises. At the same time, the volume of emissions does not depend on the volume of coal mining, but it depends on the methane content of coal seams.

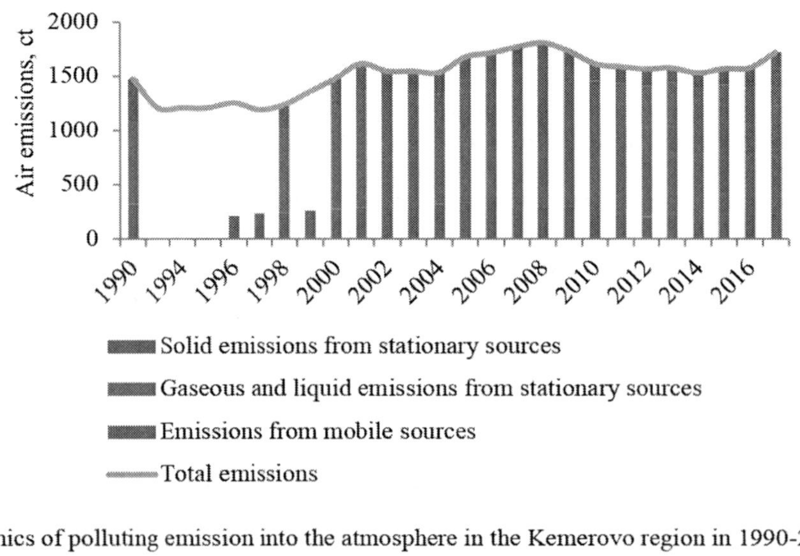

Figure 1. Dynamics of polluting emission into the atmosphere in the Kemerovo region in 1990-2017.

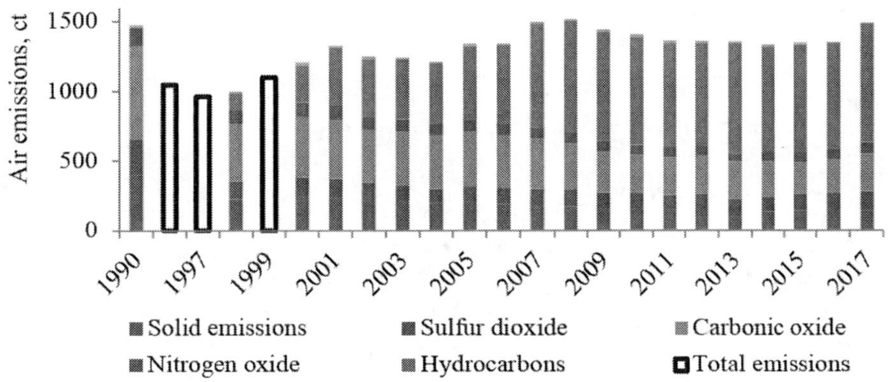

Figure 2. The structure of emissions from stationary sources into the atmosphere in the Kemerovo Region in 1990-2017.

4. Discussion

In general, characterizing the situation on the protection of atmosphere, it should be noted that in 1978 was adopted the State Standard of the USSR GOST 17.2.3.02-78 "Environmental Protection. Atmosphere: Rules for Establishing Permissible Emissions of Harmful Substances by Industrial Enterprises" (approved by the USSR State Standard Resolution of 24 August 1978 N 2329). Subsequently, this GOST was repeatedly modified. Also, various technical requirements were introduced to the newly created objects. The USSR Law No. 2353-X on the Protection of Atmospheric Air was adopted on June 25, 1980. A little later, the Law of the RSFSR on the Protection of Atmospheric Air was adopted on July 14, 1982 [1]. However, these requirements were introduced by directives, without taking into account the scientific and technical potential and economic capabilities of the country. Moreover, the failure to comply with environmental protection measures did not bear substantial responsibility, as a result of which the implementation of treatment facilities was implemented according to the residual principle [2]. In 1999, the Federal Law on the Protection of Atmospheric Air was adopted. Unlike Soviet legislation, it contains an article establishing a charge for the emission of pollutants into the atmosphere [1].

In the Criminal Code of the Russian Federation, in accordance with Article 251, air pollution that caused harm to human health can be punished for up to 2 years of imprisonment, and up to 5 years for causing death [1]. However, according to the report of Roman E. Erygin from the Main Directorate of the Ministry of Internal Affairs of Russia in the Kemerovo Region, 2 criminal cases under Article 251 of the Criminal Code of the Russian Federation were initiated as of November 16, 2018. At the same time, not a single conviction has been pronounced in the entire history of this article.

Estimating the impact of motor transport on the state of atmospheric air, we find that relative to 1990, the number of cars was 297% as of 2017 [2, 7], and the amount of emissions was 71%. Thus, with the increase in the number of cars almost in 3 times, the amount of pollutants emitted from them decreased by 29%. This situation became possible due to a number of measures taken by the Government of the Russian Federation on the phased introduction of environmental standards regulating the content of harmful substances in motor vehicle exhaust gases. In 2005, on the basis of these resolutions, the *Euro-2* environmental standard was adopted. It pushed domestic automakers to stop producing carburetor cars since 2006. Since 2016, the *Euro-5* standard is valid in Russia, according to which the standards for polluting emissions into the atmosphere are significantly tightened.

According to some publications in the media, local governments "turn a blind eye" to a number of environmental offenses. So, at the press conference on the issue of environmental pollution from the Belovo Zinc Plant and the health problems of the local population caused by this circumstance, the

IOP Conf. Series: Earth and Environmental Science **395** (2019) 012022 doi:10.1088/1755-1315/395/1/012022

position of the deputies was formulated as follows: "Yes, production is harmful, but you really want to lose another enterprise?" [8].

5. Conclusion

In sum, in the history of atmospheric air protection, two main development stages should be distinguished:

- In the Soviet period, the legal basis for environmental protection and a centralized system of monitoring the quality of the environment had been founded.
- In the Russian period, financial incentives for transiting to environmentally friendly production technologies began to be actively applied. Much more publicity on ecology and environmental protection one could observe, which also contributed to the growth of public activity.

At the same time, despite the presence of positive trends in the field of atmospheric protection for three decades, in large industrial cities of the Kemerovo region, such as Kemerovo and Novokuznetsk, air pollution is rated as "high" and "very high," respectively. The current criminal article on air pollution is not applied in practice. On the part of state authorities, there is a fear that a more stringent environmental policy will lead to stagnation of production and, consequently, an increase in unemployment.

References

[1] *GARANT system* n.d. Available at: http://ivo.garant.ru/ (Accessed 29 03 2019)

[2] Administration of the Kemerovo Region 1993 Amphosphere rehabilitation program In *Territorial Integrated Environmental Protection Program of the Kemerovo Region until 2005* (vol 3) (pp 149-197) (Kemerovo, Russia)

[3] Regional Committee of Natural Resources 2018 *Ecology and Natural Resources of the Kemerovo Region* Available at: http://ecokem.ru/doklady-o-sostoyanii-okruzhayushhej-sredy-kemerovskoj-oblasti/ (Accessed 31 03 2019)

[4] Ministry of Natural Resources of the Russian Federation State Service for the Protection of the Environment 2003 *Methodical recommendations for the assessment of the quality of the environment on the state of living organisms (assessingthe developmental stability of living organisms by the level of asymmetry of morphological structures)* (October 16, 2003 No. 460-p) (Moscow, Russia)

[5] Uporov D A, and Prilepskaya L L 2006 State and development of the chemical industry of the Kemerovo region in 1985–2005 *Vestnik of the Kuzbass State Technical University* **6-1**(57) pp 145-149

[6] Biryukova O V 2013 History of the Kuzbass coal industry restructuring (1994-2012): achievements and difficulties *Vestnik of the Kemerovo State University* **2-3**(54) pp 194-198

[7] *Federal State Statistics Service: Kemerovostat* (n.d.) Available at: http://kemerovostat.gks.ru/wps/wcm/connect/rosstat_ts/kemerovostat/ru/statistics/grp/ (Accessed 29 03 2019)

[8] Nikolaev A 2005 *Belovo Zinc Plant was ordered to live long* Available at: https://www.metaltorg.ru/analytics/publication/?id=1684/ (Accessed 09 04 2019)

IOP Conf. Series: Earth and Environmental Science **395** (2019) 012023 doi:10.1088/1755-1315/395/1/012023

Green areas as an element of sustainable development in urban settlements

E V Potapova[1*], O E Sokolova[1] and O A Barkhatova[1]

[1] Irkutsk State University, 1 Karl Marx str., Irkutsk 664003 Russia

E-mail: e.v.potapova.isu@mail.ru

Abstract. The authors propose a number of indicators of green areas and objects of gardening as proxies for sustainable development: closeness of crowns, soil sodification, a protective cover of herbaceous plants, and a one-time recreational load. The article provides a brief explanation on the inability to use some classical indicators, such as species diversity, biomass, etc. on the territories of some settlements.

1. Introduction

Currently, the well-being of humanity is increasingly determined by the quality of life in cities. More than half (52%) of the estimated 7 billion people in the world live in the urbanized space, and the growth rate of urban residents (on average by 2.2% annually) is much higher than the rate of increase in humanity [1]. After the decision was taken at the most significant United Nations Conference in Rio de Janeiro, on the transition to the path of sustainable development of all countries, the approach to assessing the quality of the urbanized environment changed somewhat. It began to depend on the degree of sustainability of the development of settlements, and subsequently, according to the "Habitat" UN program and the "Healthy Cities" WHO program on the viability. The viability of settlements is determined by the totality of their constituent elements, including plantings and green areas.

In contemporary conditions, the problem of increasing sustainability and optimizing the functioning of a landscaping facility is becoming increasingly important. According to the GOST "Greening Cities: Terms and Definitions," sustainability of green plantings is the ability of plantings to maintain the nature of functioning under the influence of anthropogenic factors [2]. Indicators are needed to identify the status and prospects of development.

As a rule, indicators of sustainable development are the numerical, measured values of a number of parameters, including the environment, which can be estimated either by their absolute value or by comparison with previously observed values. In many ways, these indicators have become indicators of the quality of the human environment, which should be achieved in a healthy, environmentally friendly settlement.

Sustainable development indicators are necessary for:

1. Justifying the decision by quantifying and simplifying;
2. Interpretation of changes;
3. Identifying deficiencies in environmental management;
4. Providing access to information for different categories of interested users;

Content from this work may be used under the terms of the Creative Commons Attribution 3.0 licence. Any further distribution of this work must maintain attribution to the author(s) and the title of the work, journal citation and DOI.
Published under licence by IOP Publishing Ltd

5. Exchange of scientific and technical information [3].

Development of criteria and indicators of sustainable development is actively going on in the world. According to the UN, the constant work on the identification and use of new indicators can reduce the subjectivity of the estimated parameters. Leading international organizations such as the UN, the World Bank, the Organization of Countries for Economic Cooperation and Development, the European Commission, and others are engaged in this. The Center for Environmental Policy of Russia and S. N. Bobylev developed accessible recommendations for identifying, applying, and classifying various indicators, as possible indicators of sustainable development. One of the most comprehensive in terms of coverage of sustainable development indicators system was developed by the UN Commission on Sustainable Development, consisting of 132 indicators, divided into four main groups: social, economic, environmental, institutional [4]. Their analysis leads to contradictory conclusions: urbanization is a global process, as indicated by a number of social indicators, most of the processes of environmental degradation are due to urbanization processes, which is reflected in both social, economic, and environmental indicators. At the same time, in indexing settlements and their conditionally natural component in the form of green areas, the latter have no weight and are not even actually represented by any indicator. This seems to be extremely strange, since the plantings of settlements fulfill the whole variety of ecosystem services.

Therefore, the purpose of the presented work is to determine one of the most important indicators of the quality of an urbanized environment: the presence and condition of plantations and green areas [5].

2. Materials and Methods

The methodology of this work can be divided into three main parts, which are combined by the system analysis. At the first stage, both domestic and foreign literature on the topic is studied and summarized. The second part is the field research in more than 200 settlements and 10 thousand green areas. And the third is an analytical and comparative analysis of the first two stages and the identification of indicators of the state, pressure, and response of plantings and green areas as elements of a sustainable development strategy for the world, country, region, etc.

3. Results

Plant communities (as well as individual species) that have a sufficiently defined and persistent connection with environmental conditions and are used to recognize these conditions are the natural biological indicators. Species diversity (usually represented as indices of diversity – Shannon, Simpson) is a classic bioindicator. But the calculations for the settlements territories have extremely large deviations due to, for example, the high dominance of some species, the local distribution of rare species, and those areas with artificial diversity. Due to the fact that during dumping, digging, and other improvement activities, at a particular point in time, a set of species can be so random and change two or three times per season. Biomass is a commonly used indicator. In a particular case, a comparison of the biomass of naturally overgrown plots, mowed areas covered by the elemental road network are also to be a mistake [6].

Field studies in each greened territory fix more than 100 different indicators, which are the characteristics of greening elements, determination of anthropogenic load, class of hemera activity, the establishment of compliance parameters for a greened territory with the requirements of town planning regulations applied for each type in the category, and many others. Some of them can be considered as indicators. According to the criteria applied to indicators of sustainable development, they must meet the following conditions:

1. Relation to sustainability (indicative / not indicative);
2. Representativeness (matching the characteristics of the type of greened territory);
3. Representativeness (single, rare, usually, often);
4. Measurability (complex, medium, light);

5. Exclusivity (condition applies only to specific areas, for most, for all);
6. Regular availability (constant, every season, every year or every 2 years, less often);
7. Easy feasibility and affordability;
8. Rapid applicability in a short time;
9. Providing an opportunity to clearly determine the direction of the changes (positive or negative);
10. An accessibility for the general public to understand;
11. An efficiency when using;
12. Significance (high, medium, low);
13. Reliability.

Out all the indicators of plantations and green areas, the authors suggest the following as the most representative indicators: a tree crown density, a level of soil sodification, a projective cover of grassy plants, and a one-time recreational load. Cultivation indicators are not extrapolated to the entire area of a recreational load is an indicator of pressure and can be considered under certain conditions for stability with the opposite sign.

In fact, all 13 conditions for the proposed five indicators are performed in the greatest manifestations. The highest values of sustainability characterize the most stable areas – urban forests. A large representativeness is predetermined by the spread of green areas, including as mandatory building elements. Representativeness is associated with a complete contingency, as a result, it is frequent, constant. Measurable - light, does not require special equipment. Availability is constant, it may decrease to the length of the growing season for some regions. Financially, it is determined only by labor hours, unambiguous fast applicability. In accordance with high rates for most criteria, indicators can be considered reliable and significant. In fact, they are understandable and effective in use, since they do not require recalculations. Another positive criterion is a small range of indices from 0 to 100% (soil sodification, as well as a protective cover of grassy plants), one from 0 to 1 (tree crowns closeness). And only one conditionally does not have a maximum size (a one-time load) [5].

The proposed indicators reflect the specific state of the greened area, and their changes or compliance with certain values can be indicators of pressure. It should be noted that each of the proposed indicators can be viewed from the point of view of the environmental, economic, social, and institutional components of sustainable development or divided into 4 environmental and 1 social.

4. Conclusion

For more than a quarter of a century, sustainable development has been high on the international agenda. Assessment of the quality of life of the urban population requires the study of all the factors determining it: economic, social, environmental, and institutional. The authors present some indicators that can be used as indicators of sustainable development of green areas and settlement estimates. Undoubtedly, further development that satisfies all the requirements for evaluating the object of the survey, including the dynamics, is necessary. Environmental adjustment of indicators of economic development and progress is required. The nature "competitiveness" in the fight against the man-made solutions should be improved.

References

[1] World Bank 2013 *World development indicators* Available at: http://data.worldbank.org (Accessed 03 02 2019)

[2] GOST 28329-89 1990 *Urban landscaping: Terms and definitions* (Moscow, Russia: Publishing Standards)

[3] Klimov A G 2005 *Features of farm organization in the forest-park part of the green zone of the city of Chita* (Dissertation of the Candidate of Agricultural Sciences) (Krasnoyarsk, Russia)

[4] Tarasova N P n.d. *Indexes and indicators of sustainable development* Available at: http://www.ustoichivo.ru (Accessed 05 02 2018)

[5] Potapova E V 2017 *Landscaped territories of settlements: structure, state, problems, risks, transformation, development indicators* (Dissertation of the Doctor of Agricultural Sciences) (Yekaterinburg, Russia)

[6] Goodman G T, and Roberts T M 1971 Plants and soils as indicators of metals in the air *Nature* **231** pp 231-287

On the issue of ensuring environmental safety of anthropogenic systems in cross-border areas of the Novosibirsk Region

E Malgin[1*], E Shemetova[1] and I Bochkarev[1]

[1] Siberian University of Consumer Cooperation, 26 Karla Marksa prosp., 630087 Russia

E-mail: malgin1954@mail.ru

Abstract. This article explores the concept of a cross-border region and analyzes sources contributing to environmental pollution. More than that, the authors focus on the features of monitoring and ensuring environmental safety of the cross-border region of Novosibirsk (Russian Federation) and Pavlodar Regions (Republic of Kazakhstan). As a result, measures and algorithms to optimize their sustainable development are suggested.

1. Introduction

The relevance of our study is determined by the practical importance of the topic related to the role played by cross-border (formed by interactions of administrative-territorial and municipal entities and their subjects) regions in the processes of cooperation and security. Globalization, on the one hand, and national identity, on the other, create contradictions that are clearly manifested in the implementation of the concept of sustainable development and environmental security in cross-border regions.

With respect to their operation, the border areas are divided into geopolitical "buffers," border "clamp regions," frontier regions, and "cross-border regions." At the present stage of research, the concept of "cross-border region" is commonly used. The concept of a "cross-border region" implies mutual border crossing and various forms of cooperation [3, 4, 5, 8, 13, 17]. V. A. Kolosov introduced this concept into scientific circulation [6, 7]. According to V. S. Korneevets, this is a "specific form of organization of society" [8, 9].

Of course, there is a problem of categorical design of the concept depending on research aspects [3, 10]. O. E. Brednikova argues that contemporary space oversteps and ignores borders, but it is formed due to the existence of the border itself [2, p. 492-497]. P. Ya. Baklanov and M. Yu. Shinkovsky believe that a cross-border strip defines a part of the territory linking different sectors of a region, and a contact zone indicates areas where interaction is most successful. These parts are necessary to understand certan characteristics of the formation of a cross-border region. [1, p. 16]. According to the definition of T. N. Kuchinskaya, "A transboundary region is a potential region: it is divided by the sovereignty of neighboring states, it has a complex of national, regional, and zonal elements with their own characteristics, it reflects their historical and cultural identity, interacts between adjacent border regions to preserve the management and development of their "living space," it is formatted by the state administrative-territorial border" [10]. According to A. N. Demyanenko, terminological

Content from this work may be used under the terms of the Creative Commons Attribution 3.0 licence. Any further distribution of this work must maintain attribution to the author(s) and the title of the work, journal citation and DOI.
Published under licence by IOP Publishing Ltd

IOP Conf. Series: Earth and Environmental Science **395** (2019) 012024 doi:10.1088/1755-1315/395/1/012024

vagueness (and even entanglement) has a very definite character, an extremely complex object of study [5].

How to combine the concept of sustainable development and ecologization of a cross-border region? For the first time, the term sustainable development was used (1987) in the report "Our Common Future" by the UN World Commission on Environment and Development as a development model based on the need to balance the solution of social, economic problems in favor of the environment. How to ensure this *balance*, environmental safety in specific *cross-border regions*, taking into account different legal and organizational conditions of border areas management? Is sustainable development relevant in general to cross-border regions? In contemporary publications on cross-border interaction, according to V. S. Molotov and other researchers, assessing the response of cross-border geosystems to different types and intensity of economic impact did not receive sufficient coverage [11].

2. Materials and Methods

Our research methods include (1) analysis of publications on this issue (conducting a literature review); (2) identifying sources of chemical pollution of ecosystems in the Pavlodar region (Kazakhstan); (3) statistical reporting: environmental conditions, an integrated assessment of the state of air, water, and soil according to the results of chemical monitoring.

The studied transboundary Novosibirsk-Pavlodar region borders with the Omsk region in the north, with the Novosibirsk region in the northeast, and with the Altai region in the southeast. It is the largest industrial region of Kazakhstan. 7.3% of the country's total industrial production is concentrated in the region. "Kazdata marketing technologies" informs about 18 large enterprises in Pavlodar and its surrounding region [16].

The concept of border cooperation of the Novosibirsk and Pavlodar regions until 2025 provides an analysis of the area and unresolved problems associated with a low level of investment cooperation, common interests in environmental protection, veterinary and sanitary well-being, mutual information on the transfer of hazardous waste by land, air and water flows, with the implementation of environmental monitoring, prevention and assistance in eliminating natural and man-made emergencies [12].

The portal "Energy.Media" and the information bulletin of RSE "Kazhydromet" *On the State of the Environment of Pavlodar Region for the Period 2014-2018* [15] noted a *high level* of air pollution in sulfur dioxide (2.4%) and nitrogen dioxide (14,6%); *an increased level* of suspended particles of PM-10, carbon monoxide, nitrogen dioxide, nitrogen oxide, ozone, hydrogen sulfide, phenol, hydrogen chloride; and a *low level* of nitrogen oxide and ozone. In the region, , the volume of industrial emissions into the atmosphere has grown by 46% over the past 10 years. The reason behind such an increase is the growing electricity generation (by 40-60%) and a simultaneous increase in emissions to 58.2 tons only at the following enterprises: CHP-1, Aluminum Kazakhstan JSC, Ekibastuz SDES-1 (state district electric station). The "Kazgidromet" has only 7 environmental posts in the region, which are designed to measure only 9 components. Since 2017, air monitoring has been carried out on 15 pollutants (suspended particles PM-10, suspended particles PM-2.5, sulfur dioxide, sulfates, carbon monoxide, nitrogen dioxide, hydrogen sulfide, ozone, phenol, chlorine, hydrogen chloride, the sum of hydrocarbons, ammonia, methane). At the same time, large industrial enterprises emit more than 130 harmful substances. How does this affect the Novosibirsk region? Let's review the repeatability of various wind directions (Table 1).

Table 1. Repeatability of various wind directions, %.

Directions	JAN	FEB	MAR	APR	MAY	JUN	JUL	AUG	SEP	OCT	NOV	DEC	Year
North	3	5	8	11	12	17	21	18	11	7	5	3	10
Northeast	4	6	8	11	8	13	12	11	7	5	5	2	8
East	8	8	8	10	10	10	10	10	9	7	7	6	8
Southeast	13	11	10	9	7	5	5	5	7	10	11	11	8
South	27	25	19	12	15	14	12	13	17	20	21	28	19
Southwest	22	21	15	13	13	10	7	8	13	20	21	22	16
West	19	20	25	21	20	15	13	17	21	22	23	22	20
Northwest	4	4	7	13	15	16	20	18	15	9	7	6	11
Calm	5	4	3	2	4	5	6	6	5	4	3	3	4

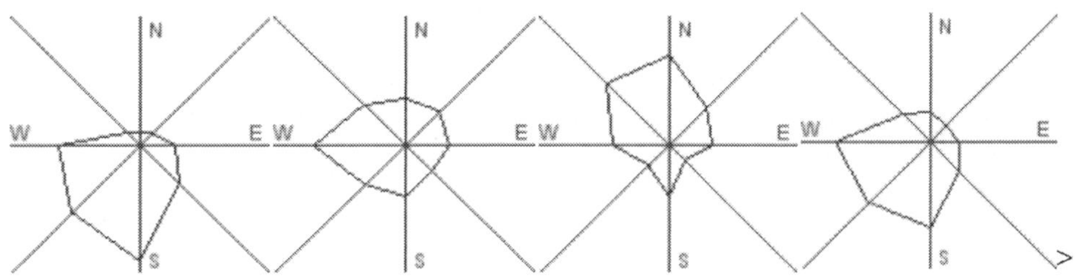

Figure 1. An average rate of wind direction for the year [16].

The contact area of the Bagansky, Karasuk, Kupinsky, Chistoozerny districts of the Novosibirsk and Pavlodar regions is 200-250 km, and the distance from the enterprises to the border of the Novosibirsk region is 150 km. An average indicator of wind direction for the year is 36% to the north and east, and these are the areas of the Omsk and Novosibirsk regions, which means that the likelihood of pollution remains. In 2018, the FGUZ Center for Hygiene and Epidemiology in the Novosibirsk Region measured a content of zinc in the soil of the Chistoozerny District, in which pollution can be characterized as dangerous [14]. On October 25, 2018, the Government of Kazakhstan made a decision on installing automated stations of atmospheric air, 14 complexes of high-precision measuring instruments of operational response, determining the content of hydrogen sulfide, mercury, radiation in the Pavlodar region. They are to be located at a distance of 3 meters from the source, where the air flow is uniform and there is no inversion. In 2015, 8 enterprises were subject to environmental analysis, but their number increased to 19 in 2019 [15].

3. Research Results

Annually, the following is emitted into the atmosphere in Kazakhstan:

1. 4.5 million tons of harmful substances;
2. 25% of this amount falls in the Pavlodar region;
3. 98% is formed on the territory of three cities – Pavlodar, Ekibastuz and Aksu (Kazkhstan), the main sources are the ferroalloy plant, "Ksp Steel," and "Aluminum of Kazakhstan";
4. The problem of air and water pollution remains relevant: violations in operation of ash dumps are made by the Eurasian Energy Corporation (EEC);

IOP Conf. Series: Earth and Environmental Science **395** (2019) 012024 doi:10.1088/1755-1315/395/1/012024

5. Water samples in the Irtysh River observe an excess of the maximum allowable concentrations of copper, zinc, manganese, and petroleum products from 1.5 to 2.5 times; the flow of the Irtysh river has decreased three times in the last 30 years; the river grows shallow; floodplain meadows suffer;
6. Eco-bioprotection technology is not actively implemented, installation of automatic monitoring systems at enterprises is at the decision stage;
7. More than 30% of the air masses per year is directed to the territory of the border regions of the Novosibirsk Region, in which an automatic system for a comprehensive assessment of the environmental situation has not yet been created. At the same time, the Pavlodar region is the only one in the republic that is exploring the target indicators of environmental quality (the second stage in 2020, and the third stage in 2023) [18].

Modernization of the wastewater treatment systems of enterprises includes installation of electrostatic precipitators of the new generation (ZUU) and boilers at power units; eco-bioprotection technology (scrubbers, emulsifiers, gas ducts). Modernization of sewage treatment plants of enterprises will give flue gas cleaning efficiency up to 99.38%. So, to ensure environmental safety of anthropogenic systems in the cross-border region of the Novosibirsk Region, we need to apply the dual purpose of V. S. Molotov: (1) Ensuring the quality of environmental conditions; (2) creatng an economy with ecological industries. They can be implemented through *"optimization of environmental management,"* on the one hand, or *"optimization of anthropogenic impacts on the environment"*, on the other [11]. The first one does the following: (1) improving an environmental management system; rationalizing a territorial distribution of economic activity (b) developing a "basin" approach. The second implies: (1) technological modernization of production; (2) environmental safety engineering; (3) environmental regulation and control.

4. Discussion

On October 25, 2018 in the Government of Kazakhstan, the hearings were held with the presence of the public and experts on environmental issues existing in the Pavlodar region, which ranked first in the country in terms of harmful emissions. According to the revealed facts of exceeding the threshold limit value (TLV) at the borders of sanitary zones at the "Maikuben-West" LLP, "EEC" JSC, and "Kazchrome" JSC, large fines were imposed on them. According to the results of these hearings, the decision was made on the installation of automated stations that would determine the content of hydrogen sulfide, mercury in ambient air, and radiological contamination at the enterprises of the region [18]. According to the results of our study, the West Siberian Administration for Hydrometeorology and Environmental Monitoring intends to improve monitoring over the cross-border environment, focusing on the Bagansky, Karasuk, Kupinsky, Chistoozerny districts of Novosibirsk and Pavlodar regions.

5. Conclusion

The problem of ensuring environmental safety of anthropogenic systems in cross-border areas remains unresolved. The basis for its implementation is to address those causes affecting the environment in the Novosibirsk-Pavlodar transboundary region, including the following: (1) a weak interregional environmental regulatory framework; (2) lacking financial resources; (3) irrational ecology of production, etc. The following tools should be applied: (1) improving the interregional environmental legislation; (2) geographical justification used in anthropogenic loads regulation; (3) a cartographic analysis of environmental management; (4) developing and applying a cross-border system of environmental quality indicators, a forecast scheme and environmental risk reduction measures; (5) providing a comprehensive automated control of environmental impact.

Thus, the activities to ensure environmental safety of cross-border territories are the implementation of means and methods of interstate regulation regarding the negative impact of economic activities on the environment, requiring a coordinated environmental policy of environmental management of cross-border territories and development of institutional mechanisms for cross-border cooperation.

6. Acknowledgments

We thank the West Siberian Administration for Hydrometeorology and Environmental Monitoring, as well as Administrations of the Bagansky, Karasuk, Kupinsky, and Chistoozerny Districts of the Novosibirsk Region for addressing a set of environmental problem and providing us with relevant information used in our research.

References

[1] Baklanov P Ya, and Ganzey S S 2010 *Border and cross-border territories of Asian Russia and neighboring countries (problems and prerequisites of sustainable development)* (Novosibirsk, Russia: Siberian Branch of the Russian Academy of Sciences Publishing House)

[2] Brednikova O E 2008 Frontierland as a social phenomenon (directions of sociological conceptualization) *Vestnik of St. Petersburg University* **4** pp 492-497

[3] Velikaya S A 2019 *The essence of the notion of a cross-border region* Available at: http://nauka-online.com/ (Accessed 02 03 2019)

[4] Volynchuk A B 2009 A cross-border region: theoretical foundations of research *Humanitarian Research in Eastern Siberia and the Far East* **4**(8) pp 49-55

[5] Demyanenko A N 2009 Sustainable environmental management in cross-border territories: theory and practice issues (on the book by P Ya Baklanov and S S Ganzey Cross-border territories: problems of sustainable environmental management) *Spatial Economics. Reviews* **3** pp 159-171

[6] Kolosov V A 2003 New limology: theoretical approaches *International Processes* **3** pp 44-59

[7] Kolosov V A Theoretical limology: new approaches Available at: http://www.intertrends.ru/three/004.htm (Accessed 02 03 2019)

[8] Korneevets V S 2010 Classification of the Russian border regions *Regional Studies* **4**(30) pp 48-53

[9] Korneevets V S 2010 Transnational and transboundary regions as specific forms of social organization (Dissertation of the Doctor of Geographical Sciences) (Saint-Petersburg, Russia)

[10] Kuchinskaya T N 2011 Cross-border region as a form of sociocultural space: in search of a cognitive model of research *Contemporary Problems of Science and Education* **6**

[11] Molotov V S 2014 Algorithm for ensuring the environmental security of cross-border territories *Vestnik of the Buryat State University* **4** *(Biology. Geography)* pp 16-22

[12] Administration of the Novosibirsk Region 2009 *Resolution "On approval of the Concept of Border Cooperation of the Novosibirsk Region until 2025 with the Administrative-Territorial Formation of the Neighboring State Pavlodar Region" (Republic of Kazakhstan)* Available at: http://www.garant.ru/hotlaw/novosibirsk/215423/ (Accessed 15 10 2014)

[13] Chernobay O L 2015 Cross-border cooperation of the Novosibirsk region with the Republic of Kazakhstan *Power* **23**(1) pp 87-90

[14] The Office of Rospotrebnadzor in the Novosibirsk Region and the Center for Hygiene and Epidemiology in the Novosibirsk Region 2017 *State report "On the state of sanitary and epidemiological welfare of the population in the Novosibirsk region in 2017"* Available at: rospotrebnadzor.ru (Accessed 02 03 2019)

[15] *RGP "Kazgidromet" of the Ministry of Energy of the Republic of Kazakhstan* n.d. Available at: kazhydromet.kz (Accessed 02 03 2019)

[16] *"KazData Marketing Technologies" Database of Kazakhstan: enterprises and organizations, import and export of Kazakhstan, production* n.d. Available at: kazdata.kz (Accessed 02 03 2019)

[17] *International Electronic Scientific Journal "Science Online"* n.d. Available at: http://nauka-online.com/ (Accessed 02 03 2019)

[18] *Pavlodarnews.kz* 2019 Available at: pavlodarnews.kz (Accessed 02 03 2019)

The importance of ecotonic habitats in the spatial organization of the Late Pleistocene ecosystems of Yakutia

A Protopopov[1], V Protopopova[2] and A Neizvestnykh[1*]

[1] Academy of Sciences of the Sakha Republic (Yakutia), 33 Lenina str., Yakutsk 677007 Russia

[2] Institute for Biological Problems of Cryolithozone of the Siberian Branch of the RAS, 41 Lenina str., Yakutsk 677007 Russia

E-mail: anna08062009@rambler.ru

Abstract. The mosaic of the existing vegetation cover of Yakutia in the Late Pleistocene is due to a combination of arid climate, as well as because of permafrost spreading and frequent fires. The combination of these factors creates necessary conditions for the formation of ecotone natural systems represented by dissipative plant communities. Sparse plant communities are characterized by the absence or weak expression of phytocenotic bonds. Their formation occurs stochastically under the absolute influence of external factors and is genetically entirely dependent on the surrounding vegetation, which acts as a donor of patient and ruderal species. Such processes were characteristic of the Late Pleistocene ecosystems of the Northern Hemisphere. A low productivity of such biotopes did not allow them to act as the main pastures for the representatives of the mammoth megafauna, apart from the saiga and bighorn sheep, but contributed to the increase in the biodiversity of the late Pleistocene ecosystems.

1. Introduction

In the 70s of the twentieth century, the concept of "zone ecotone" was introduced into scientific circulation. This concept meant a transitional state of the natural zone, characterized by the mixing of two different formations, for example, in the form of a mosaic cover [1]. One of the factors determining the existence of ecotone communities is the unstable extreme abiotic environment. Such border communities are usually formed by adding more ancient elements adapted to the microclimatic parameters, with evolutionarily new ones, for example, mixing ancient steppe groups with younger tundra. It was massively distributed in the Late Pleistocene in the Northern Hemisphere.

The mammoth faunistic complex biodiversity significantly exceeded the modern faunas of tundra, deciduous forests, taiga and steppes, which are now common in the same geographical latitudes. It is assumed that the high mosaicity of the vegetation cover due to the currently unparalleled natural and climatic conditions was the cause of such faunistic wealth.

2. Materials and Methods

The paleo-faunistic composition of the orictocoenosis of Yakutia was refined in the course of expedition studies, during which the osteological remains of large Late Pleistocene mammals were collected. In recent years, new osteological remains of *Saiga tatarica L. and Ovis nivicola Eschscholtz* have been found, it has greatly expanded the concept of their Late Pleistocene ranges. The age of the finds was determined by radiocarbon analysis. The choice of the saiga and bighorn sheep as an object of research

Content from this work may be used under the terms of the Creative Commons Attribution 3.0 licence. Any further distribution of this work must maintain attribution to the author(s) and the title of the work, journal citation and DOI.
Published under licence by IOP Publishing Ltd

due to the fact that the saiga is absent in the contemporary fauna of Yakutia, but it is widespread in the steppe zone of Eurasia, and bighorn sheep strictly adheres to the mountain habitat of Northeast Asia.

The description of modern vegetation of Yakutia was carried out by generally accepted geobotanical methods. Paleoentomological information is taken from open sources. Reconstruction of quantitative indicators of reconstructed paleofiocenoses was carried out by the method of terioindication [2, 3].

3. Results

During the expeditionary work of 2014-2015, fossil finds of bighorn sheep (*Ovis nivicola Eschscholtz*) were discovered in Khangalassky and Tattinsky districts in the territory of Central Yakutia. The interest to these finds is due to the fact that these animals do not live in these areas at the present time; and there is no mountain relief, which is a prerequisite for modern populations of bighorn sheep. The dating showed that the finds belong to the late Pleistocene. Saiga finds made in recent years on the Bolshoi Lyakhovsky Island and on the Adycha River of the Verkhoyansk District also belong to the Late Pleistocene. Fossil finds of saiga in Yakutia are rare, but they give a general idea of the range of this animal in in the Late Pleistocene to the north. More than that, the remains of the saiga have not yet been found in the Aldan-Amga interfluve and to the south (Central Yakutia) [4].

The contemporary cryoxerophilic communities in the Kolyma basin developing on higher ground with a deflated snow cover in winter and an increased summer draining of the soil surface are less productive [5]. Based on the zooindication features of the pill beetles (*Morychus viridis*) widespread in the Late Pleistocene orictocoenosis, the widespread associations formed by *Carex argunensis Turcz. ex Trev.* and *Polytrichum piliferum Hedw* in the Late Pleistocene are assumed [5]. Based on the characteristics of the modern disjunctive area of *Carex argunensis* with the presence of a relict population in the upper reaches of the Kolyma and the frequency of occurrence of residues of *Morychus viridis Kuzm. et Kor.* in the Late Pleistocene orictocoenosis, we can assume that in sartan, unproductive and rarefied communities involving this sedge were much more common than at present.

4. Discussion

Extreme climatic conditions of the Late Pleistocene of Yakutia were characterized by severe winters, low precipitation, continuous development of the cryolithozone [6].

Extremely low winter temperatures along with low snowfall, which contributed to the soil cooling, and summer desiccation to cryochrons of the Pleistocene led to the extinction of the dominant plant communities on the gentle elevations of the relief, where mountain-steppe, mountain-tundra and ruderal plants, phytocenotically unrelated to each other, occupied free ground in that time. One of the main factors for selecting plants for growing in such empty or ruderal habitats was xerophile combined with tolerance for cooling in winter in the absence of snow cover. This feature could be developed evolutionary only in mountain dry and cold enough conditions. This is similar to the microclimatic conditions of contemporary *Morychus viridis* habitats, which were characterized by low snowfall and strong winter winds, summer dryness, a wide distribution of sparse chionophobic groups in the vegetation in the stochastic combination of steppe and tundra plants [5].

Sparse plant communities are characterized by the absence or weak expression of phytocenotic links. Their formation occurs stochastically under the absolute influence of external factors and is genetically entirely dependent on the surrounding vegetation serving as a donor of patient and ruderal species adapted to the climatic and edaphic conditions, which are decisive in such dissipative communities [7].

Under the conditions of the Late Pleistocene Yakutia, such biotopes could be populated as synusia groups *Carex argunensis, Dryas punctata Juz., Selaginella rupestris (L.) Spring., Empetrum nigrum L.,* various steppe cereals, and individual wormwood, willows, grains, and haze species, creating a mixed type of vegetation.

A low productivity of such biotopes did not allow them to act as the main pastures for representatives of the mammoth megafauna, possibly other than the saiga. Since the 19th century, the saiga has been one of the main reasons for the tundra-steppe appearance of the mammoth fauna, personifying its steppe part. But contemporary studies of the ecology and biology of this animal indicate that it is not related to

IOP Conf. Series: Earth and Environmental Science **395** (2019) 012025 doi:10.1088/1755-1315/395/1/012025

the steppe, but more to semi-desert communities [8]. B. D. Abaturov and colleagues [9, 10] conducted very interesting studies on the saigas' food preferences, as a result of which it turned out that saigas try to avoid eating cereals in the presence of haze and buckwheat ones in pastures. As a result of research, it is also found that saigas prefer biotopes that have undergone pasture digression with a large number of weeds: *Kochia prostrate (L.) Scrad., Ceratocarpus arenarius L., Bassia sedoides (Pall.) Asch., Chenopodium album L.* of the haze family, *Polygonum aviculare L., Polygonum patulum M. Bieb.* from the buckwheat family, *Potentilla bifurca L. from Rosaceae*, and others. Some of these species are noted in the paleofloristic complex of the Late Pleistocene of Yakutia, such as *Polygonum aviculare, Descurainia Sophia L.*, and others (such as *Kochia prostrata* being present in the contemporary Yakut flora), and it was likely more widespread in the Late Pleistocene. Thus, a saiga with a wider distribution of ruderal communities in northeast Asia was provided with food supply to a sufficient extent. It should be noted here that *Kochia prostrata* necessarily grows in the contemporary saiga reproductive stations [8].

The saiga's remains in the Pleistocene sediments of Yakutia indicate a wide distribution of ruderal communities, composed of the haze, buckwheat, cruciferous, aster, and cereals in the Late Pleistocene. It is possible that these remains indicate being even more widespread than they are now, despite the increased anthropogenic load on the phytocenoses, which has recently led to an increase in the areas of ruderal communities of anthropogenic origin. Most likely, this is due to the fact that exogenous factors in the formation of vegetation cover prevailed over endogenous factors in the Late Pleistocene. In turn, this led to an increase in the role of stochastic communities of price-phobic ruderals in the regional natural system and, accordingly, a decrease in the role of climax communities.

The presence of saiga in the Late Pleistocene faunistic complex of northeastern Asia also indicates a low snow cover in winter. This correlates well with the hionophobicity of the *Carex argunensis* associations, possibly widespread in the Late Pleistocene, judging by the number of *Morychus viridis* in orictocoenosis and the presence of not only the saiga antelope but also the bighorn sheep and the musk ox in the fauna. The musk ox, being a typical tundra species, also prefers a zone of polar deserts with a very rarefied vegetation cover and low snow cover. That is, both saiga and musk ox are those species which depend on the height of snow cover in winter, and they are adapted to winter pastures of semi-desert type with a sparse cover. A bighorn sheep, also being a part of the mammoth faunistic complex, belongs to the same animals. In the Late Pleistocene, winters were characterized by low snowfall [6], and the areas of ruderal and wasteland dissipative communities in the Late Pleistocene of Yakutia surpassed the contemporary areas.

5. Conclusion

A wide distribution of ruderal and cryoxerophytic grass wading communities in the late Pleistocene of northeastern Asia is confirmed by the finds of saiga, *Morychus viridis*, as well as a large number of macro-remains of haze, buckwheat, and crucifers in the Late Pleistocene orictocoenosis. In complex, all these findings, as well as the presence of cohabitation of saiga and bighorn sheep, indicate a wide distribution of stochastic ruderal and cryoxerophytic grass wastelands. Such communities could form and spread due to extreme macroclimatic factors (low winter and high summer temperatures, little snow in winter), leading to cooling in winter and drying of the upper soil horizons in summer. Also, the following influenced their formation: a spread of permafrost with cryogenic processes of swelling, leading to the formation of micro and meso-positive landforms; a trampling of grass by large herd mammals (mammoths, bison), which initiated and enhanced the processes of pasture digression.

The biodiversity of the Late Pleistocene ecosystems in Yakutia was much higher than the current one. Similarly, a spatial organization of landscapes was more complex than it is at present. Contemporary microgroupings of xerophilous steppe, mountain-steppe, and ruderal plants in the Late Pleistocene were more widespread, creating an ecotonic effect and diversifying the ecological niches of some large mammals, such as the saiga and bighorn sheep. It is possible that even more numerous mammoths with bison found their feeding stations in the Late Pleistocene, presently represented by intrazonal or azonal plant communities occupying small areas.

6. Acknowledgments
This work was supported by the RFBR grant 18-45-140007 p_a and the project No. 0376-2019-0003 AAAA-A17-117020110056-0 “Fundamental and Applied Aspects of Studying the Diversity of the Plant World of the Northern and Central Yakutia.”

References
[1] Walter H, and Box E 1976 Global classification of natural terrestrial ecosystems *Vegetatio* **32**(2) pp 75-81
[2] Protopopov A V 2016 Using terioindicators in the reconstruction of the vegetation cover of the Late Pleistocene of Northern Yakutia *Science and Education* **3**(83) pp 16-20
[3] Protopopov A V 2017 *Dynamics of Yakutia's terrestrial ecosystems in the Late Pleistocene and Holocene* (Yakutsk, Russia: Alaas Publishing House)
[4] Boeskorov G G, Nogovitsyn P R, Mashchenko E N, Belolyubsky I N, Stepanov A D, Plotnikov V V, Protopopov A V, Schelchkova M V, van der Plicht J, … Solomonov N G 2016 New data on mammals of the mammoth fauna of the Middle Lena basin (Yakutia; “Lenskiye stolby” National Natural Park and adjacent territories) *Reports of the Academy of Sciences* **469**(2) pp 190-194
[5] Berman D I, Alfimov A V, Mazhitova G G, Grishkan I B, and Yurtsev B A 2001 *Cold steppes of the Northeast Asia* (Magadan, Russia: Institute of Biological Problems of the North, Far Eastern Branch of the Russian Academy of Sciences)
[6] Velichko A A 1973 *The natural process in the Pleistocene* (Moscow, USSR)
[7] Zaletaev V S 1997 *The structural organization of ecotones in the context of ecotone management in the biosphere* (Moscow, Russia: Russian Academy of Agricultural Sciences)
[8] Sokolov V E, and Zhirnov L V Ed 1998 *Saiga: phylogeny, systematics, ecology, protection and use* (Moscow, Russia: Russian Academy of Sciences Publishing House)
[9] Abaturov B D 2005 Fodder resources, food availability and viability of populations of herbivorous mammals *Zoological Journal* **84**(10) pp 1251-1271
[10] Abaturov B D, Petrishchev B I, Kolesnikov M P, and Subbotin A E 1998 Seasonal dynamics of forage resources and saiga feeding on a natural pasture in semi-desert *Successes of Contemporary Biology* **118**(5) pp 564-584

The effect of the sum of soil temperatures on the yield model of onion (*Allium cepa* L.) in the High Altai Priobye

I Gefke[1*] and S Zharkova[1]

[1] Altai State Agrarian University, 98 Krasnoarmeyskiy prosp., Barnaul 656049 Russia

E-mail: ivgefke@mail.ru

Abstract. The article presents an information-logical model of the dependence of the productivity of onion on certain meteorological and soil-physical factors. The authors conclude that meteorological factors of the growing season have a great influence on crop production, namely the hydrothermal coefficient and the sum of temperatures. Soil-physical factors (humidity and volumetric heat capacity) are also highly important.

1. Introduction

Currently, due to the intensification of agricultural production, much attention is paid to the development of the vegetable industry. As a rule, one person eats fresh vegetable products almost all year round. In addition, vegetable crops are processed, canned, dried, and frozen. One of the important problems of today's agriculture is the production of domestic high-quality agricultural products, including vegetable products, in a volume that allows the population to completely eliminate the need for food [1].

Onion is one of the most sought after and common vegetable crops in the world. Perennial by nature, onion has gone a long way in cultivation. Currently, onions are grown under production conditions through sowing, seedlings and seeds; in winter, bulbs are used for forcing green onions.

Onions are used by peoples for food all year round in fresh and processed form. This plant is a powerful antioxidant, people know its medicinal and insecticidal properties from ancient times. In terms of its biochemical parameters, it surpasses many vegetable crops. Onions are valuable for their essential oils, flavonoids, mainly glycosides. Onions contain a large group of carbohydrates: sucrose, fructose, pectin, etc. [2, 3].

Onion is cultivated everywhere, it can be successfully cultivated in the Altai region as well. However, the onion turnip production is not debugged in the region. Up to 16 thousand tons of onion turnips are required for the population of the region annually for normal consumption. On the territory of the region, about 4.5 thousand tons are grown, the missing amount of onion comes from the southern regions of the country and the countries of the near abroad.

Climatic conditions of the Altai region are considered to be sharply continental, the duration of the growing season with favorable conditions for onions is 120-130 days on average. This time is sufficient for the formation and maturation of onion. In relation to temperature conditions, onion is a relatively cold-resistant culture. Its seeds can be grown at a soil temperature of 1 ... 20°C. An optimum temperature regime for seed germination is 20 ... 22°C; under such conditions, seedlings appear already for 6-7 days. Onion shoots are relatively easy to tolerate late spring cooling, and even short frosts up to -3 °C. The best temperature for the growth of leaves is 15-20 °C. During the bulb

Content from this work may be used under the terms of the Creative Commons Attribution 3.0 licence. Any further distribution of this work must maintain attribution to the author(s) and the title of the work, journal citation and DOI.
Published under licence by IOP Publishing Ltd

formation, the outflow of nutrients comes from the leaves; consequently, the temperature of 20-25°C is necessary. It is at this temperature that the bulb successfully forms and matures [4, 5, 6, 7].

2. Materials and Methods

The studies were carried out in the conditions of stationary experience in the selection field of the laboratory of onion crops of the West Siberian Vegetable Experiment Station (a branch of the Federal Research Center for Vegetable Growing in 2004-2007). The station is located according to the climatic zoning of the region in the territory belonging to the subzone of chernozem of moderately arid and forested steppes [8]. The soil of the experimental plot is represented by leached chernozem [9].

The field production experience was laid for research. The area of the plot is 1000 m^2, there are two options in the experiment: 1 – a plot occupied with onion; 2 – a steam (control). Onions were cultivated through the seeding culture. Its rate was 1300-1500 kg / ha, the planting date was 5-10.05.

To determine the temperature in the soil profile under the bulb onions, we used a soil electrothermometer, which allowed to measure the temperature of the upper (0-20 cm) soil layer every 5 cm [10].

The obtained results were processed using the methods of information logical analysis [11].

3. Results

During its vegetative development, vegetable crops, including onion, are capable of forming a temperature regime with the characteristics of crops [12]. The task of our study was to identify such features under the planting of onions, formed in the conditions of high Altai Priobye.

During the growing season of onion culture plants, we observed the formation of temperature conditions in the soil under plantings and under control. The measurements were carried out in a 20 cm layer at different depths, this is the productive layer for the formation of the bulb and the location of 35-45% of the root mass of the plant.

The results of observations of the temperature in the active layer of the soil in the form of daily amounts at different depths in 2004 and 2006, as well as in 2007 are presented in Table 1 and 2.

Table 1. The sum of daily temperatures in the arable layer of chernozem in 2004, 2006.

	2004					
Depth,	June 9-10		June 30 - July 1		July 31 - August 1	
cm	Steam	Onion	Steam	Onion	Steam	Onion
0	181.0	156.2	183.4	176.1	145.0	142.6
5	176.8	159.1	173.8	166.6	137.5	123.1
10	156.7	160.4	168.1	162.1	122.4	114.8
15	154.8	155.8	165.0	159.5	112.0	110.6
20	151.2	153.9	162.2	156.4	104.7	104.2
∑	821.5	785.4	852.5	820.7	621.6	595.4
	2006					
Depth,	June 27-28		July 25-26		August 8-9	
cm	Steam	Onion	Steam	Onion	Steam	Onion
0	187.8	177.2	147.2	138.2	138.5	123.7
5	176.0	167.7	139.4	131.1	128.7	115.7
10	169.5	160.9	127.4	128.0	114.3	109.5
15	166.8	155.7	126.0	126.2	109.2	107.9
20	163.1	156.5	128.7	126.1	100.7	106.0
∑	863.2	818.0	668.7	649.6	577.4	562.8

Table 2. The sum of daily temperatures in the arable layer of chernozem in 2007.

Depth, cm	June 16th-17th		July 21st-22nd	
	Steam	Onion	Steam	Onion
0	155.9	135.8	185.0	175.2
5	141.1	127.4	180.1	173.9
10	127.0	124.6	175.6	170.6
15	120.7	116.4	168.9	168.8
20	115.4	112.2	162.5	164.1
∑	660.1	616.4	872.1	852.6

The data of the sum of temperatures at a specific observation hour (13.00 hours) in a layer of 0-20 cm was obtained in the same years of observations (Tables 3, 4).

Table 3. The sum of temperatures at 13.00 in a layer of 0-20 cm (2004, 2006).

Layer thickness, cm	2004					
	June 9th		June 30th		July 31st	
	steam	onion	steam	onion	steam	onion
0-20	140.6	123.7	127.5	128.7	97.1	96.4
	2006					
	June 27th		July 25th		August 8th	
0-20	164.5	159.6	132.6	131.5	86.5	84.1

Table 4. The sum of temperatures at 13.00 in a layer of 0-20 cm (2007).

Layer thickness, cm	Observation time									
	May 28th		June 16th		June 23rd		July 21st		August 8th	
	steam	onion	steam	onion	steam	onion	steam	onion	steam	onion
0-20	94.6	100.9	97.1	88.7	148.1	134.6	137.6	128.0	125.3	130.4

4. Discussion

Analyzing the data in Table 1, we can note that in 2004, the daily sum of surface temperatures of the black soil was always higher in steam than under the cover of onion. The difference in the values of the sum of temperatures was about 25°C in June, whereas it was 7°C at the end of June and 3°C at the end of July. This difference may be due to different weather conditions. Intensive warming of the air entailed a significant difference in soil temperature on the studied variants. This effect smoothed out with depth. And, in general, in the upper 20-centimeter layer in a steam, the amount of daily temperature for the growing season was more by 25-40°C than in the area occupied by onion.

Continuing to analyze Table 1, we see that during the observations in 2006, the sum of daily temperatures of the vapor field surface was higher than in 2004. At the end of June, it turned out to be 187.8°C in the first case and 183.4°C in the second case. Similar results were observed in late July and early August. Throughout the topsoil, the sum of temperatures was higher in a steam than under the onion.

In 2007, the obtained amounts of daily temperatures in the control area and under the onion indicate the trend of previous years (Table 2). So, during the growing season in a steam, they were higher as before and amounted to 660 °C and 616 °C in June. In July, the air warmed up to 30°C, this was reflected in the value of the sum of daily temperatures, making them higher than in the same periods in previous years.

Tables 3 and 4 represent the sum of temperatures of the 0-20 cm layer at 13.00 with these options in the same years of observation. Calculations showed that in 2004, the difference in the sum of temperatures in the steam and under the onion was 170°C in June, and it was practically absent in July. In 2006, the trend continues: in June is 50°C, in subsequent periods is 1-20°C. And as it turned out that in 2007, the patterns found in previous years are confirmed: the sum of temperatures in the steam and under the onion differs slightly (5-15°C).

5. Conclusion

Meteorological factors of the growing season have a great influence on crop production, namely the hydrothermal coefficient (HTC) and the sum of temperatures ($\sum T$, ^0C). Other soil-physical factors are important, such as humidity (U, %) and volumetric heat capacity (C_ρ, J / m^3K).

The effect of the sum of temperatures in 0-20 cm layer on the formation of yield of onions is presented in Table 5.

Table 5. The effect of the sum of soil temperatures on the productivity of onion.

Factor	Condition factor	Productivity, t / ha	Yield rank
$\sum T$, ^0C	<90.0	<25.0	2(3)
0-20 cm layer	90.1-110.0	25.1-30.0	3(4)
	110.1-120.0	30.1-35.0	3
T=0.4775	120.1-140.0	35.1-40.0	3
K=0.2303	>140.1	>40.1	2

Information analysis of the dependence of yield on the sum of soil temperatures showed a high reliability, which is 0.4775. A high coefficient of effectiveness of communication channels (K = 0.2303) corresponds to the information analysis. This coefficient shows the measure of dependence between phenomena.

The most significant factors (if evaluated by the magnitude of the efficiency factor of the communication channel with the yield) can be arranged in a logical series:

$$HTC > \sum T > U > C_\rho$$

The results obtained made it possible to create an informational logical model:

$$Y = HTC * \sum T * U * C_\rho,$$

where Y is the yield rank of onion crops, HTC is the yield rank, depending on the HTC, $\sum T$ is the yield rank, depending on the amount of temperature in the 0-20 cm layer, U is the yield rank, depending on soil moisture, C_ρ is the rank yield, depending on the volumetric heat capacity of chernozem.

References

[1] Litvinov S S, and Shatilov M V 2014 Vegetable-growing: a new impulse of development *Potatoes and Vegetables* **7** pp 2-4.
[2] Stolbova T M 2018 Biochemical indicators of bulb onions in the south of Western Siberia *Vestnik of the Altai State Agrarian University* **7**(165) pp 16-19
[3] Zharkova S V 2009 *Adaptive potential of the Siberian varieties of bulb onions* In Vegetable-growing of Siberia (Novosibirsk: Publishing House of the NSAU)
[4] Pivovarov V F, Ershov I I, and Agafonov A F 2001 *Onion crops* (Moscow, Russia: VNIISSOK)
[5] Vodyanova O S 2007 *Onions: napiform, shallots, leeks, perennial, wild edible and decorative, garlic* (Almaty, Kazakhstan: Book House JSC)
[6] Zharkova S V 2009 *Scientific substantiation and improvement of onion crop breeding methods (Allium cepa L., Allium ascalonicum L., Allium sativum L.) for creating varieties with high adaptability to the conditions of Western Siberia* (dissertation of the doctor of agricultural sciences) (Moscow, Russia)
[7] Zizina Ya F 2016 *Perfection of the elements of the technology of cultivation of bulb onion in the annual culture in the forest-steppe of the Novosibirsk Priobye* (Dissertation of the Candidate of Agricultural Sciences) (Tyumen, Russia: NSAU)
[8] Burlakova L M, Tatarintsev L M, and Rassypnov V A 1988 *Soils of the Altai Region* (Barnaul, Russia: Altai Agricultural Institute)
[9] Gefke I V, Zharkova S V 2018 Features of agrophysical properties of soil in the cultivation of bulb onions *Vestnik of the Altai State Agrarian University* **6** pp 79-86

[10] Bolotov A G 2002 *Measurement of soil temperature in field conditions* In A N Kupriyanov Ed *Proceedings of the II International Conference: Anthropogenic Impact on Forest Ecosystems* (pp 148-150) (Barnaul, Russia: ASU)

[11] Dospekhov B A 1985 *Methods of field experience (with the basics of statistical processing of research results)* (Moscow, USSR: Agropromizdat)

[12] Makarychev S V, Gefke I V, and Reger A I 2009 Hydrothermal regime of leached chernozem in onion cultivation under conditions of Altai Priobye *Vestnik of the Altai State Agrarian University* **1** pp 29-32

IOP Conf. Series: Earth and Environmental Science **395** (2019) 012027 doi:10.1088/1755-1315/395/1/012027

Present climate development in Southern Siberia: a 55-year weather observations record

N Kharlamova[1*], M Sukhova[2] and J Chlachula[3,4]

[1] Faculty of Geography, Altai State University, 61 Lenina st., Barnaul 656049 Russia
[2] Faculty of Geography, Gorno-Altaisk State University, 1 Lenkina str., Gorno-Altaisk 649000 Russia
[3] Department of Environmental Studies, T. Bata University in Zlin, nam. T.G.M. 5555, 762 01 Zlin, Czech Republic
[4] Institute of Geoecology and Geoinformation, Adam Mickiewicz University, Dziegielowa 27 Poznan 61-680 Poland

E-mail: harlamovageo@rambler.ru

Abstract. Siberia is a key region for study of the climate development in north-central Eurasia in terms of global climatic change. The relief creates a major orographic barrier for atmospheric streams influencing the regional altitudinal weather zonality. Systematic 55-year (1963-2017) weather observations along the 700 km N-S latitudinal transect across the southern Siberian plains and the adjoining ranges of the Altai-Sayan Mountains document progressing seasonal temperature and humidity shifts. Standard meteorological data from twenty meteorological stations in different topographical and natural settings with specific weather conditions document a rise of a regional atmospheric humidity balance reflected by the increased bulk annual precipitation rates in the mountain areas accompanied by pronounced seasonal air-temperature fluctuations. This regionally trend provides evidence speaking in favor of the strengthening climate continentality over Siberia, which is also manifested by the pronounced seasonal temperature regime with increased thermally positive and negative air temperature anomalies. A landscape response to a climate warming is particularly evident in the high mountain zone. The present thermal conditions with raised MAT contribute to the progressing melting of mountain glaciers and degradation of permafrost in the alpine zone, as well as aridization of the parkland-steppe areas that are being partly transformed into continental semi-arid to deserted steppes. The associated environmental transformations trigger shifts in the local biotopes and ecosystems, with an altitudinal expansion of taiga-forest into the alpine tundra belt and xerothermic grassland invasions in the foothills. The modern climate changes have a major bearing to the Altai-Sayans socio-economic development.

1. Introduction

Weather dynamics reflecting dominant atmospheric circulation patterns in conjunction with regional topographical conditions plays the main role in the present climate formation and climate change over a particular territory, ultimately governing a functioning of natural environments as well as all aspects of the modern civilization. The goal of our studies in southern Siberia is to research a number of regional manifestations of climate trends for the last 55 years (the period of 1963-2017). These spatially encompass the northern plains and the southern mountain ranges of the Altai and Sayans

Content from this work may be used under the terms of the Creative Commons Attribution 3.0 licence. Any further distribution of this work must maintain attribution to the author(s) and the title of the work, journal citation and DOI.
Published under licence by IOP Publishing Ltd

IOP Conf. Series: Earth and Environmental Science **395** (2019) 012027 doi:10.1088/1755-1315/395/1/012027

Mountains (altitudinal gradient of 180-4000 m asl.). The unique location in the center of Eurasia characterized by a strongly continental climate regime underlines the fundamental relevance of the systematic weather observations in this particular strategic area.

This study summarizes results of meteorological observations on directions and rates of the present climate development over southern Siberia based on archival meteorological data for the period of 1963-2017 [1]. Implications of the weather-observations are of principal relevance in environmental management of the protected natural areas of southern Siberia and for the central governmental sustainable-development planning of the Gorny Altai and Tuva Republics, as well as in East Kazakhstan [2, 3].

The monitored territory encompasses the Altai and Sayan Mountains, which forms the main mountain system of southern Siberia adjoining the Mongolian Altai by the Tavan-Bogd-Ula massive (Nairamdal Mt., 4374 m) and the Southern (Kazakh) Altai (3483 m) in the south (Fig. 1). This vast area of over 1 mil km^2 is characterized by a very diverse physiography. The Altai Mountains (Belukha Mt., 4506 m asl.) and the Western Sayan Mountains (3492 m) jointly create a major natural barrier delimiting the southern geographic limits of Siberia bordering the West Mongolian Altai, the Chinese Altai and the Southern Altai of East Kazakhstan. In the west, the continental topography gradually passes into the North Altai Plains (200-300 m asl.), representing the marginal parts of the West Siberian Lowland. The Western Sayans form the eastern continuation of the Altai Mountains, stretching for 500 km (89°-96° E). The Eastern Sayans extend 1,000 km from the Yenisei River at 92° E to the southwest end of Lake Baikal at 106° E.

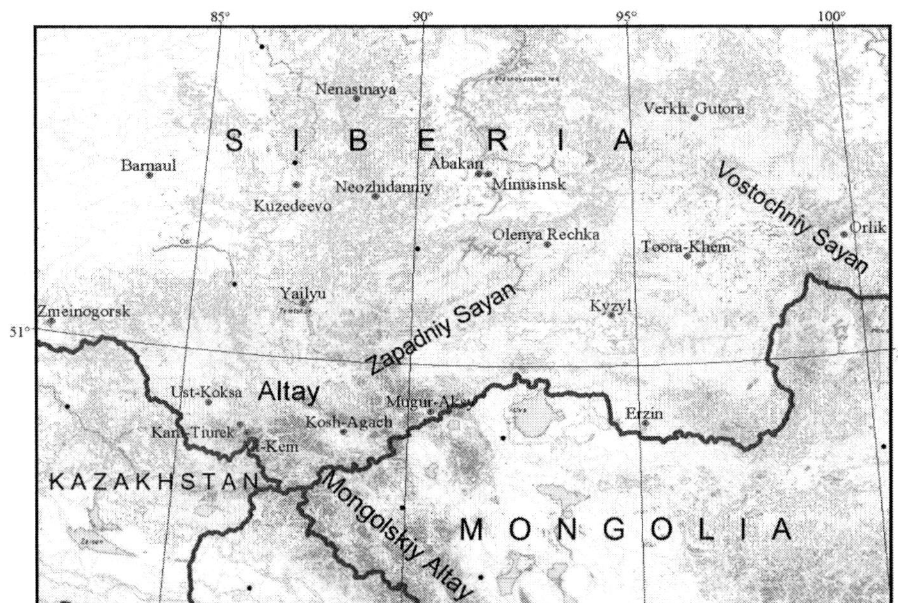

Figure 1. Geographical location of the study area in southern Siberia with the spatial and topographic distribution of the monitoring weather stations.

The present climate is strongly continental with major seasonal temperature deviations between the northern lowlands and the southern mountains. In winter, except for the highest elevations, climatic conditions in the mountains are generally less severe than in the open northern steppes, and a local microclimate prevails throughout the year in some protected locations in the Altai Mountains (the Upper Biya, Katun', Chulyshman and the lower Chuya River basins) with a relatively thin snow cover during winter [3, 4, 5]. Annual temperatures as well as precipitation rates vary greatly, reflecting particular topographic settings. Most of the annual precipitation falls on the W/NW slopes in the northern and central Altai, while the southern areas became more arid. In the Chuya Depression (1800-2000 m asl., Kosh-Agach), which is one of the most continental places in the Altai, an average July temperature is +25°C, whereas an average January temperature is -33°C, but it can occasionally drop to -60°C. Most of the local area is underlain by perennial mountain permafrost with an active thaw layer belong only 30-70 m thick. The lowland (200-500 m asl.) vegetation of southern Siberia is formed by the open steppe-parkland with mosaic birch-pine forests. Mixed taiga covers most of the foothill and the lower mountain zone (500-1500 m asl.), gradually passing at higher elevations (1500-2500 m asl.) into the alpine tundra with the Siberian pine, larch and dwarf birch. Semi-desert communities with an admixture of plant taxa characteristic of the Mongolian steppes are found in the upland depressions of the southern Altai.

2. Materials and Methods

This study summarizes results of investigations on present climate change in south-central Siberia delivered by 20 weather stations situated at different altitudes and specific topo-geographical conditions. The significance of this area is underlined by its unique position near a geographical center of Eurasia with a strongly continental climate regime and marked seasonal weather variations with low winter and high summer temperatures influenced by the Siberian cyclone/anticyclone atmospheric streams. The 20 state weather stations (WS) are positioned in diverse physio-geographic areas (of the Altai District, the Gorno Altai Republic, Khakassia, Buryatia and the Tuva Republic) following the territorial continental topographic gradient (180-4500 m asl.), providing regular daily measurements of standard atmospheric parameters (i.e. precipitation and temperature changes, wind-direction, atmospheric pressure, ground-frost and permafrost table). The analyzed summary databases include systematic meteorological data for the 1963-2017 periods.

The climate-change investigations encompass an extensive territory, geographically delimited by 49°55′- 54°45′ N and 82°10′- 99°49′E (the spatial position range of the weather stations). The N-S geographic transect of the monitored area runs over about 700 km through all the topographic and environmental zones from the northern steppe plains at the south-eastern margin of the West Siberian Lowland, across the Altai-Sayan foothills, and the low-, mid-, and high-mountains, representing the principal mountain ranges of southern Siberia adjacent to the Kazakhstan-Chinese-Mongolian boarder. The corresponding continental relief gradient rises from 183 m asl. (Barnaul WS) to 2600 m. asl. (Kara-Tiurek WS), accounting for a total of 2417 m of the vertical regional elevation difference.

3. Weather Monitoring Results

In dependence of the dominant regional climatic conditions, the thermal regime of the mountain areas is significantly differentiated. The mean seasonal winter temperature varies from -5 to -27 °C. In the low-mountain areas and foothills, a mean July temperature is around 16-19 °C, 12-15 °C in the mid-mountain zone, and 13-21°C in the basins and depressions; in the high-mountain zone it drops to 6 °C.

A maximum amount of precipitation falls on the NW slopes exposed to atmospheric stream fronts; the minimum precipitation rates in the arid intermountain basins are enclosed by mountain ranges and in other precipitation-shadowed geomorphic locations [6, 7, 8].

A gradual rise of mean annual air temperature is observed from the late 1960s and especially after 1976. A minimal MAT rise has been documented at the weather stations Kara-Tiurek, Ak-Kem, Oleniya Rechka in the high mountain areas of southern Siberia (the Western Sayans and Gorno Altai Mnts.), on average 0.2-0.3°C/10 years, whereas a maximum progressing climate warming is in the major inter-mountain depressions (Kosh-Agach, Erzin, Kyzyl and Toora-Khem) on average 0.4-0.6°C/10 years. Numbers of frost-free days are also increasing with the strengthening trend of climate continentality (Fig. 2).

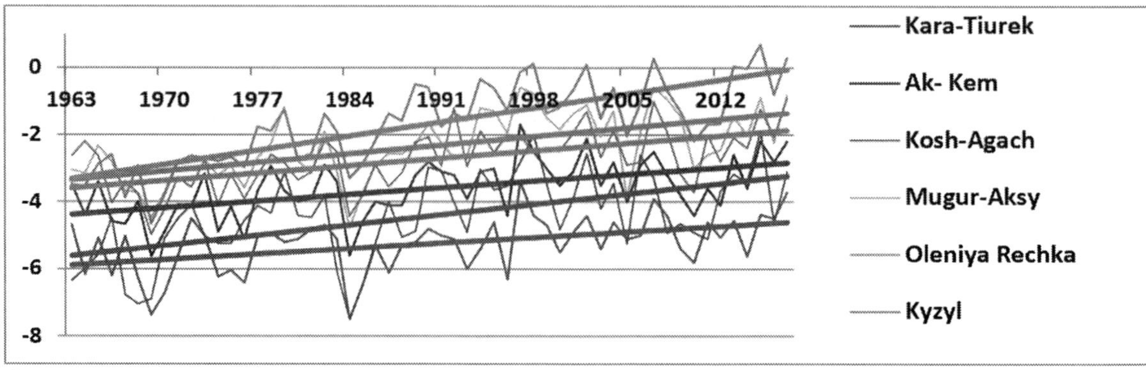

A

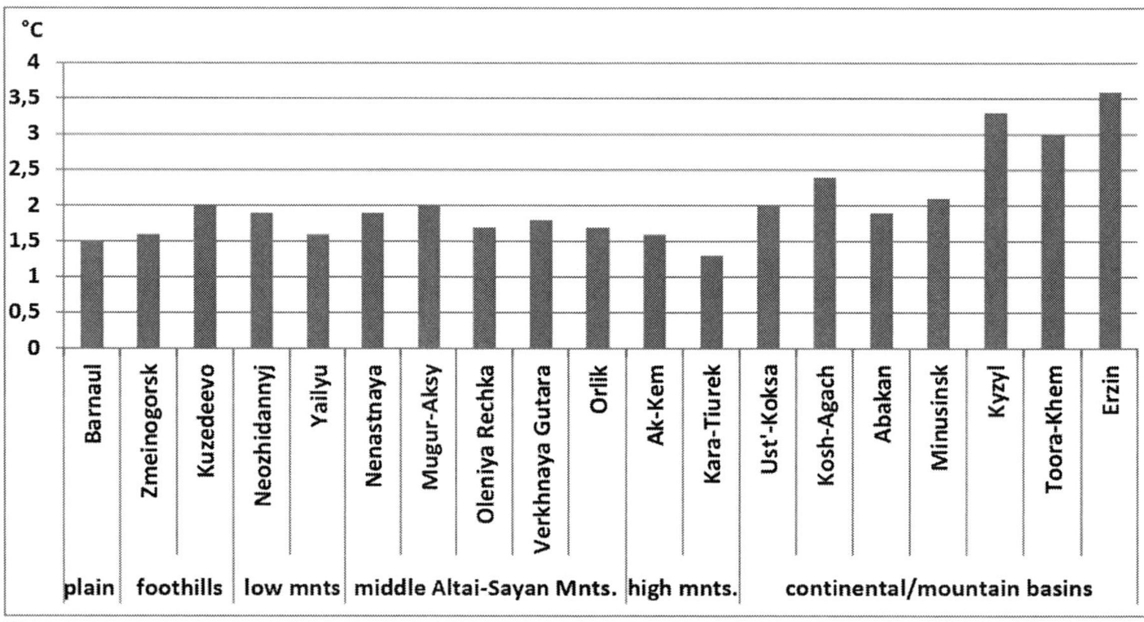

B

Figure 2. Increased temperature rate in respect to the MAT at the monitored locations (weather stations, A) following the topographic gradient from the northern plains to the southern mountain regions for the 55 year observation period (1963-2017), Altai-Sayans Mnts (Altai-Saian) and continental/mountain basins (B).

According to the mid-seasonal air temperature rates, the monitoring southern Siberian meteorological weather stations (WS) can be grouped into three zones: (a) 1st zone (SMAT 15-20°C): plains, foothills, low-mountains (500-1000 m asl.) and the inter-mountain basins; (b) 2nd zone (SMAT 10-14°C) of the mid-mountain areas (1000-2000 m asl.); (c) 3rd zone (SMAT 2-9°C) of the high-mountain areas (2000-4000 m asl.), with the highest values (8-9°C) at the Ak-Kem WS (2050 m asl.) and (5-7°C) at the Kara-Tiurek WS (2598 m asl.). A long-term (55-year) rise of the SMAT values across all geographical zones is only 0.5-1.0 °C except for slightly the higher values (2.0-2.2°C) in the mountain basins of Gorno Altai and Tuva (Fig. 2, 3).

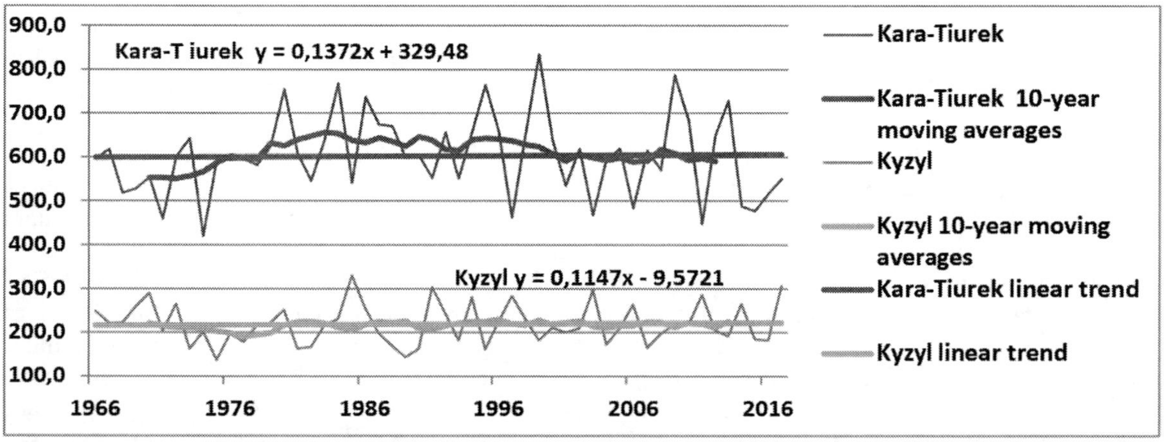

Figure 3. The variability of annual precipitation, Kara-Tiurek (2600 m, Katunsky range of Altai) and Kyzyl (626 m, Tuva intermountain basin), 1966-2017: linear trends and 10-year moving averages.

4. Discussion

Comparative analyses of the historical and the present meteorological databases provide eloquent evidence on the changing climate in southern Siberia with the trend of a strengthening continentality. Despite the significant interannual precipitation variability, a little reduction of the annual total has been observed in the Altai in 1963-2017 (Fig. 3). The mean seasonal temperature values across the monitored territory based on the 55-year observations show a differential rise by ca. 3.5 °C in winter, 1.6 °C in spring, 1°C in summer, and 0.9 °C in autumn. This seasonal temperature trend unequivocally documents a progressing climate warming in the inner Eurasia in agreement with biotic and glacial records from the Altai and Sayan Mountains [5]. This atmospheric dynamics is clearly expressed by the current environmental transformations that have direct effects on functioning of the southern Siberian ecosystems. The reported results on the present climate change in the broader Altai–Sayan Eco-Region corroborate the precipitation increase trends in the mountain areas and, on contrary, the progressing rates of aridization and desertification in the lowland steppe regions as well as in the intra-mountain basins of the southern Altai area [8].

The Altai-Sayan Eco-region that constitutes the major part of the investigated southern Siberia belongs to the most pristine natural areas in the mid-latitude boreal forest belt of Eurasia. This environmentally unique territory in the Russian-Mongolian-Kazakh border zone (Fig. 1) includes several major national parks and nature reserves. The present parkland-steppe and mountain ecosystems of southern Siberia experience progressing environmental transformations and ecology degradations as a result of the complex processes trigerred by the present global climate development. The related changes in natural (bio- and geo-) environment affect the wide mosaic of parkland-steppe, foothill and moutain ecosystems and biotopes, hosting a number of endemic and/or elsewere rare species of flora and fauna that are subject of national and international protection. Hence, systematic monitoring of the regional climate is of utmost relevance. Raising summer temperatures also contribute to the regional aridization, the mountain permafrost degradation and to an accelerated ablation of the retreating alpine glaciers, reaching a maximum rate up to 20 m per year on the Southern Altai Range [8].

5. Conclusion

The results of the 55-year (1963-2017) systematic meteorological observation in the Altai and Sayan region display significant both annual and seasonal temperature and humidity variations with a definite trend of climate warming corroborating the present climate change in the continental Eurasia, particularly winter and spring mean air temperatures have significantly increased. A minimal MAT rise is in the high mountain areas, whereas the maximum progressing climate warming has been observed in the inter-mountain continental depressions. The meteorological data also indicate an increase of regional humidity reflected by the bulk annual precipitation rates in the foothills and the mountain areas. In places of the higher topographic relief, the moisture regime and atmospheric circulation dynamics is being accentuated.

This definite climate-change trend has been recorded in conjunction with the related natural transformations throughout the entire southern Siberian territory as manifested by the expansion of grasslands in the foothills and by the rise of tree-line in the alpine mountain zone. Raising annual temperatures trigger a progressive melting of glaciers, a localized degradation of the mountain permafrost in the alpine zone (> 2,500 m asl.). The associated ecological processes include shifts in the mountain biotopes: an expansion of boreal taiga forests (now with limits at 2100-2300 m asl.) into the alpine tundra zone, a geo-botanical restructuring of xenotheric alpine meadows and concomitant biological changes in behavior of some fauna species, such as earlier arrival and a later departure of migratory birds for nestling. On the contrary, a progressing aridization of the mountain parkland-steppe areas due to the strengthened continental climate regime with high solar radiation with increased number of sunny days and a precipitation decline lead to a gradual transformation into semi-arid parkland steppes and semi-desertic steppes [9].

The implications resulting from the historical and present meteorological data analysis as well as the climate change proxies can be eventually used in the assessment of the ongoing and future climate change feedbacks to the southern Siberian environments with implications to management of natural resources, biodiversity protection and the present/perspective sustainable socio-economic development, with an optimal planning of exploitation of natural resources. The present climate change-linked economic effects may, on long-term, positively promote production in agriculture and forestry over the monitored territory due to ameliorated weather conditions and expansion of arable lands and taiga forests, respectively. On contrary, a complex evaluation of the main natural climate-triggered risks (including soil cover erosion, desertification and industrial resource exploitation) can ultimately contribute to establishment of an integrated concept of effective nature protection strategies in the Altai and Sayan Eco-Region, and an environmental hazard monitoring related to present climate change.

References

[1] VNIIGMI WDC 2018 *Data* Available at: http://meteo.ru/

[2] Kokorin A O 2011 Climate change and its impacts on ecosystems, population and economy of the Russian part of the Altai-Sayan Eco-region *Assessment Report of the World Wildlife Fund (WWF Russia)* (Moscow, Russia)

[3] Geography of Siberia 2016 *The geography of Siberia at the beginning of the XXI century: a monograph in 6 volumes - Tom 5 Western Siberia* (Novosibirsk, Russia: Academic publishing house "Geo")

[4] Sukhova M G 2010b *Ecological and climatic potential of mountain areas* (Saarbrucken: LAP LAMBERT Academic Publishing)

[5] Kharlamova N F 2013 *Evaluation and forecast of the current climate change in the Altai region:* (monograph) (Barnaul, Russia: Alt. University)

[6] Kharlamova N F 2010 Climate change trends in the Altai-Sayan region In *Climate Change and Biodiversity Conservation in the Altai-Sayan Eco-region: Papers of the International Meeting, Ust-Koksa* (pp. 10-15) (Barnaul, Russia)

[7] Chlachula J, and Sukhova M G 2011 Regional manifestations of present climate change in the Altai, Siberia. *Proceedings, ICEEA Conference on Environmental Engineering and Applications, Shanghai, China (August 19-21, 2011)* In: Li Xuan (ed) Environmental Engineering and Applications *IACSIT Press Singapore* **17** pp 134-139

[8] .N. Mikhailov 2013 *Climate change and biodiversity of the Russian part of the Altai-Sayan Ecoregion* Krasnoyarsk: City

[9] Sukhova M G, and Zhuravleva O V 2018 Changes in the temperature of air and draft in intermediary potteries of the Southeast and Central Altai *News of the Russian Academy of Sciences Geographical series* 6 pp 93-101

A granulometric composition of the Zauralye black soils

A Plotnikov[1] and A Sozinov[1*]

[1] Kurgan State Agricultural Academy named after T. S. Maltsev, Lesnikovo 641300 Russia

E-mail: savrey@ya.ru

Abstract. The article presents results of the study devoted to the granulometric composition of Zauralye black soils. The main patterns of the profile differentiation of soils in connection with the genesis and anthropogenic load on agricultural land are considered. There are signs of elution in the upper part of the soil profile when analyzing the nature of the distribution of fractions of granulometric composition. In general, a black soil profile on the granulometric composition is poorly differentiated.

1. Introduction

Currently, the rational use, protection, and improvement of soil fertility have become one of the most important problems of the mankind. Especially acute, it is placed in areas of intensive agricultural development. Numerous studies have found that morphological features, water-physical, and agrochemical properties change significantly in black soils due to prolonged agricultural use. The loss of turf and the homogenization of horizon A to the depth of the main treatment are the most significant changes in the morphology of virgin black soils after plowing. A number of scientists believe that the profile of arable Eastern European black soils varies slightly [1, 2]. However, the West Siberian black soils change when plowing to a much greater extent due to their provincial features [3-7].

The granulometric composition of contemporary soils reflects the genesis of soil-forming rocks and indicates the direction of their transformation under different conditions of agricultural use. It largely determines its structural and functional properties, being one of the main structural levels of the organization of the soil solid phase and directly participating in the formation of micro and macrostructures. The quantitative ratio in the distribution of elementary soil particles and their properties by fractions significantly affect both the aggregate level and the nature of the dynamics of the pore space [8, 9].

An optimal granulometric composition varies for different crops. The soils are light in granulometric composition, they are well aerated, but they have low water holding capacity, they are less resistant to drought, water erosion, and deflation. Medium and heavy soils have a higher coherence and moisture capacity. By comparison to light sandy soils, they are well supplied with nutrients and humus. Reserves of moisture and nutrients in these soils are able to provide good yields of agricultural crops, they have a pronounced structure and contain a sufficient number of water-resistant aggregates [1, 2, 9].

In the territory of the Kurgan region, the type of black soil is represented by the following two subtypes: leached black soils ($Ч^В$); ordinary black soils ($Ч^О$) (they occupy 2,149.7 thousand hectares, or 30.3% of the total area). In the subtype of ordinary black soils in the region, the following genus are distinguished: ordinary, alkaline, alkaline carbonate, carbonate, solodized, deep stripping.

Content from this work may be used under the terms of the Creative Commons Attribution 3.0 licence. Any further distribution of this work must maintain attribution to the author(s) and the title of the work, journal citation and DOI.
Published under licence by IOP Publishing Ltd

The share of these soil subtypes in the structure of arable land is different in zones. In the black soils northwestern zone, a share for leached sols is 44%, and it is 35.7% for ordinary alkaline soils; in the central zone, it is 33 and 45.7%, respectively. In the eastern zone, there shares are 48.3 and 68.0%, and 48.3 and 25.4% in the southern zone [3].

Ordinary and leached black soils have a loamy and clayey granulometric composition in most parts of the Zauralye forest-steppe [3, 5]. Leached black soils, light in granulometric composition, are less common.

2. Materials and Methods

The purpose of the study is to measure the granulometric composition of black soils in the process of soil formation at different levels of anthropogenic load.

The research objects included two subtypes of black soils: leached and ordinary. Research was conducted on the following experimental fields of the Kurgan Research Institute of Agriculture:

- Central and Kurgan State Agricultural Academy (Ketovsky district, central zone);
- Makushinsk (Makushinsky district, eastern zone);
- Yuzhnoye (Tselinny district, southern zone).

Related studies of soils with different agricultural loads were carried out to determine the influence of anthropogenic factors on soil formation processes. Sampling was carried out at the closest possible distance from each other so that, unlike the anthropogenic factor, the natural factors of soil formation were as close as possible.

Two cuts to the horizon of parent rock C are laid on each experimental field. Sampling was conducted on two agricultural backgrounds: virgin and arable. Samples for laboratory analysis were taken from each genetic horizon. We determined the soil granulometric composition by the pipette method (according to N. A. Kachinsky) in the laboratory.

Wild vegetation on virgin areas is mainly represented by Aster and Bean families. Agrocenoses on arable land are represented by field crop rotation (crops of continuous sowing - wheat, annual grasses).

3. Results

The results of analyzes of the granulometric composition showed that in leached black soils an increase in clay fraction occurred as a result of agricultural use. So, on the Experimental Field of the Kurgan State Agricultural Academy in the upper humus horizon on virgin soil, the content of physical clay was 29.5%; on arable land, its content increased to 33.0%; and, according to Kachinsky's classification, the soil already goes into the category of medium loamy. Fine sand and sludge fraction predominate in the composition of granulometric fractions. In this regard, the medium loamy variety has received an additional name – silty sand (Table 1).

In the Central Experimental Field, a significant weighting of the granulometric composition occurred due to an increase in the fraction of fine dust. In horizons A and AB in virgin soil, the content of physical clay was 19.4-22.6%, while its content was 32.7-33.5% in arable and subsurface horizons.

On the Yuzhnoye Experimental Field, in the upper humus horizon on virgin soil, the content of physical clay was 31.0%, its content increased to 31.2% on arable land. According to Kachinsky's classification, it is classified as medium loamy soil. The same was observed in the subsurface horizon, where the increase was 0.1%.

Thus, the content of the sludge fraction within the humus-accumulative soil horizons tends to increase regardless of the type of use. In leached black soils, it increases by 0.2%.

The granulometric composition of the ordinary saline black soils on the Makushinsky Experimental Field is represented by a heavy variety. The coarse silt (Krupnopylevataya) and sludge fraction predominate in the composition of the granulometric fractions. A heavy variety on virgin soil is silty-large-silty (ilovato-krupnopylevataya). An increase in the fine sand fraction was observed on arable land. A similar pattern is observed in the treated area in the horizon, as well as in leached black soils: an increase in the fraction of sludge and physical clay in horizon A (from 41.7 to 48.3%).

Table 1. A granulometric composition of the leached and ordinary Zauralye black soils.

Cut location	Genetic horizon	The content of granulometric fractions, % by weight of soil						
		Large and medium sand, 1-0.25 mm	Fine sand, 0.25-0.05 mm	Coarse dust, 0.05-0.01 mm	Medium dust, 0.01-0.005 mm	Fine dust 0.005-0.001, mm	Sludge, less than 0.001 mm	Physical clay, less than 0.01 mm
Leached, light loamy black soil (Experimental field of KSAA)								
Virgin lands	A (4-27)	7.0	49.9	13.6	6.9	10.3	12.3	29.5
	AB (27-40)	8.2	49.5	12.1	7.0	7.1	16.1	30.2
	B (40-62)	6.3	45.7	10.6	9.2	7.5	20.7	37.4
	BC (62-89)	10.0	49.3	11.4	4.7	7.3	17.3	29.3
	C (> 89)	11.4	56.4	8.9	2.3	7.8	13.2	23.3
Arable lands	A (0-24)	5.2	57.2	4.6	2.9	10.5	19.6	33.0
	AB (24-37)	4.6	54.5	5.6	1.7	9.6	24,0	35.3
	B (37-60)	5.3	55.0	5.5	1.6	8.7	23.9	34.2
	BC (60-85)	5.5	64.7	3.7	1.3	6.0	18.8	26.1
	C (> 85)	12.0	48.5	19.1	1.8	5.9	12.7	20.4
Leached, light loamy black soil (Central Experimental Field)								
Virgin lands	A (3-26)	6.4	43.3	27.7	2.4	8.7	11.5	22.6
	AB (26-43)	5.5	37.6	37.5	1.5	7.5	10.4	19.4
	B (43-68)	8.9	24.8	41.8	1.0	7.5	16.0	24.5
	BC (68-102)	8.4	20.0	33.8	3.5	14.5	19.8	37.8
	C (102-120)	8.3	20.5	43.6	1.4	13.5	12.7	27.6
Arable lands	A (0-24)	7.5	39.9	19.9	4.2	15.9	12.6	32.7
	AB (24-40)	6.6	36.5	23.4	6.6	13.5	13.4	33.5
	B (40-66)	8.0	21.9	24.6	3.6	15.6	26.3	45.5
	BC (66-100)	6.2	22.8	33.8	1.1	12.7	23.4	37.2
	C (>100)	7.9	21.2	38.3	1.6	16.3	14.7	32.6
Ordinary alkaline heavy loamy black soil (Makushinskoe Experimental Field)								
Virgin lands	A (5-28)	5.1	19.4	33.8	8.1	11.7	21.9	41.7
	AB (28-46)	6.2	24.1	28.0	6.6	11.7	23.4	41.7
	B (46-68)	2.7	22.3	23.6	8.1	12.1	31.2	51.4
	BC (68-80)	4.3	16.4	16.3	8.4	16.1	38.5	63.0
	C (> 80)	7.7	28.6	11.8	2.2	10.9	38.8	51.9
Arable lands	A (0-22)	3.5	26.1	22.1	9.8	11.9	26.6	48.3
	AB (22-38)	5.3	28.6	21.7	6.9	13.0	24.5	42.4
	B (38-62)	5.4	20.5	22.2	6.1	11.9	33.9	51.9
	BC (62-78)	6.1	21.6	10.1	8.4	14.3	39.5	62.2
	C (> 78)	4.0	28.3	5.6	6.0	14.9	41.2	62.1
Leached medium loamy black soil (Yuzhnoye Experimental Field)								
Virgin lands	A (4-26)	6.3	47.6	15.1	5.8	10.4	14.8	31.0
	AB (26-41)	6.7	41.6	15.7	3.5	7.8	24.7	36.0
	B (41-72)	7.6	47.9	13.2	0.8	6.3	24.2	31.3
	BC (72-93)	5.6	46.8	14.6	4.6	7.5	20.9	33.0
	C (>93)	6.7	42.0	12.3	3.9	18.1	17.0	39.0
Arable lands	A (0-24)	6.3	47.1	15.4	5.1	10.9	15.2	31.2
	AB (24-40)	4.8	45.8	13.3	6.5	4.7	24.9	36.1
	B (40-68)	5.7	49.4	11.0	7.9	4.2	21.8	33.9
	Bк (68-78)	4.5	39.1	15.4	5.7	9.5	25.8	41.0
	BC (78-90)	7.5	46.3	16.5	0.1	13.6	16.0	29.7
	C (>90)	4.5	50.4	10.9	3.2	18.6	12.4	34.2

Source: Compiled by the authors.

4. Discussion

Zonal features of soil formation are revealed in a comparative study of the distribution over the soil profile of the most dispersed and active fraction, which is the sludge fraction (particles <0.001 mm).

IOP Conf. Series: Earth and Environmental Science **395** (2019) 012028 doi:10.1088/1755-1315/395/1/012028

This study gives an idea of the most important manifestation of the soil-forming process: the nature of transformations and movements of the soil solid phase, the degree of its fixation, mobility, migration in the soil profile. That is, such a study gives an idea of the processes of elution, eluviation, and argillization of the soil mass, etc.

Signs of elution in the upper part of the soil profile are detected when analyzing the nature of the distribution of fractions of granulometric composition. In general, the black soils profile is poorly differentiated by the granulometric composition. There is a slight silt depletion in the upper part and some compaction in the middle part with a weakly pronounced maximum in the illuvial horizons. This is a physical characteristic of the soil studied.

5. Conclusion

The content of the sludge fraction within the humus-accumulative soil horizons tends to increase regardless of the type of use. In leached black soils, it increases by 0.8-4.8% and by 1.5-4.9% in ordinary alkaline.

References

[1] Kiryushin V I 2010 *Agronomical soil science* (Moscow, Russia: ColosS))
[2] Milashchenko N Z 1998 *Black soils fertility of Russia* (Moscow, Russia: Agroconsult)
[3] Egorov V P, and Krivonos L A 1995 *Soils of the Kurgan Region* (Kurgan, Russia: Zauralye)
[4] Egorov V P, and Krivonos L A 1968 Changes in the physical properties of the Zauralye black soils during cultivation *Soil Science* **11** pp 53-56
[5] Ivanyushin E A, Plotnikov A M, Sozinov A V, and Yakovlev V A 2006 *The Evolution of Zauralye black soils and measures to regulate their fertility and increase the productivity of field crops* (Kurtamysh, Russia: Kurtamyshskaya Printing House)
[6] Plotnikov A M 2012 General physical properties of leached black soils in the southern agroclimatic zone of the Kurgan region *Vestnik of the Kurgan State Agricultural Academy* **1**(1) pp 35-38
[7] Plotnikov A M 2012 *Transformation of the agrophysical properties of chernozems during fallowing in the Zauralye conditions* In Materials of the All-Russian youth conference *"Modern Problems of Soil Science and Environmental Management in Siberia"* (pp 256-267) (Tomsk, Russia: Kopi-M)
[8] Kaurichev I S 1986 *Practical work on soil science* (Moscow, Russia: Agropromizdat)
[9] Kachinsky N A 1965 *Physics of the soil. Part 1* (Moscow, Russia: High School)

IOP Conf. Series: Earth and Environmental Science **395** (2019) 012029 doi:10.1088/1755-1315/395/1/012029

Accounting for the fertility of land resources as a factor in the sustainable development of the Kursk region

T Bobovnikova[1*] and T Vorozheykina[1,2]

[1] Russian State Agrarian University – Moscow Timiryazev Agricultural Academy, 49 Timiryazevskaya str., Moscow 127550 Russia
[2] Financial University under the Government of the Russian Federation Russian, 49 Leningradsky prosp., Moscow 125167 Russia

E-mail: bt32-1@bk.ru

Abstract. The article deals with the problems of accounting for the fertility of land resources. The analysis shows that the most fertile soils (black soils) are under the threat of degradation. In the Kursk region, a negative balance of nutrient removal is observed annually. Growth in crop production is due to a decrease in soil fertility, as a result of an irrational farming system. Therefore, technologies that contribute to the preservation and improvement of land fertility should be used for the effective use of land resources in the Kursk region. The results of our study confirm the need to make adjustments to the current system of organization of agricultural production by improving the accounting system.

1. Introduction

The preservation and sustainable development of the agricultural sector is impossible without maintaining the fertility of agricultural land. This is due to the features of agricultural production and its role in ensuring national food security. 55% of the world reserves of black soil belong to Russia. The black soil is mainly located in the Central Black Earth Region, including the Kursk region, which is a cross-border territory with Ukraine. The most valuable type of land in the world must be preserved with current credentials [6], [7].

2. Materials and Methods

This paper analyzes various aspects of the fertility accounting of agricultural land using control procedures (methods and techniques), identifies the main directions of research in the field of accounting for land resources. In the framework of an integrated approach, the methods of observation, comparison, grouping, as well as the balance method were also used. The information base of the study was the official statistics of Russia and the Kursk region in particular.

3. Results

Soil is an indispensable, exhaustible, relatively renewable natural resource. Rules for the rational use of soil resources are not always followed, and soil loses its fertility as a result. Protection of renewable natural resources should be carried out through their rational use and expanded reproduction [1]. The presence and therefore quantitative and qualitative land accounting, especially of arable land, hayfields, and pastures, determines food security of any country.

Content from this work may be used under the terms of the Creative Commons Attribution 3.0 licence. Any further distribution of this work must maintain attribution to the author(s) and the title of the work, journal citation and DOI.
Published under licence by IOP Publishing Ltd

In accordance with the State Standard GOST 27593-88 "Soils. Terms and definitions," the term "soil fertility" means "an ability of soil to satisfy the need of plants for nutrients, moisture and air, as well as to provide conditions for their normal activity."

Increasing the effective soil fertility implies increasing the content of nitrogen, phosphorus, and potash compounds available to plants [2].

The legislation of the Russian Federation provides for the main directions of ensuring the fertility of agricultural lands. In accordance with these directions, the following is carried out:

- Development and implementation of federal and regional targeted fertility programs;
- State rationing of fertility;
- State registration fertility indicators, fertility monitoring;
- Mandatory confirmation of compliance with plant protection products;
- Creation of data banks in the field of fertility.

Accounting data of land assets in various departments have significant differences. But reliable, standardized, and relevant data on the state of agricultural lands is necessary for their effective management.

Historical experience shows that accounting for agricultural lands is carried out on a separate synthetic account. Some authors suggest using analytical accounting, but fertility accounting is not highlighted (Table 1).

Table 1. Reflection of land as an accounting object.

Author	Publishing year	Name and code of the synthetic account	Analytical accounts
Pestrzhetsky, I.	1864	Land account	With field numbers, a number of tithes and cost of each plot
Rainbot, P. I.	1874	Land account	No data
Popov, A.	1891	Account No 27 "Land and Forest"	No data
Utekhin, N.	1897	Land account	With types of land according to the economic plan
Golts, T.	1898	Account "Land and buildings". He offered to open a separate the land account of and the buildings.	No data
Yasenetsky, B. I.	1913	"Land" account	No data
Pizengolts, M. Z.	2002	Account 08 "Fixed assets"	1 Acquisition of land. 2. Acquisition of environmental facilities 3. Construction of fixed assets 4. Acquisition of fixed assets (except cattle) 5. Acquisition of intangible assets 6. Transfer of young animals to the main herd 7. Purchase of adult animals 8. Acquisition of fixed assets for leasing 9 Other investments
Kozmenkova, S.V., Kovanov, S.I..	2015	Account 06 "Agricultural lands," "Agricultural lands" is an article in the balance sheet	1 Agricultural grounds 2 Lands in the stage of fertility restoration 3 Forest area 4 Trees and shrubs 5 Land under water 6 Other lands

Pizengolts, M. Z. notes that the main means of production in agriculture is land. It is important to ensure accurate accounting of land and land investments. Land takes into account additional (as capital) investments in physical terms (hectares), and it is in monetary terms for the purchased land [3].

Kozmenkova, S. V. & Kovanov, S. I. highlight effective aspects of its use. They propose to disclose information on unused lands (deposits), as well as on lands in the fertility restoration stage, as this directly affects the efficiency of agricultural production. These authors single out the account 06.2, i.e. these are those lands which are in the stage of fertility restoration [4].

The account 01.6 includes information related to land fertility, stating that the last should be disclosed, i.e. the distribution of agricultural land in terms of size, humus level, nutrients (nitrogen, phosphorus, potassium), lease period. Classification of land with the above parameters should be reflected in the subaccounts of the account 01.6 (01.6.01, 01.6.02, etc.).

Soil peculiarities as the main means (PBU 6/01) of agricultural production lie in the fact that soil does not reduce, but it increases fertility with right agricultural practices, relying on fertilizers and other methods.

The Kursk region is part of the Central Federal District of the Russian Federation. The territory of the region is 29.9 thousand km^2, the population is 1115, 237 thousand people, the rural population of it is 358.357 (32%). There are 28 municipal districts. The region's territory is 4.6% of the area of the Central Federal District and only 0.2% of the total area of Russia. The Kursk region is one of the five regions of the Central Black Earth economic region, occupying 18% of agricultural land and 19% of arable land. Two thirds of the soil cover of the Kursk region is "chernozem" (high quality black soils), and one fifth is represented by gray forest soils. The total length of the borders is 1,250 km, including 245 km (20%) falls on the border with the Sumy region of Ukraine. That is, the Kursk region has the status of the state border of Russia.

The high yield of agricultural crops in the Kursk region is the result of interaction of the following number of factors: increasing the quality of seed material, introducing mineral fertilizers, using new plant protection products, introducing innovative technologies for the cultivation of crops, improving forms and methods of organizing and paying for labor, etc.

In 2017, on average in the region, 151 kg of mineral fertilizers in terms of 100% of nutrients were applied to 1 hectare of sown areas, but it was more in 1990 (193 kg were added, or 27.8% more). During the period from 1990 to 2017, the measures contributing to the improvement of farmland were practically negated. In 1990, 36.4 hectares of acid soils were saturated with lime, then it was 8.7 thousand hectares in 2014, 7.0 thousand hectares in 2017. In 1992, 53.7 thousand hectares were produced, 28.4 thousand hectares in 1993, and 2.7 thousand hectares in 1994. In 1990, 339.9 thousand tons of limestone flour was introduced, which was 3.8 times more than in 2017. In comparison with the Soviet period, this is not enough.

Recently, organizational, economic, and legislative measures to return farmland into circulation have been undertaken. To accurately determine the amount of unused land, their inventory must be carried out and put on record.

One should note the concentration of land resources in the Kursk region. So, the land area of the average agricultural organization in the region amounted to 4008.2 hectares in 2016 (more than 2 times increase in comparison with 2006). The concentration of agricultural land requires tougher state control in order to prevent the monopolization of agricultural markets and the degradation of agricultural land.

A negative balance of nutrients is characteristic of contemporary agricultural production in the Kursk region (Table 2).

Table 2. Balance of nutrients, thousand c.d.v.

Balance sheet items	Years				
	2013	**2014**	**2015**	**2016**	**2017**
COMING					
Agricultural organizations					
With mineral fertilizers	1027.3	1130.3	1297	1649	1812.4
With organic fertilizers	43.0	65.3	65.7	83.2	92.9
Non-symbiotic nitrogen fixation	123	125	130	133	132
Symbiotic nitrogen fixation	41	42	43	44	44
Flow of nitrogen with precipitation	93	93	97	100	99
Households	*97.1*	*107.6*	*113.5*	*138.2*	*135.9*
Peasant farms	*238.7*	*279.6*	*326.2*	*419.5*	*454.7*
COMING - TOTAL	1620	1777	2007	2484	2679
CONSUMPTION					
Takeaway with crop	3741	4085	3779	4835	5294
In the process of soil erosion	54	55	57	58	58
During infiltration (leaching)	108	109	113	116	116
Denitrification (gaseous losses)	57	69	76	97	107
CONSUMPTION - TOTAL	3960	4318	4025	5106	5575
BALANCE	***-2340***	***-2541***	***-2018***	***-2623***	***-2896***

Based on the data in Table 2, we can conclude that about 50% of the nutrients removed with the crop are not contributed annually. To control and ensure the preservation of fertility of agricultural land, we consider it appropriate to make an annual calculation of the balance of nutrient removal, which could be reflected in the form of the 9-AIC annual report of organizations, taking into account changes in PBU 6/01.

PBU 6/01 in terms of the restoration of fixed assets should be summarized as follows. Restoration of fixed assets can be carried out through repair, modernization, reconstruction, and restoration of land is possible only with a positive balance of nutrient removal, taking into account the soils' radioactive contamination of the territories.

World experience shows that the mechanisms of economic incentives for agricultural producers need to be used; legislative acts aimed at preserving soil fertility, in the form of granting preferential loans, subsidies for fertilizers, etc. perform the role of such mechanisms. At the same time, the economic support of those land users who consciously allow land degradation should be completely excluded [5].

Funds of municipal budgets that come in the form of land tax can also be a source of funding for soil fertility reproduction. Currently, the amount of land tax does not depend on the results of economic activities of landowners, landowners, land users, and it is established in the form of stable payments per unit of land area per year.

4. Discussion

Accounting information is one of the sources of data on the state of agricultural land (acreage, gross yield, application of mineral and organic fertilizers, etc.). The bylaws do not stipulate how to use the available information to reduce soil fertility, and how they force land users to eliminate negative consequences of their activities.

The rise of the agrarian economy will be possible only if the role of the budget in the investment process increases. That is, the competitive selection of investment projects will take place. To do so, the regulatory and legal acts that control the implementation of programs on certain parameters, including the preservation of land fertility, are necessary.

As of today, many agricultural producers use software products that provide a source of uniform data. Accounting control occurs, it allows agricultural producers to obtain the appropriate agro-economic effect.

The authors have identified and systematized the main contemporary research directions in the field of land asset accounting.

1. Fertility as a unit of the agricultural land being allocated.
2. Clarifications on the application of the system of analytical subaccounts to account 01 "Fixed assets" / subaccount 6 "Land plots and objects of environmental management" are justified.
3. A nutrient removal balance leading to deterioration of agricultural land use is analyzed.
4. Amendments concerning the restoration of land resources are proposed for PBU 6/01.

5. Conclusion

Currently, the state has actually lost the function of land management, not properly carrying out land accounting and inventory, land use development planning, land management, and control over the use of land and their protection. The data of the 2006 and 2016 agricultural census do not coincide with other sources, it cannot serve as a basic assessment of the state. In order to monitor the state of food security, affordable food, indicators characterizing the structure, quality of agricultural land, acreage should be monitored. The use of digital technologies in accounting will significantly speed up and simplify decision-making on financing and state subsidization of projects for the development of various branches of agriculture.

References

[1] Sycheva I N, Voronkova O Yu, Vorozheykina T M, Yusupova G R, Semenova A N, and Ilyin A E 2018 Directions to improve economic efficiency of regional production *European Research Studies Journal* ***XXI***(3) pp 50-61

[2] Mukhi V D Ed 2003 *Agrology* (Moscow, Russia: ColosS)

[3] Pizengolts M Z 2002 *Accounting in agriculture* (Moscow, Russia: Finance and Statistics)

[4] Kozmenkova S V, and Kovanov S I 2015 Agricultural lands as an object of accounting: historical analysis and problems of the present *International Accounting* **29**(371) pp 50-60

[5] Lukin S V, and Avramenko P M 2006 Soil fertility control in the Belgorod region *Agrochemical Vestnik* **1** p 2

[6] Bogoviz A V, Semenova E I, and Sandu I S 2019 Innovational tools of provision of food security through state support for the AIC in the conditions of the digital economy *Lecture Notes in Networks and Systems* **57** pp 334-340

[7] Bogoviz A V, Semenova E I, and Sandu, I S 2018 The problem of provision of food security through management of the AIC: Transnationalization Vs import substitution *Advances in Intelligent Systems and Computing* ***622*** pp 83-88

Phytopathogenic mycobiota of the Far Eastern species of the genus *Aristolochia* L. in the culture *in vitro*

E Demidenko[1], G Gukov[1] and S Berseneva[1*]

[1] Primorskaya State Academy of Agriculture, 44 Bluhera prosp., Ussuriisk 692510 Russia

E-mail: svshatal@mail.ru

Abstract. Microclonal reproduction of plants of the Far Eastern species of the genus Aristolochia L on explants revealed the micromycetes of various systematic groups. In microclonal propagation, the authors used various methods of sterilization of the starting material with diacid and sublime. The total period of manifestation of the pathogenic properties of micromycetes on the explants of species of the genus Aristolochia L. was 24 days *in vitro*.

1. Introduction

The family Aristolochiaceae (Aristolochiaceae Juss.) has 7 genera and more than 650 species of woody lianas and perennials, distributed mainly in the tropics and subtropics of both hemispheres. Some members of the genus Aristolochia L. attributed to the relics of the Tertiary period [14]. The genus Aristolochia L. usually occurs in tropical latitudes; in Russia, the genus Aristolochia L. is represented by five species, of which two grow in the Far East: Aristolochia manshuriensis Kom., A. contorta Bunge, and one A. clematitis L.

A. manshuriensis Kom. and A. contorta Bunge grow in the south of Primorsky Krai, outside of Russia, their range covers some provinces of North-East China, A. contorta grows on parts of Japan and the Korean Peninsula [10]. Natural populations of two representatives of the *kirkazon* genus – Aristolochia manshuriensis and A. contorta – are highly fragmented and scarce [8], which makes them particularly vulnerable, these species are rare in the flora of the Far East, listed in the Red Book of Primorsky Krai with the status of endangered (EN) (A. manshuriensis) and (VU) vulnerable (A. contorta) [6].

The weak seed productivity and absence of natural seed propagation agents among representatives of the Far Eastern flora lead to a permanent reduction in the number, including *kirkazon* (Aristolochia manshuriensis). The presence of unique medicinal properties in species of the genus leads to the extermination of rare individuals as a medicinal raw material.

The restoration of natural populations of rare and endangered plant species by the method of clonal micropropagation is a promising direction of biotechnology. Biological features of the growth and development of *kirkazon* is associated with the hindered vegetative and seed reproduction, a high degree of infection with fungal and bacterial diseases, and a single occurrence of individuals at reproductive age [5]. In this case, the condition for obtaining high-quality raw materials for the cultivation of regenerated plants is the use of a sterile culture obtained by the method of microclonal propagation.

Content from this work may be used under the terms of the Creative Commons Attribution 3.0 licence. Any further distribution of this work must maintain attribution to the author(s) and the title of the work, journal citation and DOI.
Published under licence by IOP Publishing Ltd

The micromycetes on the vegetative organs (source material) causing the death of the explant can be a serious reason preventing the introduction of plants into the culture *in vitro*.

The issues of the phytopathogenic state of the species of the *kirkazon* genus in natural populations have been studied in foreign and domestic literature. Mycobiota of the Far Eastern species of *kirkazon*, obtained by the method of microclonal reproduction, has not been studied to date.

Studies of the species diversity of parasitic fungi on explants of plants of the genus Aristolochia L. will allow optimizing methods for sterilizing the source material introduced into an *in vitro* culture.

In connection with the foregoing, the purpose of this work is to study the phytopathogenic microbiota on explants of the Far Eastern species of the genus Aristolochia L. *in vitro* and to optimize the method of sterilization of the starting material.

To solve it, the following tasks were set:

1. To identify species diversity and characteristics of the development of phytopathogenic micromycetes on explants of Far Eastern representatives of the genus Aristolochia L. in vitro.
2. To optimize the methods of sterilization of the source material introduced into the culture *in vitro*.

2. Materials and Methods

Objects of study for introduction into culture in vitro are the species of the *kirkazon* genus: Aristolochia manshuriensis, A. contorta Bunge, and A. clematitis L.

The plant material is collected in the collection of the arboretum belonging to the Branch of the Federal Scientific Center for Terrestrial Biota of East Asia of the Far Eastern Branch of the Russian Academy of Sciences at the Gornotaezhny Station named after V.L. Komarova (hereinafter referred to as "Gornotaezhny Station)".

Young shoots of the virginal age of the current year of development with one or two axillary buds (explants), which were taken in June, were used as the starting material. Studies were performed on 20 explants of each plant species in triplicate.

During the experiment, the proportion of explants that were affected by the fungi was estimated.

Sterilization of the starting material (microcurrents) was carried out in two ways. In the first case, the explants were sterilized by the standard technique, which consisted in the sequential processing of a soap-alkaline solution (15 min) and a 0.1% solution of the diacid (4 min) with 3-fold washing with sterile distilled water [5;1]. In the second, a 0.2% mercury chloride solution was used to sterilize the source material (micro-cuttings) II ($HgCl_2$) (mercuric chloride) within 4 minutes, with 3-fold washing in a sterile distilled water.

Under boxing conditions (BAVnp-01 "Laminar-S.," 1.2), our explants were placed vertically in the previously prepared sterile nutrient media. Preliminary preparation of all instruments was carried out in a glossy sterilizer (S-01). The working surface was disinfected with ethyl alcohol (70%).

Morphometric measurements of the samples were carried out in an aqueous solution using a Nikon Eclipse E200 microscope at 80–400X, with 1500X magnification. Microscopic examination was performed according to generally accepted botanical techniques [15], [3]. Latin names of mushrooms and plants, and abbreviations of authors' names are in accordance with the open database "Index Fungorum, 2018" [16].

After sterilization, the cuttings were placed vertically in the nutrient media prepared on the basis of macro and micro salts in MS and WPM with the addition of indole acetic acid (IAA) at a concentration of 1.0 mg / l and 4.0 mg / l and 2ip - 8 mg / l [4].

The nutrient media were sterilized by autoclaving at 0.8 atm for 20 minutes. Test tubes with a diameter of 10 mm and a height of 15 cm were used as culture vessels for growing microshoots.

Explants at all stages of the experiment were cultured at a temperature of +24 °C, a 16-hour photoperiod (16/8), illuminated with white fluorescent lamps with an intensity of 4 thousand lux. and 60% relative humidity [5].

3. Results

As a result of our research, taking into account the conditions of sterilization of explosives, *in vitro* cultivation of the species with the genus Aristolochia L., the micromycetes of different systematic groups were found (Table 1).

Table 1. The causative agents of fungal phytopathogenic infections during clonal micropropagation of species with the genus Aristolochia L.

Species	Phytopathogenic micromycetes	
	Sterilization diacidom	**Sterilization $HgCl_2$**
A. manshuriensis Kom.	Alternaria alternata (Fr.) Keissl., Botrytis cinerea Pers., Fusarium oxysporum Schlecht., Rhizopus stolonifer (Ehrenb.), Mucor hiemalis Wehmer., Trichothecium roseum (Pers.) Link.	Fusarium oxysporum Schlecht., Fusarium oxysporum Schlecht., Rhizopus stolonifer (Ehrenb.), Trichothecium roseum (Pers.) Link.
A. contorta Bunge.	Alternaria alternata (Fr.) Keissl., Botrytis cinerea Pers., Mucor hiemalis Wehmer.	Not identified
A. clematitis L.	Botrytis cinerea Pers., Mucor hiemalis Wehmer.	Not identified

Observations have shown that during sterilization by diacid, explants of A. manshuriensis Kom. are more susceptible to infection. At the same time, the A. contorta Bunge, which occupies a transitional position between woody and grassy species of the genus, is struck by a significantly smaller number of pathogenic mycobiota species. In the herbaceous species A. clematitis L., explants *in vitro* are susceptible to infection with only two types of mold fungi.

When using the sublimate solution, we observed the presence of only 3 types of micromycetes on A. manshuriensis explants *in vitro*: Fusarium oxysporum Schlecht., Rhizopus stolonifer (Ehrenb.), Trichothecium roseum (Pers.) Link. All these types of micro-cuttings struck the A. manshuriensis Kom.

According to literary data in grassy species and species that are transitional between grassy and ligneous plants, the main phase of the mass detection of infected explants occurs at the end of the third week of cultivation. It has been established that the main condition preceding the onset of development of fungal phytopathogenic microorganisms during microclonal propagation of plants *in vitro* is increased humidity, an insufficiently effective method of sterilizing primary explants and feeding habits of phytopathogenic micromycetes.

The total period of manifestation of phytopathogenic microflora on non-sterile micrograins and micro-shoots of species of the genus Aristolochia L. in vitro was 24 days from the initial landing.

The first manifestations of visible signs of fungal infection on the explants were observed on days 5-7 of cultivation (Fig. 1).

Figure 1. Fungal infection in the cultivation of Aristolochia manshuriensis using sterilization by diacid (on the right, the explants affected – Mucor hiemalis; Fusarium oxysporum on the left).

In each test tube with the explants, only 1 type of micromycetes was developed. After 3 days of cultivation, micromycetes appeared in 100% of the explants, the explants of all three species completely blackened and died. In the first week of cultivation, a fungal infection was observed in approximately 30% of A. manshuriensis and A. contorta explants, this figure was 2 times lower for A. clematitis. The proportion of infected explants A. contorta and A. manshuriensis for 10 days of cultivation increased in 1.3 times. A. alternata (28% of the explants out of 41.7%) was noted most often in test tubes on explants of A. contorta, 28.7% of B. cinerea and M. hiemalis were recorded. The explants for A. clematitis were M. hiemalis (10.8% of 15%) and B. cinerea (4%).

In the course of the research, it is found that the highest rate of activity of pathogenic microorganisms was observed from the 2nd week from the moment of planting until the middle of the 3rd, during these periods the percentage of infection of the microdrops sharply increased, and was 2 times higher than in the 1st week of cultivation (Table 2).

Table 2. Infectiousness of explants of Aristolochia L. species when using diacine sterilization after 24 days of cultivation.

Species	The proportion of affected explants, % by day						
	6	7	10	14	17	21	24
A. manshuriensis Kom.	28.9	30	37.8	67.1	67.5	93.3	100
A. contorta Bunge	28.3	27.5	38.3	41.7	41.2	96.7	100
A. clematitis L.	5,6	11.1	15	20	22	38.3	40

After 3 weeks of cultivation, the micromycetes appeared on almost all explants of A. contorta and A. manshuriensis, while about 60% of A. clematitis explants showed no signs of infection. After two weeks of cultivation (before the end of the experiment), B. cinerea was most often observed on the explants of A. contorta and A. manshuriensis, without obvious signs of fungal infection on the explants of A. clematitis M. hiemalis. Since phytopathogenic micromycetes developed on all explants in the absence of exchanges of pathogens and the inflow of propagules of phytopathogenic fungi from the outside, it can be concluded that the infection of the studied A. contorta and A. manshuriensis samples was initially 100%.

A slightly different picture of infection of the studied species of the genus *kirkazon* when using the sterilization of sublimate (Table 3).

Table 3. Infection of explants of Aristolochia L. species when using sterile sterile sterilization after 24 days of cultivation.

Species	The proportion of affected explants, % by day						
	6	7	10	14	17	21	24
A. manshuriensis Kom.	8.0	12.0	12.0	26.0	29.0	37.0	49.0
A. contorta Bunge	-	3.1	3.1	4.0	4.8	6.3	6.3
A. clematitis L.	-	-	1.3	1.7	2.1	2.4	2.4

The use of a more concentrated solution of sublimate as a sterilizer led to a significant reduction in the number of identified pathogenic micromycetes on *in vitro* kirkazon explants. In the case of A. manshuriensis microcurrents, the amount of cultivated material infected by internal infection averaged 50%, the similar indicators did not exceed 5-10% in the grassy species.

The main symptoms of diseases caused by pathogens of pathogenic fungi are: the appearance of dark brown or olive-black raids, consisting mainly of conidiophores and conidia of the fungus (Alternaria alternata (Fr.) Keissl.); the formation of a thick fluffy dark gray patina sprinkled with light gray flour (Botrytis cinerea Pers); a lesion in the form of pink (with shades of soft pink to crimson color) mold; and it was accompanied by the rotting of internal tissues (Fusarium oxysporum Schlecht.) sporangia of the fungus as black heads (Rhizopus stolonifer (Ehrenb), the formation of a fluffy white plaque which darkens with (Mucor hiemalis Wehmer) over time [2], [7], [9], [11], [12].

It is known that the antifungal effect is shown for species of the genus Aristolochia. Therefore, the extracts of A. debilis, A. bracteolata Linn, and A. ringens suppressed the development of Botrytis cinerea [18]; the extracts of A. indica and A. debilis affected the development of F. oxysporum [19], the extracts of A. bracteolata influenced R. stolonifer [17]; while the extracts of A. contorta affected M. hiemalis et al. [13].

According to the above, we assume that the micromycetes found as a result of the experiment are not brought outside (with the initial material), but are inside the explants tissues.

According to the data obtained, the method used for diacine sterilization of primary explants during the cultivation of A. manshuriensis *in vitro* promotes the development of pathogenic micromycetes. In some cases, a longer period of treatment with diacid leads to damage to both the outer integumentary tissues of the plant's microcurrents and the tissues on the cutting section, which entails the partial or complete die-off of the damaged plant tissues.

4. Conclusion

The species diversity of parasitic fungi was carried out on the explants of plants of Far Eastern species of the genus Aristolochia L., which manifest themselves in microclonal reproduction in vitro.

According to the obtained results, the phytopathogenic micromycetes are represented by two sections – Zygomycota and Ascomycota, 6 genera and 6 species (Alternaria alternata (Fr.) Keissl., Botrytis cinerea Pers., Fusarium oxysporum Schlecht., Rhizopus stolonifer (Ehrenb.), Mucor hiemalis Wehmer., Trichothecium roseum (Pers.) Link.

For the plants of A. manshuriensis, a description of phytopathogenic species of fungi is given for the first time. When cultivating microcurrents and Aristolochia manshuriensis microshoots, the need to improve the method of sterilization of primary explants is shown, which is due to an increase in infection of micrografts (67.5-100%) after treatment with a diacid solution.

Among those identified in this method of sterilization, the most numerous was Fusarium oxysporum Schlecht. They affected from 15 to 35% of A. manshuriensis microcurrents. The share of Fusarium oxysporum and Rhizopus stolonifer accounted for no more than 5-15% of the total number of microcurrents. The period of manifestation of visible signs of infection of the explants was approximately 23-26 days. The explants of A. manshuriensis, which were not subjected to the manifestation of infection during this period, formed shoots and partial thickening of the epithelial tissues at the base of the microcurrent. For A. clematitis and A. contorta microchips, this sterilization method is the most acceptable, since in this case the number of explants with obvious signs of damage by pathogenic microbiota does not exceed 5-7%.

The use of biotechnological methods of reproduction contributes to the conservation and sustainable reproduction of rare and endemic species *in vitro*. The obtained data on the phytopathogenic mycobiota will allow optimizing the microclonal propagation methods of Aristolochia manshuriensis and obtaining a sufficient amount of planting material in order to restore the number of natural populations.

5. Acknowledgments

We express our deep gratitude and appreciation to the staff of the Federal Scientific Center for the Biodiversity of Terrestrial Biota of East Asia, FEB RAS L.N. Yegorova, I.V. Gafitskoy, O.V. Nakonechna for consultations in determining the species composition of mycobiota and staff and dendropark of the branch of the Federal Research Center of the Terrestrial Biota of East Asia of the Far Eastern Branch of the Russian Academy of Sciences of the Gornotaezhnaya Station nmed after V. L. Komarova for the opportunity to collect the source material.

References

[1] Babikova A V, Gafitskaya I V, Koren O G, Muzarok T I, Zmeyeva V N, Pinkus S A, Akimova L A, … Barkalova O K 2013 Microcloning of decorative woody plants. In *Problems of landscaping settlements* (Vladivostok, Russia)

[2] Blagoveshchenskaya E Yu 2015 *Phytopathogenic micromycetes: educational determinant* (Moscow, Russia: Lenand)

[3] Barykina R P, Veselova T D, Devyatov A G, Dzhalilova KH KH, Ilina G M, and Chubatova N 2000 *Fundamentals of microtechnical research in botany: a reference guide* (Moscow, Russia: Publishing House of Moscow State University)

[4] Demidenko E N, Gafitskaya I V, Mikheeva (Babikova) A B 2016 On the issue of microclonal

IOP Conf. Series: Earth and Environmental Science **395** (2019) 012030 doi:10.1088/1755-1315/395/1/012030

propagation of species of the *kirkazon* genus (Aristolochia L.). In *Water and environmental problems, transformation of ecosystems in the context of global climate change: VI Druzhinin reading* (pp. 128-130) (Khabarovsk, Russia)

[5] Kataeva N V, and Butenko R G 1983 *Clonal micropropagation of plants* (Moscow, USSR: Nauka)

[6] *Red Book of Primorsky Krai: Plants* 2008 (Vladivostok, Russia: Apelsin)

[7] Melnik V A 1977 *Determinant of the fungi of the genus Ascochyta Lib.* In M A Litvinov Ed. (Moscow, USSR: Nauka)

[8] Nakonechnaya O V, Zhuravlev Yu N, Bulgakov V P, Koren O G, Sundukova E V 2014 The *Kirkazon genus in the Far East of Russia* (Aristolochia manshuriensis Kom. и A. contorta Bunge) (Vladivostok, Russia: Dalnauka)

[9] Popkova K V, Shkalikov V A, Stroykov Yu M *et al.* 2005 *General phytopathology: a textbook for universities* (2nd ed.) (Moscow, Russia: Drofa)

[10] *Vascular plants of the Soviet Far East (Vascular plants)* 1987 (Leningrad, USSR: Nauka)

[11] Teterevnikova-Babayan D N 1987 *Fungi of the genus Septoria in the USSR* In A Melnik Ed. (Yerevan, USSR: Academy of Sciences of Armenia. Institute of Botany)

[12] Braun U 1994 Studies on Ramularia and allied genera (VII) *Nova Hedwigia* **58** pp 191-222

[13] Chung Y J, Lee K S, Han S H, Kang D I, and Lee M H 2001 Antifungal and insecticidal activity of ohyang (five medicinal plants) *Int. J. Conserv. Sci.* **10**(1) pp 21-30

[14] Gonzáles F A, and Stevenson D W 2002 A phylogenetic analysis of the subfamily Aristolochioideae (Aristolochiaceae) *Botanica* **26**(98) pp 26-58

[15] Hawksworth D L 1974 *Mycologist's Handbook* (Surrey, UK: CAB International)

[16] *Index Fungorum* 2018 http://www.indexfungorum.org/ names/Names.asp. (Accessed 15 11 2018)

[17] Kavitha D, and Nirmaladevi R 2009 Assessment of Aristolochia bracteolata leaf extracts for its biotherapeutic potential *Afr. J. Biotechnol.* **8**(17) pp 4242-4244

[18] Li H Y, Guan L J, Shao S, and Ma J S 2009 Inhibition of chinese herb extracts on fruit-vegetable pathogenic fungi. *North. Hortic.* **9** p 10

[19] Liu J Y, Xue G H, Wang X L, Song M H, Wang L, and Fan H Y 2011 Study on the antibacterial activity of ten ultrasound extracts from medicinal plants *Hubei Agricultural Sciences* **9** p 26

Using natural zeolites to improve the quality of unrefined vegetable oils

L V Tereshchuk[1], K V Starovoytova[1*] and A N Crol[1]

[1] Kemerovo State University, 6 Krasnaya str., Kemerovo 650043 Russia

E-mail: centol@mail.ru

Abstract. The article is devoted to the study of the possibility of using natural zeolites as adsorbents for the purification of unrefined vegetable oils, intended for direct consumption in food, from undesirable concomitant substances. To increase their sorption capacity, various activation methods were used, such as physical and chemical. The best sorption properties were found in zeolites, activated by the complex method (heat and acid treatment). Conducted refining of unrefined sunflower oil allowed to increase its quality to meet the requirements of the state standard for the highest grade oil.

1. Introduction

The beneficial properties of sunflower oil are determined mainly by its fatty acid composition, which to a very large extent depends on the climatic conditions of the area, cultivation methods, and varietal characteristics of sunflower. Among the many cultivated varieties on the basis of the predominant fatty acid in the composition, there are traditional low-oleic acids, as well as later derived medium and high-oleic acids. The sunflower seed oil of traditional varieties contains linoleic acid most (up to 75%). In the oil of high-oleic varieties, the predominant acid is oleic, its content reaches 70%, which makes this oil closer in properties to olive. High oleic oil has a greater, compared to linoleic varieties, oxidation resistance and can be successfully used for direct consumption, for example, as a salad dressing. It should also be noted that oleic acid, which does not possess the physiological activity of essential fatty acids, enhances the action of linoleic acid, being its synergist. The nutritional value of high oleic sunflower oil, especially unrefined, is determined not only by the high content of unsaturated oleic and linoleic acids, but also by the fact that it contains the highest amount of tocopherols (738 ± 82 mg / kg) and tocotrienols (270 ± 61 mg / kg), with E-vitamin activity. Refining according to the classical scheme leads to almost complete loss of these substances, which significantly reduces the nutritional value.

Sunflower oil, intended for direct consumption in unrefined form, is obtained only by pressing from the collapsed sunflower seeds. After pressing in unrefined vegetable oils, along with the presence of free fatty acids, there are about 2% non-fatty substances belonging to the unsaponifiable fraction. They consist of phospholipids, sterols, waxes, hydrocarbons, pesticides, proteins, trace amounts of metals and pigments [1]. Most of the unsaponifiable substances are undesirable in oil, and their removal is an urgent task in the problem of improving the quality of unrefined vegetable oils. At the same time, the cleaning of oils should be strictly directed and not accompanied by the removal of useful components, which is observed during classical refining [1]. In addition to physico-chemical and organoleptic indicators, safety indicators, including the content of toxic elements such as iron,

Content from this work may be used under the terms of the Creative Commons Attribution 3.0 licence. Any further distribution of this work must maintain attribution to the author(s) and the title of the work, journal citation and DOI.
Published under licence by IOP Publishing Ltd

IOP Conf. Series: Earth and Environmental Science **395** (2019) 012031 doi:10.1088/1755-1315/395/1/012031

lead and arsenic, as well as indicators of oxidative damage (acid and peroxide number [2]) are important in assessing the quality of oils. Metals are usually found in oilseeds in trace amounts, the highest concentration in plants is reached during the growing season. For removal of vegetable oils of compounds like chlorophyll, carotene, phospholipids, peroxides, soaps in the oil and fat industry use highly active polar sorbents supplied by foreign companies, mainly natural active or chemically activated aluminosilicate bentonite bleaching clays. At present, chemisorption and acid catalysis are mainly used for these purposes [3], [4].

Recently, interest in natural zeolites has increased, and their multifaceted useful properties have been identified. Zeolite is an aluminosilicate with a skeletal structure containing voids occupied by large ions and water molecules that have significant freedom of movement, which leads to ion exchange and reversible dehydration. The zeolite lattice is constructed from silicon and aluminum-oxygen tetrahedra SiO_4 and AlO_4, which are the primary structural units and are designated TO_4, where T is a silicon or aluminum atom. The main structural element of zeolites is a tetrahedral ortho-group SiO4 of the metal type with a cation in the center surrounded by four oxygen atoms located at the vertices of the tetrahedron, where part of the Si^{4+} ions are replaced with Al^{3+} ions in order for the crystal to maintain stability, it is necessary to maintain its electrical neutrality. The negative charge of the alumina-oxygen tetrahedron is compensated by cations of alkali, alkaline-earth and other metals [5].

Despite the virtually absent literature data on the use of zeolites in the oil and fat industry, it can be assumed that zeolites can be used as adsorbents for the extraction of related substances from vegetable oils. Research on zeolites as a new sorbent in the oil and fat industry is an urgent task that requires its solution.

The aim of the work was to study the possibility of using zeolites in the oil and fat industry as adsorbents for the purification of unrefined vegetable oils (intended for direct consumption) from unwanted concomitant substances.

2. Materials and Methods

We selected the samples of high oleic sunflower oil, obtained by hot pressing, as well as natural zeolites of the Pegasus field in the Kemerovo region. To study the phase composition of zeolite rocks, an X-ray phase analysis was used. To determine the chemical composition of the zeolite tuff, an X-ray fluorescence analysis was used. The fatty acid composition of vegetable oils was determined on an Agilent 7890A gas chromatograph [5]. Determination of the color of vegetable oils was also carried out [6]. The content of free fatty acids in the oil before and after filtration through a layer of zeolites was determined by the level of acid number [7]. The content of toxic elements in oils was determined by the atomic absorption method [8].

3. Results

We were faced with the task of obtaining unrefined sunflower oil, intended for direct consumption, which meets the requirements of the standard for unrefined oil of the highest grade [1]. To solve this problem, the possibility of using natural zeolites for cleaning press oil from free fatty acids and related non-fatty substances was investigated.

The X-ray fluorescence analysis showed that the zeolite tuffs of the Pegasus field in the Kemerovo region are characterized by the following chemical composition: Na – 0.92%, SiO_2 – 66.8%, TiO_2 – 0.4%, Al_2O_3 – 13.92%, Fe_2O_3 – 3.60%, FeO – 0.05%, MgO – 1.66%, CaO – 4.92%, K_2O – 0.88%, Si_2O – traces, P_2O_5 – 0.06%, Pb – 0.002%, Cu – 0.0044%, Co – 0.00055%, Mo – 0.00036%, Zn – 0.0069%. Along with zeolites, volcanic glass, quartz, feldspar, mica, and clay minerals are present in the tuff. It is characterized by the following unit cell parameters: a=17,73 Å, в=17,82 Å, c=7,43 Å, β=116 Å, density 2,198 g/cm^3, a unit cell volume = 210 Å^3.

Before carrying out a trial purification of vegetable oil, the zeolites were dried. The dried zeolite has a free-flowing structure, which facilitates its grinding. Next, the zeolites were crushed and sieved to different fractions, with a particle size of up to 0.25 mm, 0.25–0.4 mm, 0.4–0.65 mm, 0.8–1 mm. In the course of the study, it was found out that the fraction with a particle size of up to 0.25 mm has the best adsorption properties.

The sorption capacity of zeolites is determined by the size of the free surface. Natural zeolites contain a lot of water, which covers the entire active surface and reduces its sorption capacity [9]. The zeolite fraction was heat treated at different calcination temperatures. During heat treatment, part of the adsorption water leaves, which leads to the release of the surface of the sorbent. Found that both excessive moisture content in the sorbents, and excessive dehydration greatly reduces their activity. Adsorption water is retained by the zeolite mainly by forces of a physical nature and therefore a significant part is removed at temperatures up to 350–380 °C. Chemically bound water is removed at sufficiently high temperatures. In this regard, we carried out the activation of zeolites at a temperature range from 200 °C to 700 °C.

Cleaning of sunflower oil from related substances was carried out by thermally activated zeolites by contact method with the following process parameters: the amount of adsorbent is 2.5% by weight of the oil, mixing with a paddle stirrer at a speed of 60 min^{1}, at a temperature of 80 ° C, for 30 minutes. Data on changes in color and acid numbers of sunflower oil are presented in Table 1.

Table 1. The results of filtration of raw sunflower oil by thermally treated natural zeolites of the Pegasus field.

Activation temperature, °C	**Values of indicators**			
	Color number (mg I_2)		Acid number (mg KOH / g)	
	Before cleaning	After cleaning	Before cleaning	After cleaning
200	35	30	4.67	1.34
300	35	32	4.67	1.38
400	35	33	4.67	1.50
500	35	35	4.67	1.35
600	35	35	4.67	1.35
700	35	35	4.67	1.35

However, our research also shows that the use of thermal activation only does not lead to the removal of the entire complex of related substances. The number of pigments, determined by the indicator "color number" decreases slightly (from 35 mg iodine to 30 mg iodine). The decrease in the content of free fatty acids is noticeable, since acid number significantly decreased to 1.34 - 1.38 mg KON / g. This is probably due to the fact that these free fatty acids are more easily adsorbed on thermally activated zeolites than chlorophylls and carotenoids. Thus, it can be concluded that heat treatment is an insufficient measure of the activation of the sorption capacity of zeolites. At the next stage, chemical modification of zeolites was performed.

The activations were carried out as follows: 50 g of dry ground zeolites and 250 ml of hydrochloric acid solution, taken at the rate of 50% acid by weight of dry zeolites, were placed in a 500 ml round flask equipped with reflux condenser and stirrer. The treatment was carried out at different temperatures, with stirring for different periods of time, from 30 minutes to 6 hours. The activated zeolite was filtered on a Buchner funnel, washed with distilled water heated to (40–50) °C to neutrality by methyl orange, then dried in a thermostat at (105–110) °C. After this, the sorption properties of the modified zeolite were studied. The results of the research indicators of quality and safety of the sample, purified by chemically activated zeolites are presented in Table 2.

Thus, our research clearly shows that the maximum increase in adsorption activity with respect to pigments takes place when zeolite is treated with a solution of 10% hydrochloric acid at a temperature of 80 °C for (4.5–5.0) hours, with a quantitative ratio of solid and liquid phases of 1: 4. We found that the zeolite in the activation process responds well to processing. Suspension is well transported when being settled, it quickly separates and is freely filtered.

Table 2. The results of raw sunflower oil filtration by thermally treated and chemically activated natural zeolites.

Indicators	Value before filtering	Value after filtering	Требования НД [1,11].
Transparency	Turbid, there is sediment	Transparent with a slight turbidity	Light turbidity or "mesh" is allowed.
Smell and taste	Peculiar to sunflower oil, without foreign smell and taste	Peculiar to sunflower oil, without foreign smell and taste	Peculiar to sunflower oil, without foreign smell and taste
Color number, mg iodine	35	13	No more 15
Acid number (mg KON / g)	4.67	1.1	No more 1.5
Tocopherol content, mg / 100 ml	74	71	-
Plumbum, mg / kg	Less 0.01	Less 0.01	No more 0.1
Arsenicum	Less 0.03	Less 0.03	No more 0.1
Ferrum	0.79	0.41	No more 1.5
Caesium-137, Bq / kg	2.67	1.15	No more 40 Bq / kg
Strontium-90, Bq / kg	2.55	1.11	No more 80 Bq / kg

4. Discussion

Processing of unrefined sunflower oil with zeolites that have undergone double activation allowed reducing the color number to 13 mg of iodine, which corresponds to the requirements of the state standard for unrefined highest quality vegetable oil; in addition, the content of tocopherols in the oil remained almost unchanged. It should be noted a significant decrease in the content of free fatty acids in the oil, as well as a decrease in the amount of iron and radionuclides ^{137}Cs и ^{90}Sr. Apparently, this is due to the specific ion-exchange properties of zeolites, and the presence of a large amount of silicon dioxide (66.8% of SiO2 in a content), the main component of quartz sand, a known agent used for wastewater treatment [11], [12].

5. Conclusion

Natural zeolites of the Pegassky deposit of the Kemerovo region can be successfully used in the oil and fat industry to obtain high-quality sunflower oil, suitable for both direct consumption and for food production, including baby food.

6. Acknowledgments

The authors are grateful to the staff of the laboratory of the department "Technology of food from vegetable raw materials" (KemSU), who took part in practical research.

References

[1] State Standards 2013 *Sunflower oil: Technical conditions (with amendment) – GOST 1129-2013* (Moscow, Russia)

[2] Anirudhan T S, and Suchithra P S 2010 Heavy metals uptake from aqueous solutions and industrial wastewaters by humic acid-immobilized polymer/bentonite composite: kinetics and equilibrium modeling. *Chem Eng J* **156** pp146-56

[3] Anirudhan T S, Divya L, Bringle C D, and Suchithra P S 2010 Removal of copper(II) and zinc(II) from aqueous solutions using a lignocellulosic-based polymeric adsorbent containing amidoxime chelating functional groups. *Sep Sci Technol.* **45** pp 2383-93

[4] Batistela V R, Fogaca L Z, Favaro S L, Caetano W Fernandes-Machado NRC, and Hioka N 2017 ZnO supported on zeolites: photocatalyst design, microporosity and properties. *Colloid Surface A.* **513** 20-7

[5] Anirudhan T S, Suchithra P S, and Radhakrishnan P G 2009 Synthesis and characterization of humic acid immobilized-polymer/bentonite composites and their ability to adsorb basic dyes from aqueous solutions. *Appl Clay Sci.* **43** 336-42

[6] Cobzaru C 2012 Modified zeolites: modification of natural zeolites for catalytic applications In *Handbook of Natural Zeolites* (pp 185-213) (Sharjah, U.A.E: Bentham Science Publishers)

[7] *Vegetable oils: Method for determination of fatty acid composition (GOST 30418-96)* 1996 (Moscow, Russia)

[8] *Vegetable oils: Methods for determining chromaticity (GOST 5477-2015)* 2015 (Moscow, Russia)

[9] *Vegetable oils: Methods for determining the acid number (GOST 31933-2012)* 2012 (Moscow, Russia)

[10] *Raw materials and food products: Atomic absorption method for the determination of toxic elements: GOST 30178-96* 1996 (Moscow, Russia)

[11] Technical Regulations of the Customs Union 2011 *Technical Regulations on oil and fat products* (TR TS 024/2011) (Minsk, Belarus: Customs Union)

[12] Suchithra Padmajan Sasikala, T A Nibila, Kunnathuparambil Babu Babitha, Abdul Azeez Peer Mohamed, and Ananthakumar Solaiappan 2019 Competitive photo-degradation performance of ZnO modified bentonite clay in water containing both organic and inorganic contaminants *Sustainable Environment Research* **29**(1)

IOP Conf. Series: Earth and Environmental Science **395** (2019) 012032 doi:10.1088/1755-1315/395/1/012032

Medical-geographical indicators of the Altai Krai as a cross-border region of Western Siberia

S Tokmakova[1], O Bondarenko[1*] and Y U Lunicyna[1]

[1] Altai State Medical University, 40 Lenin str., Barnaul 656038 Russia

E-mail: agmuterst@mail.ru

Abstract. As a result of a comprehensive study on the methodology of the World Health Organization, the dental incidence of children and adults in the districts of the city of Barnaul with different levels of air pollution has been studied. It is established that the ecological situation in the city is complicated by the emissions of harmful substances by industrial enterprises (nitrogen dioxide, formaldehyde, soot, dust). The structure of the indicator of the intensity of teeth caries of hard tissues in key age groups was studied, and a positive correlation was found between the intensity of dental caries in adolescents of Barnaul and the exceeded emission concentrations of the air pollutants studied by us. A high prevalence of dental anomalies in adolescents has been established. The prevalence of periodontal and oral mucosa diseases was evaluated in the examined individuals. It was revealed that this pathology did not depend on the degree of contamination of the air basin in the districts of the city of Barnaul. Analysis of the conducted clinical and epidemiological research reveals the features of dental pathology in various age groups of a large industrial city of the Siberian region. The data can be used for a differentiated approach in creating and planning programs to improve the prevention of dental health of the population, taking into account the environmental situation in the districts.

1. Introduction

The Altai Krai, like most other regions of Siberia, is characterized by a tense ecological situation, uneven density and steady trends in the health of the population. Studies of domestic and foreign scientists have established that environmental pollution is of no small importance in the violation of public health. For many years, the ecological situation has been aggravated by the functioning of the Semipalatinsk test site for nuclear devices, the concentration of large industrial enterprises polluting the air in the cities of the region and its capital. The adverse effect on the prevalence of dental morbidity has water supply, which is carried out according to the Center for State Sanitary and Epidemiological Surveillance of the city of Barnaul from sources with a low content of fluorine and microelements [1], [2], [3]. Ecology, visceral pathology and dental diseases are interdependent and interrelated. The most common dental diseases are the defeat of hard dental tissues, periodontal pathology, dentofacial anomalies, diseases of the oral mucosa, as evidenced by epidemiological studies in recent years [4], [5]. An analysis of the prevalence and intensity of oral pathology under the influence of harmful environmental factors in various districts of the city of Barnaul will make it possible to develop a prevention strategy that can be extrapolated to transboundary regions.

Content from this work may be used under the terms of the Creative Commons Attribution 3.0 licence. Any further distribution of this work must maintain attribution to the author(s) and the title of the work, journal citation and DOI.
Published under licence by IOP Publishing Ltd

2. Materials and Methods

To assess the prevalence and intensity of dental diseases, a comprehensive dental examination of the population of five districts of the city of Barnaul was carried out according to the WHO methodology in key age groups: 6, 12, 15 years, 35-44 and 65-74 years. A total of 1,250 people were examined, of which 542 were males and 708 females. The examination was carried out under standard conditions, the results (179 signs) were recorded in the assessment chart of dental status. Climate-geographical features of the regional center of Altai were studied using reference data on the climatography of the USSR and the West Siberian Hydrometeorological Center, based on the average annual data on temperature, force and wind direction, number of sunny days per year. The environmental situation was assessed according to the state of the air and water basins of the city of Barnaul. We used data from the Research Institute for Water and Environmental Problems of the West-Siberian Branch of the Academy of Sciences, the General Directorate of Natural Resources and Environmental Protection of the Ministry of Natural Resources of Russia for the Altai Krai.

To characterize the water basin, data from the center of state sanitary and epidemiological surveillance of the city of Barnaul on the condition of drinking water in the studied areas on the content of fluoride and mineral substances were used.

3. Results

In the air basin of the city of Barnaul, Altai Krai there is a high concentration (above the maximum permissible value) of carbon monoxide, nitrogen dioxide, formaldehyde, soot and dust. The highest emissions of harmful substances are registered in the Zheleznodorozhny and Oktyabrsky districts of the city of Barnaul, where the atmospheric pollution index is characterized by a high ($1 < API \leq 4$) and high degree ($4 < API \leq 16$), respectively.

In 12-year-old children with severe pollution of atmospheric air with nitrogen dioxide, formaldehyde, there are extraoral signs of damage ($r \approx 0.3$, $p < 0.05$). When contaminated with dust, soot and formaldehyde in 15-year-olds, a high intensity of caries of hard dental tissues is observed ($r \approx 0.4$; $r \approx 0.3$, $r \approx 0.4$, $p < 0.05$, respectively). These pathological changes are prevalent mainly in the Zheleznodorozhny and Oktyabrsky districts of Barnaul.

A high prevalence of dental caries was found in 12 and 15-year-old children, especially in the Oktyabrsky district (94.0%), and 100% caries susceptibility of adults at high intensity. In indicator of intensity in 6-year dental caries predominates temporary teeth, 12, 15 year old - fillings permanent teeth in adults - extracted teeth. For the first time, 12-year-old children began to observe removed permanent teeth (2.0 ± 2.0 - 4.0 ± 2.79%).

Non-carious lesions of hard dental tissues, in the form of focal and diffuse spotting (up to 50%), are not associated with the intensity of air pollution with harmful substances. Fluorosis among the population of the regional center of the Altai Krai was not identified, which is probably due to the low fluorine content in drinking water in all areas of the city (0.19-0.22 mg / l).

The prevalence of pathology of the oral mucosa and temporomandibular joint increases with age (up to 40%, 50%, respectively) and does not depend on the degree of air pollution.

Against the background of the high prevalence of periodontal diseases (84-100%), the severity of tissue damage is negligible in childhood (tartar, gingival bleeding). It increases in the middle and older age groups (periodontal pockets of different depths) and does not depend on the emissions of the air pollutants we studied in Barnaul. Despite the fact that WHO recommends at least three healthy sextants, in adults in the CPI index only ¼ - ½ sextant were on average healthy. Dental calculus was observed significantly more often in 35–44-year-olds at 1.34 ± 0.30 - 1.98 ± 0.40, compared to 6-, 12-year-olds in all areas, and in the elderly - at 0.60 ± 0, 12 - 1.28 ± 0.20 sextant. Dental anomalies in 12-, 15-year-old children are found in 62%, 59%, respectively, and are not associated with emission concentrations of harmful substances. Of the pathological signs, the crowding of the teeth on the lower jaw prevails, a deep bite. Their stomatological aesthetic index (DAI) averages 18.0 ± 0.7, which basically requires minimal therapeutic interventions.

4. Discussion

The high prevalence of dental caries among the young population of Barnaul confirmed the fact that the state of children's health is one of the most sensitive biological indicators that reflect the quality of the environment. Our results correlated with research data from ecologically unfavorable areas of the country, which attributed a large number of caries-affected teeth to a number of reasons: lack of ultraviolet rays, nutrition imbalance (seasonal changes in the diet of children), high air pollution, low fluoride content in drinking water and low levels oral hygiene. The incidence of pathology of hard dental tissues among the middle and older age urban population was high and comparable with the results in similar areas in ecology and climate and geographical location [5], [6], [7].

Our findings on the prevalence of periodontal disease at a young age are consistent with the results of studies in the European North, Krasnoyarsk Krai, and the Far East, where gum lesions were recorded in 64.0-100% of cases. Although, the author found that in the Novosibirsk region a high prevalence and intensity of diseases of the periodontal disease is observed in areas affected by atmospheric emissions from industrial enterprises. Also in the components of the CPI index, 15-year-old children of the Republic of Udmurtia have already registered periodontal pockets 4-5 mm deep. According to the researchers, the cause of high susceptibility of periodontal tissues is a low level of oral hygiene, poor quality prosthetics, multiple somatic pathology, lack of qualified dental consultation on oral care, and to a certain extent the environmental situation (impact of atmospheric emissions from industrial enterprises) [8], [9].

Although we have not identified the dependence of the prevalence of non-carious teeth lesions on air pollution, a number of authors have noted in their studies; for example, its increase due to the accident at the Chernobl NPP, which affected most Europeans.

The prevalence of diseases of the oral mucosa in Barnaul is similar to other regions of Russia, and our data correlate with the results of a number of researchers [8], [9]. Despite the fact that the etiology and pathogenesis of these diseases are not fully elucidated, the authors have no doubt that the pathological processes on the mucous membranes (candidiasis, lichen planus, diseases of the tongue and others) are the result of diseases of internal organs, disorders of metabolic processes, nervous system activity, changes immune system, adverse environmental conditions.

5. Conclusion

Thus, due to our research, a high prevalence and intensity of caries, gum disease, dental mandibular deformities, temporomandibular joint and oral mucosa has been established. The connection between the incidence of hard tooth tissues in children and the emission of harmful substances by industrial enterprises was also revealed. At the same time, the quality of medical and preventive care of various dental diseases to the population has not reached international standards and needs to be improved. The frequent occurrence of caries and periodontal diseases in children, and as a result, 100% incidence in adults and the elderly, requires the urgent universal introduction of primary preventive measures developed taking into account climate and other geographical features of the regions in order to effectively plan dental care for people of various ages. groups. Inadequate quality of treatment of hard dental tissues, as well as periodontal disease, should direct managers of the city dental service to a differentiated approach to planning preventive and curative work of major oral diseases in most areas of the regional center of the Altai Krai. The results of studying the influence of environmental factors on the dental morbidity of children can be used in the practice of sanitary doctors, as well as pediatricians and pediatric dentists in the city of Barnaul during an annual in-depth medical examination. The experience of numerous studies in other regions [2], [3], [4], p5] suggests that it is possible to reduce the prevalence and intensity of dental pathology by introducing a comprehensive prevention program that takes into account environmental indicators.

References

[1] Kolyado V B, and Dorofeyev Yu Yu 2017 Innovative technologies in the medical-demographic assessment of public health *Medical Science Bulletin* **2**(6) pp 8-10

[2] Gerasimenko N F 1993 Nuclear tests, environment and health of the population of the Altai Krai Research materials In *Assessment of radioactive contamination of the territory of the Altai Krai and radiation doses generated during the period of nuclear tests at the Semipalatinsk test site* (Volume 1, Book 3) (Barnaul, Russia: Altai Krai Administration Committee)

[3] Kolyado V B, and Puhovec I A 2006 The state of health of the population of the Altai Krai *Occupational Medicine and Industrial Ecology* **6** pp 23-27

[4] Zelenova E G, Zelenova O V, and Lukinykh L M 1999 Dental health level *International Journal on Immunorehabilitation* **12** p 117

[5] Zhumatov U Zh 1996 The influence of environmental factors on the immunological parameters of the oral cavity in children *Stomatology* **75**(4) pp 33-35

[6] Clark D C, Hann H J, Williamson M F, and Berkowitz J 1995 Effects of lifelong consumption of fluoridated water or use of fluoride supplements on dental caries prevalence *Community Dental Oral Epidemiol* **23**(1) pp 20-24

[7] WHO 1996 Monitoring global dental caries levels in adults aged 34–44 years: Oral Health programe WHO Caries Geneva pp 35-44

[8] Gazhva S I, and Gadayeva M V 2014 Comparative analysis of the level of children's dental health in the regions with different fluoride content in drinking water *Modern Problems of Science and Education* **6** p 1038

[9] Averyanova I V, Vdovenko S I, and Maksimov A L 2017 Assessment of the influence of environmental conditions on the state of the body of young men (17-19 years old) belonging todifferent ethnic groups of the North-East of Russia *Hygiene and Sanitation* **96**(8) pp 766-769

IOP Conf. Series: Earth and Environmental Science **395** (2019) 012033 doi:10.1088/1755-1315/395/1/012033

Glacier changes in the Tabyn-Bogda Ola mountain massif from 1994 to 2015

K V Lapina[1], O V Ostanin[1*] and G S Dyakova[1]

[1] Altai State University, 61 Lenina pr., Barnaul 656049 Russia

E-mail: christi.lapina@gmail.com

Abstract. In 2015, according to the Landsat and ASTER satellite imagery decoding, the contemporary glaciation of the Tavan Bogd (Tabyn-Bogda Ola) mountain massif range is represented by 82 glaciers, with a total area of 167.19 km². For the period from 1994 to 2015, the area of glaciation decreased by 3.61 km². The greatest changes occurred on glaciers of a complex-valley type with an area of more than 3 km².

1. Introduction

Modern climate change is a determining factor in the development of highland areas. Along with the arctic, high-mountain geosystems are the most vulnerable structural components of the geographic shell. The most significant consequence of contemporary climate change in the Altai is a steady trend of decreasing glaciation. The result is a reduction in water resources and, in the long term, aridization of the territory. In the continental regions of Central Asia, glacier melting is most acute, since glaciers are the main sources of river feeding and regulators of the hydrological regime of rivers in this region.

The Tavan Bogd (Tabyn-Bogda-Ola) mountain massif is the second highest after the Belukha mountain massif in the Altai-Sayan mountainous country. Within this mountainous country are the largest in area Potanin glacier. The Tavan Bogd (Tabyn-Bogd-Ola) is located on the territory of three states, such as Russia, China and Mongolia, at the junction of the Southern Altai, Sailyugem and Mongolian Altai ranges.

The study of the territory of the mountain massif is limited to the passes of Betsu-Kanas (in the west), Tsast-Daba (in the east), Postigiyn-Daba (in the south). The glaciers located in the Russian part belong to the Ak-Alakha river basin, other glaciers located in the Mongolian part feed the Tsagan-Sala-Gol and Postigin-Gol rivers. Those glaciers that are located on the territory of China belong to the Sangadyr and Poshtyk river basins.

Over the years, the research of glaciation in this area was carried out by V. V. Sapozhnikov [7], brothers B. V. Tronov, and M. V. Tronov [9], V. S. Revyakin [6], V. P. Galakhov and A. G. Redkin [1], N. N. Mikhailov, O. V. Ostanin [4], D. A. Ganyushkin [3], T. Kadota and G. Davaa [10], U. Kamp and C. Pan [11] C. Liu, G. You and J. Pu [12], and others.

2. Materials and Methods

The study of changes in the glaciation of the Tavan Bogd (Tabyn-Bogd-Ola) mountain massif was carried out by visual interpretation of Landsat and ASTER "Terra" satellite images and data from the Bing maps internet map service, providing space images with a spatial resolution of 1.5 m in a pixel

Content from this work may be used under the terms of the Creative Commons Attribution 3.0 licence. Any further distribution of this work must maintain attribution to the author(s) and the title of the work, journal citation and DOI.
Published under licence by IOP Publishing Ltd

over a given area. Data processing and calculation of the glaciers area was carried out in the software ArcGIS 10.3.1 and ENVI 5.0.

When distinguishing the boundaries of glaciers, there was both an objective and a subjective error. The first can be attributed to the spatial resolution of the images themselves, which is due to the technical characteristics of the imaging equipment and image processing technology. The resolution of Landsat-7 and Landsat-8 images is 14.25 meters per pixel, in the panchromatic channel; 28.5 meters per pixel, in the multispectral channel, and for ASTER images the resolution is from 15 meters per pixel (depending on the channel). The second group includes discrepancies resulting from the ambiguous interpretation of the position of frozen glacier tongues and the glacier boundaries in the places where snowfields adjoin them.

The main criterion for choosing space images is the condition for confidently interpreting the boundaries of glaciers, taking into account their light and shade conditions, the lack of cloudiness, and the presence or absence of seasonal snow on the surface of the glacier. The shooting date was chosen in such a way that it fell at the time of maximum glacial ablation for the region (July-August). Panchromatic and spectrozonal data were combined into one color snapshot.

In ArcGIS, all selected glaciers were assigned attribute information, including a serial number, a type, an exposure, a length, a square, a river basin.

3. Results and Discussion

The contours of 82 contemporary glaciers with an area of 167.19 km² (as of 2015) were identified in the territory of the Tavan Bogd (Tabyn-Bogdo-Ola) mountain massif. According to the results of the interpretation of satellite images, it was found that from 1994 to 2015, the glaciation of the Tavan Bogd (Tabyn-Bodo-Ola) massif changed only slightly. Changes in square glaciers range from 0.01 km² to 0.47 km². The largest change is observed in large glaciers: their square has decreased by 0.1 km² and more.

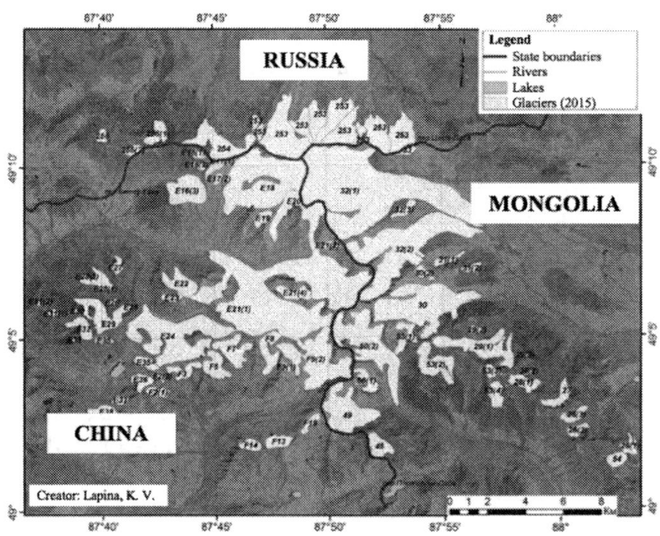

Figure 1. The positions of the glaciers of the Tavan Bogd (Tabyn-Bogdo-Ola) mountain massif for 2015 [Compiled by the author].

On the territory of Russia was allocated 14 glaciers. Numbering the glaciers of the Russian part of the Tavan Bogd (Tabyn-Bogdo-Ola) mountain massif was carried out in accordance with the Catalog of the USSR glaciers [3].

In 2015, the total square glaciation of the Russian part of the massif was 21.01 km². The largest glaciers No. 253 are a series of glacier flows merging with each other, covering dome-shaped peaks. The largest of them are glaciers No. 253 and No. 253. In 2015, their square was 4.64 km² and 3.15 km², respectively.

On the territory of Mongolia, 25 glaciers were allocated. Seventeen of them belong to the Tsagan-Sala-Gol river basin, and nine to the Postiyn-Gol river basin. The glaciers' numbering is given according to V. P. Galakhov [1]. Interpretation of satellite images showed that the large glaciers were divided into several streams and / or parts; therefore, the glacier numbers in brackets indicate the numbers of glacier streams and / or its parts into which the glacier was divided. Among the glaciers of the Tsagan-Sala-Gol river basin, crust and complex valley glaciers prevail. This basin includes the four largest glaciers, such as the Potanin Glacier No. 32(1), Alexandra Glacier No. 32(2), Granet Glacier No. 30, and Krylov Glacier No. 29(1). These glaciers in the upper reaches are closed with each other, constituting an almost continuous ice shell along the main watershed. The biggest change occurred in the area of the glacier Potanin. The area was decreased by 0.47 km² (1994-2015). The square was decreased by 0.47 km² (1994-2015).

The most elevated section with peaks exceeding 4000 m above sea level is stretched along the feeding zone of the Potanin and Alexandra glaciers. Further, the watershed turns to the east, and the height of the peaks decreases to 3,600-3,700 meters above sea level. With a decrease in the height of the mountains, the square of glaciers decreases by approximately 5 times.

Among the glaciers of the Postigiin-Gol river basin, the valley-type glaciers prevail. The largest glaciers of the basin are No. 49 and No. 50. Their square is 4.23 km² and 6.53 km², respectively. These are the eastern glacier valley exposure. During the study period the glacier No. 50(2) changed most significantly. Its square has decreased by 0.1 km². The remaining glaciers of the basin have changed slightly.

The boundaries of 42 glaciers belonging to the basins of the Sangadyr and Poshtyk rivers were identified in China. Glacier numbers were assigned in accordance with the Glacier Catalog of China [12]. The icing is most developed along the centerline of the Mongolian Altai ridge. Among the glaciers of the Sangadyr river basin, small crusty glaciers with an area of up to 1 km² prevail. The largest glacier of the basin No. E21(1), has an square of 25.29 km² (2015). From 1994 to 2006 there were no changes in the square of the glacier, and from 2006 to 2015. the square of the glacier decreased by 0.16 km². Among the large glaciers of the basin, one should also mention the glaciers No. E18 (contemporary square 9.88 km²), No. E24 (8.51 km²)б and No. E16(3) (2.72 km²). During the study period, each of them decreased by less than 1 km².

The glaciers of the Poshtyk River Basin remained almost unchanged during the study. The largest glacier of the territory is glacier No. F2(2). This is a valley glacier southern exposure. Its square for 2015 was 6.07 km². Since 1994, it has changed to 0.28 km². The remaining glaciers of the basin, by their morphological type, are crust and do not exceed 2 km² in square.

Not all glaciers have shortened lengths; but the immutability of their area does not indicate a stable state. It is possible that in the period from 1994 to 2015, there were changes in the volume, which cannot be established as a result of the interpretation of satellite images.

The glaciation of the Tavan Bogd (Tabyn-Bodo-Ola) mountain massif is mainly concentrated on the slopes of the Mongolian Altai Range from China (45.5% of the total glaciation of the massif) and Mongolia (41.9%). In Russia, the process of glaciation is 12.6%. Despite the fact that the territory of China has the most glaciers, retreats are observed more on the glaciers of Mongolia (1.89 km²), where the largest glaciers of the massif are located. The areas of glaciers in the mountain range in China decreased by 1.32 km², the glaciers of the Russian part decreased by 0.4 km².

4. Conclusion

From 1994 to 2015, based on the interpretation of satellite images and images, a decrease in the glaciation area of the Tavan Bogd (Tabyn-Bogd-Ola) massif by 3.61 km² was revealed, with half of the glaciers of the massif (40 of 82) remaining unchanged.

The materials of this work can be used to monitor changes in the area of glaciers in the region. In the future, this will provide an opportunity to obtain homogeneous data for a comparative analysis of the reaction of glaciers of various morphological types, sizes and exposures to current climate change.

References

[1] Galakhov V P, and Redkin A G 2001 Contemporary and ancient glaciation of the Tavan Bogd (Tabyn-Bogd-Ola) mountain site *Geography and Nature Management of Siberia* **4** pp 153-175
[2] Dushkin M A, and Revyakin V S 1977 *The USSR glacier catalog: Argut River Basin* (Moscow, USSR)
[3] Ganyushkin D A 2015 *Glacial complexes of the sharply continental region of the northwest of Inner Asia*
[4] Ostanin O V, Mikhaylov N N, and Arkhipov S M 2004 Changes in the glaciers of the southern and central Altai from the late 19th century and their development trends in the 21st century *Geography and Nature Management of Siberia* **7** pp 172-182
[5] Otgonbayar D 2013 *The present-day glaciation of the Mongolian Altai (on the example of the Munkh Khairkhan, Sutai ranges, the Tsambargarav mountain site)* (Barnaul, Russia)
[6] Revyakin V S, and Mukhametov R M 1993 Dynamics of the glaciers of the mountain node Tabyn-Bogdo-Ola *Glaciology of Siberia* **4**(19) pp 83-92
[7] Sapozhnikov V V 1911 *Mongolian Altai in the origins of the Irtysh and Kobdo, 1905-1909* (Moscow, Russia)
[8] Syromyatina M V, Kurochkin Yu N, and Chistyakov K V 2014 Current state and changes in the glaciers of the Tavan Bogd (Tabyn-Bogd-Ola) mountain range (Mongolia) *Ice and Snow* **3** pp 31-38
[9] Tronov B V 1949 *Essays on the Altai Glaciation* (Barnaul, Russia)
[10] Kadota T, and Davaa G 2004 *In Proceedings of the 2nd International Workshop on Terrestrial Change in Mongolia a Preliminary Study on Glaciers in Mongolia*
[11] Kamp U, Krumwiede B, McManigal K, Pan C, Walther M, and Dashtseren A 2013 *The Glaciers of Mongolia* Available at: https://instaar.colorado.edu/uploads/occasional-papers/OP61-Kamp-et-al-2013-Glaciers-of-Mongolia.pdf
[12] Shi Ya, Liu C, and Kang E 2009 Glacier Inventory of China *Annals of Glaciology* ***50(53)*** pp 1-4

IOP Conf. Series: Earth and Environmental Science **395** (2019) 012034 doi:10.1088/1755-1315/395/1/012034

Assessing the ecological and economic state of agricultural land in the border areas of the Altai Krai

T V Baykalova[1*]

[1] Altai State Agrarian University, 98 Krasnoarmeysky pr., Barnaul 656049 Russia

E-mail: tan.space@mail.ru

Abstract. The article discusses the method of integrated assessment of ecological and economic state of agricultural land territory of the Mikhailovsky District of the Altai Krai. The joint analysis of the results of remote probing data processing, cadastral information and statistical reporting allowed us to carry out an areal assessment of lands classified according to different degrees of anthropogenic transformation. The characteristics of the areas exposed to anthropogenic influences were identified, which allowed us to isolate land with a special regime of use and adjust the ecological and economic state towards a balanced and sustainable development.

1. Introduction

Currently, the concept of sustainable development has found wide application in the economic, social and environmental spheres of public activity. One of the main objectives of the program for sustainable development is the optimization of the spatial organization of rural areas based on the rational use of existing resource potential. For the successful implementation of this task in the framework of the concept of sustainable development, new approaches to the study of natural-anthropogenic systems are needed. One of such approaches is the concept of ecological and economic balance of the territory, which includes the following conditions: the organization of territories on a landscape-ecological basis; preservation and maintenance of natural and slightly modified landscapes that perform important environmental and resource-forming functions; rational use of the natural potential of the territory; development of innovation processes [1]. The study and analysis of the prerequisites that exist in a particular territory for the implementation of each of the conditions of the concept of ecological and economic balance will identify the main problems, as well as positive and negative development factors for taking measures for its sustainable development.

2. Materials and Methods

The object of research was the border area of the Mikhailovsky District, located in the south-west of the Altai Krai. The territory borders on Kazakhstan, Kliuchevsky, Uglovsky and Volchikhinsky District of the Altai Krai.

To solve the problem of assessing the ecological and economic condition of agricultural territories, in order to ensure their sustainable development, the following methods were used: analysis of the land use structure based on the land cadastre classification units and remote sensing data; definitions of integrated environmental and economic assessment; geo-information technologies for creating thematic cartographic material based on the results of the research.

Content from this work may be used under the terms of the Creative Commons Attribution 3.0 licence. Any further distribution of this work must maintain attribution to the author(s) and the title of the work, journal citation and DOI.
Published under licence by IOP Publishing Ltd

The study used, remote sensing data obtained by Aster and Landsat multi-zone satellite systems, cadastral information and statistical reporting forms No. 22-2 of the Mikhailovsky District [2, 3].

3. Results

An analysis of the joint processing of all available data shows that the total area of the Mikhailovsky District within the administrative boundaries is 311,712.59 ha. The basis of the land fund is agricultural land (54.4%). Arable land prevails in the northern part of the district and occupies 35% of the entire territory. Hayfields and pastures are spread throughout the Mikhailovsky District and account for 22%. The second largest land area of 43.1% is occupied by forest land. Forests are widespread in the southern part of the Mikhailovsky District and occupy 38% of the territory's area. Other lands with small areas include development land, under roads, under water, deposits, perennial plantings and other lands.

Given the structure of land use, as well as the availability of land of various categories, it is possible to calculate the level of anthropogenic transformation of a territory [4], [5]. The level of anthropogenic load of the study area was determined by the method of B. I. Kochurova [6]. According to the proposed method, in the case of extensive development of the territory, the analysis of the structure of land use is carried out within administrative boundaries. To determine the degree of anthropogenic load of land, expert point estimates are introduced. Each type of land receives a corresponding score. The area valuation of lands classified by varying degrees of anthropogenic transformation is shown in Figure 1 and Table 1.

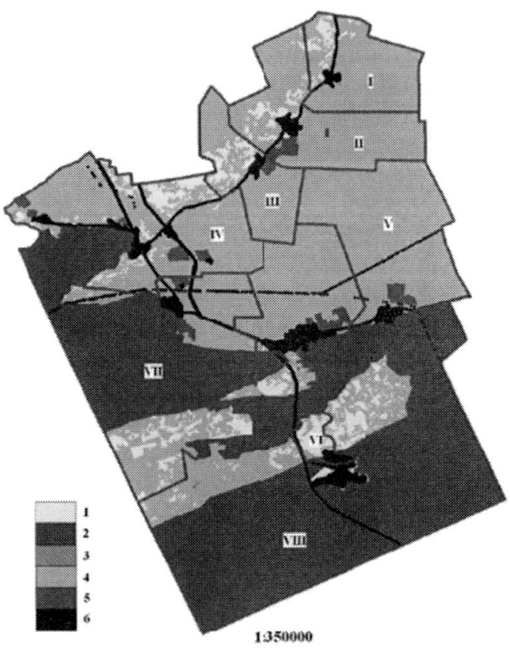

Legend: 1) Very low (conservation and unused land); 2) low (hayfields, forests used sparingly); 3) average (perennial plantations, recreational land); 4) high (arable land, areas of intensive logging, pastures and hayfields, used irrationally); 5) very high (irrigated and drained lands); 6) higher (land of industry, transport, settlements, infrastructure, disturbed land). Village councils: I – Ashchegulsky; II – Poluyamsky; III – Nazarovsky; IV – Nikolayevsky; V – Rakitovsky; VI – Malinovozersky; VII – Bastansky; VIII – Mikhailovsky.

Figure 1. The classification of the territory of the Mikhailovsky District according to the degree of anthropogenic transformation.

Table 1. Point classification of lands according to the degree of anthropogenic transformation within the administrative boundaries of the village councils of the Mikhailovsky District.

Indicator	Degree of anthropogenic transformation					
	Very low	Low	Average	High	Very high	Higher
Village Council	An1, ha	An2, ha	An3, ha	An4, ha	An5, ha	An6, ha
Ashchegulsky	1,477.77	1,165.78	19.03	13,110.88	0	248.61
Poluyamsky	1,628.26	1,356.34	0	18,967.09	808.16	424.30
Nazarovsky	2,915.67	1,054.89	12.30	9,194.79	297.78	272.70
Nikolayevsky	4,866.68	14,183.27	138.53	24,636.14	1,056.40	1,374.45
Rakitovsky	4,988.33	1,567.73	87.86	31,030.30	701.52	687.22
Malinovozersky	1,776.80	15.43	0	567.66	0	563.54
Bastansky	4,570.56	26,031.68	2.44	17,797.02	602.66	594.21
Mikhailovsky	3,843.10	88,548.12	142.81	23,671.65	889.42	3719.10

The grouping of lands according to the degree of anthropogenic load allows us to estimate the anthropogenic transformation of the territory in comparable terms [7], [8]. They are the absolute (K_a) and relative (K_o) coefficients of the ecological and economic situation of the territory, calculated using the following formulas:

$$K_a = \frac{An6}{An1}, \qquad K_o = \frac{An4 + An5 + An6}{An1 + An2 + An3}, \tag{1}$$

where $An1, \ldots An6$ proxies the numerical value of the degree of anthropogenic transformation corresponding to the point score.

The lower K_a shows a better state of the environment. In general, the ecological and economic balance of the territory is characterized by the coefficient K_o. When K_o is equal to or close to 1.0, the tension of the ecological and economic balance of the territory is balanced in terms of the degree of anthropogenic load and the potential for sustainability of nature.

Each anthropogenic impact, or their combination, has its own limit of sustainability of natural and natural-anthropogenic landscapes. The more diverse the landscape, the more stable it is. This is expressed, first of all, by a large number and uniform distribution of natural biogeocenoses, the total area of which is the ecological fund of the territory. The larger this fund, the higher the natural security of the territory and, accordingly, the stability of the landscape. Thus, the total land area with environmental and resource-stabilizing functions can be calculated using the following formula:

$$P_c = An1 + 0{,}8An2 + 0{,}6An3 + 0{,}4An4. \tag{2}$$

If we correlate the area of land P_c to the total area of the study area (P_o), then we obtain the coefficient of natural protection of the territory:

$$K_e = \frac{P_c}{P_o}. \tag{3}$$

Values of K_e are less than 0.5; therefore, it is characterized by a critical level of protection of the territory.

On the basis of the point classification of the lands of the Mikhailovsky District according to the degree of anthropogenic transformation, coefficients characterizing the ecological and economic condition of the studied territory were calculated (Table 2), and the relevant cartograms of the distributions of these indicators were constructed (Figure 2).

a b

Legend: Village councils: I – Ashchegulsky; II – Poluyamsky; III – Nazarovsky; IV – Nikolayevsky; V – Rakitovsky; VI – Malinovozersky; VII – Bastansky; VIII– Mikhailovsky.

Figure 2. The distribution of the values of the coefficients of the relative anthropogenic intensity (a) and natural security (b).

Table 2. Indicators of the main coefficients of the ecological and economic condition of the territory of the Village councils of the Mikhailovsky District.

Village councils	P_o	K_a	K_o	P_c, га	K_e
Ashchegulsky	16,022.07	0.17	5.02	7,666.16	0.48
Poluyamsky	23,184.15	0.26	6.77	10,300.168	0.44
Nazarovsky	13,748.13	0.09	2.45	7,444.88	0.54
Nikolayevsky	46,255.47	0.28	1.41	26,150.87	0.56
Rakitovsky	39,062.96	0.14	4.88	18,707.35	0.48
Malinovozersky	2,846.42	0.32	0.66	1,939.20	0.68
Bastansky	49,779.19	0.13	0.62	32,696.80	0.66
Mikhailovsky	120,814.20	0.97	0.31	84,235.94	0.70

4. Discussion

An analysis of the results shows that the territories of village administrations located in the south and west of the Mikhailovsky District have a prosperous ecological and economic condition, since their lands contain a sufficient proportion of tree-shrub vegetation.

The high degree of development of the territory in the form of arable land and pasture with insufficient land with environmental and resource-stabilizing functions in the form of forests, land in the redistribution fund, etc., determined the coefficient of relative tension greater than 1.0 in most village councils. The prosperous situation is developing in the Mikhailovsky, Bastansky, and Malinovozersky village councils, as there is a sufficient amount of forests and unused lands on the territory for their stabilization. The unfavorable situation develops in the Ashchegulsky, Poluyamsky, and Rakitovsky village councils, the coefficient of natural security is critical. Here there is an intensive use of land; therefore, measures are required for environmental stabilization. In the Nazarovsky and Nikolayevsky village councils, the coefficient of natural security is slightly above the norm. However, a high value of the coefficient of relative tension indicates an imbalance in the degree of anthropogenic load and the sustainability of nature. This situation has arisen due to the fact that in these areas, a significant proportion of arable land in the composition of land.

In general, the ecological and economic balance of the territory of the whole Mikhailovsky District is characterized by the following indicators: K_a= 0,30; K_o=0,94; P_c= 189141,37 ha; K_e= 0,61. According to the results obtained, a favorable situation remains in the territory of the district in terms of the ratio of the highest and very low anthropogenic load, thanks to a significant share of the forest cover. In general, the situation on the anthropogenic load in the general territory of the district can be considered satisfactory.

5. Conclusion

The study of the natural conditions and resources of the area allows one to conduct a successful economic activity. The modern economic system has developed in such a way that most of the land area of the district has the category of agricultural land. Extensive farm management has led to a number of environmental problems associated with the presence of high anthropogenic pressures, which in turn identified a number of negative phenomena in the territory: a pasture digression, a soil erosion, a decrease in humus in the soil, waterlogging, salinization, etc. The study of the ecological and economic condition of the Mikhailovsky District was carried out within the boundaries of village administrations. As a result of the analysis, the characteristics of areas of land exposed to anthropogenic influences were identified. This allowed us to allocate land with a special mode of use and adjust the ecological and economic state in the direction of balanced and sustainable development. Also, on the territory of the Mikhailovsky District, it is necessary to revise the nature of land use, moving to intensive management, introducing innovative technologies and projects, to eliminate and solve the growing problems of the environment.

References

[1] Orlova I V 2006 Landscape planning for balanced agricultural use *Geography and Natural Resources* **2** pp 124-131

[2] Shovengerdt R A 2010 *Remote sensing. Models and methods of image processing* (Moscow, Russia: Tekhnosfera)

[3] Baykalova T V 2011 Mapping of soil degradation processes *Geography and Nature Management of Siberia* **13** pp 26-32

[4] Morkovkin G G, Baykalova T V, Maximova N B, Ovtsinov V I, Litvinenko E A, Demina I V … Demin V A Anthropogenic transformation of arable soil of the steppe zone of the Altai Territory 2014 *Bulletin of the Altai State Agrarian University* **6** pp 43-48

[5] Istomina E A, and Cherkashin A K 2000 The use of mathematical methods and GIS technologies for functional zoning of the territory *Landscape ecology and land use planning* pp 67-71

[6] Кочуров Б И 2003 *Eco-diagnostics and balanced development* (Smolensk, Russia: Madzhenta)

[7] Vinokurov Yu I 2003 *Sustainable development of Siberian regions* (Novosibirsk, Russia: Nauka)

[8] Isachenko A G 2001 *Ecological geography of Russia* (St. Petersburg, Russia: Publishing House of St. Petersburg University)

IOP Conf. Series: Earth and Environmental Science **395** (2019) 012035 doi:10.1088/1755-1315/395/1/012035

Socio-economic patterns of development planning in rural areas and targets of land management in industrial regions

A Orlova[1], I Trapeznikova[1*] and I Tschelihina[1]

[1] Kemerovo State University, 6 Krasnaya str., Kemerovo 650000 Russia

E-mail: trapeznikova_1976@mail.ru

Abstract. The article is devoted to the analysis of the influence the coal-mining industry has on the development of rural territories, as well as identifies the tasks of land management of industrial territories. The paper proposed directions for developing programs for cross-border cooperation in the framework of resettlement of those residents living in the areas undermined by mining.

1. Introduction

According to statistics, a number of rural population in Russia decreases annually (38 million people in 2015, 37.6 million people in 2018) [6]; however, a percentage ratio of the urban and rural population in the Russian Federation does not change from 2009 (74% and 26%, respectively). This trend is also characteristic of the Siberian Federal District. In 2016, the All-Russian Agricultural Census showed a decrease in the number of agricultural organizations, as well as in the number pf peasant farms and private entrepreneurs. This trend is characteristic not only of the Kemerovo Region and the Siberian Federal District, but also of Russia in general [5].

Most of the rural settlements are dying or on the verge of extinction. After the 2014 sanctions were imposed, certain measures were taken to support the agro-industrial complex, but they has only a small effect. There are various reasons for that. For instance, budget funds do not fully reach the direct producers. More than that, peasant farms and partially current joint-stock enterprises (former collective and state farms) still dominate, and they cannot provide either a large-scale increase in quality products, or growth in labor productivity.

In the rural labor market compared to urban problems, associated with labor activity of the population, unemployment, income levels, poverty and tension in the rural labor market as a whole. Economic and social problems existing in rural areas contribute to migratory attitudes, and for the most part among young people and the able-bodied population. Monitoring of the state of rural areas in the Russian Federation reflects the main reasons for the migration of rural residents: lack of jobs, low salaries and social benefits, lack of prospects [7].

Social differences between urban and rural populations have a protracted nature and tendencies to exacerbate. Industrial regions, in fact, do not pay attention to developing agriculture and providing support for its inhabitants. In absolute terms, GRP (gross regional product) grows at the expense of other industries, while agriculture remains outside its framework. The main problems: (1) reducing the number of jobs; (2) failure of most agricultural machinery; (3) low salaries; (4) lack of infrastructure.

Rural settlements in regions with a developed coal industry also suffer from the operation of coal mining enterprises, since the main percentage of coal production is carried out by the open method,

Content from this work may be used under the terms of the Creative Commons Attribution 3.0 licence. Any further distribution of this work must maintain attribution to the author(s) and the title of the work, journal citation and DOI.
Published under licence by IOP Publishing Ltd

IOP Conf. Series: Earth and Environmental Science **395** (2019) 012035 doi:10.1088/1755-1315/395/1/012035

which leads to large-scale pollution of agricultural land and, accordingly, to a reduction in the number of rural settlements and rural territorial associations.

2. Materials and Methods

The methodological basis of the study consists of general scientific methods of cognition (retrospective and comparative analyzes, synthesis, induction, deduction), which made it possible to obtain reliable and reasonable research results. The digital data were taken from the federal and regional sites of state statistics, administrative Internet resources, as well as from the data of annual state reports on the state and protection of the Kemerovo environment. Analytical and scientific materials were obtained as a result of the study and synthesis of various theories, approaches and scientific developments.

Mining enterprises have to create a favorable social environment in response to the loyalty of the local government and residents. Those companies that do not develop and do not embody an active policy of support and development of the areas of presence, do not work with the population, stumble upon the active social discontent of citizens living in the area of activity of mining and processing industries.

Studies of social tensions show that in extractive and old industrial regions, the "environmental tensions" factor holds the greatest weight in 3rd place in the Urals region, 7th in the Siberian region and 4th in the Tyumen region. The situation developing in the territories bordering the coal-mining enterprises does not contribute to social stabilization. Therefore, as one of the solution methods, it is proposed to use agreements on mutually beneficial cooperation between cross-border regions.

3. Results

Since the start of mining operations at the mines in the Kuzbass, over 55 thousand hectares of land were disturbed, and more than 35 thousand hectares of mines, about 91 thousand hectares of land were disturbed, and this is without taking into account the influence of enterprises on the adjacent territories. At the same time, the climatic and soil resources of the region make it possible to grow various agricultural crops, from grain bread standards to vegetables in open ground. The economy of the region as a whole is insignificantly influenced by agriculture, but this industry depends on the results of providing the residents of the region with food.

The state of agricultural land area can be assessed as catastrophic. The Kuzbass coal-mining areas are particularly hardly hit. For instance, the disturbed territories of the city of Prokopyevsk make up 21%, the city of Kiselevsk 28%, the city of Mezhdurechensk 14.6%, the Belovsky District 3%, and the Novokuznetsk District 1%. The total area of waste dumps formed in open-cast coal mining is 34.3 thousand hectares. At the place of their formation, the external dumps (that is, outside the mining allotment of enterprises) occupy 22.5 thousand hectares; they are the hydraulic tailings formed by the hydraulic method of overburdening loose sediments, their area is 5.6 thousand hectares. Internal dumps, i.e. being dumped in the developed space, occupy 6,2 thousand hectares [1].

For all the time of economic and social development of the Kuzbass region, the total area of reclaimed land has been slightly more than 21 thousand hectares. And the forest reclamation dominates, covering arounf 14.3 thousand hectares. This does not even compensate for the territories newly allocated for coal mining.

Moreover, statistical data show that there is a steady change in agricultural land downward. Every year, agricultural land decreases by about 30 thousand hectares.

The proportion of the disturbed land in the Kemerovo Region is less than 1%, but the situation is complicated by the fact that most of the region's territory (60%) is mountain formations and the main anthropogenic load is formed on the flat part of the Kuznetsk Basin and the south of the West Siberian Lowland. Thus, almost 17% of the flat part of the Kemerovo region is disturbed land. These lands territorially coincide not only with deposits of hard coal, but also with the region's main agricultural zone, with its highly productive soils [2].

Until recently, the industry of recovery and utilization of large-tonnage waste of the fuel and energy industry, which is practically absent, has led to the continued accumulation of waste: by the beginning of 2017, about 8 billion tons of overburden and waste from coal enrichment processes accumulated in the Kuzbass, including 108.8 million tons and ash and slag waste more than 35.6 million tons. [2]. Overburden and host rocks are usually stored in dumps. Storage of rocks in dumps is accompanied by such negative consequences as air pollution by rock and coal dust, and during self-ignition of rocks, particularly by harmful gases (carbon oxides, sulfur, nitrogen, hydrogen sulfide, phenols), ground and surface waters, and removal of agricultural land from agricultural circulation [3].

The situation is complicated by the fact that the industry of coal mining in the Kuzbass is increasingly conducted by open pit, since it is cheaper. Coal cuts come close to the borders of settlements, barely observing the regulated boundaries of sanitary protection zone. According to modern epidemiological requirements, this is 1,000 m. Social responsibility of business is expressed in the provision of infrastructure services and priorities to residents of villages and small towns affected by coal mining, such as repairing streets, roads, lighting, creating children's sections, providing transport to schoolchildren, and so on. However, this does not help to preserve agriculture and ensure its development.

4. Discussion

One of the options for social assistance for residents of the territories under the influence of the coal industry is resettlement from problem areas, jointly funded from the federal and regional budgets and the budgets of coal companies. Local development programs that provide for resettlement are mainly aimed at relocating residents of mining allotments of abandoned mines. A situational analysis allows to consider the possibility of cross-border cooperation with those regions where agriculture is basic for the economy of the region. For example, the Altai Territory is suffering losses in agricultural production due to negative migration, the losses from which in 2017 alone amounted to 7163 people. per year. It is proposed to compensate for the losses through relocating the interested agricultural enterprises and personnel from agricultural areas disturbed by mining workings. In turn, the Altai Territory may cover the needs of the Kemerovo Region for agricultural products.

5. Conclusion

In general, to effectively address the challenges of contemporary agriculture, it is necessary to use a systematic approach, one of the options for which may be trans-regional agreements. The introduction of such social programs will contribute to the achievement of the goals stated in the Doctrine of Food Security of Russia, as well as in the State Program for the Development of Agriculture and Regulation of Agricultural Products, Raw Materials and Food for 2013-2020.

References

[1] Administration of the Kemerovo Region 2018 *State of the environment report of the Kemerovo Region in 2017* (Kemerovo, Russia)

[2] Administration of the Kemerovo Region *2017 Report on the state and protection of the environment in the Kemerovo region in 2016* (Kemerovo, Russia)

[3] Committee on Land Resources and Land Management of the Kemerovo Region 2016 *Annual report on the state of the land fund of the Kemerovo region* (Kemerovo, Russia)

[4] Federal State Statistics Service 2018 *Russian statistical yearbook 2018* (Moscoe, Russia)

[5] Federal State Statistics Service 2018 *Statistics on demograpy in Russia* Available at: http://www.gks.ru/wps/wcm/connect/rosstat_main/rosstat/ru/statistics/population/demography/# (Accessed 09 04 2018)

[6] Kemerovostat 2016 The 2016 *all-Russian agricultural census* Available at: http://kemerovostat.gks.ru/wps/wcm/connect/rosstat_ts/kemerovostat/resources/bd160c80487e87038d919db3ce167dd4/%D0%B4%D0%BB%D1%8F+%D1%81%D0%B0%D0%B9%D1%82%D0%B0.pdf (Accessed 09 04 2018)

[7] Toshchenko Zh T 2017 What is a contemporary Russian village? *Socis* **3** pp 95-96

Formation of geoinformation-cartographic support of the Altai interregion on the example of creating the international atlas "Greater Altai: Nature, History, Culture"

I N Rotanova[1*], G A Efremov[1] and S V Peremitina[2]

[1] Altai State University, 61 Lenina prosp., Barnaul 656049 Russia
[2] D. Serikbayev East Kazakhstan State Technical University, 69 Protozanovs str., Ust-Kamenogorsk 070004 Kazakhstan

E-mail: rotanova@email.asu.ru

Abstract. The article discusses the conceptual provisions used in creating the international cartographic project "Atlas of the Greater Altai: nature, history, culture" in the context of the formation of an international geo-information and cartographic support. The creation of the atlas of the Greater Altai is based on a number of methodological, conceptual, and organizational provisions, among which, first of all, the research search for new methods and approaches, the development of modern technologies for processing spatial data, cartographic modeling and cartographic research methods. Within the framework of this project, implementation of international cooperation, exchange of experience and achievements in the field of mapping, geoinformatics and GIS design is no less significant than the international cooperation, exchange of experience and achievements in the field of mapping, geoinformatics and GIS design.

1. Introduction

Creation of electronic cartographic products, including satin geo-information systems, is conditioned by the contemporary informatization and digitalization of society. The atlas "Greater Altai: Nature, History, Culture" is created as an interdisciplinary geo-information cartographic model by a team of specialists from the Altai interregion countries, based on the principles of international information interaction. The idea to create an atlas of the Greater Altai arose due to the development of scientific and intercultural communications among the countries participating in the Altai Interregion. The accumulated results of many years of work by many teams of researchers from Russia, Mongolia, Eastern Kazakhstan, Xinjiang Uygur Autonomous Region of China, in the form of separate databases, thematic maps and geo-information systems for targeted and environmental programs, combine all our knowledge gained in the product of spatial analysis of the territorial development of the Altai interregion. The widespread use and dissemination of geo-information and web technologies makes it possible to create innovative cartographic models, including geo-information atlas projects [1].

Atlases in electronic form are posted in national and international information networks. They become an important means of spatial planning and management, a source of scientific reference information, a means of learning, education, and maintaining spatial databases and user programs (interfaces).

Content from this work may be used under the terms of the Creative Commons Attribution 3.0 licence. Any further distribution of this work must maintain attribution to the author(s) and the title of the work, journal citation and DOI.
Published under licence by IOP Publishing Ltd

2. Materials and Methods

In the 21st century, a traditional geographic map is most often replaced by a geo-information model created using a geo-information system (GIS), which is based on a large amount of variable information. In turn, the last has a set of technical tools and special procedures for processing and reproducing graphic and text data, with visualization management capabilities. A serial (satin) mapping using GIS systems is being developed, which today remains one of the leading methods of the systematic geographical study of natural conditions and resources, as well as socio-economic, historical and cultural development of territories. This confirms an increase in the number of new atlases of foreign and Russian territories: from the level of the country to individual regions. The quality and scope of geographic knowledge and information concentrated in the thematic geo-information maps and complex atlases. They are the primary factors determining the efficiency of using natural resources, developing the economy, forming satellite communication and navigation systems, and substantiating the geo-political interests of states [2].

Most modern atlases cover the territories of individual countries and their administrative and territorial units. However, there are only a few examples of atlases with international regions that have a natural and historical community, but are separated by state borders. The cross-border Greater Altai is a territory for which there is a need to create an interregional atlas (i.e. a connected territory within the Altai mountain system of the border regions of Russia, Kazakhstan, Mongolia, and China).

The importance of creating an atlas is primarily due to the fact that the large transboundary region of Asia – the Greater Altai needs to create a modern, complex satin cartographic work. The satin mapping will be developed through the use of new approaches and methods, both in the subjects of thematic maps, and with the use of innovative geo-information technologies.

The main purpose of creating this Atlas is to provide the international community with an access to reliable, modern, and accurate spatial information about the cross-border region of the Greater Altai for its effective development [3].

In the process of developing geo-information support, object-oriented databases are created that meet the principles of modularity, complexity, and universality. Modularity is necessary for diverse activities, which involves the use of a single database and a cartographic basis, with the possibility of modifying the modules to solve various problems. Complexity will provide the user with information on the physical, geographical, historical, cultural and socio-economic conditions of the territory. Universality is based on the main characteristics of the geo-information software used and the possibility of creating additional application programs on its basis [4].

3. Results

The development of the atlas required the solution of a number of issues at various research levels: methodological (development of concepts and programs, priorities, architecture and functionality); scientific and methodological (development of structure, engineering and creation of GIS, construction of cartographic models in the GIS environment, etc.), information (database development, homogeneity, accessibility and reliability of mapped data), and constructive (development and construction of maps, formation of an atlas as an integrated satin geo-information system), etc. [5].

The novelty and uniqueness of the atlas "Greater Altai: Nature, History, and Culture" is predetermined on the basis of the conceptual provisions underlying it as a geo-information cartographic model, as well as the thematic content of individual original maps.

The Atlas has a multilevel structure. The Greater Altai, in the spatial aspect, is created as an integral interregion comprising from the territories of four states. The format (landscape orientation) was adopted for the atlas, and the scale of the mapping series was determined (1: 5,000,000). The inter-regional and macro-regional level includes the general maps of the Greater Altai on the following scales: 1: 4 000 000, 1: 5 000 000, 1: 8 000 000, 1:10 000 000. The local-regional level corresponds to the scale: 1: 1 000 000, 1: 2 000 000, 1: 3 000 000. A full map can be accessed (in Russian) at [6].

Unity in the atlas as a system is ensured by a set of projections and scales, a single set of cartographic generalization, a consistent system of conventional signs and a single design.

IOP Conf. Series: Earth and Environmental Science 395 (2019) 012036 doi:10.1088/1755-1315/395/1/012036

In the content and plot context, it has three main sections, corresponding to its name: "Nature," "History," and "Culture". These sections are preceded by an introductory section containing geographic information, borders of countries and border administrative and territorial units (subjects of the Greater Altai), as well as an image of the region from space.

The database structure of the atlas of the Greater Altai was created on the same principle as the structure of the atlas itself. Thematic databases are distinguished in accordance with the titular sections of the atlas: "Nature," "History," "Culture". A distinctive feature is that the databases are created separately for each of the countries of the Greater Altai: Russia, Mongolia, Kazakhstan, and China. This is due to the fact that they are formed by the author teams of each of the states. Shared databases containing aggregated data for Greater Altai are created collectively [7].

Additionally, a separate database called "Basic mapping" is being developed, which contains elements of a cartographic basis, namely: geodetic basis, hydrography, communications, settlements, administrative division, vegetation and soils, land relief and socio-economic objects. It also has an internal structure and is subdivided into 2 thematic blocks: socio-geographical and physico-geographical. The database "Base mapping" is the basis for creating base maps. The creation of base maps is given special attention, since they serve as the foundation on which thematic maps of sections are formed. For the "Nature" section, three maps were taken as basic: Physical Map, Landscape Map, and Population Settlement Map. The "Nature" section contains information about the natural conditions, ecological and natural potential, economic impact on the environment, and the quality of the environment in the Greater Altai region at the beginning of the 21st century. The quality of the environment is considered as a consequence of the economic, social and cultural development of the border administrative and territorial entities of four countries located within the Altai mountain system. The special attention is paid to the environmental policy and the network of specially protected natural territories of the Greater Altai.

For the section of thematic maps "History," as the base maps were selected the Map of the stages of development and settlement of the Greater Altai, showing the stages of formation of the territory the Map of historical areas, the Map (a series of maps) of the location of objects of historical (archaeological) heritage with a display of the location and a database containing information about artifacts reflecting the history of human society based on material evidence (monuments) of people's lives and activities. The "History" section is closely related to the "Culture" section, one of two subsections of which is devoted to the rich historical and cultural heritage of the Greater Altai, and the other one addressed the modern culture. The "Culture" section selected these base maps: Ethnic and Ethnographic Map, Map of Cultural Heritage Objects, Cultural and Educational Map [8].

4. Discussions

In the process of developing geo-information support, object-oriented databases are created that meet the principles of modularity, complexity and universality. Modularity is due to the diverse content of sections of the atlas blocks, however, it assumes the use of a single database and a cartographic base, with the possibility of modifying them within the framework of solving various problems. Complexity will provide information on the physico-geographical, historical, cultural, and socio-economic conditions of the territory. The universality is based on the main characteristics of the geo-information software used and the possibility of creating additional application programs on its basis [8].

The GIS-project of the atlas allows to conduct an analysis of the natural, economic, socio-demographic situation in the Greater Altai, it can become the basis for creating a permanent geo-information monitoring system available for its use through the Internet.

5. Conclusion

When creating an atlas as a modern GIS-cartographic project, the design of various thematic maps took into account modern and traditional approaches. The experience of creating the best samples of domestic and foreign satin cartographic works was studied. In accordance with them, approaches to the design of maps of the Atlas of the Greater Altai were determined.

The atlas "Greater Altai: nature, history, culture" creates a new image of the Altai interregion, thanks to its visibility and informativeness in displaying the main advantages of its territory. As a result of modern geo-information cartographic modeling, the atlas will contribute to the development of GIS engineering methods in order to provide information support for international projects to strengthen international relations and investment attractiveness of the Altai interregion.

References

[1] Krayukhin A N, and Pozdnyak G V 2006 Satin mapping - information base for sustainable development of the territory. In materials of the International Conference: *Sustainable development of territories: GIS theory and practical experience* (pp. 16-25) (Kaliningrad, Russia)

[2] Rotanova I N, and Repin N V 2013 The development of complex satin mapping on the example of two new atlases of Altai *Geography and Nature Management of Siberia* **16** pp 150-162

[3] Rotanova I N, Tikunov V S *et al.* 2014 International mapping project "The Atlas of Greater Altai: Nature, History, Culture" as the goundation for models of sustainable development *Geography, Environment, Sustainability* **7**(1) pp 99-108

[4] Rotanova I N, Tikunov V S, and Tishkin A A 2014 "The Greater Altai: nature, history, culture": an idea and approaches to creating *Geodesy and Cartography* **1** pp 59-63

[5] Rotanova I N, and Tikunov V S 2015 Concept and program of the Atlas "Great Altai: the Nature, History, Culture" on the basis of means of GIS In *International Geographical Union Regional Conference "Geography, Culture and Society for our Future Earth* (p. 317) (Moscow, Russia)

[6] Efremov G A, and Rotanova I N 2017 Satin geo-information mapping and its implementation on the example of atlas "Greater Altai: nature, history, culture" – Geography issues. In V M Kotlyakov et. al. Eds. *Cartography in the digital age* (pp. 98-120) (Moscow, Russia: Publishing House "Kodeks")

[7] Rotanova I N, Tikunov V S, Peremitina S V, and Efremov G A 2015 *Development of methods for constructing web-GIS cartographic models implemented in the creation of the atlas of the Greater Altai* Available at: http://conf.nsc.ru/SDM-2015/ru

[8] Rotanova I N, and Tikunov V S 2014 Historical and cultural heritage of the Altai region in the atlas "Greater Altai: nature, history, culture" In *Preservation and study of the cultural heritage of the Altai Krai* (vol. 20, pp 267-272) (Barnaul, Russia)

Heavy metals in soil and plants (*Agropyron pectiniforme Roem. Et Schult.*) of the Pavlodar region (Kazakhstan)

Yu Kanibolotskaya[1*], W Listkov[1] and N Shmidt[1]

[1] Siberian University of Consumer Cooperation, 26 Karl Marx Avenue, Novosibirsk 630087 Russia

E-mail: yu_leonova@mail.ru

Abstract. The paper presents the key results of a study on the content of heavy metals (Ti, Mn, Cr, Fe, Ni, Cu, Zn, Pb, Sr) in soil and plant (*Agropyron pectiniforme Roem. et Schult.*) samples, which are taken in the zone of influence of enterprises of the northern and eastern industrial zones of Pavlodar (Kazakhstan) as part of a study of the state of vegetation in the same territory.

1. Introduction

The city of Pavlodar City (Republic of Kazakhstan) is a large industrial center, forming the territorial industrial complex "Pavlodar-Aksu-Ekibastuz," which activities are very important for the development of the economy of Kazakhstan. So, among the leading enterprises of the considered region are the Pavlodar Aluminum Plant, Pavlodar Oil Chemistry Refinery (POCR LLP), Kazakhstan Electrolysis Plant (KEP JSC), HES-1,2,3, Aksu Ferroalloy Plant, Aksukaya Electric Generation Plant, and many other. The total number of such large enterprises is about 40. According to the development plans of the republic, the number of enterprises will increase. At the same time, the environmental situation is complex, since industrial enterprises in the region have been continuously operating for 60 years; therefore, there is pollution caused by the long periods of activity of enterprises, due to a significant number of enterprises and their emissions and discharges of pollutants.

As a result, a significant amount of emissions containing heavy metals (taking into account the specifics of existing enterprises) enters the environment of the region. The existing environmental situation is well reflected by the distribution of pollutants in the soil cover. At the same time, the rate of movement of the substance in the soil is much lower than in other environments; therefore, the composition of the soil shows not only the current (up to 5 cm in the soil layer), but also long-term pollution processes (in the deeper layers) due to the continuous operation of industrial enterprises. Vegetation, being the most "noticeable" component in the composition of ecosystems, as well as soil, accumulating emissions from industrial enterprises and acting as a buffer and detoxifier, is capable of changing the species composition of the community and/or the state of individual plants to "demonstrate" the presence of anthropogenic impact, fulfilling the role indicator [1]. Therefore, we present the results of a study of the accumulation of some heavy metals in the soils and vegetation of the territory located near the existing enterprises of the northern (HES-2,3, POCR LLP, Kazakhstan Metallurgical Company "Casting" JSC, and others) and eastern (Aluminum of Kazakhstan JSC, HES-1, KEP JSC, and others) industrial zones of Pavlodar, Kazakhstan.

Content from this work may be used under the terms of the Creative Commons Attribution 3.0 licence. Any further distribution of this work must maintain attribution to the author(s) and the title of the work, journal citation and DOI.
Published under licence by IOP Publishing Ltd

2. Materials and Methods

The object of research was the vegetation cover of the region. The study of the state of cover was carried out in dissertation research in 2006-2009, and further as part of the development of the initiative-search topic, particularly in 2011-2013, 2015-2016. The soil and plant samples for chemical analyzes were taken in 2006, 2011, 2013, 2015-2016 at various distances from the industrial areas of Pavlodar and Aksu [2-5], according to standard guidelines [6], [7].

More than 50 sites were surveyed, they were located at different distances from the leading industrial enterprises of the region. Sampling sites are confined to the predominant plant communities in which detailed geobotanical descriptions were periodically conducted, including assessing the floristic composition, vertical and horizontal structure of communities, the ecological and biomorphological characteristics of individual species (using classical methods adopted during geobotanical studies) [8].

The determination of the content of elements in soil and plants was carried out via the X-ray fluorescence analysis (annealed for soil; with a preliminary ash content for plants) and using a Spectroscan GF-1E spectrometer (Russia, 2000 g/v). The samples were analyzed in the laboratory of the Physical Technical Institute of the Ministry of Education and Science of the Republic of Kazakhstan (Alatau village, Almaty region, Kazakhstan). Comparison of the maximum permissible concentrations of elements with the concentration detected in the soil was established according to the standards of the maximum permissible concentration [9-11], the concentrations of Fe were compared with the background content.

To assess the pollution of the territory, we selected the underground and ground parts of the comb-wheatgrass (*Agropyron pectiniforme Roem. et Schult*). This cespitose grass is found in varying degrees of abundance in most of the study area, often as a dominant in the community, because it is digression-active.

3. Results

Soil samples taken at 23 sites were analyzed. These samples were located at different distances, 0.5-1-3-5-15-20 km to the north, south, east, and west of the enterprises of the northern and eastern industrial zones of Pavlodar, taking into account the rose winds and transport and walking accessibility).

Table 1 shows the maximum levels of heavy metals (Ti, Mn, Cr, Fe, Ni, Cu, Zn, Pb, Sr) from soil samples collected in these areas:

Table 1. Maximum levels of heavy metals in soil (in mg/kg).

	Soil layer		Ti	Cr	Mn	Fe	Ni	Cu	Zn	Pb	Sr
NIZ	0-5	1	**4150**	**142**	735	27578	**34**	22	**97**	27	265
		2	**0.83**	**2.84**	0,49	1.38	**0.76**	0,55	**0.97**	1.35	0,53
	10-15	1	3400	104	675	27494	**35**	22	**73**	20	260
		2	0,68	2,08	0,45	1,.53	**0.78**	0.55	**0.73**	1	0,52
EIZ	0-5	1	**4150**	112	**1005**	**31119**	32	**29**	87	**46**	**285**
		2	**0.83**	2.24	**0,67**	**1.56**	0.71	**0.73**	0.87	**2.3**	**0.57**
	10-15	1	**4500**	**113**	**885**	**27674**	34	**29**	72	**44**	**270**
		2	**0.9**	**2.26**	**0,59**	**1,.54**	0.76	**0.73**	0.72	**2.2**	**0.54**
MPC (in mg/kg)			5000	50	1500	19948* / 17970*	45	40	100	20	500

Legend: NIZ – a northern industrial zone (data of soil samples from the impact areas of HES-2-3, POCR, and others); EIZ – an eastern industrial zone (HES-1, Aluminum of Kazakhstan JSC, Kazakhstan Electrolysis Plant JSC, and others); 1 – concentration of elements in the soil (mg/kg), 2 – concentration of elements in fractions of maximum permissible concentration; **highlighted** – the highest values of each element; * – a background content Fe, mg/kg.

The results of analyzes of soil samples in the table indicate fairly high concentrations of almost all the elements considered. Vegetable samples were collected at the same sites as soil samples. The maximum concentrations (in mg/kg) of Ti, Cr, Mn, Fe, Ni, Cu, Zn, Pb, and Sr in the underground and above-ground parts of the comb-wheatgrass in Table 2.

All the considered elements accumulate more in the underground than in the above-ground parts *Agropiron pectineforme* (except for a few samples).

Table 2. The maximum content of heavy metals in *Agropiron pectineforme* (mg/kg).

	Parts of a plant		Ti	Cr	Mn	Fe	Ni	Cu	Zn	Pb	Sr
NIZ	Below	1	**1867**	850	417	**13200**	**31.7**	**7.1**	**778**	**100**	**62**
	ground	2	**2.7**	170	1.95	**1.7**	**3.6**	**1.82**	**17.3**	**12.2**	**1.8**
	Above	1	**748**	12.5	121	646.5	2.0	1.3	**120**	**5.65**	62
	ground	2	**34**	2,1	1.6	2.5	2.5	2.2	-	**20.2**	4.4
EIZ	Below	1	1449	148	509	10352	13.9	5.6	92	13.9	55
	ground	2	2.1	29.6	2.4	1,4	1.6	1.4	2	1.7	1.6
	Above	1	470	**143**	205	**3628**	**43**	**4.8**	22	4.8	**108**
	ground	2	21.4	**23.8**	2.8	**14.1**	**53.8**	**8**	-	8	**7.7**
BG	Below ground		691	5.0	213	7615	8.7	3.9	45	8.2	35
	Above ground		22	6.0	74	257	0.8	0.6	<LLQ	0.6	14

Legend: background – 50km east of Pavlodar; LLQ – a lower limit of quantification; 1 – a maximum content of heavy metals in *Agropiron pectineforme* (mg/kg), 2 – a share of the background content of the element in plant samples.

4. Discussion

The data of the analytical survey of the soil both in the zone of impact of enterprises in the eastern and northern industrial zones of Pavlodar indicate a significant excess of the maximum allowable concentration of *Chrome* (Cr), *Plumb* (Pb) and background values by *Ferrum* (Fe), *Titanium* (Ti), *zinetum* (Zn), and *Niccolum* (Ni)content in soil reaches its maximum permissible concentration; slightly more than 0.5 of the maximum permissible concentration content: *Manganese* (Mn), *Cuprum* (Cu), and *Strontium* (Sr).

Significant concentrations in soil samples of Cr and Fe are provided by the activity of the Ferroalloy Plant (Transnational Company Kazchrome JSC), and the remaining elements considered by the cumulative impact of industrial enterprises in the region, since their number and similar composition (for example, for HES) do not clearly define sources, but provide a summation of the pollution fields.

When analyzing the content of metals in the pyrex with a comb-like, it was revealed that this species is characterized by accumulation of heavy metals by an acropetal type "roots>" above the ground parts of plants. It is the territory of influence of the enterprises of the eastern and northern industrial zones of the city of Pavlodar.

The maximum concentrations of *Plumb* (Pb) and *Zinetum* (Zn) elements of the first class of danger (hazard class of danger are designated according to M. S. Panin [1]), in the above ground parts of plants registered 1 km north of the HES-2, it is 0.8 km north-east from the HES-2, Pavlodar, for the below ground parts.

Elements of the second class of danger, maximum concentrations of *Niccolum* (Ni), *Cuprum* (Cu), and *Chrome* (Cr) in above ground parts of *Agropiron pectineforme* were found at a site located 20 km west from the Aluminum of Kazakhstan JSC (12 km north from Transnational Company Kazchrome JSC); *Niccolum* (Ni) in the below ground parts – 2 km north from the HES-3; in 0.5 km north-east from the HES-3 *Cuprum* (Cu) was found; and 0.8 km north-east from the HES-2 – *Chrome* (Cr).

Elements of the third class of danger, maximum concentrations of *Manganese* (Mn) (completely) and *Strontium* (Sr) (in the above ground part) in *Agropiron pectineforme* were observed at a distance of 20 km west from the Aluminum of Kazakhstan JSC (12 km north from the Transnational Company

Kazchrome JSC); *Strontium* (Sr) in the below ground part was found 0.5 km north-west of the HES-3, Pavlodar.

The maximum content of *Ferrum* (Fe) in the below ground parts *A. pectineforme was* noted 1 km north from the HES-2; Fe in the above-ground parts and *Titanium* (Ti) completely were also noted at a site 20 km west of the Aluminum of Kazakhstan JSC (12 km north of Transnational Company Kazchrome JSC).

5. Conclusion

High concentrations of the considered elements are found in soil samples collected at a distance of 0.5-1 km from the HES-1 (eastern industrial zone) and HES-2 (northern industrial zone) of Pavlodar, as well as in the territory between the city of Pavlodar and Aksu (a site, located 20 km west of Aluminum of Kazakhstan JSC (12 km north of Transnational Company Kazchrome JSC)).

For *Agropiron pectineforme* samples, the same pattern was found. A correlation between the increased content of heavy metals in the soil and plant samples of the studied areas was also noted, i.e. the transfer of pollutants from the soil to the plant organism is possible. The fact that the metals in question are obtained from soil by plants is indicated by a significant amount of them in the roots, and much less in its above-ground parts (the aeration mechanism of the intake of pollutants is more developed in broadleaf).

Soil and plant pollution is mainly due to the proximity of sites to industrial enterprises (based on the location of sampling sites, this distance is 0.5-0.8-1 km), as well as the sum of emissions coming from enterprises of industrial zones of Pavlodar (eastern and northern) and the city of Aksu in the environment of the region.

6. Acknowledgment

The author is express gratitude to the Rector of the Siberian University of Consumer Cooperatives, Dr of Sci. (Technical), Prof., V. I. Bakaitis, for financial support in publishing this article.

References

[1] Panin M S 2002 *Chemical ecology* (Semipalatinsk, Republic of Kazakhstan: Shakarim State University of Semey)

[2] Kanibolotskaya Yu M 2012 Analysis of heavy metal content in plants (Artemisia austriaca Jacq., Agropyron pectinatum, Potentilla bifurca) and soil in Pavlodar region *Mongolian Journal of Agricultural Sciences* **6**(02) pp 101-109

[3] Kozyrenko M, and Kanibolotskaya Yu M 2012 Heavy metals in soil and plants. In Proceedings of the international scientific conference of young scientists, undergraduates, students and schoolchildren: *XIV Satpayev's readings* (pp. 182-185) (Pavlodar, Republic of Kazakhstan)

[4] Kurmanbaeva A, and Kanibolotskaya Yu M 2014 The main factors influencing the plant communities of the suburb of Pavlodar. In Proceedings of the international scientific conference of young scientists, undergraduates, students and schoolchildren: *XIV Satpayev's readings* (pp. 191-194) (Pavlodar, Republic of Kazakhstan)

[5] Berikova A T, and Kanibolotskaya Yu M 2017 Anthropogenic transformation of vegetation cover in the impact zone of the Eurasian Energy Corporation JSC and Transnational company Kazchrome JSC. In *Proceedings of the IX International Conference Toraigyrovskiye Readings* (pp. 75-78) (Pavlodar, Republic of Kazakhstan: Pavlodar State University named after S. Toraigyrov)

[6] Ussr State Committee on Hydrometeorology 1981 *Methodical recommendations on carrying out field and laboratory researches of soils at control of environmental pollution by metals* (Moscow, USSR: Hydrometeoizdat)

[7] Belyakova E E, Vasilieva N N, Grabovskaya L I *et al.* 1972 *Biogeochemical and geobotanical studies* (Leningrad, USSR: Nedra)

[8] Lavrenko E M, and Korchagin A A Eds. 1959-1972 *Field geobotany* (in 4 vol) (Moscow, USSR: Publishing House of the USSR Academy of Sciences)

[9] Kloke A Richwerte 1980 Orientierungsdaten fur tolerierbare Gesamtgehalte einger Elemente in Kulturboden *Mitteilunger VDLUFA* 1-3 pp 9-12

[10] Cabinet of Ministers of the Republic of Kazakhstan 1993 *Scientific and methodological instructions for monitoring the lands of the Republic of Kazakhstan* (Almaty, Republic of Kazakhstan)
[11] Chief State Sanitary Doctor of the USSR 1983 *Sanitary norms of permissible concentrations of chemicals in the soil SaNiP* (May 13, 1983 No. 2546-82)

Features of microclonal reproduction of species of the genus *Aristolochia L.* in various nutrient environments

E Demidenko[1] and S Berseneva[1*]

[1] Primorskaya State Academy of Agriculture, 44 Bluhera st., Ussuriisk 692510 Russia

E-mail: svshatal@mail.ru

Abstract. The paper presents a comparative analysis of the development of microshoots in primary explants of the species belonging to the genus *Aristolochia,* using various nutrient media. This allows selecting the most optimal methods for mass production of the sterile regenerated plant cultures in a shorter time for further introduction, reintroduction, and creation of industrial plantations.

1. Introduction

The problems of plant biodiversity conservation in the context of the development of biotechnology and sustainable use of biological resources were first considered in the International Convention on Biological Diversity [2]. The use of biotechnological methods offers fundamentally new opportunities for the preservation and reproduction of the gene pool of plants. When creating genetic banks *in vitro* plant species, the primary explant is preferred as a seed material [3], [5].

Domestic research in the field of microclonal reproduction of the genus *Kirkazon* confirms that for liana, such as *Aristolochia manshuriensis Kom.,* reproduction is carried out by activating the development of existing meristem [4]. It has also been established that the presence of *Cytodef and Dropp* preparations at a concentration of 0.01 mg/l in the composition of the nutrient medium stimulates the process of induction of the formation of adventitious buds, the formation of micro-shoots, and increases the multiplication factor [7], [8].

A comparative analysis of the development of microshoots in primary explants of species of the genus *Aristolochia* using various nutrient media is the purpose of the present study.

2. Materials and Methods

The following species of the genus Kirkazon are the objects of research for introduction into culture *in vitro*: *A. manshuriensis Kom., A. macrophylla Lam., A. contorta Bunge, A. fimbriata Cham. & Schltdl,* and *A. clematitis L.*

Experiments on microclonal reproduction were carried out in the period 2017-2018, in the Laboratory of Biotechnology of the Primorskaya State Agricultural Academy. After sterilization using the Butenko method [6], the cuttings of the studied types of Kirkazon were placed vertically on three variants of nutrient media.

The I nutrient medium is made on the basis of macro-and micro-salts according to MS-Murashige [10], with addition of casein hydrolyzate (80 mg/l), meso-inositol (50 mg/l), thiamine (0.5 mg/l), pyridoxine (0.5 mg/l), ascorbic acid (0.5 mg/l), kinetin (0.2 mg/l), and Indole-3-acetic acid (1.0 mg/l), pH is 4.49;

Content from this work may be used under the terms of the Creative Commons Attribution 3.0 licence. Any further distribution of this work must maintain attribution to the author(s) and the title of the work, journal citation and DOI.
Published under licence by IOP Publishing Ltd

The II nutrient medium is made on the basis of macro and micro salts according to WPM [9], with the addition of nicotinic acid (0.5 mg/l), pyridoxine (0.5 mg/l), glycine (1.0 mg/l), 2-isopentenyladenine (8 mg/l), and Indole-3-acetic acid (4 mg/l); pH is 4.51;

The III nutrient medium is made on the basis of macro- and micro-salts according to MS (half content), with the addition of thiamine (0.5 mg/l), pyridoxine (0.5 mg/l), and ascorbic acid (0.5 mg/l); pH is 4.52.

At the first stage of the experiment, in connection with a small amount of starting material, the microcurrents of the folloeing species *A. trilobata L., A. sempervirens L., A. tomentosa Lam., A. maxima Jacq., A. littoralis Parodi. and A. gigantea Mart. & Zucc.* were planted in the variant III of the nutrient medium, with the purpose of obtaining the required number of explants.

After obtaining the required number of micro-cages, in the second stage, the explants were transplanted to the (IV) and (V) variants of modified nutrient media containing macrosols according to the MS and WPM micro-salts (Woody Plant Medium). The staff of the Microclonal Reproduction Sector of Rare, Ornamental, and Agricultural Crops (A. V. Mikheeva and I. V. Gafitskaya) developed the basic medium composition [1] for the cultivation *in vitro* of rare and decorative rhododendron species and other woody and herbaceous plants.

In the IV variant of the medium, we used the Murashige macro-salts and WPM micro-salts with the addition of nicotinic acid (0.5 mg/l), pyridoxine (0.5 mg/l), glycine (1.0 mg/l), Indole-butyric acid (1.0 mg/l), sucrose (20 g/l); pH is 5,58.

The V nutrient medium is made on the basis of macrosols according to MS and micro salts according to WPM, with the addition of nicotinic acid (0.5 mg/l), pyridoxine (0.5 mg/l), glycine (1.0 mg/1), 2-isopentenyladenine (4 mg/l), and Indole-3-acetic acid (2 mg/l); sucrose (10 g/l); pH is 5.53.

The main stages of research are presented in Figure 1.

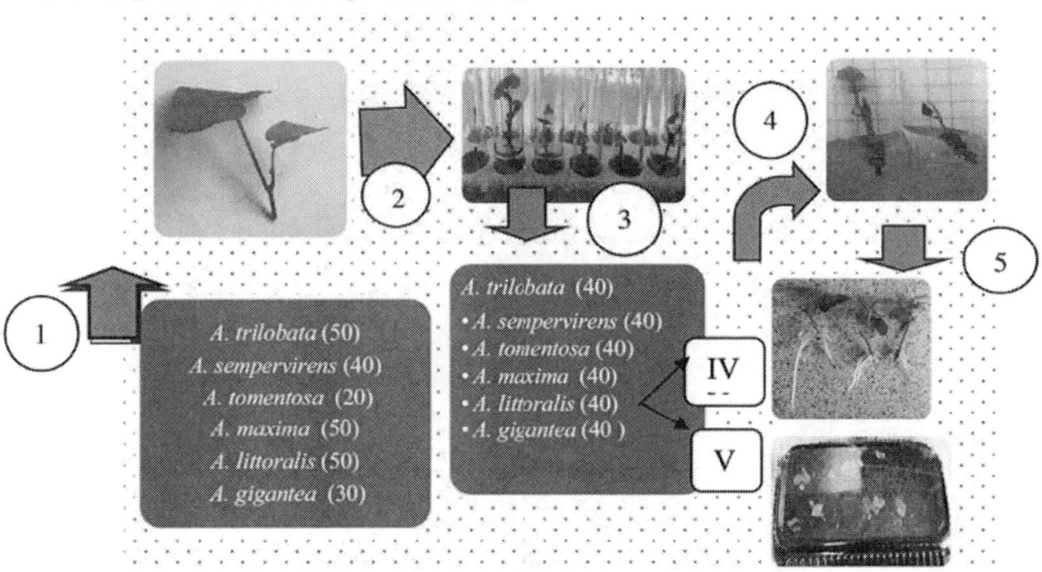

Figure 1. Diagram of the microclonal reproduction of species of the genus *Aristolochia*: (1) selecting the initial explant; (2) obtaining a sterile culture; (3) reproduction of microshoots (grafting) and transplantation of explants for IV and V variants of media; (4) the growth of the kidneys and the formation of micro shoots; (5) conversion of plants to greenhouse conditions.

3. Results

Our research (which took 4 weeks) allowed us to conduct a comparative analysis of the development of microshoots in primary explants of species of the genus *Kirkazon* (Table 1).

Table 1. Development of microshoots in primary explants of species of the genus *Aristolochia*.

Species of the genus *Aristolochia*	Environment option								
	I			II			III		
	Microshoots age, weeks	Microshoots height, cm	Number of shoots, %	Microshoot age, weeks	Microshoot height, cm	Number of shoots,%	Microshoot age, weeks	Microshoot height, cm	Number of shoots, %
A. manshuriensis Kom.	2-3	to 0.9	10%	–	–	–	2-3	to 0.5	25%
A. macrophylla Lam.	4	to 1.0	6%	–	–	–	2-3	to 0.8	5%
A. contorta Bunge	2-3	to 0.5	20%	2-3	to 0.6	45%	2-3	to 0.6	36%
A. fimbriata Cham. & Schltdl	3-4	to 0.5	40%	5	to 0.5	50%	2	to 0.5	50%
A. clematitis L.	2	to 2.5	83%	2	to 0.3	17%	2	to 3.0	83%

For grassy non-climbing species A. *fimbriata Cham. & Schltdl* and A. *clematitis* L., the highest number of shoots in all variants was observed, especially high rates were found in the mediums No. I and III. At the same time, for woody A. *manshuriensis Kom.*, A. *macrophylla Lam.*, the number of shoots is small, and in the II medium variant, the shoots did not develop. Interestingly, A. *contorta Bunge*, being a grassy liana, occupies an intermediate position in terms of lignescent vines and grassy non-climbing species on I and III medium variants; at the same time, the largest number of shoots of a species is recorded on medium II. The maximum height of the shoots (2.5–3.0 cm) was noted in A. *clematitis* L. on I and III variants of nutrient media. In the remaining cultivated species of *Aristolochia* L., this figure ranged from 0.3 to 0.9 cm (Fig. 2).

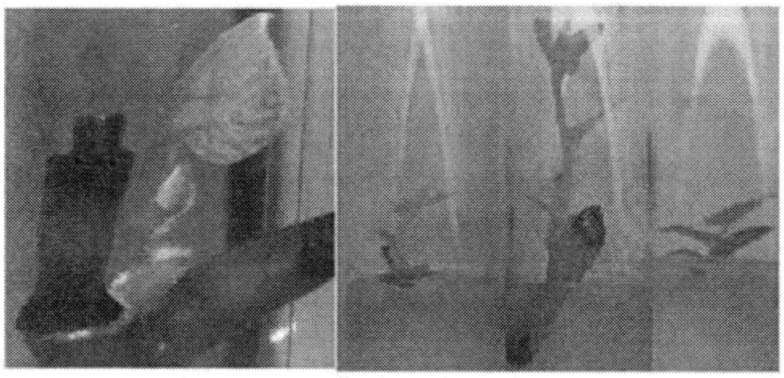

Figure 2. Micro shoots of species of the genus *Aristolochia* L.: on the right is A. *Manshuriensis Kom.*; on the left is A. *clematitis* L.

After practicing the method of introducing 4 species of the genus *Aristolochia* L. into culture, the experiment was supplemented with 6 more species in 2018: A. *trilobata* L., A. *sempervirens* L., A. *littoralis Parodi*, A. *maxima Jacq*., A. *tomentosa Sims* and A. *gigantea Mart. & Zucc*. In 2018, the cultivation period of the above-mentioned species averaged two months. The main results on the cultivation of species are presented in Table 2.

In newly introduced species, the formation of vegetative shoots bearing leaves at the base was recorded after 3 weeks. The maximum height of microshoots was noted in A. *sempervirens* L. (4.6 cm.), The minimum was for in A. *trilobata* L. (0.5 cm). In most species, elongation of micro-shoots is observed only by the end of the 6th week, disclosure of leaf blades occurs very slowly, their deformation (blackening of edges, yellowing of the entire surface) is observed in some cases. In A. *sempervirens* L., in comparison with the other studied species of the genus *Kirkazon*, a more intensive development of microshoots was noted. Preliminary results showed that the species such as A.

trilobata L. and A. *littoralis Parodi*, the formation of aerial roots is typical already at the end of the first and beginning of the second week of cultivation in 6-10% of cases.

In explants of A. *maxima Jacq.*, during the entire cultivation period, the callus developed at the base, but the differentiation of the roots did not occur (Fig. 3).

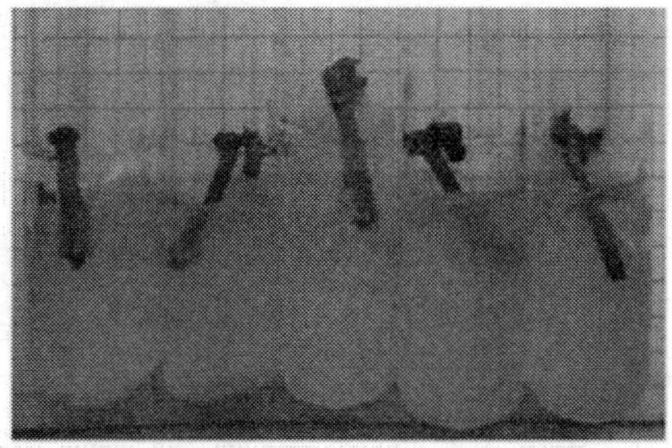

Figure 3. Microcurrents *Aristolochia maxima Jacq.*

Cultivation of explants on a hormone-free nutrient medium *in vitro* contributed to a change in morphology. This was manifested in an increase in the internal tissues, i.e., a change in their structure (gelatinous mass of dark color) and a rupture of the outer integumentary tissues.

According to the preliminary results of cultivation of this species, a repeat of the experiment is planned for a more detailed study of the response of the A. maxima Jacq explants on the composition of the nutrient medium and sending material for cytological examination. This will allow to identify exactly which plant tissues respond in this way to micropropagation *in vitro*. In some cases, there was a staining of the nutrient medium in brown or almost black color and the blackening of the microcurrents. In this case, the release of phenols by the plant into the nutrient medium contributes to this, which leads to a change in its basic properties and undesirable consequences for the explant itself.

Table 2. Cultivation results of species of the genus *Aristolochia* L.

Genus *Aristolochia*	*In vitro*	Number of microshoots, pcs; including those being rooted *in vitro*, pcs	Dimensions of microshoots, cm (3 weeks)	Dimensions of microshoots, cm (6 weeks)	Sizes of primary sheet, cm (3 weeks)	Sizes of primary sheet, cm (6 weeks)	Number of microshoots removed from the experience, pcs
A. trilobata L.	50	41(3) (82% (6%))	0.5	1.1	0.4x0.4	0.8x1.1	9 (8%)
			0.2	1.0	-	0.5x1.0	
			0.3	0.8	-	0.7x1.0	
			0.3	0.5	0.4x0.4	0.8x0.7	
			0.3	0.8	-	0.8x1.0	
			0.3	0.8	-	0.5x0.8	
			0.3	0.9	-	0.6x0.8	
			0.3	1.1	-	0.8x1.1	
	50	**41(3)**	**0.31**	**0.88**	**0.4x0.4**	**0.69x0.94**	**9**
A. sempervirens L.	40	33 (82.5%)	0.3	1.4	0.6x0.5	1.0x1.0	7 (7.5%)
			0.5	1.7	0.6x0.5	1.0x0.7	
			1.2	4.6	-	-	
			0.7	2.1	0.6x0.5	1.0x0.7	
			0.2	1.2	-	-	
			0.3	1.5	-	-	
			0.3	0.6	-	-	
			0.6	2.2	-	-	
	40	**33**	**0.51**	**1.9**	**0.6x0.5**	**1.0x0.80**	**7**

A. littoralis Parodi	50	29(5) (58% (10%))	0.5	1.3	-	-	21 (42%)
			0.3	0.9	-	-	
			0.5	1.1	-	-	
			0.9	1.6	-	-	
			0.3	0.9	-	-	
			0.4	1.4	-	-	
			0.5	1.1	-	-	
			0.8	1.8	-	-	
	50	**29(5)**	**0.53**	**1.26**	**-**	**-**	**21**
A. gigantea Mart. & Zucc	30	30 (100%)	0.9	1.0	0.4x0.4	1.0x1.1	-
			0.6	1.0	0.4x0.4	0.6x1.2	
			0.6	1.2	0.6x0.6	0.8x1.2	
			0.4	0.8	-	-	
			0.5	1.0	-	-	
			0.5	1.1	-	-	
			0.3	0.6	-	-	
			0.2	0.6	-	-	
	30	**30**	**0.5**	**0.91**	**0.47x0.47**	**0.80x1.17**	**-**
Total	170	133(8) (78% (4,7%))					37 (22%)

As a result of cultivation of *A. tomentosa Sims*, elongation of microshoots was observed only in 7 out of 20 explants (Fig. 4). Some micro shoots had a yellowish tint. The remaining 13 shoots did not show significant changes after 6 weeks of cultivation, except for staining the nutrient medium in a dark yellowish-brown tint. *Aristolochia tomentosa Sims* and A. *maxima Jacq.* woody. Bud formation was recorded in the 1st explant in the second month of cultivation. This indicates that at the time of planting in a nutrient medium, the generative kidney was already laid in this explant.

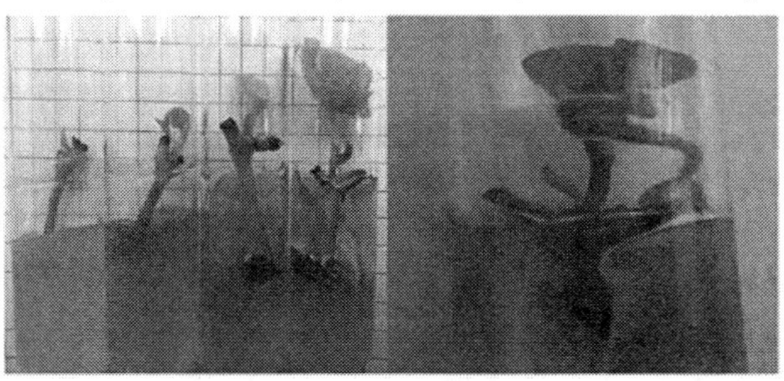

Figure 4. Micro-shoots of *Aristolochia tomentosa* L. is on the right, and the bud formation is on the left.

To continue the experiment on microclonal propagation, sterile micro shoots after preliminary grafting were transplanted into two variants of the nutrient medium. The variant IV contains substances stimulating rhizogenesis in the *in vitro* micro-vegetative shoots. And the variant V contains those stimulants that cause an intensive growth of meristem tissue and the formation of several micro-shoots on the 1st explant. The results of this part of the experiment are shown in Table 3.

The total number of microplants planted in greenhouse conditions was 31.87% of the total number of secondary explants used in the experiment. 25.63% of them rooted on the IV variant of the nutrient medium. Higher rates of rhizogenesis in microchips according to the results of the experiment were recorded in A. *littoralis Parodi* (62.5% of those used in the experiment), low results were recorded in A. *maxima Jacq.* (5%). In the secondary explants of A. *sempervirens* L. and A. *tomentosa Sims*, rhizogenesis was not recorded. In A. *tomentosa* L., a more intensive release of phenols into nutrient media and darkening of microcurrents was observed. During the growing period from January 4 to March 4, 2018, the survival rate of microplants in A. *littoralis Pparodi* and A. *trilobata* L. was 100% in greenhouse conditions. The experiment on microclonal reproduction of these species must be repeated to clarify the causes of the death of microplants A. *maxima Jacq*. and A. *gigantea Mart. & Zucc* during acclimatization to greenhouse conditions.

Table 3. The results of the cultivation of secondary explants of the genus *Aristolochia* L.

Landing options	A number of explants on the medium variant		A total number of explants, pcs
	IV	V	
	A. *gigantea Mart. & Zucc*		
In vitro	20	20	40
Closed ground	8	-	8
	A. *littoralis Parodi*		
In vitro	20	20	40
Closed ground	18	7	25
	A. *maxima Jacq.*		
In vitro	20	20	40
Closed ground	2	-	2
	A. *trilobata* L.		
In vitro	20	20	40
Closed ground	13	3	16

3. Conclusion
Comparative analysis of the development of microshoots in primary explants of species of the genus *Aristolochia* using various nutrient media allows, in a shorter time, to select the most optimal methods for mass production of sterile regenerated plant cultures, with the aim of further introduction, reintroduction, and the creation of industrial plantations. We believe that this method of breeding of species of the genus *Aristolochia* L., including A. *manshuriensis Kom.*, is be the most effective, because the technology for producing a large number of plants allows to constantly update the *Kirkazon* plantings in greenhouses, plantations, and other conditions of the artificially restored populations.

4. Acknowledgments
We express our deep gratitude and appreciation to the staff of the Arboretum of the Federal Scientific Center of the East Asia Terrestrial Biodiversity of the Far Eastern Branch of the Russian Academy of Sciences "V. L. Komarov Mountain-Taiga Station" for the opportunity to collect the source materials.

References
[1] Babikova A V, Gafitskaya I V, Koren O G, Muzarok T I, Zmeeva V N, Pinkus S A, Akimova L A, ... Barkalova O K 2013 *Microcloning of ornamental woody plants Problems of Landscaping Settlements* (pp. 10-14) (Vladivostok, Russia)
[2] Vinogradova Yu K, Gorbunov Yu N, Makridin A I, Molkanova O I, and Shvetsov A N 2005 Developing principles for the conservation and reproduction of genetic phytoresources In L P Rysin, B R Striganova, M I Shatunovskiy, V G Petrosyan (Eds.*), Basic grounds of biological resources management* (pp. 343-351) (Moscow, Russia: Fellowship of scientific publications KMK)
[3] Batygina T B, and Vasilyeva V E 2002 *Plant reproduction* (St. Petersburg, Russia: St. Petersburg University Press)
[4] Golovach A. G 1963 Aristolochia Manchu (Kirkazon Manchu) Aristolochia manshuriensis Kom. In *Green Building* (pp. 35-52) (Leningrad, USSR: Science)
[5] Ishmuratova M I, and Tkachenko K G 2009 *Grass seeds: specific features of the latent period, use in vitro for introduction and reproduction* (Ufa, Russia: Gilem)
[6] Kataeva N V, and Butenko R G 1983 *Clonal micropropagation of plants* (Moscow, USSR: Nauka)
[7] Nakonechnaya O V, Nesterova S V, and Voronkova N M 2012 Ontogenesis *Aristolochia contorta (Aristolochiaceae)* in Primorye Region *Botanical Journal* **97**(12) pp 1505-1515
[8] Nakonechnaya O V, Zhuravlev Yu N, Bulgakov V P, Koren O G, and Sundukova E V 2014 *The genus Kirkazon in the Far East of Russia (Aristolochia manshuriensis Kom. and A. contorta Bunge)*

(Vladivostok, Russia: Dalnauka)
[9] Lloyd G, and McCown B H 1981 Commercially-feasible micropropagation of Mountain Laurel, Kalmia latifolia, by shoot tip culture *Proc. Int. Plant Prop. Soc.* **30** pp 421-427
[10] Murashige T, and Skoog F 1962 A revised medium for rapid growth and bio-assays with tobacco tissue cultures *Physiologia Plantarum.* **15** pp 473-497

IOP Conf. Series: Earth and Environmental Science **395** (2019) 012039 doi:10.1088/1755-1315/395/1/012039

The state of ichthyofauna of Yakutia in modern conditions

M Tyaptirgyanov[1*], V Tyaptirgyanova[2], and T Tyaptirgyanova[1]

[1] M.K. Ammosov North-Eastern Federal University, 48 Kulakovskogo str., Yakutsk 677000 Russia

[2] Center of Hygiene and Epidemiology in the Republic of Sakha (Yakutia), 60/2 P. Alekseeva str., Yakutsk 677000 Russia

E-mail: matyap@mail.ru

Abstract. The article presents the results of many years of research on the fish fauna of Yakutia. Big changes in the general ecological situation, especially the influence of the anthropogenic factor, had a negative impact on individual structural indicators of populations.

1. Introduction

On the territory of Yakutia, almost 0.5 million rivers flow with a total length of more than 1.5 million km. Snow and rainwater and to a lesser extent glaciers, ice and groundwater take part in the feeding of rivers. Most rivers are characterized by high spring floods. Every year, rivers infuse about 900 km^2 of fresh water. The freezing season lasts 6-8 months, the thickness of the ice cover reaches 3 m. The total number of lakes with an area of 1 ha is about 709,000.

The most modern fish fauna are heterogeneous in origin; they are composed of various faunistic complexes. The distribution of fish of Yakutia is presented in the Table 1.

Table 1. Distribution of fishes in Yakutia according to their faunal complexes.

Arctic	Arcto-boreal	Boreal foothill	Boreal plain	Chinese plain	Other	Total
(n=12)	7	5	13	4	2	43
% 28	16	12	30	9	5	100

2. Material and Methods

The main objects of our research were the fishes from the waters of Yakutia, living in conditions of anthropogenic pollution. A complete general biological analysis (morphometry, size-age composition, population estimation, etc.) was carried out for all of them. From June to December 1970-2018, the collection and processing of materials were carried out. The processing of caught fish was carried out according to the methods generally accepted in ichthyology (developed by I. F. Pravdin [1]), using the methodological instructions developed by M.V. Mina [2] and Yu. S. Reshetnikov [3]. They were verified according to the latest reports "Atlas of Freshwater Fish of Russia" [4].

Content from this work may be used under the terms of the Creative Commons Attribution 3.0 licence. Any further distribution of this work must maintain attribution to the author(s) and the title of the work, journal citation and DOI.
Published under licence by IOP Publishing Ltd

IOP Conf. Series: Earth and Environmental Science **395** (2019) 012039 doi:10.1088/1755-1315/395/1/012039

3. Results

Arctic fish (28%) and arctic-boreal fish (16%), which together account for about half of the species list (44%), form the basis of the fish fauna. Typical arctic species are the coregonus (*Coregonus clupeaformis*), salvelinus (*Salvelinus alpinus alpinus*), and fourhorn sculpin (*Myoxocephalus quadricornis*). The boreal lowland complex follows them in the number of species (30%); the fish of the Boreal foothill complex (12%) is much smaller (this is the Siberian stone loach, the Siberian grayling, lenok, taimen, and malma). There are relatively many representatives of the Chinese lowland complex (9%), to which we refer the Soldatov's Gobio, Czekanowski's minnow, as well as the Amur and Lake minnow. The main range of this group of fish is located in the Amur basin (Russia) and in China. Among others, we would like to note the roach (*Rutilus rutilus*) as a representative of the Ponto-Caspian complex and catostomus (*Catostomidae*), who clearly came from the region of North America.

Most of the rivers and lakes of Yakutia are located north of the Arctic Circle in the Arctic zone; therefore, the ichthyofauna of Yakutia has features of the Arctic fauna. As we move from the European part of the Arctic to Siberia, i.e. with an increasing continental climate, the number of Arctic species increases, with a general trend of decline in the total number of species (Table 1). This is one of the characteristic features of the Arctic ichthyofauna.

Another characteristic feature of the Arctic is the low endemism of ichthyofauna at the level of families (5%) and genera (2%). So, there is not a single endemic unit for the Arctic among all fish. Only 2 species (or 5%, Yakut loach and Chersky loach) of 43 species and subspecies of Yakutia are endemic. A small number of monotypic genera (nelma, burbot) is typical for Yakutia, as well as for the whole Arctic. There are well represented species with a wide, almost circumpolar range (omul, nelma, whitefish, arctic char, pike, perch, ruff) and with a complex intraspecific structure (omul, whitefish, and charrs).

It was noted earlier that polymorphism and omnivore are the characteristic features of the Arctic ichthyofauna, and these features equally apply to Yakutia's water bodies. Properties such as burrowing in silt for the winter (two types of crucian carp, *Carassius auratus* and *C. carassius*) can be considered as one of the devices for living in these conditions. Broad whitefish (*Coregonus nasus*) spawns during freezing, and it develops in this mixture of ice and water at first. It is believed that in many whitefish fish, caviar can develop in the ice [5], [6].

There are several highly specialized fish in the food line in Yakutia. By the nature of feeding, benthophages (40%) and euryphages (29%) prevail, predators ichthyophages (19%) and zooplanktophages (10%) are relatively small, and only one detritophages species is represented (Siberian lamprey, 5%). At times, the Siberian bullhead, both smelt species and even the Siberian grayling and humpback whitefish go on to eat fish [7], [8]. The complexity of the structure of the northern ecosystems of Europe and Siberia is not achieved due to the diversity of the number of species, but due to the diversity of intraspecific forms that are energetically equivalent to independent species [9-11]. Therefore, a large intraspecific diversity of whitefish and salmon fishes is interpreted as a result of the evolution of northern ecosystems along the path of increasing the number of internal connections and stability of northern ecosystems.

According to the breeding pattern (type of spawning grounds and the method of laying caviar), all fish are divided into 4 groups. The most species lay caviar on sand (III, 26%) or on stony-pebbly soil (II, 19%), which is typical of many salmon, whitefish, grayling, lamprey, sturgeon, Lagovskoy's minnow, and Soldatov's Gobio. The majority of carps lay caviar on vegetation (IV, phytophils, 19%). A special group consists of fishes that take care of the offspring (I, 17 %). The salmon fishes (flax, pink salmon, chum salmon, loach) dig holes in the ground, where the female lays caviar, and the male fertilizes them with milt, after which the caviar are covered with soil. The cottus finsh choose a place for ther nest under the stones, the caviar is deposited in the form of masonry on the underside of the stone and then guarded by the parents (usually by males). The males of the ninespine stickleback (*Pungitius pungitius*) build a nest from the grass. Then a female lays the caviar into them, and a male protects the nest with caviar all the time until their larvae appear. The main breeding strategy of fish in the Arctic is designed so that the released larvae can find food and have time to grow in the short polar summer. Spawning, which occurs not every year or more, is characteristic of many fishes.

If one does not take into account the region of Anabar, then the specificity of this region (Khrom, Indigirka, Alazei, Kolyma) is that there is no longer Kamchatka lamprey, roach, ide, trout, small smelt. The areas of the Siberian and variegated cottus, fourhorn sculpin, perch, and ruff end there. Kolyma was the border, beyond which to the east, all sturgeon, many carp and all perch fishes did not pass. Only here, in the rivers of Kolyma, Alazei, and Indigirka, the first representative of the American fauna, *Catostomus*, appear. The malma (*Salvelinus malma*) appears in the Kolyma. The Pacific salmon spawning is also more frequent in that region, while the pink and chum salmons are in Indigirka and Kolyma).

One of the promising methods for assessing the anthropogenic effects is bioindication, when a degree of impact is assessed by the response reactions of the biological systems themselves. The content of mercury in the organs and tissues of fish varies in the most polluted areas, mostly in the diamond and gold mining industrial zones, from 1 to 3 MPC. As a result, the reduction in the number of fish species in the Vilyuy River is noted from 26 to 19. The excess of cadmium in freshwater fish in Yakutia is 1-2 MPC, while localizing mainly in the gills and liver of fish. The highest lead content was found in the Khroma River, where its content in organs and tissues was 1.0-1.7 MPC. In the Khroma River, the zooplankton biomass decreased from 80 to 4.2 mg / m^2.

4. Discussion

Thus, the water bodies of Yakutia, as well as all of Siberia (with the exception of Anadyr), belong to the Ledovitomsk province of the Siberian district, in which there are 3 districts: (a) Ob and Yenisei region, (b) Lena region (Olenek, Lena, Yana), and (c) Kolyma region (Khroma, Indigirka, Alazey, Kolyma).

The deterioration of water quality and changes have affected the structure of the whole community of freshwater ecosystems, ranging from invertebrate organisms to fish. First of all, this was manifested in the appearance of morpho-pathological anomalies in fish.

5. Conclusion

The whitefishes (*Coregonus*) are the most common species in the rivers of the Far North. It was shown that renal stone disease, pathologies in the skeleton and disturbances in the reproduction system were detected based on the analysis of changes in morpho-physiological indicators in natural water bodies. One of the measures for the biological diversity conservation is the organization of ichthyological monitoring, as well as the expansion of the geography of fishing due to the development of fish stocks in the upper and middle reaches of rivers. Another measure implies including the fish-rich lake systems and coastal areas in the fishery.

6. Acknowledgments

The authors are grateful to the Leading Researcher of the Laboratory of Lower Vertebrates of the Severtsov Institute of Ecology and Evolution of the Russian Academy of Sciences (Moscow) Professor Yu. S. Reshetnikov for his contribution to the study of contemporary problems of fish ecology of Yakutia under the conditions of powerful anthropogenic impact.

References

[1] Pravdin I F 1966 *Guidelines for the study of fish* (Moscow, USSR: Food Industry)
[2] Mina M V 1973 Growth of fish (research methods in natural populations) In *Growth of animals: Zoology of vertebrates* (Moscow, Russia: VINITI)
[3] Reshetnikov Yu S 1980 *Ecology and systematics of whitefish* (Moscow, USSR: Nauka)
[4] Yu S Reshetnikov Ed 2002 *Atlas of freshwater fish of Russia* (*vol 1-2*) (Moscow, Russia: Nauka)
[5] Chernyaev Zh A 1968 *Embryonic development of the Baikal omul* (Moscow, USSR: Nauka)
[6] Chernyaev Zh A 1971 On the possibility of developing caviar of Baikal whitefish in the supercooled state of the "pagon" *Biological Problems of the North* **42** pp 67-73

[7] Kirillov F N 1972 *Fishes of Yakutia* (Moscow, USSR: Nauka)
[8] Kirillov A F 1983 *Strategy of ecological adaptation of whitefish in extreme conditions* (Novosibirsk, USSR: Nauka)
[9] Reshetnikov Yu S 1999 *Features of functioning of freshwater northern ecosystems* In *Adaptation of animals and plants to the conditions of the Arctic seas* (pp. 159-162) (Apatity, Russia: KSC RAS)
[10] Reshetnikov Yu S 2007 *Ichthyofauna of the Arctic: Contemporary studies of the ichthyofauna of the Arctic and southern seas of the European part of Russia* (Apatity, Russia: KSC RAS)
[11] Lee D S, Gilbert C R, Hocutt C H *et al.* 1980 *Atlas of the North American freshwater fishes* (USA: North Carol. State Mus. Natur. Hist.)

Influence of the saline waters of the flowing well on the chemical composition of water and bottom sediments in the Irtysh river and its tributary (Aremzyanka river)

L V Mihailova[1], A A Chemagin[2*] and A S Aleksandrov[3]

[1] Russian Federal Research Institute of Fisheries and Oceonagraphy, 33 Odesskaya str., Tyumen 625027 Russia
[2] Tobolsk Complex Scientific Station UB RAS, 16 Yuri Osipova str., Tobolsk 626150 Russia
[3] Tyumen State University, 6 Volodarskogo str., Tyumen 625003 Russia

E-mail: chemaginaa@yandex.ru

Abstract. The article presents data on the influence of waters from the self-flowing wells in the floodplain of the Aremzyanka river of the first order inflow on the chemical composition of the water in the large transboundary Irtysh river in the south of the Tyumen region (Tobolsk District). The article shows the excess of the background of various elements in the bottom sediments of the studied rivers. The number of ions in the water of the Aramzyanka river is 2 times higher in comparison with the main river. The paper discusses the impact of saline water from a flowing well on a freshwater ecosystem, its components.

1. Introduction

The state of the water bodies of the Ob-Irtysh basin is determined by both the natural features of the hydrographic network and the economic activity. The main problems in the field of protection and use of water resources of the Irtysh river basin are their low quality and water content. At the same time, in the period of 1950-1980s, during exploration works in the territory of the Tyumen region, a large number of exploration wells were drilled. For 30 years after their decomposition, the reservoir mineralized waters through them uncontrollably pour out and gush into the basin of the Irtysh River and its tributaries, which include the Aremzyanka river [1], [2].

This right-bank inflow flows into the Irtysh River at 576 km from the mouth at the border of the following 2 areas (Tobolsk and Uvat). In the left-bank part of the Aremzyanka River on the terrace above the floodplain 9 km from the mouth, one of these wells is No. 36 RG, with a flow rate of 1000 m^3/day and a temperature of 73°C. There is a transformation of the natural state of the rivers as a result of the outpouring of saline water from such wells to the riverbed rivers' ecosystem [2], [3]. Meanwhile, no one takes into account the effect of saline water from the wells when assessing the state of large watercourses, since this issue is poorly studied, which increases the relevance of our research.

In this regard, the purpose of the work is to determine the effect of saline waters on the chemical composition of the waters and bottom sediments of the Aremzyanka and Irtysh rivers.

Content from this work may be used under the terms of the Creative Commons Attribution 3.0 licence. Any further distribution of this work must maintain attribution to the author(s) and the title of the work, journal citation and DOI.
Published under licence by IOP Publishing Ltd

2. Materials and Methods

Studies of the hydrochemical regime of the Irtysh and Aremzyanka rivers and the chemical composition of bottom sediments (BS) were carried out during the autumn low seasons in 2013, 2017, 2018. The concentration of elements in the bottom sediments was compared with background values for the south rivers of the Tyumen region [4]. River water and bottom sediments were sampled using standard methods in accordance with GOST 8.563-2009 [5] and GOST 31861-2012 [6]. For sampling of bottom sediments, a Petersen scooper bottom with a capture area of 0.025 m^2 was used. Chemical analysis of water, bottom sediments and the heavy metals content in them, and other elements were carried out in accredited laboratories of the TKNS UB RAS, FSUE "Gosrybtsentr" (Russia).

3. Results

As a result of studies on the rivers' hydrochemical regime, we found that during the period under study, the sum of the main ions in the Irtysh River in Tobolsk and Uvat districts averaged 197.68 and 190.60 mg / l, respectively (Table 1).

Table 1. The chemical composition of the Irtysh and Aremzyanka rivers in September 2013, 2017, and 2018.

Indicators	Units	Irtysh river		Aremzyanka river, Tobolsk district		MPC	Background (Tura River)
		Tobolsk district, 2013, 5 stations, n = 15	Uvat district, 2013, 2 stations, n = 6	Above the estuary, 2017, n = 2	The estuary of Aremzyanka river, 2018, n = 3		
O_2	mg / l	872 – 11.07 10.09	8.98 – 9.96 9.24	-	9.87 – 11.04 10.43	100%	100%
pH	units	8.75 – 9.14 8.97	8.78 – 9.13 8.94	7.7 – 7.69 7.7	7.44 – 7.60 7.54	6.5-8.5	6.5-8.5
Ca^{+2}	mg / l	43.26 – 61.86 48.2	38.75 – 49.38 44.07	48.1 – 68.14 58.12	40.08 – 64.13 49.43	180	-
Mg^{+2}	mg / l	5.40 – 7.72 6.01	4.83 – 6.16 5.55	20.66 – 24.5 22.48	18.23 – 20.66 19.44	40	-
$Na^{+}K^{+}$	mg / l	-	-	16.2 – 34.94 25.57	3.19 – 10.85 7.04	-	-
HCO_3^-	mg / l	96.0 – 113.53 103.65	101.20-111.0 106.10	286.8 – 280.7 283.7	183.1 – 250.2 207.5	-	-
CO_3^{2}	mg / l	1.88 – 2.55 2.28	1.93 – 3.09 2.23	-	-	-	-
Cl-	mg / l	14.70 – 16.28 15.42	15.55 – 16.14 15.88	38.29 – 38.30 38.3	28.4 – 31.9 29.5	300	14.80
$SO4^{-2}$	mg / l	20.70 – 26.33 22.93	20.6 – 26.0 22.45	3.65 – 4.56 4.11	4.56 – 7.30 6.08	100	9.60
Σion	mg / l	85.87 – 220.78 197.6	149.2 – 241.0 190.6	428.5 – 436.1 432.3	287.9 – 377.3 319.0	<1000	-
Total hardness	0.0200* ppm	1.95 – 2.79 2.18	1.75 – 2.19 1.99	4.4 – 5.1 4.75	3.5 – 4.8 4.07	-	-
N/NH_4^+	mg / l	0.53 – 0.73 0.65	0.79 – 1.12 1.05	0.74 – 0.77 0.75	0.36 – 0.77 0.50	0.40	0.35
N/NO_2^-	mg / l	0.008 -0.017 0.012	0.011 – 0.019 0.015	0.101 – 0.103 0.102	0.038 – 0.127 0.071	0.02	0.014
N/NO_3	mg / l	1.0 – 1.18 1.08	1.03 – 1.24 1.13	1.57 – 1.79 1.68	0.44 – 1.63 1.16	9.10	0.27

PO_4^{3-}	mg / l	0.052 – 0.074 0.062	0.059 – 0.071 0.65	1.10 – 1.11 1.11	0.46 – 1.04 0.9	0.1	-
Fe_{common}	mg / l	0.86 – 1.46 1.05	0.89 – 1.11 1.01	2.63 – 2.73 2.68	-	0.1	0.5
OP	mg / l	8.83 – 10.70 9.83	10.69 – 13.11 11.6	18.4 – 20.0 19.2	16.4 – 23.2 18.7	<10	-
BOD_5	mg O_2/ l	0.88 – 1.46 1.15	0.66 – 1.06 0.91	2.4 – 2.7 2.6	0.81 – 1.84 1.29	2.0	1.9

Note: Permanganate oxidability – OP, biochemical oxygen demand – BOD_5.

Moreover, both the minimum and maximum values were noted in the Uvatsky district below the confluence of the Aremzyanka river – 149.20 and 241 mg/l, respectively. The difference between them was ≈ 60%. The basis of mineralization was bicarbonates, chlorides, sulfates, calcium ions, magnesium, and sodium. Significant differences were noted in the water chemical composition of the Aremzyanka river: the number of ions exceeded this figure for the Irtysh River on average in more than 2 times.

Downstream of the well, this indicator amounted to 432.2 mg/l, a decrease in the concentration of ions occurs (dilution) in the mouth area – 319 mg / l. The most important role in the formation of the ion sum was played by bicarbonates, chlorides, calcium, magnesium, and sodium ions.

It has been shown [2] that the salt ions coming from the catchment area of the Cherkashinskaya 36 RG well into the Aremzyanka river have a significant impact on the ion composition of river waters. In turn, this leads to a change of the bicarbonate class of the calcium group with the chloride class of the sodium group, while the concentration of Na^+ и Cl^- increases significantly. These ions in the water have a high migration capacity [7], while entering the river in large volumes.

Table 2. Heavy metals in bottom sediments of the Irtysh and Aremzyanka rivers in September 2013, 2017.

Indicator, mg / kg	**Irtysh river**		**Tobolsk district**		**Plot (stream) of the water inflow into the Aremzyanka river, 2017, n = 3**	**Background value**
	Tobolsk district, 2013, 5 stations, n = 15	Uvat district, 2013, 2 stations, n = 6	Estuary, 2017, n = 3	Above the estuary, 2017, n = 2		
Ni	2.50-22.14 10.70	1.05-23.91 10.67	20.0-46.0 33.0	6.20-15.0 10.60	5.40 – 47.0 21.13	13.4
Cu	1.49-15.89 5.90	0.53-20.81 9.2	3.90-4.60 4.30	0.77 – 3.40 2.09	3.10-8.20 5.60	10.6
Zn	4.39-41.45 17.90	1.37-56.52 24.24	10.0-11.0 10.50	7.60-11.0 9.3	0.56-11.0 5.92	29.3
Cd	0.04-0.69 0.18	0.40-0.87 0.58	0.23-0.33 0.28	0.20-0.22 0.21	0.32-0.55 0.45	0.6
Pb	0.33-109.20 40.35	0.40-137.70 0.58	4.20-5.30 4.75	1.40-3.0 2.20	5.40-16.0 9.37	13.8
Mn	89.34-723.40 262.5	52.25-1159.0 436.50	309.0-331.0 320.0	65.0-121.0 92.0	234.0- 5480.0 2021.0	291.0
Fe	-	-	-	21290.0	2775.0-6529.0 4152.0	5745.0
Se	-	-	5.70-7.10 6.20	7.10-43.3 22.70	2113.0-3102.0 1561.50	13.8

Note: Background values for the bottom sediments are taken according to [4].

In the study of bottom sediments, for the south rivers of the Tyumen region, the excess of the background of the maximum levels of Ni in the stream flowing into the river Aremzyanka from the well, and the content of Ni is exceeded in 3.5 times in the river Aremzyanka, 1.8 times in the Irtysh river. The values of Mn are higher in 18.8-1.1 and 4.0-2.5 times, respectively. In the bottom sediments of the Irtysh River, the excess over the background is observed in Pb (8-10 times) and Zn (2-1.4 times) in the Uvat and Tobolsk districts (Table 2).

It should be added that the reservoir water, in addition to chlorides, which accumulate in bottom sediments, also contains fluorine, iodine, bromine, boron, strontium, and barium in concentrations significantly exceeding the MPC. At the same time, these components are not controlled by anyone either in the water or in the bottom sediments.

4. Discussion

The ongiong pollution of bottom sediments of the Irtysh river by specific toxic substances, including heavy metals, is associated with transboundary transport from overlying areas (Omsk region, Kazakhstan, PRC). Also, the pollution of bottom sediments is associated with the river fleet and the removal of pollutants from those floodplain areas where various enterprises operate, as well as the discharge of polluted wastewater in the catchment area [8].

In turn, according to [9], the composition of bottom sediments of the lower Irtysh consists in the increased content of those elements whose minerals are resistant to weathering, and floodplain soils have a high content of biogenic cadmium and zinc, alluvial soils-chromium, and nickel. It was shown [9] that the accumulation of trace elements in the bottom sediments of the Irtysh is 1.3-1.6 times more than in the soils of the floodplain, in which the tributary is located (Aremzyanka river). Anthropogenic halogenesis for watercourse functions is a problem with practically unexplored consequences [10], [11]. According to researchers, it has been shown [12] that salinization changes the fundamental physicochemical nature of the soil-water environment, increasing ion concentrations and disrupting the existing chemical equilibrium and solubility of minerals. At the same time, elevated concentrations of dissolved substances, especially sulfates, change the biogeochemical cycle of the main elements, including carbon, nitrogen, phosphorus, sulfur, iron. Also, an increase in the concentration of salts and sulfide causes physiological stress in the biota of floodplain lands, and ultimately can lead to significant changes in natural communities and related ecosystem functions [10-13]. For example, it was established [14] that the ionic stress caused by salinization has an inhibitory effect on both aerobic and anaerobic mineralization of organic substances. For the carp fish [15] and invertebrate hydrobionts [16], it was shown that as a result of changes in salinity, their behavior may change. Also, a decrease in reproductive adaptation of the plankton crustaceans to natural and anthropogenic stressors is observed [17].

5. Conclusion

Thus, our research clearly shows the difference in the chemical composition of the water and bottom sediments of the main Irtysh river and its first-order tributary (Aremzyanka river), which is affected by the mineralized iodine-bromine water of the flowing well. In the studied area of the Aremzyanka river, the number of ions, as well as the chloride content, exceeds these indicators of the Irtysh River in 2 times.

References

[1] Starkov V D, and Tyulkova L A 1996 *Geological history and mineral wealth of the Tyumen region* (Tyumen, Russia: IPP Tyumen)

[2] Svanidze I G, Moiseenko T I, Yakimov A S, and Soromotin A V 2014 The impact of man-made halogenesis on the catchment landscapes of river valleys and water systems (the south of the Tyumen region) *Water Resources* **1** pp 94-103

[3] Kovalenko A I, Aleksandrov A S, and Mikhailova L V 2017 *Ecological danger of geothermal waters of self-flowing wells to the catchment area of the Irtysh River within the Tobolsk District of the Tyumen Region* In *Modern Scientific and Practical Solutions in the Agricultural Sector* (vol 1) (pp 325-337) (Voronezh, Russia)

[4] Mikhailova L V 2006 Calculation of the standard of maximum permissible adverse effects (MPAE) on the Tura River within the Tyumen Region In *Clean water: Abstracts of reports* (pp 26-29) (Tyumen, Russia)
[5] Gosstandardt 2009 *Techniques (methods) of measurement* (GOST P.8.563-2009)
[6] Gosstandardt 2012 *Water: General requirements for sampling* (GOST 31861-2012)
[7] Nikanorov A M 2001 *Hydrochemistry* (St. Petersburg, Russia: Gidrometeoizdat)
[8] Mikhailova L V, Chemagin A A, and Medvedeva I N 2015 Retrospective analysis and the current state of the hydrochemical regime of Irtysh River in the lower reaches *Vestnik of Fishery Science* **2**(6) pp 60-75
[9] Moskovchenko D V 1998 *Oil production and the environment: an ecological-geochemical analysis of the Tyumen region* (Novosibirsk, Russia: Science)
[10] Gonçalves A L, Carvalho A, Bärlocher F, and Canhoto C 2018 Are fungal strains from salinized streams adapted to salt-rich conditions? *Philosophical Transactions of the Royal Society B: Biological Sciences* **374** 20180018 pp 1-11
[11] Gutiérrez-Cánovas C, Sánchez-Fernández D, Cañedo-Argüelles M, Millán A, Velasco J, Acosta R, Fortuño P, Otero N, Soler A, and Bonada N 2018 Do all roads lead to Rome? Exploring community trajectories in response to anthropogenic salinization and dilution of rivers *Philosophical Transactions of the Royal Society B: Biological Sciences* **374** 20180009 pp 1-11
[12] Herbert E R, Boon P A, Burgin J S, Neubauer C R, Franklin B M, Ardón K N, Hopfensperger L P, Lamers M, … Gell P 2015 A global perspective on wetland salinization: ecological consequences of a growing threat to freshwater wetlands *Ecosphere* **206**
[13] Megonigal J P, and Neubauer S C 2019 *Biogeochemistry of tidal freshwater wetlands coastal wetlands* (Elsevier)
[14] Wen Y, Bernhardt E S, Deng W, Liu W, Yan J, Baruch E M, and Bergemann C M 2019 Salt effects on carbon mineralization in southeastern coastal wetland soils of the United States *Geoderma* **339** pp 31-39
[15] Leite T, Santos J M, Ferreira M T, Canhoto C, and Branco P 2019 Does short-term salinization of freshwater alter the behaviour of the Iberian barbel (*Luciobarbus bocagei*, Steindachner 1864)? *Science of The Total Environment* **651**(1) pp 648-655
[16] Botwe P K, Carver S, Magierowski R, McEvoy P, Goonan P, Madden C, and Barmuta L A 2018 Effects of salinity and flow interactions on macroinvertebrate traits in temporary streams *Ecological Indicators* **89** pp 74-83
[17] Stănescu F Székely D Székely P Székely S and Topliceanu D 2017 Cogălniceanu Hydrobiologia: The impact of salinity on early developmental stages in two sympatric spadefoot toads and implications for amphibian conservation in coastal areas *Hydrobiologia* **792**(1) pp 357-366

IOP Conf. Series: Earth and Environmental Science **395** (2019) 012041 doi:10.1088/1755-1315/395/1/012041

Modern computer technologies in Biology: a scientific depository of phytodiversity of the transboundary territory of the Altai highland country

A Vaganov[1,2]

[1] Altai State University, 61 Lenina prosp., Barnaul 656049 Russia
[2] Sakhalin Branch of the Botanical Garden-Institute FEB RAS, 25 A.M. Gorky str., Yuzhno-Sakhalinsk 693023 Russia

E-mail: vaganov_vav@mail.ru

Abstract. In the modern period of biodiversity development, when studying the territory of the mountainous countries of the world, the active influence of such trends as genomics and digitalization is noted. Within Russia, Kazakhstan, Mongolia and China, the Altai highland country (AHC) is the highest uplift of North Asia, its biodiversity level is three times richer than it is in the West Siberian Plain. The collected material in the herbarium collections of the Altai State University (ASU, Barnaul) is more than 450 thousand herbarium leaves (ALTB Herbarium), and it is the biological repository of unique genetic material on AHC, along with other large depositories of the world (herbariums, zoological collections, etc.). The concerted actions of many countries of the world have already allowed to aggregate large data on the biodiversity of many mountain countries in order to ensure the sustainable development of these unique natural complexes. During our project, ASU scientists have developed a scientific depository with the possibility of integrating bio data into the Global Biodiversity Information Fund (GBIF). Today, data from ASU and other GBIF operators allow digital resources to be integrated across the whole spectrum of living organisms in the AHC, from genes to ecosystems. They allow linking them to issues important to science and society using GIS tools. Also, in this study, we produced a modern botanical and geographical zoning of the territory of the AHC, using open GIS systems (GeoJSON).

1. Introduction

Biological diversity is an essential component of the environment, directly and indirectly affecting the state of global ecosystems and the planet's climate. Activities for the study and preservation of living organisms should be carried out according to international standards and at the highest methodological level. As this process directly influences strategic national priorities related to health and food security. The publication of data on plant distribution within the Altai highland country (Russia, Kazakhstan, Mongolia, China) in the global biodiversity information system (GBIF.org) is one of the indicators of the country's active work in the field of biodiversity informatics at the level of contemporary standards. The Altai Region and the territory of the Altai highland country are included in the 200 key ecoregions of the world, to which the attention of the United Nations and the scientific world community are currently drawn. One of the key priorities of the UN is the conservation and study of biological diversity.

Content from this work may be used under the terms of the Creative Commons Attribution 3.0 licence. Any further distribution of this work must maintain attribution to the author(s) and the title of the work, journal citation and DOI.
Published under licence by IOP Publishing Ltd

An important basis for organizing activities for the study and preservation of biological diversity and maintaining the integration processes of data on biological diversity in the world is the *Convention on Biological Diversity* (CBD) (https://www.cbd.int). In 1995, the Convention was ratified by Russia. The Article No. 17 of the Convention postulates the need to pool the efforts of all countries to assess their own biological resources. Obviously, the study of biological diversity at the world level or the level of large regions is possible only through the concerted actions of many countries. In 2010, the Parties to the Convention adopted a *Strategic Plan for the Conservation and Sustainable Use of Biodiversity for 2011–2020*, together with targets (the so-called Aichi Biodiversity Targets – https://www.cbd.int/sp/targets). The Target No. 19 of this list aims at improving, sharing and applying knowledge, scientific data and technologies in the field of biodiversity research. This task is impossible without further digitization of biological collections and the discovery of data on the distribution of species and the dynamics of their ranges [1].

The AHC (Altai highland country) among the continental mountainous countries of Siberia (and in general in Northern and Central Asia) is at the same time both the highest contemporary uplift and the space in Siberia most favorable for the development of organic life. Due to the fact that the latitudinal and meridional stretching intermontane depressions, ridges, high-altitude plateaus, areas of typical low-hill areas and vast highlands are combined here, the relief is considered extremely complex [4], [5].

With its relatively small territorial area, the AHC is inhabited by a large number of biological objects. D. Olson and E. Dinerstein [2] rank the territory of the Altai and Sayans among the two hundred priority ecoregions of the world for global biodiversity conservation. They note that in the 200 ecoregions that make up only 10% of the land area, up to 90% of the planet's biodiversity is concentrated.

2. Materials and Methods

In December 2017, the ASU received an official registration in the GBIF network (www.gbif.org/publisher/943a5811-d56e-4c37-853d-bd64957d3833, see also the agreement with the data publisher GBIF). To date, the university has published 5 databases (Datasets GBIF). The key dataset of ASU is "Virtual Herbarium ALTB (South-Siberian Botanical Garden)" (Occurrence dataset https://doi.org/10.15468/y6xmme). The herbarium of the Altai State University (ALTB) is the core of the current scientific depository on phytodiversity of the AHC [3].

In order to unite researchers-systematists studying the flora of the AHC, in 2018, the "Flora of Altai" project was launched and posted in the Internet at the following address: www.altaiflora.asu.ru. The site is based on the Word Press platform. It is an open source content management system. For the preparation of the section containing the AHC map (altaiflora.asu.ru/kart/AHC/) with nineteen botanical and geographical areas according to R.V. Kamelin (2005), we have drawn 19 polygons using the WP Google Maps plugin. The basis for drawing polygons is topographic maps on a scale of 1: 500,000. The obtained sets of coordinates of all polygons were edited in the modern specification of creating geographic structures GeoJSON. GeoJSON is an open format for storing geographic data structures. The format is based on JSON (Java Script Object Notation); JSON is a text-based data interchange format. The AHC boundary, including 19 botanical and geographical areas (polygons) in the GeoJSON format, was obtained.

Directly, the procedure for publishing data through GBIF.org is carried out using special IPT software (Integrated Publishing Toolkit), written in Java and functioning as a server application with a visual interface accessible via a browser. The IPT running Apache web server and TomCat 7 web application services are also used. Label tables for empirical depository materials are integrated into GBIF via IPT in Darwin Core format.

3. Results

As of now (April 10, 2019), the ASU has transferred to the GBIF a total of 5075 records with a digital image of the herbarium leaf and geolocation. So far, the resulting registry with records of species locations in the AHC for the "Flora of Altai" project and the global scientific community mostly includes representatives from such plant families as in *Caryophyllaceae, Ranunculaceae, Fabaceae, Liliaceae*.

Sets of maps with accurate geolocation of the spatial distribution of rare, endemic, invasive species of economic value in the Altai Region and the Altai highland country have been obtained. A total of eight final papers with an interdisciplinary approach, using a variety of techniques from the field of information technology, GIS technology and bioclimatic modeling, have been prepared.

In the course of conducting the "Virtual Herbarium ALTB (South-Siberian Botanical Garden)" dataset, taxonomists identify new plant locations, describe new species, monitor the dynamics of habitats to obtain information about the predictive locations of promising plant species. Due to the active uploading of data on biota of AHC, research teams enter into partnership with other international organizations, such as: Catalogue of Life partnership, Biodiversity Information Standards (TDWG), International Barcode of Life project (iBOL), Encyclopedia of Life (EOL) и GEO.

4. Discussion

In the coming years, besides the increase in the number of publications in the GBIF, we also expect an increase in citation of datasets, which will eventually lead to an increase in Altai State University in the world rankings (THE, QS and Webometrics), due to the increasing interest of research teams in empirical data on AHC. As of today, the "Virtual Herbarium ALTB (South-Siberian Botanical Garden)" dataset also solves many other important problems: the study of the flora and vegetation of the AHC and individual taxa; the preservation of fragile and aging material in digital form; continuation of the publication of the printed multivolume report "Flora of Altai".

5. Conclusion

In total in our country, 38 organizations (of which only 6 universities) are the official publishers publishing data in the GBIF. At the same time, the contribution of operators of Russia to the data placed in the GBIF is no more than 0.5%, which indicates to us the enormous amount of work, given the scale of our country.

6. Acknowledgements

This work has been supported by the Russian Foundation for Basic Research (19-44-220004_ p_a).

References

[1] Filippova N V, Filippov I V, Schigel D S, Ivanova N V, and Shashkov M P 2017 Biodiversity informatics: global trends, national perspective and regional progress in Khanty-Mansi Autonomous Okrug *Environmental Dynamics and Global Climate Change* **8**(2) pp 46-56

[2] Olson D M, and Dinerstein E 2002 The global 200: Priority ecoregions for global conservation *Annals of the Missouri Botanical Garden* **89**(2) pp 199-224

[3] Vaganov A V 2018 Prospects for studying and preserving the biodiversity of the Altai-Sayan ecoregion in the context of the development of open databases *Ukrainian Journal of Ecology* **8**(2) pp 380-382 DOI: 10.15421/2018_357

[4] Kamelin R V 1998 *Materials on the history of flora of Asia* (Altai mountain country) (Barnaul, Russia: Publishing house Altai University)

[5] Kamelyn R V Ed 2005 *Flora Altaica* (Barnaul, Russia: Az Buka)

IOP Conf. Series: Earth and Environmental Science **395** (2019) 012042 doi:10.1088/1755-1315/395/1/012042

A tourist and recreational potential of the Kulunda steppe as the basis for the diversification of its economy

D Dirin[1], E Krupochkin[1] and E Rygalov[1]

[1] Altai State University, 61 Lenina prosp., Barnaul 656049 Russia

E-mail: denis_dirin@mail.ru

Abstract. This article presents the results of a comprehensive assessment of the tourist and recreational potential (TRP) of the Kulunda steppe (within the Altai Region). Evaluation methodology was used in the paper, which allowed to take into account both the natural and recreational resources of the territory, as well as the socio-economic and socio-cultural conditions for developing tourism and recreation were analyzed. Evaluation and mapping of the territory's TRP was carried out using the functionality of the GIS MapInfo and ArcGIS. Based on the selected cores of the TRP concentration, a system of tourist and recreational clusters of various specializations was designed. These clusters should become nodes of the basic recreational framework of the Kulunda steppe.

1. Introduction

The Kulunda steppe is located in the transboundary space of Russia and Kazakhstan. After the collapse of the USSR, the following typical socio-economic problems of the new border area of Russia were revealed in the Kulunda steppe: breaking economic and infrastructural ties; aboliting large agricultural enterprises; rising unemployment; reducing budget financing of agriculture and industry; lower funds for social facilities that were on the balance of collective and state farms; sharply declining living standards. The consequence of these processes was the migration outflow of the population, the growth of negative social phenomena, such as alcoholism, drug addiction, and crimes.

One of the key incentives for the sustainable development of the Kulunda steppes can be the diversification of their economies, through the introduction of new rapidly developing sectors of the economy that can not only produce products that are in demand in today's market conditions, but also create a large number of jobs and also contribute to the social development of rural areas. One of such industries is tourism and recreation. For its development, a detailed assessment of the tourist and recreational potential (TRP) of the territory is necessary, especially in order to identify the localization of recreational resources.

2. Materials and Methods

The score method developed by the authors was applied to assess the tourist and recreational potential (TRP) of the Kulunda steppe (within the Altai Region of the Russian Federation) [1]. The estimated indicators and specific weights of each of them were selected based on the method of expert assessments. Potentially significant indicators identified by experts were presented in the form of a matrix, in separate columns of which different variants of manifestation of each of them and the corresponding score in points are indicated (Table 1).

Content from this work may be used under the terms of the Creative Commons Attribution 3.0 licence. Any further distribution of this work must maintain attribution to the author(s) and the title of the work, journal citation and DOI.
Published under licence by IOP Publishing Ltd

Table 1. The system of indicators for assessing the tourist and recreational potential of the territory.

The name of indicators	Points			k weighted
	0	1	2	
Horizontal unbundling relief km / km^2	>0.8	2.5 – 0.8	> 2.5	1.2
Vertical dismemberment of the relief, m	< 300	300 – 800	> 800	1.8
Average January temperature, °C	<-25	-19 – -24	-9 – -18	1.3
Average July temperature, °C	11 – 15	16 –19	20 – 25	1.7
Average annual precipitation, mm	<300	>500	300 – 500	1.2
Duration of the period of stable snow cover, days	0 – 140	140 – 160	> 160	1.3
Duration of the period with average daily air temperature above 0 °C, days	up to 160	160 – 190	> 190	1.0
Coastline length, water bodies, km	Absent	Up to 2 km	> 2 km	1.0
Water and mineral sources, units	Absent	1	> 1	2.0
Objects of mud, units	Absent	1	> 1	2.0
Species / landscape diversity, units	1	2	> 2	1.5
Types of medicinal and food plants	Absent	1	> 1	1.0
"Red Book" plant species, units	Absent	1	> 1	1.3
Commercial animal species, units	Absent	1	> 1	1.4
"Red Book" animals, units	Absent	1	> 1	1.6
Number of nature monuments, units	Absent	1	> 1	0.9
Specially protected natural areas (coverage area, %)	Absent	Up to 50	> 50	1.1
Monuments of history and archeology, objects of cultural heritage, units	Absent	Local value	Regional and federal significance	1.4
Architectural monuments	Absent	Local value	Regional and federal significance	1.5
Objects of cultural leisure (museums, theaters, etc.)	Absent	Local value	Regional and federal significance	1.8
Objects of sports tourism, units	Absent	1	> 1	1.4
Objects of medical and health tourism are sanatoriums, etc., units	Absent	1	> 1	1.9
Trade enterprises	Absent	Point of Sale	Store Network	1.0
Catering, units	Absent	1	> 1	1.0
Equipped hunting and fishing farms, units	Absent	1	> 1	1.2
Length of the transport network by category, km	Country road	Paved road; dirt road	Railway; federal highway	1.6
Presence of railway stations, bus stations, units	Absent	1	>1	1.8
Roadside infrastructure (gas stations, service stations)	Absent	1	>1	1.5
Capacity of collective tourist accommodation facilities	Absent	Up to 20 people	> 20 people	1.8

In order to identify the patterns of spatial differentiation of the tourist and recreational potential of the Kulunda steppe, the entire study area was divided into a network of operational territorial cells (OTC). The calculations showed that the optimal form of OTC is hexagonal, with the following cell parameters: a side (edge length) is 3 km, a circumscribed circle radius is 3 km, and an area is 23.3 km^2 [2]. All primary information was presented in the form of a relational database (DB) in the environment of geographic information system (GIS) MapInfoPro.

Integral indicators of tourism and recreational potential are defined as the sum of all factors calculated from OTC. Their calculation, according to the main thematic sections and their conversion into points, was made in the MapInfo Pro GIS environment by digital layers and the developed database. The indicators selected using the "Request" and "Sampling" operations were transferred to a point system and then saved to the base.

In the future, on the basis of the constructed map, which represents a set of operational cells filled with assessment information, the isolineous map for assessing the tourist and recreational potential of the territory is being developed. In this case, the isolines reflect the patterns of territorial differentiation of the value of tourist and recreational potential.

3. Results

The result of the assessment was the map of the tourist and recreational potential of the Kulunda steppe (Fig. 1). Its analysis allows to draw conclusions about the main features and patterns of territorial differentiation of the tourist and recreational potential in the Kulunda steppe.

Climatic conditions are favorable for summer holidays. There are the sunniest days in all Siberia, and the amount of incoming solar radiation is similar to that of the southern coast of Crimea. Microclimatic conditions in the belt forests and on the adjacent steppe areas are most favorable.

The most important recreational resource of the territory is the numerous lakes with different chemical composition, many of which are characterized by favorable conditions for beach-bathing holidays (comfortable water temperature, sandy beaches, even bottom, etc.). Many salt lakes have reserves of curative mud, while fresh and alkaline lakes are excellent hunting and fishing grounds. The unique intrazonal relict formations are ribbon pine forests stretching in parallel ribbons from the northeast to the southwest along the hollows of the ancient glacier runoff. They cross the Kulundinsky plain. In addition to the favorable microclimatic conditions, the ribbon forests with their rolling relief create good conditions for active types of tourism and recreation (skiing, cycling and walking, etc.). A high biodiversity allows you to develop commercial types of recreation (hunting, fishing, picking mushrooms, berries, valuable plants).

The main "nuclei of concentration" of the tourist and recreational potential of the Kulunda steppe are the areas characterized by a combination of several types of recreational resources and favorable socio-economic factors. Such areas act as potential tourist-recreational clusters and nodes of the reference recreational framework of the territory (Fig. 2).

4. Discussion

Based on the analysis, we can state that the development of tourism and recreation can be the basis for diversifying the economy of the Kulunda steppe. Resource potential is available for this and it must be used effectively. The main types of recreational specialization in the territory in question should be mud therapy, beach swimming rest, hunting and fishing with related activities.

Based on the assessment of the tourist and recreational potential of the Kulunda steppe, the formation of the territorial tourist and recreational system of Kulunda was determined and the main nodes (tourist centers and destinations) and axis (transport communications) of its support frame were determined (Fig. 2).

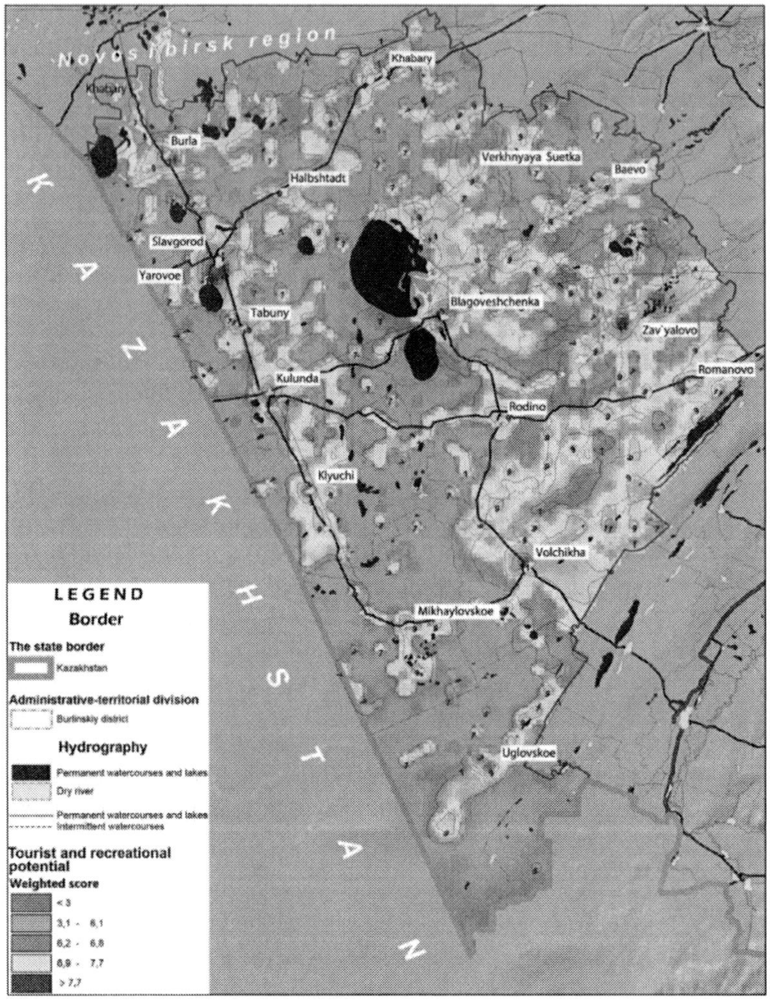

Figure 1. Map of the integrated assessment of the tourist and recreational potential of the Kulunda Steppe [compiled by the authors].

In total, there are 8 potential tourist clusters in the territory under consideration, grouped into three types by specialization.

The ozerno-borovoy beach-bathing type (No. 1) is formed on the basis of large lakes, stretched in a chain along the Barnaul and Kasmalinsky ribbon burs. The main specialization is beach swimming rest, water activities (yachting, skyting, etc.), recreational fishing (hunting, fishing, picking mushrooms and berries) and ecological tourism. This type includes one cluster (Volchikhinsky).

The ozerno-stepnoy mud treatment type (No. 2) is formed on the basis of steppe salt lakes with curative mud and brine as the main healing factors. The recreational specialization is mud therapy, balneology, climate therapy, beach swimming rest, water activities, and sometimes also commercial recreation. There are 5 clusters of this type: Yarovskaya, Zavyalovsky, Romanovsky, Mikhailovsky, amd Blagoveshchensky.

The fishing and recreational type (No. 3) is formed on the basis of the richest biological resources of hunting and fishing grounds. Allocation of the Proslaukhinsky and Burlinsky tourist clusters is proposed.

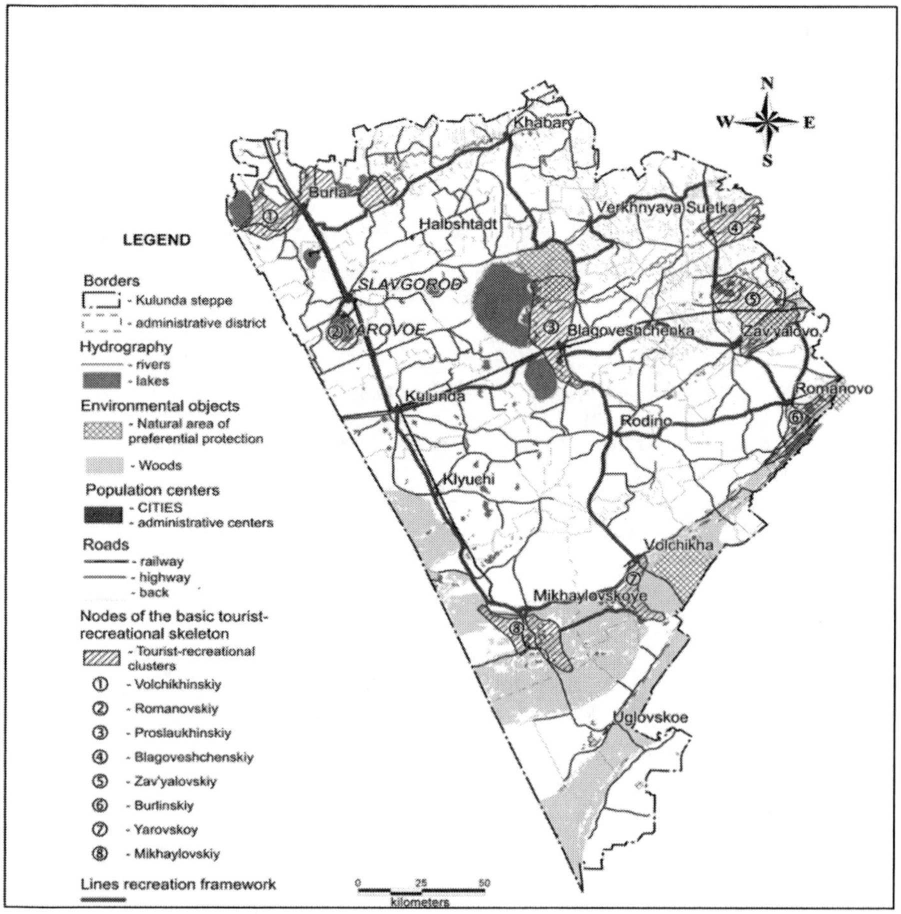

Figure 2. The reference frame of the territorial tourist and recreational system of the Kulunda Steppe.

5. Conclusion

The study leads to a number of conclusions. First, the territory of the Kulunda steppe has a tourist and recreational potential sufficient for the effective development of the tourism industry as one of the significant sectors of economic specialization. Second, the basis of the tourist and recreational potential of Kulunda consists of climatic conditions, significant balneological (curative mud and mineral waters) and biological (hunting, fishing, etc.) resources, waters and shores of numerous lakes, as well as unique ecosystems of ribbon pine burs. Also, an important factor is the relatively good infrastructure development of the territory. Third, a comprehensive assessment of the tourist and recreational potential of the territory revealed 8 promising areas for the creation of tourist and recreational clusters of various specializations. Investments in tourism infrastructure and in improving transport accessibility are necessary for their formation and development.

References

[1] Dirin D A, Krupochkin E P, and Golyadkina E I 2014 Methods of integrated assessment of the tourist and recreational potential of the region *Geography and Nature Use of Sibiria* **18** pp 64-78

[2] Krupochkin E P, Dirin D A, Dunec A N, and Rygalov E V 2016 Quantitative substantiation of the parameters of regular-cellular models as a method of numerical evaluation and GIS-mapping of territories *Polzunovskiy Vestnik* **4** pp 70-79

IOP Conf. Series: Earth and Environmental Science **395** (2019) 012043 doi:10.1088/1755-1315/395/1/012043

Unwanted appearance of invasive species in the transboundary Irtysh river basin

L S Leskovskaya[1*] and L V Mikhaylova[1,2]

[1] State Agrarian University of Northen Trans Urals, 7 Republic str., Tyumen 620005 Russia
[2] Tyumen branch of the FSBSI "VNIRO", 33 Odesskay str., Tyumen 620023 Russia

E-mail: LL-78@mail.ru

Abstract. This paper presents data on the size and weight characteristics of the Amur sleeper (*Perccottus glenii*) in floodplain reservoirs of the transboundary Irtysh river basin: Bezymyannoe Lake, Amalby Lake, Mangadaevo Lake (Yarkovsky District), and Andreevskoe Lake (Tyumen District) of the Tyumen Region.

1. Introduction

The Irtysh River is the largest tributary of the Ob River, it originates in the mountains of the Mongolian Altai and, being called as "Black Irtysh," flows until it flows into the Zaisan Lake, then it continues its way through China, Kazakhstan, Omsk, and Tyumen Regions. The area of the Irtysh basin is 1.64 million km^2, 0.52 million km^2, or about 32% of them are drainless areas [1].

The appearance of such an invasive species as the *Perccottus glenii and Dybowski* (1877) in the basin of the transboundary Irtysh river leads to undesirable consequences for the aboriginal structure of the ichthyofauna. Features of the Irtysh River basin and its floodplain reservoirs contribute to the penetration of a typical representative of alien ichthyofauna and the creation of sustainable *Perccottus glenii* populations. On the territory of the basin, a large number of dead lakes. They are well heated due to the shallow depth with stagnant water and an abundance of higher aquatic vegetation. These lakes are a typical habitat of native species, such as the silver and gold crucian carp, which are the only representatives of commercial ichthyofauna due to their high resistance to oxygen deficiency [2], [3], [4], [5], [6]. The Amur sleeper (*Perccottus glenii*) is well accustomed to these lakes and began to crowd out the local ichthyofauna [2], [6]. Because of its biological characteristics, in particular, undemanding to environmental factors of the environment, including oxygen content, and also due to the lack of a sufficient number of predators (pike, pike-perch, perch) and parasites characteristic of the natural range, the Amur sleeper expansively develops in water bodies and water courses [7-9].

The natural range of this species is the Amur region, Primorsky region, northeastern China and north Korea [10-12]. At the beginning of the twentieth century, he was introduced by aquarists to the European part of Russia, from where he actively settled in shallow reservoirs. In the basin of the Irtysh river, a "golovoshka" type of the Amur sleeper (*Perccottus Clenii* Dybowski) appeared relatively recently [13], [14].

The purpose of this work is to study the characteristics of a "golovoshka" type of the Amur sleeper (*Perccottus Clenii* Dybowski) from the water bodies of the basin of the transboundary Irtysh river.

Content from this work may be used under the terms of the Creative Commons Attribution 3.0 licence. Any further distribution of this work must maintain attribution to the author(s) and the title of the work, journal citation and DOI.
Published under licence by IOP Publishing Ltd

2. Materials and Methods

The material was collected for research on the water bodies of the Irtysh River basin in the south of the Tyumen Region. In particular, we obtained our materials in the Yarkovsky District in 2011, 2016, and in the Bezymyanoye, Amalba and Mangadaevo Lakes in 2017, as well as in the Tyumen region of Andreevskoe Lake in 2015. The lakes are characterized by a high overgrowth of the higher aquatic vegetation. Trapping was done by fixed nets with different meshes. The size and age structures were determined, along with the sex composition. Studies of ichthyological materials were carried out according to standard ichthyological methods [15], [16]. Statistical data processing was carried out according to the standard data processing method [17].

3. Results

The age composition of our samples in catches of floodplain lakes was represented by individuals from 2/2+ to 6/6+. In the Bezymyannoe and Mangadaevo Lakes, the prevailed age groups was 2/2+ to 4/4+, as well as from 3/3+ to 5/5+ in Andreevskoe and Amalba Lakes (Table 1).

Table 1. The size and age structure of the Amur sleeper in floodplain water bodies of the Irtysh River basin.

Lake, District, Date	Indicator	Age					Average Indicators
		2/2+	3/3+	4/4+	5/5+	6/6+	
Bezymyannoye, Yarkovsky, Autumn 2011	l±mx, cm	8.0±0.3	12.0±0.6	14.2±0.6	16.9	17.7	11.7±0.4
	M±mx, r	16.3±2.7	52.6±5.7	76.1±7.5	126.3	145.5	52.6±4.3
	CVl / CVm	23.3 / 83.2	32.0 / 73.3	21.9 / 49.5	-	-	34.3 / 82.1
	n, instances	26	46	25	3	1	101
Andreyevskoye, Tyumen, Autumn 2015	l±mx, cm	-	12.6±0.1	13.3±0.1	15.7±0.4	18.5±1.4	14.3±0.3
	M±mx, r	-	48.3±2.3	59.9±2.2	100.1±6.1	203.6±43.1	80.2±6.6
	CVl / CVm		4.7 / 20.5	4.9 / 18.9	9.8 / 25.8	17.7 / 51.9	15.9 / 68.1
	n, instances	-	18	26	18	6	68
Amalby, Yarkovsky, Spring 2016	l±mx, cm	10.4	11.7±0.2	13.9±0.3	15.2±0.4	17.0	13.2±0.3
	M±mx, r	34.3	47.1±2.4	82.6±5.6	107.3±9.5	154.1	71.9±4.6
	CVl / CVm	-	8.3 / 24.3	7.8 / 27.8	8.1 /26.6	-	13.9 / 45.6
	n, instances	1	22	17	9	1	50
Mangadaevo, Yarkovsky, Autumn 2017	l±mx, cm	9.7±0.04	12.6±0.1	15.2±0.07	15.9±0.49	18.7	12.4±0.36
	M±mx, r	24.5±0.34	61.3±1.6	110.3±1.43	136.9±10.7	234.24	66.1±6.41
	CVl / CVm	9.9 / 33.7	13.2 / 43.5	4.9 / 14.3	7.5 / 19.1	-	22.7/74.5
	n, instances	24	17	11	6	1	59

Note: "l" is the fishing length; "mx" is the average error; "M" is the total mass of fish; "n" is the number; "CVl" is the coefficient of variation in length; "CVm" is the mass variation coefficient.

The ratio of females and males in the samples from the lakes was as follows: in Bezymyannoe Lake, 73 specimens are females out of 101 specimens, and 28 specimens are males; in Amalby Lake, out of 50 specimens, 30 specimens are females and 20 specimens are males. That is, in both lakes, females prevailed (72 and 60%, respectively).

In Andreevskoe Lake, out of the 60 Amur sleepers being captured, 27 specimens are males and 33 specimens are females. In Mangadaevo Lake, the proportion of males was 52.5% (31 specimens), with females of 47.5% (28 specimens), respectively. That is, the sex ratio is 1: 1. (Figure 1).

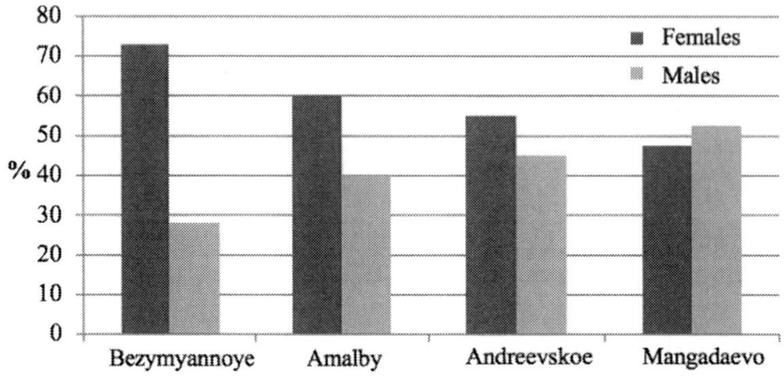

Figure 1. The percentage ratio of males and females in the named reservoirs of the Irtysh river basin.

When comparing the size and weight characteristics of the Amur sleepers from the lakes, it can be seen that length variations in all age groups are within the same limits. The variability of signs along the length in all water bodies is weak or medium (CV<10%, 10%<CV<25%) with the exception of the age group 3 - 3+ of the Bezymyannoye Lake. Mass fluctuations begin with an age group of 4/4+. By mass, the variability is medium or high (CV>25%), especially in the Bezymyannoye Lake in age groups of 2/2+ and 3/3+, where CV is equal to 83.2 and 73.3%, respectively. This is due to the large variation in weight from 13.5 to 47.8 g (2/2+) and from 27.3 to 98.2 g (3/3+).

4. Discussion

The existence in the studied reservoir of different age groups of the Amur sleepers indicates its acclimatization to the conditions of the region and the process of penetration of this invasive species into the floodplain reservoirs of the Irtysh and Ob rivers and the rivers themselves, which is confirmed by literary data [18], [19].

Comparing our data with the data of other authors [20], [21], we note that the linear and weight characteristics of the Amur sleepers inhabiting floodplain water bodies of the Irtysh river are higher than those of natural habitat.

Since 2007, the inclusion of the Amur sleeper in the field statistics of the Tyumen region indicates an increase in the number of species. Since this year, this fish is included in the forecast of the size of the allowable catch of the species in the water bodies of the Tyumen region. The Amur sleper became an ordinary representative of ichthyofauna of the Karas Lakes and such rivers as Pyshma, Tura, Iset, Irtysh [2], [9].

5. Conclusion

Thus, the naturalized in the floodplain waters of the transboundary Irtysh river basin of the Tyumen region. The main size and weight characteristics of this invasive species are relatively high compared to its natural range. A wide range of variation of these indicators indicates a high adaptive capacity, which leads to an increase in the number and inclusion in the field.

References

[1] Lezin V A 1999 *Rivers of the Tyumen region (southern regions)* (Tyumen, Russia: Vector Buk)

[2] Leskovskaya L S, Taskaeva K R, Yankova N V, and Petrachuk E S 2017 Spreading, biology and fishing for rotan *Perccottus glenii*, Dybowski, 1877 in water bodies of the Tyumen region *Agricultural and Food Policy in Russia* **3** pp 74-78

[3] *Atlas of the freshwater fish of Russia* 2002 (Moscow, Russia: Nauka)

[4] Krivoschekov GM 1953 Karas of Western Siberia *Proceedings of the Baraba department VNIORKh* **6**(11) pp 28-40

[5] Yankova N V, Niyazov N S, and Mukhachev I S 2006 Rational use of reserves of crucian carp in water bodies of the Tyumen region *Fish Industry* **5** pp 53-55

[6] Taskaeva K R, Yankova N V, Sladkova V V, and Shelyagin K E 2017 The current state of the ichthyofauna

of Bolshoy Taraskul Lakeof the Tyumen District *World of Innovation* **2** pp 79-83

[7] Elovenko V T 1985 Rotan – a new species in the waters of Europe *Fish Farming and Fishing* **11** pp 3-4

[8] Leskovskaya L S, and Mikhailova L V 2015 The spread of rotan *Perccottus glenii*, Dybowski, 1877 in reservoirs of Western Siberia *Young Scientist* **6-5**(86-5) pp 172-174

[9] Chemagin A A 2014 Spreading rotan-goloveshka *(perccottus clenii* Dybowski, 1877) in the name-lakes of the Lower Irtysh *Basic Research* **11-12** pp 2656-2660

[10] Taranets L Ya 1937 A brief guide to the fish of the Soviet Far East and adjacent waters *Izvestiya of the Pacific Research Institute of Marine Fisheries and Oceanography* **11** pp 1-148

[11] Berg L S 1949 *Freshwater fish of the USSR and neighboring countries* (Leningrad, USSR: Publishing House of the Academy of Sciences of the USSR)

[12] Nikolsky G V 1956 *Fish basin of the Amur: Results of the Amur ichthyological expedition of 1944-1949* (Moscow, USSR: Publishing House of the Academy of Sciences of the USSR)

[13] Reshetnikov A N 2001 Influence of introduced fish of rotan *Perccottus glenii* (*Odontobutidae, Pisces*) on amphibians in small water bodies of the Moscow region *Journal of General Biology* **62**(4) p 361

[14] Babueva R V 2001 The ichthyofauna of the waters of the Ob-Irtysh interfluve In *Current Problems in Hydrobiology of Siberia* pp 83-84 (Tomsk, Russia)

[15] Pravdin I F 1966 *Guidelines for the study of fish* (Moscow, USSR: Food Industry)

[16] Kafanova V V 1984 *Methods for determining the age and growth of fish* (Tomsk, Russia: Tomsk University Press)

[17] Lakin G F 1990 *Biometrics* (Moscow, USSR: High School)

[18] Chemagin A A 2014 Introductive distribution of the species of rotan-goloveshka (*Perccottus Clenii* Dybowski, 1877) in the Poymenno-Ruslovoy complex of the Lower Irtysh *Basic Research* **11-11** pp 2449-2454

[19] Zhuravlev V B 2012 Morphological and ecological characteristics of rotan (Percottus glenii) in floodplain water bodies of the upper course of the Ob *Vestnik of the Novosibirsk State Agrarian University* **2**(23) pp 28-31

[20] Nikolsky G V 1956 *Fishes of the Amur basin* (Moscow, USSR: Publishing House of the Academy of Sciences of the USSR)

[21] Kasyanov A N, and Goroshkova T V 2012 Study of morphological features in *Rerccottus glenii* (*Rerciformes, Eleotrididae*) introduced into the waters of the European part of Russia *Siberian Journal of Ecology* **19**(1) pp 81-96

IOP Conf. Series: Earth and Environmental Science **395** (2019) 012044 doi:10.1088/1755-1315/395/1/012044

The possibilities of using remote sensing data when monitoring cross-border regions

E Khlebnikova[1*] and G Simonova[1]

[1] Siberian State University of Geosystems and Technologies, 10 Plakhotnogo str., Novosibirsk 630108 Russia

E-mail: e.p.hlebnikova@sgugit.ru

Abstract. The article shows the possibility of using the results of remote sensing data processing for the purpose of geo-information support for monitoring, evaluating, and forecasting the development of cross-border regions.

1. Introduction

For the near future, the dynamics of the spatial development of individual regions of the country is connected with the domestic and foreign practice of cross-border cooperation. The intensification of cross-border interaction affects not only the economic development of the region, but also the political trends of the state. The development of global geopolitical systems requires reliable information on the impact of cross-border formations on neighboring territories.

The components of a transboundary region are the same signs that are generally characteristic of the region, namely: economic entities, geodemographic and socio-territorial communities, ecosystems, settlements.

The cross-border areas of interaction can significantly change the economic and social structure of a territory, since they are often peripheral formations of neighboring states. Effective use of the modernization process requires adequate resources, and for the effective management of these resources, objective information about the state of the region is needed. The main resources for cross-border cooperation are economic feasibility and geographical location [1]. Russia's strategy in implementing transboundary capabilities should be based not only on resource, infrastructure and institutional capabilities, but also on innovative information projects.

The recently adopted *Digital Economy of the Russian Federation* program shows the need to create a domestic digital platform for collecting, processing, and distributing spatial data to solve various political and economic problems. [2]. In accordance with the above concept, the use of modern methods of quickly obtaining objective information can be very useful for the effective management of territories and ensuring their sustainable development.

2. Materials and Methods

The presence of transboundary situations stimulates a change in ground infrastructure, contributes to the emergence of traffic flows, new settlements, objects of economic activity, which entails a change in the natural situation in the region. The diversity of the observed processes makes it difficult to monitor and analyze the situation as a whole.

Content from this work may be used under the terms of the Creative Commons Attribution 3.0 licence. Any further distribution of this work must maintain attribution to the author(s) and the title of the work, journal citation and DOI.
Published under licence by IOP Publishing Ltd

Contemporary means of remote sensing allow to obtain objective information regarding the dynamics of transboundary territories, and to carry out a comparative analysis of their development with almost any time interval and with a given accuracy of determining spatial coordinates. Images can be obtained in various fairly narrow areas of the electromagnetic spectrum with high spectral resolution and ultra-high spatial resolution. Pictures can differ in a different field of view, and in this area we can get pictures of both high detail and large coverage.

For example, Figure 1 shows pictures of one of the mineral deposits of the Trans-Baikal Region with an interval of four years (2012–2016).

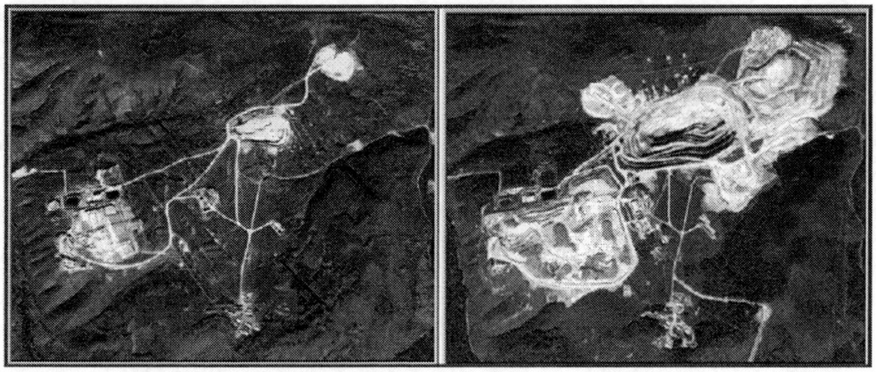

Figure 1. Mineral deposits, Trans-Baikal Region, a four-year shooting interval [3].

Figure 2 shows images of the border area of the North-East of the People's Republic of China with a time interval of seven years (2011–2018).

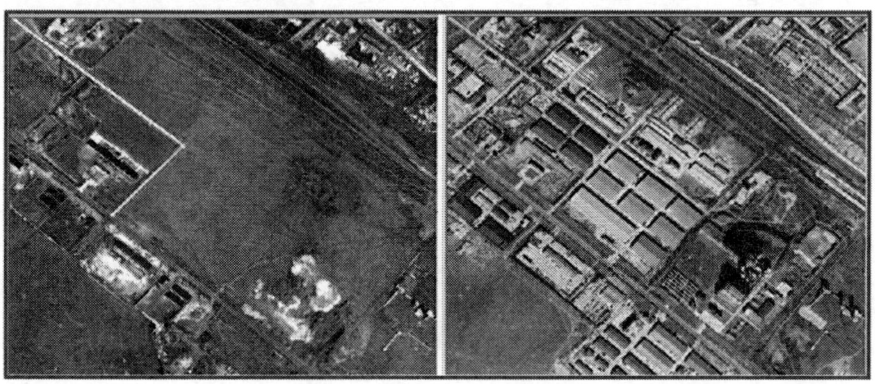

Figure 2. Border area of the North-East of the People's Republic of China, Manchuria, a time interval of seven years [3].

The given images are geolocated, they visually observe a significant change in the state of the territory; however, it is difficult to identify small tumors in this way and the quantitative characteristics of the transformation of the plots cannot be given.

For processing multi-time snapshots, various methods of automated decryption are used to increase their informativeness [4]. For example, the identification of changes in the studied areas can be detected by comparing the original images and composite images, as well as using the images converted by the principal component method, etc. The choice of methods or their combination is determined by the task, as well as by the spatial, spectral, radiometric, or temporal resolution of the satellite images used.

To detect changes in territories using different-time images, both separate image processing and the use of source images for their subsequent comparison using Change Detection algorithms are used.

The method of differentiation of multi-time images (Image differencing) is based on the elementwise subtraction of the brightness of the original images. In unchanged areas, the difference in brightness values will be close to zero. At the same time, they will have positive or negative values in the changed areas (depending on the direction of change).

A comparison of multi-time snapshots after classification (post-classification comparison) can lead to serious difficulties in comparing the results of classification associated with a mismatch of both the number of classes and their correlation with objects. However, in many cases, the processing of generalized thematic rasters shows the highest reliability of the results at regional monitoring scales.

Detection of changes in multi-temporal images converted by the principal component analysis (PCA) is based on the elementwise subtraction of the brightness values of the corresponding pixels of the main components. However, there is another way in which multi-temporal images are combined into one file, which is converted using PCA, and those components with a lower correlation are analyzed. This approach consists in the joint processing of a set of images for different dates, which are combined into a single image and processed like a multizone image.

To select objects of certain classes (vegetation, water objects, soil, anthropogenic formations, etc.), index images can be formed from multi-zone images. Obtaining index images is based on the choosing spectral channels for which the images of a given object have maximum differences in brightness. The following indexes are built on this principle: NDVI (Normalized Difference Vegetation Index), NDWI (Normalized Difference Water Index), GEMI (Global Environmental Monitoring Index), etc. [5].

3. Results

Figure 3 shows results of the comparison of the multi-temporal images presented above in order to identify changes in the state of the territories, performed by the method of differentiating multi-temporal images (image differencing).

The resulting thematic image masks superimposed on images for earlier dates, contains information on the difference in brightness of pixels having the same geographical coordinates. The magnitude of these changes can be expressed both in absolute values of brightness and in percentage. In each case, this parameter is determined experimentally and depends on the materials used and the objects under study.

In accordance with the legend in Figure 3, the changes are of a different nature, from "a significant increase" to "a significant decrease." They correspond to the appearance or disappearance of an object. In most cases, the "slight increase / decrease" of brightness values are caused by differences in shooting conditions, season, light level, etc., and they are not included in the resulting mask. The number of pixels of the image areas corresponding to the identified changes is also given in the legend. Based on this characteristic, we can determine the area of the territory that has been transformed.

Row	Histogram	Class_Names	Color
0	0	Background	
1	44553	Decreased	
2	73482	Some Decrease	
3	3756	Unchanged	
4	544744	Some Increase	
5	81277	Increased	

Figure 3. The results of the identification of changes in the state of the territories by different time shots.

The results show significant modifications in the studied objects, their nature of vegetation, the emergence of many buildings and structures, a new network of roads and communications, etc. Moreover, they are equally observed both in the territory of Russia and in the neighboring state.

Based on the information obtained from the processing of remote sensing data, we can assess trends in the development of natural and anthropogenic systems, the focus of economic activity in the region, compare changes in the Russian and adjacent territories, namely:

1. The increase in areas of residential neighborhoods in cities indicates an influx of population in the studied areas and a general economic rise.
2. Significant expansions of industrial and commercial areas indicate an increase in trade in cross-border areas.
3. Monitoring the position of communications allows one to track the disappearance or occurrence of roads, including those being not officially registered, which can reveal the illegality of certain actions.
4. Assessment of the state of forests allows to identify illegal logging and risk areas associated with the forest fires and epidemics.
5. Conducting operational monitoring of emergency situations, exciting the territory of neighboring states, ensures the effectiveness of rescue operations.

Thus, the obtained results show the possibility of operational control over the state and development of cross-border regions.

4. Discussion

The considered examples of detecting changes of a different nature have shown the possibility of on-line monitoring of territories and local objects by remote sensing methods, provided that the optimal method for decryption is chosen [6], [7], [8].

- The quality and reliability of the interpretation results is determined by the parameters of the source materials and the target task of the study.
- It is shown that the multi-temporal images obtained from space allow not only to confidently record the changes, but also their character.
- The information obtained allows us to identify the dynamics of not only the territories, but also to assess the social and economic trends associated with these changes.

The use of remote sensing data corresponds to the direction of the state policy in the field of ensuring the sustainable development of cross-border regions [2].

5. Conclusion

The conducted studies show the effectiveness of using the whole array of methods of automated processing of remote sensing data for monitoring the dynamics of the development of regions, including cross-border regions. And these studies provide the efficiency and informative assessment of not only the geophysical but also socio-economic processes occurring in these territories.

References

[1] Zykov A A 2008 A transboundary region in the system of international cooperation of Russia *Humanitarian Research in Eastern Siberia and the Far East* **2** pp 17-27

[2] GISGeo Geomarketing Spatial Analysis official site 2017 *The program "Digital Economy of the Russian Federation" is supplemented with a section on geospatial data* Available at: http://gisgeo.org/news/programma-cifrovoj-ekonomiki-rossijskoj-federacii-dopolnena-razdelom-o-geoprostranstvennyh-dannyh.html (Accessed 08 08 2017)

[3] *Google Earth Database* n.d. Available at: https://earth.google.com/web/ (Accessed 02 04 2019)

[4] Guk A P, Evstratova L G, Khlebnikova E P, Altyntsev M A, Arbuzov S A, Gordienko A S, ... Guk A A 2013 Development of methods for automated interpretation of aerospace images: Identification of changes in the state of the territories by multi-zone satellite images obtained on different dates

Geodesy and Cartography **8** pp 33-41

[5] *GIS-Lab: Geographic Information Systems and Remote Sensing* n.d. Available at: http://gis-lab.info/ (Accessed 10 04 2019)

[6] Solonko E V, and Khlebnikova E P 2016 The use of multi-time satellite images to assess the development of landslide processes in the city of Barnaul *Interexpo Geo-Siberia* **9** pp 156-161

[7] Khlebnikova E P, and Abisheva MT 2016 Features of detecting changes in engineering structures in the interpretation and analysis of space images *Interexpo Geo-Siberia* **4(1)** pp 9-14

[8] Khlebnikova E P, and Simonov D P 2014 The use of statistical analysis methods for the interpretation of multizonal space images *Interexpo Geo-Siberia* **4(1)** pp 19-23

Accumulation of heavy metals with the lichen thalli of *Cladonia rangiferina* L. at the roadside phytocenoses of the West Siberian Subarctic

E Popova[1*]

[1] Tobolsk Complex Scientific Station, Ural Branch of the Russian Academy of Sciences, 15 Akademik Yu. S. Osipov str., Tobolsk 626152 Russia

E-mail: popova-3456@mail.ru

Abstract. The article reflects the results of a comparative analysis of the content and accumulation of heavy metals such as chromium, lead, nickel, zinc in the thalli of *Cladonia rangiferina* L. in the anthropogenic phytocenoses of the West Siberian Subarctic, Russia. The next series of accumulations of heavy metals in descending their concentrations in soils is Zn> Pb> Cr> Ni. Analysis of the results showed that among the factors affecting the accumulation of elements, the most significant wa the distance from the highway. The content of most elements is significantly ($p < 0.05$) related. *Cladonia rangiferina* L. accumulates heavy metals depending on the location of the sampling point relative to the source of pollution. The coefficient of biological absorption, as well as the biogeochemical activity of the species, varies in different areas but has the greatest value in areas close to the source of anthropogenic load.

1. Introduction

The *Cladonia rangiferina* L. belongs to a group of lichens that grow well in warm and cold types of climate, but the tundra is still the ideal place. Possessing high resistance to large frosts, the plant is indispensable in the diet of many animals. In appearance, the reindeer lichen is a branched tubular sinuous overgrown, which is very soft and pleasant to the touch. However, it is very fragile when dry, it crumbles easily. The plant is of great national economic importance. As is known, it serves as one of the most important forage plants for deer in the tundra.

The when dry makes up to 1/3 of the reindeer ration. The value of moss is high nutritional value, it is rich in carbohydrates, well absorbed by deer. Playing the role of the basis of the diet for reindeer in the winter, lichens serve as a source of pollutants in trophic chains [1], [2].

Among the many environmental pollutants of technogenic origin, heavy metals are the most common and toxic. Therefore, their effect on plant objects has been actively studied for many years. Pollution of the environment near the roads leads to a change in the roadside soil [3-6].

Lichens are the edificators of many plant communities of northern phytocenoses. There is a known increased sensitivity of this group of organisms to various forms of anthropogenic impact and the absorptive capacity of thalli with respect to a number of chemical elements, including heavy metals. However, in the conditions of the North, research devoted to the issues of the distribution and accumulation of heavy metals in the environment were developed only in recent decades.

Based on the foregoing, the purpose of this work is to study the characteristics of the

Content from this work may be used under the terms of the Creative Commons Attribution 3.0 licence. Any further distribution of this work must maintain attribution to the author(s) and the title of the work, journal citation and DOI.
Published under licence by IOP Publishing Ltd

IOP Conf. Series: Earth and Environmental Science **395** (2019) 012045 doi:10.1088/1755-1315/395/1/012045

accumulation of heavy metals in the thalli of *Cladonia rangiferina* L. growing on the territory of the West Siberian Subarctic.

2. Materials and Methods

2.1. Selecting sites

To determine the degree of anthropogenic impact, observational sites with different intensities of technogenic load were selected.

Section 1. The watershed ridge near the highway is located 30 km north of the city of Novy Urengoy, Yamalo-Nenets Autonomous Okrug, Russia. The view of the natural area is the northern taiga-forest-tundra (N $66^0$05131, E $76^0$19897).

Section 2. The natural zone type is the sedge-cotton grass-sphagnum tundra-shrub tundra; a road near the town of Nadym, Yamalo-Nenets Autonomous Okrug, Russia (N $75^0$56823, E $75^0$16475).

Section 3. The natural zone type is the northern valley taiga, the highway is located 20 km from the town of Pangody, Yamalo-Nenets Autonomous Okrug, Russia (N $66^0$48420, E $76^0$24522).

Section 4. (7E, 7A, 7B). This is a highway near the village of Tazovsky, Yamalo-Nenets Autonomous Okrug, Russia (N $67^0$29358, E $78^0$44469).

2.2. Soil sampling

Soil sampling and sample preparation for quantitative chemical analysis were carried out in accordance with [7], [8]. Five samples taken from each site. Samples were taken at a distance of 10, 25, 50 meters from the roadway.

2.3. Sampling Cladonia rangiferina L.

Sampling of plant material, their preparation for quantitative chemical analysis, was carried out from areas of 20 × 20 cm, in accordance with [9], [10]. The composition of lichen syusus was divided, and a homogeneous sample without dead parts was left for analysis.

2.4. Sample preparation

Sample preparation was carried out using the speedwave MWS-2 microwave decomposition system from Perkin Elmer (USA). A soil sample of m=4.0 g was placed in a plastic tube, then HNO_3: HCl=1:3 was added. The tube was placed in a microwave oven to decompose the sample using the program recommended by the furnace manufacturer, using the following heating mode: raising the temperature to 200°C for 5 min, keeping for 5 min at 200°C, cooling to 45°C. The dissolved sample was transferred to a 15 ml tube, adjusting the volume to 10 ml with distilled water, and the analysis was performed.

For the decomposition of plant samples, it (m = 0.3 g) was placed on a tube, then was added H_2SO_4:H_2O_2=1:3. The following decomposition steps were carried out in the same manner as previously described.

Quantitative chemical analysis of the accumulation of trace elements and heavy metals Zn, Pb, Cr, Ni in soil and plant samples was carried out by inductively coupled plasma on an OPTIMA-7000DV atomic emission spectrometer (Perkin Elmer, USA). For calibration, standard solutions from Perkin Elmer (USA) were used.

2.5. Biological absorption coefficient

To assess the efficiency of absorption of heavy metals, the coefficient of biological absorption (A_x) is used. It is the content of the trace element in the ash of plant material for its content in the root zone of the soil [1]:

$$A_x = I_x / n_x \qquad (1)$$

Where I_x is the content of the *i*-th heavy metal in plant ash, mg / kg; n_x is the content of the *i*-th heavy metal in the soil, mg / kg. An index value from 1 to 10 indicates an intense accumulation of the element by the plant; from 0.1 to 1 is an average accumulation; from 0.01 to 0.1 is a weak absorption; from 0.001 to 0.01 means absence of biological accumulation of the element [11].

2.6. Biogeochemical activity of species (BCA)

The biogeochemical activity of the species expresses the general ability of the species to concentrate microelements, it is an integral characteristic allowing to quantify the ability of plants to accumulate trace elements [12].

Statistical processing of the results was carried out in the software package Statistica 10.0. Arithmetic averages (X), standard arithmetic mean (SD) errors, estimates of the significance of differences in arithmetic averages by Student's t-test were calculated.

3. Results

The concentration of chromium in the soils of these studied areas was in the range of 5.75-20.07 mg / kg (12.91 ± SD on average); 7.25-20.46 for Pb (ave. 13.86 ± SD); 0.35-12.30 for Ni (ave. 6.32 ± SD); 4.78-28.30 for Zn (ave. 16.54 ± SD). According to the average content, the studied elements are presented as the next row (in decreasing concentration): Zn > Pb > Cr > Ni.

To study the accumulation of heavy metals in plant objects, their accumulation was determined in the thalli of *Cladonia rangiferina* L. (Table 1).

Table 1. The content of heavy metals (I_x), the coefficient of biological absorption (A_x), and the biogeochemical activity of the species (BCA) in the thalli of *Cladonia rangiferina* L.

L, m	Cr		Pb		Ni		Zn		BCA
	I_x (mg / kg)	A_x	I_x (mg / kg)	A_x	I_x (mg / kg)	A_x	I_x (mg / kg)	A_x	
	Section 1								
50	7.61±0.04*	0.6	10.23±0.09*	0.5	4.03±0.03*	0.6	4.47±0.02*	0.6	2.3
25	9.32±0.08*	1.1	14.69±0.08	1.1	6.78±0.02	1.2	6.09±0.04*	1.1	4.5
10	11.36±0.07	1.6	18.25±0.09	1.3	9.12±0.03*	1.6	7.17±0.03*	1.5	6.0
	Section 2								
50	4.69±0.02*	0.5	11.23±0.09*	0.3	3.69±0.01*	0.3	6.00±0.02*	0.6	1.7
25	6.35±0.04*	1.0	12.36±0.08*	0.9	5.69±0.02	1.0	7.71±0.04*	1.1	4.0
10	9.21±0.05	1.6	16.02±0.06	1.5	8.96±0.04	1.8	12.48±0.03	1.3	6.2
	Section 3								
50	7.04±0.03*	0.7	8.36±0.06*	0.8	4.70±0.09*	0.8	6.89±0.02	0.6	2.9
25	10.36±0.04	0.8	9.36±0.07*	0.9	6.41±0.09*	0.9	7.48±0.02*	0.8	3.4
10	13.02±0.07	1.2	13.20±0.09	1.6	11.52±0.08	1.3	8.40±0.03*	1.4	5.5
	Section 4								
50	8.03±0.09*	0.4	12.69±0.08*	0.8	8.34±0.10*	0.7	8.49±0.02*	0.3	2.2
25	12.69±0.08	1.6	14.32±0.09	1.5	10.58±0.12	1.2	9.78±0.06*	0.7	5.0
10	14.78±0.09	1.9	17.36±0.09	1.9	17.39±0.10	1.9	12.89±0.07	1.6	7.3

Note: * differences from control are significant at $p < 0.005$.

Analyzing the results obtained on the content of heavy metals in the thalli of *Cladonia rangiferina L.* shows that their greatest accumulation is observed on the sections nearest to the highway (10 meters). The content of Cr in all the studied sites varied at 4.69-14.78 (9.74 ± SD on average), 8.36-18.25 for Pb (ave. 13.31 ± SD), 3.69-17.39 for N (ave. 10.54 ± SD), 4.47-12.89 for Zn (ave. 8.68 ± SD) mg / kg.

The average concentration of heavy metals in the soil significantly exceeds their concentration in the *Cladonia rangiferina* L. thalli in areas remote from the source of anthropogenic load, but their accumulation in lichens increases in areas close to the highway.

IOP Conf. Series: Earth and Environmental Science **395** (2019) 012045 doi:10.1088/1755-1315/395/1/012045

The high correlation between the metal content of *Cladonia rangiferina* L. thalli and the soil of the studied areas is noted (r = 0.95-0.99).

To characterize the intensity of the absorption of elements by *Cladonia rangiferina* L. thalli from the soil, the biological absorption coefficient (A_x), as a ratio of the metal content in the plant to its gross content in the soil, is calculated. The transfer of heavy metals from soil to lichen occurs through biological absorption of elements by biomass. The danger of pollution is higher, the greater the number exceeds 1. The highest coefficient of biological absorption is in the areas 1, 2, 3, 4 (10 meters from the highway) in relation to Cr, Pb, Ni and Zn. Also, the highest rate of biological absorption coefficient (Ax> 1) is in areas 1, 2, 4 (Cr, Ni), 1, 4 (Pb), 1.2 (Zn) (25 meters from the highway).

In all areas, the coefficient of biological absorption is from 0.1 to 1 (an average accumulation is $A_x \leq 1$) for Cr, Pb, Ni, and Zn (50 meters from the highway), with the exception of the following sections: 2 A_x = 0.9 (Ni); 3 A_x = 0.8-0.9 (Cr, Pb, Ni, Zn); 4 A_x = 0.7 (Zn) (25 meters from the highway).

The magnitude of the biogeochemical activity of the species (BCA) depends mainly on the quantity and composition of the pollutants most vigorously absorbed by plant objects, and to a lesser extent on the elements that are poorly captured. The biogeochemical activity of the species in the plots varied from 1.7 to 7.3, the highest value was noted at plots 1 (6.0), 2 (6.2), 4 (7.3) 10 meters from the highway.

4. Discussion

A number of authors note that the content of heavy metals in lichen thalli correlates with the intensity of anthropogenic activity, with vehicle emissions, first of all. The spatial distribution of elements has a direct relationship with the sources of emissions. The maximum content is observed near the highways, the minimum is in green areas [1], [13], [14]. In our studies, the biological absorption coefficient (Ax) is calculated. It showed the movement of heavy metals by bioabsorbing elements through biomass. The highest result was found at the nearest point, 10 meters from the highway.

Many researchers note that the degree of biological absorption of heavy metals by the lichenbiota is inversely proportional to their concentration in the environment. Lichens absorb heavy metals not only from the soil but also from the atmosphere, sometimes they absorb up to half of the elements they contain [2], [15], [16]. Studies have shown that in less polluted areas, an average concentration of heavy metals in the soil significantly exceeds their concentration in *Cladonia rangiferina* L thalli. In areas close to the highway, their accumulation increases.

5. Conclusion

The analysis of the obtained data allowed us to construct the next elemental series of the accumulation of heavy metals according to their decreasing concentrations in soils: Zn > Pb > Cr > Ni. The intensity of biological absorption and the magnitude of the biogeochemical activity of the species were assessed. A direct relationship between the content of heavy metals in the thalli of *Cladonia rangiferina* L. and their habitat was found.

The degree of accumulation of pollutants in the thalli of *Cladonia rangiferina* L. is determined by the distance of the latter from the point of emission. The highest concentrations of heavy metals were recorded at a distance of up to 10 km from the source, which the biogeochemical activity of the studied species confirms.

Epiphytic lichens are distinguished by the ability to accumulate various chemical elements in thalli, which serves as the basis for using them in monitoring environmental pollution. Initiated studies should be continued in the direction of determining the content of pollutants in various components of ecosystems. By increasing the content of heavy metals in lichens, roadside phytocenoses, we can judge the deterioration of the environment.

6. Acknowledgments

The article was prepared with the financial support of the Ministry of Education and Science of the Russian Federation in the framework of the theme of fundamental scientific research "Anthropogenic Transformation of Floodplain Ecosystems of the Ob-Irtysh basin" (No. AAAA-A19-119012190088-0).

References

[1] Regan S, Matwichuk L, Cloutis E, Goltz D, and Mann P 2016 Potential signatures of heavy metal complexes in lichen reflectance spectra *Int. J. Remote Sensing* **11** 21
[2] Soldatenkova Yu P 1977 *Small workshop on botany: lichens* (Moscow, Russia: Moscow University)
[3] Arunakumara K K I U, Walpola B C, Yoon M-H 2013 Current status of heavy metal contamination in Asia's ricelands *Reviews in Environmental Science and BioTechnology* **12**(4) pp 355-377
[4] Zhao Q, and Lu W 2010 Review and prospect of research on heavy metal contamination in soil - bibliometric analysis based on web of science database environmental *Science and Technology* **33**(6) pp 105-111
[5] Li W, Zhang M, and Shuai H 2006 Study on the adsorption characteristics of brown soil to copper *Journal of Soil and Water Conservation* **20**(4), pp 113-115
[6] Bashkin V N, and Kasimov N S 2004 *Biogeochemistry* (Moscow, Russia: Scientific world)
[7] National Standard 17.4.3.01-83 2008 *The Nature Conservancy. Soils. General requirements for sampling. Protection of Nature. Soils.* (Moscow, Russia: Standartinform)
[8] National Standard 17.4.4.02-84 2008 *Soils. Methods of sampling and sample preparation for chemical, bacteriological, helminthological analysis. Protection of Nature. Soils.* (Moscow, Russia: Standartinform)
[9] Kabata A, and Pendias H 2016 *Trace elements in soils and plants* (Washington, DC: CRC press)
[10] Menning W J, and Feder U A 1985 *Biomonitoring of atmospheric pollution with the help of plants* (Leningrad, Russia: Gidrometeoizdat)
[11] Lakin G F 1990 *Biometrics* (Moscow, Russia: Higher School)
[12] Novikov Yu V 2002 *Ecology, environment and man* (Moscow, Russia: Education)
[13] Tsyganov D N 1983 *Phytoindication of ecological regimes in the subzone of coniferous-broadleaf forests* (Moscow, Russia: Nauka)
[14] Ilyin V B 1991 *Heavy metals in the system soil – a plant* (Novosibirsk, Russia: Nauka)
[15] Kalinowska R, Bačkor M and Pawlik-Skowrońska B 2015 Parietin in the tolerant lichen *Xanthoria parietina* (L.) Th. Fr. increases protection of Trebouxia photobionts from cadmium excess *J. Ecological Indicators* **58** 132
[16] Brunialti G, and Frati L 2014 Bioaccumulation with lichens: the Italian experience *Int. J. Env. Stud.* **1** 115

IOP Conf. Series: Earth and Environmental Science **395** (2019) 012046 doi:10.1088/1755-1315/395/1/012046

Reconstruction of grain yield in the steppe zone of the Altai territory (south of Western Siberia) in the XIX-XX centuries on the basis of tree-ring chronologies of tape burs

N V Rygalova[1*], E V Rygalov[1] and N I Bykov[1]

[1] Altai State University, 61 Lenina prosp., Barnaul 656049 Russia

E-mail: natalia.ml@mail.ru

Abstract. The article presents the results of the reconstruction of the yield of grain crops of the economy of the Volchikhinsky district of the Altai Territory (south of Western Siberia) on the basis of an extended tree-ring chronology until 1822. Based on the restored series, the cyclical nature of cereal productivity was analyzed. In particular, the 22-, 25-, and 27-year cycles are highlighted as the most significant. The increase in the amplitude of variability of the yield series was noted by year from the middle of the 19th century to the end of the 20th century.

1. Introduction

Analysis of the dynamics of yields over a long period (several centuries) is necessary for understanding long-term trends in the production of main food crops (in particular, cereals). Grain crops are grown mainly in the steppe belt, which is also represented on the territory of the Altai Territory. This makes the region the main producer of food beyond the Urals. At the same time, the region is characterized by significant annual fluctuations in the volumes of crop production, which is caused by the frequency of occurrence of droughts (of varying intensity) on its territory. Thus, we can say that the grain yield in the Altai Territory is predetermined by the climatic conditions of the growing season.

To obtain a long series of yields, tree-ring chronology (TRC) tape burs were used. The growth of Scots pine (*Pinus sylvestris* L.) along the radius (formation of the annual ring) is the result of the productivity of the woody plant as well as the yield is an indicator of the productivity of the agrocenosis. The basis for comparing these indicators was the presence of climate as a common determining factor. It was previously established [1] that the chronology of the width of the Scots pine's annual rings are indicators of the climatic conditions of the vegetation period, which makes them suitable for the reconstruction of the climate-mediated component of the yield dynamics. Studies in this direction were previously conducted in Russia and in foreign countries [2-7].

2. Materials and Methods

The research material is represented by a series of yields (a volume of harvested grain per unit of sown area) of grain crops of farms of Volchikhinsky district (steppe zone) for the period from 1976 to 2000. The choice of a region is predetermined by its location in the steppe zone and its proximity to the belt forests. ThVolchikhinsky district specializes mainly in the production of crop products, primarily grain. Traditionally, grain crops are not irrigated in the district. There is no information about the facts of fertilization in the field at the same time. The average yield in the farms of the Volchikhinsky district

Content from this work may be used under the terms of the Creative Commons Attribution 3.0 licence. Any further distribution of this work must maintain attribution to the author(s) and the title of the work, journal citation and DOI.
Published under licence by IOP Publishing Ltd

for this period was 9.4 c / ha (the standard deviation of the indicator for the farms of the district varies from 2.4 to 3.5).

Tree-ring chronologies of the region's forest belt (Aleutian, Kulunda, Barnaul, Kasmalinskaya) were taken as the basis for the reconstruction. They are the generic, standardized (arstan-chronology) ranks of the width of tree rings. This type of standardization (arstan) allows eliminating the age component in the dynamics of tree growth, minimizing autocorrelation noise, while maintaining long-period fluctuations (unlike residual chronologies). The mean value of the TRC is 1 (since the growth indices are used in the analysis), with the standard deviation in chronologies ranging from 0.14 to 0.27. The TRC with a high standard deviation are considered sensitive, their use in dendroclimatic analysis is the most promising (these mainly include chronologies of pine obtained for the steppe zone).

The main methods used were dendrochronological and statistical. All tree-ring chronologies of pine forests were obtained by the standard technique [8]. The work used extended chronologies to increase the retrospective range of yields. The chronology of Uglovskoe-42 (1844-1997), on the basis of which a number of grain crops was reconstructed, was extended to the past for 22 years (it allowed to increase the replication of individual series at the beginning of the TRC, which increases the accuracy of reconstruction). For the extension of the DKH, samples of historic wood were used, taken from the logs of the old wooden houses in the villages of the region, located near the forests. The "floating" (relative) chronologies obtained for them were dated according to the TRC of the belt forests. In cases where TRC houses covered an earlier period of time than chronologies of forests, the latter extended into the past for this period. To extend the chronology of Uglovskoe-42, the TRC of two wooden houses, the village "Laptev Log" (Uglovsky district) were used. The cumulative duration of historical wood chronologies was from 1973 to 1822.

Restoration of yield series was carried out using the Pearson correlation method (pair correlation) and regression. The first method allowed us to identify the most significant links between the yields and the TRC, the second allowed us to reconstruct a number of yields on the basis of the obtained statistically significant links. Obtaining and processing of tree-ring series are performed in the TSAP programs and DPL software package. Statistical calculations carried out in Statistica 13.

3. Results

As a result of the correlation analysis of the grain yield series in the farms of the Volchikhinsky district with TRC tape burs, some significant connections (0.4-0.47) were identified. There are no clear geographic dependencies in the manifestation of a mediated relationship between pine growth and grain yield both in the farms and in the region average. The highest correlation coefficient (0.47, significant at $p < 0.05$) was obtained for a series of crop yields of the farm "Za vlast' Sovetov" (Vostrovo), adjacent to the Kasmalinsky forest, Uglovskoe-42 (a dry steppe zone, southern part of the Barnaul belt forest). This chronology was used as an indication basis for the retrospective recovery of yield data. The linear regression equation (1), which underlies the reconstruction, has the following form:

$$Y_i = 4{,}749 * X_i + 4{,}9855 \tag{1}$$

where Y_i is the yield value in year i'; X_i is the index of the annual increase in a year of i.

The reconstructed yield values have a smaller variation compared to the actual yield range (standard deviation is 1.4 and 3.0, respectively), which is characterized by significant annual fluctuation amplitudes. For greater clarity, the graph (Fig. 1) presents the averaged values of the observed yield (a five-year smoothing is used).

The recovered grain yield range has a well pronounced cyclical nature. Using the Fourier spectral analysis, several groups of significant cycles of the series were allocated: 22-25-29 years, 35-44 years, 88 years, also a group of 9-12-year cycles and a cycle of 170 years is allocated. The 22-25-29-year cycle is of the greatest importance (Fig.2). It is clearly manifested in the following periodicity of dry years (which determine the decline in the productivity of phytocenoses): 1963/1967, 1990/1991, and 2012.

The last year was not included in the analysis presented but was noted as a year with low grain yield throughout the region [9] and the formation of narrow rings in Scots pine.

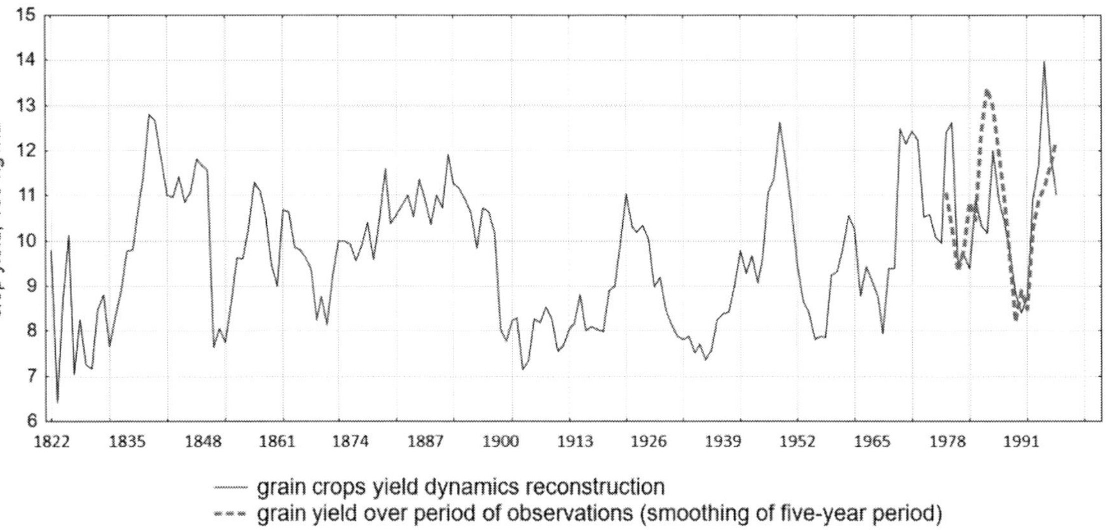

Figure 1. Reconstruction of the yield of grain crops for the economy of the Volchikhinsky district of the Altai Territory based on the tree-ring chronology.

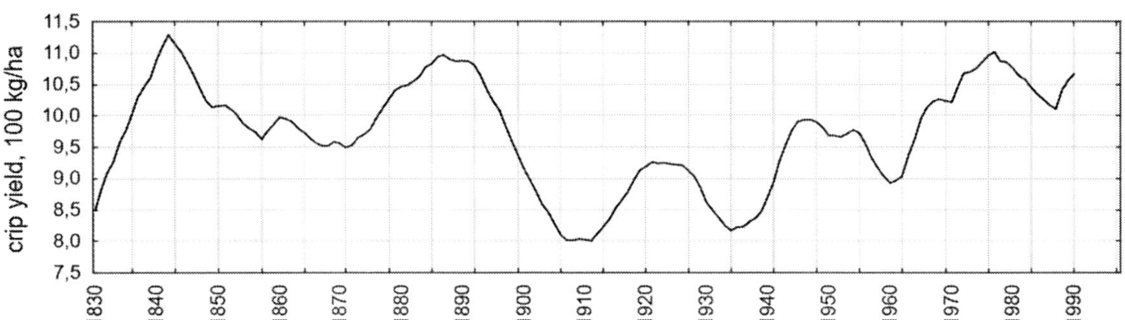

Figure 2. Smoothed (15-year averaging) recovered grain yield series.

In general, we can say that there is an increase in the amplitudes of fluctuations in the value of the yield of grain crops (Table 1) during the analyzed period. This indicates an increase in climatic instability over time, which is a sign of climate change.

Table 1. Statistical estimates of the amplitude of variation of grain yield over time.

Time period	Standard deviation value
1850-1899	1.0
1900-1949	1.18
1950-1997	1.48

4. Discussion

The study of the reconstruction of the productivity of crops based on the use of indicator data is a promising direction, especially in conjunction with the study of long-term climate dynamics. However, several points remain debatable.

- Significant correlations are important for establishing an adequate regression model for reconstructing retrospective values. In the case under consideration, the grain yield series on average showed low values with all TRC (in some cases, the correlation coefficient was higher than 0.4 with $p < 0.05$). Most likely, this is due to the fact that for the analysis, the yield values for the enlarged group of crops were taken, namely grain. Different grain crops (wheat, buckwheat, oats) show a different connection with an increase (for example, there is a weak connection between the TRC of tape forests and the buckwheat yield in the steppe districts of the region [4]), which introduces heterogeneous noise and leads to a decrease in the correlation values.

- Also, the contribution of the anthropogenic component (application of fertilizers, measures of land reclamation) to the formation of the yield volume is rather difficult to determine. But despite this, the limiting importance of climate for the productivity of agrocenoses is decisive; therefore, the reconstruction of crop yields through the climate component seems possible (in this case, the anthropogenic component will be defined as noise).

5. Conclusion

In the course of the work performed, a number of grain crops yields of the Volchikhinsky district (a steppe zone of the Altai Territory) were reconstructed on the basis of the dendrochronological method until 1822. The restored yield values are characterized by the presence of cycles of various duration. 22-29-year cycles, which are manifested in the frequency of strong droughts, are most significant. Also, there is an increase in the variation (fluctuations) of the yield values over the years from the middle of the 19th century to the end of the 20th century. In general, over the course of the 20th century, there was a general trend towards an increase in the climatic potential of the productivity of agrocenoses (with negative fluctuations local in time). To clarify long-period cycles in the reconstructed rows, further researchon the extension of the tree-ring rows is necessary. This will allow you to get the most complete picture of different-period cycles of climate change of the territory and climate-dependent processes and make it possible to forecast them.

6. Acknowledgments

The publication was prepared with the financial support of the Russian Foundation for Basic Research in the framework of the project № 18-05-00694_a.

References

[1] Malysheva (Rygalova) N V, and Bykov N I 2011 *Dendrochronological studies of tape burs in the south of Western Siberia* (Moscow: Russia, Azbuka)
[2] Bitvinskas T T, and Brukshtus V I 1990 Radial growth of trees, climate extremes and crop yields In *Problems of dendrochronology and dendroclimatology* (pp 20-21) (Sverdlovsk, USSR: Publishing House of the UB of the USSR Academy of Sciences)
[3] Vaganov E A 1989 Method for predicting grain yield using dendrochronological data *Ecology* **3** pp 15-23
[4] Rygalova N V, Bykov N I, and Plutalova T G 2014 Retrospective study of the dynamics of crop yields in Altai Territory by the method of dendrochronology *Vestnik of the Altai State Agrarian University* **10**(120) pp 43-49
[5] Babushkina E, Belokopytova L, Shah S, and Zhirnova D 2018 Past crops yield dynamics reconstruction from tree-ring chronologies in the forest-steppe zone based on low- and high-frequency components *International Journal of Biometeorology* **62**(5) pp 861-871
[6] Huhtamaa H, Helama S, Holopainen J, Rethorn C, and Rohr C 2015 Crop yield responses to temperature fluctuations in 19th century Finland: provincial variation in relation to climate and tree-rings *Boreal Environ Res* **20** pp 707-723
[7] Huhtamaa H, and Helama S 2017 Reconstructing crop yield variability in Finland: long-term perspective of the cultivation history on the agricultural periphery since AD 760 *The Holocene* **27**(1) pp 3-11
[8] Shiyatov S G, and Vaganov E A *et al* 2000 *Methods of dendrochronology (vol 1)* (Krasnoyarsk, Russia: KrasSU)
[9] Rozova M A, and Ziborov A I 2016 Correlations between the yield of spring durum wheat and the elements of its structure depending on the level of genotype productivity and weather conditions in the priobskaya forest steppe of the Altai Territory *Vestnik of the Altai State Agrarian University* **2**(136) pp 44-49

IOP Conf. Series: Earth and Environmental Science **395** (2019) 012047 doi:10.1088/1755-1315/395/1/012047

A spatio-temporal analysis of the Teniz-Korgalzhyn trench geosystems based on the Landsat and Sentinel satellite image decoding materials

E N Sagatbaev[2], E P Krupochkin[1] and A N Dunets[1]

[1] Altai State University, 61 Lenina prosp., Barnaul 656049 Russia
[2] L. N. Gumilyov Eurasian National University, 29/1 Shakarim Khudaiberdiuly, Nur-Sultan 010000 Republic of Kazakhstan

E-mail: sagatbaeve@mail.ru

Abstract. The article presents a unique research on the Teniz-Korgalzhyn trench geosystems. Using the computer-aided decoding of multi-zone satellite imagery, the space-time patterns of the structure, functioning, and dynamics of the development of landscape components of the studied area are determined. Using the functionality of contemporary instrumental GIS and interpretation algorithms, the indicators of productivity of geosystems and anthropogenic transformation are determined. Significant anthropogenic factors that do not have a ubiquitous, but local influence on the dynamics of geosystems in the context of significant changes in the ratio of moisture and productivity are established. Conclusions onthe need to control anthropogenic influence and ensure the rational use of resources of the territory are made, treating them as a factor of sustainable development.

1. Introduction

The flooded areas, as well as geosystems of runoff dispersion zones in the Teniz-Korgalzhyn lake-flow system, have a significant diversity of biota and high productivity in the Republic of Kazakhstan. A scientifically based assessment of the state, structure, and dynamics of the functioning of these geosystems is required to maintain their ecological balance and to ensure the necessary watering of wetlands.

The research purpose is to conduct a spatial-temporal analysis of the structure and dynamics of the Teniz-Korgalzhyn trench geosystems based on using the computer-based thematic decoding of space imagery. The estimated parameters of terrestrial natural geosystems include the following: slope steepness, maximum collection area, horizontal curvature, maximum curvature, illumination. The parameters of aquatic geosystems (river and lake) include the following: an organic carbon content, humidity, soil reaction, hydrolytic acidity, saturation with bases, a mobile iron content, a granulometric composition. In turn, the parameters of the anthropogenically transformed geosystems (arable land, fallow lands, downed pastures) include the spectral brightness characteristics of vegetation, which are detected from multi-zone images, as well as the elevated phytomass reserves (calculated using the NDVI index).

Content from this work may be used under the terms of the Creative Commons Attribution 3.0 licence. Any further distribution of this work must maintain attribution to the author(s) and the title of the work, journal citation and DOI.
Published under licence by IOP Publishing Ltd

2. Materials and Methods

The geosystem approach is adopted as a methodological basis. It allows one to establish the distribution of natural complexes of different rank. A geosystem is a part of a territory characterized by uniform terrain, one type or subtype of soil, and a set of plant communities with general species composition and productivity, a similar response to natural and anthropogenic influences, and resistance to them [1], [2].

To determine the boundaries of geosystems and study their dynamics, we used scenes of coatings with the Landsat and Sentinel images. These images were selected from the summer, spring and autumn seasons with the lowest cloudiness. A set of large-scale topographic maps of the region (1: 500 000 - 1: 200 000) were used in the complex with images.

Meanwhile, there is a certain specificity of working with these data. For example, the Landsat TM images are presented in series of different years, from 1982 to 2018. But there is a certain difference between them in the structure of the data itself. The Landsat 5 has the following channel structure: VIS (3), NIR (1), SWIR (2), TIR (1); the Landsat 7 is panchromatic, multispectral: VIS (3), NIR (1), SWIR (2), TIR (1), with a plan resolution of 15 m for the panchromatic channel and 30 m for the multispectral zone; the Landsat 8 is panchromatic, multispectral: VNIR (6), SWIR (2), TIR (2), with a resolution of 15 m for the panchromatic channel, 30 m for the near and middle zone of the spectrum and 100 m for the thermal zone.

To highlight the boundaries and classify the components of geosystems, all the initial data of the remote sensing data were converted into mosaic coatings developed in the ENVI 5.0 program with a spatial resolution of 30 m (1982 - 2013).

For the collection and processing of a modern data slice (2014–2018), remote sensing data from the Sentinel satellites of the 1, 2 series were used in the research.

The first Sentinel-1A radar satellite was first launched into orbit by the European Space Agency (ESA) on April 3, 2014. It became the first in the satellite constellation of the global monitoring of the environment and safety Copernicus. The Sentinel-1A is developed by the Thales Alenia Space. The C-SAR synthesized an aperture radar equipment (developed by Astrium) being installed on board. It provides all-weather and round-the-clock delivery of satellite imagery.

The Sentinel-2A, 2B satellites of the second series are equipped with an optical-electronic multispectral sensor for surveys with a resolution of 10 to 60 m in the visible, near-infrared (VNIR) and short-wave infrared (SWIR) zones of the spectrum. This ensures the display of differences in the state of vegetation, including temporary changes, and it also minimizes the impact on the quality of shooting the atmosphere. In their class, their capabilities correspond to the Landsat-7 and SPOT-5 images.

In the process of thematic processing of images, the calculations of the MNDWI and NDVI indices with the construction of index maps for the entire territory of the Teniz-Korgalzhyn system were applied. NDWI is a measure of the moisture content of soil and leaves [3]. The spectral brightness values in the green (Green) and near-infrared (NIR) spectral ranges are used for the following index calculation:

$$NDWI = (Green - NIR) / (Green + NIR),$$

where "Green" is the reflection in the green region, and "NIR" is the reflection in the near infrared region of the spectrum. NDVI (Normalized Difference Vegetation Index) is an indicator of photosynthetically active biomass on the Earth's surface. The values of spectral brightness in the red and near infrared spectral regions are used to calculate it:

$$NDVI = (NIR - Red) / (NIR + Red),$$

where "NIR" is the reflection in the near infrared region, and "Red" is the reflection in the red region of the spectrum.

IOP Conf. Series: Earth and Environmental Science **395** (2019) 012047 doi:10.1088/1755-1315/395/1/012047

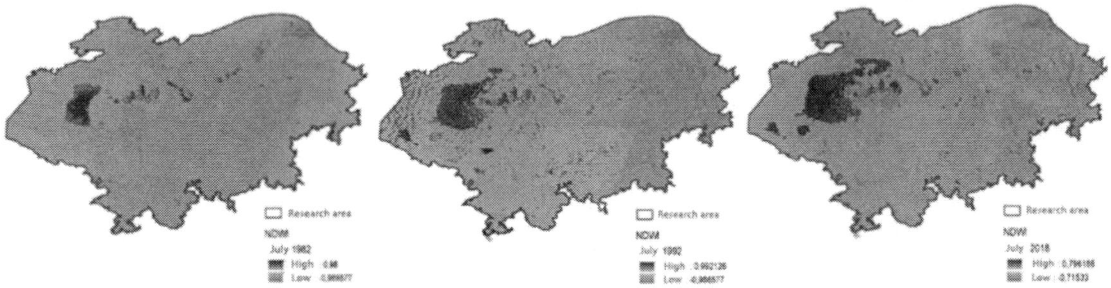

Figure 1. The moisture dynamics of vegetation cover, calculated from the values of NDWI: 1982, 1992, 2018.

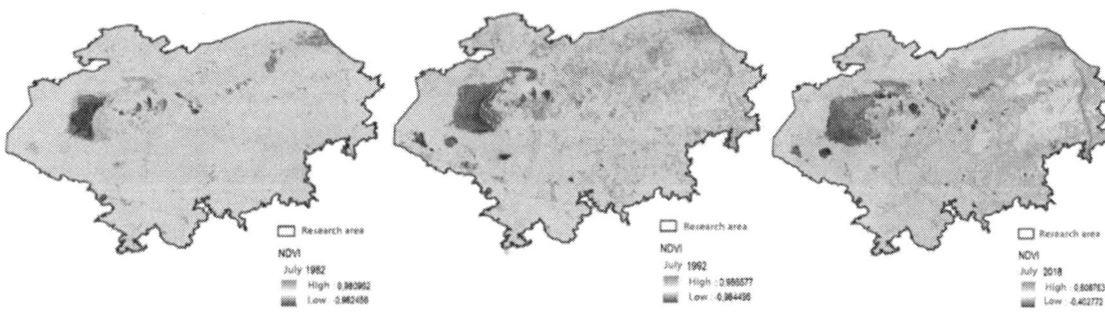

Figure 2. Vegetation intensity dynamics of vegetation, calculated from the NDVI values: 1982, 1992, 2018.

All our calculations of the indices made it possible to obtain a series of derived images, describing the degree of moisture content of geosystems as one of the agroclimatic indicators, and the intensity of vegetation as an indicator of the state (see Figures 1, 2). A visual and representative analysis of the images allowed us to establish a general trend for the increasing influence of moisture (uneven, with fluctuations in individual periods). For example, in 1992, there was an increase in moisture content not only in the floodplain plains, but also in the denudation plains with a shrub-wormwood-grass vegetation, as well as in the diluvial-proluvial plains with a shrub-wormwood-tyrsa vegetation on chestnut soils, in lakeside alluvial plains with meadows, wormwood, and fescue-wormwood factions on solonetz and meadow salt marshes with dark chestnut soils. After 20 years, there was a slight decrease in moisture content, which naturally affected the landscapes. So, for example, by 2018, we observe the concentration of moisture mainly in floodplain and lake-alluvial plains with wormwood, feather-grass and meadows on dark-chestnut soils with saline soils (Fig. 3).

The values of the NDVI index are quite strongly correlated with the distribution of NDWI; however, there are some differences. The meridional zones of average degree of moisture (this can be seen in the eastern part of the territory) practically do not manifest themselves on NDVI maps. The NDVI index values here are either medium or higher. From here, we can assume that the productivity of geosystems here depends not only on the nature of wetting, but also on other factors, such as anthropogenic influence, growth of natural / biological load on ecosystems, changes of microclimate in a less favorable direction, etc.

3. Results

In the course of the research, the following was obtained: a generalized map of the Teniz-Korgalzhyn trench geosystems; on the basis of a digital terrain model SRTM, as well as on the basis of interpretation and study of space-time changes of the Teniz-Korgalzhyn lake-flow system, a new landscape map was built.

In the classification of geosystems, three orders were identified: (1) terrestrial natural, (2) terrestrial and anthropogenically transformed, and (3) aquatic.

Natural geosystems include such areas in which the anthropogenic impact has not reached significant dimensions; moreover, they are only slightly altered by human activity. At the same time, a part of terrestrial geosystems, which have lost their natural soil and vegetation cover, is considered as terrestrial anthropogenically transformed, or rather the natural-anthropogenic geosystems. To aquatic geosystems, we assign the geosystems of the water surfaces of numerous lakes and large rivers.

Two sub-systems emit when studying the Teniz-Korgalzhyn trench: The Middle-Nura sub-geosystem and the Lower-Nura sub-geosystem, whose development is timed to the Nura river runoff within the basin, where space-time connections of channel-forming processes dominate from the source to the mouth.

The digital model of the territory was created using the materials of the SRTM survey. It allowed to identify, in combination with soil conditions, vegetation and geological and geomorphological conditions, the prevailing types of landscapes, as well as to assess their condition and degree of resistance to anthropogenic impact.

So the dominant landscapes are denudation plains with a shrub-wormwood-grass vegetation, deluvial-proluvial plains with a shrub-wormwood-tyrs vegetation on chestnut soils, lacustrine-alluvial plains, wormwood, and fescue-wormwood groups on saline and meadow salt marshes with the participation of dark chestnut soils, etc.. The main forms of relief are flat and low-slope, flat geosystems prevail in the territory, but low-slope geosystems and sandy arrays are also characteristic.

In general, it is fair to say that the patterns of distribution of geosystems of a region are determined by the following interrelated factors: climatic conditions and the geological and geomorphological structure of the territory. It is the ratio of these factors that causes a significant diversity and contrast of habitats. However, over the past few decades, as follows from the analysis of multi-temporal images, while maintaining all the prevailing landforms and its morphometric properties, other factors such as climate and soil cover have changed significantly, and anthropogenic influence has increased.

4. Discussion

Based on the integration of classical geographic research methods with remote sensing methods and the GIS tool base, new data on the state, structure, and dynamics of the functioning of the Middle-Nura and the Lower-Nura sub-geosystems of the Teniz-Korgalzhyn trench were obtained.

For the study area, a soil map at a scale of 1: 200,000 was updated; in the ArcGIS environment, a new map of geosystems on a scale of 1: 200,000 and 1: 500,000 was compiled. The structure of the legend on the presented maps is based on the classification divisions of geosystems of various ranks. So, on the basis of the geosystem map, many aspects of differentiation and areal relationships between various sub-geosystems are identified. Thus, a detailed geosystem structure of the region is determined. Here it is worth noting the important role of information obtained from the interpretation of multi-zone satellite imagery in the form of structural components of the environment, classified and recognized in images of different spatial and radiometric resolutions.

In addition to the distinguished structure of geosystems, the value indicators, namely the soil-vegetation indices (NDWI, MNDWI, NDVI), are the basis for calculating their productivity. Many side factors (such as species composition of vegetation, closeness, condition, slope exposure and surface slope angles, soil color under sparse vegetation, etc.) affect their numerical values. For example, according to Rouse et al., 1973 and Kriegler et al., 1969, if the density of vegetation cover is more than 70%, the index is moderately sensitive to changes in the soil background [4], [5]. To avoid such situations that weaken the reliability and information capabilities of remote sensing data, remote surveys are best carried out along selected key areas and routes that best cover the diversity of the studied area.

5. Conclusion

Our studies clearly show the feasibility of using, along with the classical methods of physical-geographical research, innovative methods of remote environmental surveys based on the GIS tools and computer-aided decoding algorithms.

The established spatial-temporal changes in the state and structure of the Teniz-Korgalzhyn trench

IOP Conf. Series: Earth and Environmental Science **395** (2019) 012047 doi:10.1088/1755-1315/395/1/012047

geosystems make it possible to single out the following as key factors: climatic conditions, geological and geomorphological structure, and anthropogenic influence. The latter has a direct impact on the anthropogenic transformation of geosystems, which is confirmed by a decrease in the productivity of landscapes in the floodplain part and the most densely populated areas. However, it is fair to note that this does not fit into the overall dynamics. For example, a time analysis of July images from 1982 to 2018 shows a whole inverse pattern, i.e. an increase in vegetation activity. Perhaps this is a consequence of the improving microclimate of the territory as a whole, although it does not exclude the increased anthropogenic influence in local foci. The presented methodology can be recommended for mapping and monitoring land use, monitoring changes in vegetation, water resources.

References

[1] Dzhanaleeva K M 2014 *Physical-geographical zoning of the Republic of Kazakhstan* (Almaty, Kazakhstan: EveroEvero)

[2] Dunets A N 2000 Anthropogenic transformation of geosystems on the territory of the Altai Mining District In *Mountains and man: anthropogenic transformation of mountain geosystems* (pp 54-57) (Novosibirsk, Russia: Siberian Branch of the Russian Academy of Sciences Publishing House)

[3] Xu H 2006 Modification of Normalized Difference Water Index (NDWI) to enhance open water features in remotely sensed imagery *International Journal of Remote Sensing* **27**(14) pp 3025-3033

[4] Rouse J W, Haas R H, Schell J A, and Deering D W 1973 Monitoring vegetation systems in the great plains with ERTS In *Proceedings of the Third ERTS Symposium* (vol 1) (pp 309–317) (Washington, DC: NASA)

[5] Kriegler F J, Malila W A, Nalepka R F, and Richardson W 1969 Preprocessing transformations and their effects on multispectral recognition In *Proceedings of the Sixth International Symposium on Remote Sensing of Environment* (pp 97-13) (Ann Arbor, MI: University of Michigan)

Natural radioactivity in aquatic ecosystems formed by self-flowing wells: the case of a water flow, a well "Cherkashinskaya 36-RG – stream – Aremzyanka river"

A S Aleksandrov[1]

[1] Tyumen State University, 6 Volodarskogo str., Tyumen 625003 Russia

E-mail: asaleksandov1991@gmail.com

Abstract. The article is devoted to the assessment of the radiation hazard of water and nearby bottom sediments of self-flowing hydrogeological wells, as well as the assessment of the radiation safety of personnel working with these materials for geological prospecting and national economic enterprises. The article is devoted to the study of the impact on the ecosystem of the geothermal waters of the self-flowing exploratory well "Cherkashinskaya No. b36-RG." In 1965, it was drilled to study a group of iodine-bromine water deposits. The article presents data from a survey of natural background radiation, as well as data on the natural radioactivity of bottom sediments and the volume content of radium in water. The relevance of the work is due to half a century of pollution by highly mineralized iodine-bromine waters of the Irtysh catchment as one of the largest transboundary rivers in Western Siberia. The Aremzyanka River is a tributary of the 1st order of the Irtysh River, a stream flowing from the flowing well No. 36-RG flows into this river.

1. Introduction

The Tyumen region is extremely rich in resources of underground mineral thermal waters, the reserves of which are about 3 million m^3 per day and belong to the West Siberian basin [1]. These resources were discovered almost throughout the south of the Tyumen region when searching for oil and gas fields, as well as iodine-bromine waters by exploration wells drilled in 1950-1980. Aquifers were found in the Mesozoic sediments of the Cretaceous (Apt-Cenomanian and Hotheriv-Barrem longlines) at a depth of 1-2 km, as well as in deeper Upper Jurassic sediments (more than 2 km). Many geothermal wells were drilled during the exploration in the 1950-60s of the last century. And now, they are abandoned. The total number of self-inflowing geothermal wells in the south of the region is 22, 13 of which are in the Tobolsky district [2].

There are the Cherkashinskoe and Tobolskoe deposits of thermal iodine and iodine-bromine waters in the territory of the Tobolsk district of the Tyumen region [1].

The well survey has been conducted since 2008. It shows that the manifestation of fluids is noted in 13 of the 27 suspended wells of the Tobolsk group [2], including the Cherkashinskaya well No. 36-RG, which was drilled in 1965 (Fig. 1).

Content from this work may be used under the terms of the Creative Commons Attribution 3.0 licence. Any further distribution of this work must maintain attribution to the author(s) and the title of the work, journal citation and DOI.
Published under licence by IOP Publishing Ltd

Figure 1. The well "Cherkashinskaya No. 36-RG," February 2019 [author's photo].

2. Materials and Methods

The object of the study is the area near the well No. 36-RG of the Cherkashinsky iodine-bromine water field, which is located on the II above-flood terrace of the Aremzyanka River (in the east of Tobolsk district). The river flows into the Irtysh River, on the right, at the 576th kilometer from the mouth, 0.5 km north of the Panushkova Village. The length of the river is 98 km, the basin area is 957 km^2, the riverbed is meandering, sandy-muddy, [3] the stream flowing from the well flows into the river. Mineral geothermal waters of the well No. 36-RG are opened in the depth interval of 1681-1747 m and 1797-1889 m; they are confined to terrigenous sediments (sandstones, mudstones, siltstone) belonging to the Apt-Cenomanian and Goteriv-Barrem tiers of the Lower Cretaceous [4].

The flow rate of the well is 42 m^3 / hour. Water temperature at the spout is 73 °C. At present, water is drained into a lowering of the relief and into the Aremzyanka River through a stream. The length of the stream of constant flow is 223 meters.

The study of natural radioactivity included the following: (a) studying bottom sediments in the area of the well and the stream flowing from it into the Aremzyanka River, which was conducted in the Laboratory of Radiation Ecology of Tyumen State University; (b) selecting and studying water samples for radium volumetric activity in the Testing Laboratory of "Regional Analytical Center" JSC; and also (3) measuring the level of surface radioactivity with a radiometer SRP-88N.

Table 1. Classification of industrial waste of oil and gas enterprises.

Waste category	Effective specific activity of natural radionuclides (Aeff), kBq / kg	Gamma radiation dose rate of natural radionuclides contained in waste (N), μGy / h
First	Aeff ≤ 1,5	N ≤ 0,7
Second	1,5 < Aeff ≤ 10	0,7 < N ≤ 4,4
Third	Aeff > 10	N > 4,4

The studies were conducted in January-February 2019 on three sections (Fig. 2).

In accordance with SanPiN 2.6.6.1169-02, the classification of industrial waste is introduced according to the effective specific activity of natural radionuclides (Aeff) in them. This classification is introduced for the following several reasons: (1) establishing requirements for ensuring the radiation safety of the population and employees of organizations of the oil and gas industry; (2) limits on human pollution of the environment by natural radionuclides; (3) planning the types and amount of radiation control when handling industrial waste, as well as to establish radiation and hygienic requirements for handling them. The classification is given in Table 1.

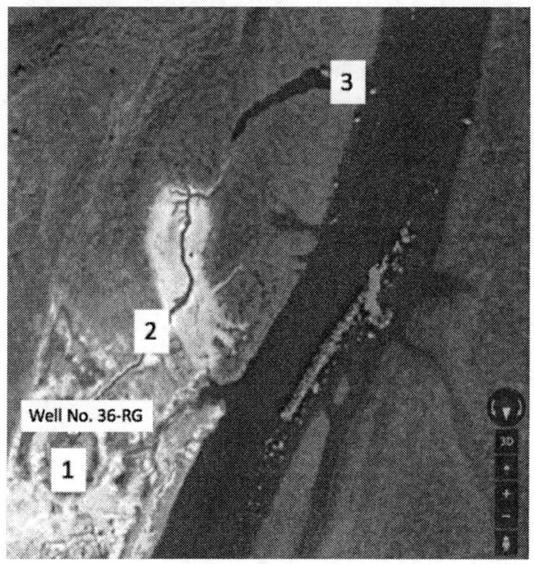

Figure 2. Sampling sites: (1) The Well No. 36-RG; (2) A creek is 90 m below the well; (3) A creek is 200 m below the well, before flowing into the Aremzyanka River.

3. Results

Analysis of bottom sediment samples in the area of well No. 36-RG of the Cherkashinsky iodine-bromine field, conducted by the Laboratory of Radiation Ecology of the Tyumen State University at the "USK Gamma Plus" spectrometric complex, showed the results given in Table 2.

Table 2. Natural radioactivity in the area of the well "Cherkashinskaya 36-RG."

The name of the indicator, Unit of measurement	A place of sampling, a type of sediment		
	Well No. 36-RG	Creek 90 m below the well	Greek 200 m below the well before flowing into the Aremzyanka River
	Clay rock	Clay rock	Clay rock
Activity ^{226}Ra, Bq / kg	18.45	13.15	22.22
Activity ^{232}Th, Bq / kg	28.00	42.70	74.50
Activity ^{40}K, Bq / kg	362.70	375.40	319.80
Activity ^{137}Cs, Bq / kg	3.52	2.30	5.36
Effective activity of the sample, Bq / kg	86.00	101.00	147.00
Volumetric activity of radium (^{226}Ra) in water, Bq / dm^3	0.68	0.31	<0.3
RA, mSv / h	0.09	0.09	0.09

The volumetric activity of radium in water was investigated at the Testing Laboratory of the Regional Analytical Center JSC; measuring the level of surface radioactivity was carried out using a SRP-88N radiometer, the results are also presented in Table 2.

The effective activity of the samples was calculated by the following formula: Aeff=ARa+1,31ATh+ +0,085Ak (1).

4. Discussion

According to I. I. Pluman [5], the content of thorium and potassium in all sedimentary rocks of the Mesozoic West Siberian Plate is almost the same, different rocks differ from each other only in the concentration of uranium, and the content of uranium increases from sandstones to clayey rocks.

Table 3. The content of natural radioactive elements in the rocks of the West Siberian plate [5].

Rock	Number of samples	Uranus, 10-4%		Thorium, 10-4%		Potassium, %	
		From / to	Average	From / to	Average	From / to	Average
Argillite (cemented clay rock)	58	1.0-5.0	2.6	3.7-13.2	7.3	1.2-4.8	2.2

5. Conclusion

The effective activity of all samples of the bottom sediments does not exceed 1500 Bq / kg. According to this indicator, the analyzed bottom sediments belong to the first category of industrial products of oil and gas enterprises according to SanPin 2.6.6.1169-02 [6]. This means that the bottom sediments near self-flowing iodine-bromine wells can be stored and disposed of as normal industrial waste.

According to SanPiN 2.6.1.2523-09 "Radiation Safety Standards NRB-99/2009" [6], the maximum content of ^{226}Ra in water is 0.68. As Table 2 shows, there is no exceedance; therefore, the water flowing today from the Cherkashinskaya No.36-RG well does not represent a radiation hazard.

The level of surface radioactivity also does not exceed acceptable values.

References

[1] Starkov V D, and Tyulkova L A 1996 *Geological history and mineral wealth of the Tyumen region* (Tyumen, Russia: "Tyumen" IPP)

[2] Konovalov I A, and Pak I V 2010 *Ecological condition of the territories in the areas of exploratory wells of the unallocated subsoil fund of the south of the Tyumen region* In *Materials of International Conference on Environment and Natural Resources Management* (pp 177-179) (Tyumen, Russia: TyumSU Publishing House)

[3] Lezin VA 2017 *The rivers and the lakes of the Tyumen region: Tobolsk and Vagai districts: an encyclopedic dictionary* (Tyumen, Russia: TyumSU Publishing House)

[4] Kovalchuk A I, Yushkova S O, and Bortnikov I V 1984 Bromine, iodine and boron in the groundwater of the Lower Eocene aquifer of the Tobolsk artesian basin *Information materials / Academy of Sciences of the USSR, Ural Scientific Center, Institute of Geology and Geochemistry named after A. N. Zavaritsky*

[5] Pluman I I 1976 Possibilities of using data on radioactivity to solve some problems of petroleum geology in Western Siberia. In *Geological structure and oil and gas potential of the south of the Western Siberia according to new data* (pp. 108-113) (Tyumen, Russia)

[6] GosStandard 2002 *SanPiN 2.6.6.1169-02. "Ensuring radiation safety in handling industrial waste with a high content of natural radionuclides at the facilities of the oil and gas complex of the Russian Federation"* (Moscow, Russia)

IOP Conf. Series: Earth and Environmental Science **395** (2019) 012049 doi:10.1088/1755-1315/395/1/012049

Features of using the land resources in the agricultural sector of the Russian Federation

A Razin[1*], R Meshcheryakova[1] and O Razin[2]

[1] All-Russian Scientific Research Institute of Vegetable Growing – Branch of the Federal Scientific Center for Vegetable Growing, Vereya, Ramenskoye district, Moscow region 140153 Russia

[2] Federal Scientific Center for Vegetable Growing, 14 Selektsionnaya str., VNIISSOK Village 143080 Russia

E-mail: vniioh@yandex.ru

Abstract. The article presents the results of a systematic analysis of land use in the world and the Russian Federation, which has a significant potential for agricultural land. The authors analyze such issues as the land reform and formation of a mixed agricultural sector; the use of farmland and arable land assigned to users; soil fertility depletion; a sharp reduction in the use of mineral and organic fertilizers; underestimation of the role of crop rotations.

1. Introduction

Today, more than 200 countries of the world are trying to overcome and get out of the economic, financial, food, and other crises. Most underdeveloped and developing countries are concerned about food shortages. The problem of providing the population with food in the world has not been solved yet. Over 50 years, from 1950 to 2000, the population of the planet grew from 2.5 billion to 6.2 billion people. The planet population reached 7.6 billion people in June 2018. According to UN experts, the planet population on average is increasing by one and a half times. Thus, it will reach 9.1 billion people by 2050. According to the UN, the number of hungry people on the planet reaches 2 billion people or every third inhabitant of the world is undernourished [1].

According to the FAO, with the growth of the world's population, global food demand will double by 2050. For this, grain yields in developing countries need to be raised by 40%, irrigation water consumption needs to be raised by 50%, to 200 million hectares of additional land, which not only is not present on the planet, but a permanent reduction of arable land per capita is underway, should be involved in the economic turnover. At the same time, due to the lack of resources for intensive agricultural production, 20% of arable land and 10% of pasture land in the world is undergoing degradation. The degree of plowing up of territories with limit values of 40% reached 75% in China, 70% in India, and more than 55% in the Russian Federation [2].

Meanwhile, according to the FAO, the humans receive 95-97% of food through agricultural production using soil resources and no more than 5% comes from the World Ocean and aquaculture. Therefore, the effective use of land resources is a determining factor for sustainable agricultural production.

The economic development of most countries in the world continues to be carried out mainly by reducing natural resources.

Content from this work may be used under the terms of the Creative Commons Attribution 3.0 licence. Any further distribution of this work must maintain attribution to the author(s) and the title of the work, journal citation and DOI.
Published under licence by IOP Publishing Ltd

The pollution of water, atmosphere, and soil has become threatening. The natural soil erosion is magnified by human activity. It has been established that about 75 billion tons of soil is removed from land as a result of wind and water erosion each year. In Europe, 46% of the soil is subject to erosion. Dust storms take out up to 20-40 tons of soil per hectare, while the natural recovery is 20-40 times lower. In Europe, the rate of soil erosion is 17 times higher than the rate of its formation, and it is not more than 17 t / ha per year in recent years.

In 2007-2009, in the world, 6-7 million hectares of land fell out of agricultural use as a result of erosion, and another 1.5 million hectares of land fell due to waterlogging, salinization and leaching annually [2].

For Europe, acidification is the most common type of soil pollution, especially in Ukraine (about 11 million hectares) and in Poland (about 10 million hectares). The deposition of nitrogen oxides and sulfur leads to a decrease in soil pH. In Central Europe, the deposition of nitrogen oxides exceeds 50 kg / ha per year, while in Central Europe and China, the deposition of sulfur oxides exceeds 10-100 kg / ha per year.

In Europe, more than 3 million hectares of land are saline. Salinization can be caused by overexploitation of groundwater in coastal areas, contributing to the infiltration of seawater, the use of saline water for irrigation. In most cases, salinization slows plant growth, reduces biodiversity, leads to desertification.

For Russia, in addition to erosion, salinization, waterlogging, deflation, desertification, waterlogging, overgrowing of fields by scrublands and shrubs and other processes leading to the loss of fertility of agricultural land and the withdrawal of arable land from agricultural circulation are characteristic. The main reasons for this are both climate change and economic activity.

According to the state report on the state of the environment in 2007, 80% of agricultural land (mainly erosion) is subject to degradation in Russia. Salinization covers up to 20% of land, 12.3% of the area falls on waterlogging, 17.8% is due to water and 8.4% on wind erosion. The most important consequence of erosion is the reduction of organic matter in the soil and the productivity of ecosystems, the reduction of biodiversity in the plant - soil microorganisms' system. It is established that over the past 100 years, the content of humus in the soil has decreased by 30% in Russia. For the period from 1979 to 2000, the area of degraded soils increased 1.6 times [3].

Thus, the humanity will have to tackle the food problem in the conditions of a growing shortage of land and water resources, global climate change, the need to preserve natural landscapes, and maintain biological diversity.

2. Land Resources: Russia in Focus

The Russian Federation has the necessary potential for the development of the agro-industrial complex, which includes extensive territories with fertile soils, rich reserves of fresh water, and territorial proximity to the largest consumption centers. At the end of the last century, 9% of the world's productive arable land, 52% of the world's black soil, 20% of the world's freshwater reserves and 8.5% of the production of mineral fertilizers accounted for Russia [4].

Russia had a significant potential for agricultural land, ranking fourth in the world in terms of its reserves (124 million hectares of arable land) after the United States (176 million hectares), India (162 million hectares), and China (143 million hectares). And it is also ranked 4th in the presence of arable land per resident (0.86 ha), after Australia (2.40 ha), Canada (1.46 ha) and Argentina (0.89 ha) [4].

As of January 1, 2013, the land area of the Russian Federation amounted to 1,709.8 million hectares. In the structure of the land fund of Russia, forest land (1121.9 million hectares or 65.6%) and agricultural land (386.1 million hectares or 22.6%) occupy the main share. The area of agricultural land in all categories of land amounted to 220.2 million hectares, which corresponds to 12.9% of the land fund of Russia. This means that only one-eighth of the entire land fund of the country is suitable for farming.

In the structure of agricultural land, the share of arable land was 55.2% of its total area, and this was one of the best indicators in the world. A high proportion of forest land provides a favorable economic environment not only in our country, but throughout the world. At the same time, the existing structure assumes a careful attitude to the existing fund of agricultural land.

Over the years of land reform, a steady trend of reducing the area of agricultural land has developed, including arable land (Table 1) [5], [6].

Table 1. Availability and use of land resources of the Russian Federation, thousand hectares (according to the results of the All-Russian Agricultural Census 2006 and 2016).

	On June 1, 2006	On June 1, 2016
Total land area	450,599.5	348,362.8
of the total area: agricultural land	165,985.1	142,659.7
including: arable land	102,140.0	94,641.1
hayfields	13,930.2	10,309.5
pastures	35,200.5	26,539.4
fallow lands	13,936.1	10,519.2
Of the total agricultural land, the following amount is actually used	125,480.3	125,031.4

Over the 10 years that have passed between the agricultural censuses of 2006 and 2016, the total land area has decreased by 22.7%, agricultural land – by 14.1%, and arable land – by 9.8%. In fact, the used area of farmland increased from 27.8% in 2006 to 35.9% in 2016.

In Russia, a total of 255 million hectares have moved to other categories from agricultural land since 2008. The area under crops was reduced from 118.4 million hectares in 1986-1990 to 77.7 million hectares in 2011–2015, or by 34.4% due to alienation and transition to other categories, including the category of waste land, also not used by the state in agricultural production (Fig. 1).

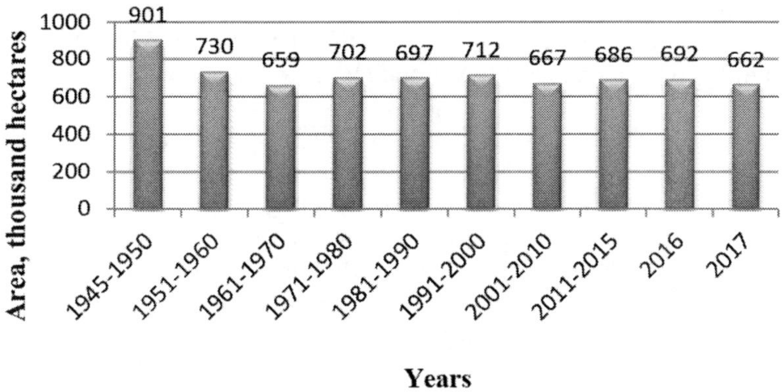

Fig. 1. Dynamics of average annual indicators of sown area of vegetable crops in farms of all categories, thousand hectares (according to Rosstat).

According to A. V. Petrikov (All-Russian Agricultural Census, 2016), as of July 1, 2016, among personal subsidiary and individual farms of citizens in rural settlements, the number of abandoned land plots and with empty houses in 17 regions ranged from 20.2% (Tula region) to 43% (Sakhalin region), in the Kamchatka region – 20.5 %%, in the Komi and Altai Republics – 20.1% of the total number of farms. Abandoned land plots of citizens and abandoned land, turned off from circulation, including fertile and floodplain vegetable-suitable land plots, are overgrown with shrubs, they are being swamped. In the Far East and Siberia, part of the unused agricultural land is already overgrown with forest.

Throughout the entire period of reforms, the overgrowing of agricultural land by forest has been observed, which is due to their lack of demand. In some regions, the voluntary refusal of the agricultural organizations and peasant farms from the lands given to them earlier due to the lack of their own resources and the lack of state support measures by the regional authorities is recorded annually. Lands of this category are often allotted for the construction of new and expansion of the territory of existing enterprises of industry, communications and transport. Large areas that belong to the peasant farms are used for unauthorized landfills. This situation testifies to the irrational use of land due to the inaction of regional authorities.

Measures to engage unused land into circulation are ongoing, but their rates remain low.

In Russia, about 30 million hectares of arable land remains unused. According to expert assessments of the specialists of the Ministry of Agriculture of the Russian Federation, about 12 million hectares of arable land can actually be put into circulation, while the rest of arable land not used in the pre-reform period has been overgrown with forest. Therefore, it is constantly drawn up and transferred to the forest fund.

The reclamation construction and restoration of soil fertility is extremely slow. The irrigated agricultural land decreased from 2375.1 thousand hectares in 2006 to 1816.5 thousand hectares in 2016 or by 23.5%. 55.2% of agricultural land from the total area was actually irrigated in 2016 (Table 2).

Table 2. The presence and use of reclaimed land of the Russian Federation, thousand hectares (according to the results of the All-Russian Agricultural Census 2006 and 2016).

	On June 1, 2006	**On June 1, 2016**
Irrigated farmland	2,375.1	1,816.5
actually irrigated lands from them	939.2	1,003.0
Drained farmland	2,405.0	1,072.6
land with actually operating a drainage network of them	1,100.4	667.7

In 2012, in the country, 489.7 thousand hectares were at the stage of land reclamation and soil fertility restoration, if necessary 1.5-2.0 million hectares per year. Large areas of land are irrigated in the Krasnodar, Stavropol, Rostov, Astrakhan and Volgograd regions. At the same time, the ameliorative condition of most of the lands of these regions is characterized as good.

Thus, the ongoing land reforms are not yet conducive to the full and effective use of existing land potential.

During the years of the Russian Empire and the Russian Federation, several land reforms were carried out [7], [8]. In the early 90s, in Russia, another attempt at a fair distribution of land was made. The economic reform in the agricultural sector of Russia began with land reform. The main task of the agrarian reform in Russia carried out since 1991 was the privatization of land and the reorganization of collective and state farms with the aim of creating effective agricultural production based on private ownership of land and personal enterprise of agricultural producers.

The adopted legal acts were aimed at changing the state land ownership to private. In the pre-reform period, the main forms of management in the agrarian sector of the country were collective and state farms, the number of which was just over 25 thousand. During the reforms, 11.9 million Russian peasants received conditional land shares (10-15 hectares per person). Collective and state farms were reorganized into various organizational and legal forms of private property: 5,400 joint-stock companies, 6,700 limited liability companies, 116,000 agricultural cooperatives, 3,200 state unitary and municipal enterprises were created out of 25.8 thousand collective and state farms (1990).

The main goal of land transformations in the Russian Federation was to ensure the rational use and protection of land as the most important natural resource, the creation of legal, economic, organizational and technological and other conditions for the reproduction and improvement of soil fertility, the preservation of rural, forest, and other lands, improvement of the natural environment, development of rural and urban settlements. Restrictions on developing personal subsidiary farms (PSF) were removed. Farm sector began to be created. In 2006, as a result of transformations, a mixed agrarian sector was formed, consisting of agricultural organizations, peasant farms and individual entrepreneurs, personal subsidiary farms.

The distribution of land and material means of production into private property due to contradictions and non-conceptual agrarian land policy of the state did not lead to an increase in business activity of millions of new owners, the flow of capital into agriculture. All these factors led to the fact that the destructive tendencies in the agrarian sphere are clearly ahead of the creative processes.

According to the 2006 Agricultural Census, 58,901 agricultural organizations were in the agricultural sector, including 27,781 large and medium-sized enterprises, 253,236 peasant farms and 31,789 of individual entrepreneurs. Only 67.6% of agricultural farms carried out agricultural activities, including 69.5% of large and medium, 49.3% of peasant farms and 66.0% of individual entrepreneurs.

In 2001-2005, the most active cessation of the activities of peasant farms and individual entrepreneurs was registered. The reform of state and collective farms led to a sharp decline in production. The main reason was the division of the material and technical base into property shares of the owners, which eventually led to a resource and production crisis. At the same time, small enterprises did not receive the equipment necessary for processing the received land plots, as well as the production facilities necessary for carrying out production activities. The absence in small enterprises of the necessary equipment, high-quality seeds, qualified specialists (agronomists, engineers, machine operators, etc.), necessary services and departments capable of maintaining the existing technological and technical level of production led to the need for the following actions: either refuse to use the land, or lease their land resources to large highly integrated holding companies and agrofirms, or join rapidly developing farms. On the basis of medium and small enterprises, with the participation of investors, large agricultural holdings were created, which had financial and technical capabilities to involve unused land resources into circulation [9].

According to the results of the agricultural census of 2016, for the period from 2006 to 2016, the number of agricultural farms decreased by 22.8 thousand (38%), the number of farms decreased to 136.6 thousand (53.5%), personal farms remained around 16.2 million (i.e. their number decreased by 20%), private households remained at a figure of 3.2 million, or three times less than in 2006.

In 1991, the transfer of land ownership rights to new owners in the amount of 213.8 million hectares was carried out, including 209,8 million hectares of agricultural farms and 4 million hectares of peasant farms and personal subsidiary farms. After 15 years, due to the withdrawal of land from agricultural land, the area of agricultural land in the agricultural farms decreased by 36% (to 133.6 million hectares), and it increased in farms to 24.5 million hectares (6 times). The weakening of state control over the level of land use, the reduction of state support to a minimum, the provision of excessive autonomy to landowners when using agricultural land led to the fact that some of the land was not used by new landowners. According to the 2006 agricultural census, the area of agricultural land used was reduced to 96.5 million hectares (by 27.7%), 23% of agricultural land was not used by large and medium-sized farms, as well as it was not used in by small enterprises (47.5%), subsidiary farms (38%), and by farms (19.3%) already by 2006.

Table 3. The area of land assigned to different users, thousand hectares (according to the results of the All-Russian Agricultural Census 2006 and 2016).

	Agricultural organizations		**Peasant farms**		**Individual entrepreneurs**	
	2006	**2016**	**2006**	**2016**	**2006**	**2016**
Total land area	402,110.6	290781.2	26349.3	37878.2	3375.8	5434.3
including farmland	133,596.8	90184.0	21918.5	35046.7	2554.7	4531.1
of which: arable land	82,268.7	64891.2	14993.7	23645.5	1734.3	3202,3
Farmland from the total area i s actually used	96,542.5	80193.1	17697.8	32368.4	2124.5	3920.3

The census of 2016 showed that in the 10 years since the previous census, the area of actually used land has increased significantly. In agricultural organizations, these lands increased from 72.3% in 2006 to 88.9% in 2016, in peasant farms and individual entrepreneurs – from 81.0% to 91.7%, respectively. At the same time, 12.2% of agricultural farms and 17.2% of peasant farms and individual entrepreneurs did not have land in their ownership. The land area over 10,000 hectares had 76.7% of agricultural organizations (the average size per organization was 6,612.2 hectares), 33.8% of peasant farms, and the individual entrepreneurs had a land area of 501 to 3,000 hectares, 30.2% more than 3,000 hectares, and 0.3% of farms possessed up to 4 hectares (with an average area per farm around1.7 hectares) (Table 3).

Depending on the regional location of agricultural organizations with a land area of up to 3 hectares, potatoes, vegetables, and melons and gourds predominate in crops.

More than half of agricultural products, including vegetable, are produced in personal subsidiary farms, although an increase in production in recent years is reduced in them. At the same time, small agricultural organizations are engaged in the cultivation of monocultures, and medium-sized and large farms are beginning to use a science-based farming system, which includes crop rotation. Sowing and harvesting complexes were effectively used only in 7.8% of medium and large farms. Agricultural and farm enterprises are engaged in entrepreneurial activities, while private farms are developed mainly for the purpose of self-employment and self-sufficiency. Therefore, the development of agricultural farms and peasant farms is of economic importance, while personal subsidiary farming has the aim of social development of rural areas.

According to Rosstat, in Russia, in the post-perestroika period, 660-690 thousand hectares are actually under vegetable crops annually, which is about 0.9% of the total crop area in the country.

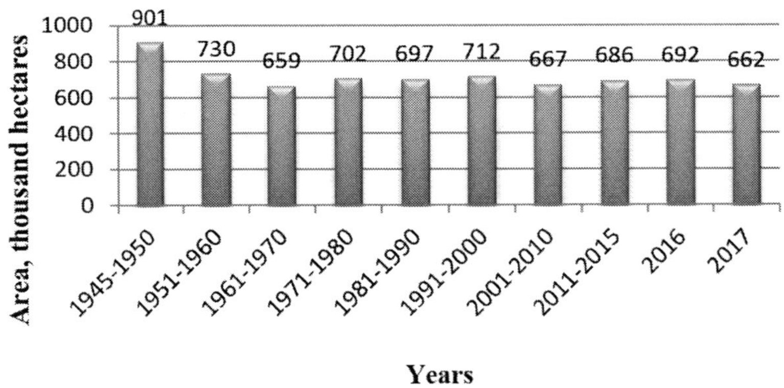

Fig. 2. Dynamics of average annual indicators of sown area of vegetable crops in farms of all categories, thousand hectares (according to Rosstat).

Since 1945, the general dynamics of the open-field vegetable crop area tends to decrease. Sown areas of vegetable crops in farms of all categories decreased from 901 thousand hectares in 1945-1950 to 697 thousand hectares in 1981-1990 and to 662 thousand hectares in 2017 (Fig. 2) [9, 10].

During the period of reforms, a sharp decline in acreage in the agricultural organizations, and an increase in the farms of the population and emerging peasant (farmer) farms occur. In 1986-1990, vegetable crops occupied 669.4 thousand hectares, 470 thousand hectares (72.2%) of them were sown by agricultural organizations, 199.4 thousand hectares (29.8%) of which were households. In 2001–2005, 118.0 thousand hectares (17.1%) of 690.1 thousand hectares belonged to agricultural organizations, 530.2 thousand hectares (76.8%) belonged to households and 41.9 thousand hectares (6, 1%) - to peasant (farmer) farms and individual entrepreneurs. From 2006 to 2016, the share of sown areas for categories of farms changed slightly, including for agricultural organizations, the share changed from 13.6% to 16.1%, for households – from 70.8% to 75.5 %, for peasant farms and individual entrepreneurs – from 10.3% to 13.5%. At the same time, from 2015 to 2017, the production of vegetables in agricultural organizations increased from 17.9% to 21.2%, in peasant farms – from 15.1% to 15.9%, but the production of vegetables decreased in households from 67,0% to 62.9% of the relevel production level in farms of all categories.

As we see, the agrarian reform was accompanied by the irrational use of land resources in some cases.

Evaluating the efficiency of farmland use, we should note that 67.6% of the arable land available is used for crops in Russia as a whole. The level of arable land use by districts of Russia ranges from

64.0% in the Far Eastern Federal District to 76.4% in the North Caucasus Federal District. The irrational use of arable land is reflected in a decrease in the general culture of agriculture, a violation of crop rotations, and cultivation technologies. The application of mineral and organic fertilizers, plant protection products from pests, diseases, and weeds were reduced. In solving the problem of preserving and increasing the volume of production, this leads to a depletion of fertility and an increase in the need for material and technical resources. However, with a lack of material and technical resources, the soil fertility decreases, and with an increase in anthropogenic impact, the soil is excessively polluted and compacted. So, the use of heavy agricultural machinery leads to soil compaction. Physical degradation leads to a decrease in porosity and permeability of soil particles, reduced access of water and oxygen to plant roots, slowed plant growth, reduced number of available habitats for soil organisms, including earthworms and microorganisms, and, ultimately, reduced productivity and biodiversity. Such physical degradation covers more than 28% of the agricultural land of European countries.

Optimal solutions in the use of land resources can be achieved by maintaining the area of arable land and the optimal levels of intensification of farming and advanced technologies.

In 2000, 144 million tons of mineral fertilizers were used in the world, and this is an average of 23 kg per person. Globally, 103 kg of fertilizer was applied on average per 1 ha. From 1980 to 2010, the use of mineral fertilizers decreased from 860 kg to 520 kg per 1 hectare in the Netherlands, while a number of countries increased their use: from 297 to 346 in the UK, from 244 to 356 in Israel, from 542 to 651 in Ireland, and from 453 to 836 kg / ha Malaysia. The growing use of mineral fertilizers in the world threatens environmental degradation, and it is dangerous for the environment and the population.

One of the guarantees of a stable growth in crop production in Russia can be the rational use of mineral and organic fertilizers. After all, our country produces 8.5% of the global production of mineral fertilizers.

In the framework of the target program "Preservation and Restoration of Soil Fertility of Agricultural Lands and Agrolandscapes as the National Achievement of Russia for 2006-2010 and for the Period up to 2012," measures to restore soil fertility were envisaged. Within the framework of the State Program, land protection measures were fully financed, and fertilization measures were partially funded. Therefore, with the plan of 12.4 million tons, 10.9 million tons were contributed in fact, 90 kg per 1 ha of arable land. The plan for the application of mineral fertilizers as a whole is fulfilled only by 88.1%. However, a sharp decline in these indicators is going on (Table 4).

Table 4. Production and application of mineral and organic fertilizers for vegetable crops (according to Rosstat).

	2015	2016	2017
Production of mineral or chemical fertilizers (in terms of 100% nutrients) million tons including the following:	20.1	20.8	22.5
nitric	8.7	9.5	10.0
phosphoric	3.3	3.6	3.9
potash	8.1	7.8	8.6
Total mineral fertilizers applied for sowing in agricultural farms (in terms of 100% nutrients), million tons	2.0	2.3	2.5
per 1 hectare, kg the total area under crops	42	49	55
vegetables and melons	166	195	198
Organic fertilizer applied total million tons	64.2	65.2	66.6
per 1 hectare, t the total area under crops	1.3	1.4	1.5
vegetables and melons	3.6	2.8	2.6

The sale of mineral fertilizers in the domestic market is no more than 23-25% produced in the country. Import of mineral fertilizers increases from 69.6 thousand tons in 2015 to 97.0 thousand tons in 2016, with exports of 15,354.9 thousand tons in 2015 and 15,504 thousand tons in 2016.

According to Rosstat, in 1972, up to 1000 kg of mineral fertilizers and more than 15 tons of organic fertilizers were applied to each hectare of a vegetable field. A sharp decrease in applied fertilizer occurred from 400–450 kg / ha during the pre-perestroika period, to 131 kg in 2006 and to 198 kg in 2017, and a sharp decrease in applied organic fertilizer occurred from 15 t / ha in 2006 to 2.6 t / ha in 2017. At the same time, in agricultural organizations, organic fertilizers were applied to only 8-9% of the cultivated area. Such amounts of mineral fertilizers do not cover 40-50% of the removal of nutrients from the soil, as a result, the soil devoid of organic and mineral fertilizers is depleted. The annual removal of nutrients from the soil is 4-5 times greater than their return. The lack of organic and mineral fertilizers causes a steady trend of soil dehumification, leads to the development of degradation processes, namely dehumification, acidification, erosion, and secondary salinization [10, 11].

According to Rosstat, the proportion of the sown area fertilized with mineral fertilizers is 33.7% in agricultural organizations, 19.5% in peasant farms and individual entrepreneurs. The state monitoring of the lands of the Russian Federation showed that 46% of the surveyed area is characterized by a very low content of organic substances.

The renewal of the fertility of vegetable-friendly soils can be carried out only by a well-thought-out, scientifically based combination of two measures, namely fertilization and proper crop rotation [10], [11].

The role of crop rotation is primarily to improve soil fertility through the following measures: regulating the regime of soil organic matter and mineral nutrients, maintaining a satisfactory structural state of the soil; preventing erosion and deflation; limiting the development of weeds; controlling phytosanitary in soils; increasing productivity and product quality of agricultural crops, including vegetables. Long-term studies of VNIIO in cooperation with vegetable experimental stations located in the main areas of commercial vegetable production on seven soil varieties with humus content from 0.6% to 6.6% and thickness of the humus horizon from 15-20 cm to 100-120 cm, showed the following. Preservation of high soil fertility in vegetable growing, optimization of their nutritional regime, improvement of physicochemical, agrophysical, and agrochemical properties of the soil is possible only with the use of organo-mineral or organo-mineral-biological fertilizer systems in vegetable-feed and vegetable- crop rotations [10], [11].

Scientists and practitioners of agricultural production are in a state of constant search and development of the most rational economically acceptable technologies for the production of agricultural products, including vegetable. Currently, a number of alternative farming systems are developed and exist in practice, including biological (ecological) farming. The farming system suggests the following: using crop rotations and fertilizers properly, such as organic, plant residues, straw, green manure, etc.; ensuring natural soil fertility, which solves the problem of nutrition; improving the uptake of nitrogen due to legumes; complete rejecting the use of mineral fertilizers and pesticides. Ensuring the ecological purity of products and the absence of environmental pollution and increasing the natural fertility of the soil is a valuable and attractive feature of biologization.

3. Conclusion and Suggestions

The state should play a major role in solving the complex problem of rational use of land resources, including with respect to arable lands. Only with the help of state regulation (land management, land monitoring, the state land cadastre, etc.), prudent management on earth can be achieved using economic impact mechanisms. Mechanisms of state control over the targeted use of land will solve the problems of entering unused land into agricultural circulation. The modern land reform does not promote the rational use of land resources. The mechanism of rational use of land resources in the agro-industrial complex, including in vegetable production, should include both the reproduction of soil fertility and the use of science-based agricultural technologies that contribute to the preservation

IOP Conf. Series: Earth and Environmental Science **395** (2019) 012049 doi:10.1088/1755-1315/395/1/012049

of soil fertility. Preservation and improvement of soil fertility in the agro-industrial complex of the Russian Federation, including vegetable growing, can be provided by a combination of the necessary agrotechnical and phytosanitary measures. Development of the unclaimed land (land shares) redistribution and their return to agricultural use is needed.

References

[1] Tulcheev V V, Lukin N D, and Yagfarov O M 2013 *The strategy for food and national security of Russia in the world economic space in the 21st century* (Moscow, Russia: Rosselkhozakademiya)
[2] Golubkina N A, Pivovarov V F, Nadezhkin S M, Loseva T A, and Sokolova A Ya 2013 *Global environmental crisis: problems and solutions* (Moscow, Russia: VNIISSOK Publishing house)
[3] Fateeva T V, Shubich S M *et al* 2012 *State National Report on the State and Use of Lands of the Russian Federation in 2012* (Moscow, Russia: "Petrovsky Park")
[4] Zhuchenko A A 1994 *The strategy of adaptive intensification of agriculture (concept)* (Pushchino, Russia: Scientific Center of the Russian Academy of Sciences)
[5] Federal State Statistics Service 2008 *Results of the All-Russian Agricultural Census 2006: Land resources, their use* (Moscow, Russia: SRC "Statistics of Russia")
[6] Federal State Statistics Service 2018 *The results of the All-Russian Agricultural Census 2016: Land resources their use* (Moscow, Russia: SRC "Statistics of Russia")
[7] Galishin E U 2012 Development of land relations in Russia and in the world *Legislation and Economics* **5** pp 28-31
[8] Varlamov A, and Galchenko S 2013 Development of land relations in Russia *Expert* **5** pp 17-21
[9] Oksanich N I 2007 *Economic sustainability of agricultural organizations* (Dissertation of Doctor of Economic Sciences) (Moscow, Russia)
[10] Litvinov S S 1998 *The scientific foundations of contemporary vegetable growing* (Moscow, Russia: Rosselkhozakademiya)
[11] Borisov V A 2016 *Vegetable fertilizer system* (Moscow, Russia: Rosinform Agrotech Publishing House)

A cost-effective evaluation of carrot resistance to *Alternaria sp.* and *Fusarium sp.*

L M Sokolova[1], M V Shatilov[1] and O A Razin[2*]

[1] All-Russian Scientific Research Institute of Vegetable Growing – Branch of the Federal Scientific Center for Vegetable Growing, Vereya, Ramenskoye district, Moscow region 140153 Russia
[2] Federal Scientific Center for Vegetable Growing, 14 Selektsionnaya str., VNIISSOK Village 143080 Russia

E-mail: vniissok@mail.ru

Abstract. Creation of forms, varieties, and hybrids of plants resistant to biotic factors is an important national economic task. It is known that the defeat of agricultural plants by fungal, bacterial, and viral diseases, as well as pests leads to a loss of 38.8% to 60.4% of the crop. The most promising way to protect plants from phytopathogens is selection, which allows to obtain varieties and hybrids of crops with complex resistance. However, the process of creating varieties and hybrids remains long and reaches ten years or more. At the same time, the degree of plant resistance to harmful organisms is often insufficient and does not always meet the requirements of production. The efficiency of the breeding process can be improved and accelerated through the use of rapid methods based on the selection in the laboratory of samples resistant to selective factor. The study was carried out in the Laboratory of Immunity at the VNIIO – Branch of the Federal State Budgetary FNCO in the Moscow region (Russia). As source material used, three samples of carrot breeding in VNIIO characterized by different resistance to *Fusarium* and *Alternaria*. The results of the experimental work showed the possibility of rapid assessment of carrots for resistance to PP fungal diseases. *Alternaria* and *Fusarium* seeds on the culture liquid filtrate at a concentration of 50% when germinating. This allows in the laboratory in a short time to evaluate a large number of breeding samples. The method of rapid assessment allows in 3.4 times to reduce the time required for the evaluation of breeding material and to reduce the consumption of seed.

1. Introduction

In recent years, the demand for sustainable hybrids and varieties of vegetable crops has increased significantly. Sustainable hybrids and varieties should become the basis of integrated protection, which is especially important during the period of application of new agricultural production technology in connection with the threat of increased damage to crops.

The susceptibility of carrot root crops to disease is the main difficulty in obtaining a consistently high crops of garden carrots, preserving marketability, especially complicates the solution to the problem of growing healthy seeds [4; 11].

Defeating of plants by wreckers occuers at all stages of growth and development; therefore, a timely identification of the first symptoms of a disease and their correct diagnostics are highly important [1].

Content from this work may be used under the terms of the Creative Commons Attribution 3.0 licence. Any further distribution of this work must maintain attribution to the author(s) and the title of the work, journal citation and DOI.
Published under licence by IOP Publishing Ltd

The genus *Fusarium* includes a number of species that are the causes of various diseases in a number of important crops, such as cereals, vegetables, etc. Some of the important species in this regard are *F. oxysporum* (*Fo*), *F. avenaceum* (*Fa*) и *F. poae (Fp)*. The most common are the fungus species of *F. oxysporum,* causing wilting disease and affecting the vascular system of plants [2]. *F. avenaceum* is a widespread species that may exist, including as a saprophyte. *F. poae* belongs to the *Sporotrichiella Wollenw* section [7; 12].

About 10 p. *Alternaria* species are causative agents of the most harmful diseases, significantly differing in pathogenicity, degree of specialization, harmfulness, sensitivity to fungicides, etc. The main problems of monitoring alternariosis in Russia are associated with the lack of modern determinants, poor use of microscopy and molecular methods to identify pathogens [8; 6]. Crop losses can reach even 40-99% [3; 14].

The main purpose of the research was to increase the efficiency of the selection process of garden carrots for resistance to *Fusarium* and *Alternaria,* based on research on the evaluation of carrot plants by various methods: germination of seeds on the filtrate of the culture medium of the causative agent of *Fusarium* and *Alternaria,* evaluation based on an artificial infectious backgrounds.

2. Research Tasks

1. To develop elements of the express-evaluation methodology of carrots for resistance to *Fusarium* and *Alternaria,* based on the germination of seeds on a filtrate of the culture medium of fungus *F. oxysporum* and *A. radicina.*
2. To evaluate carrot samples for resistance to *Fusarium* and *Alternaria* against an artificial infectious background using traditional methods.
3. To determine the economic efficiency of express-evaluation in laboratory conditions and traditional evaluation on an artificial infectious background.

3. Materials and Method

As an object of research, we used 3 samples of carrots from the VNIIO (Branch of the Federal Scientific Center for Vegetable Growing), characterized by different field resistance to *Fusarium* and *Alternaria*: stable to fungal diseases from the p. *Alternaria and Fusarium* variety "Vitaminnaya 6," the medium susceptible variety "Amsterdamskaya," and the susceptible variety "Tayfun-16."

3.1. Method of preparation of laboratory experience

The laboratory research was conducted according to methodical recommendations [10]. The pathogen cultures were sown on an agarized nutrient medium with hyphae and mycelium spores, and on liquid nutrient media with a suspension of macro- and microconidia. Pathogen cultures were cultured at 26 °C under thermostat conditions in a nutrient medium using distilled water [5]. The concentration of sucrose in the nutrient medium was 30 g/l. Accounting for the development of colonies was carried out visually on 10, 20, 30, and 40 days by increasing their size in diameter.

The suspension culture of the pathogen strain was prepared according to the Bilay method and introduced into a liquid nutrient medium.

A culture medium was obtained by growing fungal isolates in 300 ml flasks in 100 ml of nutrient medium. To obtain a culture medium, a piece of agar measuring 0.5 cm x 0.5 cm, and containing about 10^8 conidia of the fungus was placed in each flask. The flasks were placed in a thermostat at the temperature of 25 °C, for 1 month with regular shaking in a shaker. A freshly grown culture medium (turbid liquid with a specific smell) was filtered through 4 layers of gauze fabric. The resulting filtrate of the culture medium) was kept in an autoclave for 30 minutes [10].

The degree of damage to plants was determined and calculated by the formula:

$$Dp = (a \times 1) + (в \times 2) + (c \times 3) / (3 \times p) \times 100\%,$$

where Dp – a disease progression (%); a, в, c, d – a number of affected plants; 1, 2, 3 – a defeat score; p – total plants.

3.2. *Preliminary research results and discussion*

As a result of the studies, it was found that the germination of carrot seeds in a susceptible sample "Typhoon-16" on the filtrate culture medium of the fungus p. *Alternaria*, at a concentration of 70% seed germination was not observed. On the resistant variety "Vitaminnaya-6" and the medium-receptive variety "Amsterdamskaya," when germinating seeds on 70% filtrate culture medium, the seed germination was 50% and 15%. The root length of the resistant variety "Vitaminnaya-6" and the medium susceptible variety "Amsterdamskaya" did not differ significantly, – it was 10.6% and 9.6%.

When germinating carrot seeds on the filtrate of the culture medium in concentrations of 5%, 10%, and 20%, in germination and root length in all three samples – "Vitaminnaya-6," "Amsterdamskaya," and "Typhoon-16," no significant differences were revealed.

When germinating seeds on a 50% filtrate culture medium, there was a significant inhibition of germination. At the same time, differences were found in the germination of seeds of the studied varieties of garden carrots. In the susceptible variety "Typhoon-16," the root length was 29.2% less than in the resistant variety "Vitaminnaya-6," and 9.2% less than in the medium-susceptible variety "Amsterdamskaya."

The best results were obtained when germinating seeds of garden carrots on the filtrate culture medium of the fungus p. *Alternaria* at a concentration of 50%, which allowed to obtain breeding material with increased resistance to *Alternaria.* An analysis of the data indicates the possibility of express-evaluation of carrots for resistance to *Alternaria* when using a 50% filtrate of the culture medium.

As a result of the studies, it was found that when germinating carrot seeds, in a susceptible sample of "Typhoon-16" on the filtrate of the culture medium of the fungus p. *Fusarium* at a concentration of 70%, seed germination was not observed. On the stable variety "Vitaminnaya-6" and the medium-receptive variety "Amsterdamskaya," when germinating seeds on 70% filtrate culture medium, the seed germination was 60% and 35%. The root length of the resistant variety "Vitaminnaya-6" and the medium susceptible variety "Amsterdamskaya" did not differ significantly, – it was 12% and 10.7%.

When germinating carrot seeds on the filtrate culture medium in concentrations of 5%, 10% and 20%, the germination and root length in all three samples ("Vitaminnaya-6," "Amsterdamskaya," and "Typhoon-16") did not reveal significant differences.

When germinating seeds on a 50% filtrate culture medium, there was a significant inhibition of germination. At the same time, differences were found in seed germination of the studied varieties of garden carrots. In the susceptible variety "Typhoon-16," the root length was 33.1% less than in the resistant variety "Vitaminnaya-6," as well as 8.7% less than in the medium-susceptible variety "Amsterdamskaya."

The best results were obtained when germinating seeds of garden carrots on the filtrate culture medium of the fungus p. *Fusarium* at a concentration of 50%, which allowed to obtain the breeding material with increased resistance to *Fusarium.* An analysis of the data indicates the possibility of express-evaluation of carrots for resistance to *Fusarium,* using a 50% filtrate culture medium.

The results of the analysis indicate the possibility of express-evaluation of carrots for resistance to phytopathogenic fungi from pp. *Alternaria* and *Fusarium,* when using a 50% filtrate culture medium.

The duration of express-evaluation of carrots for resistance to fusarium and alternariosis in laboratory conditions, taking into account the preparatory work, is 54 days. In particular, to obtain a pure culture fungus pp. *Alternaria* and *Fusarium,* a total of 14 days is required. It takes 30 days to get a culture medium. The period of seed germination on the filtrate of the culture medium of the mushroom pp. *Alternaria* and *Fusarium*, according to the requirements for carrot culture, is 10 days.

Let us analyze the cost of this experience (Table 1). In total, 22 m/hrs were spent for carrying out experience. Despite the fact that the experiments were carried out by full-time employees, to determine the total cost of conducting the experiment, it is necessary to take into account how much of

IOP Conf. Series: Earth and Environmental Science **395** (2019) 012050 doi:10.1088/1755-1315/395/1/012050

the payroll fund was paid for the time spent on the experiment. We divided the salary before deducting personal income tax, taking into account deductions by the number of working hours in the current year. Electricity costs were calculated according to the data indicated by the manufacturers of the devices, the current tariffs of the electricity supplier and the actual amount of the devices. Calculation of depreciation was carried out by a linear way. At the same time, the annual amount of deductions was divided by the number of experiments conducted per year.

Table 1. Expenses for laboratory testing.

Expenditures	Expenses, rub.	Expenses, m/hrs
Cooking medium (mixing components)	546.12	1 m/hrs
Selective medium cooking	546.12	1 m/hrs
Autoclave sterilization	2,184.48	4 m/hrs
Experience development	8,737.92	16 m/hrs
Seeds	0.20	
Preparations	1,040	
Electrical energy	51,758.22	
Depreciation	20,600	
Total	85,413.06	

Our cost analysis showed that the fixed costs account for 24% of the total ones. In this case, payroll expenses were assigned to the workers as variables, since the costs were calculated on conducting experience for the general case, and the time spent by the employees could be used to solve other current tasks that are part of the employees' duties. In the structure of variable costs, the costs of electrical energy occupy a large part. Most of the expenses are spent on the thermostat, since according to the methodology for conducting the experiment, it was activated for 10 days continuously.

3.3. The method of laying the field experience on two artificial infectious backgrounds

To create a soil infectious background, pathogens were propagated on a grain nutrient medium. The oat grain nutrient medium for reproduction of infectious material was prepared in the following way. The oat grains were poured into flasks, poured with water in a ratio of 1: 1, and sterilized with dry steam in an autoclave under a pressure of 1 atmosphere for 1 hour. The prepared cereal medium was seeded with a clean fungus culture (14 days old), using standard microbiological methods. The flasks were kept for 14 days in a thermostat at a temperature of 20-250C, and during this time they were shaken every day to evenly distribute the infection. After 14 days, when the substrate is uniformly overgrown with mycelium, it was removed from the flasks and dried to dry completely at the room temperature for 3 days [13].

Preparing the inoculum to create a soil infectious background began 1 month before sowing the seeds. The inoculum propagated by the grain mixture was placed directly in rows (a plot size of 1.50 cm, a row spacing of 30 cm) in the surface soil layer (5-7 cm), the seeds were leveled and sown with further planting.

4. Results and Discussion

After assessing resistance against an artificial infectious background of *Alternaria*, the resistant variety "Vitaminnaya-6" had 54% of resistant genotypes with a lesion score of 0 to 0.8. The variety "Amsterdamskaya" was included in the group of low-susceptible varieties, with a score of 0.9 to 1.6 and had 78% of genotypes. The "Typhoon-16" was included in the group of susceptible with a score of 3.3 to 4, and it had only 2% of stable genotypes.

According to the *Fusarium* infectious background, the distribution of the genotypes is as follows: (a) the resistant variety "Vitaminnaya-6" has 52% of the resistant genotypes; (b) the medium susceptible variety "Amsterdamskaya" has 78% of the genotypes; and (c) the susceptible variety "Typhoon-16" has 2% of the resistant genotypes.

Thus, combining both laboratory and field experiencea, it is possible to obtain a complete assessment of samples for disease resistance.

The traditional method' duration for evaluating carrots for resistance to *Fusarium* and *Alternaria* in an infectious background is 181 days. To obtain a pure culture fungus pp. *Alternaria* and *Fusarium,* it is necessary to spend 14 days. Another 14 days are necessary for the germination of the mycelium on the substrate (oats), as well as 3 days are taken for drying the substrate. The period from sowing to final assessment for sustainability during harvesting is 150 days.

Similar to the previous analysis, we will analyze the second experiment (Table 2). In the field experiment, the working time of employees for conducting the experiment increases significantly, up to 955.5 m/hrs. This is due to the fact that it is necessary to prepare the soil for sowing and care for the plants throughout the duration of the experiment.

Comparison of the studied express method and the traditional method for evaluating carrots for resistance to alternariosis and fusariosis shows the possibility of accelerated evaluation and saving material for selection. The evaluation method based on germination, in laboratory conditions and on the filtrate of the culture medium, allows to reduce the duration and increase the productivity of the evaluation of samples compared to the traditional assessment on an artificial infectious background by 3.4 times.

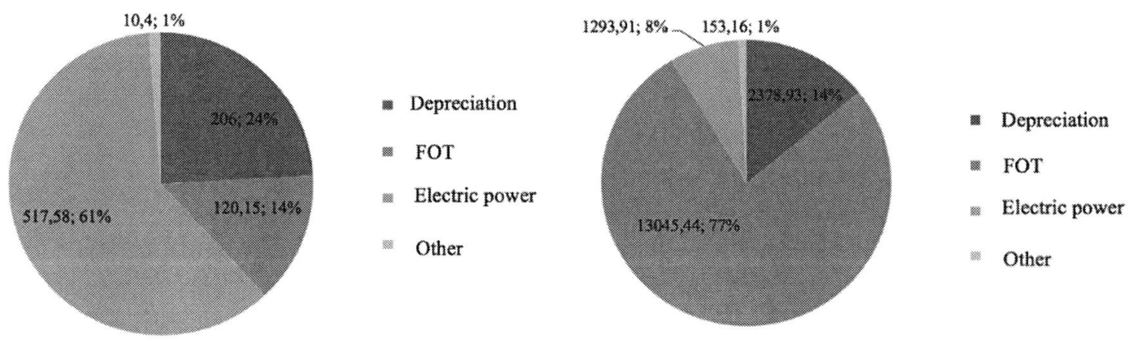

Figure 1. The structure of operating costs for the analysis of 1 sample in the laboratory and the field, RUB, %.

After analyzing the costs, we got the prime cost of 1 experience at the full possible load during the year (6 laboratory experiments and 1 in the field). It is 854.13 rub. for the analysis of each sample by the laboratory method and 16,871.43 rub. by the field method. Unfortunately, the laboratory experiments do not replace field experiments, and some studies require both options. Let's consider operating costs of each method of carrying out experience to understand what way would be less expensive, as well as to choose it in case both options are suitable for a research.

In a specific case, operating costs include all of the costs presented, other than depreciation. Operating costs for the analysis of 1 sample in the laboratory are 648.13 rub., it is 14492.51 rub. in the field experiment. Let's analyze their structure (Figure 1). The diagram shows that all indicators are significantly higher in the field experiment in absolute values. The payroll fund accounts for 77% of the costs, which means a large share of manual labor when conducting experience using this method. The multiple difference in the absolute values of the depreciation rate is due to the shorter downtime of the equipment during the laboratory experiment.

Table 2. The costs of the field experience.

Expenditures	Expenses, RUB.	Expenses, man-hours
Soil preparation and frame making	5,461.20	10
Reproduction and preparation of the substrate	4,642.02	8.5
Sowing, soil preparation	8,737.92	16
Care	494,238.60	905

Cleaning, rating	8,737.92	16
Seeds	0.4	
Preparation – Czapek's medium + alcohol	480	
Film	5,646	
Electrical energy	51,756.23	
Depreciation:	95,157.34	
Total	674,857.63	

5. Conclusion

Our analysis of the effectiveness of the two variants of the experiments showed that the laboratory experience in conducting year-round studies demonstrated a significant reduction in costs compared with the field.

References

[1] Alekseeva K L, and Ivanova M I 2015 *Diseases of green vegetable crops (diagnostics, prevention, protection)* (Moscow, Russia: Rosinformagrotech)
[2] Beckman C H 1987 *The Nature of Wilt Diseases of Plants* (St Paul MN: American Phytopathological Society Press)
[3] Ben-Noon E, Shtienberg D, Shlevin E, Vintal H, and Dinoor A 2001 Optimization of chemical suppression of *Alternaria dauci,* the causal agent of alternaria leaf blight in carrots *Plant Disease* **85**(11) pp 1149-1156
[4] Ben-Nun E 1998 Influence of the type of irrigation on al ternaria blight in carrot *Phytoparasitica* **26**(2) p 172
[5] Bilay V I 1982 *Experimental Mycology Methods* (Kiev, USSR: Naukova Dumka)
[6] Farrar J J, Pryor B M, and Davis R M 2004 *Alternaria* diseases of carrot *Plant Disease*
[7] Gagkayeva T A, Gavrilova O P, Levitin M M, and Novozhilov K V 2011 Fusarium cereal crops *Supplement to the journal Plant protection and quarantine* **5** p 112
[8] Gannibal A И, Orina A S, and Levitin M M 2010 Alternariosis of agricultural crops on Russian on Russian territory *Plant protection and quarantine* **5** pp 30-32
[9] Leunov V I, Khovrin A N, Tereshonkova T A, Sokolova L M, Gorshkova N S, and Alekseeva K L 2011 Methods of accelerated selection of dining carrots for complex resistance to fungal diseases (*Alternaria* и *Fusarium*). In *Methodical recommendations* (Moscow, Russia)
[10] Polyakov A B, Tkacheva A A, Tarasenkov I I, and Biryukova N K 2006 Production of cucumber plants with increased resistance to Fusarium wilt methods IN VITRO. In V A Borisov Ed. *Methodical recommendations* (Moscow, Russia)
[11] Schneider Y I 1972 Disease Queen of carrots *Protection of Plants* **7** pp 18-19
[12] Semenov A N, Divashuk M G, Bazhenov M S, Karlov G I, Leunov V I, Khovrin A N, Egorova A A, Sokolova L M, Tereshonkova T A, Alekseeva K L, ... Leunova V M 2016 Comparative analysis of the polymorphism of microsatellite markers in a number of species of the genus Fusarium *News Timiryazev Agricultural Academy* **1** pp 40-50
[13] Sokolova L M 2010 *Creating a starting material for dining carrot selection for resistance to Alternaria radicinia m. dr. et e, Fusarium avenaceum link.ex er* (Cand. of Sci. (Agricultural) dissertation)
[14] Vintal H, Ben-Noon E, Shlevin E, Yermiyahu U, Shtienberg D, and Dinoor A 1999 Influence of rate of soil fertilization on Alternaria leaf blight (*Alternaria dauci*) in carrots. *Phytoparasitica* **27**(3) p 193

Economic efficiency of biological protection of cucumber from root rot in a closed ground

K L Alekseeva[1], M V Shatilov[1] and O A Razin[2*]

[1] All-Russian Scientific Research Institute of Vegetable Growing – Branch of the Federal Scientific Center for Vegetable Growing, Vereya, Ramenskoye district, Moscow region 140153 Russia
[2] Federal Scientific Center for Vegetable Growing, 14 Selektsionnaya str., VNIISSOK Village 143080 Russia

E-mail: vniissok@mail.ru

Abstract: The article presents an analysis of the world and Russian production of cucumbers, as one of the most common vegetable crops in many countries, open and closed ground. Among the vegetable crops grown in protected soil, the most profitable is cucumber. The share of each district in the gross harvest of cucumbers closed in Russia is calculated, and the regions with the maximum absolute index are listed. The volume of imports to Russia is shown in terms of value and in kind, indicating the key supplier countries. In recent years, more and more demand in the market is becoming environmentally friendly vegetable products, the cultivation of which is widely used biological remedies. To obtain environmentally friendly products, it is necessary to use plant protection systems based on the use of biological plant protection products and growth regulators, which increase the disease resistance of plants in adverse conditions. To analyze the effectiveness of such means, in the course of research, the effect of the drug *Gliocladin* in combination with the growth regulator *Zircon* against the root rot was studied. On the basis of the obtained data, the economic effect of their use in the conditions of film greenhouses is calculated.

1. Introduction

Cucumber is one of the most common vegetable crops in many countries, the world production of which reached 83.8 million tons in 2017. Russia's share in world production is not yet large and it is 2.5% [4]. In Russia, the gross yield of cucumbers is close to 2 million tons, including the gross yield in protected soil amounted to 0.9 million tons (in 2017). To the level of 2016, the increase in gross yield amounted to 47.7 thousand tons (5.6%). More than 75% of greenhouse cucumbers are grown in the Volga, Central, South and Siberian districts (Fig. 1).

Content from this work may be used under the terms of the Creative Commons Attribution 3.0 licence. Any further distribution of this work must maintain attribution to the author(s) and the title of the work, journal citation and DOI.
Published under licence by IOP Publishing Ltd

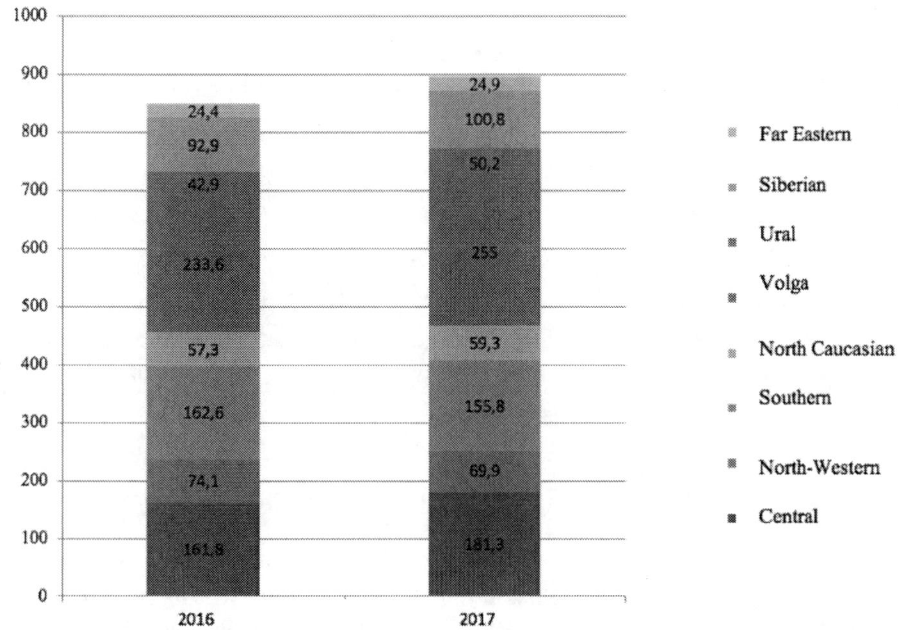

Fig. 1. Gross yield of greenhouse cucumbers in Russia, thousand tons.

In 2017, among the subjects of the federation, the following regions with the maximum collection of greenhouse cucumbers are highlighted (in thousand tons): Republic of Bashkortostan (68.9), Krasnodar Region (63.1), Republic of Tatarstan (53.1), Republic of Crimea (44.3), Moscow region (38.8), Lipetsk Region (32.4), Leningrad region (28.0), Volgograd region (27.3), Perm region (25,4), Belgorod region (22.1), Saratov region (19.9), Nizhny Novgorod region (18.8), Novgorod region (18.5), Stavropol region (16.7), Udmurt Republic (14.1), Rostov region (14.0), Republic of Mordovia (13.4), Vladimir region (11.1), Samara Region (10.9), and Yaroslavl region (10.7).

Currently, cucumber is the most common and profitable crop among greenhouse vegetable crops. The dynamics of its production remains positive, but so far insufficient to provide the population according to WHO-recommended increased standards of vegetable consumption and reduce inter-regional imbalances.

The lack of domestically produced vegetables, including cucumbers, is partially compensated by imports, which amounted to 107.3 thousand tons in the amount of $ 118 million, which is 2% more in weight and less than 5% in value to the level of 2016. In 2017, the main suppliers of cucumbers to Russia were Iran, China, Azerbaijan, Turkey, Georgia, Serbia, Morocco. It should be noted that the imported greenhouse vegetables, including cucumbers, are grown using significant quantities of agrochemicals and plant protection products, many of which are not registered in Russia. To reduce crop losses from diseases and pests, cucumber plants are repeatedly treated with pesticides (10–12 treatments per crop turnover), which reduces their nutritional value, but allows one to transport these perishable products over long distances and extend shelf life.

To obtain environmentally friendly products, plant protection systems based on the use of biological plant protection products and growth regulators increasing the disease resistance of plants in adverse conditions are important.

2. Materials and Methods

In experiments on a protected cucumber culture conducted at the All-Russian Research Institute of Vegetable Growing – a Branch of the Federal Research Center for Vegetable Growing (Moscow region, Ramensky district), the effectiveness of using the microbiological drug *Gliocladin* (d.v. *trichoderma harziannum*) in combination with growth regulators *Zircon* (d.v. *hydroxycinnamic acid*

0.1 g / l) against the root rot of cucumber studied. The work was carried out in a film greenhouse with the use of *Parthenocarpic F1 Forsazh* hybrid. *Glyocladin* was applied by watering under the root three times as follows: when transplanting seedlings and then with an interval of two weeks. The drug consumption was 60 g / ha, consumption of the working solution was 1000 l / ha. The treatment of plants with *Zircon* was carried out three times as follows: the first treatment was carried out in the phase of 2-4 true leaves, the second treatment was carried out in the budding phase, the third was carried out after 14 days. The drug consumption was 30 ml / ha, consumption of the working solution was 1000 l / ha. Our control is water, the standard is Previkur Energy, VC (watering was carried out at the root 3 times as follows: after transplanting the seedlings and then with an interval of 2 weeks). The consumption of the drug was 3 l / ha, the flow rate of the working fluid was 2000 l / ha. The repetition of the experiment was 4 times. The area of the accounting plot was 5 m^2. Standard methods adopted in greenhouse vegetable farming were used in the formulation of experiments and data processing [1], [2], [3].

3. Results and Discussion

As studies have shown, the processing of *Glyocladin* in combination with *Zircon* reduced the root rot's prevalence next to cucumber plants in the greenhouse by 78.2%. Acceleration of the timing of mass flowering on average by 3 days compared with the control, the timing of mass ripening of the fruits - by 4-5 days was noted under the influence of treatments. By stimulating the growth and development of cucumber plants and reducing the prevalence of root rot, the yield of early production increased in the greenhouse, and the total yield increased by 1.5 kg / m^2 (19.2% of the control).

To calculate the economic efficiency of using *Glyocladin* in combination with *Zircon* against the root rot of a cucumber, the following indicators were used: the cost of the *Glyocladin* biopreparation – 2,080 rubles / kg; costs of the growth regulator *Zircon* – 5,760 rubles / l; costs of the Previkur Energy chemical drug – 4,132 rubles / l; s selling price of a cucumber with a chemical protection system – 50 rubles / kg, with a biological one – 60 rubles /kg. The results of calculations are presented in Table 1.

Table 1. Economic efficiency of using the biological product Gliocladin in combination with the growth regulator *Zircon* against the root rot in cucumber plants.

Indicator	Control, without treatment	Previkur Energy, 300 ml / ha, three times (standard)	Glyocladin, 60 g / ha + Zircon, 30 ml / ha, three times
Output of standard products, kg / ha	78,000	96,000	93,000
Gain to control, kg / ha	-	18,000	15,000
Cost of additional products, rubles / ha		900,000	900,000
The cost of drugs, processing, collection, and sale of additional products, rubles / ha	-	237,188	210,893
Economic effect, rubles / ha	-	612,812	689,107

As follows from the data presented in Table 2, using the biological product *Glyocladin* in combination with the growth regulator *Zircon* against the root rot for cucumber plants is cost-effective. The economic effect amounted to 689,107 rubles / ha. In the experiment, using the drug *Previkur Energy*, the economic effect amounted to 612 & 812 rubles / ha. The results obtained allow us to conclude that the biological system of protecting cucumbers is not inferior in terms of chemical protection effectiveness. At the same time, it ensures the production of environmentally friendly products.

References

[1] Litvinov S S 2011 *Methods of field experience in vegetable growing* (Moscow, Russia: VNIIO)
[2] Dolzhenko V I Ed. 2009 *Methodical instructions on registration tests of fungicides in agriculture* (St. Petersburg, Russia: VIZR)

[3] Dospehov B A 1985 *Methodology of experimental work* (Moscow, Russia: Agropromizdat)
[4] Soldatenko A V, Razin A F, Shatilov M V, Ivanova M I, Razin O A, Rossinskaya O V, … Bashkirov O V 2018 Interregional exchange in the context of the alignment of the consumption of vegetables in subjects of the Russian Federation *Vegetables Crops of Russia* **6** pp 37-42

IOP Conf. Series: Earth and Environmental Science **395** (2019) 012052 doi:10.1088/1755-1315/395/1/012052

Monogerm table beet from the perspective of economic efficiency

S A Vetrova[1*], V A Zayachkovskyi[1] and S M Sirota[1]

[1] Federal Scientific Center for Vegetable Growing, 14 Selektsionnaya str., VNIISSOK Village 143080 Russia

Email: lana-k2201@rambler.ru

Abstract. Garden beets are one of the most popular vegetable root crops, which are grown everywhere. The existing commodity production in Russia is not able to provide the population's need for fresh vegetable products. In this regard, attracting the attention of large farms to the culture of garden beets, as one of the economically viable among the group of garden root crops, is an actual task. In connection with the intensification of agricultural production, the cultivation of single-seeded (single fruit) varieties and hybrids suitable for mechanized cultivation is an important direction in the selection of garden beets. The article provides a comparative analysis of economic efficiency in the cultivation of multi-seeded and single-seeded varieties of breeding of the Federal Scientific Center for Vegetable Growing (Russia). The calculations were carried out on tariffs and prices relevant in 2017. It is proved that in the production of single-breed varieties, the total amount of production costs decreases by 19.4%, compared with the multi-seeded standard variety. The high marketability of the products obtained in the context of lower production costs determines an increase in net income from sales, as a result, the profitability of production increases, which makes it possible to introduce single-fruited varieties in crop rotation without risking profit.

1. Introduction

Currently, along with potatoes, much attention is paid to the selection and seed production of vegetables of the "borsch" group: white cabbage, carrots and garden beets, onions. In Russia, garden beets are one of the most popular vegetable root crops and are grown everywhere. Its root crops are rich in carbohydrates, mineral salts, organic acids, and vitamins (C, B1, B2, P, PP, biotin, pantothenic and folic acid). The content of betanine and betaine (nitrogen-containing heterocyclic pigments belonging to the class of betacyanins in the roots of this crop) makes it unique and indispensable for a balanced human nutrition [4], [5], [10], [13], [14], [15], [16]. Due to its high safety, the use of root crops of garden beets for food is possible throughout the year [11].

In Russia, about 70 thousand hectares of sown area are occupied by garden beets, of which most in the Central (28%), Volga (20.4%), and Southern (14.6%) federal districts. The main producers are personal (subsidiary) households and peasant (farm) households. In the farms of the Astrakhan and Voronezh regions, due to the favorable agro-climatic conditions, when using new varieties and modern cultivation technologies, the productivity of garden beets can reach 67.3-85.5 t / ha. However, on average in Russia productivity does not exceed 30.0 t / ha [9].

Thus, the existing commodity production is not able to meet the population's need for fresh vegetable products. Actual consumption of beets, even taking into account imports, does not exceed

Content from this work may be used under the terms of the Creative Commons Attribution 3.0 licence. Any further distribution of this work must maintain attribution to the author(s) and the title of the work, journal citation and DOI.
Published under licence by IOP Publishing Ltd

70% of the recommended norm per person per year. In this regard, the urgent task is to attract the attention of large farms to the culture of garden beets, as one of the economically profitable for commercial production, among the group of root crops.

In connection with the intensification of agricultural production, an important direction in the selection of garden beets is the cultivation of single-seed (single-fruited) varieties and hybrids suitable for mechanized cultivation. Their advantage is to reduce costs during cultivation by eliminating the thinning procedure, increasing the marketability of crops, and reducing the seed rate by 30-35%. At the same time, important conditions that determine the value of the variety are: high and stable productivity, marketability, attractive appearance of the root crop (flat surface, thin axial root, small head), intense coloring of the pulp in the absence of ringing, product quality, processability, and resistance to diseases and pests, responsiveness to improved growing conditions. All this determines the need for a high level of selection work and improvement of applied technologies.

Beet gardening is characterized by multi-seedling, which occurs due to the growth of the receptacle from two to four flowers forming an inflorescence, which, after fertilization, turns into a fruit (glomerulus) containing several seeds. Multi-seed as a trait is steadily inherited in the offspring [1]. When sowing from such glomeruli-fruit, as a rule, several seedlings develop, the roots of which are intertwined and mutually suppress each other's growth. Thinning of crops, or rather creating optimal growth conditions for each plant, leads to mechanical injury to the plants and their root system. This operation requires a lot of manual labor in a limited time period for the formation of crop density [8]. This factor is the main obstacle to the mechanized cultivation of beets.

The first attempts to impart single-sprouting properties to multi-seeded beet seedlings were made in the USA in the first half of the last century by a mechanical division, but in then this did not find distribution due to a number of negative aspects [3], [12]. V. T. Krasochkin considered the hybridization of garden multi-seeded beet varieties with single-seed sugar beets to be one of the ways to create single-seeded garden beets, but this path turned out to be long, since along with the selection of single-seeded forms, it was necessary to improve the selection in the form of root crops and the quality of their pulp. The way to search for biotypes with single-seeded fruits among the testes of multi-seeded beets, with further selection of single-fruited forms in their offspring, turned out to be more effective.

The sign of single-fertility is of a recessive nature, which greatly complicates the selection work when breeding single-seeded varieties of garden beets. Single-seeded forms, as a rule, are characterized by a low level of productivity, small seeds, weak seedling of shoots, low seed productivity, extended flowering and ripening of seeds, which negatively affects their germination. In order to genetically improve single-seeded beets and increase selection efficiency, preliminary breeding of materials of various origin is carried out in breeding practice.

Breeding work in the direction of the multiplicity of garden beets in the laboratory for selection and seed production of table roots of the Federal Scientific Center for Vegetable Growing (VNIISSOK) was started in the early 60s of the last century. The first single-sprout variety of garden beet was created by the method of individual family selection from the multiple variety Bordeaux-237 and zoned since 1976.

Subsequently, the Odnorostkovaya variety, along with other populations of domestic and foreign origin, was used in breeding as a starting material for the creation of new varieties with a high level of single fertility. At the first stages, the free pollination of single-fertility forms at different levels was carried out. Then, using individual-family selection and return crosses and inbreeding, the proportion of single-seeded fruits on a plant and the proportion of single-born plants within individual families were gradually increased. Selection for high single fertility was combined with selection for productivity, which led to a gradual increase in the productivity of families with a high proportion of single fruits.

As a result of the intensive selective improvement of single-breeding populations, their stabilization by commodity, morphological and biological characteristics, single-seeded varieties of garden beets with a high level of single fertility (98-99%) were created – the Bordeaux single seed and the Dobrynya variety. For the purpose of providing valuable recommendations for producers, this paper presents a number of economic characteristics of these varieties and calculates the economic efficiency of production crops in comparison with multi-seeded varieties of garden beets.

2. Materials and Methods

The studies were conducted on the basis of the Federal Scientific Center for Vegetable Growing (VNIISSOK) in 2015-2017. The objects of research were beet varieties of garden selection VNIISSOK: Bordeaux single seed, Dobrynya. The study of varieties was carried out in accordance with the "Guidelines of the Research Institute of Plant Industry named after N.I. Vavilov on the study and maintenance of the world collection of root crops" (1968), as a standard was used variety - Bordeaux 237. The list of technological operations is presented in table 1.

Table 1. The technology of growing marketable products of garden beets (experimental field of the Federal Scientific Center for Vegetable Growing 2015-2017).

No.	Technological operation	Agricultural machinery and implements
	Application of fertilizers and processing of the soil	
1	Spring plowing	MTZ-82.1
2	Harrowing	John Deere -6130
3	Transportation and application of mineral fertilizers (nitroammophos 300kg / ha)	MTZ-82.1 + RMU (Disc Spreaders) Kverneland
4	Disc harrow	MTZ -82.1 + BDT-3 (Disc harrow)
5	Cutting rows	John Deere-5100M + Row Maker GRIMME
6	Seeding with rolling (according to the scheme 5+50+5+90)	MTZ 82.1 + Vegetable seeder Klen-1,8
	Care of plants	
7	Preparation of the herbicide solution (dual gold 1.6 l/ha)	Manual
8	Solution delivery and spraying (after sowing before germination, and in a phase of 2-3 weeks)	МТЗ-82.1+ОП-600
11	Interrow processing up to the joining rows (3 times)	John Deere-4720
*12	Weeding with thinning (double in phase 2-3 weeks and 3-5 weeks)	Manual
17	Watering (2 times)	MTZ-82.1 + sprinkling machine IRTEC
18	Weeding with hoeing	Manual
19	Trenching	MTZ-82.1
20	full cleaning with pruning	Manual
21	Loading	Manual
22	Transportation	MTZ-82.1
	Root storage	
23	Cleaning storage	Manual
24	Storage Disinfection (sulfur checkers)	Manual
25	Storage loading	Forklift truck FD-18
26	Sorting of root crops	Manual
27	Waste disposal	MTZ-82.1

* Only for multi-seed varieties of garden beet.

Root crops harvesting was carried out in the second decade of September with qualitative and quantitative accounting for the crop. The number and mass of commodity, sick, ugly, cracked and underdeveloped root crops were taken into account. For storage in the vegetable store, commercial roots were laid in containers with polyethylene inserts. During storage, observations were made of the temperature regime (optimally 4-5 °C) and humidity (95-98 %).

The economic efficiency of growing root crops of multi-seeded and single-seeded varieties of garden beets was evaluated taking into account the following indicators: productivity, production costs, gross output, net income, profitability.

Mathematical processing of experimental data was carried out by method of dispersive analysis on a personal computer, using a package of the Microsoft Exel application programs [2].

3. Results

Cultivation of garden beets can be cost-effective if there are a sufficient number of varieties and hybrids that meet modern market requirements and provide a stable crop when grown in a particular region. An important condition for increasing the efficiency of growing this crop is the gross collection and reduction of material and monetary costs for the production and sale of marketable products. Higher productivity and lower costs can be achieved by the proper selection of highly productive varieties and hybrids suitable for mechanized cultivation, responsive to fertilizer application, crop protection products, and irrigation [6]. Decrease in prime cost of the made products due to cost reduction on seeds, compensation and material resources is of great importance.

The stability of demonstration of the average value of varietal characters in a particular zone, regardless of the prevailing conditions of the growing season, is an urgent requirement for vegetable producers to ensure the return on crops. As a result of a competitive variety test conducted on the fields of the main crop rotation of the experimental production base of the Federal Scientific Center for Vegetable Growing (VNIISSOK) of the Odintsovo district of the Moscow Region in 2015-2017, it was shown that the Bordeaux 237, Bordeaux single-seeded, and Dobrynya, on average over three years, did not significantly differ in root crop productivity (Fig. 1) and were characterized by high agronomic stability on this basis (A_s > 90%). Slight variability in productivity of varieties (V = 4.9-10.4%) was determined by the difference in temperature and water regimes in the years of research. In 2016, under favorable weather conditions for the cultivation of garden beets, the single-seeded variety Bordeaux turned out to be the most fruitful (35.2 t/ha), exceeding the norm by 1.7 t/ha.

Of great importance in the formation of the commodity part of the crop when growing garden beets is the ability of biotypes of the varietal population to form root crops of a certain mass regardless of the conditions and density of the plants on a unit area. Overgrown and unformed root crops reduce the marketability, taste and nutritional quality of products. In terms of average indices, the single-seeded varieties Bordeaux and Dobrynya were distinguished by the highest marketability and stability on this basis, compared with the multi-seeded standard (fig. 1). The level of marketability of the standard over the years of research did not exceed 88%. In 2015, the Bordeaux-237 variety showed the lowest marketability (80%), when a lack of soil moisture contributed to a decrease in yield due to the large number of unformed root crops.

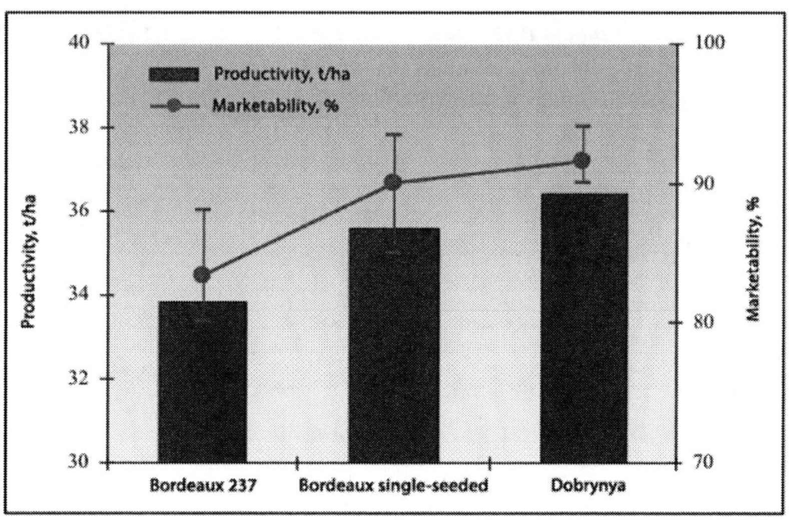

Fig. 1. Characteristics of garden beet varieties on productivity and commercialization (on average for 2015-2017).

For three years, taking into account the average productivity, a comparative analysis of the economic efficiency of growing multi-seeded and single-seeded varieties was carried out.

Production costs were calculated according to the process cards used at the Federal Scientific Center for Vegetable Growing, at the rates and prices that are relevant in 2017. Table 2 presents the list of costs per 1 ha of the sown area. The largest share (about 35%) in the cost structure for growing the multi-seeded variety Bordeaux-237 is labor costs, because the growing technology involves manual operations (Table 1).

Table 2. Cost structure in the cultivation of multi-seeded and single-seeded garden beet.

Items	Multi-seeded variety (Bordeaux 237st)		Single-seeded garden variety (Dobrynya)		Single-seeded garden variety (Bordeaux)	
	Amount, rub.	%	Amount, rub.	%	Amount, rub.	%
Payment of labour charges: incl. mechanics,	11869	7,6	12916	10,3	12428	10,3
workers	42534	27,3	26173	20,9	25183	20,9
Seeds	13357	8,6	6279	5,0	6042	5,0
Fertilizers (mineral)	22731	14,6	24737	19,7	23801	19,7
Plant protection products	3448	2,2	3753	3,0	3611	3,0
Depreciation and maintenance	11577	7,4	12599	10,0	12122	10,0
Fuel and lubricant materials	7927	5,1	8626	6,9	8300	6,9
Machinery insurance	231	0,1	251	0,2	241	0,2
Overhead costs	41849	26,9	30069	24,0	28931	24,0
Total costs	155524	100	125403	100	120658	100

When cultivating single varieties, this cost item is reduced by 3.7%, which is due to the exclusion from the technology of double manual thinning, which is paid in the amount of 14636.54 rub./ha. Also, a large share in the cost structure is occupied by the overhead costs and expenses for mineral fertilizers, regardless of the cultivated variety (Table 2). Accurate calculation of fertilizer application doses, taking into account soil fertility and crop needs, is the main way to reduce fertilizer costs.

Reducing the cost of production also depends on the cost of seeds. When growing single-fruited varieties, the cost of seeds is 5.0% of the total, which is 3.6% less than when cultivating a multi-seeded variety. Cost reduction occurs due to a decrease in the seed sowing rate from 12.5 kg / ha to 5.6 kg / ha, since the weight of 1000 single-seed seeds is less, and when sown with precision seeders, they are evenly distributed in rows.

Thus, due to the exclusion of the operation of thinning seedlings and lowering the rate of sowing seeds, there is a decrease in the total amount of production costs when cultivating single-fruited varieties of garden beets, compared with multi-seeded ones.

Cost indicators give a more accurate idea of production efficiency and cost recovery. When analyzing the economic efficiency of vegetable production, the following cost indicators are used: a gross output (t); a size of gross income and net income per unit of vegetable crops (rub./ha); prime costs per unit of output (rub./t.). The level of profitability is a general indicator of the economic efficiency of vegetable production.

Table 3. Economic efficiency of cultivation in the multi-seeded and single-seeded garden beet.

Possible options	Commercial productivity, t/ha	Production costs, rub./ha	Gross output costs, rub./ha	Prime costs, rub./t	Net income, including income tax (20%), rub./ha	Return rates, %
Bordeaux - 237st	28,2	155,524	423,300	4,574	214,221.0	137.7
Bordeaux single seed	32,0	120,658	480,600	3,389	287,953.6	238.7
Dobrynya	33,8	125,403	505,050	3,389	303,717.6	242.2

Gross income was calculated on the basis of the total productivity of a variety per hectare, the level of marketability and the selling price of 15,000 rub./t, taking into account 18% VAT. The highest productivity was obtained when growing a new single-fruit variety of the garden beet "Dobrynya," which was characterized by the highest productivity and marketability. Net income from selling products of this variety exceeded the norm by 89,496.6 rub./ha (table 3). In terms of profitability, this variety also exceeded the norm by 1.7 times.

4. Conclusion

Thus, one-seeded varieties of the garden beet breeding from the Federal Scientific Center for Vegetable Growing (VNIISSOK) is characterized by a set of traits that make it possible to obtain a stable crop. When they are grown by eliminating the operation of thinning seedlings and reducing the rate of sowing seeds, the total production costs decrease by 19.4%, compared with the multi-seeded standard variety. A high marketability of the products in the context of lower production costs determines an increase in net income from sales, as a result, the profitability of production increases. All this shows that the cultivation of single-fruit varieties of garden beets is economically viable and determines the possibility of their introduction into crop rotation without risking profit, which, on the one hand, would reduce the import share of imported seeds, and on the other, provide the Russian food market with quality vegetable products.

References

[1] Balkov I Ya 1978 *Selection of sugar beets for heterosis* (Moscow, USSR: Rosselkhozizdat)
[2] Dospekhov V A 1965 *Methods of field experience* (Moscow, USSR: Agropromizdat)
[3] Ivanchenko I P 1963 Production of single-seeded sugar beets by mechanical crushing *Proceedings of UkrNIISHOM* **1** pp 3-11
[4] Krasochkin V T 1971 Root plants *Cultural Flora*, **19** (Leningrad, USSR)
[5] Lapin A A, Bykovsky D V, Davydov Yu A, and Zelenkov V N 2007 Beetroot juice – a source of Antioxidants *Potatoes and vegetables* **6** p 27
[6] Makarets L I, and Makarets M N 2002 *Economy of production of agricultural products* (St. Petersburg, Russia: Lan')
[7] Krasochkin, V I Burenin, B I Sechkarev, *et al.* (Leningrad, USSR: VIR VASKhNIL)
[8] Roik N V 2010 The creation of single-seed varieties and hybrids of sugar beet in the Soviet Union. In S I Maletskiy Ed. *Encyclopedia of the genus Beta: Biology, Genetics and Selection of Beets* (pp. 248-265). (Novosibirsk, Russia: Sova)
[9] Soldatenko A V, Pivovarov V F, Razin A F, Shatilov M V, Razin O A, Rossinskaya O V …. Bashkirov O V 2019 Problems of production of competitive vegetable products *Russian Vegetables* **1** pp 3-7
[10] Fedorova M I 2003 Krasnobokiy doctor *Homestead farming* **3** p 43
[11] Fedorova M I, Vetrova S A, Zayachkovsky V A, and Stepanov V A 2016 Varieties of beet garden breeding VNIISSOK *Russian vegetables* **2** pp 56-59
[12] Brewbaker H E, Wood R R, and Bush, H L 1947 *Single germ seed* (New York, NY: American Society of

Sugar Beet Technologists)

[13] Escribano J, Pedreño M A, García-Carmona F, and Muñoz R 1998 Characterization of the antiradical activity of betalains from Beta vulgaris L. roots. *Phytochem. Anal.* **9**(3) pp 124-127

[14] Pialetti, M 1976 Betalains. *Chemistry and Biochemistry of Plant Pigments N.Y.: Acad. Press.*, **1** p 560

[15] Mabry T J 1980 Betalains. In E A Bell, and B V Charlwood Eds. *Secondary plant products* (vol. 8) (Berlin, Germany: Springer-Verlag)

[16] Tesoriere L, Allegra M, Butera D, and Livrea M A 2004 Absorption, excretion, and distribution of dietary antioxidant betalains in LDLs: potential health effects of betalains in humans *American Journal of Clinical Nutrition* **80**(4) pp 941-945

Analysis of the world lettuce market

M V Shatilov[1], A F Razin[1] and M I Ivanova[1]

[1] All-Russian Scientific Research Institute of Vegetable Growing – Branch of the Federal Scientific Center for Vegetable Growing, Vereya, Ramenskoye district, Moscow region 140153 Russia

E-mail: vniioh@ya.ru

Abstract. The article presents an analysis of the lettuce market (*Lactucasativa* L.) as one of the most economically important leafy vegetable crops in the world and the Russian Federation. In human nutrition, the value of lettuce is determined by those properties that are recognized as health-improving due to the high content of vitamin C, polyphenols, and fiber. The paper roves the effect of lettuce in the prevention of cardiovascular disease. It also shows the production performance of world lettuce production, including information about countries that have a significant share in global production. The authors consider the interstate exchange of lettuce products, indicating the volume of exports and imports in value and volume terms. The leading countries with the maximum indicators of import and export in value terms are identified, and the maximum and minimum prices of purchase and sale among them for the period under review are indicated. Separately, imports and exports of lettuce in the Eurasian Economic Union are analyzed, indicating the States to which the countries of the Union export, as well as the share of each member country in exports and imports. A detailed analysis of Russia's imports and exports, with allocation of the Federal districts having the greatest share, is carried out, and also the main directions with indication of the region of the country are discusses.

1. Introduction

The lettuce (*Lactucasativa* L.) is one of the most economically important leafy vegetable crops in the world [1], [2]. The healing properties of lettuce are determined by a high content of antioxidant compounds, primarily vitamin C and polyphenols, as well as fiber [3]. Recent studies on rats and humans have demonstrated the health effects of lettuce in the prevention of cardiovascular diseases [4].

Several morphotypes of lettuce are grown in production, but the most common are romaine lettuce, leafy, headed crispy, and oily ones. Growing morphotypes and varieties of lettuce depend on both market strategies and commercial requirements [5], [6].

2. Materials and Method

In the framework of the general systems approach to the study of the problem, the tools of abstract logical, comparative, economic-statistical, economic-mathematical, and computational-constructive research methods are used. The informational basis of the research includes legislative and regulatory documents of state authorities, methodological and guidance materials of the Ministry of Agriculture of the Russian Federation. The empirical base of the research is presented by the materials of the Federal State Statistics Service of the Russian Federation, the Ministry of Economic Development of

Content from this work may be used under the terms of the Creative Commons Attribution 3.0 licence. Any further distribution of this work must maintain attribution to the author(s) and the title of the work, journal citation and DOI.
Published under licence by IOP Publishing Ltd

the Russian Federation, and other ministries and departments, as well as the following: Federal Customs Service of the Russian Federation, EurAsEC Customs Union, Organization for Economic Cooperation and Development, World Trade Organization, Statistical Agency of the European Union.

3. Results and Discussion

According to the FAO, in 2017, lettuce was produced in 106 countries. We present the production indicators of countries whose share in the global gross yield is more than 1% (Table 1).

Table 1. Lettuce production figures in the world and in some countries.

A country	Sown area, ha			Productivity, t / ha			Gross yield, t		
	2015	2016	2017	2015	2016	2017	2015	2016	2017
Total	1203677	1216232	122735 8	21.63	21.93	21.89	2603299 3	26668741	2686655 7
including China	618279	622990	632972	23.59	23.86	23.95	1458374 1	14865653	1516081 8
India	170397	171200	171884	6.37	6.36	6.35	1084850	1088161	1090770
USA	105660	110800	111900	35.92	36.44	34.29	3795480	4038020	3836820
Spain	34314	35646	34508	27.03	26.09	28.29	927378	929944	976112
Italy	31329	34343	34069	20.00	21.43	21.60	626525	735967	735873
Turkey	21099	22426	22889	21.21	21.33	21.43	447492	478442	490423
Japan	21500	21600	21048	26.42	27.12	27.30	568000	585700	574646
Mexico	20289	20066	21030	21.57	21.92	22.86	437562	439831	480808
Iran (Islamic Republic)	16918	17280	16645	30.16	30.58	30.82	510289	528439	513041
Germany	14150	14278	15096	24.25	23.33	23.44	343083	333034	353883

The biggest share of the gross yield and area under crop (75%) is in China, India, and the USA. More than half of the global acreage falls on China (51.6%). During the study period, the country has seen an increase of 1 ... 2% annually. In total, 14% of the sown area is accounted for by India, it is 9.1% in the USA.

The gross yield of lettuce in China is greater than that of all other countries combined, its share in the global harvest is 56.4%. The second place is occupied by the United States (14.3%), which is due to the low (3.5 times less than the world average) yield in India, the share of which was 4.1%.

Due to the unevenness of production across countries, the surplus of grown products is exported. Thus, Spain (which has a double separation from the United States in quantitative terms) continues to increase its exports (Table 2).

Table 2. The volume of lettuce exports in natural and value indicators of the main exporting countries.

Country	2015		2016		2017	
	Amount, kg	Cost, USD	Amount, kg	Cost, USD	Amount, kg	Cost, USD
Spain	723847640	699686126	743130750	726966041	761902481	747555634
USA	336642960	527292672	335278042	469441803	331443417	508124108
Italy	74186039	153865332	77935554	150195842	73356138	167226125
Netherlands	96296885	150300081	94664196	150599317	105237386	193136857
Mexico	153030414	137906142	161658497	153308373	213993557	229969746
Belgium	40687673	61784044	39335961	58896066	39603357	60908452
China	91068678	46941834	115304202	81152450	95703289	85963701
Canada	42277278	46734367	39773112	41445012	39019670	43515950
Germany	43650139	46314934	36561685	38715797	40406537	47012737
France	21547278	35241210	21721083	34523972	21579211	42026682

A composition of the top ten countries with the maximum export in value terms did not change from year to year for the period under consideration. The minimum average selling price (0.52 ... 0.90 USD / kg) was in China, and Italy sold the most expensive products among the countries surveyed (1.93 ... 2.28 USD / kg).

The countries compensate for the lack of own production by purchasing imported products. Table 3 lists the countries that were in the top ten in terms of imports from 2015 to 2017. Canada is the leader, but there are quantitatively incomplete data in 2017, because Canada does not provided accurate information for all types of salads imported into the country.

Table 3. The volume of imports of lettuce in the natural and value indicator of the main importing countries.

Country	2015		2016		2017	
	Amount, kg	Cost, USD	Amount, kg	Cost, USD	Amount, kg	Cost, USD
Canada	282342477	440768576	279010361	386011309	87469608	425345615
Germany	262202231	370788569	248326738	361419634	244957571	397620264
USA	224182722	269457076	218349185	271039141	184392220	302882715
United Kingdom	207445083	236805444	221408467	225809881	192632853	213117184
France	88540233	101991097	90140487	108357813	83857577	109023035
Netherlands	67897024	96691478	69729155	101201748	87183690	127018509
Italy	83404441	71979732	84399875	60812391	88718877	72967335
Poland	48597471	57639221	45838145	51928060	49640830	56806774
Sweden	37157653	52265169	34261643	48767601	34510549	52791112
Belgium	27561426	50739378	29256420	53686244	24345073	46113532
Austria	27440779	49751442	27066854	49436687	24833713	53937166

Among the countries considered, the minimum average purchase price was in Italy (0.72 ... 0.86 US dollars / kg), the maximum one was in Belgium and Austria (1.81 ... 2.17 US dollars / kg).

We will consider the top five countries of the Eurasian Economic Union (Table 4).

Table 4. The volume of exports and imports of lettuce in the EurAsEC countries.

	2015		2016		2017		2018	
	Amount, kg	Cost, USD	Amount, kg	Cost, USD	Amount, kg	Cost, USD	Amount, kg	Cost, USD
EurAsEC (export)	54942	35 982	38352	26284	45337	37628	88648	55419
EurAsEC (import)	31576467	31142923	33143739	31989970	40565118	37040116	45180243	40995 035
Country share in value EAEU - total, %	export	import	export	import	export	import	export	import
Armenia	2.3	0.0	-	0.0	-	-	0.0	0.0
Belarus	10.5	59.0	26.7	47.6	-	35.7	12.5	27.6
Kazakhstan	-	2.2	-	1.7	-	1.8	-	1.8
Kyrgyzstan	-	0.0	-	0.0	-	0.0	-	0.0
Russia	87.2	38.8	73.3	50.7	100.0	62.5	87.5	70.6

The EurAsEC countries export lettuce to Belgium, Georgia, China, Lithuania, Morocco, Norway, Uzbekistan, and Ukraine. Most exports come from Russia to Ukraine.

The excess of imports over exports to the Union countries is a hundred times. More than half comes from imports to Russia. During the study period, Morocco, China, Iran were the leading countries in the supply of lettuce to the Union in different years.

According to the Customs Service, in 2016-2018, Russia exported lettuce to Belarus, China, Kazakhstan, Norway and Ukraine.

The main share of exports falls on the Central Federal District, and the North Caucasus, on the contrary, does not participate in exports of salads during the study period (Table 5).

Table 5. Lettuce export from Russia.

Federal districts	2016		2017		2018	
	Cost, USD	Amount, kg	Cost, USD	Amount, kg	Cost, USD	Amount, kg
Russia	522816.1	157333.4	168088	79632.1	253727.2	152351.1
Central	504009.4	133252.4	130447	34295	201744.9	72147
Northwestern	620.1	75	60	145	3535.7	802
Southern	18032.1	23836.0	37571	45192	48211.4	79110.5
Volga	44.2	40	-	-	-	-
Ural	-	-	10	0,1	-	-
Siberian	110.3	130	-	-	2.4	0.6
Far Eastern	-	-	-	-	232.8	291
North Caucasus	-	-	-	-	-	-

The sale of lettuce to Belarus brings in more than half of the revenue, namely, $527,209.3. The Moscow and Smolensk regions of the Central Federal District and St. Petersburg of the North-West Federal District export to Belarus. Exports to Kazakhstan amounted to $312,694.6 and was mainly from Moscow. The Omsk, Orenburg, and Tyumen regions supplied an insignificant part (less than 2 centners of products). The Southern Federal District (from the Krasnodar, Astrakhan and Rostov regions) almost fully delivered to Ukraine in the amount of $103,874.5.

The main problem for the production of lettuce in Russia is the climatic conditions. There is no opportunity to harvest year-round in the open field. Imports to Russia increased both in value and volume terms (Table 6). The main consumers are the fast food businesses. Moreover, in Russia, the market for washed and sliced lettuce is developing at a fast pace, where the assortment includes Salanova® varieties, baby-lettuce and variety types with the Knox™ attribute.

Table 6. Lettuce imports to Russia.

Federal districts	2016		2017		2018	
	Cost, USD	Amount, kg	Cost, USD	Amount, kg	Cost, USD	Amount, kg
Russia	18651346.8	27812250.8	25431731.8	36538435.8	31248931.8	41125871.7
Central	13214934.7	21711105.1	17553053.4	27671062.7	23045577	31193925
Northwestern	288727.1	260468.8	971400.4	962790.7	1261015.3	1477765.5
Southern	1116201.5	1093517.3	1759463.6	1825016.4	1480156.4	1704345
Volga	1265404.5	1252757.9	1625545.1	1659263	1268518.3	1157178
Ural	133.5	205	74829.2	88002	206055.1	240741
Siberian	814649.1	1005649.6	988924.8	1228748.4	839446.6	1034638.9
Far Eastern	1940301.9	2477160.6	2386104	3014932.6	3134842.1	4301899.5
North Caucasus	10994.5	11386.5	72411.3	88620	13321.4	15378.8

Deliveries to the Central Federal District account for about 70%. In 2016-2017, more than half of deliveries to the Central District accounted for the Moscow region, mainly (> 75%) from Iran, Tunisia, and Egypt. In 2018, the key deliveries to Moscow were from Iran, Uzbekistan, and Belarus by 85%. In the Far Eastern Federal District, imports from China to the Primorsk region account for more than 80%.

4. Conclusion

In general, the data studied are necessary to understand the current situation on the lettuce market and to evaluate the prospects for its development. Of course, the study will be useful for producers and suppliers of lettuce in the context of understanding the current market situation, industry dynamics, monitoring current industry events for building short-term and long-term development and management strategies. This study plays an important role for contractors performing services or supplying equipment or products to enterprises in the industry when developing cooperation. There is a need to have a reliable information on development in the industry and, accordingly, about the availability of effective demand.

References

[1] Soldatenko A V, Razin A F, Shatilov M V, Ivanova M I, Taktarova S V, Kuzyakin M V, ... Sokolova ES,... Bukanov V. S.2018 Problems of lettuce production in the open field and features of its cultivation in the conditions of small-scale production (the "Vesely Agronom" LLC case, Dmitrovsky district, Moscow region) *Russian Vegetables* **2** pp 55-60

[2] Soldatenko A V, Pivovarov V F, Razin A F, Meshcheryakova R A, Shatilov M V, Ivanova M I, Taktarova S V, Razin O A 2018 The economy of vegetable growing: the state and the present *Russian Vegetables* **5** pp 63-68

[3] Materska M, Olszówka K, Chilczuk B, Stochmal A, Pecio Ł, Pacholczyk Sienicka B, ... Piacente S, ... Pizza C, ... Masullo M 2019 Polyphenolic Profile In Lettuce (*Lactucasativa* L.) after $CaCl_2$ treatment and cold storage *European Food Research and Technology* **245** pp 733-744 https://doi.org/10.1007/s00217-018-3195-0

[4] Nicolle C, Cardinault N, Gueux E, Jaffrelo L, Rock E, Mazur A, ... Amouroux P, ... Remesy Ch 2004 Health effect of vegetable-based diet: lettuce consumption improves cholesterol metabolism and antioxidant status in the rat *Clin Nutri* **23** pp 605-614

[5] Ivanova M I, Kashleva A I, Alekseeva K L, and Davletbaeva O R 2017 Lettuce: diversity of species and varieties *Potatoes and Vegetables* **5** pp 22-24

[6] Ivanova M I, Kashleva A I, and Alekseeva K L 2017 Lettuce: biology and technology *Potatoes and Vegetables* **6** pp 23-25

Analyzing the risk factors for improving the Russian mink efficiency

N R Khusainova[1*], and T M Vorozheykina[2]

[1] Russian State Agrarian University – Moscow Timiryazev Agricultural Academy, 49 Timiryazevskaya str., Moscow 127550 Russia
[2] Financial University under the Government of the Russian Federation, 49 Leningradskiy prosp., Moscow 125167 Russia

E-mail: khusainova@rgau-msha.ru

Abstract. The article analyzes those factors that affect the functioning of domestic fur farms. In order to diagnose risks at the enterprises of the fur farming industry, a scientific approach using the matrix method is applied. The benefits of using this method have been studied. Using the example of fur animals for breeding mink, the risk matrix is considered. An author's vision of the classification of consequences, probabilities, and risk levels is proposed. Using an integrated approach that includes various methods for measuring risk is proposed. This will provide for the most complete picture of the risks inherent in a particular business entity. The most probable and dangerous consequences and factors affecting both domestic and international fur farms' activities are analyzed. A risk analysis at domestic fur farms should be carried out in order to prevent the occurrence of critical situations at all stages of the production process.

1. Introduction

In the conditions of modern market relations, the process of functioning of any business entity is accompanied by risk management, which is usually unsystematic or unconscious, and it is carried out in the context of individual areas of economic activity. Currently, the problem of risk management is relevant. If in the recent past, the state practically took upon itself all the risks of enterprises and organizations, then in a market economy, the situation has fundamentally changed. The business entity is forced to act independently in terms of reducing the degree of influence of business and financial risks.

For enterprises in the fur-breeding industry for the cultivation of cell breeding minks, risk management is of particular importance since this industry is most affected by risk factors (competition between the European Union and China; feed supply; weather conditions; delivery of mink animals; export of mink pelts for sale at auction, etc.) [3].

When making decisions regarding the activities of fur farming (introducing new technologies, the releasing a new product, defining an enterprise strategy), possible future developments are taken into account. However, reliable information about certain enterprise's risks is realized only in the aggregate of knowledge and skills in the field of risk management in specific business conditions. Thus, for the most efficient operation of these enterprises, methodological explanations for diagnosing risks should be developed in detail.

Content from this work may be used under the terms of the Creative Commons Attribution 3.0 licence. Any further distribution of this work must maintain attribution to the author(s) and the title of the work, journal citation and DOI.
Published under licence by IOP Publishing Ltd

2. Materials and Method

At different times, the fundamental research by domestic and foreign scientists was devoted to risk management problems. Among them, we can distinguish such authors as K. Arrow, J. Keynes, J. von Neumann, B. Milner and F. Liis, K. V. Baldin, E. M. Korolkov, and others.

In addition, at the legislative level, theoretical aspects of risk management are enshrined in the national standards of the Russian Federation, in which a risk is understood as the result of the influence of uncertainty on achieving certain goals [6], [7].

To date, a fairly wide range of practical tools used in the analysis of project risks has been developed. Research will determine their capabilities, advantages, and disadvantages. However, the methods for diagnosing risks for the operating conditions of agricultural enterprises are not well understood, which reduces entrepreneurial performance and requires further research in this industry. The purpose of this article is to study the matrix diagnosis of risks and the development of methods for constructing a risk matrix for enterprises in the fur farming industry for the cultivation of minks, as well as an analysis of factors affecting their activities.

3. Results

A risk is a result of the interaction of at least two systems, at least one of which seeks to survive and has a partial effective free will. One can manage risks by acting on one of them, or even on both at once. The relationships of these systems are partially uncertain [2].

A risk analysis is carried out to determine the likelihood of a hazard and the potential consequences of individual risks or a group of risks identified at earlier stages. Such a risk assessment is carried out after risk analysis to determine which risks are more dangerous, for which additional and possibly more comprehensive studies are needed, and which can be provided with less attention. Information collection methods may include interviewing, self-completion of questionnaires, group workshops, searching for relevant information in libraries, and chronological data processing for these purposes.

Even with reliable data, it is first advisable to conduct a qualitative analysis in order to better clarify the situation regarding the likelihood and extent of exposure to risks. After this, it is advisable to conduct an appropriate quantitative risk analysis. The need for these analyzes depends on the nature and quality of data characterizing a particular risk, the nature of the project being carried out, as well as potential consequences and the possibility of obtaining additional useful information [6].

Risk management is an integral part of ensuring the resulting return on business. Today, risk management is understood as as a process of choosing new directions of activity, levels of risk and distribution of resources (“responses to risks”) in the activities of enterprises, it is an integral part of a single process of increasing competitiveness and profitability [1].

Based on the foregoing, risk management is a component of the enterprise management process, and more precisely, all its constituent elements, such as processes, resources, technologies, etc. This management takes place by minimizing the undesirable consequences of possible events, which, if being completed, may cause the deviation of the actual results of an enterprise from planned.

We will consider the most common method for diagnosing risks, namely the matrix one. A risk matrix is a risk classification and presentation tool by ranking the consequences and likelihood of probability. Schematically, the risk matrix is a graphical and textual description of the risks of a business entity, which are located in a rectangular table.

Risk matrices allow one to determine the level of each identified risk for a subsequent decision to reduce the level of risk. Matrices include risk assessment criteria, namely, the level of damage from the implementation of the risk and the likelihood of a risk event occurring over a period of time. Each criterion is ranked from a minimum value to a maximum value. The final risk level is determined at the intersection of two criteria. For example, catastrophic risks should include those risks that are most likely and cause the maximum possible damage to the business entity during its implementation [4].

We will consider the risk matrix using the example of fur-bearing animal husbandry for growing cell breeding minks (Table 1). In the course of the analysis, the main risk factors that may arise during the implementation of the activity of fur farming, were identified. Risks were entered into the matrix by the degree of their influence and frequency of exposure, using ranking and assessment of factors. The identified factors have a significant impact on the fur farming industry and are more typical of fur animals for raising minks not only in the Russian Federation, but also internationally. The factors indicated in the matrix are manifested not only in a decrease in sales volumes of mink skins, but also in sharp jumps in prices for mink skins, which has a negative effect not only on producers, but also on end consumers.

Table 1. Risk matrix for fur mink farming .

Risk probability	Consequences		
	Minor	Medium	Significant
High probability	Adverse currency change	Decrease in consumer confidence index among consumers	Advertising campaigns for the protection of animals and the refusal to purchase clothing made from natural fur
Probably	Lack of full reporting on mink sales	Lack of advertising / own brand of domestic mink as a product of international trade	Weather conditions, namely positive temperatures, global warming
Possibly	New competitors entering the market	Raw material price increase	Implementation of mink skins below cost

Source: Developed by the author.

In the presented risk matrix (Table 1), the following classification of consequences, probabilities and risk levels is proposed:

1. The consequences:

- *Minor* (low financial losses), namely: a decrease in sales, pressure to lower prices and margins of products sold. Measures to reduce risk, such as: developing policies to promote products on the market, including pricing; search for alternative sales channels, access to Asian markets;
- *Medium* (high financial losses), namely: lack of demand for products, decrease in sales, the need to attract additional financing to maintain the functioning of the animal husbandry, decrease in motivation among buyers - manufacturers of ready-made clothes and end-users, the departure of key employees, such as livestock specialists, veterinarians, feeding services, from fur farming. Measures to reduce risk are the following: improving product quality, conducting market research, developing policies to promote products on the market, including pricing, attraction of an administrative resource to stimulate demand, constant search and training of personnel changes (profile order in institutions of secondary and higher professional education);
- *Significant* (large financial losses), namely: large losses, liquidity gaps, litigation with banks, the need to attract additional financing to maintain the functioning of the company, the elimination of fur farming. Measures to reduce risk are as follows: analysis of import substitution (search for alternative suppliers / markets for the sale of mink skins), conducting information campaigns on the need to maintain the fur farming industry, placing free cash on deposits in foreign currency, attracting investors, replacing obsolete equipment with energy-saving and knowledge-intensive.

2. Probability:

- *Possibly* means that the event has occurred no more than three times in the last ten years;
- *Probably* means that the event happens every year;
- *High probability* means that the event occurs once every six months.

3. Risk levels:

- A potentially moderate level of risk is when periodic monitoring of the level of risk is recommended;

- In case of potentially high risk, in case of economic feasibility, it is recommended that a decision is made on the development of measures to minimize risks, as well as periodic monitoring of the level of risk;
- Potentially very high risk is when immediate actions are recommended as those being necessary to reduce the risk of the target level by developing measures to minimize risks.

In this example, all situations with a high and potentially very high risk will be unacceptable and require the immediate development of measures to minimize risks. Situations with a potentially moderate risk level are acceptable; however, these situations require an increased attention of the fur management. The information presented in the risk matrix can be used to identify risk and make decision by the management of fur animals to minimize expected losses or maximize profitability.

4. Discussion

The considered risk factors are confirmed by the data of the Russian statistics service. We will analyze the factors that have the greatest impact on the activities of fur farms. According to the results of the I quarter of 2019, more than 284 thousand fur products were sold in Russia, which was 29.9 thousand less than the volume of goods sold in the I quarter of 2018 (-9.5%) [8]. The drop in comparable indicators may be due to a combination of factors. Firstly, fur coat sales are directly related to weather conditions. Warm January and February inevitably lead to lower sales. Secondly, at the beginning of 2019, a decrease in the consumer confidence index was recorded. During 2019, the population does not expect income growth; therefore, expensive purchases are delayed until "better times."

In addition, advertising campaigns to protect animal rights and refuse to purchase clothing made from natural fur are an important factor that has a significant impact. The target category of consumers from 20 to 40 years old falls under this influence, which is subject to the influence of the post-modern culture that came from the West. This is expressed not only in the refusal to wear clothes made of natural fur, but also a complete rejection of food of animal origin, namely veganism, raw food diets, etc.

The legislative prerequisites for protecting the rights of fur animals in the European livestock farms were adopted by the European Union. According to the recommendations of the European Council for the Cultivation of Fur Animals (published in 1999), it is undesirable to grow for tailoring those animals that were born in the wild and also those belonging to the species that are not able to adapt to captive life. Recommendations for the cultivation of fur animals, published in 2001 by the Scientific Committee of the European Commission on Animal Health and Well-Being, stipulate that all types of fur animals must be provided with housing that will meet all their needs on farms. In 2009, the Directive on the Protection of Animals During Slaughter was adopted, ruling that many improvements should be introduced sunce from January 1, 2013. For example, the killing of fur-bearing animals should be supervised by a competent, certified specialist. Certain specifications must be considered, e.g. carbon monoxide must be used to kill [5].

Recently, in Russia, as well as in the European Union, there has been criticism of the fur farming industry. Fighters for animal rights and people who consider the use of fur unethical and inhumane in relation to animals, oppose the use of fur products and for the closure of fur enterprises. There are also activists who are not opposed to the fur farming industry and the use of fur in general, but to they speak in favor of improving the conditions for keeping fur animals. Also, there are cases of attacks on farms and farms in order to steal animals for their subsequent release. The main international organizations against fur include the following:

- International Organization for People for the Ethical Treatment of Animals (PETA);
- Humane Society International;
- Respect for Animals (UK);
- Fur Free Alliance (FFA);
- Coalition to Abolish the Fur Trade (CAFT, UK);
- NP Center for the Protection of Animal Rights "VITA", Russian public non-profit organization (Russia, Moscow).

In addition, the cost of natural fur in the world is decreasing due to the trend to refuse natural fur set by the leading fashion houses. So, during 2018, the rejection to use the natural fur in the manufacture of clothing was announced by the leading world designers, including: Armani, Calvin Klein, Gucci, Hugo Boss, Ralph Lauren, Stella McCartney, Tommy Hilfiger, Versace, Michael Kors, Vivienne Westwood.

In addition, according to the final report of the Finnish company Saga Furs for 2018 (this is one of the largest auctions for the sale of mink and fox), it can be noted that the company's revenue fell sharply (by 20 percent) from November 1, 2017 to October 31, 2018, and their profit before tax decreased from 9.2 million euros in 2016-2017 to minus 1.7 million euros (a decrease of 10.9 million euros) [9]. This is partly explained by the many bans on fur products in fashion houses, as well as problems with the image of animals' suffering, being kept in fur farms around the world, as well as because of the slowdown in in Chinese markets. The report also states that, according to a number of experts, in comparison to the previous year, the world mink production fell by 20 percent in 2018, prices for mink and fox skins fell by 24 percent and 20 percent, respectively.

The fall in profits at global auctions clearly shows that the fur industry has been hit hard by the leading global designers who have implemented a fur-free policy; and the trade in fur skins was clearly shaken by this blow after falling prices. One of the key problems is the "problem with the positioning of the fur trade in the western market." No seller wants to spoil his/her brand with the image of a tortured animal that was kept in unsuitable conditions (tiny and cramped cages) before being gassed or electrocuted. Table 2 provides an overview of those countries that have introduced a complete ban on the keeping of fur farms, as well as restrictive regimes on the breeding of fur animals and trade in fur and fur products.

Table 2. Countries with restrictions on the breeding of fur farms.

Complete prohibition of cage fur farming	Breeding of fur animals is partially prohibited or strict conditions for their keeping are legalized	Bans on sellin fur and fur products
	Countries (cities)	
Austria (since 1998)	Switzerland (no animal farms due to strict rules for keeping animals)	West Hollywood (USA)
United Kingdom (since 2000)		Dublin and its district (Ireland)
Croatia (since 2007)		San Francisco (USA)
Bosnia and Herzegovina (since 2009)	Denmark (in 2009, only a ban on fox breeding was adopted, mink breeding is allowed)	New Zealand (ban on the import of mink fur)
Norway (more than 300 fur farms will close by 2025)		
Czech Republic (since 2019)	Italy (strict animal welfare)	
Netherlands (since 2012, final decision made in 2015)		India (since 2017, a ban on the import of mink, fox, and chinchilla fur)
Slovenia (since 2013)	Sweden (strict animal welfare)	
Germany (since 2017, the remaining fur farms have a transition period)		
Luxembourg (since 2018)		
Serbia (since 2019)		
Japan (since 2016)		

Source: Based on materials from the Fur Free Alliance, FFA, 2017 [10].

Despite the fact that Denmark, Finland, Canada, the USA, and Russia are still one of the largest producers and exporters of mink fur, the public there is fighting for animal rights [3]. The main message of environmental activists is that furs are the "luxury goods without necessity."

It should be noted that many actions in defense of animal rights are aimed at destroying the reputation of the fur trade and the destruction of this industry. For many years, false accusations spread about the fur industry, claiming that animals kept in fur farms were skinned alive to supply the fur industry with fur skins. In particular, PETA, an international organization of animal rights activists, used a video depicting live hiding of mink skin in an attempt to denigrate the fur industry and force lawmakers to

IOP Conf. Series: Earth and Environmental Science **395** (2019) 012054 doi:10.1088/1755-1315/395/1/012054

prohibit the breeding of fur animals in fur farms around the world. But in 2019, the International Fur Federation (IFF) conducted an investigation in which the actions of animal rights activists were exposed. They ordered and paid for a 2009 video showing torture over a fur-bearing animal [11]. The IFF released documentary refuting activist accusations and exposing perpetrators of major international conspiracy to destroy the fur trade reputation.

In addition to the factors arranged by animal rights activists, it is necessary to consider those factors speaking in favor of the production of clothing made of natural fur. Despite its friendly attitude to the animal world, most of the faux fur is made from modacrylic fibers, synthetic polymers. This means that, like many other products created on the basis of petroleum products, these products pollute the environment both duringits production and during washing, as well as they are not biodegradable.

5. Conclusion

Studies have shown that in a market economy, the process of a business entity's functioning is accompanied by risks that affect the profitability of the fur farming industry. The analysis of risk factors affecting fur farming was carried out using a scientific approach to diagnosing risks using the matrix method. This method does not guarantee the adoption of effective decisions in managing the activities of fur farming, although it could help to identify the severity of the consequences of risky events and a possible assessment of the likelihood of their occurrence. When analyzing risks in enterprises, an integrated approach must be used that includes various methods for measuring risk, which would allow one to get the most complete picture of risks in each case relative to the enterprise. According to the results of the study, a matrix with a classification of consequences, probabilities. and risk levels was proposed. This will allow enterprises to predict risks and make informed management decisions, preventing the emergence of critical situations at all stages of the production process.

References

[1] Baldin K 2012 *Risk management* (Moscow, Russia: UNITY-DANA)
[2] Vyatkin V 2013 *Risk management: preventive management* (Moscow, Russia: AP "Science and Education")
[3] Khusainova N, and Vorozheykina T 2018 *There is a time for all things under heaven: problems and prospects of fur farming industry in Russia* (Moscow, Russia)
[4] Korolkova E 2013 *Risk management: project risk management* (Tambov, Russia: "TSTU")
[5] Council of the European Union 2009 *Regulation (EC) on the protection of animals at the time of killing* (September 24, 2009, No. 1099/2009) (Brussels, Belgium: text with EEA relevance No L 303)
[6] Federal Agency on Technical Regulating and Metrology 2007 *The National Standard of the Russian Federation on Project Risk Management: General provisions – 2007 GOST R 52806-2007 from December 27, 2007* (Moscow, Russia: Standartinform Rossiiskoi Federatsii)
[7] Federal Agency on Technical Regulating and Metrology 2011 *National Standard of the Russian Federation Risk management. Terms and Definitions* 2011 GOST R 51897-2011 / ISO 73 Guide: 2009 from November 16, 2011 (Moscow, Russia: Standartinform Rossiiskoi Federatsii)
[8] *Fur sales decreased by 10%* 2019 Retrieved from https://news.mail.ru/economics/36998792/?frommail=1
[9] Sagafurs (n.d.) *Financial information about Sagafurs 2018* Available at: https://www.sagafurs.com/ru/%d0%be-%d0%ba%d0%be%d0%bc%d0%bf%d0%b0%d0%bd%d0%b8%d0%b8/investors/financial-information/ (Accessed 14 03 2019)
[10] "VITA" Center for the Protection of Animal Rights 2018 *Fur ban in the world* Available at: http://www.vita.org.ru/fur/law.htm (Accessed 14 03 2019)
[11] *Press release: animal rights activists paid for animal torture on video* 2019 Available at: https://profur.fi/en/ajankohtaista/artikkeli/press-release-animal-rights-activists-paid-animal-torture-video (Accessed 14 03 2019)

The effect of leaf treatment (*Solanum tuberosum* L. cv Zhukovsky) with an early leaf extract (*Amaranthus tricolor* L. cv Early splendor) on the photosynthetic indicators of potato plants and the feeding of the colorado beetle larvae

V F Pivovarov[1], E M Gins[2], and A A Baikov[1*]

[1] Federal Scientific Center for Vegetable Growing, 14 Selektsionnaya str., VNIISSOK Village 143080 Russia
[2] Lorch Potato Research Institute, 23 Lorcha str., 140051 Kraskovo Russia

Email: physiol@inbox.ru

Abstract. Processing potatoes with water extract from the amaranth of *Early splendor* variety leads to an increase in the parameter (F_M – F_T)/F_T of slow induction of leaf fluorescence up to 51%. At the same time, the effectiveness of restricting the nutrition of larvae of the Colorado potato beetle on the treated plants reaches a maximum of 64%. Laboratory experiments show a decrease in larval survival on treated plants. Due to its dual properties as a biostimulant and insecticide, as well as low toxicity to vertebrates, amaranth extracts can be recommended for integrated pest management programs.

1. Introduction

The Colorado potato beetle (*Leptinotarsa decemlineata* Say) is a serious pest of solanaceous crops in our country, especially for potatoes. The use of synthetic insecticides to combat this pest has led to resistance in virtually all classes of insecticides, including organophosphates, carbamates, pyrethroids, and neonicotinoids [5]. The development of biological methods of protection is to reduce the chemical load in the field and prevent the development of resistance to chemicals in the Colorado potato beetle [4]. The use of plant extracts is promising. They stimulate growth and lead to an increase in the photo-synthetic activity of plants [2]. At the same time, they protect them by restricting the nutrition of the larvae for the Colorado potato beetle. The evidence of luminescent research methods used in the work is that, as a rule, they make it possible to reveal changes in the photo-synthetic equipment at early stages of one or another factor, even before the appearance of characteristic morphological changes [1].

2. Materials and Method

For the processing of Zhukovsky ranniy potato variety, water extracts obtained from dried leaves of Amaranth of the Early splendor variety, were used.

The activity of the photosynthetic apparatus of potato plants at the leaf level was analyzed by recording the fluorescence and thermoluminescence curves [1], [3]. As a slow fluorescence induction parameter (SFI), the following relationship was used: (F_M – F_T)/F_T, where F_M is a value corresponding to the maximum fluorescence induction curve, and F_T is the stationary fluorescence level. As a

Content from this work may be used under the terms of the Creative Commons Attribution 3.0 licence. Any further distribution of this work must maintain attribution to the author(s) and the title of the work, journal citation and DOI.
Published under licence by IOP Publishing Ltd

IOP Conf. Series: Earth and Environmental Science **395** (2019) 012055 doi:10.1088/1755-1315/395/1/012055

parameter of high-temperature thermoluminescence, the light sum S was used to reflect an area under the TL curve in the range from + 60 to + 100 °C.

As a parameter describing the effect of treating potato leaves with amaranth extracts on the nutrition of larvae of the Colorado potato beetle, an index of nutritional inhibition was used. FDI = $(S_C-S_T)/(S_C+S_T)$, where S_C and S_T are the areas of control and treated leaves eaten [4].

3. Results

Spraying potato plants with amaranth extracts with a concentration of 2-8% led to an increase in the SFI values (Table 1). This parameter changes in proportion to the specific (in terms of chlorophyll) photo-synthetic activity of $\Delta O_2/(\Delta t \cdot \text{chlorophyll})$. The data obtained indicate an increase in the functional activity of the photosynthetic apparatus of potato plants during the treatment with phytoextracts [2].

At the same time, we observed a decrease in the light sum of the high temperature thermoluminescence $S_{[60;100]}$ in the experimental plants (Table 2). The magnitude of the temperature peak of thermoluminescence characterizes the resistance of plant tissues (chloroplast membranes) to stressors [1], [3]. Amaranth extracts have a rich antioxidant composition, including phenolic compounds [2]. Probably, exogenous antioxidants prevented lipid peroxidation of thylakoid membranes, which led to a decrease in high temperature thermoluminescence. We can assume that the resistance of chloroplast membranes will increase in the experimental plants under the stress of the action of the larvae of the Colorado potato beetle.

Table 1. Values of the SFI parameter of potatoes of the variety *Zhukovsky ranniy* variety in control and experience groups.

Day after treatment	$(F_M-F_T)/F_T$, mean ± standard error (% of control)			
	Control	Amaranth water extracts *Early splendor* variety		
		2%	4%	8%
2	0.71±0.03 (100%)	0.81±0.03 (114%)	0.89±0.04 (125%)	0.89±0.04 (125%)
5	0.73±0.03 (100%)	0.93±0.04 (127%)	1.10±0.04 (151%)	1.04±0.04 (142%)
10	0.70±0.03 (100%)	0.84±0.03 (120%)	0.99±0.04 (141%)	0.97±0.04 (139%)
15	0.72±0.03 (100%)	0.79±0.03 (110%)	0.90±0.04 (125%)	0.83±0.03 (115%)

Table 2. Values of a light sum $S_{[60;100]}$ in potato leaves of the *Zhukovsky ranniy* variety in the control and experience groups.

Day after treatment	$S_{[60;100]}$, mean ± standard error (% of control)			
	Control	Amaranth water extracts *Early splendor* variety		
		2%	4%	8%
2	35±2 (100%)	28±2 (80%)	26±2 (74%)	25±2 (71%)
5	33±2 (100%)	37±2 (82 %)	23±2 (70%)	23±2 (70%)
10	37±2 (100%)	33±2 (89%)	27±2 (73%)	28±2 (76%)
15	36±2 (100%)	36±2 (100%)	29±2 (81%)	28±2 (78%)

Treatment of potato leaves with amaranth extracts resulted in restriction of the nutrition of larvae of the Colorado potato beetle on them (Fig. 1). Such action of extracts can be associated with the presence of phenolic compounds in them that inhibit enzymes, as well as receptor sites in neuroendocrine GABAergic and dopaminergic systems of insects [5].

4. Discussion

The optimal stimulating effect on the photo-synthetic apparatus of potato plants was achieved using 4% extract of amaranth of the *Early splendor* variety, while the maximum was reached on day 5 after processing. In terms of high temperature thermoluminescence, a minimum was also observed on day 5, while an increase in the concentration of the extract to 8% did not lead to a significant increase in the effectiveness of the action. The effect decreased by 15 days, and (if necessary) the treatment can be repeated.

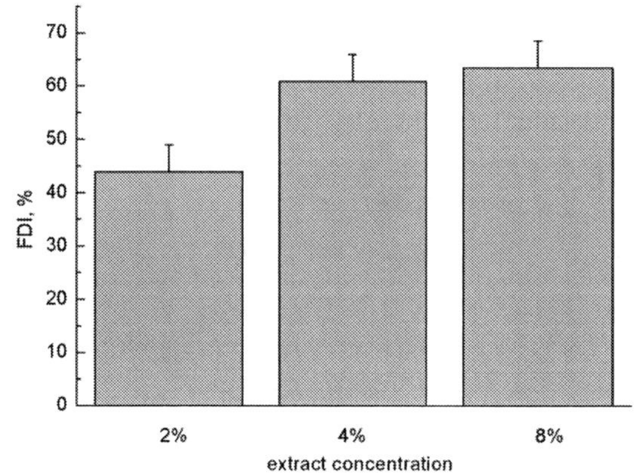

Figure 1. The effect of restricting nutrition of the larvae of the Colorado potato beetle on potato leaves treated with amaranth extracts.

In terms of the stimulating effect on potato leaves and food restriction for the larvae of the Colorado potato beetle, the use of 4% water amaranth extract of the *Early splendor* variety is the most promising. At the same time, the mortality of the larvae in the experiment with 4% extract of *Early splendor* was 53 ± 10 %.

5. Conclusion

In terms of mortality and food restriction, amaranth plant extracts cannot compete on their own with the most synthetic insecticides specifically designed to combat the Colorado potato beetle. First of all, they should be considered as biostimulants, and insecticidal properties are a bonus that should be considered when choosing a comprehensive program for pest control. The period recommended for the application of amaranth as a biostimulant for out-of-root cultivation in potatoes coincides with the period when the larvae of the Colorado potato beetle are present on the foliage, which makes it possible to increase the effectiveness of biological methods of protecting potatoes, for example, using the *Picromerus* predatory bug (*Picromerus bidens* L.).

References

[1] Baikov A A, Karavaev V A, Levykina I P, Solntsev M K, Tikhonov A N, Popov S Y, … Kvitka A Y 2013 Luminescence characteristics of strawberry leaves at early stages of injury by spider mite *Biophysics* **58**(2) pp 234-239
[2] Karavaev V A, Glazunova S A, Levykina I P, Gunar L E, Myakinkov A G, Gins M S, … Lepeshkin F D

2012 Slow fluorescence induction and productivity of barley treated with a supercritical fluid extract of amaranth *Biophysics* **57**(4) pp 502-503

[3] Pakhnenko E P, Vatsadze N S, Glazunova S A, Karavaev V A, Baikov A A, and Solntsev M K 2012 Early diagnostics of physiological state of plants under various nutrition conditions using luminescent methods *Moscow University Soil Science Bulletin* **67**(2) pp 60-64

[4] Pavela R 2010 Antifeedant activity of plant extracts on *Leptinotarsa decemlineata* Say. and *Spodoptera littoralis* Bois. larvae *Ind Crop Prod* **32** pp 13-319

[5] Wang Z, Zhao Z, Cheng X, Liu S, Wei Q, and Scott I V 2016 Conifer flavonoid compounds inhibit detoxification enzymes and synergize insecticides *Pestic. Biochem. Physiol* **127** pp 1-7

IOP Conf. Series: Earth and Environmental Science **395** (2019) 012056 doi:10.1088/1755-1315/395/1/012056

Trends in the development of native horse breeding and ways to improve its efficiency

V S Koveshnikov[1], A M Zaycev[1], and E V Slotina[2*]

[1] All-Russian Scientific Research Institute of Horse Breeding, Divovo, Ryazan region 391105 Russia
[2] Ryazan State Medical University, 9 Vysokovoltnaya str., Ryazan 390026 Russia

E-mail: e-slotina@mail.ru

Abstract. The paper analyzes the most important segment of horse breeding in the conditions of a fundamental change in the general economic situation in agriculture in general and the horse industry in particular. In a significant time aspect (1991-2018), the dynamics of the producing composition of the stud farms' breeding stock and their placement in the federal districts of the Russian Federation are rigorously analyzed. The modern livestock of horse stud mares has been categorized, included in the state breeding register and in the main breeding areas. Changes in the producing composition of the breeding stock of all breed directions are investigated. Analysis of the number of horses in the country (for 28 years) shows that the total number of factory mares decreased from 6.3 to 3.4 thousand heads or by 49.2%. The number of saddle mares increased by 42.5%, from 362 to 516 heads. It was established that during the analyzed period, all production and economic indicators of horse breeding decreased. Successfully tested in many foreign countries, a system of financial sustainability and high efficiency of horse breeding in the form of targeted support (using the national equestrian sweepstake) has not been established in the industry.

1. Introduction

During the years of developing market relations, objectively significant changes occurred in the horse breeding of Russia. They were due to fundamental changes in agricultural production. The development of privatization and multi-layered processes led to the repeated reduction of large state and collective farm agricultural enterprises and the curtailment of the main branches of productive livestock raising in them, including those in the horse-breeding sector. The development of animal husbandry in private farms and in peasant (farmer) farms only partially compensated for the decline in the number of horses. As a result, in the period 1991-2018, the horse livestock in the country decreased from 2618.4 to 1283.0 thousand heads, i.e. in more than twice. At the same time, in regional terms, horse resources have become significantly higher in Siberia and the Far East compared to the central, European districts. Negative processes in the industry could not but affect the placement and development of pedigree horse breeding, and above all, its most valuable link – the stud farm. The study analyzes the dynamics of placement of the producing breeding stock of sources with different pedigree affiliation in the region's stud farm. The paper also justifies the conditions and reserves for creating a stable economic situation in horse breeding.

Content from this work may be used under the terms of the Creative Commons Attribution 3.0 licence. Any further distribution of this work must maintain attribution to the author(s) and the title of the work, journal citation and DOI.
Published under licence by IOP Publishing Ltd

IOP Conf. Series: Earth and Environmental Science **395** (2019) 012056 doi:10.1088/1755-1315/395/1/012056

2. Materials and Methods

The research is based on the following sources: (a) the information data directory *Indicators on the Economy of Stud and Hippodromes of the USSR for 1990*, compiled by the Department of Economics VNII Horse Breeding; (b) materials of the register of breeding horse enterprises as of January 1, 2019; (c) data of annual statistical reporting on the number of horses by categories of farms in the context of regions and federal districts; (d) results of annual reporting on the horse breeding of a number of stud farms. When conducting research, the following methods were used: economic-statistical, monographic, and groupings.

3. Results

The uniqueness of horse breeding is determined by a number of biological characteristics of horses, among which their high adaptation to extreme conditions of housing occupies a special place, as evidenced by the wide distribution of the industry in different climatic zones of the country. An analysis of current trends in the development of pedigree horse breeding in Russia, which is of high relevance for practitioners, has been carried out in the stud farms, where the main breeding core of factory breeds is concentrated. Their preservation and improvement is the most important task in the tribal and mass sectors of the industry.

The development of horse breeding has historically been determined by the main tasks facing the horse-breeding industry. Until the mid-twentieth century, a horse was an important energy tool in many branches of material production, and this animal was also widely used in military units. The main purpose of the country's stud farms and breeding farms was to raise the breeding material to improve the working qualities of horses. In the second half of the last century, with an objective reduction in the role and importance of the horse as an energy vehicle, new directions for horse use began to develop, namely, meat herd, dairy, and sports and leisure. Horse breeding was reoriented to the cultivation of improvers to improve the working and productive qualities of horses, as well as horses for mass sports, tourism, rental, hippotherapy, and other goals of a social nature.

In the pre-reform period, there was a system of state leadership in horse breeding that was separate from other branches of animal husbandry. In the Union and a number of republican ministries of Agriculture, an Office of Special Horse Breeding was functioning. The most valuable tribal stud farms and racetracks were part of the All-Union Trust, which made it possible to plan and analyze the development of the industry and, above all, horse breeding.

On April 27, 1981, in order to effectively develop the horse-breeding industry, the CPSU Central Committee and the USSR Council of Ministers adopted Resolution No. 402 on *Measures for the Development of Horse Breeding*, during the implementation of which horse breeding followed the path of expanded reproduction. In the last five years of the pre-reform period (1986-1990), the number of horses in the USSR increased from 5781.5 to 6918 thousand heads (19.7%); the horse meat production increased from 213 to 270 thousand tons (26.8 %), koumiss from 31 to 35 thousand tons (12.9%). Mass equestrian sport, tourism, and horse rental were further developed.

The positive development of mass horse breeding had a positive effect on specialized horse breeding. Increased domestic demand for breeding horses led to a sharp increase in production and economic indicators of the industry. So, if in the whole country's stud farms in the five-year period before the adoption of the above-mentioned directive document (1976-1980), the profitability level from selling the breeding horses was 15.2%, it reached 59.8% in 1981-1985 and 75.2% in 1986-1990.

Unfortunately, at the end of the pre-reform period (1990), only 65 out of 102 horse breeding factories for breeding factory breeds, with the exception of regional ones operated in Russia.

Placement of producing uterine composition of the horse breeding studs in Russia in the pre-reform period in the modern federal districts is presented in Table 1.

Table 1. Placement of uterine stock of horses in the stud farms of Russia (January 1, 1991).

Federal districts	Breeding mares (aged 3 years and older)		Breeding mares per thousand horses in mass horse breeding, heads
	Number	%	
Central	1,405	22.2	4.5
Northwestern	288	4.6	4.7
Southern	1,948	30.7	9.4
North Caucasus	374	5.9	3.1
Volga	1,743	27.5	2.8
Ural	191	3.0	1.0
Siberian	276	4.4	0.4
Far Eastern	109	1.7	0.5
Total for the Russian Federation	6,334	100.0	2.5

The largest herd of breeding mares was noted in the stud farms of the Southern (30.7%), Volga (27.5%), and Central (22.2%) federal districts. In total, 80.4% of the breeding uterine stock of horses accounted for these three districts [1].

Such placement of horse breeding is due to traditional historical factors of its development. Horse factories supplying a large number of army horses for cavalry and artillery units, postal purposes and carting, as well as with the aim of improving the working qualities in agriculture and other areas of material production were created and functioned in the European territory of the country with intensive feed production and high availability of labor resources [2]. 26 out of 36 hippodromes operating in the country were created to improve the working qualities of trotting and riding horses in these regions.

By the beginning of market reforms, in the stud farms, groups of regions for raising horses of different pedigree directions developed. Thus, trotting horses (Orlovskaya, Russkaya, Amerikanskaya, Frantsuzskaya) were raised in the Central and Volga Federal Districts (79.6%) of the uterine stock. The horses of the riding breeds Arabskaya, Akhaltekinskaya, and Chistokrovnaya Verkhovaya were located in stud farms of the North Caucasus and Southern federal districts and made up 93.1% of the whole uterine stock. The breeds of seddle mares (Budenovskaya, Gannoverskaya, Golshtinskaya, Donskaya, Kabordinskaya, Trakenenskaya, and Terskaya) were grown mainly in the Southern District, 86% of all mares of this breed were grown there. According to the population of draft and trotting mares, the farms of the Volga and Central federal districts were distinguished, which had 72.4% of all the producing mares. The analysis is performed on the basis of the data, and it is presented in Table 2.

Table 2. Placement of the uterine stock of breeding horses in Russian stud farms (as of January 1, 1991).

Federal districts	Breeding types							
	Trotter		Riding		Saddle		Draft	
	Heads	%	Heads	%	Heads	%	Heads	%
Central	1041	43.5	25	6.9	45	2.3	294	18.1
Northwestern	118	5.4	-	-	100	5.1	60	3.7
Southern	60	2.5	93	25.7	1685	86.0	110	6.8
North Caucasus	-	-	244	67.4	130	6.6	-	-
Volga	864	36.1	-	-	-	-	879	54.3
Ural	60	2.5	-	-	-	-	131	8.1
Siberian	239	10.0	-	-	-	-	37	2.3
Far Eastern	-	-	-	-	-	-	109	6.7
Total for the Russian Federation	2392	100	362	100	1960	100	1620	100

During the period of market reforms (1991-2018), a sharp reduction in the number of horses from 2618.4 to 1283.0 thousand heads (by 51%) occurred in the country. In mass horse breeding, the number of horses decreased from 2,534 to 1,231 thousand heads (or 2.1 times), which objectively affected the decrease in the need of stud farms for breeding horses of factory breeds, including their breeding core.

The data from Table 3 provide an opportunity to judge the change in the number of factory females in the regional aspect.

The largest number of breeding mares is concentrated in the Volga Federal District and amounts to 776 heads (23.4%) in the North Caucasus Federal District, the livestock is 720 heads (21.2%) and the Central Federal District, and the livestock is 687 heads (20.3 %). The increasing. number of breeding mares in the stud farms of the Siberian Federal District, from 276 to 373 heads (35%), is noted, which is explained by the increased demand for breeding horses and the increase in the number of horse owners in Siberia. Currently, per thousand horses in the mass horse breeding, the number of factory mares is 2.8 heads, whereas at the beginning of 1991 there were 2.5 heads.

Table 3. Placement of the uterine stock of horses in the stud farms of Russia (January 1, 2019).

Federal districts	Breeding mares aged 3 years and older		Breeding mares per thousand horses in mass horse breeding, heads
	Heads	%	
Central	687	20.3	18.8
Northwestern	142	4.1	71.0
Southern	602	17.7	6.5
North Caucasus	720	21.2	8.6
Volga	776	23.4	3.8
Ural	91	2.3	1.3
Siberian	373	11.0	0.9
Far Eastern	-	-	-
Total for the Russian Federation	3391	100.0	2.8

A sharp decline in the number of factory breeds (by 3.2 times) in the stud farms of the Southern Federal District is due to a significant decrease in the number of representatives of the high-draft breeds (Donskaya, Budenovskaya, and Trakenenskaya), which is confirmed by the data presented in Table 4.

Table 4. Placement of the uterine stock of breeding horses in the Russian stud farms (January 1, 2019).

Federal districts	Breeding types							
	Trotter		Riding		Saddle		Draft	
	Heads	%	Heads	%	Heads	%	Heads	%
Central	489	42.7	-	-	113	11.3	85	11.6
Northwestern	50	4.3	45	8.7	-	-	47	6.4
Southern	-	-	173	33.5	429	42.9	-	-
North Caucasus	-	-	263	51.0	457	45.8	-	-
Volga	348	30.4	35	6.8	-	-	393	53.9
Ural	21	1.8	-	-	-	-	70	9.6
Siberian	238	20.8	-	-	-	-	135	18.5
Far Eastern	-	-	-	-	-	-	-	-
Total for the Russian Federation	1146	100	516	100	999	100	730	100

Currently, horses of the trotting trend, with a total reduction in their number by half, are still predominantly bred in the farms of the Central and Volga federal districts (42.7% and 30.4% of their total number). In the stud farms of the Siberian Federal District, the proportion of trotters increased from 10% to 20.8%. During the years analyzed, the number of horse mares in the horse-riding sector in the stud farms increased from 362 to 516 heads or by 42.5%; mainly, it is located in the North Caucasus and Southern federal districts and is 51% and 33.5%, respectively. The main mass of saddle mares is concentrated in the same districts, and it is 84.5%.

Unfortunately, during the analyzed period, the reduction of mares in horse-breeding factories in the draft mares from 1,620 to 730 heads or 2.2 times occurred. During the period of market reforms, the reduction in the number of mares producing composition in the stud farms was accompanied by a significant decrease in all production and economic indicators of the industry. However, the reproduction of horses has deteriorated. The business output of foals per 100 mares was 72 heads in 1990, 60 heads in 2002 and 58 heads in 2017. Implementation of the breeding horses for every 100 mares (up to 39-40 heads) fell almost one and a half times. In comparison with 1990, an average annual domestic sales of breeding horses decreased 2.4 times, and the exports fell in more than four times.

According to some economists, the horse breeding has become a deeply unprofitable industry. In recent years, annual losses range from 300 to 400 million rubles. One of the reasons for this situation is an artificial transfer of costs from other spheres of production (in a number of farms) to the horse breeding, which distorts the real economic situation.

Based on the results of research, we found that in horse breeding there are a number of reserves to improve the efficiency of technological and organizational nature [3], [4]. First of all, this is connected with the improvement of horse reproduction, since the costs of maintaining the production composition (mares and stallions) are high and range from 41% to 45% of the total annual cost of rearing. These costs are fully transferred to the litter. Consequently, at high rates of business output of foals, their cost decreases. As an example, we can estimate the cost of a Russian draft breed at birth in the Vologda and Kuedinsky stud farms for 2015-2017. In total, 23.5 million rubles were spent on the maintenance of the producing structure, which when receiving 179 foals (48 heads per 100 mares), the cost of one foal was 132 thousand rubles. If the business output of foals per 100 mares in these farms was at the level prevailing in the draft horse breeding in 1990, namely 77 heads, they would receive 286 foals with the cost price of one head 82 thousand rubles, i.e. 50,000 rubles lower.

Given the current situation with the reproduction of breeding horses, the state subsidies to breeding enterprises for the maintenance of the breeding stock using the "mare-foal" system seems appropriate [5]. In addition, stud farms incur significant additional costs for feeding horses, which is associated with the inefficient use of artificial pasture (paddocks). Our studies have found that the creation of highly productive cultivated pastures, with their rational use, reduces the cost of feeding by 1.5-2 times [6], [7].

Domestic and foreign experience testifies to the positive results of the introduction of group forms of keeping young stock of breeding horses and uterine stock, which contributes to increasing the efficiency of the breeding enterprises. Unlike the loose box technology, such methods of maintenance allow a 2-2.5 times increase in the labor productivity of horse breeders and reduce the cost of raising breeding horses by 15-20% [3].

In recent years, the costs of studs for the maintenance and hippodrome tests of horses have increased. Using the materials of the Novotomnikovsky stud farm as an example, we can conclude that the maintenance of one horse at the Central Moscow Hippodrome (tests were carried out for 2 years) cost the horse owner around 480-500 thousand rubles. Such high costs are due to insufficient use of reserves from sweepstakes and other sources of raising funds in the industry. Unfortunately, the share of revenues from sweepstakes decreased from 43.6% in 1991 to 10% in 2017. Calculations show that the effective operation of sweepstakes on previously profitable racetracks (Moscow, Pyatigorsk, Ramenskoye, by analogy with Rostovsky) makes it possible to additionally receive about 153 million rubles a year. This will allow horse owners not only to cover the costs of maintenance and testing of horses, but also to send significant amounts to other production purposes.

As the proven and successful experience of most foreign countries shows, the accelerated creation of the prize horse-breeding industry with a system of a single national equestrian sweepstake should be an important factor in improving the efficiency and sustainable economic functioning of horse breeding.

4. Discussion

At present, rational management of horse breeding should focus on the maximum use of intra-industry, organizational and technological reserves, tested in long-term practice. In the short term, the creation of functionally new production units in horse breeding, ensuring financial stability and self-sustaining profitable management of the stud farm, seems reasonable.

5. Conclusion

In the short term, the full satisfaction of the needs of mass horse breeding in pedigree improvers should form the basis of the principles of optimal placement of the industry. Also, there should be a clear program of the prize horse-breeding industry with a single system of national equestrian sweepstakes, forming the basis of the principles of optimal placement.

The leading areas for improving the efficiency of the horse breeding industry are the following: improving the reproduction of horses, rationalizing the technology of feeding and maintenance, enhancing the financial activities of racecourses, which requires considerable state support at the initial stage of formation.

6. Acknowledgments

The authors are grateful for the advisory assistance provided to Academician of the Russian Academy of Sciences, Doctor of Agricultural Sciences, Professor, Scientific Director of the All-Russian Scientific Research Institute of Horse Breeding Valery Vasilievich Kalashnikov and the Candidate 0f Agricultural Sciences, Associate Professor, Head of the Selection Department of the All-Russian Scientific Research Institute of Horse Breeding Galina Vladimirovna Kalinkina.

References

[1] Koveshnikov V S, and Yermilova V D 1991 *Indicators on the economy of stud farms and hippodromes of the USSR in 1990* (Divovo, Russia: All-Russian Scientific Research Institute of Horse Breeding)

[2] Gusev Yu P 1976 Economic appraisal and prospects for the development of horse breeding in the USSR In *Economics and organization of horse breeding* (pp 106-210) (Divovo, USSR: All-Russian Scientific Research Institute of Horse Breeding)

[3] Koveshnikov V S 2005 *Organizational, technological and economic mechanisms of rational management of horse breeding in Russia* (Dissertation in the Form of a Scientific Report for the Degree of Doctor of Agricultural Sciences) (Divovo, Russia: All-Russian Scientific Research Institute of Horse Breeding)

[4] Kalashnikov V V, Zaitsev A M, and Koveshnikov V S 2018 *Horse industry of Belt And Road countries* (Peking, China) (in Chinese)

[5] Kalashnikov V V, Koveshnikov V S, and Zaitsev A M 2011 Sectoral Target Program *Development of the Horse Breeding of the Russian Federation in 2011-2013* (Moscow, Russia)

[6] Pustovoy V F 1986 Recommendations for the creation and use of irrigated cultivated pasture-paddock for breeding horses (Divovo, USSR: All-Russian Scientific Research Institute of Horse Breeding)

[7] Kalashnikov V V, Koveshnikov V S, and Sergienko S S 1992 The current state of horse breeding in Russia and the prospects for its development in the conditions of market reform In *Ways to improve the breeding, sporting, working and productive qualities of horses* (pp 6-17) (Divovo, Russia: All-Russian Scientific Research Institute of Horse Breeding)

IOP Conf. Series: Earth and Environmental Science **395** (2019) 012057 doi:10.1088/1755-1315/395/1/012057

The life cycle and parasitic system of *opisthorchis felineus* in the Irkutsk opisthorchiasis focus

O T Rusinek[1,2*]

[1] Baikal Museum of the Irkutsk Scientific Center of the Siberian Branch of the Russian Academy of Sciences, 1 Akademicheskaya str., Listvyanka village 664520 Russia
[2] Irkutsk State University, 1 Karl Marx str., Irkutsk 664003 Russia

E-mail: rusinek@isc.irk.ru

Abstract. The article for the first time presents generalized data on the parasitic system and the life cycle of *Opisthorchis felineus* (Plathelminthes: Trematoda) in the Irkutsk center of opisthorchiasis on the river Biryus. The relatively low abundance and contagion of the first (mollusks) and second intermediate hosts (fish) clearly demonstrates that the natural and anthropogenic sources of invasion are very minor here. Human, domestic cats are infected with opistorch. However, because of the actual breaking of the chain, their eggs may not be released into the water due to objective circumstances (remoteness, small number or absence of the first intermediate owners). *O. felineus* during the implementation of its life cycle in the parasitic system can complete it as an autogenous species, without leaving the aquatic environment and as an allogeneic species, completing the life cycle in land mammals and in humans.

1. Introduction

The relevance of the study of opisthorchiasis, a pathogenic parasitic human disease, is beyond any doubt. The author has been studying the Irkutsk center of opisthorchiasis relatively since 2010. According to the results of the study of the Irkutsk focus, my colleagues and I published several research articles [11], [14], [17], [12], [15], [13], [16], [20]. According to the content, they can be divided into 3 groups: (q) those papers devoted to the study of fish and mollusks; (2) on physiological and biochemical issues; and (3) on the molecular genetic work conducted to assess the genetic diversity of opisthorchids within their Euro-Asian range.

As a result of the analysis of the obtained results, we were convinced of the need to prepare an article on life cycle assessment and the parasitic system *Opisthorchis felineus*. This is done in order to summarize the knowledge necessary for planning further work.

Research objective is to study the parasitic system of trematodes *Opisthorchis felineus* in the Irkutsk opisthorchiasis focus. First, it is necessary to analyze the life cycle and the elements of the structure of the parasitic system. Second, our task is to provide data on the structure of the parasitic system *Opisthorchis felineus* in the Irkutsk focus.

A parasitic system is a system in which the parasite is combined at all phases of its development (free-living, larval, adults) and host populations (first, second intermediate, reservoir and final) [2], [8], [1], [6]. The *O. felineus* parasitic system is complex and polynomial, since the parasite develops in it in the first and second intermediate and final hosts.

Content from this work may be used under the terms of the Creative Commons Attribution 3.0 licence. Any further distribution of this work must maintain attribution to the author(s) and the title of the work, journal citation and DOI.
Published under licence by IOP Publishing Ltd

The life cycle of the parasite reflects the size-age structure of its population. In the case of the *O. Felineus* trematode, its population consists of 7 hemipopulations (or subpopulations, or subpopulation groups): marites, eggs, miracidia, sporocysts, redias, cercariae, and metacercariae (Figure 1).

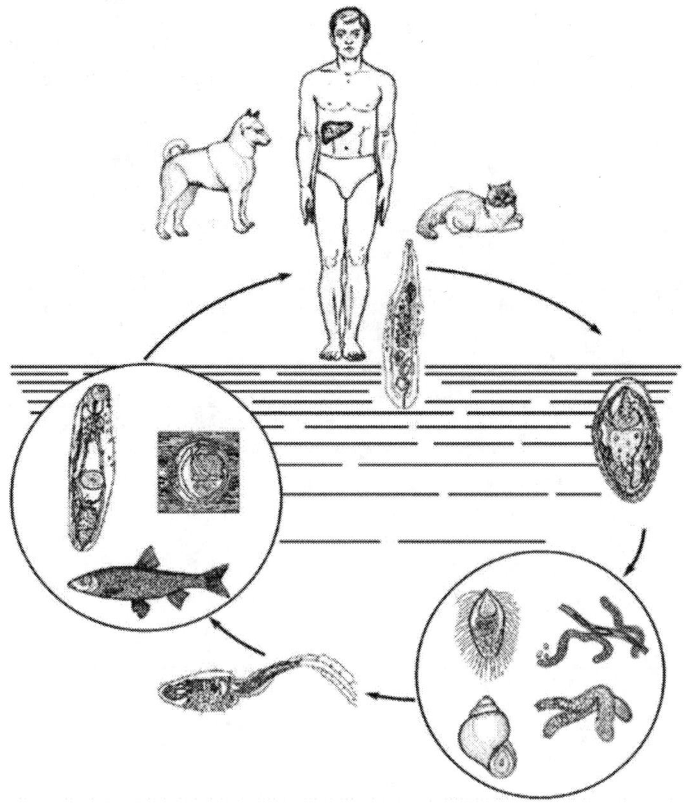

Fig. 1. Diagram of the life cycle of *Opisthorchis felineus* [3].

2. The Life Cycle of a Cat Fluke

The complex life cycle of the cat fluke was discovered by Hans Vogel, the German helminthologist. In the 1930s, he published several articles describing the life cycle of *Opisthorchis felineus* [21], [22], [24]. Later, he experimentally proved that the mollusk *Bithynia leachi* participates in the life cycle of *O. felineus* [23].

A total of 3 different generations participate in the life cycle of trematodes – one hermaphroditic (marita) and two (sporocysts and redies), consisting of parthenogenetic females. In the life cycle of a cat fluke, the stages of free life (eggs, cercariae) alternate with parasitic stages (miracidia, sporocysts, redia, metacercariae, and marites). Information on the life expectancy of individual phases of opistorchis development is somewhat different for different authors (Table 1).

Under favorable conditions, the entire life cycle of the parasite is completed within 4-4.5 months. [2]. Eggs and cercariae are the distributional free-living phases of opistorchis. It is known that at a water temperature not lower than 20ºC, up to 3.5 thousand cercariae can leave one mollusk per day [18]. The maximum number of metacercariums in old age groups of ide can reach 20-30 thousand in one fish [25].

According to various data, metacercariae in fish remain invasive for 7-8 years or more [4], 46–48 months [19], 24-26 months [18], 12-16 months [5].

Table 1. The life expectancy of individual phases of the development of *Opisthorchis felineus* (for different authors).

Phases of opisthorchis development	Pustovalova et al., 1999 [25]	Beer, 2005 [3]	Parasitic diseases ..., 2006
Egg (keeps viability in a reservoir soil)	15 months	18-24 months (according to experimental data)	5-6 months
Miracidia	-	-	-
Sporocista	-	Mollusk development time 2-2.5 months	Mollusk development time from 2 to 10-12 months
Redia	-		
Cercaria	Lifespan 60 hour	30-50 hours of active swimming	-
Metazerkaria	After 6 weeks it becomes invasive; a life time in fish from 1 year to 8 years	Up to 3 years or more	After 6 weeks becomes invasive
Marita	From the stomach, the parasite enters the liver and other organs after 3-5 hours. It becomes sexually mature after 4-5 weeks and starts producing eggs. Marita's age limit in the final host is 10-20 years.	Reaches puberty in 20-25 days	A parasite begins to produce eggs 3-4 weeks after infection of the final host; 20-25 years old – the age limit of marits in the final host

Note: - no data.

3. The Parasitic System *O. felineus*

It is obvious that the life cycle of a parasite determines the structure of its parasitic system, since in it the parasite naturally combines at all phases of its development and a host or hosts (intermediate, reservoir, final). The number of host species in different habitat conditions of the parasite may vary significantly.

The focal nature of opisthorchiasis suggests that relatively isolated territories exist in different parts of our planet, where the parasite goes through its life cycle as part of the parasitic system formed under these conditions. From the standpoint of the theory and practice of parasitic communities in all parasitic phases of development, *Opisthorchis felineus* belongs to the types of generalists, since it parasitizes in hosts of different species and genera, families and orders (Table 2).

Attention should be paid to the fact that *O. felineus* is an autogenous species (developing in the aquatic environment) up to the metacercaria phase, and in the *Marita* phase it can be an autogenous and allogeneic species depending on the habitat of the final host (Fig. 2). "Autogenous" is when the final hosts are aquatic and near aquatic mammals (otter, muskrat, water vole – the latter two rarely feed on fish but are susceptible to infection by fluke). In turn, "allogeneic" is when it completes its life cycle in land animals (bear, wolf, fox and others). It is allogeneic during the *Marita* phase and in the case of the completion of the cycle in man and domestic animals.

Thus, during the implementation of its life cycle in a parasitic system, *O. felineus* can complete it as an autogenous species, without leaving the aquatic environment, and as an allogeneic species, completing the life cycle in terrestrial mammals. The autogenous path of development is likely to be the most ancient. Allogenic, especially associated with the inclusion of humans and domestic animals, was formed in relatively recent times at the end of the Pleistocene and in the Holocene (15-10 thousand years ago) [3].

Table 2. A taxonomic status of the hosts of *Opisthorchis felineus.*

Ist intermediate hosts	IInd intermediate hosts	Definitive hosts
Gastropoda Class Squad **Family Bithyniidae** **Subfamily Bithyniinae** **Genus *Bithynia*** **Subgenus *Bithynia*** 1. Species *B (Bithynia) tentaculata* **Subgenus *Codiella*** 2. Species *B. (Codiella) liachii*	Osteichthyes Class – Bony Fish Squad Cypriniformes – carps **Family Cyprinidae** **Genus *Rutilus*** 1. Species R. *rutilus lacustris* – roach	Mammalia Class **Squad Rodentia** **Family Crecitidae** **Genus *Ondatra*** 1. Species *O. zibetica* – muskrat
Subgenus *Opistorchophorus* 3. Species *B. (Opistorchophorus) troschelii*	**Genus *Leuciscus*** 2. Species *L. leusciscus baicalensis* - dace 3. Type *L. idus* - ide	**Genus *Arvicola*** 2. Species *A. terrestris* – water vole
Subfamily Mysorellinae **Genus *Boreolona*** 4. *B. sibirica*	**Genus *Phoxinus*** 4. Genus *Ph. phoxinus* – common minnow	Family Castoridae **Genus *Castor*** 3. Species *C. fiber* – common beaver
	Genus *Abramis* 5. *A. brama orientalis* – eastern bream	**Squad Carnivora** **Family Canidae** **Genus *Canis*** 4. Species *C. lupus* – wolf 5. Subspecies *C. lupus familiaris* – dog
	6. *A. belarus* – blue	**Genus *Vulpes*** 6. Species *V. vulpes* – common fox
	7. *A. sapa* – white-eye bream	**Family Ursidae** **Genus *Ursus*** 7. Species *U. arctos* – brown bear
	Genus *Scardinius* 8. *S. erythrophtalmus* – rudd	**Family Mustelidae** **Genus *Martes*** 8. Species *M. zibellina* – sable
	Genus *Scardinius* 9. *S pelecus* – sichel	**Genus *Gulo*** 9. Species *G. gulo* – wolverine
	Genus *Culter* 10. *Culter alburnus* common bleak	**Genus *Mustela*** 10. Species *M. erminea* – ermine
	Genus *Chondrostoma* 11. *Chondrostoma nasus* – common nase	11. Species *M. vison* – american mink
	Genus *Aspius* 12. *Aspius aspius* – common asp	12. Species *M. eversmanni* – white polecat
	Genus *Barbus* 13. *Barbus barbus* – common barbel	**Genus *Lutra*** 13. Species *L. lutra* - выдра
	Genus *Carassius* 14. *C. auratus gibelio* – goldfish	**Family Felidae** **Genus *Felis*** 14. Subspecies *Felis silvestris catus* – domestic cat
	Genus *Tinca*	**Squad Artiodactyla**

	15. Species *T. tinca* – tench	**Family Suidae** **Genus *Sus*** 15. Species *Sus scrofa* 16. Subspecies *Sus scrofa domestica* – domestic pig
	Genus *Gobio* 16. Species *G. gobio* – gudgeon	**Squad Primates** **Family Hominidae** **Genus *Homo*** 17. Species *Homo sapiens* – reasonable man
	Genus *Leucaspius* 17. Species *L. delineatus* - belica	
	Genus *Blicca* 18. *B. bjoerkna* – silver bream	
	Genus *Squaliobarbus* 19. *Squaliobarbus curriculus* – mustache chub	
	Family Cobitidae – loachworm **Genus Cobitis** 20. Spined loach	

4. The Parasitic System *O. felineus* in the Irkutsk Opisthoschiasis Focus

According to the literature and our data, it can be stated that the parasitic system *O. felineus* on the Biryus river operates with the participation of 7 parasite hemipopulations, one population of *Bithynia troscheli* mollusks, four specific populations of the second intermediate hosts and three specific populations of the final hosts (Table 3). It should be understood that these are the actual data and the potential number of participants in this parasitic system may be much larger.

Taking into account the hydrological features of the Biryusa river, its comparative high water content, a low population and an infection of the first intermediate owners [7], as well as a relatively low infection of the second intermediate owners (fish) [11], it can be stated that natural and anthropogenic sources of invasion are very insignificant here. Human, domestic cats are infected with opistorch, but due to the actual breaking of the chain, eggs may not be released into the water due to objective circumstances (remoteness, a small number or absence of the first intermediate owners).

Table 3. The species composition of animals involved in the parasitic system *O. felineus* in the Irkutsk opisthoschiasis focus.

Ist intermediate hosts	IInd intermediate hosts	Definitive hosts
Bithynia (*Opistorchophorus*) *troschelii* (син.: *B. inflata*, *Boreolona sibirica*)	Dace Roach Bream Crucian Carp	Human Cat Muskrat Otter, water vole, dog can be potential definitive hosts of the parasite.

According to the classification V. D. Zavoykin, the territory of the Irkutsk opisthoschiasis focus refers to hypoendemic, since the infection of people is from 1 to 10% and no cases of the disease have been detected among children [3].

The feline flute parasitic system exists due to close ecological connections between the final, first and second intermediate hosts. In the case of the Irkutsk focus, it is necessary to continue to identify wild animals that can play the role of ultimate owners of the parasite. In addition, it is important to begin a systematic work to identify the habitats of Bithynia, i.e. the first intermediate hosts of the parasite and the study of their biology in the conditions of the Biryusa river basin.

A careful attention of various sanitary organizations should be paid to the condition and disinfection of latrines in the settlements along the banks of the Biryusa river.

In addition, it is important to conduct an ongoing outreach to the local population on the treatment of carp fish as a preventive measure against infection with opisthorch.

5. Acknowledgments

The work was done on the topic "Study of Some Ecosystem Elements of the Territory of Eastern Siberia Based on the Results of Field and Experimental Studies as a Reflection of Changes in Abiotic Environmental Factors." The state registration number: AAAA-A16-116051910013-0.

References

[1] Balashov Yu S 1991 The Meaning of the ideas of V N Beklemishev on parasitic systems and species life patterns in the development of parasitology *Parasitology* **25**(3) pp 185-195

[2] Beklemishev VN 1970 On the principles of comparative parasitology as applied to blood-sucking arthropods In *Biocenological basis of comparative parasitology* (pp 250-260). Moscow, Russia.

[3] Beer S A 2005 *Biology of opisthorchiasis causative agent* (Moscow, Russia: KMK Fellowship Association)

[4] Bliznyuk I D 1969 Experimental infection of fish by free-floating cercaria of the cat fluke *Vestnik of Zoology* **3** pp 76-79

[5] Vinogradov L I 1970 *Circulation of opisthorchiasis* Opisthorchis felineus *in Semipalatinsk region* (Dissertation of the Candidate of Biological Sciences; Almaty, Kazakhstan)

[6] Granovich A I 2009 A parasitic system as a reflection of the parasite population: concept and terms *Proceedings of the Zoological Institute RAS* **313**(3) pp 229-337

[7] Kolokoltsev M M 1988 Distribution and ecology of the Bithynia inflata mollusks of the intermediate host Opistorchis felineus in the river basin of Biryusa river *Medical Parasitology and Parasitic Diseases* **3** pp 58-60

[8] Kontrimavichus V L 1982 Modern problems of ecological parasitology *Journal of General Biology* **43**(6) pp 764-774

[9] Pavlovsky E N 1937 The Doctrine of biocenoses as applied to some parasitological problems. *News of the Academy of Sciences of the USSR, Biology Series* **4** pp 1388-1422

[10] Sergeeva V P, Lobzina Yu V, and Kozlova S S Eds. 2006 *Parasitic human diseases (protozoozy and helmitozy): a guide for a guide for doctors* (St. Petersburg, Russia: Publishing Folio LLC)

[11] Rusinek O T, and Kondratistov Yu L 2010 Study of the infection of carp fish by trematode metacercaries in the focus of opisthorchiasis (Taishet district, Irkutsk region, Russia) *Izvestiya ISU (Series "Earth Sciences")* **3**(1) pp 132-142

[12] Rusinek O T, Kondratistov Yu L, and Ivanova A I 2010 The state of the the Irkutsk opisthoschiasis focus (Biryusa river, Taishet district, Irkutsk region, Russia) *Journal of Pavlodar State Pedagogical Institute. Biological Sciences of Kazakhstan* **3** pp 96-104

[13] Rusinek O T, Kondratistov Yu L, and Rudov R V 2012 Physiological and biochemical bases for the differential diagnosis of metacercarium *Opisthorchis felineus* and *Rhipidocotyle campanula. Russian Parasitological Journal* **1** pp 85-95

[14] Rusinek O T, and Kondratistov Yu L 2012 Differential diagnosis of *Opisthorchis felineus* metacercaria from the metacharacter *Rhipidocotyle campanula* in carp fish's opisthocytosis *Veterinary* **10** pp 35-37

[15] Rusynek O T, Sitnikova T. Ya, and Kondratistov Yu L 2012 New data on the Irkutsk opisthoschiasis focus and the need for its study *Medical Parasitology and Parasitic Diseases Journal* **2** pp 15-18

[16] Rusinek O T, and Sitnikova T Ya 2012 The condition of the Irkutsk center of opisthorchiasis and issues for its further study *Proceedings of the Irkutsk State University* **3** pp 125-134

[17] Rusinek O T, and Kondratistov Yu L 2013 Differential diagnosis of *Opisthorchis felineus* metacercarium from muscle metacercaria of non-pathogenic fish for humans in determining parasitological indicators of fish safety *Vestnik of the Irkutsk State Agrarian University Named After A.A. Ezhevsky* **57**(2) pp 45-51

[18] Sidorov E G 1972 Life expectancy of *Opisthorchis felineus* and *Metorchis albidus* metacercariae *Medical Parasitology* **5** pp 611-612

[19] Fattakhov R G 1990 *The second intermediate host of opisthorchiasis pathogen in the Ob-Irtysh focus (ecology, epidemiological significance)* (Dissertation of the Candidate of Biological Sciences; Almaty, Kazakhstan)
[20] Brusentsov I I, Katokhin A V, Brusentsova I V, Shekhovtsov S V, Borovikov S N, Goncharenko G G, Lider L A, Romashov B V, Rusinek O T, Shobitov S K, Suleymanov M M, Yevtushenko A V … Mordvinov V A 2013 Low genetic diversity in wide-spread Eurasia liver fluke *Opisthorchis felineus* suggests special demographic history of this Trematode species *PLOS ONE* **8**(4) pp 1-12
[21] Vogel H. 1929 Helminthologische Beobachtungen in Ostpressen insbesondere uder *Bothriocephalus latus* und *Opisthorchis felineus Dtscfi Med. Wschr.* **55** pp 1631-1633
[22] Vogel H. 1930 Beobachtungen über die Lebenageschichte von Opisthorchis felineus in Ostpressen *Zbl. Bacter. I. Orig.* **3/4** pp 250-254
[23] Vogel H 1932 Uber den ersten Zwischenwirt und die Zerkarie von *Opisthorchis felineus* (Riv.) *Arch. Schiffs.* **36** pp 558-562
[24] Vogel H 1934 Der Entwicklungszyklus von *Opisthorchis felineus* (Riv.) nebst Bemerkungen uder die Systematik und Epidemiologie *Zoologica* **33**(86) pp 1-103
[25] Pustovalova V Ya, Stepanova T F, and Shonin A L 1999 *Opisthorchiasis* (Tyumen: Publishing House of TSMA)

Effect of superabsorbing polymers on potato yield

O A Starovoitova[1], **A A Manokhina**[2*], **M I Pekhal'skij**[1]

[1] Lorch Potato Research Institute, 23 Lorcha str., 140051 Kraskovo Russia
[2] Russian State Agrarian University – Moscow Timiryazev Agricultural Academy, 49 Timiryazevskaya str., Moscow 127550 Russia

E-mail: alexman80@list.ru

Abstract. Climate change is leading to an increase in drought, which is the largest abiotic stress factor leading to a decrease in crop yields. One of the methods in solving this problem is the use of superabsorbing polymers (SAPs) (China, USA, Canada, India, Italy, Iran, Poland, Turkey, and others). The aim of the research is to study the effect of using the superabsorbing polymers (cross-linked copolymer of potassium and ammonium salts of acrylic acid in the form of loose white granules) on the background of fractional-local application of mineral fertilizer to the crop formation. Field experience, accounting and observations were carried out in accordance with general requirements were used in our research. The studies were carried out in 2015-2017 on sod-podzolic sandy soil in the Lyubertsy district (Moscow region, Russia). When using superabsorbing polymers in an average of three years, the yield increased by 1.6...3.4 t/ha (5.5...Of 10.0%) for the studied cultivars; especially helpful was an early sort *Meteor* and a medium early variety *Irbitskiy*. It is possible to recommend the use of superabsorbing polymers at a dose of 200 kg/ha when growing potatoes in conditions of moisture deficiency.

1. Introduction

Potato is a crop of mass consumption, the cultivation of which remain high [8]. At the same time, in Russia, the population consumes mainly imported potatoes. Domestic potatoes are in little demand because of the large variance in size and shape of tubers due to the lack of specialized technologies [20].

Current climate change is projected to increase the incidence of drought, the largest abiotic stress factor leading to lower yields [16]. Numerous studies have shown that potato cultivation technology must be flexible [3],[6]. Given the large share of water use in the agricultural sector, the use of superabsorbent polymers (SAPs) is one of the strategies in this regard [16]. The influence of SAPs is studied by researchers from China [15], USA [17], Canada [21], India [18], Italy [19], Iran [22], Poland [13], Turkey [14], and others.

SAPs well retain soil moisture and mineral nutrition in it and, thus, affect the technological parameters of the soil [9], [11], [15], [17]. The use of SAPs, especially in combination with the use of fertilizers and biologics, reduces environmental pollution by retention of nutrients in the soil, leads to an optimum pH value for plants, and can improve the water holding capacity of the soil [5], [11], [12], [15].

Content from this work may be used under the terms of the Creative Commons Attribution 3.0 licence. Any further distribution of this work must maintain attribution to the author(s) and the title of the work, journal citation and DOI.
Published under licence by IOP Publishing Ltd

IOP Conf. Series: Earth and Environmental Science **395** (2019) 012058 doi:10.1088/1755-1315/395/1/012058

It is noted that the introduction of SAPs significantly increases the yield of potatoes, compared with the control groups [7], [15]. A positive effect was also obtained on the following other cultures: cherry tomatoes (*Solanum lycopersicum var. cerasiforme*) [21], Indian mustard (*Brassica juncea* L. Czern & Coss.) [18].

2. Materials and Methods

The aim of our research is to study the effect of using SAPs in the context of fractional-local mineral fertilizer application on crop formation in potato cultivation, while relying on moisture-saving technology, as well as on resource saving when using SAPs.

3. Methodology

Bookmark field experience, surveys, and observations were carried out in accordance with the requirements of the methods of field experience [1], [4].

4. Research Conditions

The studies were carried out in 2015-2017 on the experimental basis of the Lorch Potato Research Institute in the Lyubertsy district of the Moscow region.

The soil of the experimental field was characterized as sod-podzolic sandy loam, with high exchange and hydrolytic acidity (pHKCl = 4.7-4.9; Ng = 3.5-3.9 mg-eq / 100g of soil); average values of the amount of absorbed bases and the degree of saturation with them (S = 3.7-4.1 mg-eq / 100g of soil; V = 48.7-53.9%); high content of mobile phosphorus (265-356 mg / kg of soil) and average content of exchangeable potassium (127-154 mg / kg of soil); typical humus (1.7-1.9%) [10].

The scheme of the experiment in the cintext of fractional-local application of azophoska mineral fertilizer (16:16:16) with addition of potassium magnesia: 60:60:90 before planting and 30:30:30 at the 1st post-harvest tillage for crop yields. It also included the following factors: (A) varieties of different maturation groups: Meteor (ranniy), Irbitskiy (sredneranniy), Golubizna (srednespelyy), Lorkh (srednepozdniy); (B) introduction of SAPs during planting: 0 g / bush, 4 g / bush (200 kg / ha).

Aqueous superabsorbent is a crosslinked copolymer of potassium and ammonium salts of acrylic acid in the form of loose white granules.

The experiment was laid according to the scheme by the method of randomized allocation of plots. The potato precursor is grain beans. The plot area is 18.1 m^2, a number of replications is 4. The density of planting is 44.4 thousand units per hectare with a row spacing of 75 cm.

Fall plowing, made to a depth of 18-22 cm with a reversible plow, was part of the autumn soil preparation. Planting of experimental variants was carried out with ungrown tubers of the middle fraction (46… 53 mm in size according to the largest transverse diameter), with a clone potato planter into previously cut ridges, to a depth of 8-10 cm. Weed control was carried out using the following herbicides: systemic pesticide (selective herbicide, active ingredient: Rimsulfuron, active ingredient content - 250 g / kg) and systemic pesticide (selective herbicide, active substance - Metribuzin, an active ingredient content - 700 g / kg) in the recommended dose. A two-time spraying of an insecticide, namely an intestinal pesticide (with Imidacloprid as an active substance, 700 g / kg) in the recommended dose was performed against the Colorado potato beetle. During the growth of plants, chemical treatments at the recommended dose were carried out against late blight and early blight as follows: 1-3 times (depending on the conditions of the year) with a fungicide (a contact pesticide). (The active ingredients are as follows: Famoxadone (250 g / l) and Cymoxanil (250 g / l), the first was during the flowering period, then every 10-14 days. The recommended water consumption was 300 l / ha. Potato tuber harvesting was conducted in the second half of August).

In 2017, the average air temperature during the growing season was 16.2 °C, with a norm of 16.5 °C (18.6 °C in 2016 and 17.5 °C in 2015). Total precipitation during the growing season of 2017, 378.4 mm or 145.3% of the norm (260.5 mm) precipitation fell (470.2 mm or 180.5% of the norm in 2016, and 302.45 mm or 116.1% of the norm in 2015). The 2017 hydrothermal coefficient was 2.06 (wet) with a climate norm of 1.3 ... 1.4. (It was 2.16 (very wet) in 2016 and 1.46 (wet) in 2015).

5. Research Results

In 2015, the average values of soil moisture were below the optimal values in different phases of plant development (37.6 ... 61.5% of digestible nutrients / PPV in the budding-tuberization phase). The average values of soil moisture in the zone of the tuberous nest gradually decreased from 55.1 ... 55.7% during the budding stage to 34.3 ... 37.6% before harvesting. In 2016, the soil moisture values were sufficient for developing the above-ground plant mass. In 2017, for developing the above-ground mass of plants and tubers throughout the growing season, the soil moisture values were mostly optimal.

Table 1 presents data on the yield of potato varieties depending on the variety (ripeness group) and the use of SAPx. The average yield of the Ranniy Meteor variety was 33.7 t / ha under control and 37.0 t / ha with water absorbents; the Sredneranniy Irbitskiy variety — 28.0 t / ha and 30.0 t / ha; the Srednespelyy Golubizna variety - 29.6 t / ha and 31.2 t / ha; the Srednepozdniy Lorkh variety - 29.2 t / ha and 31.0 t / ha, respectively.

Table 1. Potato yield depending on the variety (factor A) and the use of SAPs (factor B), t / ha, 2015-2017

Variety	2015	2016	2017	Average	± to control	
					t / ha	%
Without superabsorbents (control)						
Meteor (ranniy)	30.3	33.6	37.1	33.7	-	-
Irbitskiy (sredneranniy)	33.7	23.4	26.9	28.0	-	-
Golubizna (srednespelyy)	34.7	28.4	25.6	29.6	-	-
Lorkh (srednepozdniy)	34.2	26.5	26.9	29.2	-	-
Average variety	33.2	28.0	29.1	30.1	-	-
With superabsorbents						
Meteor (ranniy)	34.2	34.4	42.5	37.0	+3.4	+10.0
Irbitskiy (sredneranniy)	37.1	23.6	29.3	30.0	+2.0	+7.1
Golubizna (srednespelyy)	34.8	33.1	25.7	31.2	+1.6	+5.5
Lorkh (srednepozdniy)	35.9	26.6	30.6	31.0	+1.8	+6.3
Average variety	35.5	29.4	32.0	32.3	2.2	7.2
HCP_{05}, by factor A	1.11	3.58	4.87	-	-	-
HCP_{05}, by factor B	1.34	0.69	0.74	-	-	-
HCP_{05}, common	1.84	4.17	5.72	-	-	-

6. Results and Discussion

The greatest amount of water needed by potatoes in the second half of the growing season, during the formation of tubers, falls on the period between the phases of the onset of budding and the dying of the tops. Optimum soil moisture for potatoes is 60-85% of digestible nutrients. With optimal hydration, the effectiveness of fertilizer application is manifested [2]. Soil moisture during the growing season is largely determined by the amount and distribution of precipitation and changes under the influence of certain factors [2]. Vegetation periods of 2015–2017 were characterized by uneven precipitation, which affected soil moisture.

Productivity is the main criterion for evaluating cultural activities. On average, over three years, we can note (Table 1) that when using SAPs, the yield of the studied varieties increased by 1.6 ... 3.4 t / ha (5.5 ... 10.0%). The Ranniy Meteor variety (+3.4 t / ha) and the Sredneranniy Irbitskiy variety (+2.0 t / ha) were particularly responsive, which confirms the greatest need for moisture and early varieties [2]. The Srednespelyy Golubizna and srednepozdniy Lorkh varieties were less responsive to the appliation of SAPs. The highest average yield for all varieties was obtained in a sufficiently moisture-proof (2015), and it was 33.2 t / ha in the control variant and 35.5 t / ha in the variants with the use of SAPs. In 2017, it amounted to 29.1 t / ha and 32.0 t / ha, respectively. At the same time, in very wet conditions of 2016, the yield increase on average for varieties was significant, but not so high and amounted to 28.0 t / ha and 29.4 t / ha, respectively. According to the research results, it was revealed that the use of SAPs provides a conditional net income of up to 39.1 thousand rubles / ha.

The use of SAPs when planting allows one to increase the yield of potatoes. This technology is to reduce the stresses affecting the yield of potatoes, by making superabsorbent polymers. Thus, the application of SAPs from an economic point of view can be promising [7], [9].

7. Conclusion

The application of SAPs increased the yield of the studied potato varieties of different groups of ripeness by 1.6 ... 3.4 t / ha (5.5 ... 10.0%), which makes it possible to get conditional net income up to 39.1 thousand rubles / ha.

We can recommend the use of superabsorbent polymers at a dose of 200 kg / ha when growing potatoes in conditions of moisture deficiency.

References

[1] Dospekhov B A 1985 *Field experience (with the basics of statistical processing of research results)* (Moscow, USSR: Agropromizdat)

[2] Korshunov A V, Kutovenko L N, Lysenko Yu N, and Rakhimov R L 2011 Drought lessons in potato growing *Achievements of Science and Technology of Agriculture* 3 pp 21-23

[3] Manokhina A A 2012 *Development of the technological process of planting potatoes using granular organic fertilizers* (abstract of candidate of agricultural sciences) (Moscow, Russia: MSAU)

[4] Andryushina N A *et al* 1967 *Methods of research on potato culture* (Moscow, USSR: Research Institute of Potato Economy)

[5] Smagin A V, and Sadovnikova N B 2009 *The influence of highly swelling polymer hydrogels on the physical state of light granulometric distribution soils* (Moscow, Russia: MAKS Press)

[6] Starovoitov V I 1993 Mastering technologies with regard to specific conditions *Potatoes and Vegetables* **2** p 5

[7] Starovoitov V I, Starovoitova O A, and Manokhina A A 2015 Potato cultivation using moisture-saving polymers *Vestnik of the Moscow State Agro-Engineering University named after V. P. Goryachkin* **1**(65), pp 15-19

[8] Starovoitova O A, Zhevora S V, Starovoitov V I, Oves E V, Korshunov A V, Manokhina, A A, ... Balabanov V I, ... Fedorenko V F, ... Golubev I G, ... Zvyagintsev P S, ... Zuev V V, ... Voronov N V 2018 *Competitive technologies of seed production, production and storage of potatoes* (Moscow, Russia: "Rosinformagrotekh")

[9] Starovoitova O A, Starovoitov V I, Manokhina A A 2016 Potato cultivation using water absorbents *Vestnik of the Moscow State Agro-Engineering University named after V. P. Goryachkin* **2**(72) pp 28-34

[10] Fedotova L S, Timoshina N A, Knyazeva E V, Akanova N I, Vizirskaya M M 2017 The effect of microbiological fertilizers on the productivity and quality of potatoes *Potato Protection* **4** pp 13-21

[11] Chen J, Lü S, Zhang Z, (...), Ning P, Liu M 2018 Environmentally friendly fertilizers: A review of materials used and their effects on the environment *Science of the Total Environment* **613** pp 829-839

[12] Essawy H A, Ghazy M B M, El-Hai F A, Mohamed M F 2016 Superabsorbent hydrogels via graft polymerization of acrylic acid from chitosan-cellulose hybrid and their potential in controlled release of soil nutrients *International Journal of Biological Macromolecules* **89** pp 144-151

[13] Grabin´ski J, Wyzin´ska M 2018 The effect of superabsorbent polymer application on yielding of winter wheat (*Triticum aestivum* L.) *Research for Rural Development* **2** pp 55-61

[14] Güneş A, Kitir N, Turan M, (...), Yildirim E, Avci N 2016 Evaluation of effects of water-saving superabsorbent polymer on corn (*Zea mays* L.) yield and phosphorus fertilizer efficiency *Turkish Journal of Agriculture and Forestry* **40**(3) pp 365-378

[15] Hou X, Li R, He W, (...), Ma K, Liang Y 2018 Superabsorbent polymers influence soil physical properties and increase potato tuber yield in a dry-farming region *Journal of Soils and Sediments* **18**(3) pp 816-826

[16] Khodadadi D D 2016 The effects of Superabsorbent polymers on soils and plants *Pertanika Journal of Tropical Agricultural Science* **39**(3) pp 267-298

[17] Lü S, Feng C, Gao C (...), Gao N, Liu M 2016 Multifunctional Environmental Smart Fertilizer Based on l - Aspartic Acid for Sustained Nutrient Release *Journal of Agricultural and Food Chemistry* **64**(24) pp 4965-4974

[18] Rathore S S, Shekhawat K, Rathore B S, (...), Premi O P, Kandpal B K 2018 Deficit irrigation scheduling and levels of hydrogel (SAPs) influence on productivity and economics of Indian mustard (Brassica

juncea) under semi-arid conditions *Indian Journal of Agricultural Sciences* **88**(9) pp 1474-1476
[19] Satriani A, Catalano M, and Scalcione E 2018 The role of superabsorbent hydrogel in bean crop cultivation under deficit irrigation conditions: A case-study in Southern Italy *Agricultural Water Management* **195** pp 114-119
[20] Starovoitov V I, Voronov N V, and Pavlova O A 2007 Prospects of potato growing techniques in wide rows In Haverkort A S, and Anisimov B V Eds., *Potato production and innovative technologies* (pp 246-251) (Wageningen, The Netherlands: Wageningen Academic Publishers)
[21] Suresh R, Prasher S O, Patel R M, (...), Schwinghamer T, Ehsan A M 2018 Super absorbent polymer and irrigation regime effects on growth and water use efficiency of container-grown cherry tomatoes. *Transactions of the ASABE* **61**(2), pp 523-531
[22] Wei Q 2014 Fast-swelling porous starch-g-poly (acrylic acid) superabsorbents *Iranian Polymer Journal (English Edition)* **23**(8), pp 637-643

Methods of enrichment of the qualitative composition of a beef with polyunsaturated fatty acids in ecologically unstable agrarian territories

S Shlykov[1], R Omarov[1] and O Takhumova[2*]

[1] Stavropol State Agrarian University, 12 Zootekhnicheskiy lane, Stavropol 355017 Russia
[2] Kuban State Agrarian University named after I.T. Trubilina, 13 Kalinin str., Krasnodar 350044 Russia

E-mail: takhumova@rambler.ru

Abstract. Beef is a natural carrier of beneficial n-3 PUFA (eicosapentaenoic acid (EPA, C20: 5n-3) and docosahexaenoic acid (DHA, C22: 6n-3), it is also an important source of healthy fats for human. The plant-based feeding strategies can be considered as the most appropriate and sustainable approach to increasing the content of n-3 PUFAs in beef. Taking into account the requirements for creating the pastures with high yields, productive longevity, a rapid achievement of pasture ripeness, resistance to grazing and trampling, as well as a high content of unsaturated fatty acids, a total of 3 types of herbs are selected: oat-grass + perennial sorghum + leymus. The plants were grown on the territory of the experimental pilot farm with a cut every 6 weeks and an annual fertilizer at the rate of 250 kg N per 1 hectare for 3 years. The collection of herbal material for the analysis of fatty acids was carried out in June and September 2014, 2015, and 2016. Our microscopic data showed that 2 hours after feeding on fresh grass, the protozoa had a significantly larger number of intracellular chloroplasts, which remained high and 6 hours after feeding. Therefore, data on fatty acids were correlated with the content of intraprotozoal chloroplast. These data illustrate that the absorption by protozoa of mainly plant chloroplasts occurs rapidly. More than that, the intracellular level of chloroplasts is maintained for at least 6 hours. Thus, the protozoa quickly become the main reservoir of chloroplasts and useful sources of PUFAs.

1. Introduction

The human body is able to synthesize only Omega-9 fatty acids. Omega-3 and Omega-6 fatty acids are not able to be produced in the body and must come with food, as they are also necessary for life. Relatively small differences in the structure of molecules make Omega-6 and Omega-3 acting in the human body in completely different ways. The “minus” of Omega-6 acid is a product of its metabolism. The Omega-6 metabolic by-products contribute to inflammation, blood clots, tumor growth.

The Omega-3 metabolism products have opposite effects. As soon as Omega-3 fats enter the body, they invade cells, affecting their structure and activity. Because of this, they have such a variety of useful properties. These acids improve the functioning of the heart, brain, eyes and joints; reduce the level of harmful cholesterol; are excellent antioxidants; prevent and improve eczema, allergies, asthma, Alzheimerэs disease, depression and nervous diseases, diabetes, hyperactivity of children, psoriasis,

Content from this work may be used under the terms of the Creative Commons Attribution 3.0 licence. Any further distribution of this work must maintain attribution to the author(s) and the title of the work, journal citation and DOI.
Published under licence by IOP Publishing Ltd

osteoporosis, arthrosis, cardiovascular problems, as well as more serious diseases, such as prostate cancer or breast cancer [1].

Food quality is becoming increasingly important for consumers. Determining quality is becoming more difficult for meat. It covers not only the physical aspects of meat, such as tenderness, juiciness, taste, but also includes more recent issues, such as safety, traceability, preventive properties and the working environment. Gradually, consumers become aware of the relationship between nutrition and health, especially with regard to cancer and atherosclerosis. Knowledge of these relationships has increased consumer interest in the nutritional quality of food, so that this becomes a more important aspect of food quality.

Beef is considered very nutritious and valuable food. The importance of meat as a source of protein with a high biological value (including, for example, vitamins A, B6, B12, D, E, iron, zinc, selenium) is well known. However, over the past 10-15 years, these positive qualities have often been clouded due to several negative attributes. The latter include: the perception that beef contains a large amount of fat rich in saturated fatty acids, associations between red meat and cancer, and non-nutritional issues such as detrimental effects on animal health (BSE syndrome) [2]. Intramuscular fat is the most important fat storage depot for humans. The ratio of NLC, MFA, and PUFA on average is from 0.45 to 0.48, from 0.35 to 0.45 and up to 0.05 of the total fatty acids. As a rule, the ratio of polyunsaturated fatty acids to saturated for beef is small (about 0.1). The n-6: n-3 ratio for beef is low, usually less than 3 to 1, which reflects significant amounts of useful n-3 PUFAs, especially C18:3n-3, as well as EPA and DHA. Also, beef contains CLA and, in particular, cis-9, trans-11 and trans-10, cis-12 CLA. The anticarcinogenic and anti-atherogenic effects of cis-9, trans-11, as well as the anticonvulsant effects of trans-10, cis-12 are reflected in the work of Belury M A [1, 2].

Plant-based feeding strategies are the most appropriate and sustainable approach to increasing the content of n-3 PUFAs in beef. The transformation of C18:3n-3 from feed to meat depends on two important processes: (a) an increased level of C18:3n-3 in the feed (and, therefore, in the animal); (b) reduction of the processes of biohydrogenation in the rumen.

The experience of researchers at the University of Aberystwyth shows that the grass diversity, growth stage, and methods of conservation (silage and hay, degree of wilt, etc.) affect the concentration of C18: 3n-3 [3], [4].

2. Materials and Methods

Taking into account the requirements for creating the pastures with high yields, productive longevity, the rapid achievement of pasture ripeness, resistance to grazing and trampling, as well as a high content of unsaturated fatty acids, 3 types of herbs are selected: oat-grass + perennial sorghum + leymus. The plants were grown on the territory of the experimental pilot farm with a cut every 6 weeks and an annual fertilizer at the rate of 250 kg N per hectare for 3 years. The collection of herbal material for the analysis of fatty acids was carried out in June and September 2014, 2015, and 2016 [5], [6], [9]. The content of fatty acids was determined in 1 g of the lyophilized material using heneicosylic acid methyl ester (C21: 0) as an internal standard (Sigma-Aldrich Co., St Louis, MO, USA) and one-step extraction by transesterification. Fatty acid methyl esters (FAME) were separated and quantified using gas chromatography. To evaluate the components of the variance, a linear mixed model was used. The analysis was performed using REML (Restricted maximum likelihood) analysis at Genstat (14th edition, VSN International Ltd, Hemel Hempstead, UK).

3. Results

In Table 1, for each of the three, mean and standard deviations are shown for the five essential fatty acids (C16:0, C18:0, C18:1n-9, C18:2n-6, and C18:3n-3). The number of C18: 2n-6 and C18: 3n-3 was the largest percentage of the total, and their number was the highest in the 2nd year, while three minor components tend to increase in each bevel, as a rule.

Table 1. The content of fatty acids in the test material, (mg / g-1 LM).

Year	Name	C16:0	C18:0	C18:1n-9	C18:1n-9	C18:2n-6	C18:3n-3
For all the time	Oat-grass	4.26±0.620	0.39±0.078	0.38±0.139	2.73±0.633	15.02±4.015	25.03±5.253
For all the time	Sorghum	3.92±0.693	0.38±0.103	0.46±0.270	2.65±0.846	12.09±3.708	22.95±8.706
For all the time	Leymus	4.01±0.579	0.39±0.073	0.46±0.133	2.94±0.705	11.61±2.889	22.42±6.123
1	Oat-grass	3.77±0.513	0.36±0.073	0.35±0.116	2.71±0.447	14.66±2.440	23.41±3.141
1	Sorghum	2.89±0.140	0.32±0.036	0.42±0.082	2.52±0.267	9.12±1.305	16.58±1.059
1	Leymus	3.26±0.544	0.34±0.041	0.33±0.044	2.69±0.420	10.17±2.385	18.29±3.141
2	Oat-grass	4.20±0.389	0.39±0.060	0.37±0.123	3.14±0.372	18.03±3.193	28.96±4.581
2	Sorghum	3.88±0.428	0.42±0.135	0.58±0.501	3.30±1.047	14.53±4.336	29.59±12.764
2	Leymus	3.91±0.220	0.42±0.075	0.57±0.119	3.56±0.558	13.44±2.874	27.10±8.462
3	Oat-grass	4.75±0.513	0.42±0.088	0.40±0.167	2.35±0.732	12.35±3.861	22.55±5.128
3	Sorghum	4.45±0.247	0.39±0.102	0.40±0.059	2.28±0.722	11.96±3.311	21.71±4.718
3	Leymus	4.44±0.301	0.40±0.079	0.46±0.117	2.64±0.698	11.11±2.929	21.37±3.741

Table 1 also shows that the five fatty acids are present in all samples. Analysis of the components of the dispersion using the REML model in all six cuttings demonstrated a significant influence of this factor.

Determining the interactions between plant components and rumen biotics (lipolysis and biohydrogenation) are necessary for the directional improvement of the fatty acid composition of beef and other products of ruminants.

The PUFAs are found in plant phospholipids. Phospholipids are found in large quantities in chloroplast membranes. Understanding the processes that affect the chloroplast content in the rumen may open up additional possibilities for improving the composition of beef fatty acids.

For this purpose, the *in vitro* method was used to estimate the rate of absorption, digestion, and release of chloroplasts by the biotics of the rumen. *In vitro* studies were conducted to more clearly understand the fate of chloroplasts rich in PUFAs in the rumen and, in particular, the processes between chloroplasts and rumen biotics. Intact chloroplasts were obtained from spinach leaves (Spinacia oleracea) using standard methods. The resulting chloroplasts were treated with protozoa *Epidinia spp.* and *Entodinia spp.* Intracellular and extracellular chlorophyll, as well as the content of fatty acids in protozoa were monitored with time intervals (0, 0.5, 1, 2, 4, and 8 h). The intracellular position of autofluorescent protozoa chloroplasts was evaluated using fluorescent confocal microscopy.

A number of unforeseen technical problems hindered the success of these *in vitro* approaches. At the initial stage of research, an attempt to clear the rumen's protozoa as much as possible from a larger number of intracellular chloroplasts was made, so that at the time point *0h* they only contained a very low amount of intracellular chloroplast. First of all, the protozoa fractionation was used, and then its anaerobic incubation was applied to control the time they needed to isolate intracellular chloroplasts. Protozoa density, viability and the presence of intracellular chloroplast were monitored for 24 hours. The protozoan viability steadily declined to several living cells after 6 hours, and microscopy showed that the protozoa had a lot of intracellular chloroplast even after 24 hours.

Following this, an attempt was made to replace intracellular chloroplasts rich in C18:3n-3 with chloroplasts rich in C16:3n-3 by feeding on corn after fractionation. This experiment again showed that the protozoa were not very active after 6 hours. Therefore, the removal of chloroplasts rich in C18:3n-3 and the experiment itself should be carried out during this time, which is impossible. As a result, the decision was made to receive protozoa from straw-fed cows (a low content of C18:3n-3 and chlorophyll) 1 week before the experiment and using them immediately after fractionation in experiments. The experiments were established *in vitro* with a ratio of 1:100 (1×10^4 protozoa/ml: 1×10^6 chloroplasts/ml). According to the developed experiment, the amount of protozoa that could be extracted from the rumen fluid was limited to a density of 1×10^4 protozoa/ml. Unfortunately, the use of this density did not allow to detect fatty acids and chlorophyll. Thus, due to problems of viability and density, the conclusion is made that it is impossible to control the intake/digestion/excretion by the in vitro method.

Subsequently, the decision to conduct an *in vivo* experiment was made. A total of 6 healthy animals of the Kalmyk and Kazakh white-headed breed of the Trans-Volga type (with an average live weight of 250 kg) with fistulas of the rumen and duodenum were fed with straw for 14 days. The feed samples were taken daily and weekly pooled for freezing, while grass samples were taken in the morning exclusively for the experiment (approximately 1 kg) and frozen. The selection of protozoa, planktonic and attached bacteria was carried out 1 hour before the change in diet and 2 and 6 hours after the change. The content of chlorophyll and fatty acids in each microbial fraction was analyzed, and fixed samples were monitored for the content of intracellular chloroplasts using confocal microscopy and transmission electron microscopy. The diet based on straw and feed had more ADF, NDF, C18: 1n-9 and less WSC, fatty acids, C16: 0, C18: 2n-6 and C18: 3n-6 compared to the ration on fresh grass, which was predictable.

The obtained data on fatty acids showed that protozoa were significantly enriched in C16:0, C18:0, C18:2n-6, C18:3n-6, C18:1 trans-11 content 2 hours after feeding on fresh grass. And the level of each of these fatty acids began to decrease in part over 6 hours, although the concentrations of the intraprotozoal composition C16:0, C18:0 and C18:3n-3 remained significantly higher than the values obtained 1 hour before the supply of fresh grass (Table 2).

Table 2. Fatty acid content (mg / g-1 LM) at various times before and after feeding with fresh grass.

Fatty acid	Time			SED	P
	1	2	6		
18:3n-3	0.052a	0.604b	0.511b	0.062	<0.001
18:2n-6	0.955a	2.805b	1.586a	0.298	<0.001
18:2 cis-9, trans-11	0.316a	0.298a	0.334a	0.070	0.879
18:1 trans-11	1.166a	2.792c	1.681b	0.206	<0.001
18:0	5.728b	7.479c	4.516a	0.400	<0.001
16:0	5.281a	11.625c	8.201b	0.585	<0.001

4. Discussion

Based on the results obtained (Table 1), the conclusion can be drawn that seasonal and environmental factors have a significant impact on the phenotypic variation of the fatty acid content, which in turn would require an adjustment of the fatty acid composition of the feed during the growing process.

Reducing the degree of biogenization in the rumen can increase the processes of transformation of PUFA into muscle and adipose tissue. PUFAs are rapidly hydrogenated by rumen microbiotics, which leads to the formation of SFA (mainly 18:0), but also to the formation of intermediate compounds CLA and trans monoenes (mainly trans-vaccenic acid TVA). This is one of the main reasons why the fats of ruminants are represented mainly by saturated fatty acids. Lipolysis in the rumen is a prerequisite for microbial hydrogenation (biohydrogenation) of unsaturated fatty acids. The degree to which biohydrogenation is "complete" affects the amount of SFA produced in the rumen, but also the amount of CLA and TVA. The microscopic data showed that 2 hours after feeding on fresh grass, the protozoa had a significantly larger number of intracellular chloroplasts, which remained high and 6 hours after feeding. Therefore, data on fatty acids were correlated with the content of intraprotozoal chloroplast. These data illustrate that the absorption by protozoa of mainly plant chloroplasts occurs rapidly, and the intracellular level of chloroplasts is maintained for at least 6 hours. Thus, protozoa quickly become the main reservoir of chloroplasts and then useful sources of PUFAs.

5. Conclusion

The absorption of protozoa by mainly plant chloroplasts occurs rapidly, and the intracellular level of chloroplasts is maintained for at least 6 hours. Thus, protozoa quickly become the main reservoir of chloroplasts, and then useful sources of PUFAs.

IOP Conf. Series: Earth and Environmental Science **395** (2019) 012059 doi:10.1088/1755-1315/395/1/012059

References

[1] Belury M A 2003 *Conjugated linoleic acids in type 2 diabetes mellitus: Implications and potential mechanisms.* In Sebedio J, Christie W, and Adlof R Eds. *Advances in Conjugated Linoleic Acid Research* (pp 267-281) (Champaigne, Illinois: The American Oil Chemists Society)

[2] Huws S A, Kim E J, Lee MRF, Kingston-Smith A K, and Scollan N D 2006 Fatty acid composition of rumen protozoa: a link with cholorplast ingestion? In 5th Joint RRI-INRA: *Symposium on Gut Microbiology – Research to Improve Health, Immune Response and Nutrition* (Aberdeen, UK)

[3] Baird N A, Etter P D, Atwood TS, Currey M C, Shiver A L, Lewis Z A, Selker E U, Cresko W A, … Johnson E A 2008 Rapid SNP Discovery and Genetic Mapping Using Sequenced RAD Markers *Plos One* **3** e3376

[4] Cabiddu A, Lee MRF, Scollan N D, and Sullivan M L 2009 Lipolysis and biohydrogenation of forage species at vegetative and reproductive stages of growth *Journal of Dairy Science* (E-Supplement 1) **92** pp 109-110

[5] Cabiddu A, Salis L, Tweed JKS, Molle G, Decandia M, and Lee MRF 2010 The influence of plant polyphenols on lipolysis and biohydrogenation in dried forages at different phenological stages: in vitro study *Journal of the Science of Food and Agriculture* **90** pp 829-835

[6] Huws S A, Kingston-Smith A H, Kim E J, Lee MRF, Scott M B, Tweed J K, and Scollan N D 2010 Contribution of ruminal protozoa to the duodenal flow of polyunsaturated fatty acids as a consequence of intracellular chloroplasts, in *EU Framework 6 ProSafeBeef Annual General Meeting* (Aberystwyth, UK)

[7] Huws S A, Kingston-Smith A H, Kim E J, Lee MRF, Scott M B, Tweed J K, … Scollan N D 2010 Contribution of ruminal protozoa to the duodenal flow of polyunsaturated fatty acids as a consequence of intracellular chloroplasts, in *EU Framework 6 ProSafeBeef Annual General Meeting* (Aberystwyth, UK)

IOP Conf. Series: Earth and Environmental Science **395** (2019) 012060 doi:10.1088/1755-1315/395/1/012060

The specialization areas of grain production focused on the export of products

L Silaeva[1], Zh Kulchikova[2] and E Barinova[3*]

[1] Federal Scientific Center of Agrarian Economics and Social Development of Rural Territories - All-Russian Research Institute of Agricultural Economic, 35, building 2, Highway Khoroshevskoe, Moscow 123007 Russia
[2] Kostanay Branch of the Chelyabinsk State University, 168A Borodin str., Kostanay 110006 Republic of Kazakhstan
[3] V. M. Gorbatov Federal Research Center for Food Systems of Russian Academy of Sciences, 26 Talalikhina str., Moscow 109316 Russia

E-mail: e.barinova@fncps.ru

Abstract. The country's export objectives, aimed at the growth of the Russian economy, are best addressed by highlighting the competitive advantages of its border areas. In wheat exports, as of today, Russia ranks first in the world, but its gross harvest rises in conditions of low-quality grain production. The paper analyzes the production parameters of spring wheat cultivation, explores the dynamics of its sown areas, gross yield, and yield by federal districts for the period 2001-2017. The article calculated the structure of crops and wheat production, indicators of its effectiveness depending on the quality of grain in the agricultural organizations of the country. The conditions for economic growth of individual border areas of the country are determined on the basis of the development of efficient production of strong and hard varieties of spring wheat in the regions of Western Siberia. A mechanism has been developed and proposed for creating a single large-scale West Siberian specialized zone for the sustainable production of high-quality spring wheat commercial grain with sales to the neighboring countries of the Asian continent.

1. Introduction

The globalization of the economy, significant changes in world politics are causing scientific interest in solving issues related to the functioning of cross-border regions. It is advisable to solve export tasks that are on the agenda of the country today and are of great strategic importance for ensuring the growth of the national economy in close partnership with the border regions of Russia. In this regard, we need to identify the competitive advantage of these territories, highlighting the main points of their growth, contributing to the sustainable and dynamic development of the economy, both of the regions themselves and of the country as a whole.

The favorable economic and geographical position of Russia on the world map and the colossal soil and climatic capabilities of the national territory make it possible to produce and sell competitive products on the world markets of the country. From this point of view, our close neighborhood with the countries of the Asian continent is the most economically attractive. In particular, the regions of the Siberian Federal District are able to reliably and sustainably provide the fast-growing population of the Asian continent with high-quality agricultural raw materials and food.

Content from this work may be used under the terms of the Creative Commons Attribution 3.0 licence. Any further distribution of this work must maintain attribution to the author(s) and the title of the work, journal citation and DOI.
Published under licence by IOP Publishing Ltd

2. Materials and Methods

The data of the Federal State Statistics Serves (FSSS) of Russia [7] and the indicators of annual reports of the Ministry of Agriculture of the Russian Federation [8] presented on the official websites of organizations are used as the main empirical material. The object of the research is the production of spring wheat in the country and certain regions of the Siberian Federal District. We used abstract-logical, analytical, design-constructive research methods in our work. The research was also conducted using the application software Microsoft Word and Excel.

3. Results

The world's main food grain crop is wheat, a wide variety of foods with high nutritional value, satisfying simple human food needs and gourmet requests can be produced from it. Russia occupies the leading position in the world in the export of wheat, but at the same time, the low quality of its grain and the high level of fluctuations in yield indicators do not have a positive effect on the country's economy. For example, the wheat production in the Altai, Novosibirsk, and Omsk regions of the Siberian Federal District is presented in Table 1.

Table 1. Placement of spring wheat by federal districts of the Russian Federation.

Federal districts	2001-2005		2006-2010		2011-2015		2016-2017	
	Thous. ha thous. t	%	Thous. ha thous. t	%	Thous. ha thous. t	%	Thous. ha thous. t	%
Spring wheat sown area								
Russia	15094.2	100.0	14220.1	100.0	13169.2	100.0	13318.5	100.
Central	455.2	3.0	568.6	4.0	446.6	3.4	615.7	0
Northwestern	52.1	0.3	42.4	0.3	45.0	0.3	69.4	4.6
Southern	243.4	1.6	155.7	1.1	110.4	0.8	204.0	0.5
North Caucasus	18.7	0.1	11.9	0.1	6.7	0.1	5.7	1.5
Volga	4862.2	32.2	4137.2	29.1	3751.9	28.5	3738.9	0
Ural	2141.4	14.2	2387.2	16.8	2391.3	18.2	2355.9	28.1
Siberian	7196.7	47.7	6760.0	47.5	6296.0	47.8	6184.0	17.7
Far Eastern	124.6	0.8	157.1	1.1	121.3	0.9	144.7	46.4
								1.1
Gross harvest of spring wheat								
Russia	20171.1	100.0	19684.9	100.0	17451.0	100.0	22496.5	100.
Central	829.8	4.1	1046.1	5.3	1029.7	5.9	1968.3	0
Northwestern	73.8	0.4	70.7	0.4	105.5	0.6	150.0	8.7
Southern	276.7	1.4	115.0	0.6	117.2	0.7	357.5	0.7
North Caucasus	28.1	0.1	24.1	0.1	12.9	0.1	15.2	1.6
Volga	6827.1	33.8	5242.9	26.6	4611.8	26.4	6359.6	0.1
Ural	2755.7	13.7	3192.5	16.2	3174.1	18.2	4020.2	28.3
Siberian	9245.0	45.8	9792.1	49.7	8197.7	47.0	9328.0	17.9
Far Eastern	134.8	0.7	201.5	1.0	202.1	1.2	297.7	41.5
								1.3

Located in the steppe zone, the Altai and Omsk regions are the largest producers of spring wheat in the country. They have the highest coefficient of specialization (more than 7.0) and produce over one ton of this grain crop per capita. These regions, which have a relatively large potential to meet their own needs in wheat, can provide tangible and quick returns in building up relatively cheap commodity resources of high-quality wheat, increasing its supplies to the domestic and foreign grain markets. Favorable soil and climatic conditions in the territory of Western Siberia allow us to consistently receive high grain yields of strong and hard spring wheat varieties with minimal production costs. Despite all the potentialities of the above regions, a slow but steady decline in the area under spring wheat has been observed here over the past two decades. The reduction occurs in the conditions of production of low-quality grain and unstable yield, fluctuations of which from year to year range from 8.7 to 19.3 c / ha.

The country has significant genetic and natural resources, including fertile chernozem soils, which enable stably to obtain grain of hard and strong wheat. But at the same time, the production of wheat classes 1 and 2 is 2%, and the class 3 occupies 20% of its total production (Table 2). At the same time, the 4th wheat class makes up most of its marketable volume, which significantly reduces the competitiveness of the -Russian wheat on the world market and the level of export revenues.

Table 2. The efficiency of wheat production, depending on its quality in the agricultural organizations of the Russian Federation.

Indicators	Share in implementation, %	Total cost, rub. / t	Selling price, rub. / t	Profit, rub. / t	Profitability %
		2001-2005			
Wheat, on average	100.0	1584	2062	478	30.2
Out of this: 1-2 classes	2.0	1670	2488	818	49.0
3 class	21.0	1707	2215	508	29.7
4 class and fodder	77.0	1549	2011	462	29.8
		2006-2010			
Wheat, on average	100.0	3179	4038	859	27.1
Out of this: 1-2 classes	1.8	3366	4431	1065	31.7
3 class	18.9	3326	4422	1096	33.0
4 class and fodder	79.3	3140	3937	797	254
		2011-2015			
Wheat, on average	100.0	5097	6756	1659	32.5
Out of this: 1-2 classes	1.8	5318	7171	1853	34.8
3 class	21.5	5135	6994	1859	36.2
4 class and fodder	76.7	5081	6136	1598	31.4

Due to the insignificant differentiation of prices for wheat grain by classes, its producers have no economic interest in producing high quality grain. Also, seasonal and, to a large extent, annual fluctuations in the prices of wheat grain overlap the value parameters of the quality of its grain and thus do not stimulate the production of quality products. The difference in the level of profitability of wheat of the first and second classes if compared with its grain of the third class was only 1.4 percentage points, and it was 3.4 percentage points with the fourth class and feed wheat in 2011-2015.

The natural potential of the regions of Western Siberia is able to ensure the effective functioning of the production of spring wheat in a country with high quantitative and qualitative characteristics of grain. The solution to this problem is possible by creating in these regions of Siberia a high-tech specialized zone for the sustainable production of hard and strong varieties of spring wheat using environmentally friendly and resource-saving technologies. In this regard, the use of the genetic potential of this spring crop is the most justified and least costly factor in expanding the production of spring wheat with a stable level of yield and high-quality grain. The seed growing as one of the safest ways to intensify the production of spring wheat is a biological factor in its effective functioning. It is seed production that makes it possible to expand crops under varieties of spring crops, oriented towards probable changes in weather conditions that are resistant to pests and diseases [1]. Improving the quality of grain through targeted breeding is the cheapest and most affordable way to improve the efficiency of spring wheat production in the regions of Siberia.

Sustainable and dynamic functioning of the production of spring wheat in the regions of Western Siberia is possible through the formation of a single large-scale specialized zone in the Altai, Novosibirsk and Omsk regions, with an increase in the volume of marketable grains of hard and strong varieties of spring crops. At the same time, the national interests of the state should take into account the potential growth of the regions-members of the specialized zone, moving into the direction of developing competitive production of high-quality spring wheat grain and increasing its export resources to the countries of the Asian continent. When forming the West Siberian specialized zone, the following actions are necessary:

- Considering and analyzing the already established level of distribution and specialization of wheat production, the degree of concentration of spring crops in these regions, taking into account all factors affecting the production process, including soil-climatic and socio-economic;
- Assessing the production potential of the regions, taking into account its capabilities for the storage, processing, and sale of wheat grain and finished products from it, comprehensively assessing the current state of the material and technical base and vehicles, transportation distances, and taking into account the cost of logistics;
- It is also necessary to substantiate the interregional exchange of each region, taking into account its capacity for export and the volume of wheat exports based on the calculation of the indicator of its production per capita, and the level of interregional specialization is to be defined as the volume of export of wheat on interregional exchange to the total volume of its production;
- Making a feasibility study of the project, to comprehend and evaluate the results in the final positive version of its intended justification for locating wheat production with the maximum concentration of spring crop crops in these regions, taking into account all sorts of risks and threats;
- Motivating options to intensify the production of spring wheat in order to achieve high yield and grain quality, increasing its commodity resources and increasing production sustainability.

Introducing economically viable innovative technologies, using the achievements of scientific and technological progress, ensuring the ecological balance in the cleanest regions of Russia will contribute to the rational use of production resources and the bioclimatic potential of the territory, reproduction of soil fertility, improvement of the quality of spring wheat grain and the environmental safety of the final products, which today is especially expensive in the modern world.

The creation of the West Siberian specialized zone for the sustainable production of hard and strong spring wheat will open up a significant potential for increasing the capacity to export competitive grain to nearby countries of the Asian continent. Currently, trade with China is carried out through Kazakhstan, within the framework of the Eurasian Economic Union [1]. In modern economic conditions, in conditions of ever-increasing tough competition between states and unstable political equilibrium in the world, we need to eliminate any one-sided dependence of the country in addressing the sustainable development of cross-border regions, ensuring economic interests and the national security of the state as a whole.

4. Discussion

The issues of transition to the export-oriented model of agricultural development in the country based on the rational and effective use of the potential of the regions, ensuring sustainable production and high quality of the final product, its competitiveness in world markets are actively discussed in the works of A. I. Altukhov [2], G. V. Bespakhotnyy [3], E. N. Krylatykh [4], A. G. Paptsov [5], I. G. Ushachev [6], and others.

It can be stated that the conducted scientific research allowed to summarize a part of the empirical data on the prospects for the formation and development of a rational and efficient location of the production of strong and hard spring wheat in the border regions of the country and its export to nearby countries of the Asian continent. The strategic role of creating a single specialized area for the production of spring wheat in the regions of Western Siberia is not only in building high quality commercial grain and its competitive export to the countries of the East. The creation of a single specialized area for the production of spring wheat in the regions of Western Siberia is also a fundamental basis for the development of the national territory and the cross-border regions of the country.

5. Conclusion

The formation of a large-scale specialized zone in the border regions of Western Siberia must rely not only on increasing the production of hard and strong spring wheat as a raw material resource of the country, but also on transport and logistics opportunities. These opportunities are opened when servicing infrastructure and engineering structures, accompanied by international transport corridors to the countries of the Asian continent. Such strategic points of growth will have a significant positive impact on the regions' cross-border cooperation. This is a complex of trade, economic and investment partnerships using opportunities to create the most favored conditions for the passage of products at the border.

References

[1] Altukhov A I 2017 *Improving interregional exchange in the system of territorial-sectoral division of labor in the agro-industrial production of the Eurasian Economic Union* (Monograph) (Moscow-Kostanay, Kostanay Printing House LLP)

[2] Altukhov A I 2017 On the key fundamentals of the eternal "agrarian issue" in Russia *Vestnik of the Institute of Economics of the Russian Academy of Sciences* **2** pp 188-191

[3] Bespakhotnyy G V 2017 Indicative planning in agriculture *Economy, Labor, Management in Agriculture* **2**(31) pp 2-4

[4] Krylatykh E H 2015 Scientific support of the agrarian policy: opportunities and threats *Nikonovskiye chteniya* **20-1**(20) pp 3-7

[5] Paptsov A G 2017 The role of the state in the agrarian sector of developed countries *AIC: Economics, Management* **4** pp 83-87

[6] Ushachev I G 2017 Current trends and look into the future development of the AIC of Russia *Applied Economic Research* **2**(18) pp 4-7

[7] *Federal State Statistics Service of the Russian Federation* n.d. Available at: http://www.gks.ru (Accessed 02 03 2019)

[8] *Ministry of Agriculture of the Russian Federation* n.d. Available at: http://www.mcx.ru (Accessed 02 03 2019)

Movement modes of a fruit during vibration harvesting

I Ya Fedorenko[1*]

[1] Altai State Agricultural University, 98 Krasnoarmeyskiy prospekt, Barnaul 656049 Russia

E-mail: ijfedorenko@mail.ru

Abstract. The article describes the movement of the fruit as a mathematical pendulum during branch vibrations. A non-linear numerical mathematical model was obtained. The article stated the presence of the fruit's complex movements, including non-sinusoidal oscillations, oscillations-beats, rotational modes, as well as oscillations that have a stochastic appearance. Subresonances are detected at operating frequencies $\omega \approx p\omega_0$ ($p=2,3,4,5\ldots$); at the same time, the fruit oscillates with almost its own frequency ω_0. It is established that the maximum efforts in the stem necessary for its destruction appear at a subharmonic resonance with the frequency $\omega \approx 3\omega_0$. This mode is recommended as a working one for garden vibration installations.

1. Introduction

The Altai region, the Altai Republic and a number of adjacent territories have a great potential for the development of horticulture, the ability to increase the production of fruits and berries. The revival and renovation of gardens requires the simultaneous improvement of mechanization tools for the care of fruit crops and the harvesting of finished products. In particular, the vibration working bodies of the machines show high efficiency when removing fruits. The works of Russian and Chinese scientists and designers are devoted to the development of these methods of harvesting fruits (see, for example, [1,2]). The purpose of this article is the mechanic-technological substantiation of the modes of the process of vibration harvesting of fruits based on a non-linear mathematical model. This paper continues the previously published research [3].

2. Materials and Methods

The vibration method of harvesting fruits is that the stem of a tree vibrates, and the branches fluctuate mainly in the horizontal direction (Fig. 1a). The developing inertial forces are much greater than the force of gravity acting on the fruit, making it possible to destroy the stem.

Content from this work may be used under the terms of the Creative Commons Attribution 3.0 licence. Any further distribution of this work must maintain attribution to the author(s) and the title of the work, journal citation and DOI.
Published under licence by IOP Publishing Ltd

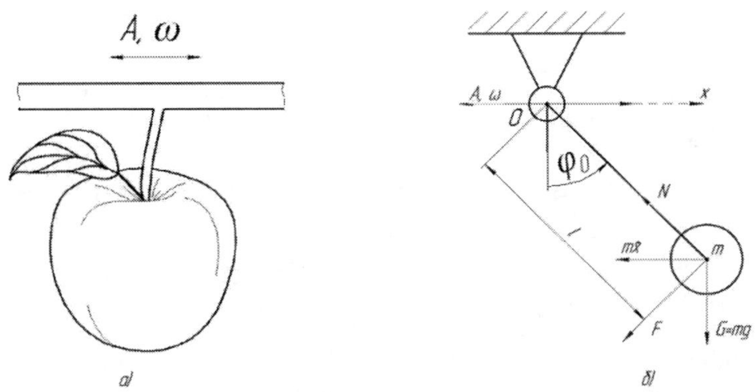

Figure 1. Physical (a) and calculated (b) diagrams of the model describing the process of the fruits' vibrational removal

We will simulate the "vibrating branch – fruit stem – fruit" system in the form of a mathematical pendulum, while we do not take into account the weight of the stem (Fig. 1b).

In a non-inertial (associated with a vibrating branch) coordinate system, the equation of fruit's oscillations, as a mathematical pendulum, has the following form [3]:

$$I\ddot{\varphi} = -k \cdot \ell \cdot (\ell \cdot \dot{\varphi})^2 \cdot sign(\dot{\varphi}) - m \cdot g \cdot l \cdot \sin(\varphi) - m \cdot l \cdot \ddot{x} \cdot \cos(\varphi), \quad (1)$$

where m is the mass of the fruit; l is the distance from the center of gravity of the fetus to the point of suspension of the stem to the branch; I is the moment of inertia of the fruit relative to the point of suspension 0 ($I = m \cdot l^2$); $k = \frac{1}{2} \cdot c \cdot \rho \cdot S$ – the air resistance coefficient (where c – is the proportionality coefficient, c = 0,3-0,4; ρ - is the air density; S is the midsection of the fetus);

$$sign(\dot{\varphi}) = \begin{cases} 1 \text{ when } \dot{\varphi} > 0 \\ 0 \text{ when } \dot{\varphi} = 0 \\ -1 \text{ when } \dot{\varphi} < 0. \end{cases} \text{ – a sign function.}$$

We divide all terms of the equation (1) by I, taking into account that $I = m \cdot l^2$; by calculating the derivative $\ddot{x}$, we get:

$$\ddot{\varphi} + \frac{k \cdot l}{m} \cdot \dot{\varphi}^2 \cdot sign(\dot{\varphi}) + \frac{g}{l} \cdot \sin(\varphi) = \frac{A}{l} \cdot \omega^2 \cdot \cos(\varphi) \cdot \sin(\omega t), \quad (2)$$

where A, ω are the amplitude and angular frequency of horizontal oscillations of the branch.

Including the following notations $\frac{k \cdot l}{m} = 2n$; $\frac{g}{l} = \omega_0^2$; $\frac{A}{l} = K$, we get the final equation of fetal oscillations transmitted from the branch:

$$\ddot{\varphi} + 2 \cdot n \cdot \dot{\varphi}^2 sign(\dot{\varphi}) + \omega_0^2 \cdot sin(\varphi) = K \cdot \omega^2 \cdot \cos(\varphi) \cdot \sin(\omega t) \quad (3)$$

where n is the damping coefficient; ω_0 is the natural frequency of the fruit's oscillation on the stem.

We have obtained the nonlinear equation of the forced fruit's oscillations as a mathematical pendulum. Recent studies of forced oscillations of a mathematical pendulum have shown its highly complex dynamic behavior [4, 5]. Moreover, the equation (3) is more complicated than the equation considered by professor E. I. Butikov in his relevant resesearch[4, 5]. This prompted us to conduct a computer simulation of the process of fetal removal, using our non-linear model. We carried out a simulation for apples of medium size, while the necessary data for calculating the parameters n, ω_0 and K were obtained from sources [1, 6]. For the numerical solution of a differential equation, we used the Runge-Kutta method with an adaptable integration step.

3. Results

The computational experiments implemented by us really show the presence of complex modes of the fruit's movement, which are not displayed by a linear oscillatory system. Non-sinusoidal oscillations, oscillations-beats, rotational modes, as well as oscillations that have a stochastic appearance belong to them. It was established that nonlinearity manifests itself even with small fruit's fluctuations (about 5°). Thus, the linear model of the forced fruit's oscillations, used in many studies, is untenable.

Subharmonic resonances, which manifest themselves at working frequencies $\omega \approx p\omega_0$ $(p=2,3,4,5\ldots)$, were also detected, while the fruit oscillates almost with its own frequency ω_0. The subharmonic resonance of the order of 1/3 is particularly strong when the frequency of fruit's oscillations is about 3 times less than the frequency of the driving force (the frequency of oscillation of the branch).

In order to identify the conditions of destruction of the stem, we consider the load acting on the stem during the vibrations of the branch to which the fruit is attached. The tension of the stem N in general is determined by the following forces: a part of gravity – $mg\cos\varphi$, a centrifugal force of inertia – $m\dot{\varphi}^2 l$, a part of the portable force of inertia – $mA\omega^2\sin\varphi$. So we have the following equation:

$$N = mg\cos\varphi + m\dot{\varphi}^2 l - mA\omega^2\sin\varphi.$$

We divide all terms of the equation by $G=mg$, we get

$$\lambda = \cos\varphi + \frac{\dot{\varphi}^2}{\omega_0^2} - \frac{A\omega^2}{g}\sin\varphi,$$

where the value $\lambda = N/G$ represents the multiplicity of an increase in the weight of an apple during vibration overloads.

Figure 2 shows the load on the stem at extreme values of the amplitude of the fruit's oscillations with an angular frequency of $\omega \approx 3\omega_0 = 27$ c^{-1}. It is noteworthy that the values of λ can take negative values. This means that during vibrations, the fruit can both stretch and shrink. Consequently, it bends and collapses.

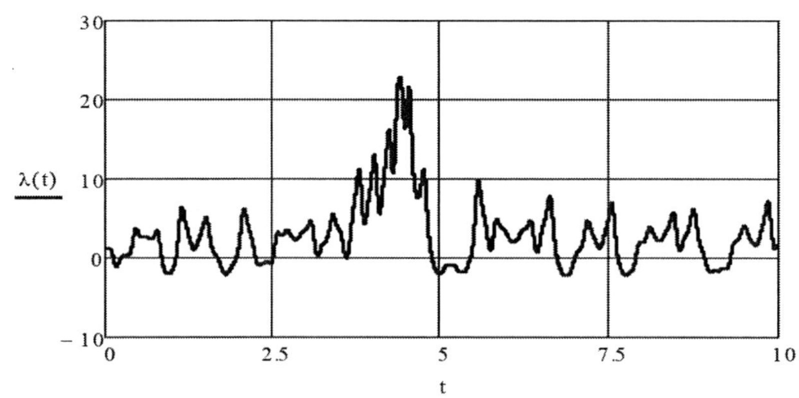

Figure 2. Developing overload λ in the fruit stem at an angular oscillation frequency ω=27 c^{-1}.

4. Conclusion

The strongest loads are tested on oscillation modes with $\omega \approx 3\omega_0$, i.e. the frequency of the forcing effect should be about 3 times higher than the natural frequency of the fetus on the peduncle. The short-term value of the overload λ must reach 20 or more units, which determines the corresponding amplitude of oscillations of the branches of the fruit tree. This ensures high efficiencies in vibratory harvesting of apples and other types of fruit. Therefore, the mode with a driving frequency is $\omega \approx 3\omega_0$. This mode is recommended as a working one for garden vibration installations.

References

[1] Mikhailova N V, Filimonova E Yu, and Levin A M 2014 *Finding methods for the mechanized harvesting and processing of sea buckthorn fruits* (Monograph) (Barnaul, RIE ASAU)

[2] Longsheng Fu, Jun Peng, Qiang Nan, Dongjian He, Yougang Yang, and Yongjie Cui 2016 Simulation of vibration harvesting mechanism for sea buckthorn *Engineering in Agriculture, Environment and Food* **9**(1) pp 101-108

[3] Fedorenko I Ya, Shestaev A V, and Shcherbakov S S 2019 *Modeling the process of vibrating fruit pick-up* In N V Gavrilets (Ed.) *Collection of the National (All-Russia) Scientific Conference: Theory and Practice of Modern Agrarian Science* (pp 254-257) (Novosibirsk, Russia: ITS "Zolotoy kolos")

[4] Butikov E I 2008 Unusual behavior of a pendulum under sinusoidal external influence *Computer Tools in Education* **1** pp 24-36

[5] Butikov E I 2011 On the movements of the pendulum under the influence of a periodic moment Mathematical modeling. Optimal control *Vestnik of Lobachevsky University of Nizhni Novgorod* **3**(2) pp 83-86

[6] Varlamov G P 1978 Machines for harvesting fruits (Moscow, Russia: Mechanical Engineering)

IOP Conf. Series: Earth and Environmental Science **395** (2019) 012062 doi:10.1088/1755-1315/395/1/012062

An analysis of the current state of vegetable production of open and protected soil in the Russian Federation

A Razin[1*], O Razin[1] and P Mescheryakova[1]

[1] Federal Scientific Center for Vegetable Growing 14 Selectionnaya str., VNIISSOK village, Odintsovo district, Moscow Region 143080 Russia

E-mail: vniioh@mail.ru

Abstract. The article analyzes the current state of vegetable production in the open and protected soils in the Russian Federation. Tomato (17.3%), cabbage (16.7%), onion (13.0%), carrots (10.4%), cucumbers (10.0%), and table beet (7.0%) occupy the largest share in the structure of vegetable crops cultures. In the structure of products of the whole agriculture, 49.5–52.8% is accounted for by agricultural organizations, 40.5–34.7% is produced by households. The bulk of commercial vegetables produced in the open ground. all categories of farms produced 14.7 million tons each, in 2016–2017. The main production of vegetables in open ground is concentrated in the Southern, Volga and Central federal districts, where about 65% of the vegetables from the total production in the Russian Federation are produced annually. The average yield of vegetables in greenhouses in the whole country is growing steadily, reaching the figure of 36.9 kg/m^2 in 2017, which is 31.8% higher than the 2013 level (28 kg/m^2). According to the Ministry of Agriculture of the Russian Federation, the level of self-sufficiency of the Russian Federation of vegetables of the protected ground was 53.4% in 2017. This indicator is 7.1% higher than the 2016 level (46.3%) and 13.0% higher than the 2015 level (40.4%).

1. Introduction

In the human diet of the Russian Federation, vegetables occupy the third place after bread and potatoes. Since vegetables are the main suppliers of phytonutrients necessary for the normal functioning of the body, the annual demand of the population of the Russian Federation for fresh vegetables is about 18.3 million tons (the current level of production is 15-16 million tons). This standard is determined in accordance with the rational norms of food consumption that meet contemporary requirements for healthy nutrition. According to the Food Security Monitoring System, the self-sufficiency of open and closed ground vegetables was 81.1% in 2017. Insufficient provision of the domestic market of the country with vegetables of domestic production is replenished with imported products from China, Turkey, Israel, Morocco, Belarus, Egypt, and other countries. Prior to the introduction of international sanctions, imports of vegetable products tended to increase, then its decline was outlined [1]. Vegetable production in the Russian Federation is inferior to industrialized countries in many respects. The vegetable production is lower by an average of 1.5-2 times. In terms of the range of products being produced on an industrial scale, vegetable production is lower by 2-3 times etc. [2].

The purpose of the study is to analyze the current state of vegetable growing in open and protected ground in the world and in the Russian Federation.

Content from this work may be used under the terms of the Creative Commons Attribution 3.0 licence. Any further distribution of this work must maintain attribution to the author(s) and the title of the work, journal citation and DOI.
Published under licence by IOP Publishing Ltd

2. Materials and Methods

Scientific studies were conducted on the basis of the first item of the FNI Program of the State Academies of Sciences for 2013–2020 and the approved research direction "Modern Economic Theory and Principles of Development of the Country's Agro-Industrial Complex in the Context of Globalization and Integration Processes in the Global Economy". In the framework of the general systems approach to the study of the problem, the tools of the abstract-logical, comparative, economic-statistical, economic-mathematical, and computational-constructive research methods are used. Legislative and regulatory documents of state authorities, methodological and instructional materials of the Ministry of Agriculture of the Russian Federation were the information basis of the study. The empirical base of the research is presented by the materials of the Federal State Statistics Service, data of the consolidated annual reporting on the financial and economic status of enterprises of the agro-industrial complex of the Russian Federation.

3. Results

The highest level of self-sufficiency in vegetables was found in the Southern Federal District (170.5%), Volga Region (88.1%), North Caucasus (76.5%), Siberian (73.6%), Ural (56.4%), Central Federal District (55.6%), and the Far East (53.1%). The lowest level of provision with vegetables is noted in the regions of the North-West Federal District (35%). For full provision of the population with domestic vegetables, it is necessary to additionally produce about 3 million tons of vegetable products. Such countries as China (25%), Turkey (14%), Israel (10.5%), Morocco (8.6%), Belarus (8.3%), Egypt (7.8%) are the main countries-suppliers of vegetables in Russia. Before the introduction of international sanctions, the import of vegetable products tended to increase, then its decline was evident. Almost twice less imported vegetables are imported into the country compared to previous years. Thus, the vegetable imports to Russia amounted to 3,000 thousand tons, in 2013, 3,380 thousand tons in 2014, 2,607 thousand tons in 2015, 1,724 thousand tons in 2016, 1,864 thousand tons in 2017. Vegetable imports to Russia amounted to $10.4 billion, with a total weight of 12,575 thousand tons, for the period 2013–2017. Tomatoes (42% of the value of imported vegetables), onions and garlic (12%), cucumber (10%), other vegetables (27%) were the bulk of imported vegetables [3]. Vegetable exports amounted to 4348 thousand tons worth $1.56 billion. Exports of Russian tomatoes increased almost 4 times, exports of domestic cucumbers increased by 29%, cabbage - by 12.6%, from the beginning of 2018 [4].

Vegetable production in the Russian Federation is inferior to industrialized countries in many respects, including yield, which in Russia is on average 1.5–2 times lower than in Western Europe, the USA, China. Russia, on the yield is in 57th place. The range of products produced on an industrial scale in the Russian Federation is very limited. If most countries in the world use more than 30–35 species of vegetable crops in industrial vegetable production, Russia uses no more than 12–15 species. 84% of the total harvest of vegetables accounted for the 6 main types of vegetable crops: cabbage (23%), tomato (18%), onion (13%), cucumber (12%), carrots (11%), table beet (7%). The share of other vegetables is not more than 16% of the total harvest of all vegetables [6].

Provision of the population with certain types of vegetable products is determined by the rational allocation of vegetable production on the territory of the Russian Federation and the species composition of cultivated crops. In the structure of vegetable crops, tomatoes (17.3%), cabbage (16.7%), onion (13.0%), carrots (10.4%), cucumbers (10.0%), and table beet (7.0%) are the largest share.

In Russia, 660–690 thousand hectares are under vegetable crops annually, which is about 0.9% of the total crop area in the country, the gross yield is 16.1–16.4 million tons, with an average yield of 22.5 –23.6 t/ha. In the structure of products of all agriculture, the share of agricultural organizations is 49.5–52.8%, the share of households of the population is 40.5–34.7%, the share of peasant (farmer) farms and individual entrepreneurs is 10.0–12.5 %. In the country as a whole, in recent years, a steady increase in agricultural production in agricultural organizations and peasant (farmer) farms has been observed, a decrease is observed in household farms. The same trend is observed in the production of vegetables. The main production of vegetables is concentrated in household farms, their share was 62.9% of the gross collection in 2017. And these farms had 478.9 thousand hectares in their use, or 70.9% of the total area of the country's vegetable field. In the course of the reforms, the vast majority of agricultural land became the individual property of workers and retirees of reorganized state and collective farms, citizens - members of garden cooperatives, owners of household plots [3]. Vegetable production (in % of total production in farms of all categories) increased in agricultural organizations from 17.1% to 21.2%, in peasant (farmer) farms was from 11.4% to 15.9%, and the vegetable production decreased in households from 71.5 to 62.9% for the period from 2010 to 2017.

The production of vegetables in agricultural organizations increased from 2.5 million tons to 3.5 million tons as a result of an increase in the sown area of 90.0 thousand hectares to 95.2 thousand hectares and an increase in yield from 276.0 c/ha to 285.7 c/ha for the period from 2012 to 2017. Modernization of old, construction and commissioning of new modern greenhouses also contributed to an increase in the production of vegetables in this category of farms.

In the Russian Federation, over the past 2 years, a positive trend in vegetable production has been observed. In all categories of farms, the gross yield of vegetables was obtained at the level of 16.3–16.4 million tons in 2016–2017, or 111.5% to the level of 2012. The bulk of commercial vegetables was produced in the open ground. All categories of farms produced 14.7 million tons each in 2016–2017. The main production of vegetables in open ground is concentrated in the Southern, Volga and Central federal districts, where about 65% of vegetables are produced annually from the total production in the Russian Federation. The main production of vegetables in the open ground in the country is concentrated in household farms. The share of open-field vegetable production in household farms is about 70% of the total gross collection. Gross harvest of open-ground vegetables in farms of all categories amounted to 14.7 million tons in 2017, or 112.5% of the 2016 level [3, 7, 8].

A comparative analysis of modern production and economic indicators showed greater stability of large-scale production in conditions of unfavorable economic situation in the vegetable industry and the country. The analysis was performed at the VNIIO in 12 large specialized farms and 12 farms of various specializations in the Moscow and Saratov regions. In small-scale farms, the average annual rates are minimal and varied over the sown area from 60 to 101 hectares, gross yield from 1.6 to 3.1 thousand tons, and crop yields from 17.8 to 38.4 t/ha, In large-scale farms, the average annual rates fluctuated respectively within 396-498 hectares (19.0-32.2 thousand tons, or 47.0-67.0 t/ha). The limit of these indicators in large-scale farms reached the level of 610 hectares in terms of area in 2016, 38.4 thousand tons in gross yield, 77.3 tons/hectare in yield. The profit per 1 hectare of sown area was 77-88 rubles / ha in large specialized farms, 17-34 rubles / hectare in small-scale farms. The average annual profitability was 27-28 and 11.7-20.1%, respectively [8, 9].

In the all-Russian structure of production cost of 1 quintal of outdoor vegetable production, the share of certain types of costs consists of: wages (19%), seeds and planting material (16%), the maintenance of fixed assets (14%), fertilizers and plant protection products (12%), petroleum products (7%), electricity (3%), general and other expenses (19%), on average for the years 2008–2011. The share of mechanisms in the production cost is so small that it is not separately allocated.

At the same time, in large-scale production with the latest technologies in vegetable production, the share of expenditures on machinery and equipment is 33.6% of the total cost, and the salary is 17%. The share of machine and tractor fleet costs in the cost structure of vegetables of the *borshchevaya* group was 26.2% for cabbage, 31% for beetroot, 33.2% for onions, and 44% for carrots.

In the Russian Federation, in recent years, the production of vegetable products has not decreased. The growth in gross harvest of vegetables is relatively steady and increasing (14.6 million tons in 2012 and 16.4 million tons in 2017), an the production profitability fluctuates from 9 to 36%, over the last 10–15 years.

Over 700 hectares of modern high-tech complexes built and put into operation, for the period from 2014 to 2017. State support provided for in the State Program helped to achieve this. As a result, the industry is experiencing an investment boom. The Russian government considers the development of the protected ground industry as a priority for the development of the agro-industrial complex, which provides the population of the country with fresh vegetable products during the off-season period. According to forecasts, the state program of developing the agro-industrial sector provides for an increase in the area of winter greenhouses to 4.7 thousand hectares. The production of domestic greenhouse vegetables increased to 1,720 thousand tons per year by 2020. To fulfill this indicator, about 820 thousand tons of vegetables should be additionally produced, which would require the construction of 1.6 thousand hectares of greenhouses with an average yield of 50 kg/m^2. It is assumed that the demand of the country's domestic market for cucumbers will be fully satisfied. Additionally, the demand of the country's domestic market for tomatoes will be 70–80% satisfied by the end of 2020 [10, 11, 12].

In the structure of vegetable production in the Russian Federation, the products of protected soil have a small share (9.6%). The main producers of greenhouse vegetable products are agricultural organizations. Agricultural organizations grow 52.1% of the total volume of greenhouse vegetables, while household farms has a slower rate (45.4%). The role of peasant (farmer) farms in the production of greenhouse vegetables is insignificant (2.5%). The main production of greenhouse vegetables is concentrated in the following 4 federal districts: Volga, Southern, Central, and Siberian. The share of these districts in all protected soil products is 75.1%.

Winter greenhouses prevail in the production of vegetable. In agricultural organizations, their area is 21,425,000 m^2 (or 68%), the area of spring greenhouses is 9,132 thousand m^2 (29%), the area of insulated soil and greenhouses is 961,000 m^2, or 3% of the total area of protected ground. In winter greenhouses, the highest yield of vegetable crops is 34.4 kg/m^2, it is 8.4 kg/m^2 in the spring greenhouses, and 1.4 kg / m2 in the greenhouses in the insulated soil. In the structure of the production of greenhouse vegetables, the share of cucumbers is 54.4%, and it is 39.7% of tomatoes. The rest of the crop (pepper, radish, green, eggplant, mushrooms) have a small proportion (5.8%). Agricultural organizations produced 922.2 thousand tons of greenhouse vegetables in 2017, which is 13.3% higher than the level of 2016 (813.6 thousand tons) and 49% higher than the level of 2013 (615.0 thousand tons).

In recent years, in the Russian Federation, the production of greenhouse vegetables shows stable growth dynamics in the vast majority of federal districts. The new 5th generation greenhouses are built using the latest technology and equipment; they are not inferior to the best foreign analogues. Due to this trend, the yield in new greenhouses reaches an indicator of 60 kg/m^2, it is 100 kg/m^2 in a photoculture. In general, in the country, the average yield of vegetables in greenhouse complexes is growing steadily, reaching 36.9 kg/m^2 in 2017, which is 31.8% higher than the 2013 level (28 kg/m^2). According to the Ministry of Agriculture of the Russian Federation, the level of self-sufficiency of the Russian Federation of greenhouse vegetables was 53.4% in 2017, which was 7.1% higher than the 2016 level (46.3%) and 13.0% higher than the 2015 level (40.4%) [9].

4. Conclusion

The agricultural organizations supply most of the vegetable products to the country's agri-food market, although the main producers of vegetables are the household farms. The agricultural organizations sold 2,443 thousand tons, or 36.7% of the products in all categories. The households sold 2,410 thousand tons (36.2%, respectively, in 2016). Also, the farms play a significant role in providing the population with vegetables. Their share is 27.1% of the volume sold by the households of all categories. The agricultural organizations and farms have high-quality production. In agricultural organizations, the level of marketability of vegetable production amounted to 79.4%, it was 75.9% in farms. In households, the level of marketability of vegetable production continues to be low, amounting to 23.3%. The Russian households grow vegetables for their own consumption and sell only a small surplus on the food market.

References

[1] Razin A, Taktarova S, and Semenov V 2018 *Innovative and investment development of vegetable growing MATEC Web Conf.* **212** doi: 10.1051/matecconf/201821207010

[2] Soldatenko A V, Pivovarov V F, Razin A F, Meshcheryakova R A, Shatilov M V, Ivanova M I, Taktarova S V, ... Razin O A 2018 Vegetable production economics: state and modernity *Vegetables of Russia* **5**(43) pp 63-68 (doi: 10.18619/2072-9146-2018-5-63-68)

[3] Federal State Statistics Service (Rosstat) (2012 ... 2017) *Indicators of agricultural production in Russia in 2012, 2013, 2014, 2015, 2016, 2017* (Moscow, Russia: Main Interregional Center of Rosstat)

[4] Export and import of Russia by goods and countries (n.d.) Available at: http://ru-stat.com (Accessed 25 03 2019)

[5] Soldatenko A V, Razin A F, Shatilov M V, Ivanova M I, Razin O A, Rossinskaya O V, and Bashkirov O V 2018 Interregional exchange in the context of the alignment of the consumption of vegetables in subjects of the Russian Federation *Овощи России* **6**(44) pp 41-46 (doi: 10.18619/2072-9146-2018-6-41-46)

[6] Litvinov S S, Razin A F, Ivanova M I, Meshcheryakova R A, and Razin O A 2016 State, problems, prospects and risks for the development of vegetable production in Russia under sanctions *Potatoes and Vegetables* **2** pp 25-29

[7] Razin A F, Meshcheryakova R A, Razin O A, Taktarova S V, and Telegina G A 2018 Analysis of the economic efficiency of vegetable production in the Russian Federation *Agro-Industrial Technologies in Central Russia* **1**(7) pp 68-75 (doi: 10.24888/2541-7835-2018-7-68-75)

[8] Razin A F, Shatilov M V, and Razin O A 2018 Vegetable production in the gross domestic product of Russia *Economics of Russian Agriculture* **7** pp 10-11

[9] Ministry of Agriculture of the Russian Federation 2016 *National report on the progress and results of implementation in 2015 of the State Program for the Development of Agriculture and Regulation of Agricultural Products, Raw Materials and Food Markets for 2013–2020* Available at: https://gov-news.ru/ (Accessed 25 03 2019)

[10] Council of Federation 2018 *Materials of the Round Table of the Committee of the Council of the Federation on Agrarian and Food Policy and Environmental Management on the topic "Ways of increasing self-sufficiency of vegetables in the Russian Federation"* (Moscow, Russia).

[11] Muravyev AK 2010 Protected ground of the Russian Federation: state, problems, prospects, ways of development *Greenhouses of Russia* **1** pp 24-25

[12] Litvinov S S, Nurmetov R J, Alekseeva K L, Razin A F, and Selivanov V G 2015 *Facilities, technologies and technical means for the production of vegetables in greenhouses* (Moscow, Russia: Methodical recommendations, publishing house Rosinformagrotekh)

IOP Conf. Series: Earth and Environmental Science **395** (2019) 012063 doi:10.1088/1755-1315/395/1/012063

On increasing the sorption capacity of semi-cokes with acid regenerates of water-desalting ion-exchange plants

A Gorelkina[1], I Timoshchuk[1*] and N Gora[1]

[1] Kemerovo State University, 47 Stroitelei blvd, Kemerovo 650056 Russia

E-mail: ecolog1528@yandex.ru

Abstract. In modern conditions, the problem of environmental protection has become one of the decisive factors determining the further development of mankind, along with technological and economic aspects. The approach is most effective when disposing the waste from one production provides an effective solution to the problem of environmental protection in other industries. The article investigates the possibility of using the acidic waste generated from the operation of ion exchange plants at power plants to increase the sorption capacity of carbon sorbents for phenol by changing the chemistry of the surface of coal. It has been established that an increase in the sorption capacity of active carbon ABG provides additional adsorption and increases the efficiency and cost-effectiveness of wastewater treatment from phenol.

1. Introduction

Currently, environmental protection is one of the significant factors influencing the further development of the whole mankind. In recent years, problems related to the pollution of the hydrosphere have worsened. The discharge of untreated or insufficiently treated wastewater containing toxic organic compounds, including phenols, into reservoirs leads to a decrease in biodiversity and even the disappearance of life in ecosystems. Development and introduction of low-waste and non-waste technological processes into practice is one of the ways to solve the problem of environmental protection and resource conservation. These technological processes ensure the utilization of wastewater from one production to increase the efficiency of another production.

Phenols are priority pollutants in the aquatic environment. They enter water bodies with wastewater from wood-chemical, aniline-dyeing, oil-refining, chemical industries (production of pesticides, organic dyes, varnishes, explosives, synthetic resins, plastics). The concentration of phenol in wastewater can reach 10 g/dm^3. In addition to phenols of technogenic origin, natural waters contain low molecular weight phenols, which are formed during the biochemical decomposition and transformation of organic substances (humic and fulvic acids). Phenol is a common protoplasmic poison and is toxic to all cells. Phenol leads to disruption of the gastrointestinal tract, liver and kidney diseases, it also is the cause of cancer [1].

Methods of phenol extraction from wastewater used in practice have a number of disadvantages: significant costs of chemicals, heat and electricity are required; the use of expensive equipment, high temperatures are often accompanied by the formation of secondary pollution. The solution to the problems of wastewater treatment is the development and widespread introduction into practice of new, reliable and highly economical technological processes.

Content from this work may be used under the terms of the Creative Commons Attribution 3.0 licence. Any further distribution of this work must maintain attribution to the author(s) and the title of the work, journal citation and DOI.
Published under licence by IOP Publishing Ltd

For wastewater treatment of organic compounds, adsorption on activated carbons (AC) is often an effective method. Active carbons are quite effective and inexpensive materials. A huge variety of carbon materials exists, differing in raw materials and activation method. Today, a new class of cheap sorbents has appeared - semi-cokes.

Earlier in our research [2], [3], we found that semi-cokes have low sorption capacity. Increasing the sorption capacity of the adsorbent with respect to the extracted component can be achieved by changing the structure and state of the surface of the adsorbent when using modifiers of various types [4]. Solutions of organic reagents, bases, mineral acids, for example, sulfuric acid, are used as modifiers.

Sulfuric acid is the main component of wastewater generated at power plants (SDPP, HPP) using ion exchange resins for cleaning large volumes of water and their repeated regeneration. The volume of acidic waste from the operation of ion-exchange units on cation exchanger is about 20-25% of the volume of desalted water. Basically, their composition includes sulfuric acid, the average concentration of which is 0.5 mol / dm^3 and Ca^{2+} (average concentration of 0.016 mol / dm^3).

The purpose of the study is to identify the fundamental possibility of solving the problem of environmental pollution by reducing the amount of wastewater from ion-exchange plants used in power plants through the use of acidic wastewater, with the purpose of increasing the sorption capacity of AC in phenol.

2. Materials and Methods

As objects of study, the semi-coke of brand ABG is used in our research. The ABG is a coal-gas carbonized carbon sorbent, which is selected. It is obtained from coal raw materials (semi-coke) by the manufacturer CJSC "Karbonika-F", Krasnoyarsk. This semi-coke is produced from the coal mined in the Kansko-Achinsky open pit. The unique technology of its receipt provides low costs.

The study of phenol adsorption from aqueous solutions in the concentration range from 0.001 to 1000 mg / dm^3 was carried out in static conditions using the AC ABG.

The ABG was additionally treated with acidic regenerants of ion-exchange units at a ratio of the mass of the sorbent in grams to the volume of the regenerate (in cm^3, 1:5, 1:10, 1:15) for 6-12 hours, followed by washing with 3 times the volume of demineralized water. The content of phenol in aqueous solutions after adsorption was determined by the method of molecular absorption spectroscopy.

3. Results

According to the obtained experimental data on the phenol adsorption from the AC ABG solution, the adsorption isotherms were built (Figure 1) and the values of phenol adsorption parameters from aqueous solutions were calculated: using the theories of monomolecular adsorption (Freundlich and Langmuir equations), the theory of volume filling of micropores (Dubinin-Radushkevich equation), and the generalized theory of the polymolecular adsorption developed by Brunauer, Emmett, and Teller (table 1) for industrial and modified sorbent wastewater from ion-exchange plants of thermal power plants.

It is established that the presence of Ca^{2+} salts in the regenerate does not affect the efficiency of the treatment of AC and the subsequent adsorption of organic substances. The precipitate of $CaSO_4$ does not fall out with this content of components in the regenerate, since the value of the solubility product is not reached ($PR(CaSO_4) = 2{,}25*10^{-4}$).

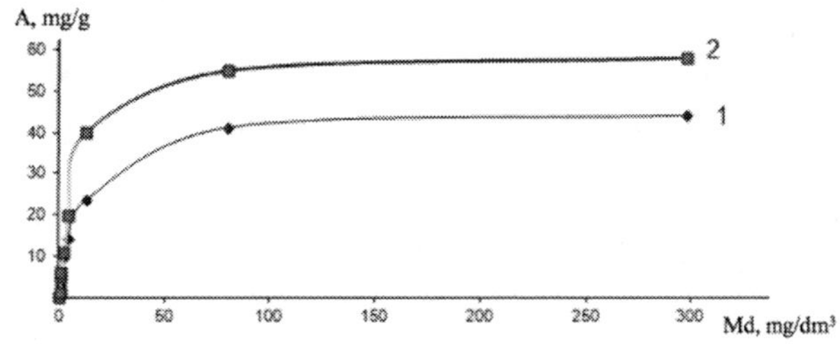

Figure 1. Adsorption isotherms of the phenol AC ABG industrial (1), modified by acid regenerates of a thermal power station (2).

Table 1. Parameters of phenol adsorption from AC ABG aqueous solutions.

Type of equation		ABG	ABG (H_2SO_4)
Freundlich	1/n	0.3536	0.344
	b, mmol / g	0.069	0.098
Langmuir	a_m, mmol / g	0.524	0.733
BET	a_m, mmol / g	0.524	0.714
	-Q, kJ / mol	14.734	13.987
Dubinin-Radushkevich	a_{max}, mmol / g	1.064	1.330
	E_0, kJ / mmol	13.445	12.950
	W, dm^3/ kg	0.0935	0.1140

On the sorbent ABG treated as described, the adsorption of phenol was carried out from aqueous solutions under static conditions. The effect of the ratio (AC mass: a regenerate volume) on the sorption capacity of AC with respect to phenol is presented in Table 2.

Table 2. The effect of the ratio (AC mass: regenerate volume) on the AC sorption capacity.

No.	The ratio of m (g): Vp (cm^3)	Increasing the sorption capacity of AC,%
1	1:5	16
2	1:10	33
3	1:15	35

The effect of the duration of the process on the sorption capacity of the AC with respect to phenol is established: treatment of active carbon with an acid regenerate for 6 hours increases the sorption capacity by 33%, 8 hours by 48%, 10 hours by 54%, 12 hours by 55%.

4. Discussion

The presented adsorption parameters (Table 1) show that surface modification allows to obtain an adsorbent with satisfactory characteristics close to active coal (AG-3, AG-OV-1, SKD-515) widely used for wastewater treatment and at drinking water treatment stations.

The magnitude of the limiting adsorption volume W suggests that the adsorption of phenol obeys the bulk mechanism of filling the pores. The values of the characteristic energy indicate that the sorption of phenol is mainly in the micro- and mesopores of the adsorbent. The adsorption isotherm of phenol AC suggests the physical nature of adsorption (Table 1). The calculated value of the heat of adsorption confirms the conclusion about the physical nature of the interaction of phenol with the ABG surface.

The increase in the adsorption capacity for phenol is established. This is due to the appearance on the surface of the sorbent of new oxygen-containing functional groups (OFG), which are additional adsorption centers. Because of this, the amount of phenol interacting through a specific adsorption mechanism due to hydrogen bonds between OFG sorbents and the OH group of phenol increases. The degree of purification increases.

To determine the effect of modifier concentrations on the adsorption of organic substances, the adsorption of phenol from aqueous solutions was studied on the initial and acid-modified samples. It is established that with an increase in the concentration of sulfuric acid of more than 2 mol / dm^3, the adsorption capacity of the modified samples with respect to phenol decreases. The combination of the obtained results allows us to conclude that the modification of carbon sorbents with acid allows one to increase the adsorption of phenol by 1.5 times. As a result of the research, the following conditions for the disposal of acid regenerates of water desalting plants were chosen: the treatment of industrial active carbons with acidic regenerate at a ratio of mass AC in grams to the volume of regenerate in cm^3 1:10 for 10 hours, followed by washing the volume of water desalinated by ion exchange no less than 3 times.

The studied semi-coke has the following characteristics: the possibility of increasing the adsorption capacity, low production costs and transportation costs, minor costs for modification, and the possibility of effective utilization. (This semi-coke does not require regeneration due to the fact that the spent sorbent can be used to produce marketable products at the "Koks" OJSC, Kemerovo). We can conclude that in the Kuzbass region of Russia, the use of semi-coke to extract phenol from natural and waste waters is advisable.

5. Conclusion

Treatment of activated carbons with acid regenerate of water desalting plants allows utilization of acid regenerate of ion-exchange plants, reducing the volume of wastewater.

Treatment of sorbents with acid regenerates of water desalting plants leads to the formation of secondary adsorption centers on the surface of active carbons. They are able to interact with functional groups of organic substances due to a specific adsorption mechanism, providing them with additional adsorption, increasing the adsorption capacity of active carbons and the efficiency of wastewater treatment from phenol.

Thus, two tasks are simultaneously solved: an increase in the efficiency of purification of some industrial wastes (wastewater containing phenol) due to the utilization of another waste (acid regenerate of ion-exchange plants).

6. Acknowledgments

The study was carried out in the framework of the State task "Initiative Research Projects" 19.4713.2017/БЧ

References

[1] Guseva T V, Molchanova Ya P, and Zaika E A 2007 *Hydrochemical indicators of the state of the environment* (Moscow, Russia: Forum: INFRA-M)

[2] Krasnova T A, Timoschuk I V, Gorelkina A K, and Dugarjav J 2017 The choice of sorbent for adsorption extraction of chloroform from drinking water *Foods and Raw materials* **2** pp 189-196

[3] Krasnova T, Kirsanov M, Belyaeva O, and Timoshchuk I 2018 Adsorption processes in solving environmental problems in water sources of Urban Territories Development *MATEC Web of Conferences* **170** 04005

[4] Tarkovskaya I A 1981 *Oxidized coal* (Kiev, USSR: Naukova Dumka)

Ensuring energy security: global trends and Russian practice

L P Goncharenko[1], E G Lisovskaya[1] and N V Sedova[1*]

[1] Plekhanov Russian University of Economics, 36 Stremyanny lane, Moscow 117998 Russia

E-mail: nadseva@mail.ru

Abstract. The article describes the latest global trends in the field of energy security and energy efficiency. The experience of developed countries in energy efficiency, as well as experience and dynamics in the Russian Federation, are analyzed.

1. Introduction

Now, energy is able to meet the demand that humanity imposes on it, but the ever-growing need for energy resources, accompanied by rising energy costs, environmental degradation and geopolitical instability, are pushing countries to switch to energy-saving technologies and energy efficiency. Success in this area varies depending on the area and country. In general, statistics suggest that if no energy saving programs were implemented, the world economy would use 12% more energy during the period from 2000 to 2016, which is comparable to the total demand of the European Union countries [1], [3].

The International Energy Agency believes that improving energy efficiency is one of the key reasons for reducing greenhouse gas emissions into the atmosphere. The transition to renewable energy and alternative fuels managed to outweigh the negative effect of economic growth on greenhouse gas emissions.

2. Materials and Methods

A methodical approach to the development of proposals for the implementation of energy efficiency policies based on the optimization of production capabilities and market requirements of the region is used in the work. The studies are based on an understanding of the development trends of the international energy system and the determination of the parameters of the problem sectors of the energy-saving technologies market. They are based on conducting a comprehensive analysis of the state of the international energy security system, identifying existing problems and developing proposals for optimizing production capabilities and taking into account the needs of the energy market in the region.

3. Results

The introduction of energy-saving technologies had a positive effect not only on the environment. In terms of energy security, many countries have significantly improved their positions. Countries such as Germany and the UK are now reporting a higher ability to cope with peak loads on the grid and less dependence on energy imports [2].

On average in the world, consumers also benefited from improved energy efficiency. Households now save between 10% and 30% of energy costs, and this effect is noticeable not only in developed but also in developing countries, such as China, where this indicator is 25%.

Content from this work may be used under the terms of the Creative Commons Attribution 3.0 licence. Any further distribution of this work must maintain attribution to the author(s) and the title of the work, journal citation and DOI.
Published under licence by IOP Publishing Ltd

IOP Conf. Series: Earth and Environmental Science **395** (2019) 012064 doi:10.1088/1755-1315/395/1/012064

The industrial sector of the economy throughout the world is actively adapting innovative approaches to energy management and introducing them into production. From 2000 to 2016, energy use per unit of economic output has decreased by 20%. Moreover, a further decrease in this indicator is projected as the introduction of new production facilities that meet energy efficiency standards occurs, and old plants and factories drop out. The combined use of financial incentives and government coercion has led to large-scale adaptation of energy management systems, which have helped enterprises save about 10% of their energy costs.

The field of construction and housing and communal services also began to change after other sectors of the economy. New buildings are equipped with technologies to reduce their consumption of energy resources. However, in this area, energy efficiency has not yet been able to overcome the growth of space. And there is great potential for improving the situation through the introduction of new and tightening of existing construction standards.

The automotive market is on the verge of significant change. For many years, government policy in this market focused exclusively on passenger cars, however, now, a policy of encouraging more efficient use of fuel in trucks is being developed. There is an increase in sales of electric cars in the car market. Despite the slowdown in 2016, this growth gives hope for the gradual ousting of gasoline engines by electric ones, although this trend is under threat. Because more often, consumer demand is growing for less energy-efficient car models: large passenger cars and SUVs.

The development of energy-saving technologies continues to experience an increasing inflow of investments from year to year. The leaders in investment growth are China (24%) and European countries (30%) [4].

Energy service companies continue to demonstrate their expansion in the market. The market for their services was estimated at $ 26.8 billion in 2016. 60% of which are in China with Europe and the USA, they are in second and third place with shares of 20% and 10% respectively.

Regulators are gradually finding new ways to create a market mechanism for the efficient allocation of energy efficient technologies, which increases the investment attractiveness of the industry and increases competition in it.

4. Discussion

Despite these positive changes in energy efficiency, there are a number of factors causing concern to regulators and international organizations. Thus, 68% of total energy use is not subject to energy efficiency programs. The share of energy resources used outside of these programs is growing annually, but in recent years, this growth is achieved more by continuing to comply with existing standards than introducing new ones. The tightening of standards is also slowing, although some states, such as Denmark and Germany, continue to raise energy efficiency standards at a stated rate. Overall, the energy efficiency improvement program index of the International Energy Agency showed the following. There is a slowdown in the improvement and dissemination of government energy efficiency programs: its growth in 2016 was below the average over the previous 15 years [5].

In Russia, the dynamics in the field of energy efficiency is significantly different from the world. Russian Ministries and Enterprises use in their reports indicators that are very different from those used in the member countries of the International Energy Agency. So, in Russia, just the specific use of the final energy per conventional unit of production is estimated, whereas in world practice, the final use of energy per unit of gross value added is taken to be measured. In some areas of the economy, the relevant agencies have announced plans for the transition to global measurement standards, but this has not happened yet. In Russia, at the moment, a number of extremely important economic indicators for calculating the energy efficiency of the industry / sub-sector of the economy do not have accounting indicators. As an example, we can cite the fact that in the field of transport there is no information on the consumption of energy resources by cars, measurements are carried out only on regular bus transport of people and freight traffic. The data cited by the Ministry of Energy of the Russian Federation with reference to experts indicate that in the Russian Federation there is a decrease in the consumption of energy resources due to modernization and a slowdown in the decline in GDP [6].

On the one hand, a comparison between world and Russian practice is not possible due to the use of different indicators and units of measurement. On the other hand, we can look at the components of the state policy of the Russian Federation in the field of energy efficiency. Since 2012, work on creating an effective management system in this area has been underway. The result of this work was the inclusion of energy efficiency indicators in 5 of 13 sectoral government programs [7]. At the level of subjects, 63% of targeted state programs contain energy efficiency indicators, and each region has at least one such program. The state information system "Energy Saving" was installed at 94% of budget institutions to assess energy consumption. A system of financial incentives for the development and implementation of energy-saving technologies, including simplified access to financial resources and tax incentives, was created. The Ministry of Energy has resorted to cooperation with the Ministry of Culture and partners from the private sector to hold the festival *#VmesteYarche*, to increase public awareness of energy saving methods and available technological solutions in this area.

5. Conclusion

Russia is actively developing international cooperation on energy efficiency, in particular, the country became a member of the *Global Construction Alliance* initiative held in the framework of the Paris Climate Conference in 2015. The Ministry of Energy identifies the following areas for further work on energy efficiency:

1. Continuing the practice of decommissioning inefficient heat generating capacities with specific fuel consumption well above average;
2. The introduction of effective mechanisms to regulate the heat market, aimed at encouraging an increase in the share of heat sources operating in the combined generation mode in heat load;
3. The introduction of advanced energy-efficient technologies in construction and overhaul in the public and residential sectors, as well as a large-scale transition to energy-efficient LED lighting in both internal and external lighting.

Both in Russia and in the world, work on improving energy efficiency is carried out in three key sectors: in industry, construction and housing and utilities, as well as in transport.

To measure industrial energy efficiency, the International Energy Agency uses two indicators: energy intensity and energy performance. The first indicator is calculated as the final energy consumption per unit of gross value added, and the second as value added per unit of energy consumed. Thus, the transition to energy efficient production is the task of maximizing energy performance, accompanied by minimizing the energy intensity of production.

6. Acknowledgments

This article was prepared as part of the project section of the government contract as requested by the Ministry of Science and Higher Education of the Russian Federation on the subject formulated as 'Development of Methodological Principles and Organizational Economic Mechanism of Strategic Management of Economic Security in Russia'. (Assignment No. 26.3913.2017/4.6).

References

[1] World Energy Outlook 2017 *Energy efficiency and renewable energy outlook* (Paris, France: OECD Publishing)

[2] European Parliament and of the Council 2012 *Energy Efficiency Directive (October 25, 2012*No. 2012/27/EU, on energy efficiency, amending Directives 2009/125/EC and 2010/30/EU and repealing Directives 2004/8/EC and 2006/32/EC) Available at: http://eur-lex.europa.eu/LexUriServ/LexUriServ.do?uri=OJ:L:2012:315:0001:0056:en:PDF (Accessed 02 03 2019)

[3] European Parliament and of the Council 2017 EU Regulation On energy labeling (July 4, 2017 No. 2017/1369) Available at: http://eur-lex.europa.eu/eli/reg/2017/1369/oj (Accessed 02 03 2019)

[4] Zindler E, and Locklin K 2016 *Mapping the gap: the road from Paris: finance paths to 2°c future* (NY, NY: CERES: Bloomberg New Energy Finance)

[5] Lauf E, Pillen N, and Stenzel B 2007 *Energy-efficient construction in Germany* (Berlin, Germany)

[6] Ministry of Energy of the Russian Federation 2014 *Order of the Ministry of Energy of Russia "On approval of the methodology for calculating the values of target indicators in the field of energy saving and energy efficiency, including in comparable conditions"* (June 30, 2014 No. 399) Available at: http://www.garant.ru/products/ipo/prime/doc/70609922/ (Accessed 02 03 2019)

[7] State Duma of the Russian Federation 2009 Federal Law *"On energy saving and energy efficiency improvement, and on amendments to certain legislative acts of the Russian Federation"* (November 23, 2009 No. 261-FZ) *Rossiyskaya Gazeta* **5050**

The environmental monitoring and management of urban greenery using GIS technology

L Trubina, O Nikolaeva[1*] and P Mullayarova[1]

[1] Siberian State University of Geosystems and Technologies, 10 Plakhotnogo str., Novosibirsk 630108 Russia

E-mail: onixx76@mail.ru

Abstract. The main provisions of using GIS technologies for the environmental monitoring of urban green spaces (UGS) are described. The main factors of anthropogenic negative impact on the UGS are characterized. The paper considers the necessity of using modern methods of the Earth remote sensing and geospatial mapping for the inventory, assessment, and monitoring of environmental conditions of UGS. The principles of geoinformation modeling of UGS are characterized. The definition of UGS geospatial model is given, the tasks solved with its help are listed, the considered parameters of UGS are provided by the authors. The main stages of UGS geospatial modeling are described. A fragment of digital greening plan of Novosibirsk is presented. The paper concludes about the importance of geospatial modeling and mapping for the effective organization of environmental monitoring of UGS.

1. Introduction

The urban green spaces (UGS) are an important component of the urban environment, ensuring environmental safety and environmental comfort of the population [1]. At the same time, they are subjected to constant anthropogenic pressure, which is expressed in the following phenomena:

- Accumulation of pollutants in plant tissues due to the pollution by industrial plants and road transport of the urban air basin and soil;
- Deterioration of plant nutrition due to the deposition of dust, soot, and other suspended substances on the aboveground parts of plants;
- Degradation of plant root systems as a result of soil compaction under the vibration effect of road transport;
- Violation of the integrity of tree crowns and soil cover as a result of industrial and recreational activities (construction and road works, lawn trampling, illegal parking, etc.).

The optimal state of the UGS is an important factor contributing to the improvement of the ecological condition of the city. Therefore, taking into account the wide range of impacts to which they are exposed, the condition of the UGS should be monitored and regularly assessed from a geoecological point of view. This will provide the city management bodies with relevant and reliable information on the state of the UGS and, making it possible to optimize care activities.

The environmental monitoring of the UGS should include the following actions [2, 3]:

Content from this work may be used under the terms of the Creative Commons Attribution 3.0 licence. Any further distribution of this work must maintain attribution to the author(s) and the title of the work, journal citation and DOI.
Published under licence by IOP Publishing Ltd

- Country-wide inventory of the UGS, providing for the collection of information on individual landscaping objects (trees, groups of shrubs, flowerbeds and lawns), the totality of which forms the UGS;
- Assessment of the ecological condition of the UGS in general and of individual landscaping objects (especially high-growth vegetation) within their borders;
- Formulation of care recommendations for existing the UGS;
- Planning of the new UGS taking into account their desired location, species composition recommended for this climate, geographical features of the area, regulatory requirements for the territorial planning of cities, financial and aesthetic value of different plant species.

The officially adopted method of inventory of green plantings is marked by methodological and technological obsolescence, since it does not provide for the use of modern capabilities of computer equipment and technology [4].

2. Materials and Methods

The modern technologies of Earth remote sensing (ERS) and geo-information mapping provide all the possibilities for collecting highly detailed information that is spatially referenced on any objects of the earth's surface. These technologies provide opportunities for collecting information including on the boundaries of the UGS and the location and condition of individual trees within their limits. Relying on the modern achievements of photogrammetry and digital cartography allows us to formulate the following principles of inventory, assessment and mapping of the UGS:

- Comprehensive use of freely distributed data of ERS to determine the main spatial, qualitative and quantitative parameters of the UGS and individual landscaping objects (trees, shrubs, flower beds, etc.) within their borders;
- Systematization, analysis, and modeling of the collected data in the GIS environment, taking into account the available geospatial information about other objects of the urban environment (buildings, engineering structures, communications, etc.);
- Assessment of the ecological state of the UGS and landscaping facilities within its borders, carried out on the basis of the results of the inventory of the UGS on materials of ERS and field surveys;
- A visual representation of the results of the inventory and assessment of the ecological status of the UGS in the form of digital cartographic materials that provide usability with the new knowledge of urban greenery, including the implementation of visual analysis that does not require special technical and technological support.

Based on the above principles, the structure and the sequence of creating the geo-information model of the UGS were developed, which lays the informational basis for monitoring and rational management of the state of the UGS.

3. Results

The geo-information model of the UGS is an ordered set of features reflecting the current state of the UGS and individual landscaping objects within its borders. It provides the following tasks [3, 5]:

- An inventory of all individual landscaping objects (to the level of individual trees and shrubs), including the determination of their spatial position;
- Creation, development, and updating of the database, which contains detailed semantic information about the qualitative and quantitative parameters of individual landscaping objects;
- Assessment of the ecological status of individual landscaping objects on the basis of the collected data, documenting the assessment results in the form of reports;
- Making recommendations for the care of each specific object of landscaping and for the green area as a whole, planning the placement of new UGS and adjusting the area and structure of existing spaces.

In the process of geoinformation modeling of the state of the UGS, the collection of information not only about the UGS itself but also about the urban environment containing them is important to ensure. Information on engineering objects can be obtained from digital maps used by municipal urban management bodies. In addition, information on the development and cadastre of urban areas can be specified and verified using the Rosreestr Public Cadastre Map.

For a more complete understanding of the causes of negative environmental changes in the UGS, their condition should be considered in the context of the general environmental situation of the city. However, it should be noted that the data of state monitoring of the state of the environment conducted by the Roshydromet are determined on a limited number of fixed and mobile posts [6]. And the collection of information on the pollution of the atmosphere by emissions from road transport is not provided for by the current governing documentation and can be carried out only with the involvement of non-governmental organizations and funds, which will lead to higher cost of work [7-9]. The current program of state monitoring of soil pollution with heavy metals provides only for sampling in industrial areas [10].

The general structure of the database of the UGS geoinformation model includes the following main blocks:

- Data on the urban environment (buildings, driveways, engineering communications, etc.);
- Data on the ecological status of the territory (air pollution, pollution of soil and snow cover of the UGS);
- Data on UGS and individual landscaping objects within their borders.

The following information is recorded in the last block of the database [3,5]:

- For lawns: a lawn type, an area, dominant grass types, a total projective cover;
- For trees: a type of planting (includes a tree), a registry number of the tree, a breed, the age, ` diameter, its height, a quality condition;
- For shrubs: a type and number of shrub planting, a type and number of bushes in planting, the age, the height of bushes, their length and width of planting, a quality condition;
- For flower beds: a type of flower bed, the perennial culture, an area, a quality condition.

The general technological sequence of creating a geo-information model of the UGS includes the following steps:

- Receiving freely distributed materials for ERS on the territory within which the inventory and assessment of the UGS is performed;
- Formation and filling of the database of the UGS geo-information model and the creation of vector thematic layers reflecting the UGS and objects of landscaping within their borders.

The main part of the work consists in deciphering the UGS and individual landscaping objects using ERS data. Decoding allows one to set parameters such as the spatial position and height of trees and shrubs, the configuration and area of lawns and flower beds, the degree of disturbance of the projective cover of the lawn. Information about the species of trees and shrubs, the nature and degree of disturbance of their crowns, forest pathological characteristics are established on the materials of panoramic street photography [4]. The UGS geo-information model is supplemented with information on the environmental situation, either by importing digital layers or statistical databases that already contain environmental information, or by creating an appropriate database and digital layers "from scratch" associated with it.

- Assessment of the ecological state of the UGS and individual landscaping objects (especially trees) within its borders. The assessment is performed in several stages using a specially developed mathematical apparatus. [11];
- Displaying the results of the inventory and assessment of the ecological condition of the UGS and individual landscaping objects in the form of digital landscaping schemes, which are the final product provided to specialists in the field of greenery management.

A Fragment of a similar scheme created in Mapinfo GIS on a landscaped area along the street Titova (Novosibirsk), is shown in Figure 1.

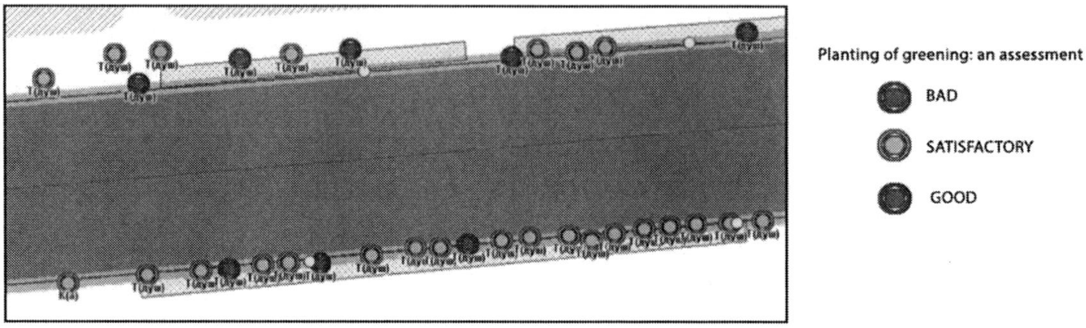

Figure 1. A fragment of a digital landscaping scheme (Titova street, Novosibirsk, Russia).

4. Discussion

The environmentally and economically efficient design of UGS involves a comprehensive analysis of the spatially distributed data on the characteristics of the placement of green spaces and indicators of their structure and condition. However, an analysis of modern processes for managing urban greening [12–15] revealed that the inventory of a state-owned plant is often carried out to a limited extent due to a shortage of man-hours and financial support. As a matter of fact, the urban management authorities identify and control several "hot spots", which are the largest and most visited greening sites.

The implementation of the proposed UGS geo-information model will ensure the collection and comprehensive analysis of information on the UGS condition throughout the city while reducing the volume of field work. Its content occurs both due to the exchange of information with the existing digital cartographic sources (on-duty cadastral maps, plans, etc.), as well as by obtaining information about UGS with freely distributed ERS data. This ensures the formation of information arrays, sufficient for the effective management of the UGS, by office methods and a visual representation of the information obtained on digital maps. The developed technological sequence of geoinformation modeling and mapping of the UGS ensures the creation of cartographic support of measures for monitoring the condition and caring for the UGS.

5. Conclusion

In the modern era, the effective management of urban greenery is impossible without the use of geo-information and computer technologies. To optimize the process of managing the UGS, a geo-information model has been proposed, which is the main tool for the inventory and monitoring of the UGS. The geo-information models created on street gardening sites and the Monument of Glory Square in Novosibirsk have established themselves as an effective means of systematizing and visualizing information on the state of the UGS. They can serve as a reference tool in the development of measures for the care of greenery in specific areas.

References

[1] Trubina L K 2012 *Methodological aspects of the environmental assessment of urbanized areas* In Interexpo GEO-Siberia-2012: a collection of materials of the International Scientific Conference *"Remote sensing of the Earth and photogrammetry, environmental monitoring, geo-ecology"* (pp 200-203) (Novosibirsk, Russia: SSGA)

[2] Trubina L K, Baranova E I, and Chagina G S 2013 Geoinformational mapping and inventory of green spaces In Interexpo GEO-Siberia-2013: a collection of materials of the International Scientific Conference *"Remote sensing of the Earth and photogrammetry, environmental monitoring, geo-ecology"* (pp 82-86) (Novosibirsk, Russia: SSGA)

[3] Trubina L K, Mullayarova P I, Baranova E I, and Nikolaeva O N 2014 Some approaches to the

IOP Conf. Series: Earth and Environmental Science **395** (2019) 012065 doi:10.1088/1755-1315/395/1/012065

geoinformation mapping of greenery In Interexpo GEO-Siberia-2014: a collection of materials of the International Scientific Conference *"Remote sensing of the Earth and photogrammetry, environmental monitoring, geo-ecology"* (pp 68-74) (Novosibirsk, Russia: SSGA)

[4] Mullayarova P I 2017 On the need to improve the methods of inventory of urban green spaces In Interexpo GEO-Siberia-2017 XIII International Scientific Congress: International Scientific Conference *"Remote sensing of the Earth and photogrammetry, environmental monitoring, geo-ecology": a collection of materials in 2 vols.* (pp 180-185) (Novosibirsk, Russia: SGUGIT)

[5] Trubina L K, Nikolaeva O N, and Mullayarova P I 2017 Inventory of urban greenery with GIS tools *Vestnik of SGUGIT* **22** pp 107–118

[6] *Method of calculating concentrations in the air of harmful substances contained in emissions of OND-86 enterprises* Available at http://docs.cntd.ru/document/1200000112 (Accessed 25 03 2019)

[7] Regulations on the state observation network for building a computerized system for monitoring emissions of motor vehicles 2017 *Fire Safety: Problems and Prospects* **1** pp 467–468

[8] Yakimov M R, Petrov V Yu, Petukhov M Yu, and Esipova M Yu 2004 Environmental problems of the automobilization of large Russian cities In *Environmental issues: a collection of scientific papers* (pp 22-30) (Vienna, Austria: Publishing House of the Vienna Technical University)

[9] Nedre A Yu, Azarov V N, and Nedre Yu A 2012 Use of summary calculations of the levels of air pollution when choosing town-planning solutions as part of optimizing the urban transport scheme *Internet Vestnik of VolgGASU (Ser.: Polythematic)* **2** Available at http://vestnik.vgasu.ru/attachments/NedreAzarovNedre-2012_2(22).pdf (Accessed 25 03 2019)

[10] *Organization and procedure for the observation of soil pollution by toxicants of industrial origin RD 52.18.718-2008* Available at http://docs.cntd.ru/document/1200081126 (Accessed 25 03 2019)

[11] Mullayarova P I, Nikolaeva O N, and Trubina L K 2018 Geoecological assessment and mapping of the state of green areas of special purpose *Vestnik of SGUGIT* **23** pp 262-274

[12] Kovalchuk, AG, Sokolov, RA, Bukharina, IL, and Vedernikov, KE 2017 On the organization of the management of the green fund of the city *Biological Sciences* **1** pp 8-13

[13] Teodoronsky V S 2010 *Landscaping populated areas: urban planning foundations* (Moscow, Russia: "Academy" Publishing Center)

[14] Filimonenko L A 2006 *Engineering improvement of urban areas and transport* (Chelyabinsk, Russia: SUSU Publishing House)

[15] Varzareva V G, Trusheva N A, Peredelsky N A, Fedorovskaya M G, Sazonets N M, and Udzhukhu M I 2016 Problems of greening the cities of Southern Russia on the example of Maykop *Actual Problems of the Forest Complex* **44** pp 154-159

IOP Conf. Series: Earth and Environmental Science **395** (2019) 012066 doi:10.1088/1755-1315/395/1/012066

An investigation of the process of producing the snow carbon dioxide

E N Neverov[1*], I A Korotkiy[1] and E V Korotkaya[2]

[1] Department of Thermal Refrigeration, Kemerovo State University, 6 Krasnay str., Kemerovo 650043 Russia
[2] Department of Analytical Chemistry and Ecology, Kemerovo State University, 6 Krasnay str., Kemerovo 650043 Russia

E-mail: neverov42@mail.ru

Abstract. Currently, the Russian and foreign scientists are actively involved in the development of technologies for the processing and utilization of carbon dioxide. At the Kemerovo State University, the technologies for processing and application of carbon dioxide in the food industry have been developed. One of them is based on throttling the CO_2 coater in the generator. In order to determine the optimal operating modes of the generator, throttling is performed through washers with different throttle holes and their configuration. Also, throttling is performed with different angles of cones of snow formers at different temperatures of liquid carbon dioxide, which allows to reveal the influence of the above factors on the process of carbon dioxide snow formation. Based on the analysis of the obtained dependencies, we can conclude that the largest volume of snow weighing 0.25 kg was obtained using a cone with an angle of 12° and a temperature of minus 10°C. In percentage terms, 31.25% of the CO_2 consumed liquid phase was reworked into snow. The additional lowering of the temperature of the liquid carbon dioxide, as well as the choice of the optimum, for a given temperature, the angle of the bell, have a significant impact on the volume of the snow carbon dioxide produced. However, in the process of choosing the temperature control parameters, possible additional economic costs of creating conditions for lowering the temperature of liquid carbon dioxide should be considered.

1. Introduction

In recent years, the Russian and foreign scientists have been actively involved in the development of technologies for the processing and utilization of carbon dioxide. This corresponds to the agreement beginning to operate from 2020, the purpose of which is to limit the growth of the average temperature on the planet by 2°C.

The Ministry of Natural Resources and the Ministry of Economic Development are developing a methodology for reporting greenhouse gas emissions for enterprises. Until the end of 2019, those companies and enterprises whose emissions exceed 150,000 tons of carbon dioxide, an equivalent for the year, will have to ensure the transfer of emissions reporting to the Rosprirodnadzor agency. By the beginning of 2020, such information will already provide all production with annual emissions of more than 50,000 tons. After that, a data verification and reporting system will be created and financial impact instruments introduced (such as tax charges).

Content from this work may be used under the terms of the Creative Commons Attribution 3.0 licence. Any further distribution of this work must maintain attribution to the author(s) and the title of the work, journal citation and DOI.
Published under licence by IOP Publishing Ltd

Thus, in the future, the demand of industrial enterprises for the methods used for utilizing greenhouse gases, including carbon dioxide produced by alcohol processing enterprises, is expected. The promising areas for the use of waste carbon dioxide: chemical and petrochemical industries; pharmaceutical and biotech; food industry and agriculture; medicine; metallurgy and engineering; laboratory research; pulp and paper industry; electronics and ecology [3].

2. Materials and Methods

At the Kemerovo State University, a technology for the carbon dioxide processing in the food industry has been developed. To study the process of obtaining snow-like carbon dioxide, a generator-snowmaker was developed. A simplified diagram of which is presented in Figure 1. Its principle of operation is based on throttling the liquid CO_2 phase and achieving a pressure lower than at the triple point, since the snow produced in such conditions is not to be wet and loose [2].

The process of throttling the snow-like CO_2 in the generator-snow skimmer occurs through throttle washers having different diameters of the bore. In addition, the angle of the snow former cone and the temperature, at which the throttling is performed, change.

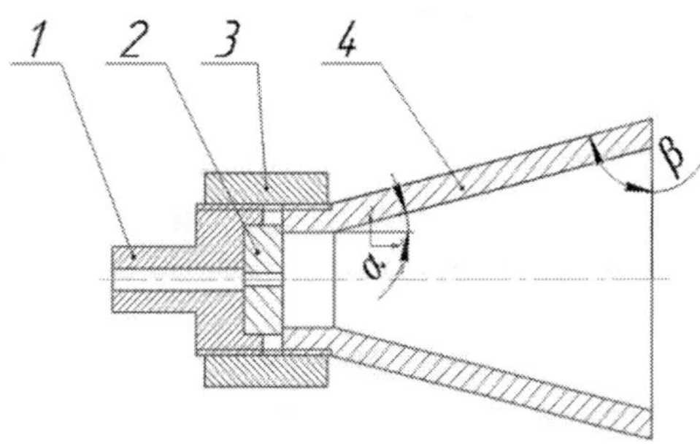

Figure 1. The working drawing of the generator – a snow former.

All elements of the generator "snow former" are made of stainless steel. The fitting (1) on the one hand joins the pipeline supplying liquid carbon dioxide. On the other hand, it has a guide recess for centering the throttle washer (2). Along the nozzle axis, the through hole is located, while the connecting nut (3) is made with a turnkey outer profile. A throttle washer with an outside diameter of 20 mm is made with a through hole. The diameter of the hole and its type were chosen experimentally to ensure maximum snow output with a minimum consumption of the CO_2 liquid phase [1].

In the manufacture of the cone "snow former" (4), the design of generating elements used in carbon dioxide fire extinguishers were considered. The cone angle was selected experimentally to ensure maximum snow yield with a minimum consumption of the liquid carbon dioxide phase.

The CO_2 throttling was carried out through the washers with a choke hole having a diameter of 1 mm. The process of snow formation occurred in cones with angles (α) from 3° to 16°, which made it possible to reveal the effect of a cone angle on the process of carbon dioxide production in the solid phase. Geometric data cones are presented in Table 1.

Table 1. Cone geometry.

No	d, mm	D, mm	l, mm	α,°	β,°
1	42	50	80	3	87
2		60		6	84
3		65		8	82
4		70		10	80
5		75		12	78
6		85		16	74

The first series of experiments was carried out at a temperature of liquid carbon dioxide of 20±1 °C. The results of the dependence of the amount of the snow-shaped CO_2 obtained on the angles of the cone 3÷16° are presented in Figure 2.

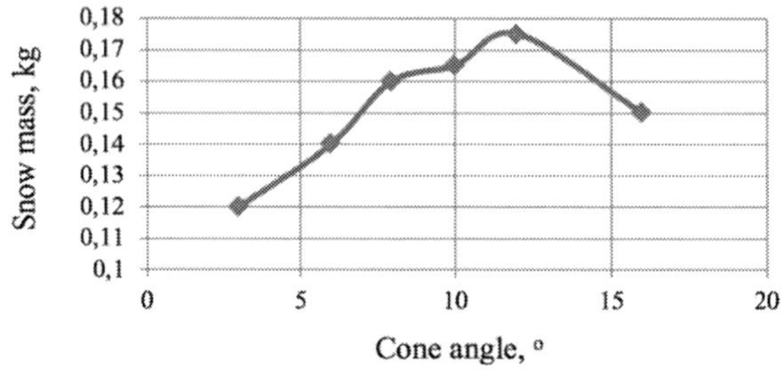

Figure 2. Characteristics of the release of snow at a temperature of liquid carbon dioxide of 20±1° C.

Analyzing the dependence presented in Figure 2, we note that the process of obtaining snow-like carbon dioxide in the socket - the snow former is divided into several stages. The first one is connected with the Bernoulli equation, since the CO_2 flow passes through a washer with a throttle orifice 1 mm in diameter, where the velocity of the gas or liquid increases (dynamic pressure increases – $\frac{pv^2}{2}$). But at the same moment, the static pressure decreases, it follows that with a decrease in the flow cross section due to an increase in velocity (a dynamic pressure), and a static pressure p drops.

Further, the stream enters the cone and with an increase in the angle of the cone, the snow former, an increase in the mass of the formed snow is observed to 0.175 kg, the angle of the cone being 12^0. Thus, an increase in pressure in front of the cone leads to an increase in speed along the entire cone - the snow former. The flow velocity in the expanding part decreases, and the pressure increases. The optimum angle of the cone is 12^0. At this angle, the maximum pressure occurs, allowing one to convert the maximum amount of carbon dioxide into a snow-like state.

The ratio of the fluid flow rate to the snow-shaped CO_2 output (Fig. 3) is 25% (0.175 / 0.7 kg / kg), while the optimum angle of the cone is 12^0.

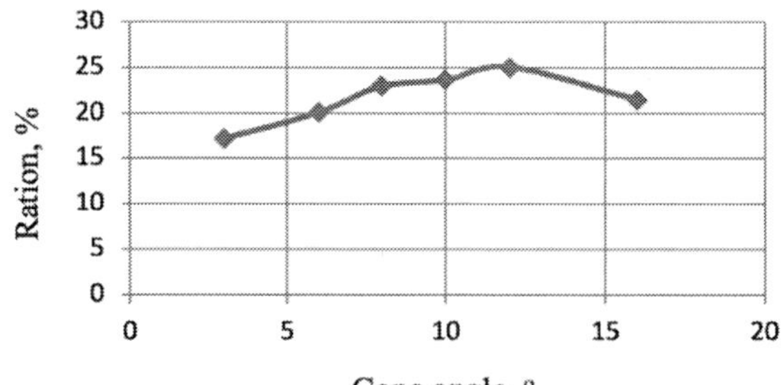

Figure 3. The ratio of liquid consumption to the output of the snow-shaped CO_2 in %, at a temperature of liquid carbon dioxide of 20±1°C.

The following experiments were carried out at a temperature of liquid carbon dioxide of 10±1 °C. With an increase in the angle of the cone-snow former, an increase in the mass of the resulting snow to 0.18 kg is observed, the optimum angle of the cone being 12^0.

The ratio of fluid flow to the snow-shaped CO_2 output is in this case 30% (0.18 / 0.6 kg / kg). Lowering the temperature of liquid CO_2 before throttling increases the output of snow-like carbon dioxide.

At a temperature of liquid carbon dioxide of 0±1°C, the ratio of liquid consumption to the output of snow-like CO_2 is 30.9% (0.17 / 0.55 kg / kg). The optimum angle of the cone was also 12°.

The following group of experiments was carried out at the temperature of liquid carbon dioxide minus 10±1 °C. The results of the dependence of the amount of the snow-shaped CO_2 obtained on the angles of the cone 3÷16° are presented in Figure 4. Optimal in this experiment is also a cone with an angle of 12°.

3. Conclusion

Characteristics of the output of snow-like carbon dioxide at different cone angles are presented in Figure 4.

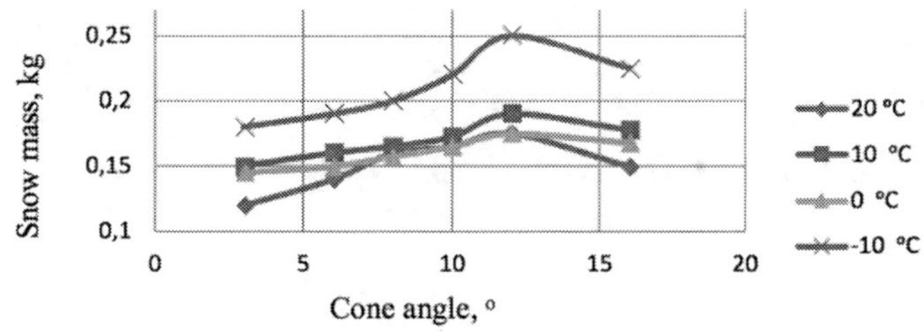

Figure 4. Characteristics of the snow emission of CO_2 at different cone angles.

Our analysis of dependences shows that the greatest amount of snow was obtained at a temperature of minus 10°C (0.25 kg), using a cone with an angle of 12°. The ratio of fluid flow to the snow-shaped CO_2 output is 31.25%.

Thus, lowering the temperature of liquid carbon dioxide before throttling allows one to increase the output of snow-like CO_2. And choosing an optimal angle leads to an additional increase in the yield of snow-like carbon dioxide. However, when choosing the conditions of temperature control, it is necessary to take into account the economic aspect of the issue associated with the cost of lowering the temperature of liquid carbon dioxide.

References

[1] Buyanov O N, Gorokhov, A A, and Neverov, E N 2005 Study of the operation of a generator of carbon dioxide snow dosing *Vestnik of the International Academy of Cold* **4** pp 20-21

[2] Pimenova T F 1989 *Basic properties and applications of dry ice in the food industry. Overview* (Moscow, Russia: AgroNIITEIMMP)

[3] Neverov E N, and Grinyuk A N 2018 *Use of carbon dioxide for transport of rabbit meat* In *Advances in Engineering Research* **151** pp 825-829

Changes in the content of low-molecular antioxidants and carbohydrates in *pinus silvestris, rosa acicularis, ledum palustre* in the context of the transition to winter dormancy in Yakutia

S Kuzmina[1*] and M Soloveva[1]

[1] M. K. Ammosov North-Eastern Federal University, 48 Kulakovsky str., Yakutsk 677013 Russia

E-mail: sskuzminar@bk.ru

Abstract. The accumulation of low-molecular antioxidants and carbohydrates in three wintering plants of Yakutia in the summer-autumn period (July, August, September) was studied. Determining the content of the sum of low-molecular antioxidants (LMAO), the sum of flavonoids, ascorbic acid, and the sum of carbohydrates was conducted spectrofotometrically in the needles of *pinus silvestris L.* and annual shoots of *ledum palustre L., rosa acicularis Lindl.* The research shows that in general a decrease of temperature (a transition from summer months to autumn months) increases the content of LMAO in the three examined plants. The study also shows that an increase of the LMAO amount in the content descends to a greater extent due to the accumulation of ascorbic acid. In the dynamics of accumulation of flavonoids during the studied period, the maximum was noticed in August and followed by a decline in September. In the dynamics of the content of carbohydrates (reducing sugars), as well as the LMAO, was revealed a gradual increase in the period from July to September. The wide fluctuations in the content of the sum of LMAO, ascorbic acid, and flavonoids with decreasing temperature were found in the *rosa acicularis Lindl,* as well as in the sugars in the needles of *pinus silvestris L.* Thus, the revealed dynamics of accumulation of LMAO in wintering plants (fraction of flavonoids, ascorbic acid) in the period of preparation for the transition to winter dormancy did not show the presence of an explicit dependence on the accumulation of the main cryoprotectants of plants, i.e. carbohydrates.

1. Introduction

Many experimental studies have shown the protective role of antioxidants in the oxidation chain of lipids as an explanation of some aspects of the nature of plants frost resistance. Thus, with increasing frost resistance by tempering plants, the content of phenolic compounds increases [1]. Also, resistance to low temperatures is characterized by a high accumulation of ascorbic acid and intensive oxidation-reduction processes. In this regard, it should be noted that in the needles of various tree species vitamin C content more in the winter months, when it is lowered [2]. According to A. D. Egorov [3], in the needles of the Yakut pine (*pinus silvestris L.*) is found that the maximum accumulation of ascorbic acid occurs in winter than in other seasons of the year.

The purpose of this work was to determine the dynamics of accumulation of low-molecular antioxidants and carbohydrates in the wintering plants in the summer-autumn period.

Content from this work may be used under the terms of the Creative Commons Attribution 3.0 licence. Any further distribution of this work must maintain attribution to the author(s) and the title of the work, journal citation and DOI.
Published under licence by IOP Publishing Ltd

2. Materials and Method

The Republic of Sakha (Yakutia) is a territorial entity of the Russian Federation. The entire territory is located in the continuous distribution of permafrost zone. The climate of Central Yakutia can be characterized by sharp continentality, which causes great contrast of the seasons of the year. The average annual air temperature is -10.3°C, the average monthly temperature is -43.2°C to -45.2°C in winter (December, January).

The most important period for the plant's life with an active temperature of 10°C and above lasts 95-98 days, from May 28th to September 4th. The summer period is short, hot and dry, includes sharp changes in daily temperatures. The average monthly temperature in July, August, and September is +18.8°C, +14.9°C, +5.7°C [4].

The object of the study was the 3 species of plants, *pinus silvestris L.*, *ledum palustre L.*, *rosa acicularis Lindl.* The pine pinus (*pinus silvestris L.*), growing in the bearberry-pine forest 7 km to the west of Yakutsk (62°N, 130°E), is a coniferous evergreen tree of the *Pinaceae* family, up to 50 meters high. The prickly wild rose (*rosa acicularis Lindl.*) is a prickly bush of the *Rosaceae* family up to 2 m high. The Swamp Ledum (*ledum palustre L.*) is an evergreen shrub of the heather family *Ericaceae*, reaching 50-90 cm in height. For the study at the end of the July, August, September months were collected: the needles of the second year of Scots Pine and the annual shoots of the Swamp Ledum, Prickly wild rose. Determination of the amount of LMAO, ascorbic, acid and flavonoids was carried out by the spectrophotometric method in ethanol extracts [5].

The determination of the amount of carbohydrates (reducing sugar) was carried out according to Bertrand [6]. Statistical data processing was carried out using the BIOSTAT and Microsoft Exel software. The data in the tables are presented in the form M±m, where M is the mean, m is the error of the mean. The reliability of the differences was assessed by the Student t-test, $p<0.05$.

3. Results

In general, the results of the studies indicate that a decrease in temperature (the transition from summer months to autumn ones) increases the content of low-molecular antioxidants in the three investigated plants. Results of the research:

1. The content of LMAO in the stalk of the *ledum palustre* increased by 29.6% ($p<0.05$) from July to September, *pinus silvestris* by 16.8% ($p<0.05$), and in the stem of *rosa acicularis* by 19.6% ($p<0.05$) (Table 1).
2. The increased content of LMAO in a greater degree descends due to the accumulation of ascorbic acid: in the stalk of *ledum palustre*, the content of ascorbic acid increased almost in 2 times ($p<0.05$), 38.5% ($p<0.05$) in *pinus silvestris*, and by 54% ($p<0.05$) in the stalk of *rosa acicularis*.
3. The maximum accumulation of the flavonoids dynamics content was in the period from July to September. The maximum accumulation was observed in August: in the stem of the *ledum palustre*, the content of flavonoids increased 2 times from July to August ($p<0.05$); it decreased 2.2 times ($p<0.05$) further in September; in needles of *pinus silvestris*, the contest of flavonoids increased in August by 1.3 times ($p<0.05$) and decreased in September by 1.7 times ($p<0.05$); in the stem of *rosa acicularis* in August, there was a tendency to increase by about 20% and a decrease in September by 20% (Table 1).
4. In the dynamics of carbohydrate content, as well as the LMAO, a gradual increase is observed from July to September: 9.5% in the stem of the *ledum palustre*, by 28% ($p<0.05$) in *pinus silvestris*, and by 22% ($p<0.05$) in the stem of *rosa acicularis* (Table 2).

Table 1. The content of low-molecular antioxidants in wintering plants in the summer-autumn period.

Defenition parametres	Months	*Ledum palustre L.*	*Pinus silvestris L.*	*Rosa acicularis Lindl.*
Amount of LMAO, µg/g	July	590±26,4	308±16,4	769±32,5
	August	698±12,3*	349±15,3*	880±11,2*
	September	765±6,4**	360±8,2**	920±12,5**
Ascorbic acid, µg/g	July	92,2±1,2	130,5±8,9	110,8±8,7
	August	180,5±6,2*	180,4±4,4*	180,4±4,4*
	September	180,6±9,4**	180,8±8,9**	170,8±8,7**
Flavonids, mg/%	July	1,78±0,4	1,89±0,3	1,24±0,2
	August	3,63±0,6*	2,39±0,2*	1,54±0,1
	September	1,62±0,3**	1,43±0,1**	1,23±0,1

The statistical significance of the differences in comparison with the July data * p <0.05, with the August data ** p <0.05.

Table 2. The content of carbohydrates (mg / g) in plants in the summer-autumn period.

Plants	July	August	September
Ledum palustre L.	21,5±1,4	23,3±3,2	23,2±2,5
Pinus silvestris L.	45,3±2,3	56,6±2,2*	58,6±1,5*
Rosa acicularis Lindl.	18,7±2,5	20,1±1,8	22,2±1,3*

The statistical significance of differences in comparison with July data * p <0.05.

4. Discussion

A.D. Egorov explains the periodicity of content change of ascorbic acid in pine needles by depending on the season of the year, by shifting the balance between the processes of synthesis and disintegration, oxidation and reduction. As the temperature of the medium decreases in autumn, the rate of irreversible oxidation of ascorbic acid decreases and the processes of its recovery and formation conversely increase, which entails a rapid increase in the total content of vitamin C in needles [3]. In our opinion, this can also be associated with an increase in the LPO intensity due to cryopreservation of cell membranes and the need to create a reserve of LMAO.

The study on the features of the accumulation of flavonoids in some species of plants in Yakutia shows that, in general, this process is natural and closely related to the developmental phases and growing conditions of plants [7]. The general tendency of increasing the accumulation of flavonoids during the transition of plants to reproductive processes and under favorable temperature and humidity conditions of growth has been established. Probably, this regularity is associated with a general increase in the intensity of metabolism of the plant organism. According to the obtained general dynamics of the accumulation of flavonoids in three studied plants, we can assume that from August to September (after the end of the reproductive processes), the content of flavonoids decreases due to a deduction in the intensity of photosynthetic processes, respiration, mineral nutrition, and the processes of movement of the starting materials and the activity of enzymes for the flavonoids synthesis.

In order to determine the interaction between the accumulation of low-molecular antioxidants and the accumulation of basic cryoprotectants, we also studied the dynamics of the total carbohydrate content using the same plants in the same time period.

5. Conclusion

The research finds that in the three wintering plants that we examined the total content of LMAOand ascorbic acidand, the amount of carbohydrates increases with the transition from July to September.

At the same time, the largest fluctuations in the content of LMAO, ascorbic acid, and flavonoids are observed in ledum palustre, when carbohydrates differences were found in *pinus silvestris*. In general, the revealed dynamics of LMAO accumulation (fraction of flavonoids, ascorbic acid) in

IOP Conf. Series: Earth and Environmental Science **395** (2019) 012067 doi:10.1088/1755-1315/395/1/012067

wintering plants in the transition preparation period to winter rest did not show the presence of an explicit dependence on the accumulation of the main cryoprotectants of plants, i.e. carbohydrates. In our opinion, the reason for this is an increase in the importance of LMAO with a decrease in the ambient temperature in protection from the intensification of LPO, i.e. the subordination to the "principle of complementarity" of the individual functioning groups of cryoprotectants.

References

[1] Mitteler R 2002 Oxidative stress, antioxidants, and stress tolerance *Trends Plant Sci.* **7** pp 405-409
[2] Maximov O B, and Gorovoy P G 1995 Antioxidant activity of leaves and bark of trees shoots and shrubs of flora of Primorsky Krai *Plant Resources* **31**(1) pp 61-69
[3] Egorov A D 1954 *Vitamin C and carotene in the vegetation of Yakutia* (Moscow, USSR: Publishing House of the Academy of Sciences of the USSR)
[4] Gavrilova M K 1973 *The climate of central Yakutia* (Yakutsk, USSR: Yakut Book Publishing House)
[5] Ermakov A I, and Arasimovich V V 1987 *Methods of biochemical research of plants* (Agropromizdat)
[6] *Molecular-genetic and biochemical methods in modern plant biology* 2011 (Moscow, Russia: BINOM - Knowledge Lab)
[7] Kuzmina S S, Popova A S 2001 Dependence of the accumulation of flavonoids and ascorbic acid in the wild plants of Yakutia from the conditions of growing *Science and Education* **1** pp 81-87

IOP Conf. Series: Earth and Environmental Science **395** (2019) 012068 doi:10.1088/1755-1315/395/1/012068

Planning the forest transport systems based on the principles of sustainable development of territories

I M Enaleeva-Bandura[1*], R N Kovalev[2] and R A Chernykh[1]

[1] Reshetnev Siberian State University of Science and Technology, 31 Krasnoyarsky Rabochy Av., Krasnoyarsk 660037 Russia
[2] Ural State Forest Engineering University, 37 Siberian Tract, Ekaterinburg 620032 Russia

E-mail: melnikov1978@inbox.ru

Abstract. The article identifies a new method of dynamic modeling in the design of the transport system in the forest fund (TSFF), which is based on economic and mathematical modeling and fuzzy logic tools. The combination of the indicated methods is designed to reduce the disadvantages of their use and increase the benefits. The article substantiates the choice of assessing the forecast level of the impact of risks on the activities of forestry enterprises (the method of expert assessments), using the methodological tools of fuzzy logic. The indicated method makes it possible to take into account a large variety of risk factors of the internal and external environment. At the same time, methodological aspects of fuzzy logic make it possible to formulate a quantitative assessment of qualitative indicators. The article substantiates the choice of tools for economic and mathematical modeling in order to state the design problem of the planned TSFF. Since the indicated method enables the formalization of the functioning of the timber transport system in the given conditions. The article presents a developed model that correctly takes into account the influence of risk factors when planning a TSFF, through the combination of fuzzy logic methods and economic and mathematical modeling. The advantages of the developed model include: considering the multivariance of material flows, vehicles, points of overload, etc.; automated processing of input parameters and effective data; using the model for forecasting, i.e. the possibility of deriving a fuzzy estimate of the efficiency of the timber transport system by identifying cause-effect relationships between the modeling object and the influence of risk factors on its functioning.

1. Introduction

The presence of an optimally developed transport infrastructure is a necessary condition, without which sustainable development of territories is impossible [1, 6]. Planning a transport system in the forest fund (TSFF) is a multi-criteria task, since the cost of its construction and operation, environmental, social, and technical efficiency are contradictory [9-12]. Proceeding from this, the planned TSFF and the territory of the forest fund proposed for development should be considered as a single system, i.e. in synergistic relationships between the designated components. In turn, this system is influenced by a number of risk-forming factors.

Content from this work may be used under the terms of the Creative Commons Attribution 3.0 licence. Any further distribution of this work must maintain attribution to the author(s) and the title of the work, journal citation and DOI.
Published under licence by IOP Publishing Ltd

IOP Conf. Series: Earth and Environmental Science **395** (2019) 012068 doi:10.1088/1755-1315/395/1/012068

2. Objects and Research Methods

Risks in the forest industry is the probability of losses due to changes in the economic, environmental, social, and technical condition of the forest industry [7, 8]. The risks shown in Figure 1 have the greatest impact on forestry enterprises.

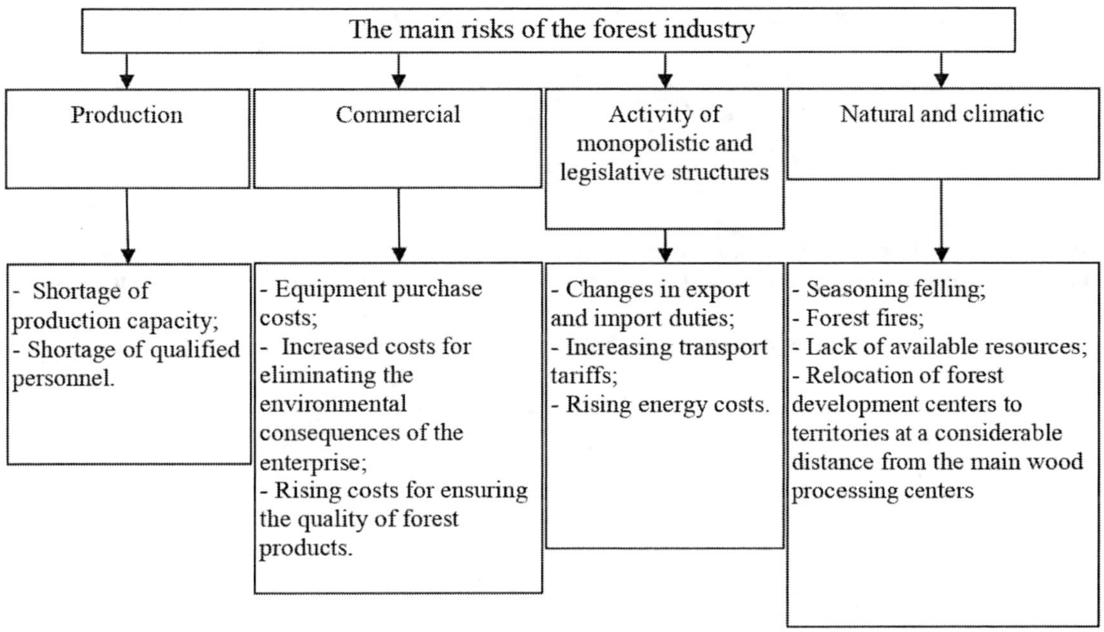

Figure 1. Risk classification in the forest industry.

Since the main criterion for evaluating any system is its effectiveness (the ratio of results to costs), we propose to predict the effectiveness of the planned TSFF's functioning, taking into account environmental and economic indicators and risk factors accompanying the project. Qualitatively, this can be expressed as reaching the maximum of all potential incomes, taking into account the influence of risk-forming factors.

3. Results and Discussion

Taking into account the analysis of sources [2-4; 15-19], quantitatively, the objective function of the efficiency of the TSFF E_{ef}^{n} can be the total value of the resulting functional E_{ef}^{n} and *£(T)*, which is a predicted value, determining the degree of influence of risk-forming factors on the effectiveness of planning a TSFF for the period of forest fund development *(T)*.

The final functionality E_{ef} in the first approximation is the ratio of net present value (NPV), obtained as a result of the development of the forest fund during the period of development of the forest fund T, to the total cost of construction and operation of the TSFF (*S*). It shows the amount of net profit per ruble of total costs over the time periods *t*. In this case, the mathematical model of the stated optimization problem takes the form:

$$\begin{cases} \mathrm{E}_{ef}^{n} = \mathrm{E}^{ef} - £(T);\ \mathrm{E}^{ef} = \mathrm{HTC} \cdot S^{-1} \to \max;\ £(T) = \sum_{b=1}^{B} g_b(t); \\ £(T) = \sum_{t=0}^{T} £(t);\ £(T) \to \min;\ b = 1,\dots B. \end{cases} \quad (1)$$

The coefficients $g_b(t)$ given in expression (1) should be determined by the method of expert assessments by forest industry specialists [8]. Taking into account these factors, the forecast of the degree of influence of one or another risk factor on the value of the current gross income at points in time *t* can be carried out. In the mathematical model, only those types of risks (B) are included, which, according to experts, have the greatest impact on the TSFF effectiveness [13, 14].

It should be noted that, according to some sources [5, 8], the forecast of the influence of risk-forming factors on the TSFF planning project for the period of forest fund development is short (from 3 to 5 years). Consequently, the predicted value, which determines the degree of influence of risk-forming factors on the TSFF planning project during the forest fund development period *£(T)*, is the sum of *£(T)* = T over time periods (*t*).

The net present value obtained as a result of the development of the territory of the forest fund through the implementation of the TSFF (for periods of time *t*) is determined by the following expression:

$$HTC = DT - S \tag{2}$$

where D_T – the current gross income from the development of the forest fund, for a period of time *t*, ruble; *S* – planned total costs as a result of the development of the forest fund, for periods of time *t*, ruble.

The current gross income as a result of the development of the forest fund for periods of time *t* is determined by the expression (3):

$$\begin{cases} \mathrm{D}_T = (1+e)^{-T}\cdot\sum_{i=1}^{m}\sum_{j=1}^{n}\sum_{k=1}^{K}\left[C_i(t)+C_{ij}^{TP}(t)\right]\cdot Q_{ijk}(t)-\sum_{t=0}^{T}£(t)\rightarrow \max; \\ £(t) = g_v(t)+g_G(t)+g_w(t)\rightarrow \min; \\ \mathrm{D}(t) = \sum_{i=1}^{m}\sum_{j=1}^{n}\sum_{k=1}^{K}\left[C_i(t)\cdot(1\pm g_v(t))+C_{ij}^{TP}(t)\cdot(1\pm g_G(t))\right]\cdot Q_{ijk}(t)\cdot(1-g_w(t)); \\ (i=1,...m),(j=1,...n),(k=1,...K),(t=0,...T). \end{cases} \tag{3}$$

where C_{ij} is the selling price of 1 m^3 (or other units of forest resources) by types of forest resources from the *i*-th forest plot, $i \in \{1,...,m\}$, (hectare, allotment, m^2), to the *j*-th warehouse (near-land plot, to the consumer), $j \in \{1,...,n\}$, rub.; $C_{ij}^{TP}(t)$ – transportation costs per unit of output, rub. at time t; $Q_{ijk}(t)$ – a total volume of forest resources from the *i*-th forest area (quarter, section, m^2), to the *j*-th warehouse to be harvested and transported by the *k*-th type of transport at time *t*, $k \in \{1,...,K\}$, m^3 (or other units of forest resources); *e* – a discount factor; *T* – a period of the forest fund development, years; *t* – time from the moment of evaluation to the moment of resource procurement, $t \in \{1,...,T\}$, years; $g_w(t)$ is the coefficient for assessing the impact of the resource factor on the volume of (shipped) products, as well as the effect of the risk of short delivery or delivery of products of inadequate quality at the time *t*; $g_v(t)$ is the coefficient for assessing the influence of the factor of shadow economic relations, as well as the effect of inflation on the cost of (shipped) products at time *t*; $g_G(t)$ is the coefficient for assessing the impact of the legislative and monopolistic factors on transportation costs at time *t*; *D(t)* is the current gross income from the development of the forest fund, for periods of time t, corrected for the influence of risk-forming factors (*B*), rub.

The planned total costs (maximum allowable costs associated with logging, reforestation, construction and operation of the TSFF for the period of development of the forest fund) are determined by the expression (4):

$$S = \sum_{t=0}^{T}(\mathrm{O}_z + \mathrm{O}_v + P_{ij}^{'}\cdot\Delta t(t))\cdot(1+e)^{t}\cdot(1+(1+e)^{-T})+\sum_{t=0}^{T}£(t)\rightarrow \min \tag{4}$$

where O_z – reduced costs for the creation and operation of TSFF, rub.; O_v – regulatory costs for the reproduction and protection of forests, guaranteeing their restoration in clearings, growing up to the age

IOP Conf. Series: Earth and Environmental Science **395** (2019) 012068 doi:10.1088/1755-1315/395/1/012068

of maturity, protection and protection, rub.; P'_{ij} stands for deadening the forestry asset (not sales) from not developing the forest fund due to the absence of LDS, per 1m^3 hectare, land plot, m^2) at time t, rub.; Δt – the period of non-development of the forest fund. P'_{ij} - those costs that include the potential income lost due to the postponement of income from forest use plus the cost of removing income from future production cycles for a period of time Δt.

When implementing the proposed mathematical model for assessing the effectiveness of the TSFF, the following limitations should be considered:

1. Payback project planning:

$$\sum_{t=0}^{T}(\text{O}_z+\text{O}_v+P'_{ij}\cdot\Delta t(t))\cdot(1+e)^t\cdot(1+(1+e)^{-T})+\sum_{t=0}^{T}£(t)\leq \text{O}_{max} \quad (5)$$

where O_{max} is the financial capabilities of the enterprise.

2. The distance between the i-th forest site and the j-th raw material storage (according to the source [4], the effective shoulder of forest resource removal from the i-th forest area to the j-th raw material warehouse should not exceed 120 km) determines the accessibility of forest sites:

$$L_{ij}\leq 120 \quad (6)$$

where L_{ij} is the distance between the i-th forest site and the j- th raw materials storage, km.

3. Natural non-negativity of freight traffic:

$$L_{ijk}(t)\geq 0,\ i=1,...m;\ j=1,...n;\ t=1,...T;\ k=1,...K. \quad (7)$$

4. The requirement of continuous, inexhaustible forest management, taking into account the influence of risk-forming factors:

$$\sum_{t=0}^{T}\sum_{i=1}^{m}\sum_{j=1}^{n}\sum_{k=1}^{K}\begin{Bmatrix}Q_{ijk}(1-g_w)(t+1)_l\cdot\left[C_i(1\pm g_b)+C_{ij}^{TP}(1\pm g_G)\right](t+1)_l- \\ -\text{O}_z(t+1)_l-\text{O}_v(t+1)_l-P'_{ij}\cdot\Delta t(t+1)_l\end{Bmatrix}-$$
$$-\sum_{t=0}^{T}\sum_{i=1}^{m}\sum_{j=1}^{n}\sum_{k=1}^{K}\begin{Bmatrix}Q_{ijk}(1-g_w)(t)_l\cdot\left[C_i(1\pm g_b)+C_{ij}^{TP}(1\pm g_G)\right](t)_l- \\ -\text{O}_z(t)_l-\text{O}_v(t)_l-P'_{ij}\cdot\Delta t(t)_l\end{Bmatrix}\geq 0, \quad (8)$$

where l is the number of the economic section (breed), $l=1,...,L.$

Based on the short-term efficiency forecasting of the forest transport systems functioning, the final functionality should be adjusted for the influence of risk-forming factors $£(T)$ each time period (t).

4. Conclusion

The proposed dynamic model of forecasting the impact of risks on the forest transport system of forest industry enterprises is able to provide: a comprehensive assessment of external and internal environment factors that influence risks in order to predict the level of certain types of risk of forest industry enterprises; a wide range of indicators in order to comprehensively study the influence of environmental factors on sustainability and risks when planning a TSFF; application of expert methods in assessing the influence of risk factors; using the method of fuzzy inference to obtain the final predicted values of the level of risk; the possibility, using qualitative values of the indicators, to obtain a quantitative result.

References

[1] Kovalev R N, and Gurov S V 1996 *Planning of transport systems of forest enterprises in a multi-purpose forest management* (Ekaterinburg, Russia: Ural. State Forestry Academy) p 250

[2] Kovyazin V F, and Romanchikov A Yu 2018 *Notes of the Mining Institute* **229** pp 98-104

[3] Kovyazin V F, and Romanchikov A Yu 2015 *Notes of the Mining Institute* **216** pp 232-237
[4] Elderstein Yu M 2003 *Modeling and optimization of production processes in the forest and woodworking industry* (Krasnoyarsk, Russia: SibGTU)
[5] Tyrtyshnyy N N 2013 *Thesis of Candidate of Economic Sciences* (Rostov-na-Donu)
[6] *Report of the World Summit on Sustainable Development* 2000 (Johannesburg, South Africa)
[7] Seraya O V, and Dunaevskaya O I 2012/2013 *Veda a vznik: Materialy IX mezinarodnivedecko – prakticka konference* (Prague, Chech Republic: Education and Science) pp 20-23
[8] Kobalinskiy M V 2006 *Thesis of Candidate of Economic Sciences* (Krasnoyarsk) pp 42-63
[9] *Indicators of sustainable development: guidelines and methodologies. Third Edition. Economic & Social Affairs* 2007 (New York, United Nations) p 93
[10] *From transition to transformation: sustainable and inclusive development in Europe and Central Asia* 2012 (New York, Geneva) p 156
[11] Fücks R 2014 *Intelligent Wachsen. Die grüne Revolution.* (Berlin: Hanser) p 362.
[12] Duwe S 2015 *Governing the transition to a green economy: doctoral thesis* (Freie Universitët Berlin) p 292
[13] Broman G I, and Robert K-H 2017 *Journal of Cleaner Production* **140** pp 17-31
[14] Detz P V 1979 *Allgemeine Forstzeitschrift* **34** pp 141-167
[15] Romanchikov A 2014 Cadastral Valuation of Forestlands Using Statistical Analysis of Taxation Parameters *55 Konferencija Studenckich Kol Naukowych Pionu Gorniczego* (Krakow: AGH) pp 10
[16] Romanchikov A 2015 Forest inventory data as an instrument for cadastral valuation of forestlands in North-West Russia *Scientific Reports on Resource Issues* (Freiberg: Technische University Bergakademie Freiberg) pp 360-363
[17] Kovyazin V, Romanchikov A, and Belyaev V 2014 Use of forest inventory data as a new method for cadastral valuation of forestlands in North-West Russia *Forestry Studies* **61** pp 69-78
[18] Kovyazin V, Romanchikov A and Belyaev V 2014 Taxation indices of forest stand as the basis for cadastral valuation of forestlands *IOP Conference Series: Earth and Enviromental Science* **21** 012026
[19] Kovyazin V, Romanchikov A, and Belyaev V 2015 Comparative analysis of forest lands cadastral appraisal estimated with regards to wood and food resources *IOP Conf. Series: Earth and Environmental Science* **27** 012039.

IOP Conf. Series: Earth and Environmental Science **395** (2019) 012069 doi:10.1088/1755-1315/395/1/012069

Adsorption of a pyridine-phenol mixture from aqueous solutions with preheated active carbons

O Belyaeva[1], N Golubeva[1*] and E Belyaeva[1]

[1] Kemerovo State University, 47 Stroitelei blvd, Kemerovo 650056 Russia

E-mail: golnadya@yandex.ru

Abstract. The article investigates the fundamental possibility of using preheating of the activated carbon in air and in an inert atmosphere to increase its adsorption capacity during adsorption extraction from aqueous solutions of a mixture of phenol – pyridine. It is found that such a modification affects more adsorption of pyridine than of phenol. At the same time, heating in an inert atmosphere turned out to be more efficient.

1. Introduction

In large industrial regions, a high level of environmental pollution, including the hydrosphere, actualizes the task of developing and putting into practice technological processes with closed water-circulation cycles. In this case, the ingress of insufficiently treated wastewater into the population's water supply sources can be excluded.

With wastewater from pharmaceutical, metallurgical and coke-chemical plants, toxic compounds such as pyridine and phenol can be released into water bodies. In addition, a low molecular weight of phenols can be formed during the biochemical decomposition of humic acids. Biochemical methods used in practice for wastewater treatment containing pyridine and phenol are not effective enough. The concentration of organic components in the effluent after the treatment plant may exceed permissible values in 1.5-2 times [1]. An effective method of purification and purification of wastewater with a low concentration of organic components is adsorption using carbon materials.

The capacity of industrial activated carbons (AC) in relation to highly soluble and water-miscible organic compounds (the latter includes pyridine) is low due to the low selectivity of adsorption. Since the energies of the interaction of the solute and solvent with the adsorbent are close [2]. To increase the adsorption selectivity, we need to either increase the adsorption of the organic component due to the specific interaction of the sorbate with the surface groups of the sorbent or reduce the adsorption of water. We can achieve a similar effect by modifying the AC with vapors and solutions of acids, bases, oxidizers, and reducing agents of various types.

Earlier we established [3] that the modification of AC with acid solutions, which increase the adsorption of phenols from aqueous media, drastically reduced the adsorption of pyridine. One of the ways to change the chemical state of the surface of an activated carbon is a heat treatment. At the same time, when heated in air, an increase in the surface oxygen compounds (SOC) of various types (acidic and basic) occurs, and during a heat treatment in an inert environment (for example, hydrogen or nitrogen), the decrease in SOC (predominantly acidic) occurs [4-5]. Previous studies also demonstrate that heating at temperatures above 120 C leads to a noticeable increase in the adsorption of pyridine

Content from this work may be used under the terms of the Creative Commons Attribution 3.0 licence. Any further distribution of this work must maintain attribution to the author(s) and the title of the work, journal citation and DOI.
Published under licence by IOP Publishing Ltd

from aqueous solutions, whereas the adsorption of phenol varies slightly, and heating AC in a nitrogen atmosphere at 1,000°C reduces water adsorption [5-7].

The purpose of this study was to clarify the possibility of using the preliminary thermal treatment of AC to increase the adsorption of the phenol-pyridine mixture from aqueous solutions.

2. Materials and Methods

The objects of our study are the granulated industrial active carbons of the grades AG-OV-1 (designation A-1) and SKD-515 (designation S-1) (manufactured by the Sorbent LLC., Perm, Russia). They were previously washed with distilled water from dust fractions and dried to air-dry condition. Then, one part of the AC prepared in this way was subjected to heat treatment in air at 250 ± 1.0°C for 5 hours (reference A-2 and S-2 for AG-OV-1 and SKD-515, respectively). And another part of the AC was subjected to heat treatment in air at 1000 ± 5.0°C for 2 hours under nitrogen atmosphere (designation A-3 and S-3 for AG-OV-1 and SKD-515, respectively). All modified samples were cooled in a desiccator and stored tightly sealed.

The adsorption of phenol and pyridine was carried out from aqueous solutions in the concentration range from 0.010 to 2.00 g/l with a ratio of organic components of 1:1. The equilibrium concentration of pyridine was determined by its own absorption at 256 nm; it was after reaction with 4-aminoantipyrine at 490 nm for phenol. In selected conditions, the components do not interfere with each other's definition.

3. Results

Experimental adsorption isotherms are shown in Figure 1.

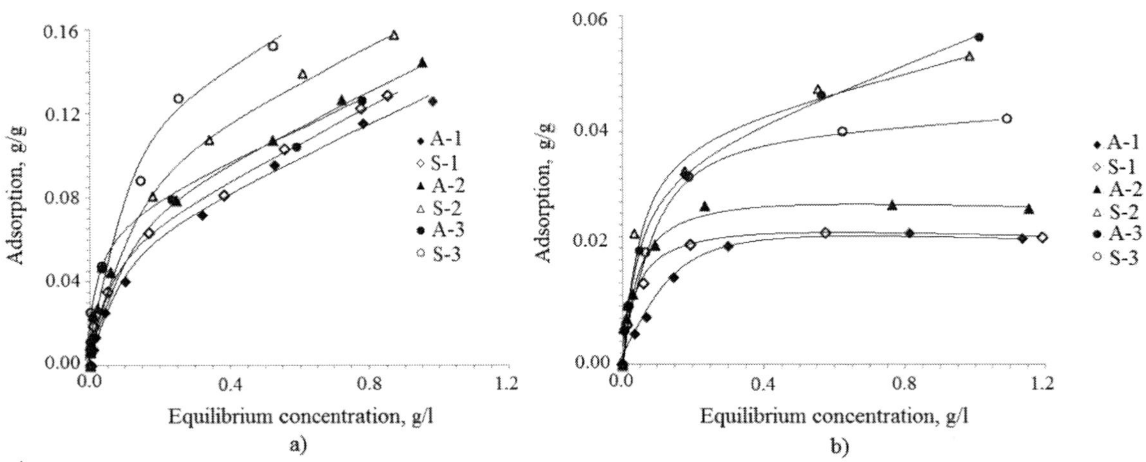

Figure 1. Adsorption isotherms of phenol (a) and pyridine (b) from an aqueous solution with the joint presence (ratio 1:1) of active carbon samples.

The obtained adsorption isotherms were analyzed using the models of polymolecular adsorption (BET equation) and volume filling of micropores (Dubinin-Radushkevich equation), modified for the case of adsorption from highly soluble and miscible substances from solutions. It was established that both models with high correlation (R2> 98%) describe the adsorption behavior of the studied organic components. Adsorption parameters calculated using the BET equation are presented in Table 1.

Table 1. Parameters of phenol and pyridine adsorption from an aqueous solution with the joint presence (ratio 1:1) of active carbon samples.

Brand AC	q_{max}, mmol / g	$-Q_L$, kJ / mol	$-Q_S$, kJ / mol	R^2
		Phenol		
A-1	0.89	16.93	8.88	99.29
A-2	0.90	19.39	9.27	99.50
A-3	0.97	20.70	8.62	99.67
S-1	0.75	19.37	9.72	99.45
S-2	0.98	20.66	9.48	99.19
S-3	1.07	20.95	10.25	98.36
		Pyridine		
A-1	0.11	19.04	13.48	99.99
A-2	0.27	20.34	10.96	99.74
A-3	0.62	17.25	6.25	99.71
S-1	0.21	20.15	11.24	99.99
S-2	0.53	19.16	7.29	99,68
S-3	0.51	18.10	4.61	99.87

The magnitude of the limiting adsorption volume (W_0), calculated by the Dubinin-Radushkevich equation, is shown in Figure 2.

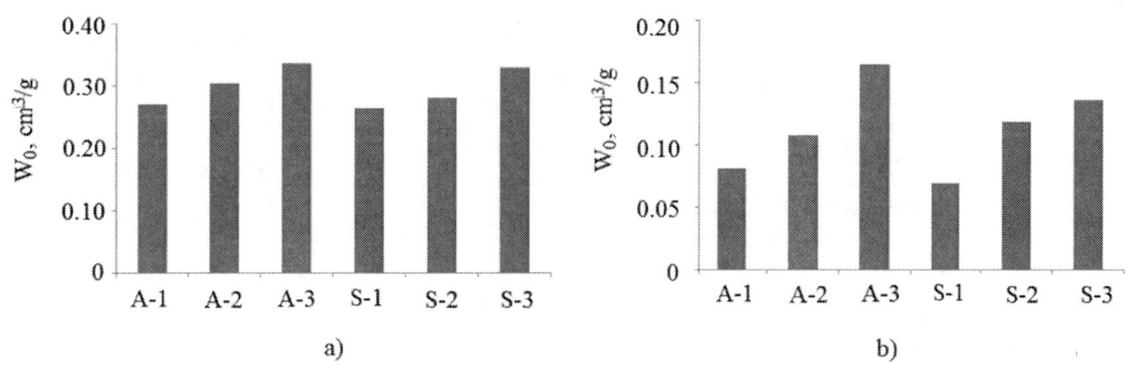

Figure 2. The limiting adsorption volume occupied by phenol (a) and pyridine (b) when extracted from an aqueous solution with the joint presence (ratio 1:1) of active carbon samples.

4. Discussion

Studies have shown that heating the samples of industrial AUs in air at 250°C practically does not change the maximum monolayer adsorption (q_{max}) for phenol for A-2 (the change in q_{max} is 1%, which is comparable to the error in determining). While the growth of q_{max} for S-2 reaches 30%. The adsorption of pyridine from a mixture with phenol increases significantly: q_{max} increases by more than 2 times for both A-2 and S-2. The change in the value of the limiting adsorption volume (W_0) (Figure 2) for phenol is the opposite. For A-2, its growth is 12.5%, and 6.2% for S-2. In the case of pyridine, W_0 increases by 32% and 71% for A-2 and S-2, respectively. Probably, this behavior is associated with the competition of molecules of organic substances among themselves, and water for the primary adsorption centers of specific adsorption, the number of which increases with this method of processing AC [5]. In this case, all components of the solution can become secondary adsorption centers for each other. The calculated heat of adsorption (Q_L) indicates a high contribution of a specific interaction (hydrogen bond) in the formation of a monolayer of both phenol and pyridine on the surface of the AC data (Table 1). When a subsequent layer of organic components is formed, the specific interaction fraction decreases (Q_S does not exceed 11 kJ / mol) [8].

Preheating AC in a nitrogen atmosphere (samples A-3 and S-3) also leads to an increase in the adsorption of organic components from their mixture. The maximum adsorption of the monolayer for phenol on A-3 and S-3 increases by 10% compared with A-2 and S-2. For pyridine, this parameter (q_{max}) is increased by 5.6 and 2.4 times if compared to the industrial AC for A-3 and S-3, respectively. The limiting adsorption volume of AC heated in this way is increased by ~ 24% for phenol and ~ 100% for pyridine (Figure 2). As in the case of A-2 and S-2, a monolayer of phenol and pyridine on A-3 and S-3 is formed due to the specific interaction of components with the surface of an AC (Table 1). Formation of the next component layer for pyridine occurs only due to dispersion interaction [8].

5. Conclusion

Studies have shown that the heating of AC can be used to increase the adsorption of a mixture of phenol-pyridine from aqueous solutions. Preliminary heat treatment both in air and in nitrogen atmosphere allows to increase the adsorption from the mixture worse than pyridine, a sorbed component. The nature of changes in the parameters of the adsorption of phenol and pyridine from aqueous solutions when co-present depends on the transformations of the chemical state of the surface of the AC during the heating. And also, the nature of changes in the parameters of the adsorption of phenol and pyridine from aqueous solutions when co-present depends on the redistribution of the ratio of micro- and mesopores and their availability.

References

[1] Kagasov V M, and Pimenov I V 1991 Cleaning and use of sewage of the coke-chemical enterprises *Coke & Chemistry* **3** pp 62-63

[2] Koganovskii A M, Klimenko N A, Levchenko T M, & Roda I G 1990 *Adsorption of organic substances from water* (Leningrad,USSR: Chemistry)

[3] Krasnova T, Kirsanov M, Belyaeva O, and Timoshchuk I 2018 Adsorption processes in solving environmental problems in water sources of Urban Territories Development *MATEC Web of Conferences* **170** 10 p

[4] Barton S S, Evans M J B, Halliop E, MacDonald J A F 1997 Acidic and basic sites on the surface of porous carbon *Carbon* **35**(9) pp 1361-1366

[5] Belyaeva O V, Krasnova T A, Gladkova O S 2015 Effect of the thermal treatment conditions of granulated active carbons on their properties *Solid Fuel Chemistry* **49**(3) pp 196-200

[6] Krasnova T A, Kirsanov M P, Belyaeva O V, & Shishlyannikova N Yu 2004 Effect of the thermal treatment of active carbons on their ability to adsorb pyridine from aqueous solutions *Russian Journal of Physical Chemistry* **78**(8) pp1344-1346

[7] Belyaeva O V, Krasnova T A, Anikina A V 2010 Effect of preliminary heat treatment of activated carbons on adsorptive extraction of aniline and phenol *Chemistry forSustainable Development* **18**(6) pp 761-765

[8] Steed J W, and Atwood J L 2000 *Supramolecular chemistry* (John Wiley & Sons)

Prospects for using the method of separation freezing for whey processing

I Korotkiy[1], E Korotkay[1*] and E Neverov[1]

[1] Kemerovo State University, 6 Krasnaya str, Kemerovo 650000 Russia

E-mail: korotkayael@mail.ru

Abstract. The use of cryoconcentration allows one to preserve the biological value and organoleptic properties of food products. This paper presents the results of studies on the concentration of whey by separation freezing. General patterns of separation freezing of whey in a capacitive crystallizer, the influence of the rate of ice formation, and the mass fraction of dry substances on the formed amount of whey protein concentrate are analyzed. The chemical composition of the initial serum and the resulting concentrates are determined. The mass fraction of protein in whey protein concentrate was 92% of the total dry matter. In the lactose concentrate, the dry matter content was 62.5%, the amount of pure lactose – 55.8 g / 100 g concentrate.

1. Introduction

In recent years, because of imposing the food embargo by the Russian government in response to sanctions from Western countries, an increase in the production of import-substituting dairy products is observed, including in cottage cheese, cheese, and cheese products [1]. In turn, this led to an increase in the amount of secondary raw materials, namely whey. Unfortunately, in our country, according to [2, 3], only 30-40% of the whey obtained is subjected to industrial processing, in contrast to the countries of Western Europe (80%) and the USA (90%). Cheese whey makes up the majority of the whey processed in our country – 20-30% [1, 3]. The final product of its processing is often a dry product, its insignificant use is due to specific organoleptic and physico-chemical parameters. A portion of the whey produced is used to fatten livestock. A significant amount of serum is discharged into sewage, which leads to severe environmental pollution (annual average BOD5 -is 50–60 g O2 per 1 liter, and it is 50.5-54 g O2 per 1 liter for COD [2, 4-6]). In this regard, the development and introduction of new technologies for the processing of secondary dairy raw materials is relevant for many enterprises, because their development and implementation meets the main directions of the country's development, which are indicated in the concept of the long-term socio-economic development of the Russian Federation for the period up to 2020 [7].

According to A. G. Khramtsov [2], six methods are known today that allow partially solving the problem of whey complex processing. These include: coagulation, centrifuging, canning, enzymatic methods, membrane methods and electrodialysis, cryoconcentration, thickening and drying. Cryoconcentration includes two main technological stages: (a) formation of a mixture of ice crystals and whey concentrate; (b) separation of water-ice suspension. The use of separation freezing allows to separate whey proteins and lactose separately. Unfortunately, this method of concentration and processing of whey receives insufficient attention, despite the fact that the undoubted advantage of the

Content from this work may be used under the terms of the Creative Commons Attribution 3.0 licence. Any further distribution of this work must maintain attribution to the author(s) and the title of the work, journal citation and DOI.
Published under licence by IOP Publishing Ltd

low-temperature concentration technology is the maximum preservation of the properties of the raw materials.

In this regard, the study of the possibility of using separative freezing for the processing of whey is very relevant.

2. Materials and Methods

As an object of research, we used curd whey that meets the requirements of GOST R 53438-2009 "Milk whey. Technical conditions." The separation patterns of the whey freezing were studied on an experimental setup, the design of which is protected by the patent of the Russian Federation [8].

The mass fraction of protein in the studied samples was determined by the Dumas method on the analyzer of protein nitrogen "Rapid N Cube", the fat content – according to the GOST R ISO 2446-2011 "Milk. Method of Determining Fat Content", lactose amount – according to the GOST R 54667-2011 "Milk and Milk Processing Products. Methods for Determining the Mass Fraction of Sugars". The content of dry substances was determined using the arbitration method in accordance with the GOST 3626-73 "Milk and Dairy Products. Methods for Determining the Moisture and Dry Matter."

3. Results

At the first stage of research, the chemical composition of whey in grams per 100 g of the product was determined: water – 92.8 ± 0.5; dry substances – 7.2 ± 0.5; fat – 0.21 ± 0.01; proteins – 0.97 ± 0.03; carbohydrates – 5.20 ± 0.05, minerals – 0.82 ± 0.03. Since the mass fraction of fat in the test serum was insignificant, this indicator was not taken into account in the course of further studies.

Separation freezing of whey was carried out in a 3.5 liter capacitive type crystallizer with an internal heat exchange surface in the temperature range of the heat exchange surface from minus 2 to minus 8°C in steps of 2°C. The resulting concentrate was removed and subjected to thermocoagulation for 10 minutes at a temperature of 90-95°C for a more complete isolation of protein. After that, the concentrate together with the precipitate was centrifuged for 25 minutes at a frequency of 4,200 rpm, the supernatant was removed by decantation, the precipitate was filtered. The amount of protein obtained, depending on the temperature of the heat exchange surface and the duration of freezing, is presented in Figure 1.

At the next stage of experimental studies, selecting the temperature differential was made in the range from 0.5 to 3°C, in increments of 0.5 degrees. The time of whey freezing corresponded to the maximum amount of protein precipitated at the corresponding temperature of the heat exchange surface: minus 2°C (240 min), minus 4°C (180 min). The research results are shown in Figure 2.

After removal of the whey protein concentrate, the dry matter content in the remaining solution was 9.6%, 7.3% of them were carbohydrates (lactose), 1.1% were mineral substances, and 0.5% were protein. This solution was re-frozen for 8 hours at a temperature of minus 5 ± 0.5 °C. This operation allowed to obtain an eutectic solution with a lactose content of 49%. To remove mineral substances and to obtain pure lactose, the eutectic solution was subjected to separation freezing at temperatures of the heat exchange surface from minus 5°C to minus 8°C, with a step of 1 degree. Duration of concentration varied in the range from 30 to 600 min, with a step of 30 min. During the experiment, the chemical analysis of the composition of the liquid phase and the ice fraction on the content of lactose and minerals was carried out. The dependence of the lactose content in the liquid and solid phases on the time of freezing at a temperature of minus 6°C is shown in Figure 3.

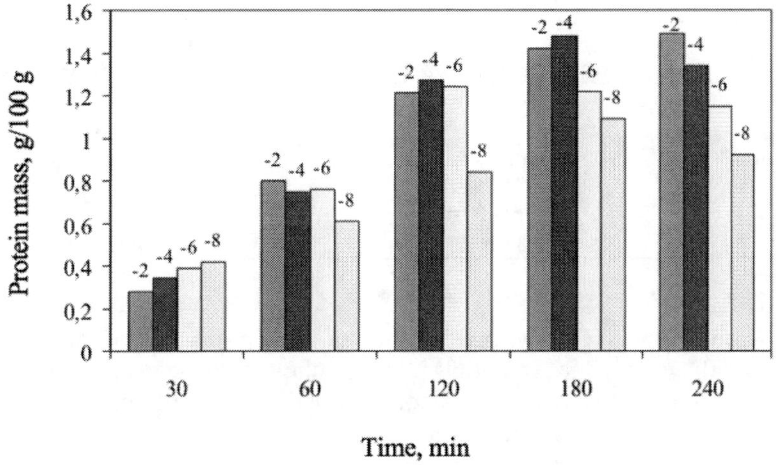

Figure 1. The dependence of the amount of protein isolated from the temperature and time of freezing.

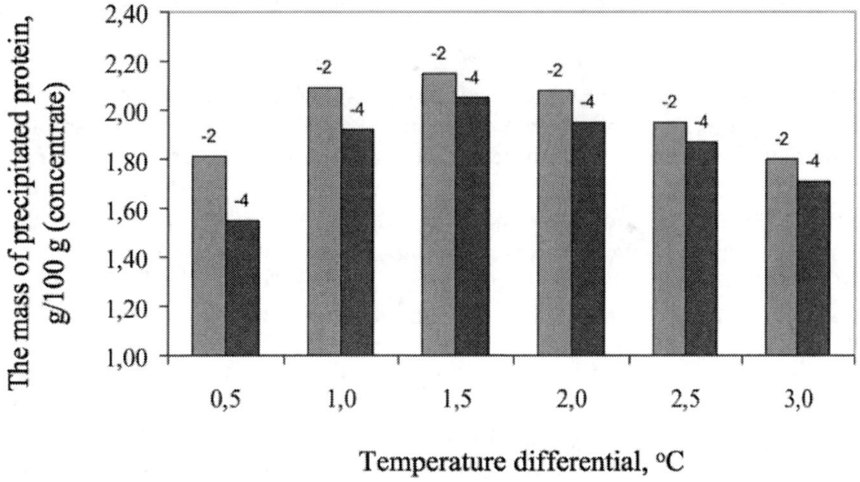

Figure 2. Dependence of the amount of protein deposited on the temperature differential.

The lactose concentrate contained about 56% of lactose and 5.7% of mineral substances. The average chemical composition of the components separated by freezing from whey is given in Table 1.

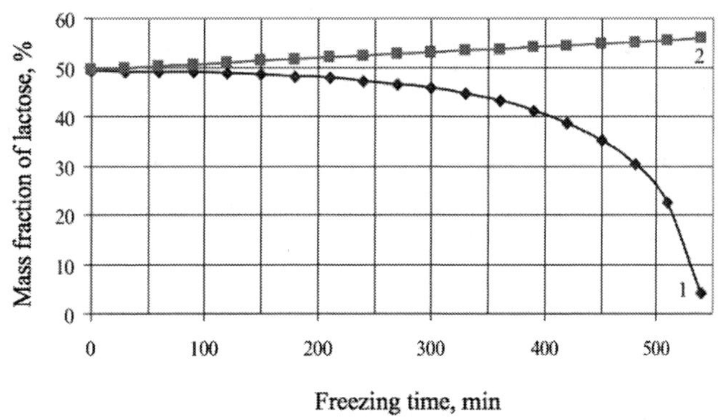

Figure 3. The lactose content in the liquid (1) and solid (2) phases depending on the time of freezing.

Table 1. The average chemical composition of protein and lactose concentrates.

Component	Content, g / 100 g	
	Protein concentrate	**Lactose concentrate**
Water	68.37±0.5	37.51±0.5
Dry matter	31.63±0.5	62.49±0.5
Fat	0.03±0.01	0.28±0.01
Proteins	29.14±0.03	0.63±0.03
Carbohydrates	2.11±0.05	55.83±0.05
Minerals	0.31±0.03	5.75±0.03

4. Discussion

During the separation freezing of whey, precipitation of protein occurred at the bottom of the working capacity of the crystallizer. Sludge formation is due to the simultaneous action of the following two factors: the change in serum acidity and, as a consequence, the loss of aggregative and sedimentation stability of the system, as well as the effect of non-stationary temperature field caused by the temperature differential of the refrigerating machine.

Based on the experimental results (Figure 1), we noted that the maximum amount of protein (1.5 g / 100 g of concentrate) precipitates after 240 minutes at a temperature of the heat exchange surface minus 2°C, while about 35% of all moisture is frozen. At a temperature of minus 4°C, the time corresponding to the precipitation of the maximum amount of protein was 180 minutes, the amount of frozen moisture is about 50%. When the temperature of the heat exchange surface decreases to minus 6 and minus 8°C, the seizure of protein precipitate by the crystallization front was observed. Therefore, further studies were carried out at temperatures of the heat exchange surface – minus 2°C and minus 4°C.

As a result of experimental studies on the effect of the temperature differential on the amount of protein precipitated (Figure 2), we found that the largest amount of protein (~ 2%) precipitated at a temperature of minus 2°C, with a temperature differential of 1.5 °C. At a temperature of minus 4°C, the dependence of the amount of precipitated protein on the temperature differential is more gentle, while the degree of loss of dry substances was greater than at minus 2°C.

After removal of whey proteins, the remaining concentrate contained a significant amount of lactose, which can also be extracted by freezing. To remove a complex of minerals, the eutectic solution of lactose was subjected to repeated freezing. During crystallization of the eutectic solution, the formation of lactose crystals occurs primarily, and the solution is saturated with minerals. Better separation of lactose and minerals was observed at a temperature of minus 6°C. At a lower temperature, the rate of ice formation was too high, which led to an increase in the content of mineral salts in the solid phase. The efficiency of separation of components is significantly reduced after 540 minutes; therefore, it is not advisable to carry out further freezing. After this time, the amount of frozen product was 87%.

The resulting whey protein concentrate contains water (68.4%) and dry matter (31.6%). The protein content in the whey protein concentrate was 29.1%, which is consistent with the data on cryoconcentration of whey presented in the work [6]. The proposed method of concentration allows to obtain a concentrate of whey proteins with a higher content of dry matter, compared with the concentrate (with the mass fraction of dry matter being at 21.8%) described in the work [9]. In addition to the whey protein concentrate, the lactose concentrate was also obtained with a solids content of 62.5%, with 90% of them being pure lactose. Organoleptic characteristics of the obtained concentrates corresponded to their characteristic properties.

5. Conclusion

The use of separation freezing for the processing of whey allows one to increase business profits, to reduce the amount of waste contaminating wastewater, and to obtain biologically valuable

IOP Conf. Series: Earth and Environmental Science **395** (2019) 012070 doi:10.1088/1755-1315/395/1/012070

concentrates of whey proteins and lactose. These concentrates can be used to expand the range of existing and develop new foods.

References

[1] Zolotareva M S, Volodin D N, Topalov K V, Evdokimov I A, and Chablin B V 2016 About the processing of whey and the introduction of the best available technologies *Milk Processing* **7** pp 17-19

[2] Khramtsov A G 2011 *Phenomenon of whey* (St. Petersburg, Russia: Profession))

[3] Shchetinin M P 2013 Production and processing of whey in Russia and the Altai region *Polzunovsky Vestnik* **4**(4-4) pp 80-84

[4] Sviridenko Yu Ya, Kravchenko E F, and Yakovleva O A 2008 The use of whey and local wastewater treatment *Dairy Industry* **11** pp 58-60

[5] Agrovestik 2016 *Perspective directions of whey processing* Available at: https://agrovesti.net/lib/industries/dairy-farming/perspektivnye-napravleniya-pererabotki-molochnoj-syvorotki.html (Accessed 02 03 2019)

[6] Gavrilov G B 2004 *Modern aspects of whey processing by membrane methods* (Kemerovo, Russia: Kuzbassvuzizdat)

[7] Government of the Russian Federation 2008 *Order of the Government of the Russian Federation: The concept of long-term socio-economic development of the Russian Federation for the period up to 2020* (November 17, 2008 No. 1662-p) Available at: http://government.ru/info/6217/ (Accessed 02 03 2019)

[8] Korotkiy I A, Fedorov D E, Gushchin A A (2018) *Patent № 2654867 (Russian Federation), MPK A23L 3/00: Separating freezing device for liquid foodstuffs № 2016147301 Stated on December 1, 2016; Published on May 23, 2018* Available at: https://findpatent.ru/patent/265/2654867.html (Accessed 02 03 2019)

[9] Sánchez J, Hernández E, Auleda J M, and Raventós M 2011 Freeze concentration of whey in a falling-film based pilot plant: Process and characterization *Journal of Food Engineering* **103** pp 147-155

[10] Bogoviz A V , Bugai Y A, and Minenko A V 2019 Economic analysis of effectiveness of the existing tools of state support for entrepreneurship in the AIC in the digital economy *Lecture Notes in Networks and Systems* **57** pp 789-794

IOP Conf. Series: Earth and Environmental Science **395** (2019) 012071 doi:10.1088/1755-1315/395/1/012071

The burnt solid in adsorption treatment of water from heavy metal ions

O V Salishcheva[1], U V Tarasova[1*] and N E Moldagulova[1]

[1] Kemerovo State University, 6 Krasnaya str., Kemerovo 650000 Russia

E-mail: onh@kemsu.ru

Abstract. The process of adsorption of ions of heavy metals Cu^{2+}, Pb^{2+}, Zn^{2+}, Cd^{2+} from aqueous solutions with the mineral adsorbent burnt solid was investigated. Kinetic data on the adsorption process were obtained. Parameters of adsorption equilibrium using the Langmuir and Dubinin–Radushkevich equations have been calculated. The efficiency of extracting heavy metal ions upon modification of the adsorbent and in the area of low concentrations is shown.

1. Introduction

Development of the industrial sector makes it necessary to solve the problems of industrial wastewater treatment and waste disposal. The deteriorating environmental situation forces to tighten the requirements for waste dumping and wastewater treatment in business and public enterprises. Discharge of wastewater containing petroleum products, phenols, nitrogen compounds and heavy metals, as well as surfactants and other toxic compounds into the sources of centralized water supply exacerbates the situation with quality water supply. In the Kemerovo region, the chemical composition of the Tom River is formed under the influence of pollutants from wastewater from the enterprises of metallurgy, coke chemistry, as well as the coal and fuel and energy industries, agricultural enterprises and transport, and additionally through natural processes. The main pollutants of water bodies are still petroleum products, phenols, and heavy metal salts. Surface and underground waters are most intensively polluted with heavy metals in the areas of large-scale coal mining and metallurgical enterprises in the cities of Novokuznetsk, Leninsk-Kuznetsky, Belovo, Prokopyevsk, Kiselevsk, and Yurga. Heavy metal compounds form a group of toxicants. High toxicity, an ability to accumulate in human body, exerting harmful effects even in relatively small concentrations, determine the anthropogenic impact on the environment and on people.

Much attention is paid to the study of sorption purification of aqueous media from heavy metal ions using porous materials. Adsorption purification is more efficiently used as an additional purification after effluent treatment with chemical reagents by precipitating agents. Mineral adsorbents of natural origin are also very promising [1, 2]. The developed specific surface area and high absorption capacity with respect to liquids and solutes are characteristic of porous materials. They can serve as excellent carriers for fixing on the surface of various compounds when modifying them. Interest in modified mineral sorbents is caused by their specific properties: an inorganic carrier gives a sorbent such properties as high equilibrium rate, chemical resistance to aggressive media, mechanical strength; a modifier provides selectivity and completeness of ion binding [3]. The study focused on kinetics of adsorption processes allows one to determine the time required to achieve adsorption equilibrium and

Content from this work may be used under the terms of the Creative Commons Attribution 3.0 licence. Any further distribution of this work must maintain attribution to the author(s) and the title of the work, journal citation and DOI.
Published under licence by IOP Publishing Ltd

to obtain the parameters required for engineering calculations of these processes. The purpose of this paper is to study the process of adsorption of heavy metal ions on mineral adsorbents.

2. Materials and Methods

We used model solutions with variable content of heavy metal ions Cu^{2+}, Pb^{2+}, Zn^{2+}, Cd^{2+} as objects for our research. A mineral adsorbent is a burnt rock with a composition of 68.2% in SiO_2, 15.5% in Al_2O_3, which was obtained from the "Dalnie Gory" deposit in the city of Kiselevsk, Kemerovo region. It was formed as a result of thermal processes caused by the self-ignition of coal waste heaps in the worked-out spaces and waste dumps. After burning out an organic part in the rock, a mineral content increases up to 84.6%. The burnt rock of various fractions with a grain size of 1.5 mm, a grain shape factor of 2.1, and a porosity of 52–60% is used. It is modified with a sodium hydroxide solution with pH=12 and temperature of 60 °C.

Our research on kinetics of ion adsorption was carried out at the temperature of 25 °C on unmodified and modified with alkali (pH = 12) samples of adsorbents. Determining the equilibrium concentration of copper, cadmium, zinc, and lead ions in solutions was carried out by the complexometric method using Trilon B. To determine the adsorbent pore volume, we used the pycnometric method (also known as the "molecular probe" method) [4, 5]. To study the adsorption of heavy metal ions under static conditions, we used model solutions of their salts with variable concentrations. We found the value of adsorption by calculating a difference of the equilibrium concentrations of metal ions in an aqueous solution before and after adsorption, taking into account the volume of this solution and the mass of adsorbent [6]. We calculated the degree of purification as the ratio of the amount of adsorbed metal ion to the amount of metal ion in the initial solution. The Langmuir and the Dubinin-Radushkevich equations were used for reaching an analytical description of adsorption isotherms and calculating the parameters of adsorption equilibrium.

3. Results

Based on the experimental data obtained, the kinetic curves of the adsorption of cadmium, zinc, copper, and lead ions on the unmodified adsorbent were constructed (Fig. 1). Adsorption equilibrium is established within 40 minutes.

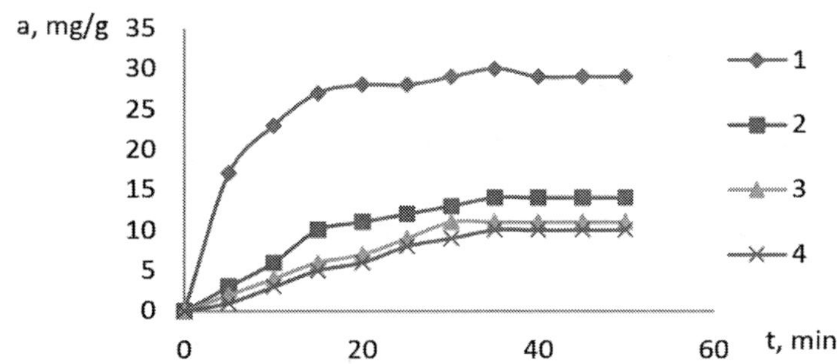

Figure 1. Kinetic curves of adsorption of metals ions Pb^{2+} (1), Cd^{2+} (2), Zn^{2+} (3), Cu^{2+} (4) from aqueous solutions on the burned rock (t=25°C).

The limiting adsorption capacity of the burned rock was Pb^{2+} 32.25 mg / g, Cd^{2+} 22.15 mg / g, Cu^{2+} 11.21 mg / g, Zn^{2+} 10.01 mg / g. To increase the number of surface adsorption centers, we processed the material under study with an alkali solution. Alkaline modification led to an increase in adsorption, the limiting adsorption capacity values of the treated burned rock were obtained at Pb^{2+} 46.08 mg / g, at Cd^{2+} 31.05 mg / g, at Cu^{2+} 18.32 mg / g, at Zn^{2+} 16.88 mg / g.

According to the research results of adsorption of heavy metal ions from aqueous solutions of natural and modified forms of adsorbents, adsorption isotherms were obtained, which are the main criterion for evaluating adsorption properties of the adsorbents under study and determining the dependence of adsorbent activity on concentration of adsorbate in equilibrium. Figure 2 shows the adsorption isotherm of lead ions.

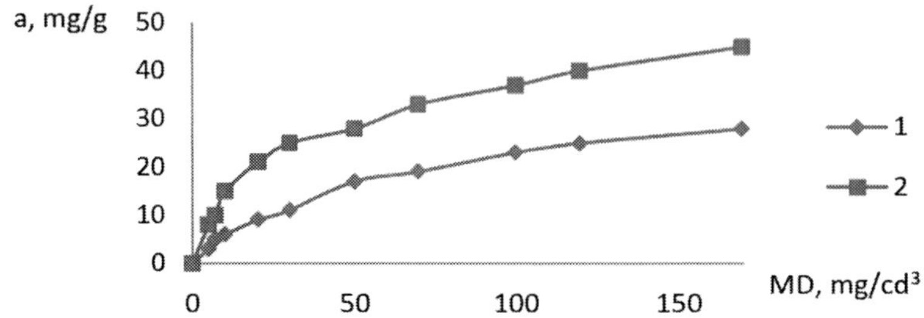

Figure 2. Adsorption isotherm of lead ions (II), samples of burnt rock without modification (1) and being modified (2).

Figure 3 shows the adsorption isotherms of lead ions in the coordinates of linear form (Langmuir equation).

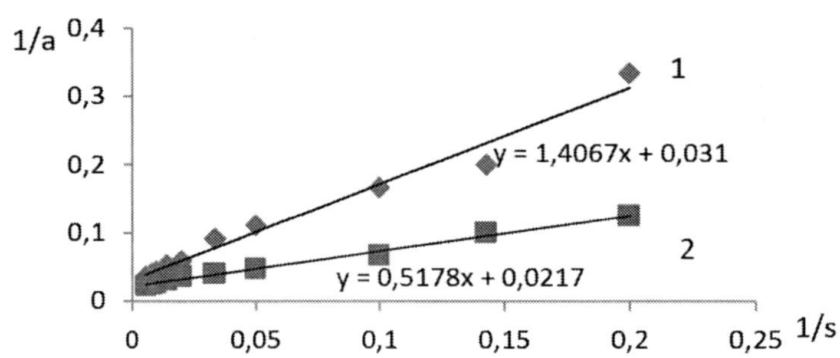

Figure 3. A linear view of the adsorption isotherms of lead ions (II), samples of burnt rock without modification (1) and being modified (2).

Based on the obtained experimental data, the parameters of adsorption of metal ions were determined (Table 1).

Table 1. Parameters of adsorption of heavy metals from aqueous solutions on the burned rock.

Ion / Adsorbent	Cleaning degree, %	Langmuir equation		Dubinin-Radushkevich equation		
		a_{max}, mg/g	K	a_{max}, mg/g	E, kJ / mol	W, cm^3/g
Pb^{2+} / burnt rock	49.5	32.25	0.031	32.6	1.31	0.036
Pb^{2+} / modified burned rock	87.5	46.08	0.021	45.5	2.03	0.050
Cd^{2+} / burnt rock	50.5	22.15	0.018	22.68	0.38	0.019
Cd^{2+} / modified burned rock	77.5	31.05	0.027	35.13	1.38	0.022
Cu^{2+} / burnt rock	55.5	11.21	0.091	11.62	0.81	0.011
Cu^{2+} / modified burned rock	84.0	18.32	0.121	19.05	2.79	0.016
Zn^{2+} / burnt rock	48.4	10.01	0.022	09.98	0.78	0.016
Zn^{2+} / modified burned rock	67.5	16.88	0.031	17.05	2.73	0.021

To determine the limiting stage of the diffusion process and establish the model of the samples' porous structure, the dependence of the degree of achieving the adsorption equilibrium (γ) and the dimensionless kinetic parameter (T) on the duration of mixing solutions (t) was studied. The equations determining the mass transfer coefficient and describing the external diffusion kinetics of absorption of solutes were used for calculations [7, 8]. The mass transfer coefficient (β) characterizes the adsorbate absorption rate by a unit volume of the adsorbent and is found according to the formula:

$$\beta_{\Pi} = \frac{dC}{dt \cdot (C_t - C_p)} \quad (1)$$

The value of T, proportional to the time of the process, is found according to the formula:

$$T = A \cdot \beta \cdot t \quad (2)$$

According to the linear dependence T = *f*(t), the adsorption of ions is limited by the process of external mass transfer for 5-10 minutes. Further deviation from the straight line shows that as the sorbent is developed, internal diffusion becomes a limiting stage in this hydrodynamic regime. More than that, values of the external diffusion mass transfer coefficients for the unmodified burnt rock were 0.004-0.008 min^{-1}, as well as 0.009-0.03 min^{-1} for the modified burnt rock.

Formation of a porous structure in the burnt rock took place as a result of an uncontrolled natural process. The volume of the adsorbent porous space falling on the micropores, determined by the "molecular probe" method, showed that micropores with a diameter of 0.35 nm make the greatest contribution to the porous space. The total volume of the porous space of the burned rock was 0.20 cm^3/g. The pore volume with a diameter of 0.35 nm was 38% of the total pore volume.

4. Discussion

The processing of adsorption isotherms of heavy metals according to the Dubinin-Radushkevich equation made it possible to calculate the following adsorption parameters of adsorbents: (a) a limiting value of the adsorption value in micropores; (b) a volume of the limiting adsorption space; (c) an energy value, which is the energy characteristic of the burning rock as a microporous adsorbent. It is known that low values of adsorption heat (up to 40 kJ / mol) and characteristic energy are typical for physical adsorption. For the studied adsorbents' samples, the value of the characteristic energy (E) is in the range of 0.38–2.8 kJ / mol, which indicates the physical nature of adsorption. The magnitude of the characteristic adsorption energy for modified adsorbents is 1.5–3.6 times higher than for unmodified ones, which indicates that, all other things being equal, the adsorption of heavy metals on modified adsorbents is more energy-efficient than on unmodified ones. Values of the limiting adsorption volume, calculated by the Dubinin-Radushkevich equation during the adsorption of heavy metal ions, are less than the total volume of micropores if being calculated by the "molecular probe" method. At the same time, incomplete filling of micropores with sizes of $0.35 < d < 0.67$ nm with heavy metal ions is observed. Thus, micropores with a size of less than 0.35 nm make the largest contribution to the adsorption of heavy metal ions ($r = 0.091 - 0.133$ nm).

Adsorption capacity of the modified adsorbent exceeds the capacity of the unmodified in 1.5-2 times. Therefore, we can assume that the extraction of heavy metals with a modified adsorbent occurs with the predominant participation of the donor groups of the modifier fixed on its surface and the silanol groups of the matrix.

For the practical use of adsorbents, in the process of water treatment, an important characteristic is the degree of extraction of the adsorptive. Studied adsorbents most effectively extract heavy metal ions in the area of low concentrations.

5. Conclusion

The obtained experimental data showed the possibility of using aluminosilicate mineral materials for the extraction of heavy metal ions from wastewater, the modification of which increases the degree of ion extraction. The obtained values of the degrees of purification of aqueous solutions with mineral adsorbents showed that the studied adsorbents most effectively extract heavy metal ions in the region of low concentrations. It is more efficient to use adsorption purification with the help of burnt rock as additional purification after treating effluents with chemical reagents with precipitating agents. Studies on the absorptive capacity of aluminosilicates of industrial origin, which are waste products of various industries, make it possible to search for directions of recycling industrial wastes that are not efficiently stored near enterprises and are harmful to the environment [9].

References

[1] Anurov S, Belevich A, and Yattara B 2018 Wastewater treatment from heavy metals with mineral absorbers *Contemporary High Technologies* **2** pp 9-14

[2] Shachneva E, Archibasova D, Magomedova E, and Zukhayrayeva A 2014 Extraction of copper, zinc and cadmium from aqueous solutions of oil fields by a sorbent obtained on the basis of the flasks of the Astrakhan region *Geology, Geography and Global Energy* **53**(2) pp 93-104

[3] Tsvetkova A D and Akayev O P 2010 The use of natural and industrial aluminosilicates in the process of adsorption of heavy metals ions *Vestnik of the Kostroma State University named after N. A. Nekrasov* **16**(2) C 18-20

[4] Karnaukhov A 1999 *Adsorption, texture of dispersed and porous materials* (Novosibirsk, Russia: "Science", Siberian Enterprise of the Russian Academy of Sciences)

[5] Adamova L, and Safronov A 2008 *Sorption method for studying the porous structure of nanomaterials and the specific surface area of nanoscale systems* (Ekaterinburg, Russia: USU named after A. M. Gorky)

[6] Gelfman M 2005 *Workshop on colloid chemistry* (St. Petersburg, Russia: Lan)

[7] Ruckenstein E, and Berim G 2010 Effect of solute-solute and solute-solvent interactions on the kinetics of nucleation in liquids *Journal of Colloid and Interface Science* **342** pp 528-532

[8] Parfitt G 1993 *Adsorption from solution at the solid* (New York, NY: Academic press)

[9] Gelfman M, Shevchenko T, and Tarasova Yu 2004 *A method of obtaining a sorbent for the purification of solutions from heavy metals* (Patent 2224143 RF МПК7 B 01 J 20/16)

Enrichment of food products with local berry raw materials for the residents of border areas

G Rybakova[1], G Gulenkova[1*] and G Chepeleva[1]

[1] Siberian Federal University, 79 Svobodny prosp., Krasnoyarsk 660041 Russia

E-mail: pochta_galiny@.ru

Abstract. The article deals with topical issues of forming a range of functional foods with useful properties in the border areas. Also, the article considers the technological possibilities of introducing berry raw materials into traditional food products.

1. Introduction

Border territories of the constituent entities of the Russian Federation in the Concept of their development are recognized as the most important resource of the country in many strategic aspects [1], [7]. As part of the development of agricultural production in these regions, the state declares "the creation of cooperatives for the harvesting and processing of wild plants, ... berries"; "the creation of enterprises for the production of ... food supplements based on biologically active substances"; the production cooperation aimed "at the processing of raw materials; the creation of enterprises for the extraction and processing of natural resources"; "organization of wholesale-distribution centers in the ... border areas for the sale of agricultural products, collecting agricultural products ..., its storage and marketing (in the future, it is possible to create processing sites on the basis of these centers with the release of finished food products);" one of the promising areas [1]. It is not by chance that in a number of key state documents aimed at ensuring the security of the state and its citizens, attention is directed to such areas of development. On the one hand, this is connected with the economic problems indicated in the documents in the regions remote from the center. For example, in relation to the border areas, the following is noted: their low share of participation in the formation of the gross regional product; weak activity of small and medium enterprises; low employment rates of the working population, a high level of migration; inefficient use of the natural resource potential of border areas. In such a situation, the proposed activities will contribute to the solution of regional economic problems.

On the other hand, these activities will allow solving issues related to ensuring a decent quality and standard of living of citizens, impossible without the realization of their right to safety, health and a healthy lifestyle, which are also part of the country's security [4-6]. In this regard, the efforts of many Russian scientists are aimed at finding new, non-traditional types of food raw materials and developing recipes for the preparation of fortified or specialized food products based on it. One of the promising areas of scientific research in this aspect is the study of the possibility of increasing the nutritional value of products by including local ingredients in the recipe, which are the sources of vitamins, minerals and other biologically active components.

Content from this work may be used under the terms of the Creative Commons Attribution 3.0 licence. Any further distribution of this work must maintain attribution to the author(s) and the title of the work, journal citation and DOI.
Published under licence by IOP Publishing Ltd

2. Materials and Methods

An information-logical model of technological processes of production and determination of the quality of the obtained enriched products based on the introduced varieties of sea buckthorn (krushenovidnaya) was taken as the object of study.

The area of distribution of wild species is extensive and includes a large part of Europe and Asia. Sea buckthorn in Siberia forms the largest thickets in the Altai Region, the Republics of Tyva and Buryatia, it is widely represented in the border areas. This makes it a promising raw material for use in food. The organization of its collection and processing will allow not only to solve the problem of enrichment of traditional products for the population, but also the task of attracting local labor resources for this, developing small or medium-sized enterprises. In the future, we can talk about improving the efficiency of using the natural resource potential of border areas. On the example of introduced varieties, the work showed a consistent information-logical model and research procedure, including the steps from studying the technological and biochemical characteristics of the raw materials and formulating the formulation using mathematical models, to producing and determining the consumer properties of the resulting enriched products. This model was considered as a system.

3. Results

The target state of the system of the object under consideration from the standpoint of a systematic approach to the overall goals of the global level, as recorded in strategically important state documents in the field of safety and ensuring the quality of life of the population, including border areas, adds goals of a lower order. They set the main vector of scientific research, including the following: conducting a study of the biochemical composition of the sea-buckthorn (krushinovidnaya) fruit of various regions; developing approaches and technologies for obtaining enriched food products based on it; identifying approaches to increase compliance with the safety and quality of new products.

The main factor generating the system under study is the contradiction between the situation in which the use of non-traditional types of local plant raw materials has a high potential for solving a number of problems (implementation of state policy in the field of healthy nutrition, environmental management, expanding the range of enriched products) and the lack of experimental justification for such a direction. Given the richness of the properties and composition of the fruit of sea buckthorn (krushinovidnaya), the presence of demand for enriched natural foods, the use of its fruits to produce such products is very important.

Analysis of the possibilities of food fortification through the use of local raw materials, the study of the range, biological features and biochemical composition of sea buckthorn fruits as raw materials for the production of puree are properties of the system. The properties of the system under study are specific biological, biochemical, chemical, and physicochemical parameters of the quality and safety of raw materials. Development of a model and substantiation of effective technological processes for obtaining puree from sea-buckthorn fruits, taking into account the biochemical parameters of raw materials (the content of dry substances, sugars, acids), is an element of connection between the studied parameters. This allows us to consider this network of interactions as a structure from the standpoint of a systematic approach.

Elements of the management system for the assortment and quality of fortified food products include the following: conducting a study of the morphological properties and chemical composition of fruits from different regions, identifying the promising varieties for industrial processing; developing a technological scheme for the production of puree, analyzing compliance with the safety and quality of raw materials (sea-buckthorn fruit) and processed products (natural puree, puree with sugar), analyzing the safety and quality of yogurt enriched recipes; conducting a mathematical modeling of process optimization.

The group of system-dependent factors for the submitted research related to the enrichment of food products is indicated by the state in the basic documents. It includes economic, managerial, and social factors, including the following: solving economic and migration problems, which were mentioned at the beginning of the article, issues of employment, providing the diet with essential nutrients.

4. Discussion

The general scheme for developing an innovative product using sea buckthorn is already described in our literature published in Russian [2], [3]. In total, 8 varieties of sea-buckthorn (krushinovidnaya) fruits during consumer maturity, growing in different climatic regions, were taken for research. At the first stage of the study, we determined the chemical composition, morphological features, their dependence on the growing area, the influence of the technological characteristics of the sea-buckthorn (krushinovidnaya) fruits on the production processes and the functional properties of new processed products with the addition of sea-buckthorn fruits. Also, we determined the quality indicators of the products based on vegetable raw materials, their compliance with the requirements of regulatory documents. The findings are also published in the sources [2], [3]. Average long-term data on sea buckthorn yields show that most varieties in the conditions of the Altai, Krasnoyarsk regions, and Khakassia have similar productivity. It was shown that all the samples of sea buckthorn studied had a balanced fatty acid composition. This characterizes the sea buckthorn fruit as a valuable food raw material. The fruits of different ecological groups differ in the accumulation of the following vitamins and minerals: vitamin C, carotenoids, and nicotinic acid. Its Vitamin E content is highest in the samples from the Republic of Tuva. On the basis of accumulation of macro and microelements, the samples introduced in the Krasnoyarsk Region and the Republic of Khakassia stand out. The total ash content of the sea-buckthorn fruit is determined by the biological characteristics of the variety. Microbiological indicators (the content of non-spore-forming microorganisms, mold fungi and yeast), sea buckthorn fruits meet the requirements of SanPiN 2.3.2.1078.

Technological processes for the production of processed products from sea-buckthorn fruits are presented in the form of an information-logical model, where the links and intensity of transients are indicated. For example, λ_{VI} is the intensity of the flow from link V to link I. In [3], the processes of the technological line were presented in the form of a model. As a result of a computational experiment, at different values of the intensities of the processes, we found that solving the Cauchy problem for a system of differential equations depending on the parameters x, y, z is a model of the dynamics of technological processes (where x is the dry matter content,%; y is the sugar content,%; z is the content of acids in the puree from the fruits of sea buckthorn,%).

The proposed method of scientific and practical research allowed us to create the structure of the technology for obtaining functional products from sea buckthorn fruits (natural puree, puree with sugar). The developed analytical model of technological processes for the production of sea-buckthorn puree made it possible to calculate the optimal values of the regime parameters for natural puree, and puree with sugar, respectively: 46.985 and 63.85% is the solids content; 4.102 and 25.546% is the sugar content; 1.024 and 0.964% is the acid content. Laboratory experiments showed that the values obtained differ from the theoretical values by no more than 5%.

The use of sea-buckthorn puree as a functional supplement made it possible to obtain new yogurt formulations that satisfy the daily need for β-carotene, vitamin E by 60% and ascorbic acid by 70%. The doses of the supplement are determined: 7 and 10% for natural puree and sugar puree, respectively. Yoghurt is enriched with carbohydrates, fatty acids, and micronutrients. The payback period of the proposed technology is 2 years [2], [3].

5. Conclusion

Optimized technological parameters for the production of sea-buckthorn puree provide not only high-quality products with biological value for consumers, but they also contribute to the creation of new efficient production. The expected payback period is 2 years. During this time, the profit will exceed the initial cost of producing this product [2], [3]. The rationality of introducing the production technology of sea-buckthorn fruit puree into practical activity has been confirmed under the conditions of individual entrepreneur N. M. Popovenko (Chernogorsk, Russia). Enriched yogurts based on sea-buckthorn fruit puree have high tasting characteristics; they meet the requirements for physico-chemical quality indicators and have a high biological value.

IOP Conf. Series: Earth and Environmental Science **395** (2019) 012072 doi:10.1088/1755-1315/395/1/012072

Making puree from the sea buckthorn fruits had a positive impact on the physico-chemical and structural-mechanical characteristics of the finished product. Yoghurts with 7 and 10% supplementation of puree from the main raw materials had the best indicators on the accumulation of vitamins [2], [3]. A new product allows to rationalize the nutrition of the population with balanced products. It also makes it possible to expand the range of fermented milk products on traditional equipment without additional costs for instrumentation of the technological process.

6. Acknowledgments

The authors are grateful for the contribution to the Candidate of Biological Sciences, Associate Professor of Goods Management and Examination of the Siberian Federal University Leontyev Vladimir Mikhailovich, as well as to the Doctor of Pedagogical Sciences, Professor, Head of the Department of Commodity Research and Examination of Goods of the Siberian Federal University Krotova Irina Vladimirovna.

References

[1] Government of the Russian Federation 2015 *Order of the Government of the Russian Federation on approval of the concept for the development of border areas of the subjects of the Russian Federation constituent of the Far Eastern Federal District and the Baikal Region* (October 28, 2015 N 2193-p) Available at: http://docs.cntd.ru/document/420313859 (Accessed 09 04 2019)

[2] Menyaylo L N, Baturina I A *et al* 2015 *The scientific basis for the formation of a range of food products with desired properties: Technologies for production and processing of vegetable raw materials* (Krasnoyarsk, Russia: Siberian Federal University)

[3] Gulenkova G S, Demakova E A *et al* 2016 *A systematic approach to managing the range and quality of products* (Krasnoyarsk, Russia: Siberian Federal University)

[4] Government of the Russian Federation 2008 *Order of the Government of the Russian Federation: The concept of long-term socio-economic development of the Russian Federation for the period up to 2020* (November 17, 2008 N 1662-p) Available at: http://docs.cntd.ru/document/902130343 (Accessed 09 04 2019)

[5] Presidential Executive Office 2015 *Presidential Decree "On the National Security Strategy of the Russian Federation"* (December 31, 2015 N 683) Available at: http://docs.cntd.ru/document/420327289 (Accessed 09 04 2019)

[6] Government of the Russian Federation 2010 *Order of the Government of the Russian Federation on the approval of the Fundamentals of the state policy of the Russian Federation in the field of healthy nutrition of the population for the period up to 2020* (October 25, 2010 N 1873-r) Available at: http://docs.cntd.ru/document/902242308 (Accessed 09 04 2019)

[7] Bogoviz A V, Tufetulov A M, and Chepik D A 2018 The mechanism of activation of the process of import substitution in the agro-industrial complex for provision of food security *Advances in Intelligent Systems and Computing* **622** pp 71-76

IOP Conf. Series: Earth and Environmental Science **395** (2019) 012073 doi:10.1088/1755-1315/395/1/012073

Prediction of the space-time state of systems using geospatial data for assessing technological risks

T Yu Bugakova[1*]

[1] Siberian State University of Geosystems and Technologies, 10 Plakhotnogo str., Novosibirsk 630108 Russia

E-mail: kaf.pi@ssga.ru

Abstract. The article deals with the tasks of mathematical modeling and forecasting of the space-time state of systems using geospatial data for assessing technological risk. Predicted functions of changes in the space-time state of the system, obtained by the exponential smoothing method, are constructed. The article provides an example of options for changing the space-time state of an object based on the exponential smoothing method.

1. Introduction

In the twentieth century, the intensive development of new technologies led to the outstanding results in all spheres of the man-made technogenic activities, creating an unprecedented potential and real threats to humans. Many emergency situations happen in the world every year. Sources of accidents and disasters are geodynamic processes and improper operation of man-machine systems consisting of hardware, computers, software and personnel actions [1]. In the aggregate, geodynamic and man-made systems are complex systems, the main security problem of which is the inability to completely eliminate the risk of emergency situations in such systems and the need to minimize this risk. Therefore, the main objective of studying the problem of man-made risk assessment is not to determine the damage to an object as an already accomplished fact, but to prevent a dangerous situation, search for managing the space-time state (STS) of the system, ensuring that the risk is reduced to a minimum.

2. Materials and Methods

The achievement of the stated goal is based on the principles of the system-target approach and is carried out as a result of solving the following set of interrelated tasks:

1. Development of a mathematical model for determining the spatial-temporal state of the system according to geodetic data for the assessment of technological risk.
2. Development of a forecast model of the space-time state of the system, the ability to predict scenarios of changes in its state to reduce risk and select alternatives.

The initial data for modeling STS are the time series of coordinates of the set of control points of the object (system) under investigation, obtained from the results of repeated cycles of geospatial measurements.

The model of the dynamic system is adopted as a formal model of the object [2].

Content from this work may be used under the terms of the Creative Commons Attribution 3.0 licence. Any further distribution of this work must maintain attribution to the author(s) and the title of the work, journal citation and DOI.
Published under licence by IOP Publishing Ltd

$$X \overset{\varphi}{\Rightarrow} S \overset{f}{\Rightarrow} Y, \tag{1}$$

Where X is the set of input signals; Y is the set of output signals; S is the system state space; φ is the display of the transition of the system from state to state as a result of the flow of input information; f is the system output display.

The task of determining the STS system is reduced to a meaningful finding of the elements of the model (1). The array of the height coordinates of the control points of the object was taken as the source data for solving this task, i.e. the state of the object at the moment t_j is determined by the height coordinates of the points. Consequently, the set X consists of scalar functions:

$$H_i = H_i(t) \tag{2}$$

The state space of an object is defined as the Cartesian product of all elements of this set. The dimension of space n is equal to the number of control points.

Each cycle of observations with a number j in the state space has a corresponding point, the radius vector of which

$$\overline{H}(t_j) = \sum_{i=1}^{n} \overline{k_i} \cdot H_i(t_j), \tag{3}$$

where $\overline{k_i}$ are the ort-vectors of the basis of n, a dimensional state space.

Thus, the function φ is a mapping that puts the phase point of the state space in accordance with the set of input signals $H_i(t_j)$. This point represents the state of the system in the loop with the number j. The set of points whose radius vectors are determined by the vector function (3) in each observation cycle forms a phase trajectory in phase space, which is an explicit function of coordinates and time characterizing the change in the state of the system from cycle to cycle. An important step in the assessment of technological risk is the prediction of changes in the space-time state of systems.

There are various options for solving this problem. One of the simplest and quite effective methods for its solution is the exponential smoothing method [3], [4], [5].

We assume that in the prediction interval, the evolution of the state of the system looks like this:

$$y = A + \eta, \tag{4}$$

where η is a random stationary uncorrelated process with zero expectation. Let the value A occasionally change in steps. The magnitude of the change in A and the moment of change are unpredictable, and the time interval during which the value of A remains unchanged significantly exceeds the interval between our observations. Under these assumptions, the smoothed observation function is:

$$S_t = A \cdot y_t + (1 - A) \cdot S_{t-1}. \tag{5}$$

3. Results and Discussion

As a result of research, the predictive functions of the change in the space-time state of the system, obtained by the exponential smoothing method, are constructed (Figure 1).

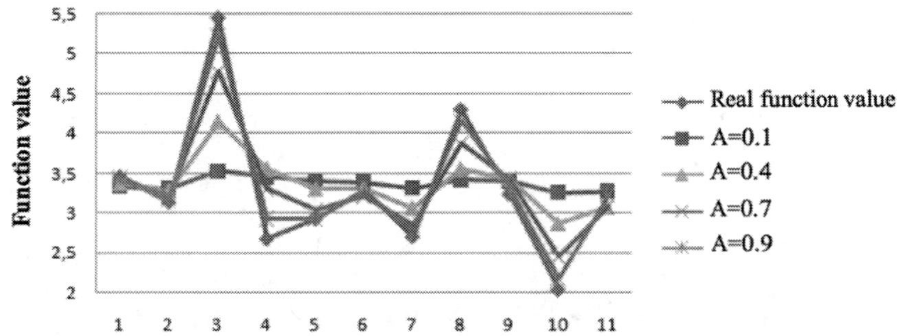

Figure 1. Predicted functions of changes in the space-time state of the system, obtained by the exponential smoothing method.

Here, $A \in [0,1]$ is the smoothing constant, S_t is the smoothed value y, referred to the moment t; y_t is the state of the object at the moment t.

We assume that the "degree of stochasticity" (entropy of the process) is known, and it is determined by the probability value $P \in [0,1]$ (Figure 2).

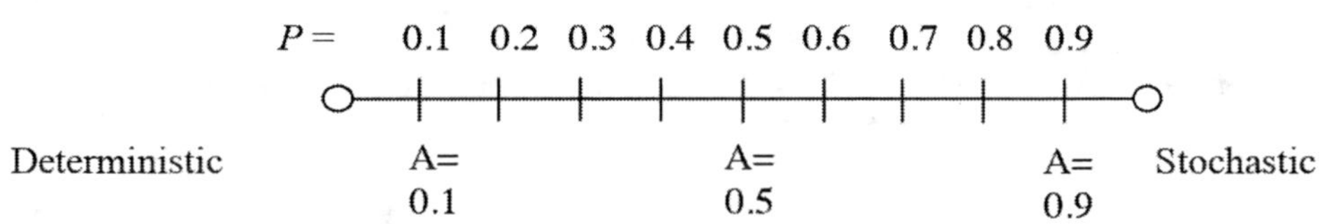

Figure 2. Probability P of the origin of random events.

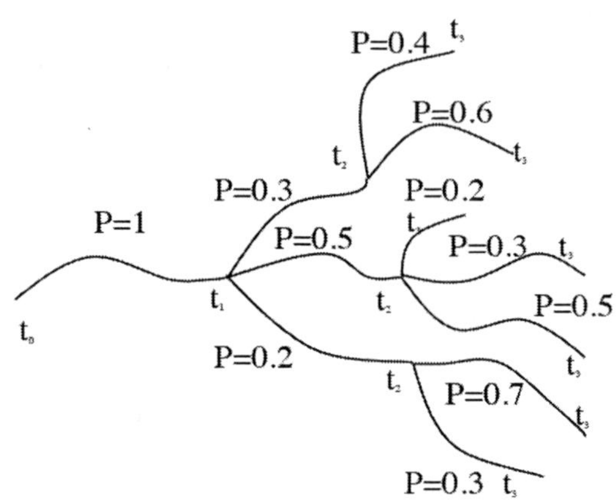

Figure 3. An example of options for changing the space-time state of an object based on the exponential smoothing method on the interval t_i=*0..3*.

The most important characteristic in this model is the smoothing constant *A*, the largest by which the forecast is carried out. The closer the value of this parameter is to 1, the greater the likelihood of the occurrence of random events, and vice versa. Therefore, the choice of the value *A* depends on the value of the probability *P*. The change in the space-time state is determined by the function, where the state of the object at a fixed time *t* is represented by a point.

At each time interval t_i, where *i=0..n*, the value of *P* can change and then for prediction the value of *A* needs to be changed.

To assess the technogenic risk, we need to know not only the spatial-temporal state of the system, but also possible scenarios for its change, as well as the probability distribution of scenarios [3] (Figure 3).

Thus, knowing the degree of "stochasticity" of the process, we can choose the optimally "safe" strategy for changing the space-time state of an object.

4. Conclusion

Determining the future of the STS system is always based on some plausible hypotheses, the reliability of which is unknown, as a rule. To reduce the measure of uncertainty when choosing the most useful solution, the concept of risk is introduced, which is characterized by the magnitude of the damage due to the choice of decision.

References

[1] Belov P G 2003 *System analysis and modeling of hazardous processes in the technosphere* (Moscow, Russia: Akademiya)

[2] Bugakova T Yu 2005 *Assessment of the stability of the states of objects from geodesic data using the phase space method* (Dissertation of the Candidate of Technical Sciences) (Novosibirsk, Russia)

[3] Bugakova T Yu, and Vovk I G 2012 Mathematical modeling of the space-time state of systems by geometric properties and the assessment of technological risk using the exponential smoothing method *Vestnik of the Siberian State Academy of Geodesy* **4** pp 47-58

[4] Bugakova T Yu 2011 On the issue of risk assessment of geotechnical systems based on geodetic data *Interexpo Geo-Siberia* **1**(1) pp 151-157

[5] Vovk I G, and Bugakova T Yu 2010 *The theory of determining the technogenic geodynamic risk of the spatial and temporal state of technical systems* In *Proceedings of the V International Scientific Congress "GEO-Siberia-2010"* (pp 21-24) (Novosibirsk, Russia: SSGA)

Multi-agent modeling of the control system of space-time states of man-made objects

T Bugakova[1], A Sharapov[1*] and I Knol[1]

[1] Siberian State University Geosystems and Technologies, 10 Plakhotnogo str, Novosibirsk 630108 Russia

E-mail: sharapov_artem@mail.ru

Abstract. The article discusses the development of a multi-agent system for solving problems of determining, controlling, and predicting the space-time states (STS) of objects. A generalized scheme of a multi-agent approach to solving the problem of determining the STS of an object is given. The article describes an algorithm "decision trees" on the example of the decomposition of the object in order to monitor and predict the STS of the structural parts of the object.

1. Introduction

Currently, the development and implementation of automated monitoring systems (AMS) for the safety of man-made objects is actively underway. Mainly, it relates to especially dangerous, unique, technically-complex construction objects, representing a threat to human life and health, and the environment. The relevance of such work is confirmed by the requirements of the Government of the Russian Federation on the need to develop a single basic system for monitoring critical facilities.

The main advantage of automated systems is the automation of the process of collecting data from the sensors installed in the object's body, which allows determining the state of an object with any degree of discreteness at any time of the day, in any weather conditions. However, the data processing algorithms incorporated in these systems are not unified, they are not capable of solving the complex problems of determining the space-time state of the entire object as a whole [1-4].

2. Materials and Methods

Each object is unique in its own way, it has its own design features, it is subject to the influence of various external factors, often with elements of uncertainty. Therefore, the task of creating such a system of control STS arises, which would work in conditions of uncertainty, make decisions based on available data, be able to adapt to changes in goals, have the ability to make management decisions independently. If the uncertainty is higher, the decision-making processes are more distributed in nature and the more unplanned events happen. In turn, it lowers efficiency in the existing systems, which are unable to make decisions independently and automatically rebuild to changes in the environment.

A multi-agent technology can solve this problem. They are implemented on the basis of multi-agent systems, using a large flow of information flows, when working with a large amount of data of diverse nature.

Content from this work may be used under the terms of the Creative Commons Attribution 3.0 licence. Any further distribution of this work must maintain attribution to the author(s) and the title of the work, journal citation and DOI.
Published under licence by IOP Publishing Ltd

3. Results and Discussion

An multi-agent system is based on the interaction between intelligent agents capable of communicating with each other and making decisions together. Each agent includes solving a problem of analysis, forecasting and diagnostics, identifying hidden dependencies and supporting optimal decision making based on the methods and algorithms incorporated in it. Similarly, each agent in the system is an "expert" in his field who is able to solve a specific task, thereby contributing to the achievement of a common goal. Figure 1 shows the enlarged structure of a typical intelligent agent. At the entrance, an agent is given the characteristics of the environment, and at the output, information about the state of the environment and the option of optimally solving the task set for him are obtained. The solver is a decision procedure that is implemented as a simple algorithm [8].

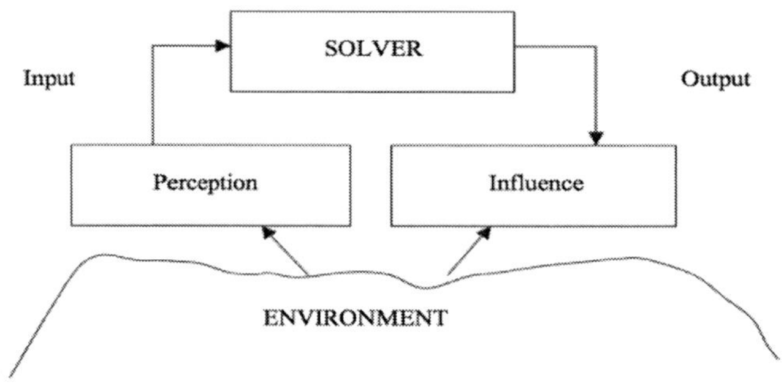

Figure 1. The integrated structure of an intelligent agent.

To solve the problem of determining the space-time state of a man-made object, the authors developed a prototype multi-agent system. The system consists of a subordinator agent, an integrator agent, executing agents, and communication agents. The general scheme of interaction between agents is shown in Fig. 2 [9].

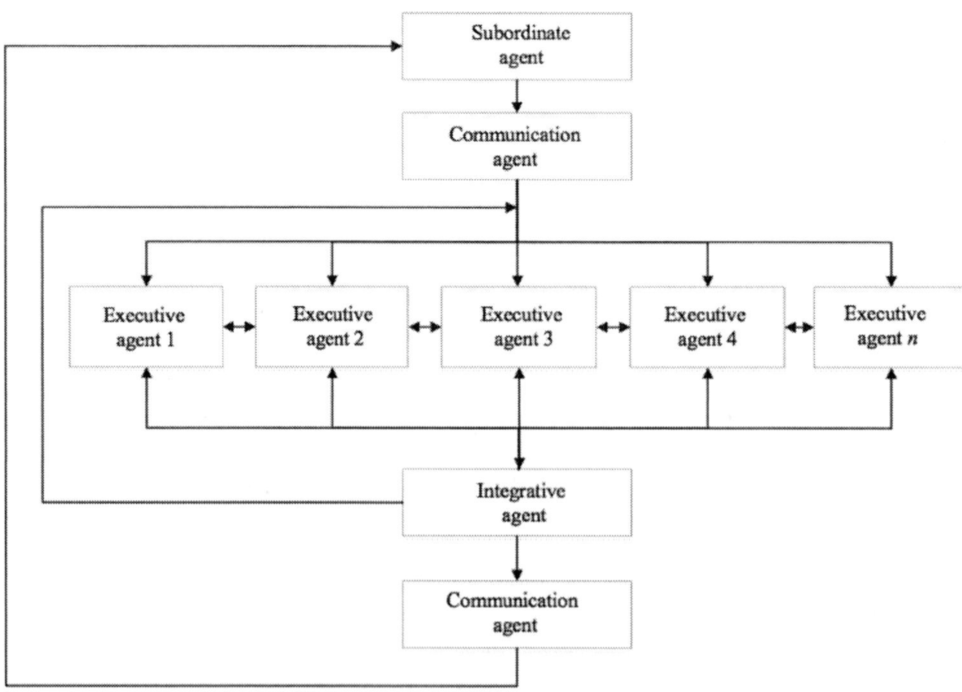

Figure 2. Interactions between intelligent agents.

In the course of work on the design of a prototype multi-agent system for monitoring the state of man-made objects, the following structural blocks were identified:

- The block of data collection, which provides for the collection, storage, transportation of data, information about the studied man-made object;
- Analytical units, providing analysis of incoming data;
- The block of user interaction, providing the user with all the necessary information about the object [5-7].

An important part of the multi-agent state-of-the-art object monitoring system is the analytics unit, implemented as a set of intelligent agents, analyzing incoming state data using various algorithms and data processing methods. When analyzing data, such algorithms as an "event tree", an algorithm for choosing optimal solutions, algorithms for working with graphs, etc. are appropriate to apply. Figure 3 shows a generalized scheme of a multi-agent approach to solving the problem of determining the space-time state of an object.

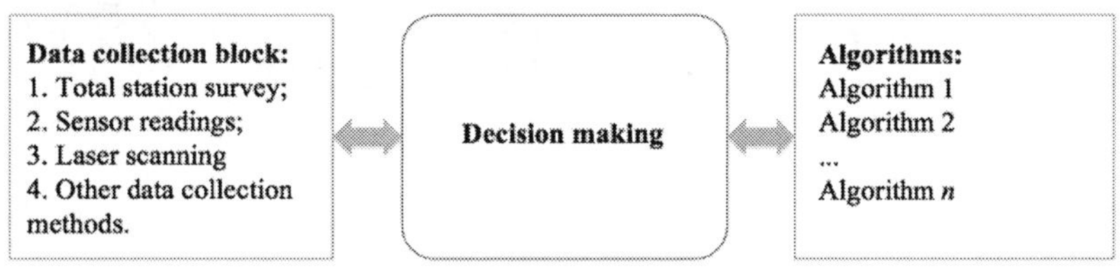

Figure 3. A generalized scheme of a multi-agent approach to solving the problem of determining the STS of an object.

First, the system requests all the necessary information about the object under study from the data acquisition unit. After that, the selection of algorithms is performed to determine the STS of the object. Further, the system is able to make a management decision.

One of the actual data mining algorithms is the "decision trees" algorithm. "Decision trees" is a way of representing rules in a hierarchical, sequential structure, where each object has a single node that provides a solution. Decisive trees reproduce logical schemes that allow one to get a final decision on the classification of an object using answers to a hierarchically organized system of questions. Moreover, the question asked at the subsequent hierarchical level depends on the answer received at the previous level. We consider this algorithm on the example of a multi-agent decision-making system about the decomposition of an object in order to determine and predict the STS of the structural parts of the object. (Fig. 4) [3].

We need to note that almost any man-made object is subject to deformation. Deformation is formed as a result of uneven exposure of external factors, which leads to the structuring of the object into parts (blocks). Each structural part under the influence of external factors may have a different direction of movement in space and a different speed. In this case, the system's tasks are to detect the boundaries between the blocks and determine the STS of each of them.

The graph presented in the form of a "decision tree" corresponds to Fig. 4. The tree includes a root node, which contains the following question: "Is the object under study as a whole in a state of relative equilibrium or not?" The ultimate goal of a multi-agent system is not only to determine the STS of an object, but also to choose the most optimal solution, selecting the strategy of developments that would provide the object with a state of "relative equilibrium".

The object is in the state of "relative equilibrium" when the deviation of the state S from the original S0 does not exceed the permissible value E. It is expressed by the following inequality:

$$|S - S0| <= E. \qquad (1)$$

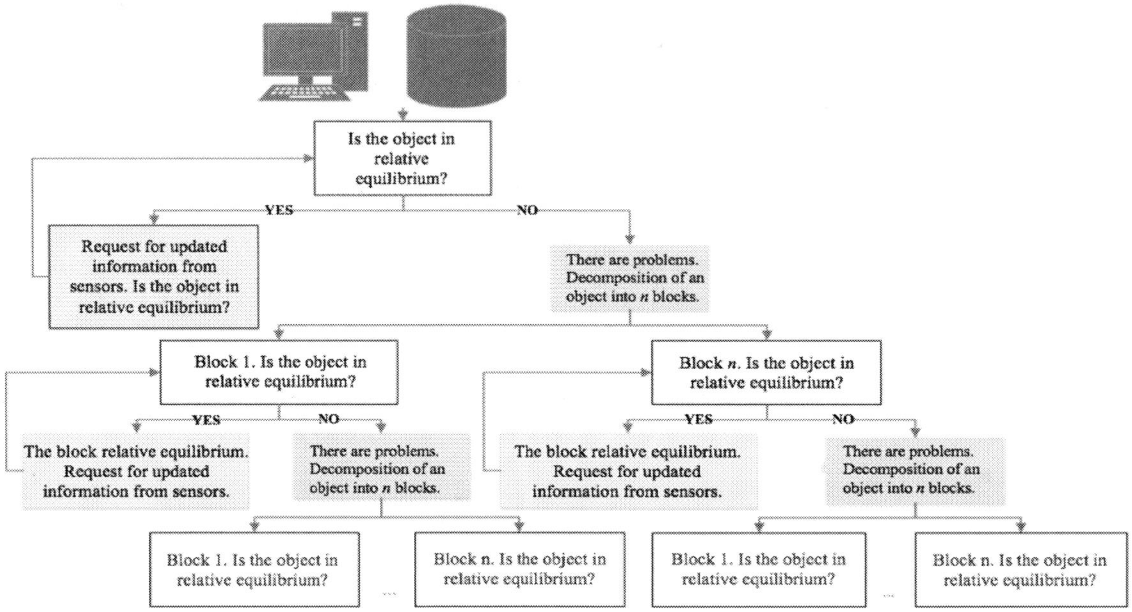

Figure 4. The object decomposition algorithm for determining and predicting the STS of the structural parts of the object.

The system makes a request to the database stored on the server, after which the data processing procedure is performed using algorithms that detect the violation of inequality (1). The top of the first level "implies" that there are answers to the question that are contained in it. Answers to the question are marked as outgoing ribs. Depending on the selected answer, transition along the corresponding edge to the top of the next level is possible. If the answer that the structural decomposition of the object under study was required was obtained (provided that the inequality (1) is violated), then according to the scheme, the block is divided into additional subblocks in order to study the object.

In turn, each sub-block becomes a top at the appropriate level, so the analysis is carried out as to whether the subblock under study in more detail is in a state of "relative equilibrium." If there is a response that there are no deviations from the norm in the subblock, a system requests additional information from the database, asks to update the data for further verification. The division into structural sub-blocks is limited by the number of sensors initially installed on the object. The more sensors there are on the object, the more accurately it becomes possible to identify the problem area, thereby increasing the time for troubleshooting. This algorithm allows one to identify the structural parts of the object and determine their STS, which contributes to the operational prevention of an emergency [8]

The advantage of using a "decision tree" algorithm is that a sufficiently clear classification model is formed, which gives high prediction accuracy [10]. When developing a multi-agent system, we took into account the fact that the user needs information about the state of the object at any stage of the system. Communication agents solve this problem; they are responsible for collecting, processing, storing, transferring data to other agents and provide user and system interaction. Thus, the user has the opportunity to obtain information on the parameters of the state of a man-made object that are specifically interested in his parameters.

4. Conclusion

In conclusion, we can say that the development of multi-agent systems is a promising direction that solves the existing problems of objectivity, timeliness. and accuracy of determining the state of man-made objects. High automation and intelligent decision making will significantly reduce the risk of man-made emergency situations.

IOP Conf. Series: Earth and Environmental Science **395** (2019) 012074 doi:10.1088/1755-1315/395/1/012074

References

[1] Gavrilova T A, and Khoroshevsky V F 2000 *Knowledge base of intelligent systems* (St. Petersburg, Russia: Piter)

[2] Rygalov A Yu, and Kubarkov Yu P 2012 Application of multi-agent systems in the electric power industry *Proceedings of the Kola Scientific Center of the Russian Academy of Sciences* **1**(4) pp 102-105

[3] Bugakova T Yu, and Sharapov A A 2016 *Application of a multi-agent approach for determining the space-time state of man-made systems* In XII International Forum *"Interexpo GEO-Siberia-2016"* (pp. 189-194) (Novosibirsk, Russia: SSUGiT)

[4] Karpik A P 2004 *Methodological and technological bases of geoinformation support of territories* (Novosibirsk, Russia: SSGA)

[5] Evgenev G B 2000 Multi-agent systems of computer engineering activity *Information Technology* **4** pp 2-7

[6] National Open University INTUIT n d *Management based on multi-agent systems* Available at: http://www.intuit.ru/studies/courses/4115/1230/lecture/24081 (Accessed 02 04 2019)

[7] Russell S, and Norvig P 2007 *Artificial Intelligence: contemporary approach* (Moscow, Russia: Williams)

[8] Bugakova T Yu 2011 *On the issue of risk assessment of geotechnical systems based on geodetic data* In *Proceedings of the VII International Scientific Congress "GEO-Siberia-2011"* (pp 151-157) (Novosibirsk, Russia: SSGA)

[9] Bugakova T Yu 2015 Modeling changes in the space time state of engineering structures and natural objects according to geodetic data *Vestnik SSUGiT* **1**(29) pp 34-42

[10] Bugakova T Yu, Shlyakhova M M, and Knol I A 2016 Structural decomposition of an object using mathematical modeling and subsequent visualization based on WebGL In *Proceedings of the XII International Forum "Interexpo GEO-Siberia-2016"* (pp 142 –147) (Novosibirsk, Russia: SSUGiT)

A robotic system for determining the state of a man-made object in space and time

T Bugakova[1], A Sharapov[1] and I Knol[1*]

[1] Siberian State University of Geosystems and Technologies, 10 Plakhotnogo str, Novosibirsk 630108 Russia

E-mail: ivanknol@mail.ru

Abstract. The article proposes a conceptually new, intellectually oriented approach with a higher degree of automation to the monitoring of man-made objects and the prediction of the space-time state of man-made objects. The paper describes an algorithm of functioning of the multi-agent system for determining and controlling the space-time state of a man-made object is completed. The article describes the development of a robotic stand, which is a prototype of a multi-agent system.

1. Introduction

The space-time state of an object (STS) is a position of the object as a whole or its structural parts with respect to a fixed coordinate system. A man-made object is an object, or a collection of objects created by man in the process of production and economic activities and interacting with the environment (buildings, engineering structures, bridges, tunnels, etc.).

Each man-made object (MMO) has its own design features that must be taken into account when monitoring and controlling its space-time state [1]. Operational safety of man-made objects depends on the quality of monitoring and forecast of STS.

According to GOST R 22.1.12-2005, the man-made objects should be equipped with structured systems for monitoring engineering structures, coupled with automated systems of dispatching services and EDDS to prevent the occurrence and elimination of emergency [2].

Existing automated monitoring systems (AMS) operate on the basis of various algorithms. One of the important tasks in the development of AMS is the selection of effective algorithms for determining and predicting the STS of the object.

The paper proposes a conceptually new intellectually oriented approach with a high degree of automation to the monitoring of man-made objects to determine and predict their space-time state, based on the creation of the space-time state of a man-made object.

2. Materials and Methods

An intelligent agent is a simulation model of a certain active element whose state and behavior in the presence of a specific goal changes in direct proportion to the state and behavior of the environment, as well as the agents existing in it [4].

The MAS assumes the functioning of agents to achieve intermediate and primary goals [5]. The main purpose of creating the MAS in the published work is to determine and predict the space-time state of man-made objects to prevent accidents and emergencies.

Content from this work may be used under the terms of the Creative Commons Attribution 3.0 licence. Any further distribution of this work must maintain attribution to the author(s) and the title of the work, journal citation and DOI.
Published under licence by IOP Publishing Ltd

IOP Conf. Series: Earth and Environmental Science **395** (2019) 012075 doi:10.1088/1755-1315/395/1/012075

A multiagent modeling of systems with the presence of actively flowing processes is a new concept in the field of geodesy and geoinformatics. This system is aimed at the consistent use of models and methods of natural and artificial intelligence for software research, state determination (with forecast function) and system behavior in a specific environment.

3. Results

It is assumed that building structures and the foundation of a building or structure must be of such strength and stability that during construction and operation, the threat of harm to human life or health, property, and the environment does not arise as a result of the following actions:

1. Destruction of individual supporting building structures or their parts;
2. Destruction of the entire building, structure, or part thereof;
3. Deformation of an unacceptable value of building structures, a base of the building or structure, as well as the geological massifs of the adjacent territory;
4. Damage to a part of a building or structure, engineering networks or engineering systems as a result of deformation, displacement or loss of stability of load-bearing building structures, including deviations from verticality [6], [7].

Knowing possible space-time changes of a MMO, we can create a list of intermediate states of a multi-agent system of a MMO. We will highlight the following states:

1. A translational motion: the movement of the entire man-made object, conventionally taken as an absolutely solid body along some vector;
2. A rotational motion (torsion roll man-made object);
3. A relative movement: a technogenic object changes its state by moving its structural parts (blocks) along multidirectional vectors with different speeds.

The MMO's structural blocks are identified as a result of the MMO decomposition procedure based on specified algorithms [8] We consider one of them, which is the algorithm of the nearest neighbors. The essence of this algorithm is to determine the set of control points of the object, the coordinates of which have changed with respect to the values determined in the zero era of time. Based on this information, we can determine the boundaries of the structural division of MMO into sub-blocks, i.e. deformation zones.

With the emergence of a new block, new characteristics of this block arise, namely the motion vector, the speed of movement, the axis of rotation, the angular velocity of rotation, the compression ratio. Thus, if we create software analogies of these structural elements (MMO blocks) and establish the laws of interaction between them and the environment, it would be possible to predict a change in the STS of the MMO. Fig. 1 shows an algorithm of the software package based on the multi-agent approach.

In this algorithm, M1(x,y,z) is a cloud of points resulting from the operation of complex multifunctional sensors (CMS) with an update resolution T; M2(x,y,z) is a cloud of points obtained as a result of surveying geodesic stamps with update resolution T*24; M3(x,y,z) is a cloud of points obtained as a result of a laser scanner survey with the incremental update T*24*30.

Independent surfaces that are the basis for generating a single digital 3D model are built on the basis of sets of points. At the next stages, automated addition of characteristics to points and surfaces based on data from CMS, processing a single cloud of MMO points by decomposition and clustering algorithms, is carried out to identify new MMO blocks.

As a result of the identification of new autonomous structural elements, the MMO generates a software analogy (agent) of this structural element with the addition of characteristics with the subsequent determination of dependencies between these agents.

The final stage is the derivation of correlation graphs for agents and a predictive 3D model with the display of structural blocks.

IOP Conf. Series: Earth and Environmental Science **395** (2019) 012075 doi:10.1088/1755-1315/395/1/012075

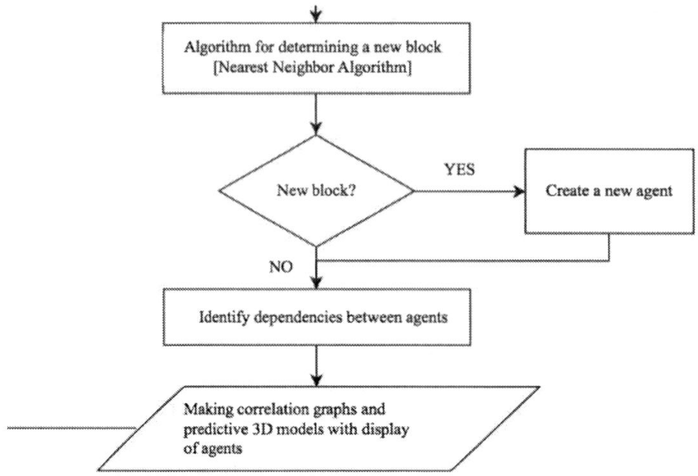

Figure 1. An algorithm of the program based on the multi-agent approach.

As part of the research, a platform for obtaining initial data is the development of the Center for Engineering and Robotics (TSIIR) of the Siberian State University of Geosystems and Technologies (SSUGiT). This is a robotic stand for determining and controlling the space-time state of a man-made object [9-13].

We are currently involved into creating integrated multi-functional sensors that would allow to determine the coordinates not only of a point in space, but to control some cloud of points of its environment. These sensors are planned to be tested on the basis of SGUGiT in a specially prepared anechoic chamber, without interference from radio and mechanical waves.

Development and implementation of the MAS for the control of the STS (of MMO) is a complex and lengthy process. Therefore, a multi-agent system modeling should include the following step-by-step processes:

1. Creating a 3D-model of the object;
2. Installing the integrated multifunctional sensors on the MMO;
3. Acquiring all necessary object data from sensors with an incremental data refresh *d*.
4. Installation of geodesic marks;
5. Obtaining data on the object by the tacheometric (theodolite, leveling) survey of geodesic marks with the resolution of updating data *d * 24*;
6. Getting a point cloud based on laser imaging;
7. Selecting an algorithm for data processing, based on the methods of the systems approach (decomposition, aggregation) and decision making:
8. Determining the frequency of receipt of geospatial data;
9. Decisiding on further action;
10. A software implementation;
11. A web interface development.

Fig. 2 shows the diagram of the web application of the MAS, based on the principle of "client" - "server." In that Figure, "A1..An" is the spatial and semantic data obtained by complex multifunctional sensors installed on the object under study (*n* is the number of blocks (structural parts of the object) obtained as a result of decomposition, *d* is the resolution of data update). "B1..Bn" is the spatial data obtained as a result of geodetic measurements based on tacheometers, theodolites, and levels (*n* is the number of geodetic marks (data refresh resolution is *d*24*), the point cloud of the studied technogenic object is a set of spatial coordinates obtained as a result of geodetic measurements based on laser scanners (data refresh resolution is *d*720*). "Algorithm № 1 .. n" stands for the data processing algorithms.

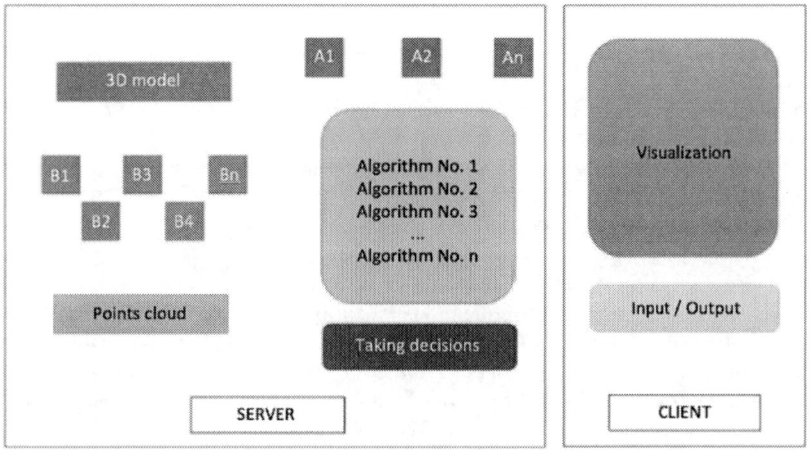

Figure 2. A scheme of the web application (MAS).

4. Discussion

Heterogeneous information may come in to determine the STS, the functions that allow to supplement or adjust the data processing algorithms are built into the system. Algorithms for data mining and "machine learning" and algorithms of "decision trees" are included in the decision block. Decision trees are a way of representing rules in a hierarchical, sequential structure, where each object has a single node that provides a solution. The advantage of using data mining algorithms is that a sufficiently clear classification model, which has a high prediction accuracy, is implemented in the "machine learning" process [14-17].

It is assumed that the system being developed will be for multi-users, with the ability to connect several man-made objects, which is why the Internet was chosen to organize the connection between the client and the server.

The server is written in Java and PHP. The multifunctional interface of a client part of this web application is implemented on the basis of various libraries of the JavaScript language (in particular, WebGL) and the synthesis of HTML5 and CSS3.

The interface allows one to display graphs of dependencies between the structural parts of the object (blocks), obtained as a result of the decomposition process, and also allows one to visualize the current 3D model MMO.

5. Conclusion

The developed robotic system is endowed with the following functionality:

- Determining an offset model of the MMO in the plane OXY;
- Determining a free movement of the MMO model in space, relative to OXYZ and the parameters of this movement;
- Determining the deformation of the MMO model (integral, differential), the definition of the structural elements of the object.

As part of continuing to work with this project, we have planned to implement the following functions:

- Modeling various factors influencing a man-made object;
- Conducting an intellectual analysis of the influence of external factors on the change in the MMO state ;
- Determining the criteria for the transition of the MMO from the "safe" state to the "dangerous" one, based on the analysis of the correlation of external factors and object states;
- MAS testing at a real object.

References

[1] Russian Federation 2004 *Town Planning Code of the Russian Federation* (December 18, 2006 N 232-FZ) (Moscow, Russia)

[2] Gosstandard 2005 *The national standard of the Russian Federation GOST R 22.1.12-2005* (Moscow, Russia: IPK Publishing Standards)

[3] Ulybin A V, and Arzamastsev A A 2010 Multi-agent approach in simulation modeling *Vestnik of the University of Tambov* **15**(5) pp 1470-1471

[4] Luger G F 2003 *Artificial intelligence: strategies and methods for solving difficult problems* (Moscow, Russia: Vilyame)

[5] Bugakova T Yu, Shlyakhova M M, and Knol I A 2016 Structural decomposition of an object using mathematical modeling and subsequent visualization based on WebGL In *Proceedings of the XII International Forum "Interexpo GEO-Siberia-2016"* (pp 142-148) (Novosibirsk, Russia: SGUGiT)

[6] Ariskin M V, and Boldyrev S A 2015 *Security basics of buildings and structures* (Penza, Russia: PGUAS)

[7] Zhuravlev Yu I, Ryazanov V V, and Senko O V 2005 *"Recognition"": Mathematical methods, software system, practical applications* (Moscow, Russia: Fazis)

[8] Gavrilova T A, and Khoroshevsky V F 2000 *Knowledge base of intelligent systems* (St. Petersburg, Russia: Piter)

[9] Evgenev G B 2000 Multi-agent systems of computer engineering activity *Information Technology* **4** pp 2-7

[10] National Open University INTUIT n.d. *Management based on multi-agent systems* Available at: http://www.intuit.ru/studies/courses/4115/1230/lecture/24081 (Accessed 02 04 2019)

[11] Lafore R 2013 *JAVA data structures and algorithms* (St. Petersburg, Russia: Piter)

[12] Bugakova T Yu, and Sharapov A A 2016 *Certificate of state registration of computer program No. 2016661521 "MAS PVS TO", October 12, 2016, a prototype of a multi-agent system for monitoring the space-time state of man-made systems* (Moscow, Russia)

[13] Bugakova T Yu 2016 Interactive control of the space-time state of man-made objects using the WebGL technology *Vestnik SGUGiT* **36** pp 114-120

[14] Sibrikov S G 2009 *Man-made systems and environmental risk* (Yaroslavl, Russia: YarSU)

[15] Mendel I D 1988 *Cluster analysis* (Moscow, Russia: Finance and Statistics)

[16] Bugakova T Yu 2015 Modeling changes in the space-time state of engineering structures and natural objects according to geodetic data *Vestnik SGUGiT* **1**(29) pp 34-42

Economic assessment of the development potential of mushroom production in the Russian Federation

N Devochkina[1*], R Nurmetov[1] and A Razin[1]

[1] All-Russian Scientific Research Institute of Vegetable Growing – Branch of the Federal Scientific Center for Vegetable Growing, Vereya, Ramenskoye district, Moscow region 140153 Russia

E-mail: vniioh@yandex.ru

Abstract. The article assesses the issue of further development of the industrial mushroom production in the Russian Federation in order to fulfill the objectives of the State Program for the Development of Agriculture and Regulation of Agricultural Products, Raw Materials, and Food Markets for 2013-2020. It was developed in accordance with the principles of ensuring food independence of Russia, sustainable development of rural territories of the Russian Federation, and financial sustainability of agricultural producers. Conducting an analysis of the state of the world mushroom industry, general trends in the organization of industrial enterprises, and the choice of modern technological solutions, which are the basis for the intensification of production and its economic efficiency. The article assesses a level of self-sufficiency of the Russian Federation in mushroom products on the basis of a comparison of the consumption levels of cultivated mushrooms in the developed European countries, the USA, as well as East and Southeast Asia.

1. Introduction

There are economically justified prerequisites for the rapid and effective development of industrial mushroom production in the Russian Federation. In 2016-2017, the demand for fresh mushrooms amounted to about 115,000 tons. Out of them, 100,000 tons of them are imports, which consumes more than $400 million. In addition, more than 20,000 tons of compost for growing mushrooms are imported annually. At the same time, Russia has gained practical experience in growing mushrooms in previous years, when large industrial mushroom complexes were built in the Moscow, Leningrad Region, Chuvash, Tatarstan Republics, Voronezh [1]. There is a powerful resource base in various regions of the country. The main components for preparing the substrate are available in sufficient quantities, namely cereal straw and chicken manure, which is the basis for the production of compost for the cultivation of mushrooms. The decisions taken on the food embargo also favor the development of the Russian mushroom industry, which significantly reduces imports.

The course on import substitution is declared in the Russian Federation since 2014. According to statistics, the level of self-sufficiency in mushroom products is only 6.4%. In recent years, positive dynamics of the construction of contemporary greenhouse and mushroom production complexes have been noted. Projects implemented in the field of vegetable and mushroom greening have a high investment attractiveness due to the use of intensive technologies provided by contemporary technological equipment and high economic indicators. The Program for the Development of Vegetable

Content from this work may be used under the terms of the Creative Commons Attribution 3.0 licence. Any further distribution of this work must maintain attribution to the author(s) and the title of the work, journal citation and DOI.
Published under licence by IOP Publishing Ltd

Growing and Mushroom Production until 2020 has been adopted at the state level. Its goal is to implement significant government support for the industry. High boundaries are indicated in this program, the implementation of which must be done in the near future [2]. According to the data of the Food and Agriculture Organization of the United Nations, Russia occupies about 22 place in the structure of world production of mushroom products, and only 46 in terms of per capita consumption only. The per capita consumption of cultivated mushrooms in developed countries ranges from 2 to 4.5 kg. In Russia, this figure is about 1 kg per person. Thus, the main target of the program is to achieve an annual volume of mushroom production of about 143,000 tons [3]. The domestic market capacity of mushroom products in financial terms is about 25 billion rubles. Thus, a unique situation is created, allowing to increase the production of fresh mushroom products, rich in protein and environmentally safe, produced in specialized cultivation facilities, and almost completely replace imported products of domestic.

2. Materials and Method

In conducting an economic analysis of the potential for the development of domestic mushroom production and modern technological systems for the production of edible mushrooms, the most widely used in the global production of mushrooms and recommended for introduction, the following methods are used: economic and statistical, methods of systems and comparative analysis, mathematical modeling and planning [4], [5]. The study is based on the collection and analysis of information and statistical data of the Rosstat, an international organization FAO (Food and Agriculture Organization of the United Nations), both of which have a reliable information in the field of global production of various types of food. We also used data from the Republican Production and Scientific Association "Greenhouses of Russia," the assessment of trends in the development of world industrial mushroom production and the state of the domestic sector of mushroom producers. Our research was conducted in the following three stages: (1) Analyzing the level of world edible mushrooms production and the rate of its development over the past 20-25 years; (2) Assessment of the state of the domestic mushroom industry; (3) A business-innovative model of the perspective development of the domestic mushroom industry based on the use of the latest organizational and technological system is justified and proposed.

3. Results

The study was conducted in three stages. At the first stage, our analysis of the world champignon production was carried out. The production of champignons as a type of economic activity in the developed countries is allocated to an independent branch of agricultural production. In the world, in 2015-2017, the production area of cultivated mushrooms was about 27.5 thousand hectares. This is a huge area, allowing to achieve production volumes at the level of about 11 million tons of mushroom products. According to the FAO, only China produced about 7.8 million tons of various types of mushrooms in 2017. Over the past 35 years, the mushroom production increased in 10 times (about 1.5 million tons in 1980, about 11 million tons in 2017). In this structure, champignons hold the championship (more than 80%), but other types of mushrooms such as shiitake and oyster mushroom are becoming increasingly popular due to their nutritional and medicinal properties. About 15 other species of various cultivated mushrooms are produced in the world. The largest regions of mushroom production are also Europe. In the European countries, 1,300,000 tons of champignons are grown annually. About 70% of them are sold fresh, and 30% are canned or frozen). Poland was distinguished in terms of active growth in the production of fresh mushrooms, reaching a level of 260 thousand tons, almost equaling the Netherlands.

The USA is the largest producer of mushrooms in North America. The annual volume of champignon production is relatively constant and amounts to 420,000 tons per year. Most champignons are sold fresh in the domestic market.

Canada is the second largest producer of champignons in North America. The annual volume is about 100,000 tons. All fresh mushrooms enter the domestic market and the US market in small quantities.

One of the fastest growing mushroom industries in the world works in Australia. Over the past 10 years, production in this country has grown by 20% and reached 75,000 tons.

Table 1. World ranking of countries in terms of production of cultivated mushrooms.

No	Country	Production volume, t	No	Country	Production volume, t
1	China	7,800,000	15	Turkey	40,200
2	USA	420,000	16	Hungary	32,000
3	Italy	683,000	17	India	30,000
4	Netherlands	300,000	18	Belgium	29,450
5	Poland	260,140	19	The Republic of Korea	26,200
6	Spain	197,000	20	Vietnam	24,000
7	France	120,000	21	South Africa	18,800
8	United Kingdom	100,000	**22**	**Russian Federation**	18,000
9	Canada	100,000	23	Lithuania	15,800
10	Australia	75,000	24	Ukraine	15,000
11	Germany	72,000	25	Romania	14,500
12	Ireland	70,000	26	Portugal	12,000
13	Japan	65,600	27	Taiwan	10,530
14	Indonesia	40,900	28	Cyprus	11,500

The production of champignons in Europe is fairly stable, although a tendency to a decrease in consumption is expected. On average, consumption in Europe is 2.0 kg per person per year.

In China, the production structure of cultivated mushrooms is changing rapidly. Large modern facilities with a full set of technological equipment are replacing small primitive farms. At the same time, the growth in production remains unchanged, but in the near future, its decline is possible due to the reorientation of the Chinese mushroom industry to the large-scale cultivation of traditional for the local market mushrooms (shiitake, enoki, eringi, black tree fungus, etc.). In China, the domestic consumption of champignons is small and amounts to only 0.2 kg per person per year, with a total consumption of other types of mushrooms up to 14 kg per person per year.

The North American mushroom industries are stable and not prone to drastic changes. In the US, champignon consumption is small and amounts to only 1.5 kg per person per year. In Canada, it is much more – about 3 kg per person per year. In recent years, the Australian mushroom industry has experienced a slow growth, as the market is close to saturation. Today, Australia has the highest per capita consumption of 3.2 kg per person per year.

At the second stage, assessment of the state of the domestic mushroom industry is carried out. In Russia, industrial mushroom production is an integral part of the greenhouse vegetable growing industry. In Russia, mushroom enterprises were built on imported and domestic equipment based on the use of targeted public investment in the period from 1976 to 1995. The enterprises used the Dutch technology of cultivation of champignon of the full technological cycle. They included the following: workshops for preparing the substrate and coating material, when the cultivation of the main product, namely the champignon fruit bodies, was carried out. The payback rate for agricultural enterprises of this type was 7-8 years. In 1980, as part of the mushroom complex with a useful area of 0.25 hectares, a plant for the production of planting material of edible mushrooms was purchased to import, the project was implemented in "Zarechye" CJSC of the Moscow Region (51 km of the Moscow Ring Road). It provided the existing of 9 complexes in the country with of mushroom mycelium, as well as private entrepreneurs, focused on the production of oyster mushroom, which was grown mainly in adapted premises. The production capacity of a plant made it possible to produce up to 300 tons of mycelium annually, which ensured the cultivation of at least 12-15 thousand tons of mushrooms of various kinds. At present, the Moscow mycelium plant has ceased to exist. Therefore, in the near future, new construction of an industrial plant for the production of quality planting material would be appropriate to implement. The plant can exist as an independent production, as well as as part of the mushroom production complex, which would allow production control of the products to be carried out.

In Russia, the growth rate of the production of edible mushrooms is extremely slow. First of all, this is due to the high capital intensity of industrial mushroom production complexes. Previously built mushroom enterprises completely ceased their production activities for a number of objective reasons. The first reason is the full or partial depreciation of cultivation structures and outdated technologies (for example, "Moskovsky" Agrokombinat CJSC). A number of built champignon greenhouses work on imported compost, which is supplied from Lithuania or Poland, and this does not guarantee the operation stability of these enterprises. A centralized compost production is absent in Russia.

Analysis of the state of the domestic branch of mushroom production showed that the enterprises built earlier almost fully developed their resources, being in operation for more than 25 years. However, the experience of new construction of mushroom complexes in a market economy is absent.

In the international practice of the agribusiness, mushroom production is one of the most economically efficient industries. To date, sanctions imposed on foodstuffs from the EU countries did not affect the Russian mushroom market and the Phase 3 compost market. Polish mushrooms are supplied to Russia under the guise of products from third countries. Imported compost is not a sanctioned product.

The main influence on the market for fresh mushrooms and compost is rendered by the ruble exchange rate against the dollar and the euro, as it directly determines the cost of imported products on the Russian market. Compared with the previous year, the cost of imported compost and mushrooms in rubles increased by 20%.

The ruble devaluation also affects the budgets of new projects. Due to the absence in Russia of the production of a complex of machines for the production of compost and mushroom cultivation, they must be purchased via import. Their ruble value also increases in proportion to the depreciation of the ruble. Evaluating the prospects of import substitution equipment and the purchase of planting material of mushrooms for the production of compost for imports before 2020, we can conclude that it is not yet technically possible.

In the third stage, a model of organization of the industrial mushroom industry was developed. This model justified the following two types of production organization: expansion of mushroom production through the organization of enterprises for the centralized preparation of compost and the construction of new complexes with a full technological cycle and significant production volumes from 3000 tons per year and more.

According to the Agroinvestor Publishing House (July 2017), the Ministry of Agriculture of the Russian Federation proposed to support the development of mushroom production, including it in the State Program of Agricultural Development for 2013-2020. The corresponding draft government decree posted on the portal <regulation.gov.ru>. This confirms the interest of the state in accelerating the development of mushroom production in the Russian Federation. Real actions in this direction are also carried out by the regional authorities of Kaluga, Belgorod, and Tula regions. Regional support for new construction projects in Moscow, Samara, and Penza regions was announced.

4. Conclusion

The proposed business innovation model for the development of domestic mushroom production was the basis for the development of the *Concept for the Development of Domestic Mushroom Production for the Period 2013-2020* and the development of agricultural policy measures laid down in the *State Program for the Development of the Industry*.

References

[1] Devochkina N L, Nurmetov R D, and Dolgikh L I 2012 Industrial mushroom production is an effective resource for the development of the Russian agro-industrial complex *Potatoes and Vegetables* **1** p 21

[2] Soldatenko A V, Devochkina N L, Razin A F, Razin O A, and Nurmetov R D 2018 Industrial mushroom production as an innovative direction of economic activity in the field of the agro-industrial complex of the Russian Federation *Russian Vegetables* **3** 89-92

[3] Devochkina N L, Dugunieva L G, Razin A F, Ivanova M I, and Nurmetov R D 2018 Investment attractiveness of the industrial mushroom production *Economics of Agriculture of Russia* **11** pp 52-59

[4] Behrens W 1995 *A manual for the preparation of industrial feasibility studies* (Moscow, Russia: "Interexpert" CJSC)

[5] Razin A F 2004 Modeling innovative production of competitive products *Bulletin of the Moscow Academy of Labor Market and Information Technologies* **11** pp 108-115

IOP Conf. Series: Earth and Environmental Science **395** (2019) 012077 doi:10.1088/1755-1315/395/1/012077

Winter hardiness of *Iris barbatus* L. in the climatic conditions of the nonchernozem zone of Russia

G D Levko[1*], V A Kharchenko[1] and I T Ushakova[1]

[1] Federal Scientific Center for Vegetable Growing, 14 Selektsionnaya str., VNIISSOK Village 143080 Russia

E-mail: gennadylevko@yandex.ru

Abstract. The article presents a comparative analysis of such an important trait that is economically valuable for vegetative reproduction in the bearded iris as the degree of winter hardiness of rhizomes. Economic efficiency for the industrial production of planting material is also calculated. The degree of winter-hardness was determined by the ratio of the number of plants left before winter to the number of plants that grew up in early spring. As a result of research, 84 varieties of domestic and foreign selection were assessed, among which a total of four groups were identified with varying degrees of winter hardiness. Observations have shown that the irises of domestic breeding and time-tested foreign varieties, which have been cultivated for a long time in central Russia, are most resistant to overwintering. The economic efficiency of growing planting material was 144% per unit area (1 ha).

1. Introduction

Winter hardiness in rhizomatous perennial plants is an important sign of resistance to the complex of adverse wintering factors. Winter hardiness is the ability of plants to withstand the complex effects of the environment during the winter and early spring periods. This includes the following: an ability to resist spring burns, low temperatures, decay, soaking, an ability to withstand frequent temperature extremes. Bearded irises are ephemeroids and xerophytes. They are the hybrids of frost-resistant irises of diploids, for example, iris Pallida, and thermophilic tetraploids – these are Cypriot irises, from which they inherited genes and modifications that contribute to their frost resistance at low temperatures in the winter dormancy period. These are the dysmutase enzyme production genes that support lipid membrane fatty acids in an unsaturated state, which enables them to maintain their resistance to frost [1], [2], [3], [4].

Currently, the global range is more than 80,000 species, varieties and hybrids. Irises of domestic breeding and time-tested foreign varieties that have long been grown in the middle lane are the most winter hardy. For most irises, 30 cm of snow is sufficient protection from frost. The most vulnerable are the novelties of the world selection of high irises. The potential of their winter hardiness and methods of shelter have to be established empirically [5], [6]. Earlier experiments have found that varieties of bearded iris with a higher total content of antioxidants (overproduction) are winter hardy by 100%, and varieties of bearded iris with a reduced total content of antioxidants (deficient) have a decreased winter hardiness. Thus, the total content of antioxidants in rhizomes can serve as a potential marker of resistance to stress [7]. The purpose of the study is to assess the winter hardiness of the rhizomes of

Content from this work may be used under the terms of the Creative Commons Attribution 3.0 licence. Any further distribution of this work must maintain attribution to the author(s) and the title of the work, journal citation and DOI.
Published under licence by IOP Publishing Ltd

bearded iris during the first to third years of vegetation and to calculate the economic efficiency of growing.

2. Research Methodology

The degree of winter hardiness was determined by the ratio of the number of plants left before winter to the number of plants grown up in early spring (expressed as a percentage). All studied varieties were distributed in the following four groups with varying degrees of winter hardiness: (1) not winter-hardly (0%), (2) poorly winter-hardly (up to 25% and from 26 to 50%), (3) moderately winter-hardly (51-75%) and (4) winter-hardly (from 75 to 95%, and more). Phenological observations were carried out according to the methodological recommendations [8]. Profitability was calculated as the ratio of costs to profit.

3. Results and Discussion

During the spring inspection, observations of the manifestation of such an important trait that is economically valuable for vegetative propagation in bearded iris as the degree of rhizomes' winter-hardiness were followed by plants planted in 2015, annually. As a rule, the regrowth of variety samples began at the end of April (from April 20 to April 30). So, 80 varieties (out of 84 varieties) that went through the winter in 2017 survived (i.e. 95%).

Of the varieties of foreign selection, 7% of them were not winter-resistant, 20% were poorly winter-resistant, 19% were moderately winter-resistant, and 34% were winter-resistant. In varieties of domestic breeding, in the four groups listed, the highest indicator was recorded in the group of winter-hardy varieties, constituting 56% (Fig. 1). On average, the overwintering percentage of domestic varieties was 72.9%, and 73.1% of foreign varieties (Table 1). Poorly winter-resistant plants among the Russian varieties turned out to be "K Yubileyu Ilicha" (22%), "Mechta" (40%), "Udacha" (33%). Not winter-resistant varieties of the foreign breeding were "Bay Fog," "Cinnamon Girl," "Satin Satan," "Pink Romance," "Royal Intrigue" (0% overwintering), and the "Henry Shaw" variety was poorly winter resistant (25 %).

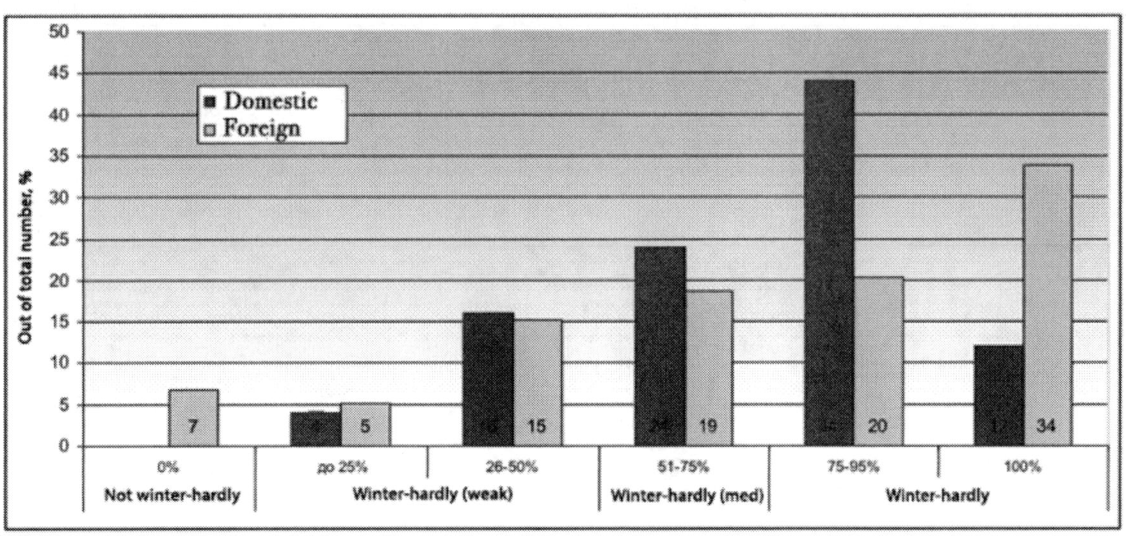

Fig. 1. Distribution of variety samples of bearded domestic and foreign iris breed by winter-resistance groups.

The variability of the "plants' overwintering" characteristic for the entire set was significant (Cv = 42.7%), and it was 1.5 higher for foreign varieties than in domestic ones (Cv = 41.1% and 28.5%, respectively).

Table 1. The level of "overwintering" in bearded iris (2015-2018).

No	Variety	Rhizomes wintered, %
	Domestic varieties	
1	Amethyst of VNIISSOK	70
2	Belosnezhka	80
3	Belyy of VNIISSOK	50
4	Belyy Karlik	70
5	Belyanchik	100
6	Bordyurnyy Belyy	67
7	Bordyurnyy Siniy	90
8	Bordyurnyy Chernyy	80
9	Viktor Dryagin	90
10	Gvardeyskiy	78
11	Graf Tolstoy	90
12	Drevniy Rim	80
13	Zharok	80
14	Irlev	89
15	K Yubileyu Ilicha	22
16	Lunnaya Sonata	100
17	Marshal Pokryshkin	100
18	Mechta	40
19	Nadezhdy	72
20	Olimpiyskiy	80
21	Pervenets	50
22	Ryzhik	64
23	Sirenevyy Venochek	90
24	Udacha	33
25	Fioletovyy Nizkoroslyy	63
	Foreign varieties	
26	Batik	100
27	Arpege	90
28	Back in Black	70
29	Bay Fog	0
30	Black tie Affair	80
31	Blushing Pink	70
32	Brook Flowers	100
33	Cascade Springs	80
34	Celebration Song	50
35	Champagne Music	50
36	Charm City Choice	86
37	Cinnamon Girl	0
38	Cosmic Dance	86
39	Cracker	100
40	Crinoline	100
41	Dark Side	63
42	Christmas Angel	50
43	Dusky Challenger	63
44	Edith Wolford	100
45	Everything Plus	100
46	Fiesta Time	100
47	Five Star Admiral	100
48	Flash Fire	100

IOP Conf. Series: Earth and Environmental Science **395** (2019) 012077 doi:10.1088/1755-1315/395/1/012077

49	Foreign Statesman	50
50	Ginger Swirl	100
51	Green Quest	80
52	Gypsy Caravan	100
53	Gypsy Jewels	17
54	Heather Hawk	100
55	Henry Shaw	25
56	Heritage Lace	50
57	Invitation	50
58	Jitterbug	100
59	Maraschino	75
60	Mary Frances	100
61	Morning Hymn	100
62	Oritam	100
63	Pink Romance	0
64	Proud Tradition	100
65	Purple Pepper	100
66	Renaissance Fair	75
67	Ribbon Round	80
68	Rosette Wine	60
69	Royal Crusader	60
70	Royal Intrigue	0
71	Ruffles and Lace	100
72	Rustic Cedar	100
73	Satin Satan	0
74	Siva Siva	80
75	Spartan	75
76	Stairway to Heaven	80
77	Stepping Out	70
78	Sultry Mood	83
79	Superstition	50
80	Timeless Moment	83
81	Queen of Angels	83
82	Vigilante	100
83	War Chief	50
84	War Sails	70
	All varieties	**72.96**

Calculating the economic efficiency of growing planting material of bearded iris for industrial reproduction (2015-2018) is presented in Table 2. From this table it is clear that profitability is 144% with a high degree of recoupment when growing planting material of iris bearded commercially.

Table 2. Calculation of the economic efficiency of growing planting material bearded iris for industrial reproduction (2015-2018).

No	Costs	Amount, rub.
1	Wages	41,898
2	Taxes (30,2%)	12,653
3	Fuels and lubricants	6,039
4	Spare parts	72
5	Depreciation	8,748
6	Auto insurance	176
7	Utilities	2,921
8	Overhead	41,898
	Total:	114,405

9	Planting material	8,925
10	Fertilizers, pesticides	472
11	Peat	590
12	Water	571
	Total:	124,963
13	Stroring	13823
	Total: cost	138,786
14	Cost of 1 tuber bud	4,09
15	Realization (revenue)	339,150
16	Profit	200,364
17	Profitability	144%

4. Conclusion

As a result of research, 84 varieties of domestic and foreign breeding were assessed, among which four groups with different degrees of winter hardiness were identified. Irises of domestic breeding and time-tested foreign varieties, which have been grown for a long time in central Russia, are the most resistant to overwintering. The economic efficiency of the growing planting material was 144% per unit area (1 ha).

5. Acknowledgments

We express our deep gratitude to the leading economist of the Federal State Budgetary Institution "Federal Scientific Center of Vegetable Growing" N. P. Khovaeva for assistance in economic calculations and the doctor of biological sciences L. I. Belyakova for the valuable information included in the article.

References

[1] Blazhek M, and Blazhkova U 2011 *The gene pool of garden group irises are high bearded irises. Search, identification and preservation of the gene pool of high bearded irises* In Materials of the II Moscow International Symposium on the Iris genus: *"Iris-2011"* (pp 31-44) (Moscow, Russia)

[2] Hattenberger P F 1980 Frost resistance of irises *Floriculture* **8** pp 34-35

[3] Borodich G S, and Belyakova L I 2018 *Features of agrotechnics of bearded irises in connection with their physiology* (Minsk, Belarus: Kolorgrad)

[4] Igonina E V 2013 Winter resistance of irises *Irises of Russia. Annual Bulletin* pp 35-45

[5] Rodionenko G I 2002 Irises (St. Petersburg, Russia: "Diamant" LLC, Agroprom izdat)

[6] Cayeux 2014 *Iris lover's catalogue* (Moscow, Russia)

[7] Levko G D 2016 The influence of the total content of water-soluble antioxidants in rhizomes on winter hardiness of garden iris varieties (*Iris hybrida* L.) *Russian Vegetables* **1** pp 76-81

[8] Dryagina I V 1991 *Cultivation of garden bearded irises: methodical recommendations* (Moscow, Russia: All-Russian Academy of Agricultural Sciences, All-Russian Research Institute of Plant Cultivation and Seed Breeding)

IOP Conf. Series: Earth and Environmental Science **395** (2019) 012078 doi:10.1088/1755-1315/395/1/012078

Expenses for producing the winter garlic planting material (*allium sativum* L.) *in vitro*

M Azopkova[1]**, I Muravieva**[1] **and A Polyakov**[1*]

[1] All-Russian Scientific Research Institute of Vegetable Growing – Branch of the Federal Scientific Center for Vegetable Growing, Moscow region, Ramenskoye district, Vereya, 140153 Russia

E-mail: vniioh@yandex.ru

Abstract. The article calculates the costs incurred at different stages of cultivation of winter garlic in an *in vitro* culture. It is shown that when using the developed technique (10,000 plants with a cost of 30.3 rubles per 1 plant), a regenerant can be obtained for 270 days of cultivation.

1. Introduction

Currently, the solution of practical problems of breeding is largely determined by the effectiveness of involving modern biotechnological methods that can be successfully used at all stages of the selection process, including the creation of genetic diversity, obtaining stable genetic lines, the identification of genotypes and their reproduction [6].

Sowing campaign garlic (*Allium sativum* L.) is a valuable vegetable and medicinal crop that is widely used in human nutrition and the pharmaceutical industry [2]. Garlic is a vegetatively propagated plant. In this regard, it is subject to numerous viral, bacterial, fungal and other infections that are transmitted to offspring, which leads to lower yields, loss of quality, keeping quality, and often degeneration of varieties [1], [4].

Numerous studies have shown the possibility of obtaining a healthy planting material using *in vitro* technologies. Therefore, improving the planting material of this type of cultivated plant is a necessary condition in contemporary technologies of garlic seed production.

2. Materials and Methods

Unripe air bulbs of garlic winter varieties Gladiator served as a research material [5]. All-Russian Scientific Research Institute of Vegetable-Growing – Branch of the Federal State Scientific Institution Federal Research Center for Vegetable-Growing has developed a method for producing the *in vitro* planting material for winter garlic (*Allium sativum* L.) [3]. To calculate the plant production costs according to this method, the following input data were used: (1) wages with a specialist's charges (40,000 rubles per month), (2) costs of 1 kW / hour (4.66 rubles), (3) an average cost of 1 l nutrient medium (600 rubles).

According to this method, a step sterilization of 7-day-old garlic inflorescences (using 1% potassium permanganate solution at exposure time 15-20 minutes, 70% ethanol solution for 30 seconds and 1.0% sodium hypochlorite solution) takes 20 minutes and allows one to get from 92.6% to 100.0% sterile explants.

Content from this work may be used under the terms of the Creative Commons Attribution 3.0 licence. Any further distribution of this work must maintain attribution to the author(s) and the title of the work, journal citation and DOI.
Published under licence by IOP Publishing Ltd

IOP Conf. Series: Earth and Environmental Science **395** (2019) 012078 doi:10.1088/1755-1315/395/1/012078

3. Results and Discussion

In accordance with the technological process, at the first stage, the isolated inflorescences are placed on a nutrient medium containing macro- and microelements, vitamins and organic compounds according to the recipe N6 [7], as well as kinetin at a concentration of 1.0 mg / l. At this stage, the duration of cultivation averaged 30 days, and the cost of wages with charges and payment of electricity amounted to 1,739.95 rubles (Table 1).

Table 1. The cost of obtaining 10 000 plants (regenerants) of garlic *in vitro*.

Indicators	Cultivation stages				Total costs
	Introduction *in vitro*	Breeding shoots	Rooting and growing shoots *in vitro*	Adaptation of regenerants to *ex vitro* conditions and their growing	
Nutrient medium	N6, kinetin 1 mg / l sucrose 3%	N6, kinetin 1 mg / l sucrose 3%	N6, PAA 0,1mg / l, sucrose 1%	Peat tablets Jiffy 24 mm	
The duration of cultivation, days	30	30	30	150	270
Expenses:					
including labor, man-hour	26	125.5	125.5	126	402
electricity, rub.	1216.95	4862.46	4748.90	83280.01	94108.32
salary with charges, rub.	523	2522.5	2522.5	2532.6	8080.2
materials, rub.		81111.48		119700	200811.48
TOTAL costs, rub.	**303.000**				
Per 1 regenerant, rub.	**30.3**				

The resulting shoots of garlic are transplanted to a nutrient medium N6, containing sucrose at a concentration of 10 g / l, PAA (0.1 mg / l). After 10-14 days, the roots and additional leaves begin to appear in the garlic regenerants.

To adapt to the new gas composition to the air with less humidity and infectious load, holes are made in the lids of the cultivation vessels, which expand as the plants adapt. This period usually lasts 3-7 days. Then, plants with a well-formed root system are removed from the cultivation vessels and planted in peat tablets. Within 10-14 days, the plants are kept in a humid chamber at a temperature of + 18 ... + 20 °C, and they are opened daily for aeration.

The cost of materials for the entire period of *in vitro* cultivation is 81,111.48 rubles, or 26.8% of the total costs, and includes the cost of the nutrient medium, consumables, and ethanol.

The stages of obtaining regenerants and their rooting are 30 days each, and the cost of wages with charges and electricity for this period is 14,656.36 rubles, or 4.8% of the total costs.

Using the described techniques provides for 81.5–94.3% of garlic regenerated plants adapted to *ex vitro* conditions. To produce well developed plants and bulbs, they are grown within 3-4 months. At this stage, costs amount to 205,512.61 rubles, or 67.8% of the total costs.

4. Conclusion

When using the described *in vitro* technology, the period of obtaining 10,000 winter garlic plants takes 270 days, and the cost of 1 plant (regenerant) is 30.3 rubles.

References

[1] Kokareka N N, and Pleshakova T I 2013 The onion and garlic viruses: diagnosis and prevention *Potatoes and Vegetables* **6** pp 13-14

[2] Pivovarov V F, Ershov I I, and Agafonov A F 2001 *Onion crops* (Moscow, Russia: VNIISSOK)
[3] Polyakov A V, Azopkova M A, and Muravieva I V 2018 *Obtaining in vitro planting material of winter garlic (Allium sativum L.). Methodical recommendations* (Moscow, Russia: Polygraph-Biznes LLC)
[4] Polyakov A V 2015 *Garlic is a multifunctional component of nutrition* In Proceedings of the III International Scientific Conference with Elements of a Scientific School for Young People: *"Quality and Environmental Safety of Food Products and Industries"* (pp 150-154) (Tver, Russia: TSU)
[5] Polyakov A V 2013 Gladiator – a new sort of winter garlic *Potatoes and Vegetables* **9** pp 31-33
[6] Polyakov A V 2005 *Obtaining regenerated vegetable crops and their reproduction* in vitro (Moscow, Russia: All-Russian Scientific Research Institute of Vegetable Growing of the Russian Academy of Agricultural Sciences)
[7] Chu C C, Wang C C, Sun C S, Hsu C, Jin K C, and Chu C Y 1975 Establishment of an efficient medium for another culture of rice trough comparative experiments of the nitrogen sources. *Scientia Sinika* **65** pp 559-568

IOP Conf. Series: Earth and Environmental Science **395** (2019) 012079 doi:10.1088/1755-1315/395/1/012079

Vertical vegetable growing: creating tomato varieties for multi-tiered hydroponic installations

I Balashova[1*], S Sirota[1] and Ye Pinchuk[1]

[1] Federal Scientific Center for Vegetable Growing, 14 Selektsionnaya str., VNIISSOK Village 143080 Russia

E-mail: balashova56@mail.ru

Abstract. The relevance of vertical vegetable growing is due to the need to ensure an uninterrupted supply of food and reduce the cost of food production in terms of moving the majority of the population of the Earth into urban space. The market of vertical vegetable growing can be characterized as capacious and fast-growing. In the Russian Federation, this market is still developing. The main vegetable crops are not represented in it. The purpose of the study was to create tomato varieties for the vertical vegetable growing (multi-tiered hydroponic plants). For this purpose, a new approach to the breeding process was necessary, in which the variety would be considered as an element of the technological chain. A collection of 692 tomato samples served as our research material. Research methods included the following: creating a model of a new form of tomato for multi-tiered hydroponic structures; analyzing the genetic basis of the characteristic "short stature" in *Solanum lycopersicum* L.; conducting an individual selection from sporophyte variety populations; testing selected samples in a multi-tiered narrow-rack hydroponics installation; selecting the most large-fruited and productive samples; transferring new varieties to the State Variety Testing (GSI), and analyzing the economic efficiency of cultivating new varieties on multi-tiered hydroponic installation. As a result, in the Federal Scientific Center for Vegetable Growing, the first varieties of tomato vertical vegetables such as "Natasha" (Patent of the Russian Federation No. 9060) and "Timosha" (Patent of the Russian Federation No. 9059) were obtained. Our analysis of the economic efficiency of tomato cultivation on multi-tiered narrow-rack hydroponics demonstrated the convincing advantages of such a method of tomato cultivation as compared to the traditional cultivation method.

1. Introduction

Vertical vegetable growing is an innovative direction in the technological development of protected ground. Its relevance is due to the need to save energy, nutrient substrate and water in cnditions of the growing world population and the growing urban space. The so-called "urban farms" means growing plants in vacant areas in an urban environment, for example, in underground boxes, on rooftops, in closed greenhouses. At the same time, less space is used, and the consumption of energy and water is significantly reduced, which distinguishes these technologies from traditional farming methods [1]. In 2015, the volume of the vertical vegetable market amounted to 1.15 billion dollars. By the end of 2016, a 25% growth was observed, and an increase in the market volume to 6.0 billion dollars is expected by 2025 [2]. The main market segment (42%) is occupied by the countries of the Asia-Pacific Region (China, India, Japan, Taiwan, Singapore, South Korea), in which the shortage of acreage and water for irrigation is already felt. Then the countries of Europe (30%), namely United Kingdom, Germany,

Content from this work may be used under the terms of the Creative Commons Attribution 3.0 licence. Any further distribution of this work must maintain attribution to the author(s) and the title of the work, journal citation and DOI.
Published under licence by IOP Publishing Ltd

France, Belgium, the Netherlands follow, and then the countries of North America (21%), namely Canada, the USA and Mexico follow. 7% of the market is occupied by other countries [3]. It should be noted that the main volume of the market are green crops and strawberries. The main vegetable crops of protected ground (tomato, cucumber, sweet pepper) are represented to a small extent [4]. The domestic market of vertical vegetable growing is in the process of formation.

Tomato ranks first in the world among vegetable crops on cultivated areas in open and protected ground, the area occupies 3.7 million hectares with a yield of 45-65 t / ha. The People's Republic of China holds the lead in the areas occupied by the tomato crop, the area is 974 thousand hectares. The annual production of fresh tomatoes in China is 25 million tons. The Russian Federation ranks sixth in terms of the area under this crop (142 thousand hectares) and occupies 11th place in the production of fresh tomatoes per year (1.82 million tons) [5]. Therefore, we chose the cultivation of tomato for our analysis, focusing on the technology of vertical vegetable growing. In the State Register of Breeding Achievements Being Approved for Use in 2011, 1,630 registered samples (varieties and hybrids) of tomato were found [6]. But among them there was not a single dwarf variety / hybrid that, according to morphological characteristics (a plant height of 35-50 cm), would be suitable for growing on multi-tiered hydroponic plants.

The purpose of the study is to analyze the dwarf tomato varieties adapted for cultivation on multi-tiered hydroponic plants, according to the technology of vertical vegetable farming.

Research objectives: (1) creating a variety / hybrid model for multi-tiered narrow-rack hydroponics; (2) analyzing the genetic basis of short stature in tomato according to the literature; (3) selecting undersized samples from the collection of varietal populations of the laboratory of new technologies using sporophyte selection; (4) testing the selected tomato samples on a multi-stage narrow-rack hydroponics installation and selecting the most productive samples; (5) analyzing the economic efficiency of cultivation of new tomato varieties intended for multi-tiered hydroponic plants.

2. Materials and Methods

The research material was the following collections of tomato *Solanum lycopersicum* L., collected by the employees of the Laboratory of New Technologies of the Federal Scientific Center for Vegetable Growing in different years of research:

- Collection from the Kazakh Research Institute of Potato and Vegetable Production (Kainar village, Alma-Atinskaya region) – 48 samples;
- Collection from the Ukrainian Institute of Vegetable and Melon-Growing of the Ukrainian Academy of Agrarian Sciences (UAAS) (Kharkiv) – 42 samples;
- Collection of tomato marker mutants from the Institute of Ecological Genetics of the Academy of Sciences of Moldova (Kishinev) – 518 samples;
- Collection of the West Siberian Experimental Station of the All-Russian Research Institute for Vegetable Growing (Moscow Region) – 27 samples;
- Collection of samples of tomato from the Republic of Sri Lanka – 57 samples.

A total of 692 tomato samples from the above collections were examined. In addition, samples of the $F_2 – F_7$ generations, isolated from the variety populations in the selection process, were studied. Our research methods included: (1) creating a model of a new form of tomato for the multi-tiered narrow-rack hydroponics; (2) an analytical review of the literature on the genetic basis of dwarf tomatoes; (3) an individual sampling from varieties of sporophytes in the rassadnom branch of a greenhouse with a polycarbonate coating type of the Richelle French company.

Seedlings were grown in standard cassettes / blocks of a mineral wool in the seedling compartment of the greenhouse. As the root environment, peat nutrient substrate / blocks of mineral wool were used. Tomato seeds, pre-treated with a 0.1% $KMnO_4$, solution, were sown in peat-filled cartridges / corks for mineral wool blocks. With adventing the 5th true leaf, the plants were planted in the pots / blocks of a mineral wool and placed on tables in the seedling section of the greenhouse. When growing seedlings, a previously developed nutrient solution was used [7].

With the advent of buds, the plants were set up on a multi-tiered narrow-rack hydroponics unit. First, testing the selected low-growing specimens on the installation of a five-tier narrow-rack hydroponic takes place. The design of such five-tiered narrow-rack hydroponics is a truncated pyramid with a base of 900 cm x 200 cm and a height of 250 cm. Trays from the stainless steel settle down on the iron basis in 5 tiers. Above trays are closed by a cover from the same material with holes for installation of pots with plants. The tray depth is 10 cm, its width is 15 cm. The site of automated solution supply through the pipes of "FITO" NPO is located at the end of the pyramid. The feeding of the nutrient solution into the trays is done automatically, the reverse flow of the solution is performed by gravity. The plants were grown either in 0.5 l pots filled with peat: perlite (1: 1), or in blocks of a mineral wool. The density of installation of plants on the structure of the MUG is 5 plants per 1 meter. The feeding mode of the nutrient solution is as follows: the feeding time is 5-10 minutes with an interval of 40 minutes. For growing the adult plants, a nutrient solution was used, which was also developed earlier [7]. The light sources were the lamps "DNa3-400" ("Reflaks" LLC). When growing tomato plants, the air temperature was + 22 / + 24 °C during the day and + 18 / + 20 °C at night; the relative air humidity within 50-60% was maintained. The duration of the light period is 16 hours / day. Second, we also relied in biometric measurements. The productivity of plants and the fetus mass was determined by the gravimetric method. Third, the dispersion analysis of research results was performed according to [8].

3. Research results

3.1 Creating a model of a new form of tomato for multi-tiered narrow-rack hydroponics

It was carried out on the basis of classical works on the genetics and biology of tomato [9-11], as well as on the basis of our own experimental data [12], [13].

Table 1. Models of the new forms of tomato for multi-tiered narrow-rack hydroponics.

	Small-fruited variety		Medium-fruited variety	
	Feature	Value	Feature	Value
1	*Dwarfism - plant height*	*30-35 cm*	*Dwarf - superdet – plant height*	*45-50 cm*
2	Productivity	450-500 g / plant	Productivity	850-1,000 g / plant
3	Maturation term (shoots - ripe fruit)	78-82 days	Maturation term (shoots - ripe fruit)	78-82 days
4	Brush	Difficult compact	Brush	Simple elongated, intermediate
5	Laying brush	After the second true leaf	Laying brush	After the second and third true leaf
6	Stem	Thick, dense	Stem	Thick, dense, popping after 4-5 inflorescences
7	Forming	2-3 stalks	Forming	Without forming
8	Leaf	Smooth / slightly corrugated, horizontal / hanging	Leaf	Smooth / slightly corrugated, horizontal / hanging
9	Mass of 1 fruit	10 - 25 g	Mass of 1 fruit	35-55 g
	A number of fruits on the plant	20-35 pieces	The number of fruits on the plant	25-35 pieces
10	Tolerance	To low light, to abiotic and biotic stresses	Tolerance	To low light, to abiotic and biotic stresses

3.2 Analysis of the genetic basis of short stature in tomato

The sign "short stature" in tomato is controlled by a group of *d*-genes, which are localized in the long arm of chromosome 2, control the biosynthesis of brassinosteroids, and are represented by 11 alleles [14], [15]. The degree of expression of a characteristic depends on environmental conditions [16]. A selectively useful property associated with the *d*-genes is their manifestation in the early stages of plant development, namely, before flowering. Given the heterophase alternation of generations in higher plants, which implies the existence of asexual generation – diplophase (sporophyte), originating from zygotes before the onset of meiosis in sporocytes, and the sexual generation – haplophase (gametophyte) [17]; the individual selection of stunted plants from populations can be conducted by sporophyte in the seedling compartment of the greenhouse. In addition, it is known that the *d*-genes are localized on chromosome 2 next to the genes controlling early maturity [10], [14]. With clutches, they can be inherited at the same time.

3.3 Selecting short-growing and early-ripe samples from the collection of varietal populations of the laboratory of new technologies using sporophyte selection

Taking advantage of the expression of the *d*-genes, we conducted a selection for the sporophyte in the seedling compartment of the greenhouse. A total of 8 short and early maturing individuals from 692 collection samples of tomato were selected. Their productivity parameters were further studied on a five-tier hydroponic installation. Thus, we managed to speed up the selection process 3 times. The results of our studies, processed statistically, are presented in Table 2. The principle of sporophyte selection is shown using 10 samples from each collection.

Table 2. Analysis of plant height and sporophyte selection in the short and early ripening tomato individuals for multi-tiered hydroponic plants. Greenhouse "Richelle" FSBI VNIISSOK. I and II turnovers, 2012.

No.	Sample	Plant height, cm		The period of "shoots-maturation", day		Sample	Plant height, cm		The period of "shoots-maturation", day	
		1	2	3	4		1	2	3	4
					I turn					
St	*Funtik*	*46*	*St*	*110 St*		*Funtik*	*46*	*St*	*110*	*St*
	Tomato Marker Mutant Collection					*West Siberian Experiment Station Collection*				
1	Mo 331	68	-22	111	-1	1C	75	-29	*96*	*+14*
2	Mo 411	91	-45	*106*	*+4*	*2C*	*52*	*+8*	*94*	*+16*
3	Mo 504	70	-24	122	-12	*3C*	*42*	*+4*	*103*	*+7*
4	Mo 579	82	-36	118	-8	4C	73	-27	*95*	*+15*
5	Mo 609	93	-47	123	-13	*5C*	*51*	*+5*	*87*	*+23*
6	*Mo 663*	*58*	*-12*	*102*	*+8*	6C	82	-36	*98*	*+12*
7	Mo 753	90	-44	123	-13	*11C*	*36*	*+10*	105	+5
8	Mo 782	88	-42	123	-13	411K	77	-31	109	+1
9	*Mo 825*	*41*	*+5*	123	-13	*4K*	*33*	*+13*	*103*	*+7*
10	Mo 940	94	-48	119	-9	*5K*	*30*	*+16*	*91*	*+19*
		HCP_{05}	16	HCP_{05}	7		HCP_{05}	16	HCP_{05}	7
					II turn					
St	*Funtik*	*38*	*St*	*110*	*St*	*Funtik*	*38*	*St*	*110*	*St*
	Collection of the Republic of Sri Lanka					*Collection from the Kazakh Research Institute of Potato and Vegetable Production*				
1	31П	116	-78	*93*	*+17*	36A	90	-52	103	+7
2	32П	104	-66	*96*	*+14*	37A	110	-72	*94*	*+16*
3	33П	98	-60	98	+12	38A	112	-74	101	+9
4	34П	118	-80	107	+3	39A	136	-98	107	+3
5	35П	130	-92	105	+5	40A	108	-70	107	+3
6	36П	72	-34	*94*	*+16*	41A	126	-88	*92*	*+18*
7	37П	86	-48	100	+10	42A	106	-68	*96*	*+14*
8	38П	70	-32	105	+5	43A	72	-34	100	+10
9	39П	108	-70	107	+3	44A	70	-32	105	+5
10	40П	90	-52	113	-3	45A	108	-70	107	+3
		HCP_{05}	20	HCP_{05}	13		HCP_{05}	20	HCP_{05}	13

Samples from the collection of the Ukrainian Institute of Vegetable and Melon-Growing of the Ukrainian Academy of Agrarian Sciences (UAAS) (Kharkiv, Ukraine) were already very tall at the seedling stage; therefore, none of them were selected for subsequent breeding. Samples from the collections of the Republic of Sri Lanka and the Kazakh Research Institute of Potato and Vegetable Production were also all tall, although a number of samples from these collections subsequently showed early maturity.

But the key feature that determines the possibility of cultivation on multi-tiered hydroponic plants is a short stature controlled by the group of *d*-genes. Therefore, when selecting for a sporophyte, we preferred this particular feature. The most productive was the selection from the collection of the West Siberian Experimental Station of the All-Russian Research Institute for Vegetable Growing. A total of 6 samples from 27 showed a short stature and early ripeness. The samples Mo663 and Mo825 were selected from the collection of tomato marker mutants of the Institute of Ecological Genetics of the Academy of Sciences of Moldova. The selection of these samples is of great value also because their genome is guaranteed to contain a group of *d*-genes. Summarizing the analytical studies of the height of plants by the sporophyte, we conclude that selection, by the sporophyte, at an early stage of plant development, could significantly (86 times) reduce the amount of material needed for further breeding work. As a result, 8 out of 692 samples were taken. This made it possible to more fully analyze the selected valuable material in a multi-tiered hydroponic installation.

3.4 Testing the selected dwarf samples on the installation of a five-tier narrow-rack hydroponics.

The installation of a five-tier narrow-tier hydroponic was installed in the "Richelle" greenhouse at the VNIISSOK in June 2013. The site of automated solution supply was kindly provided by the firm "FITO" NPO. We carried out the first round of studies ofn the productivity of the selected plants in the seedling section of the greenhouse in 2012. In subsequent years, all experiments were carried out on the installation of a five-tier narrow-rack hydroponics. In 2014, experiments were carried out in the first and second turns, in 2015 were in the first, second, and third turns. The experimental results, processed statistically, are presented in Table 3.

Table 3. Analyzing the productivity and average weight of the fetus in low-growing tomato samples, previously selected for sporophyte. Selection of productive samples. Greenhouse "Richelle" VNIISSOK. 2012-2015.

No.	Sample	Productivity, g			Average weight of 1 fruit, g		
		2012 (F_1)	2014(F_2–F_3)	2015(F_4–F_6)	2012(F_1)	2014(F_2–F_3)	2015(F_4–F_6)
1	*Funtik - St*	*221*	*Komnatnyy 187*	*Komnatnyy 199*	*53*	*Komnatnyy 7.4*	*Komnatnyy 8.5*
2	Mo 663	187	185	-	*55*	*48.0*	-
3	Mo 825	213	178	-	25	*23.0*	-
4	2C	310	180	-	28	*20.0*	-
5	3C	249	173	-	11	8.0	-
6	5C	303	215	-	20	*18.0*	-
7	11C	304	206	-	12	*11.0*	-
8	*4K*	*431*	*383*	*454*	11	*10.8*	*10.7*
9	*5K*	*358*	*369*	*501*	10	*10.0*	*10.3*
	HCP_{05}	129	54	68	16	2.2	1.8

It is quite difficult to select from the varieties of a self-pollinator, such as *Solanum lycopersicum* L., because these varieties are more or less homogeneous. However, by productivity, we were removed to select 2 small-fruited specimens with red and yellow fruits. In 2015, these samples were transferred to the State Commission of the Russian Federation for Testing Varieties. In 2017, two patents were obtained for them, particularly for the tomato varieties "Natasha" (red-fruited) and "Timosha" (yellow-fruited). These are the first tomato varieties that were created at the Federal Scientific Center for Vegetable Growing specifically for multi-tiered hydroponic plants using sporophyte selection technologies.

IOP Conf. Series: Earth and Environmental Science **395** (2019) 012079 doi:10.1088/1755-1315/395/1/012079

3.5 Analyzing the economic efficiency of new tomato varieties

The creation of tomato varieties for multi-tiered hydroponic plants allowed us to analyze the economic efficiency of the technology of vertical vegetable growing, especially in comparison with the traditional technology of cultivation of tomato culture in film greenhouses. The results are presented in Table 4. Such an analysis clearly shows the advantages of vertical vegetable production systems over traditional tomato cultivation technologies.

Table 4. Analyzing the economic efficiency of the vertical vegetable growing technology, in comparison with the traditional growing tomato technology in film greenhouses.

Indicators	Traditional technology	Five tier narrow-rack hydroponics
Number of crop turns	1-2 ordinary / 1 extended	4 turns: I turn – 19.01-9.05 II turn – 27.03-28.07 III turn – 16.06-28.09 IV turn – 16.08-14.12
Productivity, kg / m^2	45-55	93-128
Selling price, rubles	70-71	70-71
Costs, rub / m^2	2 520	2,532
Revenue, rub / m^2	2 864	8,145
Estimated profit, rub / m^2	344	5,613
Profitability, %	25	222

4. Discussion

By 2050, approximately 85% of the 9 billion people on Earth will live in cities. More than that, food and water shortages will threaten a growing world population [1]. The problem solution is seen in the creation of the so-called "urban farms," for which the vertical vegetable farming technologies are very relevant. They are already being developed and used by the leading countries of the world. Moreover, in the industrial district of Khabarovsk (Russia), the Japanese company "Mirai" built an agricultural complex equipped with modern hydroponic complexes. The peculiarity is the prohibition of using plant materials modified with microorganisms and chemical methods of pest control. With the commissioning of this complex, the region will be able to increase its self-sufficiency in vegetables and herbs from 10 to 60% [18].

In order to promote the technology of vertical vegetable growing in the protected soil of the Russian Federation, the project "Creating dwarf tomato varieties for high-bay hydroponic plants" started in VNIISSOK in 2011. Tomato is the leading open and protected ground vegetable crop in the world. However, the State Register of Breeding Achievements approved for use in the Russian Federation did not contain tomato varieties and hybrids that would meet the requirements for cultivation on high-bay hydroponic plants. We had to create these varieties "from scratch". Ancestral forms of tomato *Solanum lycopersicum* L. are tall [9], [11].

Among the cultivated varieties of tomato, there are indeterminate and determinate forms, as well as ordinary and standard forms by habitus type [19]. We could not find any dwarf forms among them. Therefore, we began our scientific search by creating a model of a such tomato shape that would be adapted for multi-tiered hydroponic plants [7]. The key characteristic of this model is dwarfism. And dwarfism in tomato is controlled by a group of mutant *d*-genes [9].

In connection with this, we began to select the necessary forms for us in the collection of tomato marker mutants, which was maintained for a long period in the VNIISSOK Laboratory of Gamet Selection. The expression level of a group of *d*-genes depends on environmental conditions, but under favorable conditions, the sign of "short stature" manifests itself in the early stages of plant development [14], which allows selection by sporophyte [20]. Our pre-breeding selection was built on this technology, which made it possible to significantly reduce the volume of the studied material before conducting basic research on the installation of a five-tier narrow-rack hydroponics. The technology was named "target breeding" and was recognized internationally [21], [22]. Using this technology in pre-breeding, we created the tomato varieties "Natasha" and "Timosha," designed specifically for multi-tiered hydroponic plants in the vertical vegetable production system. In 2017, VNIISSOK received their patents. Obtaining varieties for hydroponic plants allowed us to analyze the economic efficiency of the technology of vertical vegetable growing, which convincingly demonstrated the advantages of innovative technology of vertical vegetable growing compared with the traditional technology of growing in film greenhouses.

5. Conclusion

The innovative technology of vertical vegetable growing is the technology of the future. It requires a different approach to the selection process, when the selection process itself is considered as a technology for obtaining new plant forms, which includes several stages: (1) creating a model of a plant's future shape; (2) pre-selection (pre-breeding); (3) defining the breeding strategy; 4) selection process; 5) testing new plant forms in various technological and ecological environments; 6) seed production of new plant forms.

This technology should be not only efficient, but also economically beneficial, i.e. at each stage, it should also include an economic component. It is known that the selection process is long and difficult. And the reducing volume of the studied material is most advantageous to carry out in the early stages of the study, in pre-breeding, which allows to significantly reduce the time to obtain new forms, and at the same time to give the most complete description of their productivity and adaptability in further production tests. For this, the key parameter is always useful to highlight which should be sought during the breeding work. The remaining parameters willfully or unwittingly correlate with this "key". This principle, formulated by us as the principle of "target breeding" [22], allowed us to create (from a scratch) a number of new tomato varieties for contemporary technologies of vertical vegetable farming in a relatively short time.

References

[1] *News of high technology, science and technology and the most interesting from the world of gadgets and the Internet* 2019 Available at: http://prohitech.ru (Accessed 03 04 2019)

[2] Global Industry Report, 2014-2025 2017, Report ID: IVR 1-68038-797-1

[3] *Vertical farming market analysis report by offering (lighting, climate control), by growing mechanism (hydroponics, aquaponics), by structure, by fruits, vegetables & herbs, and segment forecasts, 2019 – 2025* Available at: http://www.grandviewresearch.com/industry-analysis/vertical-farming-market (Accessed 03 04 2019)

[4] *Markets and markets* 2019 Available at: http://www.marketsandmarkets.com/Marker-Reports/vertical-farming-market-221795343.html (Accessed 03 04 2019)

[5] *Food and Agriculture Organization of the United Nations (FAO)* 2005 Available at: www.fao.stat.org (Accessed 03 04 2019)

[6] Ministry of Agriculture of the Russian Federation 2017 *State Register of Breeding Achievements Approved for Use* (Moscow, Russia)

[7] Sirota S M, Balashova I T, Kozar E G, Mitrofanova O A, Autko A A, and Dolbik M A 2014 *Greenhouses of Russia* **3** pp 58-62

[8] Dospekhov B A 1985 *Methodology of field experience* (Moscow, USSR: Agropromizdat)

[9] Zhuchenko A A 1973 *Tomato genetics* (Kishinev, Moldova: Shtiintsa)

[10] Tanksley S M, and Mutschler M A 1989 *Genetics map* (Moscow, Russia)

[11] Pivovarov V F, Mamedov M I, Bocharnikova N I 1998 *Solanage cultures: tomato, pepper, eggplant, physalis* (Moscow, Russia: VNIISSOK)

[12] Balashova I T, Suprunova T P, Ursul N A, Vedadevage Sunila Pradipa Kumari, Guzhov Yu L 2008 The after-effects of low-temperature stress in the early stages of development of the starting material and F_1 hybrids of tomato *Agricultural Biology* **3** pp 62-71

[13] Shatilo V I, Balashova I T, Ursul N A, Tkachenko O B, Kondratyeva V V, and Pivovarov V F 2009 Photoregulation of tomato epigenetic resistance to fungal diseases *Agricultural Biology* **3** 118-123

[14] Zhuchenko A A, Balashova N N, Korol A B, Samovol A P, Grati V G, Kravchenko A N, … Dobryanskiy V A, … Smirnov V A, ... Bocharnikova N I 1988 *Ecological and genetic basis of tomato breeding*

(Kishinev, Moldova: Shtiintsa)
[15] *Tomato locus* n. d. Available at: http://solgenomics.net/locus/428/view (Accessed 03 04 2019)
[16] Bocharnikova N I 2011 *Genetic collection of tomato mutant forms and its use in breeding and genetic research* (Moscow, Russia: VNIISSOK)
[17] Gulyaev G V, and Malchenko V V 1975 *Glossary of terms on genetics, cytology, breeding, seed production and seed science* (Moscow, Russia: Rosselkhozizdat)
[18] *Komsomolskaya Pravda* n.d. Khabarovsk Available at: http://m.hab.kp.ru/ (Accessed 03 04 2019)
[19] Brezhnev D D 1955 *Tomato* (Moscow; Leningrad, USSR: Selkhozgiz)
[20] Pivovarov V F, Balashova I T, Sirota S M, Kozar E G, and Pinchuk E V 2013 Improving the selection of sporophyte in order to accelerate the selection of tomato forms for the narrow-rack hydroponics technology *Agricultural Biology* **1** pp 95-101
[21] Balashova I T, Sirota S M, Balashova N N, Kozar E G, Pinchuk E V 2014 *The heritability analysis main tomato traits for the special hydroponic technology (on narrow benches)* (Avignon, France)
[22] Balashova I T, Sirota S M, Kozar E G, and Pivovarov V F 2016 Target tomato breeding for special hydroponic technology In *Abstracts of 20th EUCARPIA Congress* (Zurich, Switzerland)

IOP Conf. Series: Earth and Environmental Science **395** (2019) 012080 doi:10.1088/1755-1315/395/1/012080

The effectiveness of using the growth regulators of a new generation in the production of garlic winter planting material

A Polyakov[1,2], T Alekseeva[1,2] and A Razin[1]

[1] All-Russian Scientific Research Institute of Vegetable Growing – Branch of the Federal Scientific Center for Vegetable Growing, Vereya, Ramenskoye district, Moscow region 140153 Russia
[2] Moscow Region State University, 10a Radio str., Moscow 105005 Russia

E-mail: kaf-bosh@mgou.ru, vniioh@yandex.ru

Abstract. The use of quality planting material is an important factor for the production of agricultural products, because it allows to realize the biological potential of the cultivated plant to a greater extent, to obtain a high-quality crop. Garlic is an exclusively vegetatively propagated crop. Due to the low breeding rate, garlic is difficult to reproduce in a short time in volumes that would fully meet the demand for planting material. In this regard, a large number of imported, often poor-quality planting material of this culture comes to the domestic market. Therefore, the issue of providing a domestic manufacturer with a quality planting material is acute. We have established the effectiveness of contemporary growth regulators in the planting of garlic. The article discusses the issues of increasing the economic efficiency of the production of winter garlic planting material using the growth regulators of the last generation. The article shows the high efficiency of using the composition of silicon organic growth regulators, including Boratran, Krezacin, and Asyak, in a concentration of 0.015% in winter garlic plantings, which allows to increase the yield of bulbs by 2 t / ha. The economic efficiency of using the preplant treatment of clove and plants during the growing season with growth regulators allows to obtain additional profit, amounting to 178 thousand rubles per hectare.

1. Introduction

Garlic is an important culture in the human diet. It is used to prevent the development of cancer and diseases of the cardiovascular system [1], [3], [4]. The medical intake rate of garlic is 3-4 kg / year per person. According to FAOSTAT (2017), garlic production in Russia is at the level of 260 thousand tons, while the need for this crop in the country is about 430 thousand tons annually. The agro-industrial complex of the country is not able to meet the demand of the population and industry in this culture in full; therefore, the imported products enter the Russian market. China is the world's largest garlic importer. In China, more than 20 million tons of garlic are grown annually, which is about 79% of the world production (Table 1) [1], [2], [3].

Content from this work may be used under the terms of the Creative Commons Attribution 3.0 licence. Any further distribution of this work must maintain attribution to the author(s) and the title of the work, journal citation and DOI.
Published under licence by IOP Publishing Ltd

Table 1. Main countries producing garlic [9].

Country	Production, t	Sown area, ha	Productivity, t / ha	The share of world production, %
China	22,216,965	820,101	27.1	78.9
India	1,693,000	321,000	5.8	6.0
Egypt	274,668	12,607	21.8	1.0
Russia	258,455	27,445	9.4	0.9
Bangladesh	425,401	66,259	6.4	1.5
USA	232,000	13,360	17.4	0.8
Ukraine	185,830	21,500	8.6	0.6
Brazil	120,897	10,588	11.4	0.4
World	28,160,584	1,577,780	17.8	

In Russia, the production of garlic is limited by the lack of quality seed and planting material. In the State Register of Breeding Achievements, 74 varieties of winter garlic are registered [5], but most of them "exist only on paper," and the high-quality domestic planting material cannot be purchased in an amount sufficient for industrial production. Because of this, the market of this culture is filled with imported low-quality products. The seed propagation method is absent in varieties of garlic sown, listed in the state register of the Russian Federation, in connection with which its reproduction is carried out by vegetative means, namely cloves, single-clove bulbils, aerial bulbils. The main method of reproduction of garlic is cloves. In total, 5-9 cloves usually form in winter garlic bulbils. Those cloves weighing at least 3-5 g are suitable for landing. A high yield of bulbils can be guaranteed to be obtained from such chives for the next year, subject to agrotechnical requirements. However, not all cloves in the bulbil meet this condition, some cloves have a smaller mass, some can be fused; therefore, the reproduction rate of cloves remains very low. To overcome this limiting factor, aerial bulbils can be used. From 40 to 300 aerial bulbs formed in the inflorescences of garlic planting. In this case, the more bulbils formed in the inflorescence, the smaller the mass of one bulbil. When using aerial bulbils weighing less than 0.1 g, the process of obtaining multi-chives bulbils is often three years. The traditional cultivation of garlic from aerial bulbils is as follows:

Aerial bulbils ⇨ Small single-clove bulbils ⇨ Large single-clove bulbils ⇨ Bulbils divided into cloves

The production of garlic according to this scheme hinders the filling of the market with planting material of this culture [3]. Therefore, the use of contemporary growth regulators (aimed at increasing the mass of aerial and single-clove bulbils), which are accompanied by an increase in yields and a reduction in the period of obtaining multi-clove bulbils and aerial bulbils for one year, is a relevant and promising direction of research.

Within the framework of this research, we studied silicon organic growth regulators (such as Energy M, Kufecin, Boratran, Krezacin, Asyak, Hermatranol, and Silacin) in various combinations. Registered in 2008, the drug "Energy M" is the most widely distributed and well-known of these regulators. This drug has shown its high efficacy on sunflower, root crops, cabbage, and other vegetable crops [6], [7]. The remaining drugs are the latest developments of the "Flora Si" LLC.

2. Materials and Method

Studies were conducted in the Department of Biotechnology and Innovation Projects of the All-Russian Research Institute of Vegetable Growing – Branch of the Federal Scientific Center for Vegetable Growing (Moscow Region, Ramensky District, Vereya Village) in open ground conditions (Table 2).

Table 2. Characteristics of the soil experimental plot.

Indicator	Measurement units	Amount
Humus content in the topsoil 0-20 cm contained: humus	%	3.41-3.44
The amount of absorbed bases	mg-eq. / 100 g of soil	46-50
Mobile phosphorus (according to Chirikov)-	mg / 100 g of soil	22.78-24.62
Exchange Potassium (according to Maslova)	mg / 100 g of soil	10.38-17.88
Hydrolytic soil acidity	mg-eq. / 100 g of soil	0.72-0.92
Neutral environment reaction – pH_{KCl}		6.7
Capillary soil moisture	%	43-44.0
Hygroscopic moisture	%	8.25
A specific soil weight of the arable layer is 0-25 cm	g / cm^3	2.61

To increase soil fertility and improve the mineral nutrition of plants, the following was introduced into the soil:

- Potassium chloride at a dose of 50 g / m^2 was added to the main treatment;
- During the growing season, plants were fed as follows: the first dressing was made with the ammonium nitrate (30 g / m^2) in the beginning phase of the intensive leaf growth, and the second dressing was made with potassium sulfate (50 g / m^2) 2 weeks after the first.

The cloves and plants of the garlic winter varieties "Gladiator" served as the material for research [8]. The study design included the triple treatment of plants with solutions of growth regulating substances. The following compositions were used in the experiment: Kufecin together with Energy M 0.01%; Boratran together with Krezacin and Asyak 0.015%; Hermatranol together with Krezacin and Asyak 0.015%; Silacin together with Krezacin and Asyak 0,015%; Silacin with Asyak 0,015%. In the control variant, the plants were treated with water. The repetition of the experiment was 4-fold, the area of one plot is 1 m^2, the sowing of a private is the distance between the lines of 25 cm; 10 cm between plants in a row; and the length of a row is 1 m^2.

3. Results and Discussion

Studies have revealed a high efficiency of use of the composition, including Boratran, together with Krezacin and Asyak, used in a concentration of 0.015%. In this embodiment, the yield of bulbils was 1.58 kg / m^2, which is 14% higher than the control, and the average mass of the bulbils was 40.4 g, which was 12.5% higher than the control.

Table 3. The economic efficiency of the use of growth regulators of the new generation in the production of winter garlic planting material.

Economic indicator	Standard technology	Using growth regulators	Deviation
Bulbil yield, t / ha	13.8	15.8	2
Bulbil sales price, rubles / kg	200.0	200.0	0
Gross income from bulbils selling, rubles	2,760,000.0	3,160,000.0	+400,000.0
The cost of production, rubles, Including the following:	1,128,800.0	1,350,800.0	+230,000.0
Preplant treatment and processing during the growing season of Boratran together with Krezacin and Asyak in a concentration of 0.015%	-	60,000.0*	+60,000.0
Production cost per ton of received bulbils, rub.	44,457.4	42,627.3	-1830.0
Taxes (UAT, 6%)	97,900.0	108,600.0	+10,700.0
Profit, rub.	1,631,200.0	1,809,200.0	178,000.0
Profitability, %	144.5	133.9	-10.6

* A regulator cost is 3000 rubles per 50 g.

IOP Conf. Series: Earth and Environmental Science **395** (2019) 012080 doi:10.1088/1755-1315/395/1/012080

Evaluation of the effectiveness of this technique showed that it allows one to get 15.8 t / g of garlic bulbils, which is 2 tons more compared to standard technology (Table 3). Thus, the use of growth regulators has led to higher production costs. However, a large profit from sales of products covered them. The use of Boratran in combination with Krezacin and Asyak in a concentration of 0.015% made it possible to increase profits by 178 thousand rubles / ha.

4. Conclusion

Pre-treatment of cloves and double treatment of garlic plants with a solution of growth regulators of a new generation, including Boratran, Krezacin, and Asyak in a concentration of 0.015%, allows to increase the yield of bulbils by 2 t / ha, and the weight of the bulbils grows up to 40 g, making it possible to get additional 178 thousand rubles / ha.

References

[1] Kapustina L 2016 Features of the cultivation of garlic in the world *Vegetable Growing* **9** pp 33-35

[2] Litvinov S S, and Shatilov M V 2015 *Efficiency of the Russian vegetable growing (analysis, strategy, forecast)* (Moscow, Russia: All-Russian Research Institute of Vegetable Growing)

[3] Polyakov A V 2014 The most important issues of the development of garlic in the Russian Federation In S S Litvinov Ed *Ecological problems of contemporary vegetable growing and the quality of vegetable production* **1** (pp 437-443) (Moscow, Russia: All-Russian Research Institute of Vegetable Growing)

[4] Nikolić, V. D., Ilić, D. P., Nikolić, L. B., Stanković, M. Z., Stanojević, L.P., Savić, I. M., … Savić, I. M. 2012 The synthesis and structure characterization of deoxyalliin and alliin *Advanced Technologies* **1**(1) pp 38-46

[5] Ministry of Agriculture of the Russian Federation 2017 *State register of breeding achievements approved for use: Plant varieties (vol 1)* (Moscow, Russia)

[6] Petrichenko V N, Loginov S V, and Turkina O S 2014 The influence of growth regulators and micronutrients on metabolic processes in table root vegetables *Agrarian Russia* **8** pp 10-12

[7] Petrichenko V N, and Loginov S V 2010 The study of silicon organic drug Energia-M *Agrochemical Vestnik* **2** pp 22-24

[8] Polyakov A V 2013 Gladiator is a new sort of winter garlic *Potatoes and Vegetables* **9** pp 19-21

[9] FAOSTAT 2019 *Statistical data* Available at: http://www.fao.org/statistics/en/ (Accessed 03 04 2019)

The economic benefits of the production of double haploid for selection of white cabbage

A I Mineikina[1*] L L Bondareva[1] and E A Domblides[1]

[1] Federal Scientific Center for Vegetable Growing, 14 Selektsionnaya str., VNIISSOK Village 143080 Russia

Email: vniissok@mail.ru

Abstract. An important task of any state in the world is to ensure food security. The Russian Federation is the third largest cabbage producer in the world. In the conditions of the modern market, for ensuring successful competitiveness, selection must promptly respond to consumer requests. To reduce some stages of the selection process, we used the modern biotechnological method of culture of isolated microspores *in vitro*. This method allows one to get homozygous lines in the first generation. In the Federal Scientific Center for Vegetable Growing, the protocol for obtaining doubled haploid cabbage cultures has been optimized and successfully applied in practice. However, for expediency of implementation of the obtained lines in the selection and seed-growing process, it is important for us to know the economic efficiency of production. The paper presents the data on the basic costs of obtaining lines of white cabbage with traditional and modern biotechnological methods of selection. The article provides a comparative assessment of their cost-effectiveness in the selection of white cabbage. It is shown that the use of the biotechnological method in combination with the classical method of selection allows to obtain pure lines with 100% homozygosity for 3 years. At the same time, economic costs of producing lines, when using doubled haploid plants of white cabbage, are reduced in 2 times.

1. Introduction

According to FAO, for 2017, global cabbage production is 71.451 million tons, and the cultivation area is 2.513 million hectares. The areas under cabbage cultivation in the Russian Federation amounted to 105.8 thousand hectares, and the production volume amounted to 3.531 million tons. In comparison to 2016, the area decreased by 6.1%, respectively, the production volume decreased by 2.4%. The same trend is observed in the world. Based on this, there is the task of improving the efficiency of production of cabbage crops in smaller areas, including by reducing production costs. To a greater extent, plant breeding can more effectively increase food production in smaller areas and is often more environmentally friendly today [1].

Among the plants of the *Brassica* genus, the most common vegetable crop is white cabbage (*B. oleracea* L. convar *capitata* (L.) Alef. var. *capitata* L. *f. alba DC*). Its popularity is explained by the whole complex of biological and economically useful properties [2]. Selection of white cabbage is mainly focused on the creation of F_1 hybrids, differing from the variety populations with high yields, evenness of plants in terms of ripening and quality of productive organs. The most difficult, time consuming and lengthy step in this process is the creation of constant parental lines, which takes from 7 to 14 years to receive [3], [4].

Content from this work may be used under the terms of the Creative Commons Attribution 3.0 licence. Any further distribution of this work must maintain attribution to the author(s) and the title of the work, journal citation and DOI.
Published under licence by IOP Publishing Ltd

Recent advances in biotechnology are a valuable and powerful tool for improving the efficiency of breeding and reducing the time to create clean lines of agricultural plants. Among biotechnological methods, the culture of isolated microspores *in vitro* has long been recognized as effective for improving breeding material. The basis of the technology is the ability of microspores to switch from the gametophytic path of development to the sporophytic one under the influence of stress factors such as elevated temperature, high osmotic pressure, etc. Under appropriate conditions, such microspores begin to develop in a sporophytic type, forming embryoids [5]. The first successful microspore culture studies in vegetable crops of the *Brassicaceae Burnett* family were carried out in the early 1980s [6]. Later, a basic rape seed microspore culture protocol was developed, which serves as the basis of DH-technology for plants of the *Brassica* L. genus [7]. Then the culture of microspores began to be used for different types of cabbage cultures [8], [9], [10], [11], [12], [13], [14]. At the Federal Scientific Center for Vegetable Growing, the protocol for creating doubled haploid plants in vegetable crops from the *Brassicaceae Burnett* family is optimized [15].

Despite obvious achievements of scientists in the field of biotechnology, the practical experience of using the obtained doubled haploid plants in domestic breeding is extremely limited [16]. At the Federal Scientific Center for Vegetable Growing, research work on the production of DH-plants of cabbage cultures has been going on for over 20 years, and the following plants have been created and are registered in the State Register of Breeding Achievements: *DH-line broccoli "BR1-1"* (Breeding achievement patent No. 7144); *Dobrynya F1 kohlrabi cabbage* hybrid, created on the basis of DH-lines (Breeding achievement patent No. 8866) [18]. A hybrid of white cabbage *Natali F1* is undergoing state variety testing.

It should be noted that the feasibility of using modern methods in breeding is determined both by their economic efficiency and time costs. Therefore, an important and crucial step in the introduction of DH-technology in the selection of white cabbage is the determination of the economic benefits of the production of doubled haploid lines.

To solve this goal, the following is necessary:

1. To optimize the classical selection and seed-growing scheme for creating F_1 hybrids of white cabbage due to the introduction of DH-lines;
2. To calculate the economic costs of plant production *in vitro* (cultures of isolated microspores);
3. To conduct a comparative analysis of the economic and time use of the classical and modern biotechnological method in the selection of white cabbage.

2. Method

The work was performed at the Federal Scientific Center for Vegetable Growing using classical and contemporary biotechnological methods for white cabbage breeding from 2014 to 2017. The starting material was the selection of cabbage samples from the white laboratory of selection and seed production of cabbage cultures and regenerated plants obtained in the culture of isolated microspores *in vitro* in the laboratory of biotechnology.

White cabbage plants were grown according to the method of field experiment according to Dospekhov [17]. Planting was carried out by the seedling method according to the scheme 70x50 cm, the breeding area was 175 m^2. Plants were selected in the stage of technical maturity of the heading according to the main approbation characteristics. Then, the selected queen cells were planted in 5-liter vegetation vessels with a nutrient mixture consisting of two parts of sod land, one part of humus and peat with the addition of a ammonium nitrate phosphate fertilizer (10 g per vegetal vessel) and dolomite flour. Then, the queen cells were placed on vernalization in a refrigeration chamber with a temperature of 4-6 °C for 60 days. At the end of vernalization (at the stage of acclimatization), the vegetation vessels with plants were placed in the vegetation chamber to receive the peduncle. Within 15-35 days, the temperature was gradually increased from + 8 °C to 16-18 °C, with a mode of 16 hours per day / 8 hours per night and under illumination of 9000 l [4].

Obtaining of regenerated plants by the *in vitro* culture of isolated microspores was carried out according to the methodological recommendations [15]. Plants, in which the leaves and root system were normally developed, were transferred to vegetation vessels filled with a mixture of peat and perlite (7:3). To adapt plants to *in vivo* conditions, they were covered with perforated plastic cups. Regenerant plants were grown under the same conditions as donor plants.

The main economic indicators that make up the cost of producing pure lines of white cabbage were determined, guided by the technological map of cultivation of white cabbage in specific conditions of the Moscow region (according to the data of the Economic Department of the Federal Scientific Center for Vegetable Growing), the production of plants doubled haploids was estimated, and highly valuable observations were collected.

3. Results

3.1 Determination of the time spent on creating pure lines of white cabbage

The use of doubled haploids in the selection of white cabbage allows for more efficient selection of the desired genotypes (by expanding the spectrum of genetic recombinant forms). Since the lines of doubled haploids are 100% homozygous, their selection can be carried out in the first generation. In classical breeding, to obtain an aligned homozygous line, 6 generations of inbreeding are required to be carried out, which can take up to 12 years for cabbage with a two-year development cycle. Thus, the time spent on creating a clean line of white cabbage is significantly reduced (Fig. 1).

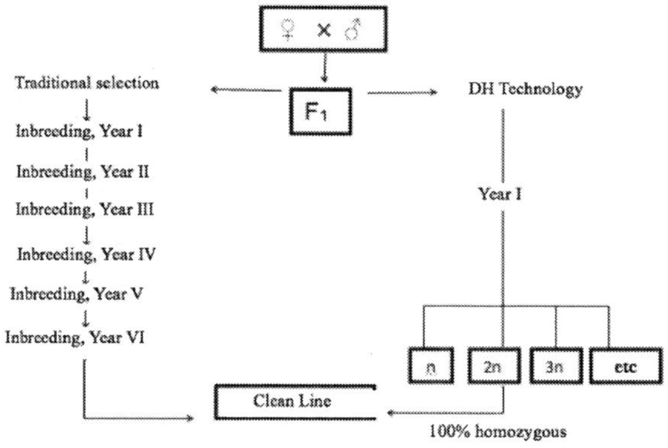

Fig. 1. Diagram of ways to create a linear material of white cabbage.

To speed up individual stages of the selection process, in our work to create homozygous lines, artificial climate chambers with a given light and temperature conditions are used. The use of these climate chambers in breeding work allows us to translate the cultivation of biennial crops into a one-year cycle. The use of artificial climate chambers makes it possible to reduce the selection process by two times under controlled conditions [4].

For the use of white cabbage lines, in breeding, an important step is to determine the ability of the line to produce heterotic offspring in one or another hybrid combination. Thus, taking into account the reduction of individual stages due to the artificial climate chambers, the duration of the selection process for creating and evaluating pure lines of white cabbage will look like this:

1. Sowing of seeds and study of source material, selection of queen cells, vernalization: spring-autumn X - the start of breeding work;
2. Production of doubled haploid lines in an isolated microspore culture (production, adaptation, vernalization at the "stem" stage): winter-autumn X + 1;
3. Duplication of doubled haploid lines, production of hybrid combinations, evaluation of the line combining ability: winter-autumn X + 2.

According to the presented scheme, it takes 3 years to get lines of white cabbage with 100% homozygosity, which can be included in the selection scheme for creating an F1 hybrid.

3.2 Determination of the economic costs of production of white cabbage lines by various methods

To determine the economic benefits of producing white cabbage lines, we compared the traditional method of white cabbage breeding for 12 years in a two-year cycle and for 6 years by converting to a one-year cycle through artificial climate chambers. Also, we compared the traditional method of white cabbage breeding for 12 years in a two-year cycle and for 6 years by shortening the stage of creating a pure homozygous line by introducing DH-technology.

In agriculture, the cost of cultivating any crop, including white cabbage, is determined by drawing up a flow chart, into which major expenses are made necessary during the entire period of crop cultivation. Since white cabbage is a biennial crop, costs must be taken into account over two years of the full cycle of plant development. In the first year, the plants are grown up to the technical ripeness phase of the head and laid for storage and vernalization. In the second year, all necessary testes are grown, and one gets the seeds.

Since we had to compare 3 methods of creating a clean line of white cabbage, we would adhere to the "only difference" principle. At the initial stage of growing the queen cells, conditions were the same for all methods. The collection nursery consisted of 500 plants of white cabbage, covering an area of 175m^2. The intensity of selection was 10%, so 50 queen cells were deposited. The Federal Scientific Center for Vegetable Growing has a storage facility with a storage space for vegetable crops of 864 m^3, while our plants occupied 1.5 m^3. After storage, 10% of the plants were culled, so 45 plants were grown in greenhouses on an area of 16m^2 (Table 1).

Table 1. Costs for cultivation of cabbage plants by the traditional method in a 2-year cycle (direct costs are provided according to the data of the economic department of the Federal Scientific Center for Vegetable Growing).

No	Expenditure	Unit cost	Amount	Cost, rub.
	1 year (in conditions of open ground)		*500 plants (area of 175 m^2)*	
1	**Growing seedlings**		40 days	
1.1	Wages	120.01	40 man-hours	4800.40
1.2	Cassettes for seedlings	70.00	10 cassettes of 60 cells	700.00
1.3	Priming	3.00	10 cassettes (36 l)	108.00
1.4	Fertilizers	250.00	1 dressing for the period	250.00
1.5	Other	40 days (depreciation, water, protection, if necessary)		500.00
2	**Growing queen cells**		*125 days*	
2.1	Wages	120.01	Total scope of work (tractor drivers and workers)	38479.60
2.2	Fuel	40.00	When using the MTZ-80 and John Deere tractor (3.8 liters)	152.43
2.3	Fertilizers	155.00	Minerals - 10.5 kg	1627.50
2.4	Means of protection	1173.00	2 treatments of 35 ml	82.11
2.5	Watering	20.00	10.5 liters for treatment and watering for 3 months	4326.00
2.6	Depreciation	12618.00	MTZ-80 and John Deere	220.82
2.7	Seeds	25000	On the whole area of2.45 g	61.25
3	**Storage of queen cells**	730.46	From October to April (6 months), 1 container (1.5 m^3)	4382.76
4	**Total for the year**			55690.87
	2 year (in protected ground)		*45 plants (16 m^2)*	
1	Wages	1713.60	In the period from April to September (6 months)	13386.64
2	Fuel	40.00	When using the Mitsubishi MT-180D mini-tractor (12.6 liters)	506.25
3	Watering	20.00	Based on 115.8 liters of water	2315.25
4	Consumables	100.00	Insulators, labels, garter rope	4500.00
5	Depreciation	45.30	On the occupied area	724.80

6	Fertilizers	155.00	They are worn by hand under each plant (4 times per period)	459.38
7	**Total:**	**Per year**		**21892.31**
		One hole cycle		**77583.18**
		For six cycles		**465499.08**

Table 2. Estimated costs for the production of doubled haploids of white cabbage by the method of *in vitro* culture of isolated microspores*.

No	Main expenses	Unit cost	Amount	Amount, rub.
1	*Climate chamber*			
1.1	Light	4.80	120 days (16 hours a day, power 600 W, 6 lamps)	33177.6
1.2	Plant care, man-hours	120.01	140 man-hours	16801.40
1.3	Priming	3.00	50 plants 5l = 250l	750.00
1.4	Fertilizers	103.00	4 kg	412.00
1.5	Vegetation vessels	60.00	50 on 5l	3000.00
2	*Laying microspores*			
2.1	Nutrient medium	512.34	1l Nitsch	512.34
2.2	Utensils for cooking environment	240.00 660.00	2 glasses 2 bottles	480.00 1320.00
2.3	Cultivation dishes	55.88	50 cans with lids	2794.00
2.4	Autoclaving	4.80	1 hour (power 3 kW)	14.40
2.5	Electrical costs laminar	4.80	6 hours (power 330 W)	9.50
2.6	Work, man-hours	960.04	2 man-days	1920.08
3	*Cultivation of microspores*			
3.1	Thermostat, energy costs	4.80	30 days (power 330 W)	1140.48
3.2	Work, man-hours	120.01	4 man-hours	480.04
4	*Embryoid cultivation*			
4.1	Nutrient medium	334.75	MS: 1,5l in cups, 10l * 2 in cans	7197.13
4.2	Utensils for cooking environment	240.00 660.00	2 glasses 2 bottles	480.00 1320.00
4.3	Cultivation dishes	10.00 55.88 26.48	50 Petri dishes, 500 cans with lids 50 Falcon tubes Other laboratory materials	500.00 27940.00 1324.00 2000.00
4.4	Autoclaving	4.80	5 hours (power 3 kW)	72.00
4.5	Electrical costs – Laminar – Shelving	4.80 4.80	40 hours (power 330 W) 720 hours (power 36 W, 14 lamps)	63.36 1741.82
4.6	Work, man-hours	120.01	5 person-days	4800.16
5	**Total**			**110250.31**

* For genotypes whose embryoids range is from 5 to 20 embryoids per Petri dish (6 mm in diameter).

To determine the cost of creating regenerated plants obtained in an *in vitro* culture of isolated microspores, the cost estimates for the main expenses were calculated based on obtaining 50 doubled haploid lines of white cabbage, according to the principle of the only difference for all methods (Table

2). At the same time, it is necessary to take into account that the average costs are based on the production of DH-plants in an isolated *in vitro* microspore culture, since the release of embryoids would strongly depend on the responsiveness of a particular genotype when the technology is used.

Further, we carried out vernalization of the obtained plants in the "stem" stage and hybridization in an artificial climate chamber with given light parameters for 5 months (Table 3).

Table 3. The costs of hybridization of white cabbage plants in an artificial climate chamber (direct costs according to the Economic Department of the Federal Scientific Center for Vegetable Growing).

No	Main expenses	Unit cost	Amount	Amount, rub.
1	*Climate chamber*			
1.1	Light	4.80	150 days (16 hours per day, power 600 W, 6 lamps)	41472.00
1.2	Plant care, man-hours	120.01	175 man-hours	21001.75
1.3	Priming	3.00	50 plants 5l = 250l	750.00
1.4	Fertilizers	103.00	4 kg	412.00
1.5	Vegetation vessels	60.00	50 on 5l	3000.00
1.6	Total			66635.75

To determine the economic efficiency of various methods of breeding white cabbage, a comparative assessment of the cost of obtaining 50 plants of white cabbage was carried out according to the principle of the only difference (Table 4).

As a result of the calculation, the obtained data showed that when creating lines for F_1 hybrids of white cabbage, the most cost-effective is the use of contemporary biotechnological method. Since due to the reduction of some stages of the selection process, financial costs are reduced. As well as the use of artificial climate chambers is advisable, since the time to get the lines is reduced in 2 times, which contributes to the rapid entry of hybrids into the market. Consequently, this increases the implementation period and profits.

Table 4. Comparative assessment of the cost necessary for creating lines of doubled haploids of white cabbage by various selection methods.

Indicators		Traditional selection method (inbreeding)		Biotechnological method (*in vitro* culture of isolated microspores 1 year)
		12 years	6 years, sing artificial climate chambers	
Costs of cultivation, rubles / ha	Open ground	334145.22	334145.22	55690.87
	Protected ground	131353.86	-	-
Cost of hybridization in an artificial climate chamber, rubles / 50 plants		-	399814.50	66635.75
Costs necessary for doubled haploid lines production, rubles / 50 plants		-	-	110250.31
Total costs, rub.		465498.86	733959.72	232576.93
Costs for 1 plant, rub.		9309.98	14679.19	4651.54

4. Conclusion

By calculating the cost of creating pure lines of white cabbage, the economic benefit of using the modern biotechnological method of culture of isolated microspores *in vitro* when creating white cabbage hybrids is proved. At the same time, the production time of hybrids is reduced from 12 to 6 years, and financial costs are reduced in 2 times.

References

[1] Meenakshi B, Barsha T, Gouri S S, and Anil K J 2017 *JPP* **6** p 384
[2] Krasheninnik N V 2010 *Gavrish* **2** p 16
[3] Pivovarov V F, and Startsev V I 2006 (Moscow, Russia: VNIISSOK Publishing House)
[4] Bondareva L L, and Razin O A 2014 *Russian Vegetables* **4** p 37
[5] Shmykova N A, Shumilina D V, and Suprunova T P 2015 *Vavilov Journal of Genetics and Breeding* **19**(1) p 111
[6] Lichter R 1982 *Pflanzenphysiol* **105** p 427
[7] Pechan P M, and Keller W A 1988 *Physiol Plant* **74**, p 377
[8] Lichter R 1989 *Plant Breeding* **103** p 119
[9] Takahata Y, and Keller W A 1991 *Plant Sci.* **74** p 235
[10] Duijs J C, Voorrips R E, Visser D L, and Custers J B 1992 *Euphytica* **60** p 45
[11] Cao M Q, Li Y, Liu F, and Dore C 1994 *Plant Cell Rep* **13** p 447
[12] Zhang W, Qiang F, Xigang D, and Manzhu B 2008 *Sci. Hortic.* ***117*** p 69
[13] Winarto B, and Teixeira da Silva J A 2011 *Plant Cell Tiss Org Cult* **107** p 305
[14] Yuan S X, Su Y B, and Liu Y M 2012 *Plant Cell Tiss Organ Cult* **110** p 69
[15] Domblides E A, Shmykova N A, Shumilina D V, Zayachkovskaya T V, Mineykina A I, Kozar E V, … Akhramenko V A, … Shevchenko L L, … Kan L Yu, … Bondareva L L, ... Domblides A S 2016 (Moscow, Russia: VNIISSOK Publishing House)
[16] Monakhos S G 2015 *Potatoes and Vegetables* **9** p 38
[17] Dospekhov B A 1985 (Moscow, USSR: Agropromizdat) p 350
[18] Domblides E, Shmykova N, Zayachkovskaya T, Vjurtts T, Kozar E, Kan L, Romanov V … Domblides A 2017 In Materials *"Agrosym 2017": VIII International Scientific Agriculture Symposium*

Economic efficiency of growing domestic hybrids of sweet peppers in the steppe zone conditions

M I Mamedov[1*], S F Gavrish[2] and O N Pyshnaya[1]

[1] Federal Scientific Center for Vegetable Growing, 14 Selektsionnaya str., VNIISSOK Village 143080 Russia
[2] Research Institute of Vegetable Crop Selection, 11/1 2ya Khutorskaya str., Moscow 127287 Russia

Email: mubaris-mamedov@yandex.ru

Abstract. One of the main factors for increasing the efficiency of production of sweet peppers is the introduction of new high-yielding varieties and hybrids adapted to the growing conditions in order to fully realize their potential productivity. Every year, the leading breeding institutions of the country and foreign companies supply a large number of new hybrids to the market for subsequent introduction into production of the most productive ones. The main criterion for evaluating the effectiveness of varieties and hybrids is the annual economic effect of their introduction into production. The purpose of our research was to test domestic heterotic pepper hybrids suitable for production in industrial volumes, clarify the varietal cultivation technology for the steppe zone, and establish the economic efficiency of their cultivation. The work shows the economic efficiency of growing domestic hybrids of sweet pepper *F_1Victor, F_1Medok F_1Lekar, F_1Aleksiy, F_1Erivan* using advanced cultivation technology. The article shows that domestic hybrids with a planting period of March 15-20 and plant density of 72 thousand / ha provided a yield of 73-78 t / ha, which was 12-20% higher than the *F_1Vedran* standard. The calculations showed that the profitability of production is determined by the productivity of a particular hybrid. Among the studied samples, the most profitable (more than 30%) are *F_1Lekar, F_1Aleksiy*, which had a yield of 77-78 t / ha. High profitability was observed in *F_1Victor* (28.9%) and *F_1Erivan* (27.2%) hybrids, profitability is slightly lower in *F_1Medok* (23.8%). In general, domestic hybrids have a 2.3-3.1 times higher profitability compared to the standard, which indicates the high efficiency of their cultivation in industrial production.

1. Introduction

Pepper is one of the most valuable vegetable crops, having a universal use in the food, pharmaceutical, perfumery and cosmetic industry, traditional medicine due to the rich biochemical composition of fruits [1]. In the face of increasing environmental stress, increasing the resistance of the human body to stress becomes one of the main problems of science [2]. One of the main ways to solve this problem is to provide the population with vegetable products, where the main role belongs to pepper as a multivitamin product. Continuous demand and growth in pepper consumption requires an increase in varietal diversity and yield growth. A variety characterized by high productivity potential combined with resistance to biotic and abiotic stressors with improved biochemical composition, high technological qualities of the products obtained is the key to providing the industry with this valuable product. At the same time, it is

Content from this work may be used under the terms of the Creative Commons Attribution 3.0 licence. Any further distribution of this work must maintain attribution to the author(s) and the title of the work, journal citation and DOI.
Published under licence by IOP Publishing Ltd

important for specific hybrids to have a variety cultivation technology, in which the cultivation of pepper will be less expensive and more profitable.

There are 786 varieties and hybrids of sweet peppers in the State Register of Breeding Achievements Approved for Use in 2019, 646 of them are domestic breeding achievements and 140 are foreign products. Every year, the leading breeding institutions of the country and foreign companies supply a large number of new hybrids to the market for subsequent introduction into production of the most productive ones.

The purpose of our research is to test domestic heterotic pepper hybrids suitable for production in industrial volumes, to clarify the varietal cultivation technology, and to establish the economic efficiency of their cultivation.

2. Materials and Method

In 2016-2017, on the basis of the "Agrosvit" Research and Production Association (Kherson region), a comparative variety testing of the following sweet pepper hybrids was conducted: *F_1Victor, F_1Medok, F_1Lekar, F_1Aleksiy, F_1Erivan* [3], [4]. A hybrid of sweet pepper *F_1Vedran* (Enza Zaden) was used as a standard. It is grown in industrial vegetable production in the south of the country. The introduction of new varieties and the technology of their cultivation primarily based on the calculation of economic efficiency. For the economic evaluation of domestic varieties and their agrotechnological passport, we used the following system of indicators: (1) the yield of products from 1 ha in physical and monetary terms, (2) the cost of 1 ton of products, (3) the conditional net income and profitability of production [5], [6].

3. Results and Discussion

For the study of domestic varieties, the technology of cultivation of pepper using fertigation was used, it is used in the economy with specifying the time of sowing and plant density. As a result of studying various dates for sowing seeds, the period from March 15 to 20 is recommended for growing seedlings (cassette technology) for the steppe zone. With earlier plantings, the plants outgrow and age in advance, which leads to the loss of the first ovaries and a decrease in yield. Study of various planting patterns, such as 90 + 50 x 20 cm (72 thousand plants / ha) and 90 + 50 x 25-30 cm (55 thousand plants / ha), made it possible to establish that the best yield results were obtained when growing peppers with a planting density of 72 thousand plants per 1 hectare. With such planting density, an average yield was more than 70 t / ha, which was 50% higher compared to another study.

Analysis of the cultivation data of domestic pepper hybrids showed that the profitability of production is directly dependent on crop yields. The final useful effect from the use of means of production and living labor, the return on total investment shows the economic efficiency of their production. It is known that the level of costs and their combination, expressed in the cost of production, are the most informative components of the criterion of economic efficiency of agricultural enterprises, along with profit and profitability [7]. The results of our research show that the following indicators account for the largest share in the cost of growing peppers: items for labor costs – 33.80%, applying mineral fertilizers and means of protection – 14.19%, and also costs for seeds and planting material – 29.58% (Fig. 1).

As a result of various testing techniques, it was noted that the domestic hybrids of sweet pepper have a sufficiently high genetic potential of productivity and adaptive properties for their effective cultivation. Features of growth and development of the studied pepper hybrids directly affected their productivity. According to the results of the analysis, domestic hybrids formed a sufficiently high yield (73-78 t / ha) and exceeded in productivity the standard *F_1Vedran* hybrid by 12-20%. The calculations showed that pepper as a whole is a profitable crop, the profitability of its production is determined by the productivity of a particular hybrid (Fig. 2). *F_1Lekar* and *F_1Aleksiy* hybrids are the most profitable (more than 30%) among the studied hybrids. They had a yield of 77-78 t / ha. A high profitability was observed in the *F_1Victor* (28.9%) and *F_1Erivan* (27.2%) hybrids, the *F_1Medok* hybrid has a slightly lower profitability (23.8%).

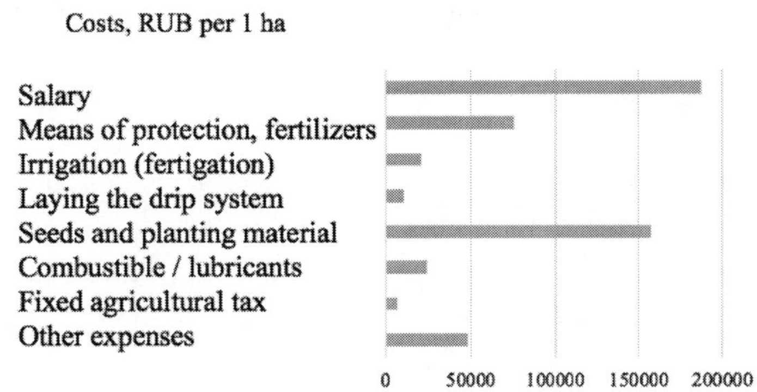

Fig. 1. The cost structure for the production of sweet pepper.

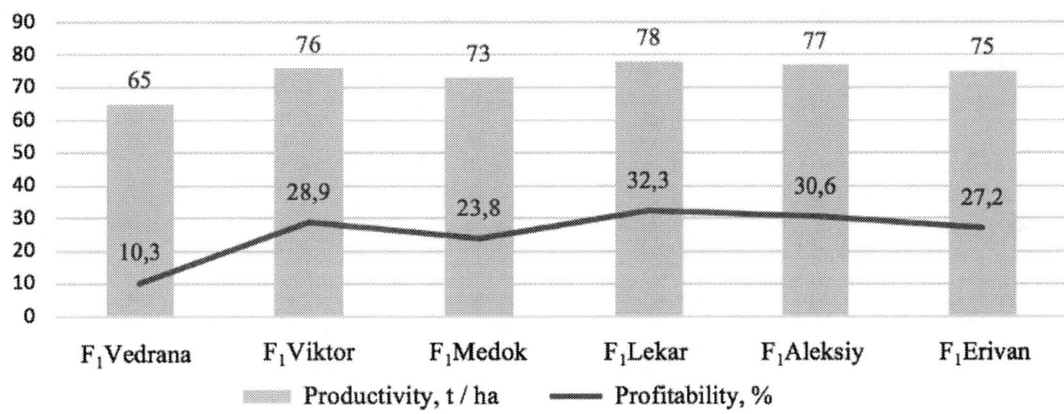

Fig. 2. Dependence of the profitability of production from the yield of hybrids.

In general, the domestic hybrids have a higher profitability (in 2.3-3.1 times) if compared to the standard. This indicates a high efficiency of growing domestic sweet pepper hybrids in industrial production.

4. Conclusion

Based on studies of the economic efficiency of growing sweet pepper hybrids, it was revealed that all tested hybrids (*F_1Victor, F_1Medok, F_1Lekar, F_1Aleksiy, F_1Erivan*) are promising for industrial production in open ground conditions, because they provide high yields and profitability of production.

References

[1] Gish R A 2017 *Pepper culture* (Krasnodar, Russia: KubSAU)
[2] Borisov V A, Litvinov S S, and Romanova A V 2003 *The quality and keeping quality of vegetables* (Moscow, Russia: Mytishchinskaya interdistrict printing house)
[3] Litvinov S S 2011 *Methods of field experience in vegetable growing* (Moscow, Russia: Rosselkhozakademiya)
[4] Ministry of Agriculture of the Russian Federation 2015 *Methods of state variety testing of agricultural crops, fourth release (potatoes, vegetables and melons)* (Moscow, Russia: Kolos)
[5] Polunin G A et al 2007 *Guidelines for determining the overall economic effect of the use of the results of research and development in the agro-industrial complex* (Moscow, Russia: RAAS)
[6] Razin A F, Bukharov A F, Ivanova M I, Bukharova A R, and Razin O A 2017 *The basic principles for calculating the economic efficiency of new varieties of vegetable crops* In *Current and New Trends in Plant Breeding and Seed Production of Crops* (pp 167-169) (Moscowm Russia)
[7] Soldatenko A V, Pivovarov V F, Razin A F, Shatilov M V, Razin O A, Rossinskaya O V, ... Bashkirov O

V 2019 Problems of competitive vegetable production *Russian Vegetables* **1** pp 3-7 Retrieved from https://doi.org/10.18619/2072-9146-2019-1-3-7

Economic justification for applying instrumental methods of seed quality control

F Musaev[1*], V Pivovarov[1] and S Beleskyi[2]

[1] Federal Scientific Center for Vegetable Growing, 14 Selektsionnaya str., VNIISSOK Village 143080 Russia
[2] Scientific Research Institute of Storage Problems of the Federal Agency of State Reserves, Moscow, Россия

Email: vniissok@mail.ru

Abstract. The basis of food independence of any country is a reliable seed system, and seed control is a concern of the state in many developed countries. The standard methods used for analyzing the quality of seeds do not meet today's requirements for seed production, they are laborious, time-consuming, and also uninformative. The contemporary level of scientific knowledge requires the use of new instrumental methods that are highly informative. We have developed an instrumental X-ray method for assessing the quality of vegetable seeds. It is highly informative, quick and easy to perform, and preserves the material being analyzed. The article compares and economically analyzes the widely used standard morphometric method and the implemented instrumental method of X-ray analysis of seeds. It turned out that the use of instrumental methods for seed quality analysis is economically and energetically justified. The method of X-ray analysis of seeds favorably differs in energy efficiency and the period of performance from the standard method. Introduced instrumental method for visual analysis of radiographs is slightly inferior to the standard in labor costs, but the process is significantly accelerated with automatic analysis. The method is automated, which allows to avoid errors in assessing the quality of seeds associated with the operator's subjectivity. The article calculates the benefit from introducing the X-ray method in the seed control case. It is shown that over a ten-year lifetime of the X-ray diagnostic unit, all the costs of organizing the laboratory pay off even with a relatively small workload and begin to make a profit.

1. Introduction

A reliable seed system is the basis of any country's food independence. In most developed countries, including the Russian Federation, seed control is the responsibility of the state. Unfavorable soil and climatic conditions in most regions of our country do not contribute to the widespread industrial seed production of heat-acceptable crops. The main problem of the industry is the quality of the produced seeds, which do not meet the requirements of contemporary agriculture. Therefore, a constant search for favorable backgrounds for seed production is necessary [1].

Seed production of vegetable crops has a special specificity associated with the huge species diversity, the increased heat demand of most crops, the biennial and long-term development cycle of a number of species. The quality requirements of vegetable seeds, due to their high productivity, is also increased.

Content from this work may be used under the terms of the Creative Commons Attribution 3.0 licence. Any further distribution of this work must maintain attribution to the author(s) and the title of the work, journal citation and DOI.
Published under licence by IOP Publishing Ltd

The traditional, standard methods of analyzing the quality of seeds do not meet today's requirements for seed production, because they are laborious, time-consuming, and also uninformative. The modern level of development of agricultural technologies, as well as the expansion of scientific knowledge requires the use of new instrumental methods in seed control, characterized by high information content and speed of execution [2].

The seed quality analysis is a rather laborious process and requires highly skilled operator. It is necessary to minimize the share of manual labor in this process through the use of instrumental methods. It will also increase the objectivity of the assessment being conducted.

There are many instrumental methods for influencing seeds to improve their sowing qualities [3], [4], [5], [6], [7], [8], [9], [10]. However, only a few are engaged in studying the internal structure of seeds in connection with their economic and biological significance [11], [12], [13], [14].

In the long-term (2016-2018), a joint work of employees of the Federal Scientific Center for Vegetable Growing, the Agrophysical Research Institute, St. Petersburg State Electrotechnical University and the Scientific Research Institute of Storage Problems of the Federal Agency of State Reserves was carried out with the purpose of developing a new instrumental biophysical method for analyzing the quality of vegetable seeds.

2. Materials and Method

The object of research was various methods of analyzing the quality of seeds, namely standard (morphometric) and instrumental (biophysical). The material for research was the different-quality seeds of various 25 types of vegetable crops.

Morphometric analysis of seed quality was carried out by the standard method GOST 12038-84 (Interstate Standard-Methods for Determining the Germination of Agricultural Crops Seeds) [15].

An instrumental analysis on the quality of seeds was carried out by a new biophysical X-ray method of vegetable seeds [16].

We have conducted a comparative economic analysis of various methods for assessing the quality of seeds.

The standard laboratory method for determining seed germination GOST 12038-84 adopted as the traditional method of analyzing the quality of seeds.

The method of X-ray analysis of seeds was considered in the following two versions: (1) visual analysis of radiographs; (2) automatic analysis of radiographs. The following works were taken into account as an example: (1) for a standard technique, we used a certified seed testing laboratory of the Federal Scientific Center for Vegetable Growing; (2) for the X-ray method, works conducted in the laboratory of X-ray analysis of seeds of the Department of Electronic Devices of the Saint Petersburg Electrotechnical University "LETI."

3. Results and Discussion

The X-ray analysis of vegetable seeds is still carried out for research purposes, and the cost of work can only be calculated. The entire analysis process consists of the following three stages: (1) sample preparation; (2) setting, shooting, recognizing images; and (3) visual analysis of radiographs on a computer monitor.

Table 1. The cost of analyzing seeds of vegetables and melons in the Testing Laboratory of the Federal Scientific Center for Vegetable Growing, March 2018.

Type of work	Standard	Cost, rub.
Complete analysis	-	1841
Clean and waste seed	GOST 12037-84	779
Germination	GOST 12038-84	814
Viability	GOST 12038-84	336
Humidity	GOST 12041-82	142
Weight of 1000 seeds	GOST 12042-80	106
Pest population	GOST 12045-97	142

IOP Conf. Series: Earth and Environmental Science **395** (2019) 012083 doi:10.1088/1755-1315/395/1/012083

The process of sample preparation is the most time-consuming technique in the whole process of seed analysis. At the same time it does not require special qualifications of the worker and can be performed by a laboratory assistant, trainee, etc. It is a method of regularly applying seeds on a thin sticky tape. Depending on the type, size of seeds, the preparation of one sample (100 seeds x 4 times) takes 20 minutes.

A radiographic survey of seeds is carried out by an electrical engineer. Shooting one sample (4 or 8 pictures) takes 10-20 minutes, and an image recognition on the scanner is also included here. More time is spent on setting up a plant, which depends on the size of the seeds being analyzed. If seeds of one size group are removed, then the need for reconfiguration disappears, accelerating the process.

The X-ray analysis is the final stage and is carried out by a qualified x-ray operator. The visual analysis is carried out on a computer monitor, the results are plotted in a worksheet. An experienced specialist will need about 30 minutes to analyze one sample.

Total, the visual X-ray analysis of one seed sample, together with sample preparation and shooting, takes 60-70 minutes (Table 2).

With automatic analysis, a computer performs the work of a radiobiologist in a few seconds. Therefore, the time for sample preparation and filming is taken into account, which totals 30-40 minutes.

In the X-ray analysis of seeds, labor costs and analysis duration almost coincide (Table 2). With the standard method of analysis, regardless the amount of labor, the analysis duration is associated with the type of seed and can be 7-14 days, even up to 21 days for individual species (GOST 12038-84).

So, the X-ray analysis differs from the standard method of less energy and speed of execution. The method for visual analysis of radiographs is inferior to the standard in labor costs, but with an automatic analysis, the process is significantly accelerated.

Table 2. Economic and energy comparison of standard and instrumental methods.

Methods	Labor costs, man / hour	Energy consumption, kW / h	Duration of analysis
	Standard method		
GOST 12038-84	0.8 – 1.0	4 – 7	7-14 days
	Instrumental (radiographic) method		
Visual radiography	1.0 – 1.2	0,5	1 - 1.2 hours
Automatic radiography	0.5 – 0.7	0,3	Less than 1 hour

We considered the economic effect of the use of the X-ray method of the current version in the current laboratory, taking into account current costs. Now, we will calculate the benefits in organizing a new laboratory of seeds radiography, taking into account the payback of equipment. The cost of analyzing a seed sample on a radiograph is calculated using the following formula:

$$\mathbf{C = (P_{eq}/XT) + P_{em} + S}$$

where **C** is the cost of analyzing a single sample; $\mathbf{P_{eq}}$ is the cost of equipment; **X** is the workload of the laboratory (a number of analyzes per year); **T** is the estimated service life of the equipment; $\mathbf{P_{em}}$ is the cost of consumable materials and electricity; **S** is the remuneration of employees, taking into account the complexity and complexity of work.

Now, we will enter our data into the formula: $\mathbf{P_{eq}}$ is 3.8 million rubles, (when purchasing several X-ray complexes; **T** is 10 years; $\mathbf{P_{em}}$ is 15 rub.; **S** – [0.5 man / hour technician (rate 15000 rub / month) = 90 rub.] + [0.25 man / hour engineer (rate 30,000 rub / month) = 90 rub.] = 180 rub.; **X** – it depends on the number of seeds received for analysis and the frequency of their quality control, ranges from 50 (in low-loaded research laboratories) to 2000 (in laboratories that carry out commercial orders) analyzes per year.

We give an example of calculating the cost of X-ray analysis of a single seed sample, when the laboratory is loaded at 50 analyzes per year:

$$\mathbf{C} = [3\,800\,000 / (50 \text{x} 10)] + 15 + 180 = 7795 \text{ rubles.}$$

The cost of determining the seed germination of a seed according to the standard method GOST 12038-84 is 814 rubles (Table 1). Obviously, with a low frequency of analysis, the laboratory of X-ray analysis of seeds cannot benefit. It is necessary to increase the number of tests for the payback of expensive equipment. Analysis of the data in Table 3 shows that with an increase in the "load" of the laboratory (100 ... 200 ... 300 ... 500 tests / year), the cost of each analysis decreases, and already when conducting 700 tests a year for 10 years, the cost of each analysis is 738 rubles. This is slightly lower than the cost of determining seed germination by the standard method. Therefore, in order to pay off laboratory equipment with a 10-year service life, it is necessary to carry out at least 700 X-ray analyzes of seeds per year or 2-3 analyzes per day, which is technically easy to do. When the laboratory has 1500-2000 tasks in a year, we can not only pay back the equipment, but also achieve tremendous savings and work with profits (Fig. 1).

Table 3. Comparative economic analysis of instrumental and standard techniques, RUB.

Number of tests per year	Expenses (for 10 years)		Costs, X-ray analysis of one sample	Annual savings
	Radiography P_{eq} + [XTx (P_{em} +S)]	GOST 12038-84 XT x 814		
50	3 897 500	407 000	7795	- 3 490 500
100	3 995 000	814 000	3995	- 3 181 000
200	4 190 000	1 628 000	2095	- 2 562 000
300	3 858 000	2 442 000	1286	- 1 416 000
500	4 775 000	4 070 000	955	- 705 000
700	5 165 000	5 698 000	738	533 000
1000	5750 000	8 140 000	575	2 390 000
1500	6 725 000	12 210 000	448	5 485 000
2000	7 700 000	16 280 000	385	8 580 000

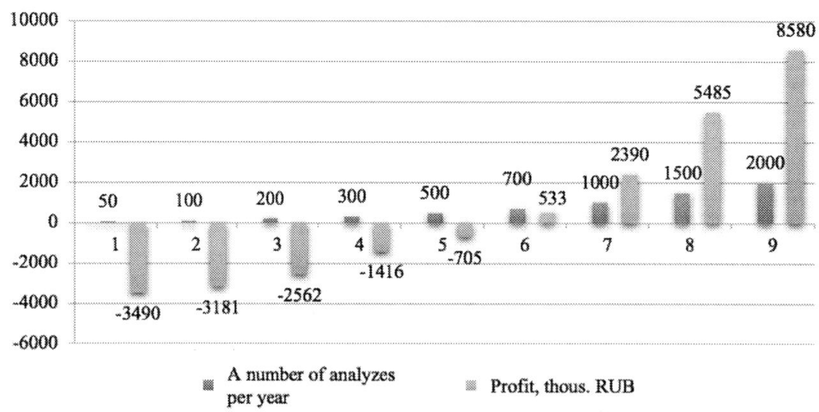

Fig. 1. Economic benefits from using instrumental methods for analyzing the quality of seeds.

Despite the higher cost of X-ray diagnostic equipment, the costs pay off with its rational use, and for a 10-year service life of the plant, the method makes a huge profit.

4. Conclusion

Obviously, the use of instrumental methods for analyzing seed quality is economically feasible both by reducing manual labor and speeding up the analysis process. When applying the instrumental analysis in the automatic mode, the reliability of such assessment is also increased due to the exclusion of the subjective factor associated with the qualification or operability of the operator.

References

[1] Pivovarov V F, and Musaev F B 2017 Ecological and geographical orientation of seed production of

vegetable crops *Proceedings of the Kuban State Agrarian University* **67** pp 185-189

[2] Arkhipov M V, Tyukalov Yu A, Priyatkin N S, Gusakova L P, Potrakhov N N, and Beleskyi S L 2017 Physical and technical basis for obtaining competitive seed material *Tavricheskiy Vestnik of Agrarian Science* **4**(12) pp 42-49

[3] Gordeev Yu A 2012 *Methodological and agrobiological bases of pre-sowing bioactivation of crop seeds by low-temperature plasma flow* (Dissertation of the Doctor of Agricultural Sciences) (Saint-Petersburg, Russia)

[4] Sterkhova T N 2012 Reserves for increasing the productivity of seeds of vegetable crops *Vestnik of the Moscow State Agroengineering University named after V. P. Goryachkin* **1**(52) pp 17-18

[5] Zuev N A, Rudobashta S P, Zueva G A, and Zotova E Yu 2012 Stimulation of onion seeds by pulsed infrared heat treatment *Vestnik of the Moscow State Agroengineering University named after V. P. Goryachkin* **2**(53) pp 9-12

[6] Sigacheva M A, Pinchuk L G, and Gridina S B 2013 Preseeding ozonization of seeds as a factor influencing the grain quality of spring wheat *Vestnik of the Altai State Agrarian University* **3** pp 21-24

[7] Kostin V I, Dozorov A V, and Isaychev V A 2018 On the issue of stimulation of agricultural plants under the action of physical and chemical factors during seed treatment *Vestnik of the Ulyanovsk State Agricultural Academy* **42**(2) pp 67-77

[8] Soproni V D, Hathazi F I, Arion M N, Molnar C O, and Bandici L 2009 Aspects regarding the adapting and optimization of mixed drying systems microwave-hot air for the processing of agricultural seeds In *PIERS* Proceedings (pp 210-213) (Beijing, China)

[9] Ragha L, Mishra S, Ramachandran V, Bhatia M S 2011 Effects of low-power microwave fields on seed germination and growth rate *Journal of Electromagnetic Analysis and Applications* **3** pp 165-171

[10] Bykovsky Yu A, Shaymanov A A, and Fefelova S V 2015 Perspectives of Q2 machine use for estimation of seeds sowing quality *Potatoes and Vegetables* **8** pp 24-25

[11] Bukharov A F, Baleev D N, Ivanova M I, Bukharova A R, and Razin O A 2018 Embryo morphometry as an element of the dill seed quality testing system *Proceedings of the Kuban State Agrarian University* **72** pp 63-66

[12] Zheludkov A G, Beleskyi S L, and Potrakhov N N 2016 A comprehensive solution to the problems of automating an X-ray method for analyzing the quality of seeds and grain of cereals *Bakery Products* **5** pp 58-61

[13] Arkhipov M V, and Potrakhov N N 2016 *Microfocal X-ray of seeds in plant breeding, seed production and plant protection* In *IV Luga Scientific Readings. Contemporary Scientific Knowledge: Theory and Practice* (pp 53-55) (Luga, Russia)

[14] Arkhipov M V, Priyatkin N S, and Zhuravleva E V 2014 *A comprehensive approach to studies of the problem of inhomogeneity of seeds by biophysical methods.* In Science. Information. Spirit XVIII International Scientific Congress: bioelectrography (pp 19-20)

[15] GOST 12038–84 1985 *Crop seeds: methods for determining germination* (Moscow, Russia: Standards Publishing House)

[16] Musaev F B, Potrakhov N N, and Arkhipov M V 2016 *Radiography of vegetable seeds* (Saint-Petersburg, Russia: Saint Petersburg Electrotechnical University "LETI")

IOP Conf. Series: Earth and Environmental Science **395** (2019) 012084 doi:10.1088/1755-1315/395/1/012084

Economic efficiency of obtaining carrot lines using classical and biotechnological methods

T S Vurtz[1], E A Domblides[1] and A V Soldatenko[1]

[1] Federal Scientific Center for Vegetable Growing, 14 Selektsionnaya str., VNIISSOK Village 143080 Russia

E-mail: vniissok@mail.ru

Abstract. Garden carrot is a traditional food in Russia. Based on the consumption rate of 10 kg per year per person, 1,469 thousand tons of carrots per year must be produced for the population of the Russian Federation. To meet the needs of the population in fresh and processed carrot products, the development of new highly productive and resistant varieties and hybrids is necessary. The traditional production of linear material and hybrids in the breeding of garden carrots is a laborious, lengthy process due to the 2-year cycle of plant development, self-incompatibility, and inbreeding depression. On average, 16-20 years can be spent on creating a hybrid of carrots with the successful course of the selection process. Using biotechnological methods, completely homozygous lines can be obtained in 1 year and a wide variety of starting material can be created for inclusion in the breeding process. The article presents an analysis of economic efficiency and a comparative assessment of the production of garden carrot lines using classical breeding and biotechnology methods. Calculations are based on the rates being used at the Federal Scientific Center for Vegetable Growing. It is shown that the economic costs of obtaining lines when using doubled haploid plants of the garden carrot are reduced in almost five times. To increase the efficiency of the breeding process, biotechnological methods need to be embedded in classical breeding schemes.

1. Introduction

Garden carrot is a valuable, indispensable for rational human nutrition vegetable culture, characterized by a high content of vitamins, biologically active substances, mineral salts, phytoncides, essential oils and dietary fibers necessary for the normal functioning of the human body. Also, it is used for medicinal purposes and in the perfume industry [1]. To meet the needs of the population in fresh vegetables, it is necessary to produce 17.9 million tons of vegetables. In the Russian Federation, 66.3 thousand hectares are occupied under crops of carrots and turnips (according to the FAOSTAT statistics database, 2018). The difference between demand and production is made up for by imported products from the countries of near and far abroad. In comparison to 2017, carrot imports decreased by 10.6% in 2018. The main exporters to Russia are Israel, China, Belarus, Egypt.

150 varieties and 150 hybrids of garden carrots are in the State Register of Breeding Achievements for 2019. Most of them are represented by foreign companies. Manufacturers prefer F_1 hybrids, which are most aligned, plastic and, as a result, technologically advanced compared to varieties. Their cultivation to the maximum eliminates manual labor, in connection with which, the time and cost are reduced.

Content from this work may be used under the terms of the Creative Commons Attribution 3.0 licence. Any further distribution of this work must maintain attribution to the author(s) and the title of the work, journal citation and DOI.
Published under licence by IOP Publishing Ltd

IOP Conf. Series: Earth and Environmental Science **395** (2019) 012084 doi:10.1088/1755-1315/395/1/012084

Table carrot is a cross-pollinated plant, its varieties represent a population of biotypes that are close to each other in morphological and economic characteristics. The traditional production of linear material and hybrids in the breeding of garden carrots is a laborious, lengthy process due to the 2-year cycle of plant development, self-incompatibility, and inbreeding depression. Carrot plants bloom and produce seed offspring only in the second year of life (*in vivo*). Root crops need to go through the vernalization stage for 3-5 months at a temperature of 2-3 °C during their winter storage.

Currently, one of the main ways to obtain carrot hybrids is the use of forms with cytoplasmic male sterility (CMS). Three types of lines are necessary to create a hybrid *ms line A*, which has a CMS, line B - a fixer of a CMS at line A, and line C, which is a pollinator of line A. These lines are created by the classical selection methods by inbreeding (forced self-pollination) and selection for 6-7 generations. This is a laborious and complex breeding work that requires a sterility control in the interaction of nucleus genes with cytoplasmic sterility genes. On average, 16-20 years can be spent on creating a carrot hybrid with a successful selection process [2]. Over this long period of time, the renewal of the fleet of agricultural machinery occurs, phytopathogens overcome the resistance of plants and consumer demands are changing. Thus, traditional breeding lags behind the technical development of the agro-industrial complex, does not keep pace with changes in consumer preferences and the evolution of harmful organisms. In addition, it is very important that the lines that make up the basis of the hybrid are aligned, have high combining ability in terms of yield, quality and adaptability, which cannot be achieved with 100% probability using only classical methods of creating linear material.

The key stage, which takes the longest time in the breeding scheme for creating a hybrid of canteen carrots, is to obtain clean lines, then the inclusion of doubled haploids in the breeding scheme (DH-technology) can reduce the time for creating a hybrid to 4-6 years. Obtaining doubled haploids *in vitro* significantly accelerates the creation of completely homozygous lines in one generation; while increasing the accuracy and efficiency of the selection process, allows you to quickly find the desired combination, reduces the time to create lines and hybrids. The development and implementation of doubled haploid technology has significantly changed the plant breeding process in the world. Nowadays, this innovative and effective technology is included in the breeding programs of many globally important crops such as corn, wheat, barley, rape, sugar beets, peppers, and cucumbers [3]. Every year, in the world, a significant number of new plant varieties obtained using haploid technologies are recorded. Today, the leaders in the field of haploid technologies are the EU countries, Australia, Canada, the USA and China [4]. We are aware of a registered patent for the production of doubled haploids of agricultural crops of the Apiaceae family in November 2006 (WO 2006/125310). In Russia, such studies are conducted only in a small number of scientific institutions. We are aware of only one registered variety of garden carrots, namely the Sonata, created with the participation of VNIISSOK (now Federal Scientific Center for Vegetable Growing) on the basis of DH-lines (Copyright certificate No. 34702 of 11.29.2000).

Doubled garden carrot haploids can be obtained using the method of gynogenesis (culture of dustless ovules) and androgenesis (anther culture, culture of isolated microspores).The complexity of the application of the anther culture method is associated with the uncertainty of the result, since embryoids and callus can develop both from microspore cells and from anther somatic tissues. The same problem exists when using the method of cultivating unfertilized ovules (gynogenesis) of carrots. However, this method is the only way to obtain doubled haploids from sterile plants. The most promising is the production of DH-plants in a culture of isolated microspores. This excludes the presence in the in vitro culture of somatic tissues, which cast doubt on the haploid nature of the resulting plants. Success in using the method of culture of microspores garden carrots was achieved only in recent years. The authors of [5] first reported the formation of multinucleated structures and callus in a culture of isolated microspores. The authors first received DH-carrot plants and presented their protocol [6], [7]. At the Federal Scientific Center for Vegetable Growing, we have created an effective technology for producing doubled haploids in the culture of isolated microspores *in vitro* [8]. The use of this technology and *in vitro* dustless ovules culture (for sterile samples) makes it possible to significantly reduce the time required to obtain homozygous lines. In this regard, we conducted a study comparing the economic efficiency of obtaining clean garden carrot lines using classical and biotechnological methods in 2017-2018.

2. Materials and Method

The studies were carried out on the basis of the Federal Scientific Center for Vegetable Growing in 2017-2018. Calculating the costs of creating a line of garden carrots was carried out taking into account labor resources, as well as the cost of materials and fixed assets on average for 2017-2018. We collected and analyzed experimental data on labor and material resources at all stages of the selection process when creating lines.

We compared the following two methods of creating clean lines of garden carrots in the conditions of the Moscow region: (1) in the open ground and using unheated greenhouse; (2) using a biotechnological method of an *in vitro* culture of isolated microspores.

We used samples of carrots from the collection of the laboratory of genetics and cytology and the laboratory of table root vegetables of the Federal Scientific Center for Vegetable Growing. Donor plants were grown in open ground conditions with the observance of agrotechnical measures for the production of testes. Root crops, after vernalization in a refrigerating chamber at a temperature of + 4 ... 6 °C for 3-3.5 months, were planted in a climate chamber at around-the-clock temperature of 19 °C, illumination of 9000 lux, ans a light mode of 16 hours a day / 8 hours a night.

For the introduction into culture, the selection of buds was performed using cytological study of the stages of development of microspores. A differential staining technique [9] and an Axio Imager A2 microscope (Zeiss, Germany) were used to determine the stage of development of microspores. The buds collected from plants were sterilized for 30 seconds in 96% ethanol, then for 5 minutes in a 50% aqueous solution of the Belizna preparation, with adding0 1 drop per 100 ml of Tvin-20. Then, the buds were washed three times in sterile distilled water. The cultivation of microspores was carried out on ½ NLN medium with a sucrose concentration of 13% and MS in Petri dishes 6 cm in diameter according to the developed procedure [8].

3. Results

The cost of breeding one variety or hybrid varies significantly and depends on the methods and technological receptions, technical equipment and skills of workers. The high cost of foreign seeds (France, Holland, China) is associated with high costs for the creation of hybrids. They include the cost of selected varietal samples, which is about 70% of the total costs.

The criterion for the effectiveness of methods for obtaining clean lines is the acceleration of the process with minimal expenditure of material and labor resources. Accounting for specific indicators reflecting the impact of various factors on this process is necessary. When evaluating performance, attention should be paid to the features that influence the final results.

To identify the economic effect, knowledge of total labor costs, which provided a specific result, is necessary. The same result can be achieved with different costs and time. For a comparative assessment of economic efficiency, we considered direct variable costs. They include the following: material costs for seeds, fertilizers, plant protection products, combustible and lubricating materials, electricity, maintenance of machinery and other material costs.

In the collection of variety samples of the source material, we studied the economically valuable traits and carried out the search and selection of plants of interest for inclusion in the breeding process (August - September). When calculating, we took into account the costs of growing 1000 plants in a collection nursery (open ground), occupying an area of 30 m^2. The selection efficiency for further work is 25% (i.e. 250 plants). Losses during storage (October - April) amounted to 10%, and we planted 225 root crops in the greenhouse in April. When using the trilinear hybrid creation scheme, for 6 cycles (seed-roots-self-pollination / inbreeding-seeds), at least three aligned lines (line A, B, C) should be provided after evaluation by economically valuable attributes.

Economic evaluation for the year was carried out by comparing the cost of the occupied area and was calculated for one unit. To determine the costs, flow charts on the culture of garden carrots were made on the basis of prices prevailing in 2017 - 2018. Analysis of the impact of costs on the cultivation of plants in open ground shows that the main part of them falls on the share of wages, namely 37,453 rubles (Table 1).

Table 1. Costs for obtaining lines in open ground and unheated greenhouse (6 generations of inbreeding).

No	Main expenses	Unit cost, rub	Amount, pieces	Sum, rub
	Open ground. The period (calendar days) of 120 days, the area of 30 m^2, 1000 plants			
1	Wages, rub.	37453.00		37453.00
2	The cost of fuels and lubricants	40.00	MTZ-80, John Deere	26.13
3	The cost of mineral fertilizers, rub.	155.00	Mineral fertilizers - 1.8 kg	279.00
4	Plant protection products	1227.00	3 treatments of 0.6 ml	56.10
5	Depreciation	12618.00	MTZ-80, John Deere	37.90
6	Seeds	25000.00	On the whole area, 6 g	150.00
7	Storage costs	730.46	Within 6 months (October-April) 1.5 m^3 (container)	4382.76
Total for 1 year				**42384.89**
	*Protected ground, area 40 m^2, 225 plants**			
1	Wages, rub.	4284.00	In the period from April to September (6 months)	27935.00
2	The cost of fuels and lubricants	40.00	When using a mini-tractor Mitsubishi MT-180 D	1265.00
3	Watering	20.00	As needed	5788.00
4	Consumables	100.00	Insulators, labels, twine	22500.00
5	Depreciation	45.30	On the occupied area	1812.00
6	Fertilizers	155.00	On the 40 m^2 (4 feeding)	1488.00
Total for the second year				**60788.00**
Total for 2 years				**103172.89**
Only 6 cycles				**619037.30**

* A sample of 225 plants was taken to simplify calculations; it also allows one to evaluate and maintain the necessary (at least three) to create a hybrid lines.

Since carrots are a biennial crop, the total costs consist of the costs of growing plants in open and protected ground. Using classical breeding methods, at least 6 generations of inbreeding are required to create a line for biennial cultures, and this is at least 12 years. For 6 cycles, the economic costs of carrying out inbreeding on the minimum number of samples (to create a subsequent hybrid) amount to 619,037.30 rubles. At the same time, it is necessary to take into account that even after 6 generations of inbreeding, 100% evenness of lines cannot be guaranteed.

One solution to this problem is to use the biotechnological method of *in vitro* culture of isolated microspores, which allows one to reduce the time to create clean lines compared to traditional methods. To determine the necessary costs for the creation of homozygous carrot lines in an *in vitro* culture of isolated microspores, an estimate of labor costs and material resources at current prices on the basis of 50 doubled haploid carrot lines were obtained. According to the results of our research, costs amounted to 129,063.60 rubles (Table 2).

In addition, there are differences in the responsiveness of a particular genotype, which affect the final yield of doubled haploids. The calculations made are given for samples whose responsiveness in a microspore culture is up to 10 embryos per 1 cultured Petri dish 6 mm in diameter. For poorly responsive genotypes, larger studies need to be carried out, which would entail additional costs for modifying the standard technique. However, a significant reduction in the time interval for creating a fully homozygous line to one year increases the economic benefit. The important point is that in doubled haploids, all recessive signs are visible and not disguised under the influence of dominant genes, in connection with which selection is greatly simplified and its accuracy increases.

Table 2. Costs for obtaining a line of garden carrots by using a microspore culture *in vitro*.

No	Main expenses	Unit cost, rub	Amount	Amount, rub
1		*Laying microspores*		
1.1	Nutrient medium	850.00	1 l NLN	850.00
1.2	Utensils for cooking medium	240.00 660.00	2 glasses 2 bottles	480.00 1320.00
1.3	Cultivation dishes	10.00	Petri dishes - 50 pieces	500.00
1.4	Autoclaving	4.80	1 hour (power 3 kW)	14.40
1.5	Electric costs - Laminar	4.80	6 hours (power 330 W)	9.50
1.6	Work, man-hours	120.01	4 man-hours	480.04
2		**Microspores cultivation**		
2.1	Energy costs - thermostat	4.80	60 days (330 W)	2280.96
2.2	Work, man-hours	120.01	4 man-hours	480.04
3.		***Embryoid cultivation***		
3.1	Nutrient medium	1000.00 850.00	MSm - 2 l (with hormones) NLN – 2 l	2000.00 1700.00
3.2	Utensils for cooking medium*	240.00 660.00	2 glasses 2 bottles	480.00 1320.00
3.3	Cultivation dishes	8.00 12.00	Petri dishes - 50 pieces Test tubes * PB2-16x150 - 500 pieces	400.00 6000.00
3.4	Autoclaving	4.80	5 hours (power 3 kW)	72.00
3.5	Electric costs - Laminar	4.80	40 hours (power 330 W)	63.36
3.6	Electric costs - racks	4.80	1800 hours (power 36 W, 14 lamps)	3110.40
3.7	Work, man-hours	120.01	5 man-days	4800.16
4		**Cultivation of regenerated plants**		
4.1	The soil	5.00	50 plants / 5 l = 250 l	1250.00
4.2	Light	4.80	90 days (16 hours / day, power 600 W 6 lamps)	24883.20
4.3	Plant care, man-hours	120.01	110 man-hours	13201.10
4.4	Fertilizers	250.00	Top dressing 4-5 times for the period	500.00
4.5	Vegetation vessels	60.00	50 pieces on 5 l	3000.00
5	Vernalization	4.80	90 days - 2160 hours Refrigerator (power 105 W)	45360.00
6	Other expenses	1000.00	1 laying	1000.00
7	Growing plants in the greenhouse to seed	50.00	The complex of necessary expenses	13508.44
		Total		**129063.60**

4. Conclusion

As a result of calculations, it is clearly shown that, in terms of the cost of production and a significant reduction in the time to create lines, biotechnological methods are the most cost-effective. They allow to significantly reduce the duration of the selection process for 8 - 12 years. To create competitive new domestic hybrids of garden carrots, biotechnology methods need to be embedded in traditional breeding schemes.

References

[1] Marshak M S 1967 *Dietary nutrition* (Moscow, Russia: Meditsina)
[2] Leunov V I 2011 *Table root vegetables in Russia* (Moscow, Russia: Tovarishchestvo nauchnykh izdaniy KMK)
[3] Touraev A, Forster B P, and Jain S M 2009 *Advances in haploid production in higher plants* (Berlin, Germany: Springer)
[4] Dunwell J M 2010 *Plant Biotechnology Journal* **8**(4) pp 377-424
[5] Matsubara S, Dohya N, Murakami K, Nishio T, and Dore C 1995 *Acta Hortic* **392** pp 129-137
[6] Górecka K, Kowalska U, Krzyżanowska D, and Kiszezak W 2010 *J Appl Genet* **51** pp 141-147
[7] Li J-R, Zhuang FY, Ou Ch-G, Hu H, Zhao Z-W, and Mao J-H 2013 *Plant Cell Tiss. Organ Cult* **112** pp 275-287
[8] Vurtz T S, Domblides E A, Shmykova N A, Fedorova M I, Kan L Yu, and Domblides A S 2017 *Russian Vegetables* **5** pp 25-30
[9] Alexander M P 1969 *Stain technol* **44**(3) pp 117-122

IOP Conf. Series: Earth and Environmental Science **395** (2019) 012085 doi:10.1088/1755-1315/395/1/012085

Economic efficiency of the raw materials production for obtaining a natural food dye from Amaranth

V F Pivovarov[1], M S Gins[1*] and V K Gins[1]

[1] Federal Scientific Center for Vegetable Growing, 14 Selektsionnaya str., VNIISSOK Village 143080 Russia

E-mail: anirr@bk.ru

Abstract. The paper presents the economic results of the raw materials production of fthe ood red-violet dye, enriched with antioxidants from the leaves of two varieties of amaranth (Valentina, Don Pedro) breeding of the Federal Scientific Center for Vegetable Growing (Russia). The amaranth plant leaves are used both as a raw material for the production of food-grade natural dye for coloring confectionery products, beverages, and so can be used as part of salads, soups, sauces, enriching them with biologically active substances (in particular, ascorbic acid and phenolic compounds), giving them both original taste and color. The article analyzes the economic efficiency of growing amaranth plants in an open and protected ground, depending on additional mineral fertilizer to obtain a natural food dye. Calculated in prices of 2018, the economic efficiency indicators for both products and resources proved to be satisfactory. Costs per hectare amounted to 77 thousand rubles for cultivation in open ground conditions and 225 thousand rubles for cultivation in a film greenhouse. The cost of raw materials varied from 550 to 880 rubles per kilogram under these conditions. The gross profit ranged from 15 to 125 rubles / ha, profitability ranged from 14 to 81%.

1. Introduction

The emerging consumer market of healthy food products requires creating coloring ingredients that are natural in origin, which is confirmed by the following marketing research results, in particular a share of vegetable dyes in food products is growing [1], [2]. As a rule, dyes of natural origin on the market are obtained from raw food sources, such as carrot, beetroot, etc., the use of which is more appropriate for the intended purpose [3]. In this regard, in recent years, more and more attention is paid to the search for alternative, natural sources of raw materials that are promising both in terms of the content of pigments and in nutritional and biological value, ensuring consumer properties of the products. The variety of red-painted species of amaranth with a high level of accumulation of water-soluble antioxidants allows to use vegetable raw materials for food purposes, to obtain a dye with a natural red-violet pigment, namely Amarantin [4]. The group of authors has developed a multicomponent concentrate, in which a dry leaf of amaranth variety "Valentina" is used as extractive substances. The resulting concentrate provides a harmonizing taste and physiological effects on the organs and systems of the body, allows you to activate the body's metabolism, strengthens the body's defenses from the negative effects of the environment and allows you to expand the range of multicomponent concentrates with preventive action [5], [6]. Obtaining the maximum amount of amaranth dry leaf with minimum cost is necessary to produce large quantities of multicomponent concentrate.

Content from this work may be used under the terms of the Creative Commons Attribution 3.0 licence. Any further distribution of this work must maintain attribution to the author(s) and the title of the work, journal citation and DOI.
Published under licence by IOP Publishing Ltd

2. Materials and Method

Amaranth plants were grown in open ground on sod-podzolic soil at the greenhouse base of the Federal Scientific Center for Vegetable Growing in the Moscow Region. Production experiments using conventional methods in vegetable farming were carried out [7]. Amaranth leaves, collected from three weeks before flowering and a week after flowering, were dried to a residual moisture content of 10% in a drying cabinet at 80 ° C.

The economic efficiency of growing amaranth in an open and protected ground in the control and experience (with the introduction of mineral supplementation at a dose of 70 kg / ha of active substance at a wholesale price of 22 thousand rubles per ton) was determined on the basis of charts of growing amaranth according to the method of determining economic efficiency [8], [9]. Realization cost of amaranth leaf mass was calculated according to the selling prices of the Federal Scientific Center for Vegetable Growing. The table shows the average values.

3. Results and Discussion

Amaranth leaves contain protein (15–20%), including a balanced composition of essential amino acids, a complete set of vitamins B, E, C, carotenoids, flavonoids (rutin, quercetin, etc.), simple phenolic compounds, hydroxycinnamic acids. The content of ascorbic acid in amaranth leaves is in the range from 8 to 33%, also there are biologically active substances with antioxidant activity are present (carotenoids, flavonoids, methionine, selenium, red amaranth pigment *betacinin amarantine*). When cultivating amaranth varieties "Valentina" and "Don Pedro," as a raw material for the preparation of the dye, it is optimal to collect their leaves starting from the 8th order (three weeks before flowering), at the moment of maximum accumulation of biologically active substances. Earlier, it was shown that the yield of amaranth leaf biomass was highly dependent on doses of nitrogen fertilizer [10].

The yield of the marketable leaf "Valentina" in the open ground was 100 kg / ha in the control without applying additional fertilizing, and the yield was 140 kg / ha in the experiment with introduction. However, when it is grown in the greenhouse, the yield of the commodity raw material was 300 kg / ha in the control, and the yield was 350 kg / ha with the introduction of top dressing.

In comparison with the control, the cost was 1.2 times lower. Indicators of economic efficiency are presented in Table 1.

Table 1. Economic efficiency of cultivation of amaranth varieties for the production of natural food raw materials using mineral supplements in open and protected ground conditions of the Moscow region.

Growing conditions	Experience Options	**Indicators**					
		Production costs, thousand rubles / ha	Amount of leaf biomass, kg / ha	Costs, rub / kg	Product price, rub / kg	Gross profit, thousand rubles / ha	Profitability, %
Valentine							
Open ground	Control	75	100	750	1000	25	33
	Top dressing	77	140	550	1000	63	81
Greenhouse	Control	220	300	733	1000	80	36
	Top dressing	225	350	643	1000	125	56
Don Pedro							
Open ground	Control	75	90	830	1000	15	20
	Top dressing	77	120	640	1000	43	56
Greenhouse	Control	220	250	880	1000	30	14
	Top dressing	225	290	780	1000	64	28

In a protected ground, when growing the Valentine's amaranth variety at a high cost (220 thousand rubles), the cost of these products corresponds to the level of being grown in open ground, but the amount of leaf biomass is 3 times higher. The option of growing in a greenhouse with the addition of top dressings was also distinguished by a high profit from the sale. In comparison to open ground, the cost is reduced by about 1.2 times.

In the "Don Pedro" variety, the yield of finished commercial raw materials of the open ground (also taking into account the collection of leaves two weeks before flowering) averaged 90 kg / ha in control (without top dressing), and the yield was 120 kg / ha in the experiment (using top dressings, respectively).

It should be noted that in greenhouses, the yield of leaf biomass was 250 kg / ha in the control and the yield was 290 kg / ha with the addition of top dressing at the "Don Pedro" variety.

4. Conclusion

It has been shown that when grown in both open and protected soil conditions of the Federal Scientific Center for Vegetable Growing (Moscow region, Russia), with the initial wholesale price of 1 thousand rubles per 1 kilogram, both varieties were profitable when grown both in open ground and in the greenhouse. At the same time, the variant with the use of mineral top dressing when grown in the open ground is most cost-effective in the cultivation of the "Valentine" leaf amaranth variety for obtaining natural dye with a profitability of 81%. In this variant, cultivation of the "Don Pedro" variety shows profitability up to 56%. The maximum yield of leaf biomass when growing both species of amaranth is achieved in a film greenhouse using top dressing, and reaches 350 kg per hectare, in the case of "Valentine" variety.

References

[1] Derkanosova N M, Andropova I I, Gins V K, and Lupanova O A 2017 The use of natural dye from amaranth in confectionery *Confectionery and Bakery Production* **11-12**(173) pp 28-31

[2] Elliot D C 1993 Betacyanin decolourizing enzyme in *Amaranthus tricolor* L. seedlings *Phytochemistry* **22**(2) pp 383-387

[3] Gins V K, Gins M S, Derkanosova N M, and Lupanova O A 2016 The potential of amaranth as a raw material source of confectionery dyes *New and Unconventional Plants and Prospects for Their Use* **12** pp 472-474

[4] Shelamova S A, Lupanova O A, Derkanosova N M, and Zaitseva I I 2015 *Microbiological studies of confectionery with a natural dye* In *Problems of Identification, Quality and Competitiveness of Consumer Goods* (pp 280-284) (Kursk, Russia: Southwest State University)

[5] Filonova G L, Oganesyants L A, Soboleva O A, Golovina T A, Kononkov P F, Gins V K, Pivovarov V F, Bezgin V M, Postnikov S V, Fedorova O V, ... Gins M S 2013 *Patent 2482753 Polycomponent* (Moscow, Russia: Rospatent).

[6] Filonova G L, Soboleva O A, Golovina T A, Kononkov P F, Gins V K, and Gins M S 2015 Amaranth of the variety "Valentina" is a promising source of raw materials for health drinks *Beer and Drinks* **4** pp 10-14

[7] Litvinov S S 2011 *Methodology of field experience in vegetable growing* (Moscow, Russia: Rosselkhozakademiya)

[8] Gins M S, Gins V K, Kononkov P F, Pivovarov V F, Romanova E V, Osokin I E, ... Gins E M 2017 *Technology of growing vegetable (leaf) amaranth* (Moscow, Russia: RUDN)

[9] Razin A F, Bukharov A F, Ivanova M I, Bukharova A R, and Razin O A 2017 *The basic principles for calculating the economic efficiency of new varieties of vegetable crops* In *Current and New Trends in Plant Breeding and Seed Production of Crops* (pp 167-169) (Moscow, Russia)

[10] Gins M S, Gins V K, Kononkov P F, Pivovarov V F, and Gins E M 2016 Elements of the technology of growing amaranth leaf biomass with a high content of antioxidants to produce functional prophylactic products *Russian Vegetables* **4**(33) pp 90-95

Efficiency of the newest sterile substrate production technology for oyster cultivation

A Soldatenko[1], N Devochkina[1], and M Ivanova[1*]

[1] Federal Scientific Center for Vegetable Growing, 14 Selektsionnaya str., VNIISSOK Village 143080 Russia

Email: ivanova_170@mail.ru

Abstract. The article conducted production tests and developed parameters of the innovative technology for substrate preparation under sterile conditions for the cultivation of oyster mushrooms. Oyster mushroom is the second largest industrial mushroom culture, which belongs to the group of wood-destroying mushrooms. World production of oyster mushrooms is about 1.5 million tons per year. Its cultivation is practiced in Italy, France, Belgium, the Netherlands, Japan, and Taiwan. Fruit bodies of oyster mushrooms have high nutritional and medicinal value. Biological features of a culture allow its production under conditions of both specialized cultivation facilities and in adapted premises, thereby solving the problem of their effective use. The industrial mushroom production in the Russian Federation is growing due to the construction of industrial complexes for mushroom cultivation. According to experts, the expansion of the cultivation of oyster mushrooms with the use of contemporary technologies and technological equipment is a very promising business for our country. The total annual production of oyster mushrooms does not exceed 4-5 thousand tons, which does not even provide 5-6% of the required domestic demand. In Russia, mushroom producers, farmers, and subsidiary farms are constantly looking for a solution to the problem of acquiring high-quality raw materials, organizing the full technological cycle of production, choosing the technology for preparing the substrate and growing the fruit bodies of the mushroom that would ensure the economic efficiency of the production of fruit bodies of mushrooms. When growing oyster mushrooms in small enterprises, the following simplest methods of processing raw materials are used: soaking in hot water without temperature control and without optimal processing time, respectively. The quality of the substrate prepared in this way is very low, and the level of the resulting crop does not provide cost-effective results, which leads to the cessation of the mushroom production. Therefore, the latest technological and technical solutions to virtually solve the issues of introducing new equipment for the production of high-quality substrate are relevant and important. This will allow to obtain stable yields and effectively lead the production of significant low-calorie fresh protein products, namely oyster mushroom bodies.

1. Introduction

The Southeast Asian countries, such as China, Taiwan, South Korea, Indonesia, the Philippines, and India, unlike the European countries, the United States and Canada, prefer to grow xylotrophic mushrooms (destroying wood or growing on cellulose-containing materials, such as shiitake, winter mushroom, etc.). Until the mid-70s of the twentieth century, the main method of growing wood-destroying mushrooms, including oyster mushrooms, was an extensive method of cultivation, which consisted in the fact that the cutting of wood was infected with the mushroom mycelium. Plantations

Content from this work may be used under the terms of the Creative Commons Attribution 3.0 licence. Any further distribution of this work must maintain attribution to the author(s) and the title of the work, journal citation and DOI.
Published under licence by IOP Publishing Ltd

IOP Conf. Series: Earth and Environmental Science **395** (2019) 012086 doi:10.1088/1755-1315/395/1/012086

were located in the natural growing conditions, and the harvest of fruit bodies was completely dependent on climatic factors. Such results were not comfortable for the producers, since the yield and efficiency of such production were extremely low. The first experience in the production of champignon and oyster mushrooms was the experience of Hungary, in the "Duna" cooperative. In the 1950s, a specialized laboratory for the cultivation of edible mushrooms was organized. From that moment, the active development of oyster mushroom cultivation technologies began. In the mid-60s, the sterile oyster mushroom cultivation technology on various plant lignin cellulosic substrates was developed. The substrate was treated with steam in autoclaves at elevated pressure and temperature over 120 °C. The sterile method was the first industrial method of growing oyster mushroom, and it was patented in 1966 [2]. The sterile technology was considered highly expensive and not widely used in production for a long time. However, in the 90s of the twentieth century, in connection with the start of the production of autoclaves for the sterilization of bulk materials in Southeast Asia, in China and Korea, the development and testing of methods for sterilizing the substrate for the production of wood-destroying mushrooms began. In 2016, the latest technology was introduced in the Russian Federation, in the commercial enterprise "Aprel" LLC (Tula), on the basis of which the full technological process of producing the substrate under sterile conditions was worked out and the cultivation of oyster mushroom bodies took place [3], [4].

2. Research Methodology

When conducting our research, along with technological research, the methods of economic analysis, mathematical modeling of the production process and business planning of production were used to analyze the cost of production, to estimate the payback of costs and economic efficiency of production [5], [6]. The production process involves using a full technological cycle, divided into the following 5 independent technological zones-departments: substrate preparation; its heat treatment; planting mycelium in the substrate and germinating the mycelium in the substrate in tanks in a dedicated process zone; growing and harvesting mushroom bodies in three cultivation chambers with a loading capacity of up to 20 tons on multi-tiered racks. The latest sterile technology for preparing a substrate based on the husks of sunflower seeds is applied with the processing of substrate containers in an autoclave (steam sterilizer) at a temperature of 121°C for 3 hours. The mass of the autoclave loading is 5.3-5.5 tons. Growing oyster mushroom fruit bodies is carried out in plastic containers with a substrate mass of up to 1.4-1.6 kg on multi-tiered racks. A series of experiments was carried out to assess the productivity of the substrate with various doses of dietary supplements and the cost of production was calculated.

3. Results and Discussion

Technological equipment for the cultivation of oyster mushrooms is an industrial line in the complete delivery. It incorporates the following: a mixer in which the mechanized mixing of the initial substrate materials, their moistening, transportation to the hopper with dispensers for packaging the substrate mass in the tank is carried out. The substrate in containers (up to 4,000 pieces) from the stuffing zone is moved to an autoclave, where it is subjected to severe sterilization on multi-tiered racks in boxes at 121 °C for 3 hours under the overpressure of 1.2 atm. / 5.5 tons of substrate (they are processed simultaneously. After autoclaving, the tanks are cooled and moved to the pre-box of the sterile zone. Sowing oyster mushroom mycelium is carried out in a clean sterile area, namely the box. Inoculated (seeded) containers are then transferred to the incubation zone (germination). The so-called multi-zone cultivation technology is applied. It involves separating the process into several zones with different microclimate parameters in accordance with the requirements of the crop and at least 8 crop turns per year in the production areas where the harvest is taking place.

Analyzing the results of the production survey revealed the need to optimize the technological parameters, change the mode of ventilation of the premises (to reduce evaporation from the surface of the substrate unit and the fruit bodies). This made it possible to increase the yield by optimizing the microclimate parameters and the ventilation mode and amounted to 3.1% (from 17.88% to 21.1% of the substrate mass under other equal conditions, namely the density, humidity, and substrate composition). In terms of value, this increase was as follows: at a yield of 178.8 kg per 1 ton of substrate (selling price: 120 rubles / kg – 21456 rubles), 211 kg – 25320 rubles. The increase amounted to 32.2 kg or 3864 rubles per 1 ton of substrate.

Also, the substrate mass in the box was reduced, the seeding rate was increased to the maximum possible value (2.4%), the substrate density was reduced to 0.5 kg / dm^3, which made it possible to increase the productivity of the substrate at the lowest cost for its preparation. The result of technological changes is the acceleration of the process of vegetative growth of mycelium in the substrate in containers for 3-4 days (20-21) instead of 24 days, output in 98% of blocks with a yield of 25 to 27%, in 2% of blocks with a yield of 32% substrate mass, a block weight was reduced from 1.7 kg to 1.2 kg (Table 1).

Table 1. The economic efficiency of the oyster mushrooms production, depending on a yield.

Crop of mushrooms with 1 ton of substrate, kg	Gross yield from 1 ha per year, t	Production costs, thousand rubles	Proceeds from sales of products, thousand rubles	Net income, thousand rubles.	Costs for 1 kg of mushrooms, rub	Production profitability, %
180	780	90840	70200	93600	116.46	3.03
200	884	95648	79560	106080	108.199	10.90
220	988	100002	88920	118560	101.216	15.65
240	1092	103200	98280	131040	94.50	26.27
260	1196	105000	107640	143520	87.79	36.68
280	1300	108900	117000	156000	83.76	43.25
300	1404	112200	126360	168480	79.91	50.16

4. Conclusion

The proposed technology of oyster mushroom cultivation using the sterile substrate preparation method, an optimized mass of substrate capacity (1-1.2 kg) and a seeding rate (2.4% mycelium of the substrate weight in a container) allows to obtain a yield of at least 25% of the substrate mass. Improved elements of the technological process in the complex can significantly increase the yield of products by 7-9% and achieve a level of profitability of at least 30%.

References

[1] Devochkina N L, Nurmetov R D, and Dolgikh L I 2012 Industrial mushroom growing is an effective resource for the development of the Russian agro-industrial complex *Potatoes and Vegetables* **1** pp 21-22

[2] Dudka I A 1978 *Industrial cultivation of edible mushrooms* (Kiev, USSR: Naukova Dumka)

[3] Soldatenko A V, Devochkina N L, Razin A F, Razin O A, and Nurmetov R D 2018 Industrial mushroom production as an innovative direction of economic activity in the field of the agro-industrial complex of the Russian Federation *Russian Vegetables* **3** pp 9-92

[4] Devochkina N L, Mukienko S V, Nurmetov R D, and Pryanishnikova L N 2019 Innovative technology for preparing the substrate in sterile conditions for the cultivation of oyster mushroom *Potatoes and Vegetables* **2** pp 17-19

[5] Devochkina N L, Dugunieva L G, Razin A F, Ivanova M I, and Nurmetov R D 2018 Investment attractiveness of the industrial mushroom production *Economics of Agriculture of Russia* **11** pp 52-59

[6] Behrens W 1995 *Manual for the preparation of industrial feasibility studies* (Moscow, Russia: Interexpert CJSC)

IOP Conf. Series: Earth and Environmental Science **395** (2019) 012087 doi:10.1088/1755-1315/395/1/012087

Economic and agri-environmental efficiency of applying fertilizers and growth regulators in broccoli cultivation

V A Borisov[1], I A Lysenko[1] and O A Razin[1]

[1] Federal Research Center of Agrarian Economy and Social Development of Rural Areas – All-Russian Research Institute of Agricultural Economics, 35, building 2 Khoroshevskoe shosse, Moscow 123007 Russia

E-mail: vniioh@yandex.ru

Abstract. The highest yield (18.6 t / ha) and the best economic effect (320 thousand rubles / ha) with good product quality was obtained on alluvial soils of the Moscow Region with the complex use of the Maraton F_1, hybrid, mineral fertilizers $N_{120}P_{60}K_{150}$ for broccoli in combination with top dressing humic fertilizer "gumistar" (3 l / ha) and microfertilizer "tenso-cocktail" (0.5 kg / ha). The developed plant nutrition system provides an increase in the content of dry matter, sugars, and vitamins in products with a slight increase in nitrates.

1. Introduction

Broccoli cabbage is one of the most valuable vegetable crops due to its excellent nutritional and healing properties, high-quality vegetable protein, ascorbic acid, a complex of other vitamins and mineral salts [1, 2].

However, in our country, the technology of cultivation of broccoli is not developed enough, the yield level of this crop rarely reaches 10-12 t / ha, and broccoli tops often crumble, which reduces the commodity properties of products [3, 4]. According to H. Michalik [5], the level of nitrates in broccoli heads ranges from 540 to 3430 mg / kg, which exceeds permissible levels. According to other data [6, 7], these fluctuations do not exceed 200-500 mg / kg NO_3. There are actually no experimental data on this issue in Russia.

2. Material and Methodology

Research on the use of fertilizers and growth regulators for broccoli cabbage hybrid Maraton F_1, which is zoned in Russia since 2006, conducted in 2012-2014. Studies were carried out on alluvial meadow soils of the floodplain of the Moscow River, containing 3.4–3.8% of humus, with a humus horizon thickness of 40–60 cm, and a soil reaction close to neutral (pH 5.5-6.1). The soil is sufficiently supplied with movable P_2O_5 according to Chirikov (176-191 mg / kg) and has an average supply of exchangeable K_2O according to Maslova (128-148 mg / kg), which is generally characteristic of the soil conditions of vegetable farms of the Moscow region.

The scheme of experiments included options with the following fertilizers: nitrogen-phosphorus, nitrogen-potassium, and phosphorus-potassium fertilizers in the form of urea (46% N), double superphosphate (43% P_2O_5) and potassium chloride (56% K_2O), a microelement fertilizer "tenso-cocktail" of the "Hydro-Agro Rotterdam" company (B-0.52%, Ca-2.57%, Fe-3.84%, Mg-2.57%, Mo-0.13%, Zn-0.53%, Cu-0.53), a biocompost "Bioud-compost" from the "Eco-Agti" LLC; with the

Content from this work may be used under the terms of the Creative Commons Attribution 3.0 licence. Any further distribution of this work must maintain attribution to the author(s) and the title of the work, journal citation and DOI.
Published under licence by IOP Publishing Ltd

content of organic matter> 30% used in a dose of 5 t / ha, a biohumus called "Gumistar" of the "Green-Peak" company is obtained using Californian worms and used in a dose of 3 l / ha; a zeolitic fertilizer from the Khotynetsky deposit of the Oryol region is used at a dose of 100 kg / ha, as well as a natural plant growth regulator "Zircon" is used at a dose of 250 ml / ha.

Agrotechnical cultivation of broccoli was generally accepted for the Central region of the non-chernozem zone. Broccoli seedlings were grown in film greenhouses for 30 days, the seedlings were transplanted in early June according to a 70 x 40 cm pattern. Mineral, organic fertilizers, zeolite were applied before planting seedlings, and the growth regulators Gumistar and Zircon, as well as the tenso-cocktail, were used 7-10 days after planting by manual spraying of plants. The area of the experimental plot was 16.8 m^2, the repetition was 3-4 times. Broccoli harvesting was done selectively as the heads mature.

3. Research Results

The results of the calculation of the yield of broccoli (Table 1) revealed a rather high efficiency of mineral fertilizers in a dose of $N_{120}P_{60}K_{150}$. For 3 years, the yield increase was 56%, and nitrogen fertilizers increased plant productivity by 21%, phosphate fertilizers – by 25%, and potash fertilizers – by 29%. The biocompost increased the yield of broccoli by 51%, which was slightly lower than when using the estimated dose of mineral fertilizers.

The use of zeolite on NPK background allowed to increase the yield of broccoli by 9%, tenso-cocktail – by 34%, humistar – by 38%; a higher efficiency was observed with using the growth regulator Zircon (by 45%). The combined use of the estimated dose of mineral fertilizers $N_{120}P_{60}K_{150}$ for spring cultivation in combination with using the growth regulator humistar and complex microfertilizer tenso-cocktail ensured the best result in the experience and the greatest yield of broccoli. For 3 years, the average yield on this variant was 18.6 t / ha, or an increase of the broccoli yield in 216%.

Table 1. Productivity and quality of broccoli depending on the use of fertilizers and growth regulators.

Options	Productivity of broccoli heads		Quality broccoli heads			
	t / ha	%	Dry matter, %	Amount of sugars, %	Vitamin C mg %	Nitrates (NO3), mg / kg
No fertilizer	8.6	100	11.2	1.8	74.8	138
$N_{120}P_{60}$	10.9	127	12.5	2.3	75.5	261
$N_{120}K_{150}$	11.3	131	11.8	2.4	87.0	270
$P_{60}K_{150}$	11.6	135	12.8	2.6	81.8	203
$N_{120}P_{60}K_{150}$	13.4	156	13.0	2.4	81.6	172
Biocompost (5t / ha)	13.0	151	12.4	2.2	90.7	166
NPK+ biocompost	14.8	172	12.5	2.2	83.8	200
NPK+ gumistar	16.7	194	13.1	2.4	82.8	191
NPK+ zircon	17.3	201	13.0	2.4	87.4	193
NPK+ zeolite	14.2	165	13.0	2.4	78.8	242
NPK+ tenso-cocktail	16.3	190	13.0	2.6	79.8	199
NPK+ gumistar + tenso-cocktail	18.6	216	13.6	2.4	89.5	204
$HCP_{0.5}$ t/ha 1.26-2.28 S_{x1}%=2.8-3.3						

Broccoli is a diet product; therefore, product quality is very important. It should be noted that fertilizers and plant growth regulators generally improved the biochemical quality of broccoli heads. The dry matter content in broccoli from 11.2 to 13.6%, the amount of sugars from 1.8% to 2.6%, and the vitamin C from 74.8 to 90.7 mg% were noted. Top yield options for experience provided a significant improvement in product quality.

The nitrate content in broccoli heads was slightly increased under the influence of fertilizers and plant growth regulators, but only slightly. The highest NO_3 concentration is noted in variant $N_{120}K_{150}$

(270 mg / kg), which is far from the MPC for cabbage vegetables (900 mg / kg for early production and 500 mg / kg NO_3 for late production). Therefore, ecologically, the developed system of application of fertilizers and growth regulators is absolutely safe and can be used in the production of broccoli.

Table 2. Economic efficiency of using the fertilizer, zeolite, and growth regulators for broccoli (2013-2015).

Options	Productivity of heads, t / ha	The cost of production, thousand rubles / ha	Net income, thousand rubles / ha	Profitability level, %	The cost of 1 tons of products thousand roubles
No fertilizer	8.6	256.9	18.3	7	29.9
$N_{120}P_{60}$	10.9	271.7	77.1	28	24.9
$N_{120}K_{150}$	11.3	269.1	92.5	34	23.8
$P_{60}K_{150}$	11.6	266.7	104.5	39	23.0
$N_{120}P_{60}K_{150}$	13.4	272.6	156.2	57	20.3
Biocompost (5t / ha)	13.0	285.9	130.1	45	22.0
NPK+biocompost	14.8	301.0	172.6	57	20.3
NPK+ gumistar	16.7	273.9	260.5	95	16.4
NPK+ zircon	17.3	277.3	276.3	100	16.0
NPK+ zeolite	14.2	280.0	174.3	62	19.7
NPK+ tenso-cocktail	16.3	274.1	247.5	90	16.8
NPK+ gumistar +tenso-cocktail	18.6	275.2	320.0	116	14.8

The economic efficiency of using fertilizers and plant growth regulators under broccoli cabbage (Table 2) showed that the cost of producing broccoli products slightly increased (from 257 to 301 thousand rubles / ha). However, they are more than compensated by an increase in net income, an increase in the level of profitability of production (from 7 to 116%), a decrease in the cost of 1 ton of broccoli cabbage heads (from 29.9 thousand rubles to 14.8 thousand rubles), which is 2 times actually. This demonstrates a high economic efficiency of the developed fertilizer system.

4. Conclusions

When growing broccoli cabbage on the alluvial meadow soils of the Central Non-chernozem Zone of the Russian Federation, mineral fertilizers in a dose of $N_{120}P_{60}K_{150}$ provided an increase in yield of 56%, while potash fertilizers, then phosphorus and nitrogen fertilizers were in the first minimum. The efficiency of the biocompost at a dose of 5 t / ha was close to the NPK level.

Plant nutrition with growth regulators (gumistar, zircon) and microelements (tenso-cocktail) during the growing season made it possible to increase the yield of broccoli by 34-45%.

A complex application of NPK with a gumistar and tenso-cocktail allowed to obtain the highest yield increase (116%, non-fertilized control). Fertilizers and growth regulators increased the content of dry matter, sugars, and vitamin C in broccoli, and the nitrate content was 2-3 times lower than the MPC.

Calculating economic efficiency of using fertilizers and growth regulators in the cultivation of broccoli revealed a twofold decrease in the cost of production and an increase in net income from the developed system of fertilizers from 18.3 to 320 thousand rubles / ha and an increase in the profitability of production from 7 to 116%.

References

[1] Pivovarov V F 2006 *Russian vegetables* (Moscow, Russia: VNIISSOK)
[2] Rabinovich A M, and Borisov V A 2008 *The healing properties of vegetable and spicy aromatic plants of Russia* (Moscow, Russia: "Arnebiya" LLC)
[3] Borisov V A, Litvinov S S, and Romanova A V 2003 *The quality and storage of vegetables* (Moscow, Russia: All-Russian Research Institute of Vegetable Growing)
[4] Soldatenko A V, Pivovarov V F, Razin A F, Meshcheryakova R A, Shatilov M V, Ivanova M I,

Taktarova S V, ... Razin O A 2018 The economy of vegetable growing: the state and the present *Russian Vegetables* **5**(43) pp 63-68

[5] Michalik H 1989 Azotany w warzywach *Biul. Warz.* **1** pp 115-119

[6] Terbe I A, Zsoldos L, and Patocs I 1987 Zöldségnövények nitráttartalma. In *Kerteszeti es elelmiszeripari egyetem kiadvanyai* (vol. 1) (pp. 125-13) (Budapest, Hungary)

[7] Santamaria P 2006 Nitrate in vegetables: toxicity content, intake and EC regulation *J. Food Agric* **86** pp 10-17

IOP Conf. Series: Earth and Environmental Science **395** (2019) 012088 doi:10.1088/1755-1315/395/1/012088

A perspective method of garlic enrichment (*Allium Sativum* L.) with selenium

A Polyakov[1,2*], T Alekseeva[1,2] and A Soldatenko[1]

[1] All-Russian Scientific Research Institute of Vegetable Growing – Branch of the Federal Scientific Center for Vegetable Growing, Vereya, Ramenskoye district, Moscow region 140153 Russia
[2] Moscow Region State University, 10a Radio str., Moscow 105005 Russia

E-mail: vniioh@yandex.ru

Abstract. Obtaining high-quality products is one of the most important tasks of agricultural enterprises of all forms of ownership. Quality indicators are not only organoleptic indicators, but first of all, the chemical composition of the products manufactured. Quality food should contain the necessary macro-and micronutrients in a dose that meets the needs of the human body in them. In this regard, the development of methods for obtaining products containing microelements, in the quantity necessary to ensure the normal functioning of a person is a particularly important area at present. We have found that preplant treatment of chives and vegetative plants with a 0.1% solution of potassium selenite in combination with SAS and DMSO makes it possible to produce bulbs in which the content of selenium exceeds the control by 29.4 and 145.3 times. A similar treatment of cloves and plants with sodium selenite leads to an increase in the selenium content in the bulbs to 8.0 mg / kg and 7.3 mg / kg, or 169.8 and 157.2 times in comparison with the control. The proposed method of enrichment of garlic with selenium allows to improve the qualitative composition of this culture and increase its economic significance.

1. Introduction

Allium sativum L. is a widely cultivated crop. Worldwide, garlic production is more than 28 million tons per year, more than 20 million tons of them are produced in China.

Numerous studies currently being conducted are aimed at finding and studying effective methods for obtaining high yields of garlic. The search for such techniques is associated with the high labor intensity of growing this crop. The vegetative method of breeding garlic does not allow to quickly fill the market with planting material and leads to the fact that all planting material must be completely updated every 3-5 years.

Unfortunately, there are very few studies related to the development of methods for improving product quality. An important indicator of quality is the mineral and organic composition of products. However, both the absence of components harmful to human health and the availability of useful elements should be considered. Therefore, a lot of attention should be paid to the methods of saturation of garlic and other crop production with individual essential elements and, in particular, with *selenium*.

Content from this work may be used under the terms of the Creative Commons Attribution 3.0 licence. Any further distribution of this work must maintain attribution to the author(s) and the title of the work, journal citation and DOI.
Published under licence by IOP Publishing Ltd

Selenium has a similar structure of the atom with sulfur, so in some organic compounds it is able to replace it. The biological activity of such selenium-containing organic compounds is higher than that of the initial ones containing sulfur, which is of great importance for human health [5].

Garlic, enriched with selenium, exhibits a high degree of biological activity, namely, it protects against the toxic effects of heavy metals – As, Cd, Hg, Sn, etc., and it also has antioxidant, anti-inflammatory, and antibacterial effects [1]. The recommended daily intake of selenium depends on the gender and age of a person and ranges from 30 µg to 70 µg [2]. The aim of our study was to find effective ways to enrich garlic with Selenium.

2. Materials and Methods

Studies were conducted in the Department of Biotechnology and Innovative Projects of the All-Russian Research Institute for Vegetable Growing – Branch of the Federal Scientific Center for Vegetable Growing (Moscow Region, Ramensky District, Vereya Village) in open ground conditions.

The material for the study was the cloves and plants of garlic (winter Gladiator variety) [3].

The study design included preplant treatment of the cloves and double treatment of vegetative plants with sodium selenite and potassium selenite solutions. The concentration of our solutions was 0.1%. Selenites dissolved in DMSO or SAS.

The cloves were soaked in a solution of selenites for 30 minutes; with subsequent treatments, the plants were sprayed with solutions as follows: they were sprayed in the phase of the onset of intensive leaf growth for the first time and the second time, and then they were sprayed after 3 weeks. The experiment was laid in the fourfold repetition; the accounting area of one plot was 1 m^2; the distance between the rows was 25 cm; between the plants in a row – 10 cm; the length of one row is 1 m. The flow rate of the working fluid was 100 ml / m^2.

3. Results and Discussion

Our studies conducted earlier [4] in greenhouses clearly shows that Selenium from solutions for non-root and root treatment is used unequally effectively (Table 1).

Table 1. The selenium content in garlic samples, mg / kg.

Experience option	Non-root top dressing	Root top dressing
Control	0.040±0.006	0.040±0.005
Sodium selenite	**0.470±0.057**	**0.200±0.024**
Potassium selenite	**0.580±0.069**	**0.100±0.013**
Zinc selenite	0.070±0.010	0.030±0.004

It was established that the root and non-root top-dressing of plants with sodium, potassium and zinc selenites was very effective and allowed to significantly increase the selenium content in the bulbs (2.5 - 14.5 times). At the same time, the content of selenium in the control samples was 0.040 mg / kg, and for non-root treatment with potassium selenite was 0.580 mg / kg. Studies have shown that non-root top dressing with sodium and potassium selenite was 2.3–5.8 times more effective than root. So, when using sodium and potassium selenite with a non-root top dressing, the content of selenium in the bulbs was 0.470 mg / kg and 0.580 mg / kg, and with the root was 0.200 mg / kg and 0.100 mg / kg, respectively.

Based on these results, further work on the enrichment of garlic with selenium was carried out using non-root top dressing. To facilitate the penetration of solutions into plant tissue, surfactants, namely DMSO and SAS were added to the studied solutions.

Studies have shown that the use of 0.1% solutions of sodium selenite and potassium selenite, dissolved in DMSO or SAS, did not have a significant effect on plant height, average weight, yield of bulbs and inflorescences (Table 2).

Table 2. Effect of Se-containing preparations on the growth and yield of winter garlic bulbs.

Experience option	Plant height	Average bulbs weight		Productivity bulbs		The average mass of inflorescences		Yields aerial bulbils	
	cm	kg	%	kg / m^2	%	g	%	kg /m^2	%
Control	103±7	31.6	100.0	1.230	100.0	8.8	100.0	0.359	100.0
Na_2SeO_3+DMSO	104±10	31.3	99.1	1.221	99.3	8.0	90.9	0.312	86.9
Na_2SeO_3+ SAS	110±12	32.3	102.2	1.260	102.4	8.5	96.6	0.331	92.2
K_2SeO_3+ DMSO	97±10	29.9	94.6	1.166	94.8	8.8	100.0	0.343	95.5
K_2SeO_3 + SAS	97±8	32.0	101.3	1.248	101.5	8.5	96.6	0.331	92.2

The chemical analysis of the grown bulbs showed that the use of solutions of sodium selenite and potassium selenite contributed to an increase in the selenium content in comparison with control. It was noted that the use of sodium selenite was more effective in comparison with potassium selenite. Thus, the use of potassium selenite in combination with SAS contributed to an increase in the selenium content in the cloves by 29.4 times, and in combination with DMSO by 145.3 times in comparison with the control. The selenium content in these variants was 1.3 mg / kg and 6.8 mg / kg, respectively. At the same time, the use of sodium selenite in combination with DMSO and SAS solutions increased the selenium level by 157.2 and 169.8 times, and the selenium content was 7.4 and 8.0 mg / kg, respectively (Table 3).

Table 3. Selenium content in winter garlic cloves when using selenites, μg / kg.

Processing option	Selenium content	
	mg / kg	excess ratio
Control	0.047.0±0.007	-
Sodium selenite + DMSO	7.390±0.740	157.2
Sodium selenite + SAS	7.980±0.800	169.8
Potassium selenite + DMSO	6.830±0.680	145.3
Potassium selenite + SAS	1.380±0.140	29.4

It should be noted that earlier, the researchers could not get such a high content of selenium in the bulbs. So, N. A. Golubkina and co-workers [1] found that with the joint application of sodium selenate and liquid fertilizers Fertika lux (N – 1.6; P_2O_5 – 2.0; K_2O – 2.7; Fe – 0.01; B – 0.002; Cu – 0.001; Mn –0,01; Mo – 0,002; Zn – 0,001%), the total dose of sodium selenate was 75 mg / m^2. The introduction was carried out in parts for 2.5 months. With each addition, 25 ml of Fertika fertilizer was added to 7.5 mg of sodium selenate and the solution was adjusted to 10 liters per 1 m^2. As a result of the experiment, the content of selenium (powder obtained from bulbs) increased 16.7 times in the final product in comparison with the control and amounted to 1.5 + 0.1 mg / kg of dry weight.

The proposed method showed high efficiency. It allows one to increase the content of selenium in the chives of garlic to 8.0 mg / kg, which will allow to fill the daily need of the body in this element with an average daily intake of garlic 10 g.

4. Conclusion

The developed effective method of enrichment of garlic with Selenium allows to improve the qualitative composition of this crop and increase its economic significance.

The proposed method, based on preplant treatment of the teeth and the subsequent double treatment of plants with 0.1% sodium selenite solution with SAS, allows to obtain cloves containing up to 8.0 mg / kg of selenium.

References

[1] Golubkina N A, Seredin T M, Koshevarov A A, Shilo L M, Baranova E V, and Pavlov L V 2018 Garlic powder enriched with selenium *Trace Elements in Medicine* **19**(1) pp 43-50

[2] Gorelikova G A 2008 *Theoretical and practical aspects of the development of food products enriched with selenium* (Kemerovo, Russia: Kemerovo Institute of Food Technology)

[3] Polyakov A V 2013 Gladiator is a new sort of winter garlic *Potatoes and Vegetables* **9** pp 19-21

[4] Polyakov A V, Loginov S V, Alekseeva T V, and Petrenko D B 2018 *Garlic is a source of organic selenium* In the collection of the International Conference of Young Scientists, Graduate Students, and Students on *Science for the Benefit of Mankind - 2018* (pp 45-49) (Moscow, Russia: MGOU)

[5] Ip C, and Ganther H E 1992 Comparison of selenium and sulfur analogs in cancer prevention *Carcinogenesis* **13** pp 1167-1171

IOP Conf. Series: Earth and Environmental Science **395** (2019) 012089 doi:10.1088/1755-1315/395/1/012089

An innovative technology for natural raw materials processing and phytocomplex production of a functional purpose

Sai-Suu S. Sadi[1*], M I Kalimullin[1] and V M Poznnyakovskiy[1]

[1] Kemerovo State Agricultural Institute, 5 Markovtseva str., Kemerovo 650056 Russia

Email: saysuu.sadi@mail.ru

Abstract. The main consumer vector of the food sector is positioned with functional products. Issues of providing the body with a complete, balanced diet are the determining factor in the formation of human health. Addressing women's health issues is of particular relevance in the social policy of the state, since the health and development of the future generation as a whole depends on women's health. The science-based prescription formula and innovative technology of the biologically active complex in the form of dietary supplements on a plant-based basis have been developed. Clinical trials on a representative group of women with premenstrual syndrome, conducted as evidence of its effectiveness. The developed phytocomplex contains synergistically active ingredients for nutritive and metabolic support of the female body during the period of hormonal imbalance. Phytoestrogens contained in herbal ingredients stimulate tissue receptors during menopause, triggering a response aimed at normalizing impaired functions. An innovative method of matrix tableting in the manufacture of dietary supplements provides a prolonged effect of biologically active substances with adjustable speed. The article shows that the developed form of dietary supplements is an effective treatment for premenstrual syndrome. In combination with the psychotropic drug *Grandoxin*, the dietary supplement leads to a significant improvement in all indicators of psychological testing, namely a decrease in the degree of depression, reactive anxiety, and a significant decrease in the indices of vegetative disturbances. Joint therapy has a more pronounced therapeutic effect. It completely suppresses or weakens the clinical manifestations of premenstrual syndrome. The developed formula of dietary supplement is well tolerated by patients and does not cause any side effects.

1. Introduction

The need of the contemporary market for healthy foods determines the need to develop innovative technologies for the processing of agricultural raw materials. The development of healthy food products is one of the key areas of contemporary nutrition. Threpsology is the science of nutrition for a healthy and sick person [1]. The basis of the production of such products is the use of centuries-old experience of scientific and traditional medicine [2]. Priority attention is paid to the development of specialized products, including dietary supplements, which are among the most effective and affordable methods of correcting nutrition and health, occupy an important place [3].

The development of new forms of specialized products with a direct influence on the maintenance of women's health is of particular relevance. A person's health and longevity are determined by balanced, wholesome nutrition of a woman during periods before pregnancy, pregnancy and lactation [4].

Content from this work may be used under the terms of the Creative Commons Attribution 3.0 licence. Any further distribution of this work must maintain attribution to the author(s) and the title of the work, journal citation and DOI.
Published under licence by IOP Publishing Ltd

IOP Conf. Series: Earth and Environmental Science **395** (2019) 012089 doi:10.1088/1755-1315/395/1/012089

Today, food ingredients that have a direct effect on the metabolic processes of the female body are sufficiently studied [5], [6]. Their role in the correction of metabolic disorders and discomfort in different periods of life, especially in adolescence and menopausal periods, is studied.

2. Materials and Methods

Raw materials, semi-finished products, and prepared dietary supplements were used as materials. Generally available and special methods of researching the quality, safety of raw materials and finished products, their effectiveness and functional orientation were used.

Clinical studies were conducted on a representative group consisting of 108 patients with a premenstrual syndrome. The main group consisted of 54 patients aged 30-42 years who received supplements of 2 tablets per day with meals, in combination with the vegetotropic drug *Grandoxin* 50 mg 3 times a day. The control group took only *Grandoxin* in the specified amount. Patients were examined for all psychological test scores (Beck test depression, Spielberger test reactivity were studied). All patients were examined gynecologically, organic lesions of the female genitalia were excluded, an ultrasound examination of the abdominal organs, an ECG were performed. Field tests were performed at the General Surgery Clinic of the Siberian State Medical University (Tomsk, Russia) under the guidance of Doctor of Medical Sciences, Professor O. S. Popov. Our research was approved by the Committee on Human Ethics.

3. Research Results

The qualitative and quantitative composition of the prescription formula of a specialized product, based on the characteristics of raw ingredients and their active principles, is scientifically substantiated. Innovative dietary supplements tableting technology was developed. Clinical trials of dietary supplements were carried out to find evidence speaking in favor of effectiveness and functional orientation of the developed product.

3.1 Recipe

Formulation of innovative phyto formulas in the form of dietary supplements has been developed. Its ingredients exhibit synergistic properties in relation to the correction of metabolic disorders during periods of the menstrual cycle, preparation for pregnancy, various infectious and somatic diseases, and menopause. The composition of 1 tablet weighing 500 mg includes (mg): calcium carbonate - 100; Vitamin Premix - 49.2, which also includes vitamins (vitamin B3 - 5, vitamin B2 - 0.4, vitamin E - 2.75, vitamin A - 0.37, vitamin B5 - 2.1, vitamin D3 - 0.0025, vitamin B6–0.5, vitamin B1 - 0.5, vitamin H - 0.05, vitamin B12 - 0.00009); hops (cones) - 27.5; sodium ascorbate - 28.1 (vitamin C - 25); zinc oxide - 3.75 (zinc - 3); magnesium oxide - 25; manganese asparginate - 3.5 (manganese - 0.48); potassium iodate - 0.0625 (iodine - 0.037); sodium selenite - 0.0375 (selenium - 0.0171); chromium picolinate - 0.1 (chromium - 0.012); Alfalfa (herb) - 25; repyashok agrimony (herb) - 25; choline bitartrate - 15; inositol - 15; Angelica Sinensis - 15; potassium chloride - 10; iron sulfate - 10 (iron - 3.7); magnolia-vine (fruits) - 7.5; blessed thistle - 7.5; Damiana - 7.5; silicon dioxide - 5; lecithin - 5; lemon bioflavonoids of lemon - 5; bromelain - 5; L-cysteine - 4; L-methionine - 4; sublimated garlic - 4; ginseng (root) - 4; Cayenne pepper - 4; Ginkgo biloba extract - 2.5; para-aminobenzoic acid - 1.65; gammalinoleic acid - 1.5; octacosanol - 1; royal jelly - 1.

3.2 Production technology

The production technology consists of the following successive stages.:

- *Raw material preparation.* Grinding of plant materials in a hammer mill, disinfection on the installation of the fluidized bed, sieved through a sieve with a mesh size of 1 mm is carried out;
- *Preparation of the mixture for granulation.* Joint dosing and pre-mixing of prescription ingredients with 1.5 kg of MCC M 12 (microcrystalline cellulose) occurs.

Then we proceeds with sifting the mixture through a vibrating screen with a hole diameter of 1 mm, grinding in a hammer mill, re-mixing occurs. Lumps and foreign inclusions should not be mixed. Mixing is carried out in a V-shaped mixer at the rate of 100 kg per 1 hour. The mixture homogeneity is checked by pressing the pestle on the surface; in this case, lumps and foreign inclusions should not be observed.

- *Preparation of wet and dry granules (extrusion).* For wet mixture granulation, a humidifier, namely 5% starch paste in the amount of 25%, is used. Granulation is performed on a horizontal granulator (extruder) press machine. The color uniformity of each granulate is checked, after which it is sent for drying with a temperature of the drying cabinet (65 ± 5 ° C) to a residual moisture content of 4-6%;
- *Preparation of powder mixture.* Preparation is carried out by joint dosing of components. Sifting the powder mixture through a 1-mm vibrating screen, grinding in a hammer mill and re-sieving to avoid lumps and foreign inclusions is carried out. Mixing components is done in a V-shaped mixer for 1 hour at the rate of 100 kg of the mixture. Checking uniformity, the presence of lumps and foreign inclusions is done by pressing a pestle on the surface of the mixture;
- *Getting a mixture for tableting.* Dusting mixture and regranulate are placed in a V-shaped mixer and stirred at the rate of 100 kg per 1 hour. Quality control is carried out in accordance with the requirements of technical documentation by sending it to a testing laboratory;
- *Tabletting and dusting.* The mixture tableting takes place on a rotary tablet machine brand. The process is followed by controlling an average weight of the tablets every 30 minutes by weighing 20 tablets and individual specimens, weighing a similar amount. Deviations should not exceed ± 5%. The appearance is checked every 60 minutes by analyzing 10 tablets (the surface should be smooth and durable in the absence of dimples and sticking, chips, hillocks and lamination. Finished tablets are dedusted);
- *Film coating.* Suspension of dry film coating mixture is being prepared. For this purpose, the calculated amount of water were charged to the reactor-homogenizer and small portions, while running the stirrer, dry film coating mixture is added (stirred for 15 min), then it is homogenized for 10 minutes and filtered through a nylon filter with holes 0,315-0,45 mm. The reactor-homogenizer is connected to the installation for applying a film coating and spraying up to a mass of film coating of 3.5%; the weight of half-product is produced. The shelf life of the film coating solution is 24 hours at 20–25 ° C and 7 days at 4–6 °C.
- *Evaluation of appearance and quality.* Examination of the finished product in appearance, its transfer to the testing laboratory for compliance with the technical documentation requirements. After receiving the research report, the finished product is placed in a container with the name, batch number, date of manufacture, operator signature on the label;
- *Packing, packaging, and storage.* 3 product packages are sent to a collection of arbitration samples. Packing and packaging is carried out according to the documented procedure.

3.3 Clinical trials

As a result of monitoring patients with premenstrual syndrome, improvement in clinical well-being was observed in the treatment process from the third day (namely, sleep improved, and irritability, tearfulness decreased, aggressiveness and headaches decreased), heart pain, fainting, hot flashes were absent, blood pressure returned to normal, dizziness decreased significantly. Tolerability of complex therapy was good; no side effects were noted.

Table 1 presents the dynamics of subjective and objective symptoms in patients with premenstrual syndrome of the main group *and* the control group. The data in Table 1 indicate that the dynamics of subjective indicators was more pronounced and reliable in the main group than in the control group.

In the subjects of the main group, the frequency of complaints on weakness, fatigue, irritability decreased in 5 times; while dizziness, anorexia, headache decreased in more than 5 times. In single patients of the main group, migraine-like pain was noted, "hot" flashes significantly decreased. Analyzing the objective symptoms of patients in this group, we noted a significant decrease in the number of patients with high blood pressure, pulse lability, and subfebrile conditions.

Table 1. Dynamics of subjective and objective symptoms in patients with premenstrual syndrome of the main and control groups.

Complaints	Group 1 (main)				Group 2 (control)			
	After treatment	%	After treatment	%	After treatment	%	After treatment	%
Weakness, fatigue	54	100	10	20	54	100	20	40
Excessive sweating	53	98	8	15	53	98	16	30
Bulimia	10	20	-	-	5	10	-	-
Anorexia	20	48	2	4	20	48	10	20
Headache	54	100	10	20	54	100	22	50
Migraine-type headache	10	20	1	2	8	18	4	9
Dizziness	54	100	8	15	54	100	27	50
Heart pain (cardialgia)	25	48	10	20	25	48	15	30
Blood pressure (increase)	20	40	5	9	25	48	15	18
"Hot" flashes	40	80	10	20	35	70	18	35
Numbness of the arms and legs	27	50	5	9	20	40	10	20
Pulse lability	20	40	10	20	25	48	12	24
Increased libido	10	20	1	2	8	18	4	9
Low libido	20	40	5	9	10	20	8	15
Subfebrile condition	25	45	5	9	20	40	10	20
Depression	20	40	8	15	10	20	8	18

4. Discussion

The functional orientation of a specialized product is associated with the implementation of the physiological functions of phytoestrogens contained in extracts of ginseng, Damiana, Angelica Sinensis and hops.

Phytoestrogens are similar in chemical structure to estrogens; therefore, they are physiologically accessible to receptors in the body of women, thus playing an important role in correcting premenstrual syndrome. Phytohormones increase anti-stress function, affects the endocrine system, inducing the synthesis of its own estrogens, namely progesterone. In turn, the normalization of progesterone levels smoothes the symptoms of premenstrual syndrome, providing good physiological and psychological well-being.

During menopauses, phytoestrogens provide a balance of calcium in bones, lower cholesterol levels. Separate properties of phytoestrogens provide a balance between excitation and inhibition of the nervous system, positively affecting the emotional state of a woman.

Included in the dietary supplements vitamins, minerals and amino acids control the normal level of

IOP Conf. Series: Earth and Environmental Science **395** (2019) 012089 doi:10.1088/1755-1315/395/1/012089

metabolism in the female body. In general, research results confirm these conclusions.

The innovativeness of technological solutions lies in the production of matrix tablet forms, a feature of which is the presence in the tablet of three types of granules with active substances that are released in the gastrointestinal tract in a certain order. Such a system is characterized by a prolonged nature of targeted release of biologically active substances with adjustable speed.

5. Conclusion

The use of dietary supplements in complex treatment in patients with premenstrual syndrome leads to clinical improvement of vegetative functions and all indicators of psychological testing. No side effects were noted, good tolerance to diet therapy is shown. Dietary supplements can be recommended in the complex treatment of premenstrual syndrome within a month. According to the doctor's prescription, a further administration of the biocomplex is possible for 2-3 months in a maintenance dose, namely, 1 tablet per day.

The technology was tested at the "Art Life" enterprises, with the inclusion of dietary supplements in the Federal Register of the Russian Federation and the organization of mass production. The stability of the quality characteristics andcompetitiveness of a specialized product are guaranteed by the implementation of the System of Management in accordance with the standards of the ISO 9001, 22000 series and GMP rules.

6. Acknowledgment

The authors are grateful to the technology corpus of the company "Art Life" and the Department of General Surgery of the Siberian State Medical University for their advice and assistance in the work.

References

[1] Gerasimenko N F, Poznnyakovskiy V M, and Chelnakova N G 2017 Methodological aspects of a complete, safe diet: the value in maintaining health and efficiency *Person. Sport. Medicine* **17**(1) pp 79 -86

[2] Tutelyan V A, and Nechayeva A P Eds. 2014 *Food ingredients in the creation of modern food* (Moscow, Russia: DeLi)

[3] Poznnyakovskiy V M, Chugunova O V, and Tamova M Yu 2017 *Food ingredients and dietary supplements* (Moscow, Russia: INFRA-M)

[4] Ancheva I N 2016 Functional nutrition during pregnancy *Nutrition Issues* **85**(4) pp 22-28

[5] Lukoyanova O L, Borovik T E, Baturin A K, Starovoitov M L, and Lebedeva U M 2016 Feeding women during periods of pregravid preparation, pregnancy and lactation *Questions of Contemporary Pediatrics* **15**(6) pp 625-630

[6] McKenna E, Hure A, Perkins A, and Gresham E 2017 Dietary supplement use during preconception: the Australian longitudinal study on women's health *Nutrients* **9**(10) pp 1119

IOP Conf. Series: Earth and Environmental Science **395** (2019) 012090 doi:10.1088/1755-1315/395/1/012090

Innovative beverage production technologies based on Amaranth biomass

V F Pivovarov[1], M S Gins[1*] and V K Gins[1]

[1] Federal Scientific Center for Vegetable Growing, 14 Selektsionnaya str., VNIISSOK Village 143080 Russia

Email: anirr@bk.ru

Abstract. The paper presents research results aimed at finding plant raw materials, food biologically active components and reviews the possibility of their organic interaction to a form nutrient phytocomplex to create functional beverages. Using the principles of food combinatorics, we formulate the criteria of safety and usefulness of extracts, the formulation of health drinks using edible red-violet dye enriched with antioxidants from leaves and seeds of the amaranth Valentina variety. The biochemical and mineral composition of the created drink is studied.

1. Introduction

Natural drinks show the most rapid growth in the market of functional foods. The development trend of the domestic consumer market of non-alcoholic beverages indicates the expansion of their range using natural ingredients in the composition of beverages, including those made from vegetable raw materials. Activation of the creation and production of domestic health drinks can be one of the factors of the state program *Basics of the State Policy in the Field of Healthy Nutrition of the Population of the Russian Federation for the Period up to 2020*.

Non-alcoholic beverages, as well as other products demanded by the consumer, should be tasty, healthy, and safe. Scientifically based composition of the plant mixture to obtain the extract and on its basis of the concentrated form, which is the main component of the product being created is determined in the process of their development [7, 9].

The purpose of this work was to develop a beverage with a preventive effect, with a manifested antioxidant effect, activating the metabolism and increasing the body's resistance to adverse factors. This is relevant as part of government programs aimed at improving quality and increasing life expectancy.

The main raw materials for the creation of natural drinks are herbal mixtures, usually including from three to six items [8, 9].

2. Material sand Method

In the process of obtaining a concentrate for a drink, the herbal mixture was used, the main component of which is the leaf biomass of the introduced plant of the Valentina amaranth variety [3]. To harmonize the functional properties and taste-forming perception of the PDA, the vegetative part and black currant berries are included in the composition of the plant mixture. Scientifically based selection of plant mixture, containing a certain mass ratio of the above components, contributes to the

Content from this work may be used under the terms of the Creative Commons Attribution 3.0 licence. Any further distribution of this work must maintain attribution to the author(s) and the title of the work, journal citation and DOI.
Published under licence by IOP Publishing Ltd

formation of a targeted action of the phytocomplex of natural nutrients. Determination of free amino acids was carried out on a KNAUER liquid chromatograph (HPLC) ("KNAUER Wissenschaftliche Geräte GmbH", Germany), with a diode-matrix detector equipped with an automatic information collection and processing system [2]. Cations and anions were determined on an ion chromatograph of the Model 861 Advanced IC- "MCS" 853 Metrohm CO2 Supressor by Metrohm AG ("Metrohm" - Switzerland). The method is based on chromatographic separation of cations in a sample and their subsequent detection with a conductometric detector (Chromatography column for Metrosep C 4150 / 4.0 cations [10].

3. Results

Amaranth leaves contain the betacyanine-amarantine amaranth red pigment, protein (15-20%), balanced in composition of essential amino acids, a full set of vitamins B, E, C, carotenoids, flavonoids (rutin, quercetin, etc.), simple phenolic compounds, hydroxycinnamic acids, ascorbic acid [2, 10].

Vegetative part and black currant berries have functional properties. It is known that 100 g of berries of this culture contain up to 4000 mg of ascorbic acid, up to 17% sugar, organic acids (tartaric, succinic, citric, malic, nicotinic), pectic substances, tannins up to 0.62%, vitamins B1, B6, E, P, K, glycosides, essential oils, macro- and microelements.

Black currant leaves also contain a wide range of macro- and microelements, ascorbic acid, phytoncides, carotene, essential oil, as well as compounds with P-vitamin activity.

The complex of nutrients of berries and leaves of black currant in the human body is able to exhibit properties that improve the state of the cardiovascular system, tonic, enhance appetite, activate the functions of the gastrointestinal tract [4].

The extract was enriched with the products of hydrolysis of non-starch polysaccharides from amaranth seeds with easily digestible substances of carbohydrate and protein nature. Hydrolysis products formed in the process of biocatalysis contribute to the formation of functional and technological properties, positively affect the creation of a complex of nutrients that form organoleptic properties with the manifestation of a soft balanced taste.

4. Discussion

The results of scientific and experimental work to determine the criteria for the process of bioconversion made it possible to optimize the size of the plant particles, including fruit raw materials, in a ground mass. The particle size in the vegetable mixture is in the range of 3-5 mm. The duration of the extraction process is 12 hours at a temperature of 46 °C / 48 °C. A filtered clear extract is the main ingredient in the drink. The drink obtained in this way has a moderately pronounced taste and aroma perception characteristic of the used components of the herbal mixture.

In the complex of nutrients of the drink, amino acids are present, they are in the composition of all protein substances in the human body as the main component, and also are part of the vitamins and other compounds important for the body. 10 amino acids in the human body are not synthesized [5]. Amino acid composition was determined in the developed drink.

In total, 100 g of the drink include the following mass concentration of amino acids, mg: aspartic acid (122.8), glutamic acid (19.2), asparagine (38.0), histidine (58.8), serine (27.2), glutamine (3.2), arginine (16.0), glycine (8.8), threonine (6.4), alanine (127.2), tyrosine (62.4), valine (27.2), methionine (344.0), tryptophan (30.0), isoleucine (26.8), phenylalanine (4.8), leucine (62.0), and lysine (27.6).

Minerals in the form of organic and inorganic salts play an important role in the body. They are part of the cells and intercellular fluids, normalize metabolic processes, participate in the enzymatic and hormonal activities of the body, regulate the tension formed in the nervous system and muscle tissue, depending on the characteristics of salt metabolism. The constancy of the blood pH and other biological systems in the body is maintained by salts [1, 6].

Mass concentration in 100 g of the drink is represented by the following minerals, mg: Na (13.6), K

(1160.4), Mg (42.4), and F (30.4).

5. Conclusion

As a result of the research, the scientifically based criteria of the basic technology were developed. Namely, we focused on the extraction and technological process of obtaining a natural beverage based on the extract from the plant mixture, including leaves and berries of the "Valentina" amaranth variety.

In the complex of natural nutrients contained in the Debyut Amaranta non-alcoholic beverages, extractive (including biologically active_ substances of the main component of the PDA, which is Valentina' s variety amaranth leaf (0.38 g per 100 cm^3 of drink), prevail.

Scientifically-based decisions in the process of performing work on the possibility of using new types of plant raw materials in the technologies of creating new functional beverages contribute to the expansion of the raw material base, improving the process activation technology and qualitative development of organoleptic indicators and the biological value of the finished product.

References

[1] Gins M, Gins V, Momyleva S, Kulikov I, Medvedev S, Kononkov P, ... Pivovarov V 2018 Mineral composition of amaranth (Amaranthus L.) seeds of vegetable and grain usage by ARHIVBSP selection *Potravinarstvo. V.* **12**(1) pp 330-336

[2] Gins M S, Gins V K, Motyleva S M, Kulikov I M, Medvedev S M, Pivovarov V F, ... Mertvischeva M E 2017 Identification of metabolites with antioxidant properties in the leaves of vegetable amaranth (*Amaranthus tricolor* L.) *Agricultural Biology* **52**(5) pp 1030-1040

[3] Gins M S, Pivovarov V F, Gins V K, Kononkov P F, and Derkanosova N M 2014 Scientific support of innovative technologies in the creation of functional products based on vegetable crops *Russian Vegetables* **1**(22) pp 4-9

[4] Latushkin V A, Shestakov V K, and Kulikov I M 1997 Environmental assessment of black currant varieties in the Kirov region *Fruit and Berry Growing in Russia* **4** pp 137-142

[5] Oreschenko A V 1999 *Food combinatorics and human genetic health* (Moscow, Russia: Pishtepromizdat)

[6] Pilat T L 2002 *Dietary supplements. Theory, production, application* (Moscow, Russia: Avvallon)

[7] Strelkov V N, Burmistrov G P, and Filonova G L 2013 *Technology of special food products and research of consumer properties* (Pyatigorsk, Russia: RIA-KMV)

[8] Filonova G L, Gernet M V, Kovaleva I L, and Litvinov E A 2013 Ultrasound and biocatalysis is a radical link in the technology of extracts from vegetable raw materials *Beer and Drinks* **3** pp 18-21

[9] Filonova G L, Kovaleva I L, Komrakova N A, Shcherbakova V V, Nikiforova E V, Osipova V P, and Grishkovsky B A 2012 Food combinatorics in the technology of multicomponent concentrates using vegetable raw materials and drinks based on them *Beer and Drinks* **4** pp 22-25

[10] Filonova G L, Soboleva O A, Golovina T A, Kononkov P F, Gins V K, and Gins M S 2015 Amaranth of the Valentina variety is a promising source of raw materials for health drinks *Beer and Drinks* **4** pp 10-14

Improving the technological process for the mini-tubers production of the original seed of Jerusalem artichoke

A A Manokhina[1*], O A Starovoitova[2] and V I Starovoitov[2]

[1] Russian State Agrarian University – Moscow Timiryazev Agricultural Academy, 49 Timiryazevskaya str., Moscow 127550 Russia
[2] Lorch Potato Research Institute, 23 Lorcha str., 140051 Kraskovo Russia

E-mail: alexman80@list.ru

Abstract. The purpose of the research is the development and improvement of the technological process of production of mini-tubers of the original seed Jerusalem artichoke to increase the quantitative yield of the standard seed fraction "original seed Jerusalem artichoke" in accordance with the requirements of GOST R 55757-2013. The development of the technological process of production of the original seed Jerusalem artichoke was carried out with the introduction into the culture and reproduction through meristem; cuttings of plants from test tubes were carried out for their further reproduction; then, plants from test tubes were planted in propagatory specially made at the Institute installation for aerogydromassage the method of obtaining the minitubers. Introduced into the culture in vitro and propagated ten varieties (hybrids) of Jerusalem artichoke. In the production 1000 in vitro micro plants, the costs amounted to 76.5 thousand RUB From one plant turned out more than 15 pieces of size more than 10 mm when growing minitubers in aerohydroplane propagator. The proposed methods can increase almost 2 times the volume of production of superelites of standard quality and increase profitability by 29%, for practical implementation with the greatest efficiency they can be implemented on the basis of modern well-equipped high-tech enterprises that are specialized in the original seed Jerusalem artichoke.

1. Introduction

Jerusalem artichoke is a tuberous perennial plant of the sunflower family, which originates from North America and is also known as wild sunflower or Jerusalem artichoke. This plant is characterized by good tolerance to frost, drought and poor soil, strong resistance to pests and plant diseases. For centuries, it has been cultivated due to the fact that it was an edible tuber and had healing properties [1, 2]. It is an important crop that is used as food, animal feed and energy culture. However, there was little basic research on its breeding, and the lack of elite varieties seriously limited its industrial use [3, 4, 5].

In the middle zone of the Russian Federation, biological seeds of Jerusalem artichoke, despite abundant flowering, ripen rarely, so this culture is propagated mainly by tubers.

The purpose of the research is the development and improvement of the technological process of production of mini tubers than the original seed Jerusalem artichoke to increase the quantitative yield of the standard seed fraction "original seed Jerusalem artichoke" in accordance with the requirements of GOST R 55757-2013.

Content from this work may be used under the terms of the Creative Commons Attribution 3.0 licence. Any further distribution of this work must maintain attribution to the author(s) and the title of the work, journal citation and DOI.
Published under licence by IOP Publishing Ltd

IOP Conf. Series: Earth and Environmental Science **395** (2019) 012091 doi:10.1088/1755-1315/395/1/012091

The analysis and mastering of the technological process for obtaining the starting material for the production of Jerusalem artichoke superelite are preliminarily carried out. For high yields, Jerusalem artichoke needs to use for planting only a new, benign regeneration material.

2. Materials and Methods

Experience shows that during the reproduction of Jerusalem artichoke in large volumes, the quality of the produced seeds may arise [3, 6, 7], therefore the development of the technological process of production of the original seed Jerusalem artichoke was carried out with the introduction into culture and reproduction through the meristem. According to the results of the assessment of clones, during the growing season in 2015, only healthy tubers were selected in the nursery and after germination, isolation, meristems were introduced into test tubes. In 2015-2016, grafting of plants from test tubes was carried out for their further reproduction. In 2016, plants from test tubes were planted in propogators - these are specially made installations at the institute for the aero-hydroponic method for obtaining mini-tubes.

The technological process of obtaining in vitro microtubes includes the cultivation of microplants according to the standard technology of clonal micropropagation with subsequent induction of tuberization. The most optimal for these purposes is Murashige-Skoog medium with a mineral base with the addition of growth regulators and a high concentration of sucrose for micro tuberization [6].

During the growing season, the air exchange and the maintenance of the optimum temperature were ensured by the combined coating with an anti-mosquito screen and ultrasil. Plant protection was carried out against Septoria and powdery mildew and was achieved by spraying of systemic and contact fungicides 3-5 times (Kurzat R, Shirlan, Skor, Infinito, Akrobat MTs).

3. Research Results

The process of clonal micropropagation consisted of four stages: 1 - introduction to the culture *in vitro* of plant material; 2 - microproduction itself, due to which the maximum number of arylones (microshoots) is achieved; 3 - rooting of propagated shoots with their subsequent adaptation to soil conditions; 4 - growing plants in greenhouse conditions or in aero-hydroponic propagators and preparing them for planting in the field.

Ten varieties (hybrids) of Jerusalem artichoke introduced into culture *in vitro* and multiply. In the laboratory of bioengineering of the All-Russian Research Institute of Potato Farm, work on the introduction of Jerusalem artichoke into tissue culture was carried out. In the Russian Federation, such varieties of Jerusalem artichoke as Skorospelka, Diyeticheskiy, Podmoskovnyy, Nakhodka, etc. were obtained in test tubes for the first time [8]. With the introduction of Jerusalem artichoke in culture *in vitro*, 4-6 cuttings can be obtained from a single plant. It is noticed that the upper cuttings root better [9, 10].

Material costs for the production of 1000 in vitro microplants amounted to 76.5 thousand rubles, including: the source material – 57.5 thousand rubles (75%); remuneration of labor – 8.2 thousand rubles (11%); electricity costs – 3.5 thousand rubles (5%); equipment depreciation costs – 3.1 thousand rubles (4%); utility costs, consumables, phytotron depreciation, overhead – 4.2 thousand rubles (5%). The cost price of one micro plant was 76.54 rubles.

When grown in vitro culture, 1-2 tubers (average of 1.5 tubers) measuring 5–9 mm in diameter are usually formed from a single plant. With the aero-hydroponic method of growing mini-tubers, the plants are fed by supplying the nutrient solution to the roots in the form of an aerosol saturated with oxygen. The modified Knop solution is used as a nutrient medium. The concentration of salts in the medium may vary depending on the timing and phases of the growing season and must be maintained within 0.9-2.2 mg / l; pH of the medium is 5.9-6.3. Periodic feeding of the nutrient solution is carried out at certain intervals. The solution is regularly completely replaced and, if necessary, replenished with an adjustment of the content of macro- and microelements.

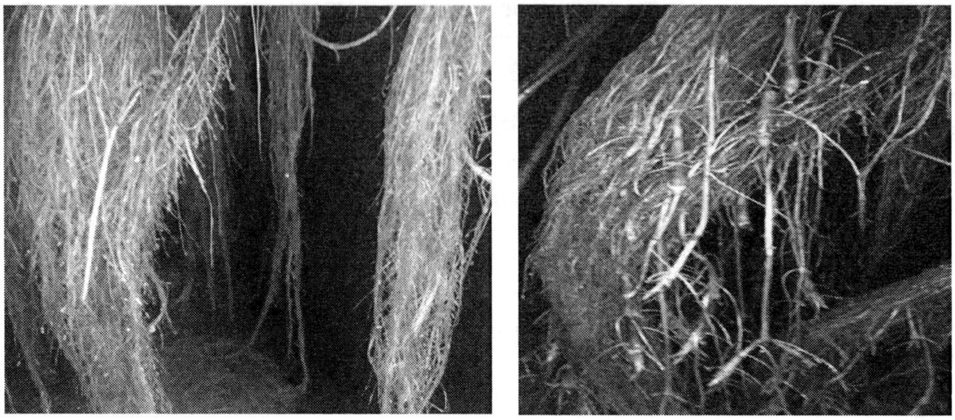

Fig. 1. The supply of the nutrient mixture and tuberization in the promoter (aerohydroponic installation).

During the growing season of plants, the correct choice of the spectrum of light and photoperiod is of great importance [11, 12, 13, 14]. The most optimal mode of plant lighting is achieved through the use of DNAT400 lamps and LED 80 LED lamps (SYLVANIAGROLUX 400W) - 2 DNAT400 + 4LED80 and provides a 14-hour photoperiod.

Plant lighting and feeding of the nutrient solution are regulated using timers that are tuned to the necessary modes. Indoors, the relative humidity of the air must be maintained, which is 50-55%.

For the aerohydroponic production method of mini-tubes, various types of installations are used where plants are fed by feeding a nutrient solution to the roots in the form of an aerosol saturated with oxygen. When grown in propagators, collection is carried out manually, gradually, as the tubers mature and reach a standard size (mass of tubers is 7-10 g). The yield from one bush was more than 15 pieces with a size of more than 10 mm when growing minitubers in aero-hydroponic propagators. The storage of mini-tubers of Jerusalem artichoke is carried out at a temperature of -1°C / + 4°C and air humidity of 95-85% [15].

4. Results and Discussion

Of great importance for the development of Jerusalem artichoke seed production and obtaining high-quality planting material is the use of *in vitro* microclonal propagation technology.

With clonal micro propagation of Jerusalem artichoke, the following is achieved: the recovery of plants from fungal and bacterial pathogens, viral and mycoplasmal infections; obtaining genetically uniform planting material; getting a high breeding rate. 10-15 thousand plants can be obtained in six months. Work can be carried out throughout the year, since the development and growth of plants *in vitro* is practically independent of seasonal changes. Also, the need for large areas required for growing planting material is reduced [9], [11], [15].

The highest costs for the production of in vitro microplants are the production of the source material, which is 75% of all costs, wages – 11%, and energy costs – 5%.

With proper environmental management and using special methods of *in vitro* tuberization, there is the possibility of increasing a number of tubers of acceptable size (over 9 mm in diameter). Such microtubers are quite suitable for direct planting in the ground in spring-summer greenhouses or tunnels of covering synthetic material (ultrasil, lutrasil, spanbond, etc.) with the condition of strict adherence to phytosanitary and agrotechnical measures that exclude the possibility of new infections by pathogens from outside.

The best conditions for the planting of Jerusalem artichoke in spring-summer aero-hydroponic propagators are created from mid-April to mid-May.

The possibility of a more productive use of laboratory equipment and cultivation facilities by obtaining *in vitro* microtubers (during the autumn-winter season (September-January) and clonal propagation of microplants to the required volumes during the winter-spring season (January-May), with subsequent cultivation of mini-tubers in a controlled environment under the protection of insect vectors of infections, including aerohydroponic structures and tunnels of light covering materials) allows for a 29% increase in profitability, as well as an increase in the production of standard-quality super-elites on almost 2 times.

5. Conclusion

The proposed techniques for practical implementation with the greatest efficiency can be implemented on the basis of modern, well-equipped high-tech enterprises that specialize in original Jerusalem artichoke seed production.

The process of clonal micropropagation with the introduction into culture in vitro of plant material; Namely, micropropagation, rooting of propagated shoots with their subsequent adaptation to soil conditions, growing plants in greenhouse conditions or in aerohydroponic propagators and preparing them for planting in the field, allows more productive use of cultivation rooms and laboratory equipment. This is done by obtaining *in vitro* plants of Jerusalem artichoke during the autumn-winter season (September-January) and clonal propagation of micro-plants to the required volumes during the entire winter-spring season (January-May), followed by cultivation of mini-tubers in a controlled environment under protection from insect vectors infections (including aerohydroponic structures). This makes it possible to ensure an increase in profitability by 29%, and also to increase the production volume of standard-quality super-elites in almost 2 times.

6. Acknowledgments

The authors are grateful to the researchers of the Institute, namely, Shmyglya Irina Valentinovna and Khutinaev Oleg Soslanbekovich, who actively assisted in our research.

References

[1] Szewczyk A, Zagaja M, Bryda J, Winiarczyk S, and Andres-Mach M 2019 Topinambur – New possibilities for use in a supplementation diet *Annals of Agricultural and Environmental Medicine: AAEM* **26**(1), pp 24-28

[2] Starovoitov V, Starovoitova O, Aldoshin N, and Manohina A 2017 Jerusalem artichoke as a means of fields conservation *Acta Technologica Agriculturae* **20**(1) pp 7-10

[3] Dunin M S 1935 Diseases and pests of Jerusalem artichoke *Proceedings of the VNII of Grain Crops* **VI**(1)

[4] Lv S, Wang R, Xiao Y, Zeng J, and Zhao C 2019 Growth, yield formation, and inulin performance of a non-food energy crop, Jerusalem artichoke (*Helianthus tuberosus* L.), in a semi-arid area of China *Industrial Crops and Products* **134** pp 71-79

[5] Fukano Y, Guo W, Noshita K, Hashida S, and Kamikawa S 2019 Genotype-aggregated planting improves yield in Jerusalem artichoke (*Helianthus tuberosus*) due to self/non-self-discrimination *Evolutionary Applications* **12**(3) pp 508-518

[6] Starovoitov V I, Starovoitova O A, Khutinaev O S, Biryukova V A, Shmyglya I V, Manokhina, A A, ... Baranov V V 2016 *Guidelines for the Typical technology of large-scale production of original seeds of Jerusalem artichoke* (Moscow, Russia)

[7] Kays Stanley J, and Nottingham Stephen F 2007 *Biology and chemistry of Jeruslaem artichoke: helianthus tuberosus L.* (Boca Raton, FL: CRC Press)

[8] Starovoitov V I, Starovoitova O A, Aldoshin N V, and Manokhina, A A 2018 Technology and mechanization of cultivation of jerusalem artichoke healthier *Research in Agricultural Engineering* **64**(3) pp 151-156

[9] Somda Z C, McLaurin W J, and Kays S J 1999 Jerusalem artichoke growth, development, and field storage. II. Carbon and nutrient element allocation and redistribution *Plant Nutr.* **22** pp 1315-1334

[10] Starovoitov V I, Starovoitova O A, Chernikov D V, Manokhina A A, Shmyglya I V, Boyko Yu P, Boyko V V, and Feoktistov A N 2012 *The method of growing plants of Jerusalem artichoke patent for invention RUS 2534350* (Moscow, Russia)

[11] Martirosyan Yu Ts 2014 Aeroponic technology in the primary potato seed production - advantages and prospects In Materials of the International Scientific-Practical Conference on *Potato Farming: Biotechnology Methods in Potato Breeding and Seed Production* (pp 175-179) (Moscow, Russia)

[12] Zubr J 1991 Performance of different Jerusalem artichoke cultivars in Denmark (1982-1984) In G Gosse

and G Grassi Eds., *Topinambour (Jerusalem Artichoke)* (pp 43-51) (Luxembourg, LUX: Commission of the European Communities)

[13] Mordashin, I S, and Lobastova E Yu 2014 An effective method for accelerated reproduction of healthy potatoes *Potatoes and Vegetables* **5** pp 23-24

[14] Khutinaev O S, Yurlova S M, and Anisimov B V 2012 Features of hydroponic cultivation of mini- and microtubers on installations KD-10 and "Minivit" In *Potato Growing* (pp 125-131) (Moscow, Russia: All-Russian Research Institute of Potato Farming)

[15] Sobrado M A, and Turner N C 1986 Photosynthesis, dry matter accumulation and distribution in the wild sunflower *Helianthus petiolaris* and the cultivated sunflower *Helianthus annuus* as influenced by water deficits *Oecologia* **69** pp 181-187

IOP Conf. Series: Earth and Environmental Science **395** (2019) 012092 doi:10.1088/1755-1315/395/1/012092

Dependence of tuber fluorescence on potato storage duration

V I Starovoitov[1], A A Manokhina[2*] and Zh Zh Allayarov[2]

[1] Lorch Potato Research Institute, 23 Lorcha str., 140051 Kraskovo Russia
[2] Russian State Agrarian University – Moscow Timiryazev Agricultural Academy, 49 Timiryazevskaya str., Moscow 127550 Russia

E-mail: alexman80@list.ru

Abstract. Fluorescent control methods are very sensitive and able to capture the ultra-weak reactions of tubers to lighting. However, they have not been developed due to the high sensitivity and complexity of the analysis of diverse information from natural samples. An important biological feature of tubers is repair, i.e. an ability to renew the cover tissue in places of mechanical damage. To assess the quality of tubers, natural, and wound periderms, a spectrofluorimeter was used. Optical measurements were performed. The spectra were recorded at a low voltage on a photovoltaic multiplier with a slit width of monochromators – 5 nm, in which the object fluorescence was illuminated by light with a variable wavelength. Results of the spectral reaction to these excitations were recorded. Tubers of potato varieties of different maturity groups were used for research. The relationship between the intensity of fluorescence of vitamins (max 468 nm) and flavins (max 520 nm) most fully reflects the state of tubers in different periods of storage. The aim of the study was to assess the quality of tubers by applying the fluorescent method to create a database for the development of portable devices that do not destroy the control of tubers during their storage.

1. Introduction

Effectiveness of a study focused on the quality of potatoes depends largely on the possibilities of instrumental assessment of parameters of tubers during storage. Optical methods of non-destructive express quality control of tubers are promising, but not fully studied for the diagnosis of potatoes. These methods allow one to monitor the status after diagnosis and, if necessary, make changes in the process of storage. Fluorescent control methods are very sensitive and capable of capturing the super-weak reactions of tubers to lighting; however, they have not received proper development due to the high sensitivity and complexity of analyzing diverse information from natural samples [1], [2], [3]. The non-destructive method allows one to enhance the quality control of potatoes during storage, minimizing both temporary and food losses [4], [5].

Reparation, as the ability to renew the integumentary tissue in places of mechanical damage, is an important biological feature of tubers. The course of wound reactions in potatoes is closely related to the respiration of tubers. The importance of this feature increases due to the extensive mechanization of harvesting and post-harvest handling, as a result of which macro- and micro-cracks of tubers, cuts, dents, occur [6], [7]. A well-formed wound periderm protects the tuber from evaporation and penetration of microorganisms. It contains a large amount of the following compounds of phenolic and glycoalkaloid nature: caffeic acid, scopoletin, solanine, chakoin, and other phytoalexins [6], [8].

The work of K. Ryan and his collegues made the most significant contribution to understanding the processes of a wound repair. They showed that mechanical injuries of tomato and potato leaves

Content from this work may be used under the terms of the Creative Commons Attribution 3.0 licence. Any further distribution of this work must maintain attribution to the author(s) and the title of the work, journal citation and DOI.
Published under licence by IOP Publishing Ltd

systematically expressed proteinase inhibitor (PI) genes. They were recognized as a marker of wound signaling. PIs function as protective proteins that protect plant tissue from the effects of mechanical wounds, as well as damage caused by insects [9], [10].

The purpose of the study is to assess the quality of tubers using the fluorescent method to create a database when developing portable devices that do not destroy the control of tubers during storage.

2. Materials and Methods

The Spectropluorimeter "MPF-44A" (Sweden) was used to assess the quality of natural and wound periderm tubers. Optical measurements were performed. The spectra were recorded at low voltage on a photomultiplier with a monochromator slit width of 5 nm, where the object was illuminated with variable wavelength to excite its fluorescence; and the results of the spectral response to these excitations were fixed. Tubers of different varieties of potatoes in groups of ripeness used for research. Studies were conducted on the basis of the Lorch Potato Research Institute and the Carcinogenesis of Blokhin Russian Cancer Center (Russia).

3. Research Results

The relationship between the fluorescence intensity indicators of vitamins (max 468 nm) and flavins (max 520 nm) most fully reflects the state of tubers during different periods of storage (Table 1).

Table 1. Dynamics of values of fluorescence intensity ratios B (I468 nm / I520 nm) in the post-harvest period λ_B= 368nm (for two years). Greenhouse Wellness.

Variety	Post-harvest period									
	Cleaning		The end of the treatment period		The end of deep peace		The end of forced rest		Correlation	
	A μ	B	A μ	B	A μ	B	A μ	B	$K_{A,B}$	$K_{A,B}$
Early varieties	15.8	1.60	24.3	1.77	29.3	2.06	30.5	1.86	0.87	0.78
Medium early	23.5	1.61	25.84	1.75	30.1	1.96	32.3	1.79	0.90	0.89
Middle late	26.3	1.58	33.0	1.79	35.3	2.10	37.7	1.96	0.93	0.88
Late varieties	6.8	1.51	34.3	1.84	39.7	2.03	40.3	1.94	0.99	0.91

Some notes: λ_B is the wavelength of fluorescence excitation; *A* is the thickness of the periderm; μ is the micron; *I* is the fluorescence intensity; nm – nanometer; $K_{A,B}$ is the correlation between the thickness of periderm and spectral characteristics before the tubers emerge from deep dormancy; $K_{A,B}$ is the correlation between periderm thickness and spectral characteristics during the entire storage period.

The analysis of changes in the shape of the fluorescence spectral bands during the healing process was carried out in young tubers devoid of peel, either partially or from the entire surface (peel of the peel), of tubers cut in half. Comparison of the intensity values of the luminescence intensity of tubers without mechanical damage showed that in undamaged tubers, the luminescence during the treatment period during ripening smoothly increases, and in mechanically damaged ones, these indicators tend to align with those of intact (whole) tubers (Table 2). Depending on the variety, the dynamics of the values of fluorescence intensity ratios B ($I_{468\,nm}$/ $I_{520\,nm}$) have a temporary length of the treatment period, healing and entry into deep rest in 2 to 4 weeks.

IOP Conf. Series: Earth and Environmental Science **395** (2019) 012092 doi:10.1088/1755-1315/395/1/012092

Table 2. Dynamics of the fluorescence intensity ratio values during the treatment period of mechanically intact tubers and tubers with mechanical wounds (Nevskiy variety).

Time after injury (days)	Fluorescence intensity ratio values $I_{468\ nm}/I_{520\ nm}$; λ_B=368 nm; t=20°C		
	Control (intact tubers)	Peel skin (from the whole surface)	Sections (cut in half)
Before injury	1,58	1,60	1,64
Immediately after injury	1,58	0,86	1,06
3	1,60	1,14	1,31
6	1,63	1,27	1,40
9	1,63	1,34	1,46
12	1,69	1,41	1,52
15	1,73	1,52	1,60
18	1,74	1,63	1,65
21	1,78	1,75	1,72

Note: Nm is a nanometer; I is the fluorescence intensity; λ_B is the fluorescence excitation wavelength.

4. Discussion

As can be seen from Table 1, during the treatment period, morphological and optical parameters changed quite quickly; due to the formation of durable peel, the activity of chemical processes takes place in the tissues.

In the phase of deep and forced rest of the studied varieties, rather stable indices of luminescence of vitamins with insignificant discrepancies are characteristic of the fluorescent characteristics of potatoes.

The time corresponding to the occurrence of tubers in a state of deep dormancy is characterized by a slowdown in the increase in the luminous intensity ratio of these components, which occurs within 2-4 weeks and depends on the group of ripeness of the variety.

In the deep rest phase, the natural periderm thickening slows down. During this period, the growth of the values of this optical indicator begins to weaken noticeably, and with the beginning of the awakening of the ocelli, it completely stops. With the onset of forced dormancy, despite the fact that the periderm thickness continues to grow slowly, this indicator begins to gradually decrease, which is associated with the opening of the lentils, due to increased respiration activity and the appearance of microcracks on the aging peel, which increased the fluorescence yield of substances in the subcutaneous layer. The mechanisms of wound repair are activated, both directly at the sites of injury and in tissues remote from the site of local exposure, as a result of which the plant's protective system responses are induced. A mobile signal that transmits wound stress information throughout the plant or its organ is involved in the induction of systemic protection. Currently, it has been established that the main components of the mobile signal are the 18-amino peptide, called systemin, as well as jasmonic acid (JA), and possibly some oxylipins, which appear during lipoxygenase oxidation. It has been established that JA can be transported along the phloem to remote areas of the plant. The Plasmodesma system allows JA to reach the target cell, notifying the plant of a stressful situation and including the expression of PI genes [11], [12], [13], [14].

The established relationship between the fluorescence of potatoes and the duration of its storage makes it possible to monitor the condition in the surface tissues of tubers and to evaluate the physiological state of the latter in the post-harvest period by the shape of the spectral bands using the indicator $I_{468\ nm}/I_{520\ nm}$. This can be used as a component of indicators of the state of potatoes during storage.

When fluorescence of the damaged areas of tubers was excited by light with a wavelength λ_B=386 nm, the obtained spectral bands resembled undupened tubers in their form. At the same time, they had a much higher yield of fluorescence in the region λp=490-550 nm, than the young unripe tubers. Thus, the lower limit of the fluorescence intensity variation values with the flavin maximum λp=520 nm was always higher for them than the upper variation limit for non-mature tubers. It is not possible to identify

IOP Conf. Series: Earth and Environmental Science **395** (2019) 012092 doi:10.1088/1755-1315/395/1/012092

a mechanical damage with a very thin peel of undigested young tubers, which formed wound periderm (healed). Varietal features did not affect this.

In connection with the participation in the healing process of mechanically damaged tubers of various types of tissue (hypoderm, vascular bundle area, the core), the formation of various forms of wound periderm is possible.

The study of a batch of tubers with a diameter of 5-6 cm, in order to determine the magnitude of the discrepancy between the fluorescence intensity values at their registration at points $I_{468\ nm}$ / $I_{520\ nm}$ depending on the depth of the lesion, showed the following. Minor discrepancies are observed only in the surface layer of the tuber, depending on the basal or apical zone.

The established relationship between the values of a "peak" fluorescence intensities at wavelengths λp=468 nm and λp=520 nm, obtained when the fluorescence is excited by light with a wavelength λ_B=386 nm, on the quality of the covering tissue can serve as a criterion for evaluating the healing of mechanical damage, physiological maturity of tubers and their readiness for long-term storage.

As noted, the freshly harvested and mechanically damaged potato tubers can form new covering tissues. A wound periderm is formed at the site of injury, which is impregnated with a wax-like substance, which prevents microorganisms from entering the tuber. This is how a mechanical barrier is formed. In addition to the mechanical barrier, a chemical barrier appears. In the wound area under conditions of actual harvesting of potatoes, infection of damaged tubers is possible [15], [16]. Phytoalexins are formed in response to contact with microorganisms, they are absent in healthy tissues and occur in them after being affected by disease [10], [11], [12], [17]. Phytoalexins have antibiotic properties and are able to suppress the development of microorganisms. The faster and larger quantities are formed, the more resistant this species is to pathogenic microorganisms. As it is stored, the ability of tubers to produce phytoalexins decreases, which reduces their resistance to disease.

Such research on the biochemical mechanisms of species immunity and their timely assessment should help to identify resistance of a plant as a whole.

5. Conclusion

The relationship between fluorescence λ_B=368 nm of potatoes and the duration of its storage allows one to monitor the state of the surface tissues of tubers by the shape of spectral bands using I_{468nm} / $I_{520\ nm}$, as well as to evaluate the physiological state of the tubers in the post-harvest period. This can be used as a component of indicators of the state of potatoes during storage.

Dependence of a "peak" fluorescence intensity ratio at the wavelengths λp=468 nm and λp=520 nm, obtained by excitation of fluorescence with the light of a wavelength λ_B= 386 nm, on the quality of the covering fabric can serve as a criterion for assessing the healing of mechanical damage, physiological maturity of tubers, and their readiness for long-term storage.

Further improvement of plant disease detection systems can be achieved by combining multisensory spectral and florescent methods.

References

[1] Metlitsky L V, Ozeretskovskaya O L, and Salkova E G 1984 *Biochemistry of plant immunity, rest, aging* (Moscow, Russia: Nauka)

[2] Starovoitov V I, Bashilov A M, and Anderzhanov A L 1987 *Automation of quality control of potatoes, vegetables and fruits* (Moscow, Russia: Agropromizdat)

[3] Starovoitov V I, Mikhalev A E 1993 Optical properties of potato tubers and carrot and beet root crops *Achievements of Science and Technology of Agriculture* **4** pp 35-36

[4] Bianchi G, Scalzo R L, Testoni A, and Maestrelli A 2014 nondestructive analysis to monitor potato quality during cold storage *Journal of Food Quality* **37**(1) pp 9-17

[5] Xue J-P, Wang X, Zhang A-M, He J-Q, and Chang L 2008 Effects of shading on photosynthetic characteristics of Pinellia ternata leaves *Zhongguo Zhongyao Zazhi* **33**(24) pp 2896-2900

[6] Polevoy V V, and Salamatova T S 1991 *Physiology of plant growth and development* (Leningrad, USSR: Leningrad University Press)

[7] Velikanov L L, Levkina L M, Prokhorov V P, and Sidorova I I 1985 *Infectious plant diseases: physiological*

and biochemical bases (Moscow, Russia: Agropromizdat)
[8] Gusev S A, and Starovoitov V I 1989 *Post-harvest handling and storage of potatoes* (Moscow, Russia: Moskovskiy rabochiy)
[9] Parfenov I A 2012 *Proteinase inhibitors from potato tubers and their role in the plant defense system* (Abstract of the Dissertation of the Candidate of Biological Sciences) (Moscow, Russia: A. N. Bakh Institute of Biochemistry, RAS)
[10] Dyakov Yu T 2012 *Fundamental phytopathology* (Moscow, Russia: Krasand)
[11] Vasyukova N I, and Ozeretskovskaya O L 2009 Jasmonat-dependent protective signaling in plant tissues *Plant Physiology* **56**(5), pp 643-653
[12] Vasyukova N I, Ozeretskovskaya O L 2009 Jasmonate-dependent defense signaling in plant tissues *Russian Journal of Plant Physiology* **56**(5) pp 581-590
[13] Tarchevsky I A, Karimova F G, Grechkin A N, and Moukhametchina N M 2000 Influence of (9Z)-12-hydroxy-9-dodecenoic acid and methyl jasmonate on plant protein phosphorylation *Biochemical Society Transactions* **28**(6), pp 872-873
[14] Korableva N P, Platonova T A, Dogonadze M Z, and Evsimina A S 2002 Brassinolide effect on growth of apical meri stems, ethylene production, andabscisic acid content in potato tubers *Biologia Plantarum* **45**(1), pp 39-43
[15] Glinka E M, Bulantseva E A, Salkova E G, and Protsenko M A 2003 Influence of BIPG on the intensity of oligouronide formation *Plant Physiology* **50**(2), pp 260-264
[16] Revina TA, Gerasimova NG, Kladnitskaya GV, Chalenko GI, and Valueva TA 2008 The influence of proteinase inhibitors of potato tubers on the growth and development of phytopathogenic microorganisms *Applied Biochemistry and Microbiology* **44**(1), pp 101-108
[17] Panina Ya S, Vasyukova N I, and Ozeretskovskaya O L 2003 Inhibition of activity of catalase from potato tubers by salicylic and succinic acids *Applied Biochemistry and Microbiology* **397** pp 307-310

Sustainable development and social and environmental risks of industrial regions of Siberia

E Sugak [1*]

[1] Reshetnev Siberian State University of Science and Technology, 31 prospect behalf of the newspaper "Krasnoyarsky Rabochiy", Krasnoyarsk 660037 Russia

E-mail: sugak@mail.ru

Abstract. The high investment potential of the regions of Siberia is substantially leveled by the low infrastructural potential and the high level of investment risks, primarily social and environmental risks. Consequently, the fundamental basis for ensuring their sustainable development and increasing investment attractiveness is the optimization of social and environmental risks. And the main indicators of their sustainable development are the indicators characterizing the environmental safety and social and environmental risks of the population. To assess the socio-environmental risks of harm to the health of the population of the region, taking into account its specific features, we propose to use data mining methods and neural network models, along with classical methods.

1. Introduction

The main task of modern society is to ensure sustainable development in order to create favorable conditions for economic growth and improve the quality of life of the population. However, the Russian Federation is a very heterogeneous spatially distributed system of national and territorial entities. This peculiarity necessitates a differentiated approach to the selection of the main indicators of sustainable development of territories and regions, taking into account the specifics of their socio-economic development and the characteristics of regional programs and development strategies.

Among the eight federal districts, the Siberian Federal District takes the 2nd place in terms of the territory, 3rd in terms of the population, 5th in the fixed assets, 2nd in mining, 3rd in the production of energy resources. The Siberian Federal District provides more than 10% of the total gross product, but it occupies the last places in many social indicators, quality of life, and the human development index.

The imbalance between the economic and social development of the Siberian regions significantly reduces their investment attractiveness and poses a serious threat to sustainable socio-economic development.

To solve the problem of equalizing the economic and social development and investment attractiveness of the regions, the most critical components of their investment potential and investment risk must be identified.

2. Materials and Methods

According to the international rating agency "RAEX-Expert RA", the ranks of investment potential and risk regions of the Siberian Federal District are significantly different (Tables 1 and 2) [1]. By the difference of ranks, the regions can be divided into three groups conditionally (Table 2):

Content from this work may be used under the terms of the Creative Commons Attribution 3.0 licence. Any further distribution of this work must maintain attribution to the author(s) and the title of the work, journal citation and DOI.
Published under licence by IOP Publishing Ltd

- Regions where the investment potential roughly corresponds to the level of risks – Novosibirsk and Omsk regions, Altai region, and the Republics of Tyva and Altai;
- Regions with an investment potential rating being significantly lower than the investment risk rating. At the same time, in the Tomsk region, the average rating of the investment potential is combined with a high rating of investment risks; and in the Republic of Khakassia, both indicators are quite low;
- Regions with an investment potential rating being significantly higher than the investment risk rating. At the same time, in the Krasnoyarsk region, the Kemerovo and Irkutsk regions, a high rating of the investment potential is combined with a low rating of investment risks, and in the Republic of Buryatia and the Transbaikal region, both indicators are quite low.

This classification allows one to identify both the most attractive and the most problematic regions [2]. In addition, the classification should be taken into account when developing regional development plans and determining the main indicators of their sustainable development.

For example, we consider it obvious that a natural way to increase the investment attractiveness of the industrial regions of the Siberian Federal District — the Krasnoyarsk region, the Kemerovo and Irkutsk regions — is to reduce investment risks. Analysis of the components of investment risk shows that the most critical for these regions are the social and environmental risks (Table 2) [2].

Consequently, the fundamental basis for ensuring the sustainable socio-economic development and increasing the investment attractiveness of industrial regions of Siberia is to reduce social and environmental risks. And, accordingly, the indicators characterizing the socio-environmental risks of the population are considered the main indicators of their sustainable development [2-4].

Table 1. Investment potential of the regions of the Siberian Federal District.

Place (SFD)	Rank (RF)	Region	Ranks of components of investment potential								
			Labor	Consumer	Industrial	Financial	Institutional	Innovative	Infrastructure	Natural resource	Tourist
1	7	Krasnoyarsk region	14	14	15	11	13	16	78	1	9
2	15	Novosibirsk region	13	16	20	17	9	6	49	40	32
3	16	Kemerovo region	17	18	16	19	26	37	53	4	36
4	18	Irkutsk region	20	23	18	15	20	20	74	7	13
5	26	Altai region	24	24	33	28	30	13	44	23	25
6	30	Omsk region	26	21	17	24	21	27	60	43	64
7	48	Tomsk region	37	60	44	50	45	10	82	37	79
8	53	Republic of Buryatia	55	58	65	59	65	63	79	10	15
9	54	Transbaikal region	64	52	64	57	62	74	70	9	43
10	77	Republic of Khakassia	75	74	67	73	70	81	67	32	78
11	82	Tyva Republic	80	80	83	80	82	78	84	31	72
12	83	Altai Republic	82	82	84	83	81	83	81	60	50

Table 2. Investment risks of the regions of the Siberian Federal District.

Place (SFD)	Rank (RF)	Region	Ranks of investment risk components						Difference of ranks of potential and risk
			Social	Economic	Financial	Criminal	Ecological	Managerial	
1	19	Novosibirsk region	44	13	9	39	41	50	-4
2	28	Tomsk region	64	35	19	27	58	12	+20
3	29	Omsk region	42	8	21	73	56	59	+1
4	32	Altai region	55	40	23	31	42	58	-6
5	43	Krasnoyarsk region	57	21	28	44	78	43	-36
6	52	Irkutsk region	68	26	17	59	72	61	-34
7	56	Kemerovo region	33	66	54	33	73	32	-40
8	64	Republic of Khakassia	60	60	61	48	52	55	+13
9	68	Republic of Buryatia	75	19	75	69	67	45	-15
10	74	Transbaikal region	71	70	73	80	75	53	-20
11	77	Altai Republic	78	73	81	41	57	10	+6
12	85	Tyva Republic	85	48	83	64	62	78	-3

Features of the ecological situation in Siberia due to the nature of the environmental impact of industry, transport, utilities, and agriculture [5]. The change in the natural environment has led to the emergence of a number of environmental problems in the region. It has also has caused deterioration of living conditions an increase in the number of diseases of the population and an increase in social tension. The main factors determining the degree of environmental hazard in the regions of the Siberian Federal District are the metallurgical industry, oil production and refining, the coal mining and gas industry, engineering and electric power industry [5]. Therefore, for industrial regions of Siberia, the main problem is air pollution (Table 3) [6].

Table 3. Rating of regions in terms of pollution and environmental impacts.

Place (SFD)	Rank (RF)	Region	Ranks of the components			
			Air pollution	Water pollution	Ecosystem disturbance	Ecosystem protection
1	3	Tyva Republic	26	9	17	20
2	4	The Republic of Buryatia	24	18	22	23
3	5	Irkutsk region	59	27	19	58
4	12	The Republic of Khakassia	70	7	36	21
5	24	Krasnoyarsk region	88	28	25	49
6	29	Transbaikal region	16	19	20	75
7	35	Tomsk region	77	31	21	70
8	49	Altai region	43	21	67	68
9	53	Novosibirsk region	74	22	45	52
10	59	Omsk region	71	36	54	71
11	63	Kemerovo region	82	61	42	9

IOP Conf. Series: Earth and Environmental Science **395** (2019) 012093 doi:10.1088/1755-1315/395/1/012093

However, at present, there are practically no methods for quantitative, statistically reliable assessment of technogenic social and environmental risks of harm to the health of the population of a particular region [3, 7]. As a rule, existing methods are based on large-scale long-term and expensive biomedical and epidemiological studies. And these methods do not take into account the factors specific to a particular region — geographical and climatic conditions, the level of industrial and socio-economic development, the particular social structure and way of life of the population, the level of development of the health care system, and others [3, 8]. Only a combination of classical methods and methods of data mining can make it possible to fully assess the influence of harmful environmental factors on the health of the population of a region [3, 8-10].

The use of artificial neural networks (ANN) seems to be promising when building a "dose-effect" relationship. They allow the development of highly efficient computer systems for predicting and analyzing mortality and morbidity with changes in environmental factors [3, 9-11].

For the assessment of environmental risks, neural network models describing the influence of harmful factors on the health of the population of Krasnoyarsk and the Krasnoyarsk region are built [3, 7-10]. Concentrations and emissions of major pollutants into the atmosphere were used as indicators of the environmental state. Primary morbidity, mortality, and life expectancy data were used as the indicators of health [3, 9-11].

3. Results

Calculations show that the neural network models satisfactorily describe the initial data – the error for various indicators of public health ranged from 0.4 to 4.7% [3, 9-11]. The results obtained allow us to determine the dose-effect relationships for individual components of the environment. And it provides an opportunity to create a comprehensive quantitative, statistically reliable methodology for the intellectual analysis and forecasting of social and environmental risks of the population of a particular industrial region using the available databases for monitoring emissions and pollutant concentrations and the state of public health [3, 9-11].

When new data is available, the resulting models can be easily corrected by the "complementary training." With appropriate "retraining" and customization, the models can be adapted to the conditions of any other industrial region of Russia, taking into account its specific features.

Interpretation of the obtained results will allow to assess the real or estimated socio-economic efficiency of implemented or planned technical, technological, administrative, and managerial decisions in the industrial and socio-economic development of an industrial region and an environmental management. Also, the interpretation of the results will make it possible to substantiate programs for the socio-economic development of the region, to develop regional standards for environmental quality.

4. Conclusion

Thus, the transition to the sustainable development of industrial regions of Siberia makes it necessary to include social and environmental factors in the system of basic socio-economic indicators of the region. This can be achieved through the development and adaptation of indicators of environmental safety at the national and regional levels. These indicators would reflect the influence of environmental factors on the health status of the population - social and environmental risks.

The currently existing traditional macroeconomic indicators that usually measure the success of the economic and social development of regions and the economic well-being of the population do not reflect the progress towards the achievement of sustainable development goals. This is due to the fact that the growth of these indicators can be based on man-made nature-intensive development. This creates prerequisites for a sharp deterioration in the economic and social indicators of the region in the future, in the event of depletion of natural resources and environmental pollution.

To assess and predict the social and environmental risks of the population of an industrial region, taking into account its specific features and the state of the environment, modern information technologies and data mining methods should be used, along with classical methods [3, 10, 11].

References

[1] Expert RA 2016 *Rating of investment attractiveness of the regions in 2015* Available at: http://www.raexpert.ru (Accessed 04 02 2019)

[2] Sugak E V 2017 Investment attractiveness and social and environmental risks of the Krasnoyarsk region *Nauka Krasnoyarya* **6**(4-2) pp 146-151

[3] Sugak E V 2014 Modern methods of environmental risk assessment *European Social Science Journal* **5**(44) pp 427-433

[4] Udartseva O V 2009 Environmental safety as a criterion for sustainable development of Western Siberia *Vestnik of RUDN* **4** pp 93-96

[5] Tsalikov R Kh, Akimov V A, and Kozlov K A 2009 *Assessment of Russia's natural, man-mad,e and environmental safety* (Moscow, Russia: FGU VNII GOCHS (FTS))

[6] Karanina E V, and Vershinina N A 2016 Ecological risks of Russian regions *Vestnik of Moscow University of Finance and Law* **2** pp 21-38

[7] Belskaya E N, Sugak E V, and Brazgovka O V 2016 Calculation and prediction of the individual death risk of the population of an industrial region *Safety in the Technosphere* **5**(4) pp 18-22

[8] Onishchenko G G, Zaitseva N V, May I V *et al* 2014 *Health risk analysis in the strategy of state socio-economic development* (Perm, Russia: Publishing House of the Perm National Research Polytechnic University)

[9] Sugak E V, Okladnikova E N, and Kuznetsov E V 2008 Computational and informational technologies for analysis and assessment of social and ecological risks *Ecology and Industry of Russia* **8** pp 24-29

[10] Sugak E V, Kuznetsov E V, and Nazarov A G 2009 Information technologies for assessing environmental safety *Mining Information and Analytical Bulletin (Scientific and Technical Journal)* **18**(12) pp 39-45

[11] Potylitsyna E N, Lipinskiy L V, and Sugak E V 2013 Use of artificial neural networks to solve applied environmental problems *Modern Problems of Science and Education* **4** pp 51-58

IOP Conf. Series: Earth and Environmental Science **395** (2019) 012094 doi:10.1088/1755-1315/395/1/012094

The concept of formation of agricultural food systems in the border regions on the principles of network-centric structures

V A Kundius[1] **and I V Kovalyova**[1*]

[1] Altai State Agrarian University, 98 Krasnoarmeyskiy prospekt, Barnaul, Russia

E-mail: irakovaleva20051@rambler.ru

Abstract. The theoretical and practical approaches to the organization of the territorial economy in agricultural regions on network-centric principles are considered in this paper. Their effectiveness has been proven within the sinter theory in the conditions of market economy and planned economy. The effectiveness of horizontal linkages in clusters of commercial complex provides the opportunity to conduct economic activities, mutually beneficial, efficient use of resources usage cross-border territories. Exploring the resource potential of Large regions of Altai enabled us to identify opportunities to significantly increase the production of the ecologically clean certified meat and dairy livestock products with special properties and contains useful microelements, vitamins, competitive not only in the domestic but also in global markets. This is due to the problem of ensuring healthy nutrition as well as recruiting in world popularity, slow food movement, based on the principle of ecological compatibility of natural products consumed. Thus, the eco-production is able to become a "growth pole" for the rural territories.

1. Introduction

One of the areas of agricultural policy aimed at ensuring the sustainability of ecological agriculture (including livestock) is the development of cooperative forms to increase production volumes of livestock products, providing the people of the region and other territories with environmentally-friendly food products. However, the development of Rural Studies show that the production, supply, and marketing cooperatives and its other forms are not sufficient in themselves to modern conditions. There is therefore a need for cooperation and integration, providing a positive impact on the development of the territories.

2. Materials and Methods

The resource potential of the studied cross-border territories, including Altai Krai, Altai Republic, and Mongolia, in the production of ecologically pure meat and meat products, on the basis of historical economic ties these regions [9] is enormous. Additionally, there are many natural objects included in the UNESCO list of world natural heritage [1]. These products are uniquely characterized by the world community as environmentally friendly.

Taken into account the fact that the territories of Altai Krai and Altai Republic, as well as in Mongolia, authentic homemade breeds adapted to local conditions. Their presence is the proof of careful attitude to the local genetic resources, the success of breeding and selection work. The largest research specialist in the preservation of the pets' gene pool Y. Stolpovskogo states that the availability of local breeds is the most important objective condition for the production of

Content from this work may be used under the terms of the Creative Commons Attribution 3.0 licence. Any further distribution of this work must maintain attribution to the author(s) and the title of the work, journal citation and DOI.
Published under licence by IOP Publishing Ltd

environmentally friendly and traditional foods. In the transboundary region, 14 species new species have been genetically bred [4]. In turn, as a result of many years of breeding work, there are the genetically bred sheep in Mongolia ("Uzjemchin" and "Bayaz") [4]. These local breeds are different with their exclusive endurance and pure genes. It is clear that the natural pets' elite gene pool acquires special significance in the context of increasing diseases in cattle fed industrially. According to the results of biochemical research, the beef from the mountains of Altai Krai, Altai Republic, and Western Mongolia is environmentally friendly. Only natural feeds are used in that region, without elements of agricultural chemistry and pesticides. Mongolia currently contains 50 million only cattle, a quarter of which is consumed by China due to its high quality [5].

Our studies on the basis of the Main Central Scientific-Industrial Veterinary Radiological Laboratory were focused on the justification of ecological purity and nutritional value of meat grown in farms of the Altai Republic, Altai Krai, and Zavhanskogo aimak and demonstrated high ecological safety and high content of useful micro and macro elements, vitamin, and acid preparations.

3. Results and Discussion

We have three samples of meat: (1) the first sample (beef, frozen in quarters) is collected in the Altai Krai (Petropavlovsky District); (2) the second sample (beef, frozen in quarters) is from the Altai Republic (Ongudaysky region, IP "Sajdanov, A. B."); (3) the third sample (beef stew) is from Ulaanbaatar, Armos, Mongolia. In the future, it is necessary to conduct research and laboratory complex studies on the qualities of the Mongolian and Russian meat livestock. Special significance is given in the research on content in meat of cattle, yaks, sheep vitamins and minerals, positively affect the human body. It is important to conduct experiments on the therapeutic effect of the high-mountainous Altai meat. It is planned to develop research in cooperation with scientists from Medical University on the appropriateness of this meat to supply patients undergoing heavy operations. There is every reason to believe that we get here very important scientific results of great social and economic importance for the development of recreational areas, tourism in Altai. Some positive results have been achieved in the medical and therapeutic practice in the Altai region.

It is worth mentioning that the processes of cooperation and integration in the development of livestock production and processing of cross-border territories of Altai and Western Mongolia have long historical roots. In Soviet times, meat sfrom Western Mongolia were almost entirely processed in the Altai Krai and the East Kazakhstan region (at the Semipalatinsk and Bijskom meat-processing plants). In the 90s years, customs and veterinary barriers of the Russian-Mongolia actually blocked the possibility of exporting Mongolian meat on territory of the Altai region and Republic of Altai.

In turn, this has led to substantial investments of 15-20 years restructuring the Mongolian herds and exacerbated the problems in animal husbandry. At the end of the 90s and in the early years of the twenty-first century, the production of high-quality is done in the cities of Kobdo and Bayan-Ulgii by the entrepreneurs from Gorno-Altaisk. Lack of effective marketing, logistics, and purely technologic reasons do not allow make their businesses profitable.

For the gradual resumption and development of the high-value production based on their food and curative properties of meat and products of its processing in modern conditions, it is necessary to form agri-food systems based on the principles of cross-border regions and network-centric structures in the form of territorial innovation economic clusters. In particular, the formation of cross-border territories cluster meat will solve the complex laws and customs problems associated with Mongolian meat import conditions on the Russian territory. Also, it is necessary to look for transferring the processing equipment and agricultural machinery on the territory of Mongolia. In particular, it is highly advisable to use the veterinary expertise and technology developed in Russia to improve the organization of sanitary-epidemiological and veterinary control in Mongolia.

Forming a cluster would accelerate solving the problems of innovative development, transport, production, and social infrastructure. A large role in these processes is assigned. There should be a successful co-op experience in Sweden and Finland, where major amounts of meat and dairy products produce just cooperatives. Given the predominantly family nature of Mongolian livestock, its cooperation in reconstruction of traditional Russian markets could provide a significant economic multiplier effect. The establishment of procurement, consumption and, possibly, credit cooperatives in the Western Mongolia will require a qualitative economic advice. To solve this problem, all interested parties should quite seek assistance form the capable economists-Altai agrarians, in our opinion. It is helpful to organize a Cooperative Advisory Centre, which would coordinate economic cooperation in the cross-border regions. The large meat processing enterprises of the Altai region and Republic of Altai become the core of such a cross-border cluster.

Within our project, we have developed a number of theoretical and practical approaches to the territorial economy of border regions on the network-centric principles. Their effectiveness has been proven in the conditions of market economy and planned economy. The effectiveness of horizontal linkages in clusters of commercial complex provides the opportunity to conduct economic activities, even in cases where they, in terms of classical investment analysis, appear to be completely unprofitable. A formation of cluster agglomeration, possibly in the form of a formalization of economic partnership structure prevailing at that time in a specific territory. In a cluster structure, in addition to agricultural producers, processors, infrastructure organizations will include scientific organizations: Altay State Agrarian University, Gorno-Altaysk Research Institute of Aagriculture, Research Institute of Water and Ecological Problems of the Russian Academy of Science, Altay Regional Institute of Ecology, and others.

Clustering economy enables a comprehensive look at public policies for regional development, increase productivity, efficiency and competitiveness, cluster members. Additional opportunities for innovative development, in general, increased the levels of employment and quality of life [3]. In addition, we justified factors of the feasibility of forming of territorial economic clusters based on principles of network-centric agri-food structures (fig. 1).

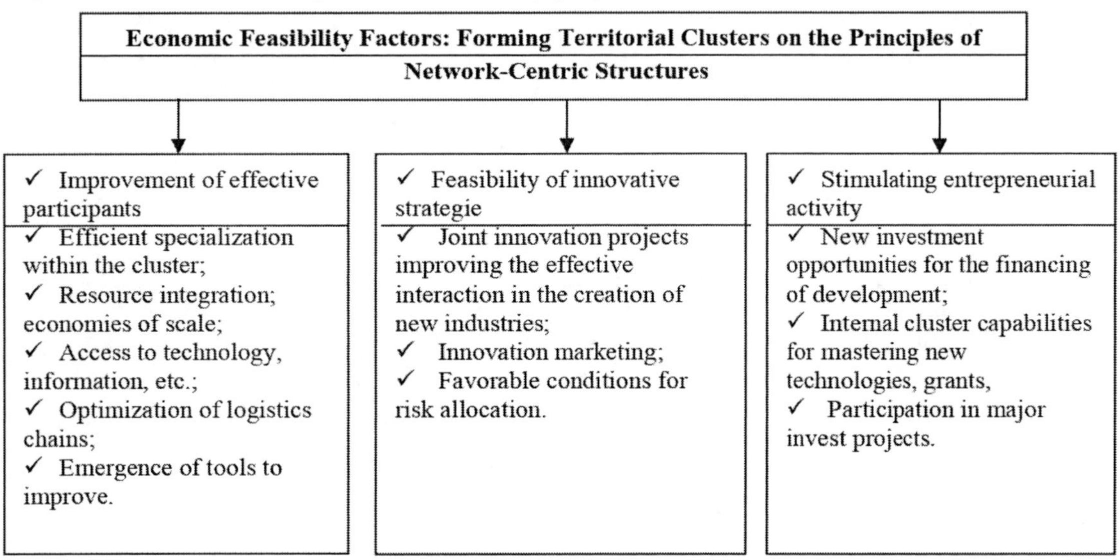

Figure 1. Factors of economic feasibility.

4. Conclusion

Our research clearly shows that the economic feasibility of developing the territorial economic clusters based on the principles of network-centric structures evolves could have other expansion opportunities, adapting to specific situations. It is a systematic interaction in the integrated entities induces additional effect.

References

[1] Shishina M. Ed. 2013 *Altai transboundary: procedures for international integration and sustainable development* (Moscow, Russia: Institute of sustainable development of the public Chamber of the RUSSIAN FEDERATION, Center for Environmental Politics of Russia).

[2] Popkov J V Ed. 2010 The Great Altai: locus of self-organization of the Eurasian world *In Eurasian world: values, constants, self-organization* (pp. 353-384) (Novosibirsk, Russia: Nonparel)

[3] StrategPlann website n.d. *The cluster approach is the basis of improving the competitiveness of the regional economy* Available at: http://www.strategplann.ru/stati-po-ekonomike/klasternyj-podhod-osnova-povyshenija-konkurentosposobnosti-regionalnoj-ekonomiki.html

[4] Cjendsurjen C, and Cjendsurjen L 2008 Genetic structure of populations of ecological types of Mongolian ovecpo loci In *Agrarian science: a collection of articles* (pp. 133-136). Barnaul: Rio AGAU

[5] Agro-industrial portal "AgroXXI" 2013 *Organic meat will be brought from Mongolia* Available at: https://www.agroxxi.ru/zhivotnovodstvo/novosti/yekologicheski-chistoe-mjaso-privezut-iz-mongolii.html

[6] *Ministry of Food, Agriculture and Light Industry of the Republic of Mongolia* (n.d.) Available at: http://www.mofa.gov.mn/

[7] Kundius V, and Kronevald O 2014 Resource potential production of the ecologically pure products in cross-border regions of Altai and Mongolia (based on the results of the veterinary-sanitary examination) *Herald of the Altai State Agrarian University* **11** pp 156-161

[8] Kundius V 2015 The utilization of natural resources of the Altai and Himalayas in the production eko-product beef cattle In *Materials First Russian-Indo-Mongolian Seminar "Altai-Himalayas"* (pp. 110-117) (Barnaul: Altai Fund – 21)

[9] Kundius V, Glotko A, Galkin D, Kashirskih P, and Gartman A 2015 *Potential environmentally friendly livestock production through developing cooperation in cross-border territories of the Greater Altai* (Barnaul, Russia: Altai Fund – 21)

IOP Conf. Series: Earth and Environmental Science **395** (2019) 012095 doi:10.1088/1755-1315/395/1/012095

The use of crisis management in ensuring sustainable food production by organizations of border areas

Yu Gorbunov[1*], A Gorbunova[1] and I Solskaia[2]
[1] Altai State University, 61 Lenin prospect, Barnaul 656049 Russia
[2] Irkutsk State Transport University, 15 Chernyshevskogo str., Irkutsk 664074 Russia

E-mail: barnaul-gorbunov@mail.ru

Abstract. The article substantiates the relevance of the practical application of the concept of sustainable development and production. The peculiarity of doing business in the border areas is considered in details. Factors affecting sustainable food production by organizations in border areas are also analyzed. The role of crisis management in ensuring sustainable production is reviewed.

1. Introduction

The concept of sustainable production emerged within the framework of a broader theory. The theory of sustainable development determines the need for a transition to a qualitatively new level of economy. This is the level of economy that "satisfies the need of the present but does not jeopardize the ability of future generations to meet their own needs" [1]. The theory is developed in response to the challenges posed by the increasing pressure on the environment from the growth of the world's population, the rapid pace of development of environmentally dirty, high-cost industries, leading to a catastrophic depletion of natural resources. Within the framework of the "sustainable development" concept, the models of production and consumption of goods should be qualitatively changed. In particular, the concept of sustainable production arose. On the one hand, it is aimed at saving and restoring nature; on the other hand, it focuses on increasing the volume of material benefits being created and more fully meeting the needs of the population, improving the quality of every person's life. Finding a balance between these contradictory trends is achieved through the intensive use of all types of resources, the widespread use of their renewable types, powerful protection and restoration of the environment. [2] Concepts of sustainable development and production are widely used in research of both domestic and foreign scholars [3, 4, 5, 6].

Ensuring sustainable production is particularly important in the border areas, which are the parts of the territories of the administrative-territorial units of the state, with administrative border coinciding with the state border line [7]. This particular importance is determined by the fact that, on the one hand, sustainable production should provide an economic basis for unconditional compliance with the border regime, the procedure for protecting the state border, compliance with customs, sanitary and other types of control, the procedure for entry and exit of citizens, which is strategically important for the country. On the other hand, the socio-economic development of most border regions is at a lower level than the socio-economic development of Russia's internal regions [8].

Solving the issue of ensuring sustainable production in the border areas, the researchers identify the following development factors: (1) cross-border cooperation, which provides additional access to commodity, financial, and transport flows in the local market; (2) price differences for industrial and

Content from this work may be used under the terms of the Creative Commons Attribution 3.0 licence. Any further distribution of this work must maintain attribution to the author(s) and the title of the work, journal citation and DOI.
Published under licence by IOP Publishing Ltd

consumer goods, which determine the expansion of the sales market for some and the production of cheaper goods and services for others; (3) the ability to import cheap labor and reduce production costs; (4) interstate development programs for cross-border territories [9, 10, 11]. In our opinion, modern scientists do not pay enough attention to crisis management, which is able to largely ensure sustainable production in the border areas. A. T. Zub notes, "Crisis management is a set of organizational measures to prevent the development of the crisis, minimize the damage caused by the crisis and post-crisis restoration of the normal functioning of various socio-economic systems" [12]. In our opinion, it is precisely crisis management that is especially important for ensuring sustainable production in border areas, which economic situation is most often characterized as a pre-crisis or crisis situation.

The purpose of this research work is to study the role of crisis management in ensuring sustainable food production by organizations of border areas. The research object is the economic processes occurring in the Agricultural Production Cooperative "Kolkhoz Rakitovsky" located in the Mikhailovsky district of the Altai region.

2. Materials and Methods

The research methodology is based on a systematic approach to the objects and processes under study. In addition to general economic research methods such as vertical and horizontal financial analysis, synthesis, specific research methods were used. These specific methods are used in crisis management, as provided for by the Government Decree on Approval of the Rules for Conducting Financial Analysis by an Arbitration Manager, Order of the Ministry of Finance of the Russian Federation on Approval of the Procedure for Determining the Value of Net Assets, Order of the Ministry of Industry and Energy of Russia and the Ministry of Economic Development of Russia on Approval of Guidelines for Drawing up a Plan (Program) for Financial Recovery [13, 14, 15].

3. Results

In the course of the study, the economic activity of the agricultural production cooperative "Kolkhoz Rakitovsky" is studied, which is located in the Mikhailovsky district of the Altai region and has a border with the Republic of Kazakhstan. This organization produces food of both plant and animal origin. The production cooperative is one of the leading farms of the region. Consequently, the supply of food for the border area depends largely on the level of its sustainable production.

The average annual production capacity of a production cooperative is:

- Milk production – 1,070 tons;
- Meat production – 90 tons;
- Horse rearing – 35 heads;
- Wheat production – 2,500 tons;
- Sunflower production – 870 tons;
- Oat production – 180 tons;
- Barley production – 190 tons;
- Buckwheat production – 200 tons.

Since 2015, due to unqualified management, unfair performance of their duties as a manager, the activities of the cooperative have been unprofitable (Table 1).

Table 1. Analysis of the financial results of the organization, thousand rubles.

An indicator	2015	2016	2017	2018
Total assets (liabilities)	166,439	157,283	153,700	146,867
Adjusted non-current assets	66,240	63,949	41,835	42,563
Current assets	100,199	93,334	111,865	104,304
Liquid assets	114	863	106	406
The most liquid current assets	82	624	56	56
Receivables	6,308	5,820	3,151	8,774

Own funds	91,912	97,599	83,363	72,380
Debtor's obligations	74,527	59,684	70,337	74,487
Debtor's long-term liabilities	43,793	38,503	29,330	28,630
Debtor's current liabilities	30,734	21,181	41,007	45,857
Net revenue	47,000	12,644	32,997	73,644
Gross revenues	47,000	12,644	32,997	73,644
Average monthly revenue	5,222	4,215	3,666	6,137
Net income (loss)	-1,996	-595	-6,947	-25,731

In 2018, the proceedings commenced on the case of declaring the debtor bankrupt, the monitoring procedure was introduced. The debt on the register of creditors' claims amounted to more than 47 million rubles. When the company was on the verge of closure, food production in the border area was jeopardized. The bankruptcy of the cooperative would lead to the loss of work for its employees, the deterioration of the socio-economic situation in the region.

As part of crisis management, we have developed a financial recovery plan for this organization. A feature of developing such a plan was that the enterprise was located in the border area and was subject to the influence of a mass of specific impacts that are not found in territories remote from the state borders. The implementation of the financial recovery plan should lead to the achievement of the financial indicators presented in Table 2.

Table 2. Projected financial performance of the organization, thousand rubles.

Indicators		March 2019 – February 2020	March 2020 – February 2021
Meat production	Income	8,331	13,523
	Costs	7,896	13,561
Milk production	Income	22,598	22,598
	Costs	12,173	12,173
Rearing horses	Income	1,100	500
	Costs	189	155
Crop production	Income	44,260	145,089
	Costs	20,938	20,938
Rental property (revenue)		152	152
Land lease (expenses)		5,559	6,670
Other expenses		2,209	2,209
Total revenue		76,441	181,862
Total expenses		48,964	55,706
Revenues less expenses		27,477	126,155
Single agricultural tax		1,079	4,041
Net profit		26,398	122,115

The implementation of the financial recovery plan allows to restore food production in the border area, to preserve 132 jobs, to prepare the base for the creation of an additional 12 jobs, to create conditions for payments to the budget and extrabudgetary funds in 2019 and 2020 more than 33.4 million rubles.

4. Discussion

The creditors of the production cooperative, employees of the tax inspectorate, the court, where the bankruptcy case of the organization under investigation is considered, acknowledge the data obtained as a result of the study as reliable. Therefore, we can talk about the objectivity and completeness of the study.

The results of the work confirmed the conclusions made by previous researchers about the special complexity of doing business in the border areas, often leading enterprises to bankruptcy. Considering the issue of preventing the bankruptcy of these organizations, following other researchers, we highlight such factors as the favorable opportunities for cross-border cooperation, the difference in prices for similar goods, and intensive traffic flows through the territory. But at the same time, we believe that the use of favorable factors, the prevention of negative impacts will be much more effective when applying crisis management. It is the crisis management that allows one to comprehensively study the current situation, make a reliable forecast of its development, comprehensively use all the methods of influence on the managed object, using the whole system of favorable factors and obtaining a synergistic effect, and thus prevent the bankruptcy of organizations and ensure sustainable production.

5. Conclusion

The research results clearly show the following:

- High relevance of practical application of concepts of sustainable development and production;
- The presence of objective factors that impede economic activities, including food production, in the border areas;
- The relevance of using the crisis management in ensuring sustainable food production by organizations of border areas.

References

[1] World Commission on Environment and Development 1987 *Report of the World Commission on Environment and Development: Our Common Future. World Commission on Environment and Development* Available at: http://www.un-documents.net/our-common-future (Accessed 03 04 2019)

[2] Donella H Meadows, Dennis L Meadows, Jorgen Randers, and William W 1972 *Behrens III* (New York, NY: Univers Books)

[3] Gorbunov Yu V, and Sokolova O N 2015 *Theory and practice of using university research projects by industrial enterprises in shaping the mechanism for their sustainable development* (Barnaul, Russia: Publishing house of Altai State University)

[4] United Nations Statistics Division Development Data and Outreach Branch 2018 *The Sustainable Development Goals Report* Available at: https://unstats.un.org/sdgs/report/2018 (Accessed 03 04 2019)

[5] Bezgodov A 2016 *Planetary project: from sustainable development to managed harmony* (St. Petersburg, Russia: Peter).

[6] Akayev A A, Korotaev A V, Malinetsky G G, and Malkov S Yu Eds 2017 *Modeling and forecasting global, regional, and national development* (Moscow, Russia: Librokom Book House)

[7] Academician (n.d.) *Border dictionary* Available at: http://border.academic.ru/943/Приграничная_территория (Accessed 03 04 2019)

[8] Government of the Russian Federation 2019 *Spatial Development Strategy of the Russian Federation for the Period up to 2025* Available at: http://static.government.ru/media/files/UVAlqUtT08o60RktoOXl22JjAe7irNxc.pdf (Accessed 03 04 2019)

[9] Kupriyanov S V, Stryabkova E A, and Gerasimova N A 2018 Features of the economic development of border regions *Scientific Vedomosti (Economy. Computer Science Series)* **45**(1) pp 5-14

[10] Shvakov E E, Lobova S V, and Kozhevnikova A A 2018 Conditions and factors of developing the border areas of the Altai region *Fundamental Research* **12(2)** pp 288-292 Available at: http://fundamental-research.ru/ru/article/view?id=42392 (Accessed 03 04 2019)

[11] Damdyn O S 2016 Analysis of the development possibilities of border and cross-border trade and economic relations of the Republic of Tyva *Young Scientist* **1** pp 342-345 Available at: https://moluch.ru/archive/105/24643/ (Accessed 03 04 2019)

[12] Zub A T 2017 Situational planning in crisis management: a recipe for success *Public Administration. Electronic Messenger* **64** Available at: https://cyberleninka.ru/article/n/situatsionnoe-planirovanie-v-antikrizisnom-upravlenii-retsept-uspeha (Accessed 03 04 2019)

[13] Government of the Russian Federation 2003 *Resolution on Approving the Rules for Conducting Financial Analysis by an Arbitration Manager"* (June 25, 2003 No. 367) Available at: https://base.garant.ru/12131539/ (Accessed 03 04 2019)

[14] Ministry of the Russian Federation 2014 *Order on Approval of the Procedure for Determining the Value of Net Assets"* (August 28, 2014 No. 84n) Available at: https://base.garant.ru/70771060/ (Accessed 03 04 2019)

IOP Conf. Series: Earth and Environmental Science **395** (2019) 012095 doi:10.1088/1755-1315/395/1/012095

[15] Ministry of Economic Development of Russia 2007 *Order of the Ministry of Industry and Energy of Russia on Approval of Guidelines for Compiling a Plan (Program) for Financial Recovery"* (April 25, 2007 No. 57/134) Available at: http://economy.gov.ru/minec/documents/doc1180526791125 (Accessed 03 04 2019)

An innovative component of the integration process of the resource potential of the regions of the Siberian Federal District

M Butakova[1*], O Sokolova[1] and O Oskina[1]

[1] Altai State University, 61 Lenin Ave., Barnaul 656049 Russia

E-mail: son-555@mail.ru

Abstract. The article contains the results of a study of the resource and innovation potentials of the regions of the Siberian Federal District as the basis for their integration. The authors develop the idea of the necessity and expediency of using the innovation component in the interaction of regions with different potential.

1. Introduction

In the conditions of increasing competition between organizations, regions and entire states, the search for sources of their development, determining the competitive advantages of subjects, plays an important role. One of such sources is the processes of integration of resource potentials of regions, allowing to provide synergistic effect from interaction. In a considerable part of empirical research, when analyzing domestic and foreign experience, the high significance of integration interactions is noted, and the problems constraining the development of integration interactions are highlighted [1-3]. As of today, a differentiated assessment of the regions, taking into account the resource, innovation potential as the basis of the integration processes is important.

The problem of finding sources of development that can provide competitive advantages to the subjects is of particular relevance for cross-border regions. These regions are threatened with penetration of finished products and capital into their labor markets by neighboring regions of neighboring countries [4].

2. Materials and Methods

The research conducted by domestic and foreign scholars in the field of regional economics, interregional interaction, vertical and horizontal integration, and innovative development formed the methodological basis of the study. In order to identify the development features of the resource and innovation potential of the regional economies of the Siberian Federal District of the Russian Federation (SFD of RF), the authors used the methods of economic and statistical analysis. Official statistics served as the source of primary data used for analysis (Rosstat, 2017). The authors analyzed the possibilities of effective integration on the materials of the Siberian Federal District, consisting of 12 subjects, which are significantly different in size of territories and indicators of socio-economic development. To compare the regions, the authors relied on characteristics that reflect their resource and innovative potential. The authors believe that the analysis of the selected characteristics will make

Content from this work may be used under the terms of the Creative Commons Attribution 3.0 licence. Any further distribution of this work must maintain attribution to the author(s) and the title of the work, journal citation and DOI.
Published under licence by IOP Publishing Ltd

it possible to substantiate possible approaches to the integration of the resource potential of the regions, taking into account the innovation component.

3. Results

In the course of the study, the authors analyzed the regional structure of the resource potential of the Siberian Federal District. Table 1 presents the subjects of the Siberian Federal District of the Russian Federation are presented in descending order of their share in the structure of the main characteristics of the resource potential.

Table 1. Regional structure of the resource potential of the Siberian Federal District, 2017, in %.

Subjects	Mining, million rubles	Electricity production, billion kWh	Production of agricultural products, million rubles
Kemerovo region	38.5	11.6	8.3
Krasnoyarsk region	23.7	31.7	12.5
Irkutsk region	19.5	23.0	10.7
Tomsk region	7.7	1.8	5.0
Altai region	0.3	3.5	21.8
Novosibirsk region	2.2	6.5	14.1
Omsk region	0.1	3.3	15.7
The Republic of Khakassia	2.3	12.1	2.6
Transbaikal region	3.5	3.4	3.9
The Republic of Buryatia	1.1	3.0	2.6
Tyva Republic	1.0	0.0	1.0
Altai Republic	0.2	0.0	1.9

Source: Calculated by the authors according to the Federal State Statistics Service [11].

As the main characteristics of the resource potential, such areas as mining, electricity production, and agricultural production were studied. This made it possible to assess the level of provision of the regions with raw materials, energy and food resources. In our opinion, it is this approach, without calculating the integral coefficient of the resource potential, that shows the diverse aspects of the regional structure of the resource potential of the SFD. In 2017, the Kemerovo and the Krasnoyarsk Regions played a dominant role in the extraction of minerals; the Krasnoyarsk and the Irkutsk regions led the production of electricity; the Altai region was the leader in the production of agricultural products.

Further, the authors analyzed the regional structure of the innovation potential of the Siberian Federal District. Table 2 presents the subjects of the SFD RF in their decreasing order according to the share in the structure of the main characteristics of the innovation potential.

As the main characteristics of the innovation potential, the following indicators were analyzed: a number of students enrolled in undergraduate, specialist, graduate programs; a number of organizations performing research and development; internal research and development costs. In our opinion, the selected indicators, not exhausting the whole range of characteristics of the innovation potential, reflect its main components. In 2017, the Novosibirsk region was ranked first in the share in the structure of all analyzed characteristics of the innovative potential of the SFD. The Krasnoyarsk and Tomsk regions immediately follow the leader, with a high value of shares in individual characteristics of the innovative potential of the Omsk and Irkutsk Regions.

The development prospects of the Siberian Federal District are largely determined by the development strategies of its constituent entities. The authors showed and analyzed previously their main provisions [5]. The key drafters of the development strategies of the subjects are unanimous in the opinion that the future of the regions is impossible without the innovation component. The Novosibirsk, Tomsk and Krasnoyarsk regions can become the driver-forming regions in terms of the use of innovative potential.

The analysis made allows to conclude that it is necessary and expedient to use the innovation component in the interaction of regions with different resource and innovation potential. The integration of the economic entities of the mining and processing industries, agriculture, science and education of the regions of the SFD creates additional impulses for their progressive development.

Table 2. Regional structure of the innovative potential of the SFD, 2017, %.

Subjects	Number of students enrolled in undergraduate, specialist, graduate programs, thousand people	Research and development organizations	Domestic costs of research and development, million rubles
Novosibirsk region	18.2	25.8	31.9
Krasnoyarsk region	13.9	14.7	23.8
Tomsk region	10.7	13.4	20.8
Omsk region	14.5	8.5	8.9
Irkutsk region	12.6	9.8	6.2
Altai region	9.2	7.2	2.6
Kemerovo region	9.3	6.6	3.3
The Republic of Buryatia	4.5	4.7	1.3
Transbaikal region	4.2	3.4	0.6
The Republic of Khakassia	1.5	3.6	0.1
Tyva Republic	0.8	2.1	0.4
Altai Republic	0.5	2.1	0.1

Source: Calculated by the authors according to the Federal State Statistics Service [11].

4. Discussion

In the contemporary scholarly literature, a number of researchers pay serious attention to the issues of ensuring the efficient use of the territories' resource potentials. N. D. Rodionova [6], A. G. Polyakova, I. S. Simarova [7] studied the spatial connectivity of regions and the forms of their interaction. From the foreign researchers focused on the forms of realization of inter-regional economic interactions, the works of E. M. Bergman, E. J. Feser [8, 9] are of particular interest. S. S. Gubanov [1], S. A. Kozhevnikov [2], Yu. A. Volkova, N. P. Dragun [10] investigated the essence and types of integration. These authors note the existence of different approaches to explaining the nature and essence of vertical integration in neoclassical and neoinstitutional economic theory, draw attention to the emergence of a broader interpretation of integration and related new forms of integration interactions.

Appeals to the subject and content of scientific discussion in the theory of spatial development on the problems of interregional interaction allow to justify the author's position on the need and desirability of regional integration. Comparing the "soft" and "hard" forms of integration, we consider the "soft" forms of regional integration to be the most acceptable. They are based solely on the contractual relationship between its members (network organizations, clusters, associations, consortia). At the same time, they do not deny the feasibility of forming, in some cases, technological value chains using "hard" forms of vertical integration.

The authors tested a methodical approach to assessing resource and innovation potentials as a condition for the integration of regions on the materials of the Siberian Federal District. The results of the approbation give grounds to consider the integration of the resource and innovation potential of the regions expedient in order to increase the competitiveness of the subjects of the SFD and Russia as a whole.

5. Conclusion

The authors reviewed the theoretical aspects of the integration processes of the regions and analyzed the level of resource and innovation potentials of the SFD subjects. Based on the foregoing, the

IOP Conf. Series: Earth and Environmental Science **395** (2019) 012096 doi:10.1088/1755-1315/395/1/012096

authors show the capabilities of the studied regions to integration interactions to improve the efficiency of the using the regional resources and ensuring their competitiveness and the competitiveness of the national economy.

Author's research confirms the hypothesis of the expediency of integrating regions with different species and level potentials for the development of their territories.

6. Acknowledgments

The study was carried out with the financial support of the Russian Foundation for Basic Research in the framework of a research project № 18-010-00135 "Conditions for Expansion and Assessment of the Level of Regional Integration in Order to Ensure Economic Security (on materials of the Siberian Federal District)".

References

[1] Gubanov S 2009 Towards a policy of neo-industrialization of Russia *The Economist* **9** pp 3-20
[2] Kozhevnikov S A 2016 Formation of technological value chains in the form of vertical integration *Territorial Development Issues* **3**(33) pp 1-14
[3] Butakova M M, and Sokolova O N 2018 Interregional economic cooperation: a systematic approach *Economy. Business. Banks* **2**(23) pp 10-21
[4] Novoselov A S 2014 *Regional and municipal management of socio-economic development in the Siberian Federal District* (Novosibirsk, Russia: IEEPP SB RAS)
[5] Butakova M M, Sokolova O N, Zaitseva N A, Larionova A A, Povorina E V, and Kushch E N 2018 Assessment of the background for interregional economic cooperation *Espacios* **39**(22) pp 1-13
[6] Rodionova N D 2015 The economic space of the region in the network interaction of the subjects of the regional innovation system *Vestnik of the Rostov State Economic University (RINKH)* **1**(49) pp 131-138
[7] Polyakova A G, and Simarova I S 2014 The conceptual model of regional development management, taking into account the level of spatial connectivity *Economy of the Region* **2** pp 32-41
[8] Bergman E M, and Feser E J 1999 *Industrial and Regional Cluster: Concepts and Comparative* Available at: http://www.rri.wvue du/WebBook/Bergman-Feser/contents.htm
[9] Feser E J 1998 Old and new theories of industry clusters *Clusters and Regional Specialisation: On Geography, Technology, and Networks* **8** pp 18-40
[10] Volkova Yu, and Dragun N 2016 Integration of enterprises in the market: the problem of choosing its shape and type *Vestnik of GSTU named after P. O. Sukhoi* **3** pp 101-110
[11] Federa State Statistics Service (2019) *Data on science and innovation* Available at: http://www.gks.ru/wps/wcm/connect/rosstat_main/rosstat/ru/statistics/science_and_innovations/science/ (Accessed 27 03 2019).

IOP Conf. Series: Earth and Environmental Science **395** (2019) 012097 doi:10.1088/1755-1315/395/1/012097

Regional food security: a concept and an assessment

M Semikolenova[1*], T Rudakova[1,2] and N Shavkunova[3]

[1] Altai State University, 61 Lenin str., Barnaul 656099 Russia
[2] Polzunov Altai State Technical University, 46 Lenin str., Barnaul 656038 Russia
[3] Novosibirsk State University of Economics and Management, 56 Kamenskaya str., Novosibirsk 630099 Russia

E-mail: semikmn@mail.ru

Abstract. The article analyzes the approaches to the definition, criteria, and threshold values of regional food security. Based on the analysis of indicators of food security of the Altai region, the main directions of the regional food policy are justified. The research shows that regional food security determines the level of security and sustainability of both the region as a whole and individual households. The main criteria for assessing food security are food independence, physical and economic accessibility of food in the region

1. Introduction

In the conditions of the modern dynamically developing world, the problem of providing food products is quite acute. According to the Western experts, food production should increase substantially by the middle of the century in order to feed a more populous and rich world [1]. A similar problem occurred in previous years, when food supply has more than tripled since the 1960s. However, given the expected adverse effects of climate change and the degradation of natural resources, concerns about the possibility of maintaining this trend exist [2].

The development of food safety strategies is the subject of scientific research and research by many scientists and government organizations. The Food and Agriculture Organization of the United Nations (FAO) considers food security as a guarantee of the constant access of the population of the planet, country, region to food in the amounts necessary to maintain an active and healthy lifestyle [3].

2. Materials and Methods

Thus, in international practice, three levels of food security are recognized: global, national, and regional. An assessment of global food security is alarming. For the third year in a row, world hunger has been rising. The absolute number of undernourished people, those facing chronic food shortages, increased to almost 821 million in 2017 from about 804 million in 2016. According to FAO [3], the food security situation has deteriorated from 45% of the population to 50% in 2018 compared with 2017: about 2.1 million people lack food; more than 320 thousand people have acute food shortages.

The problem of ensuring national food security is very relevant in the Russian Federation. Of the estimated 113 countries in the world, Russia ranks only 42nd in the global food security index. The strategy for ensuring national food security is developed by the Doctrine of Food Security of the Russian Federation [4] and consists in ensuring the population of the country with safe agricultural, fishery products, and food to the extent necessary for an active and healthy life.

Content from this work may be used under the terms of the Creative Commons Attribution 3.0 licence. Any further distribution of this work must maintain attribution to the author(s) and the title of the work, journal citation and DOI.
Published under licence by IOP Publishing Ltd

3. Results

This study is devoted to the development of approaches to the assessment of regional food security on the example of the Altai region. The effectiveness of managing and addressing food safety issues can be studied using a comprehensive index of regional food security. This indicator is based on the ratio between the levels of the gross regional product, the volume of agricultural production, the consumer price index for food, and the price index of agricultural producers.

Table 1. Assessment of the level of food security of the Altai region.

Indicators	2013	2014	2015	2016
Consumer Price Index For Food Products	107.4	115.4	112.4	104.3
Volume Index of Gross Regional Product	104.9	100.4	100.5	99.0
Agricultural Production Index	122.7	88.7	107.1	112.5
Producer Price Index of Agricultural Products	102.7	114.1	108.5	101.8
Food Security Level	0.852	0.866	0.777	0.812

The data in Table 1 indicate a decrease in the level of food security in the Altai region to 0.812 in 2016. This is mainly due to the outpacing rate of prices compared to the change in the physical volume of the gross regional product.

At the next stage, we evaluate the indicators of the socio-economic development of the region that could affect the level of food security in the Altai region.

Table 2. Population dynamics of the Altai Region in 2010-2016, thousand people.

Year	Total population, thousand people	Population of rural areas, thousand people	Growth rate, %	
			Total number	Rural population
2010	2,430.8	1,110.7	100	100
2011	2,417.4	1,093.4	99.4	98.44
2012	2,407.2	1,079.6	99.6	98.74
2013	2,398.7	1,067.7	99.6	98.89
2014	2,390.6	1,057.5	99.7	99.04
2015	2,384.8	1,049.2	99.8	99.21
2016	2,371.2	1,041.5	99.4	99.27

Over the entire period under review, a clear downward trend in the population has been observed. So, 2 371.2 thousand people inhabited the Altai region, in 2016, which was 59.6 thousand people (2.5%) less than in 2010. Together with the total population, we must consider the dynamics of the rural population of the Altai region (Table 2), since this indicator has a strictly negative direction. That is, the number of rural residents is falling on a federal scale. As a result, agricultural enterprises are declining.

This situation indicates a decrease in prospects for citizens in the rural areas. Since there is the decrease in the number of the rural population in the period under review by 6.23%, the decrease in the availability of social infrastructure occurs at a higher rate. The number of hospitals has decreased by 97.1 % since 2000, while the number schools has lowered by 42.6%.

Simultaneously with the reduction in the total number of people inhabiting the Altai region, the economically active population also decreases. 1,059.6 thousand people were employed in the economy in 2016, which was 86.1 thousand people (7.5%) less than 2010 [5].

For a complete picture of the standard of living in the Altai region, we need to pay attention to the indicators given in Table 3.

Table 3. Indicators characterizing the living standard of the population of the Altai region in 2010-2016.

Years	Per capita cash income of the population		Real money incomes of the population, in % to the previous year	The subsistence minimum (average per capita)		Population with money incomes below the subsistence minimum, in % of the total
	Rub. per month	Growth rate, %		Rub. per month	Growth rate, %	
2012	13,617.2	108.94	103.6	5,147	107.9	20.6
2013	15,979.1	117.34	109.0	5,672	110.2	17.6
2014	18,434	115.36	106.4	5,877	103.61	17.0
2015	20,985	113.84	98.9	6,647	113.1	17.9
2016	21,485	102.38	95.2	8,363	125.81	17.6

The data in Table 3 confirm the main trend prevailing in the economy of the Altai Territory. The growth rate of average per capita incomes of the population does not change much, and so does the minimum subsistence level. We should note the positive dynamics of indicators associated with a decrease in the proportion of the population with incomes below the subsistence minimum. However, it is also worth paying attention to the fact that along with a gradual increase in per capita incomes, the indicator of real incomes of the population has a negative tendency.

Therefore, taking into account the aforementioned trends in 2017 than in 2014, in the agricultural products market, residents of the Altai region will be able to afford much less products or lower quality products.

Table 4. Analysis of the availability of food for the population of the Altai region.

Product Categories	Food consumption per capita per year, kg						
	WHO standards	RF norms	2012	2013	2014	2015	2016
Meat and meat products	78	73	73	74	73	69	70
Milk and dairy products	405	325	335	335	335	330	326
Potatoes	117	90	145	145	145	142	145
Vegetables and melons food	139	140	103	102	102	101	103
Sugar	47	24	38	39	39	37	37
Vegetable oil	9.1	12	13.5	13.8	14.3	15.0	14.5
Eggs	291	260	335	335	336	338	340
Bread products	117	96	167	166	165	160	158

The data in Table 4 confirm the previously made conclusions. For the population of the Altai region, protein and the most expensive products (meat products), vegetables are less affordable. The consumption of the listed food products is below the rational consumption norms established by the order of the Ministry of Health of the Russian Federation dated August 19, 2016 No. 614 and the consumption standards recommended by the World Health Organization (WHO). Thus, in terms of the consumption of meat and meat products, the region occupies only 45th place among the subjects of the Russian Federation.

The level of food independence in the region is an important characteristic. The Altai region belongs to the agro-oriented regions. The studied region ranked 8th in terms of the volume of agricultural production among all the subjects of the Russian Federation in 2016. 57% of the total agricultural products accounted for crop production. The rest is produced by livestock organizations.

The agro-industrial complex of the Altai region in all the main categories of food products with a significant supply meets the needs of the population. The Altai region is a surplus region in terms of food security. The region produces a significant share of industrial products in Russia and the Siberian Federal District. Altai enterprises occupy the 1st place in the country for the production of flour, cheese and cheese products, dry whey, butter, breakfast cereal, dietary supplements to food, according to the results of 2016. In the production of cereals, the Altai region is in 2nd place in Russia (in the production of buckwheat and oatmeal holds the 1st place), is in 3rd place pasta. It is the only region of Siberia and the Far East that produces sugar from sugar beet.

The index of industrial production in agriculture was 112.5% in total and 105.3% for food products in particular by the end of 2016 to the level of 2015. The regional growth rate of production of agricultural and food products exceeded the Russian level by more than 2 times.

4. Discussion

The issue of regional food security is not so straightforward, and it is even considered controversial by some authors. Although at present, many economists do agree that the emphasis on regional food security is necessary [6]. And this is due to fundamental differences in the territorial situation, the climatic situation, and the economic potential of the regions of the Russian Federation. From our point of view, the allocation of food security at the regional level is fully justified and advisable to ensure food independence, accessibility and food safety in certain administrative-territorial associations. We believe that it is at the regional level that the conditions and prerequisites for ensuring national food security are laid.

In the economic literature, there are different approaches to the definition of food security in the region [7, 8]. We understand regional food security as a level of organization and economic development of a separate regional space that ensures the physical and economic accessibility of food products necessary for an active and healthy lifestyle of the population of the region.

The concept of regional security is complex from different perspectives. From the point of view of the organizational approach, the assessment of food security should take into account the need to consolidate the efforts of various sectoral, organizational, and legal entities in order to meet the food needs of the population of the region.

On the other hand, it is necessary to understand that the main guarantee of an acceptable level of regional food security lies in the effective development of the key industry, which is agriculture. Covering food needs with own production of the agricultural and food industry ensures food independence of the region, including from external factors.

Socio-economic development of the region, the increase in incomes of the population, providing the possibility of acquiring food necessary for life, have a decisive influence on the indicator of the economic availability of food products.

Thus, we can relate to the components of regional food security: food self-sufficiency; food availability for the population; effective management of agricultural and processing industries; compliance with environmental and quality standards of food produced.

5. Conclusion

Regional food security determines the level of security and sustainability of both the region as a whole and individual households. The main criteria for assessing food security are food independence, physical and economic accessibility of food in the region. Despite the fact that these indicators and their criteria are developed at the national level, they require rethinking from the point of view of a particular region.

The main directions of the food policy of the Altai region is the balance of production and processing of agro-industrial complex, the introduction of innovative agricultural technologies and the modernization of agricultural equipment, improving the standard of living of the population of the region.

References

[1] Fouré J, Bénassy-Quéré A, and Fontagné L 2013 Modelling the world economy at the 2050 horizon *Economics of Transition* **4** pp 617-54

[2] Godfray H C J, Beddington J R, Crute I R, Haddad L, Lawrence D, Muir J F, Pretty J, Robinson S, Thomas S M, and Toulmin C 2010 Food security: the challenge of feeding 9 billion people *Science* **327** pp 812-8

[3] *Food and Agriculture Organization* Available at: http://www.fao.org/about/en/ (Accessed 02 04 2019)
[4] *Doctrine of Food Security of the Russian Federation* Available at: http://ab-centre.ru/news/doktrina-prodovolstvennoy-bezopasnosti-rossiyskoy-federacii (Accessed 12 03 2019)
[5] *Federal State Statistics Service* Available at: http://www.gks.ru/wps/wcm/connect/rosstat_main/rosstat/ru/statistics/publications/catalog/doc_1138623506156 (Accessed 20 03 2019)
[6] Bondareva G S 2015 Food security of the region: criteria and threshold values *Scientific Almanac* **9** pp 90-100
[7] Dudin M N, and Lyasnikov N V 2015 Food security of regions in the national and economic security of the state *Regional Economic: Theory and Practice* **6** pp 2-11
[8] Smirnov V V 2015 Food security of the Russian Federation regions *National interests: priorities and security* **3** pp 29-42

Development of the region's pig production market based on the cluster approach

A Mamaeva[1*] and G Vinokurov[1]

[1] Irkutsk State Agrarian University named after A.A. Ezhevsky, 1/1 Molodezhny pereulok, Molodezhny settlement, Irkutsk region 664038 Russia

E-mail: tera2805@mail.ru

Abstract. The pork production market is part of the food market, and increasing its competitiveness is a priority both in managing the agro-industrial complex in general and in the region in particular. Cluster interaction in the regional market of pig production is an important direction of its development, ensuring a full exchange of goods of appropriate quality between producers and consumers of pork production. In the proposed study, based on the analysis of statistical information, as well as surveys of consumers of pig production in the Irkutsk region (conducted in 2016), recommendations aimed at developing a regional production base for pig breeding were proposed. The development of a regional production base is proposed by expanding the existing industries, as well as by improving the breeding base of pig production in the region, creating a new model of cluster interaction between producers and consumers in the market under study. The formation of a regional pig cluster will increase the production of pork, open new channels for the sale of manufactured products, as well as start exporting to other regions.

1. Introduction

The problem of interaction in the regional market of pig production between producers and consumers is of particular relevance in modern conditions. First, pig production is one of the main products on the food market in the meat product segment, it is used by processors to make various types of goods. Secondly, more than 60% of pig production is made by industrial complexes, and the rest of the production is distributed between peasant (farmer) farms and households. Third, the existing steady demand for these products stimulates the emergence of a large number of intermediaries in the market and causes an increase in product prices. The market for pork products is an integral part of the agro-industrial complex of both the region and the country as a whole. Solving these problems is part of the regional AIC policy. As one of the objectives of this policy is the formation of an effective competitive agri-food production, contributing to food security, as well as integration into the global agri-food market [1].

In the Irkutsk region, the manufacturers of pig products are forced to act independently in the search for suppliers of equipment, feed, breeding animals, as well as channels for the sale of manufactured products. The reduction of intermediary lines with the appropriate construction of a pig-breeding cluster in the Irkutsk region will lead to a gradual increase in production volumes and self-sufficiency in the region. In addition, the creation of territorial production clusters is an innovative direction for the development of the region [13, 2].

Content from this work may be used under the terms of the Creative Commons Attribution 3.0 licence. Any further distribution of this work must maintain attribution to the author(s) and the title of the work, journal citation and DOI.
Published under licence by IOP Publishing Ltd

IOP Conf. Series: Earth and Environmental Science **395** (2019) 012098 doi:10.1088/1755-1315/395/1/012098

The purpose of the study is to develop recommendations for the formation of a pig breeding cluster in the Irkutsk region, based on the analysis of statistical data and the study of consumer demand for pig production.

2. Materials and Methods

We conducted a study of statistical information for the entire territory of the Irkutsk region. A comparative analysis of data with indicators of the Russian Federation, as well as a study of more than 1,500 consumers of pig production throughout the region were conducted. Consumer research was conducted from spring to autumn 2016. From 2013 to 2017, the collection and analysis of statistical information were carried out. As statistical information, the data of the Federal State Statistics Service of the Russian Federation, the Territorial Body of the Federal State Statistics Service for the Irkutsk Region, as well as the Ministry of Agriculture of the Irkutsk Region and agricultural producers for 2005-2016 were used.

Data processing was carried out using the Microsoft Excel software using the "Data Analysis" package. Indicators of correlation between consumption of pork in the Irkutsk region and its production in the region, price indices for pork and other types of meat, as well as imports of pork from other regions were determined. The probability of increasing the level of consumption of pork, as well as its production in the region and the import of pork from other regions were evaluated.

3. Results

A study of statistical information in the field of pig production in the Irkutsk region showed fluctuations in the quantity of products produced, as well as in the structure of its production. The main producers were households as of 2005. The agricultural producers using industrial production technologies were the main producers by 2016. In addition, over the past three years, production in the whole region is decreasing.

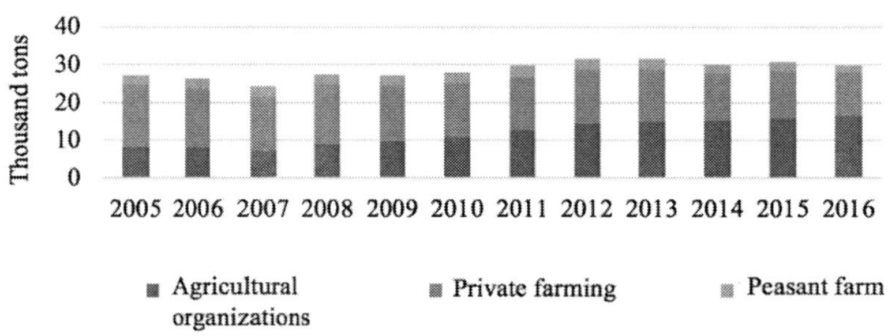

Figure 1. Production of pork for slaughter in slaughter weight by categories of farms in the Irkutsk region for 2005-2016 [5].

Thanks to a comparative analysis of data on the level of consumption of pork in the Irkutsk region and in the whole of the Russian Federation, as can be seen from Table 1, we determined the level of self-sufficiency of the population in the region of pork, which does not exceed 55%. This indicates a high level of import of pig products to the region. In addition, the level of meat consumption per person both in Russia and in the region is lower than the biological consumption rate, which is 75 kg per year [6, 4, 8].

Table 1. Consumption of pork per capita and the level of self-sufficiency of the population of the Irkutsk region pork for the period 2005-2016 [5].

Years	Meat consumption Russia per 1 person per year, kg	Including pork, kg	Consumption of meat in the Irkutsk region per 1 person per year, kg	Including pork, kg	Production of pork in the Irkutsk region per 1 person per year, kg	The region's self-sufficiency in pork,%
2005	55	15	53	16	10	61.8
2010	69	21	62	21	11	54.6
2015	74	24	71	24	13	53.4
2016	74	25	70	24	14	51.6
2016 to 2005, %	134.5	166.7	132.1	150.0	140.0	93.5

During the study period, the pork consumption in the region increased by 1.5 times; at the same time, the growth of pork production over the same period was only 40%. Differences in growth rates of pork production and consumption lead to a decrease in the region's self-sufficiency. In accordance with the Food Security Doctrine, the level of self-sufficiency in meat should be at 85%. As of 2016, the self-sufficiency of the Irkutsk region in pig meat is below the standard by 33.4% [9, 12].

The correlation analysis of the obtained data allowed to identify the degree of influence on the consumption of pig production in the Irkutsk region of such factors as the population of the region, the average per capita income of the population, the production of pork, its import from other regions, as well as the index of consumer prices for pork. Additionally, we conducted a regression analysis of the data obtained to build a model for predicting pork consumption in the region, depending on the identified factors. A two-factor model for forecasting consumption depending on pork production and import from other regions, as well as a one-factor model for forecasting demand depending on the average per capita incomes of the region's population, were built.

According to the results of forecasting, the production of pork in the Irkutsk region should be 25.5 ± 1 thousand tons by 2019. The forecast value of the regional production of pork is in the range from 25.1 to 25.9 thousand tons. Import of pork from other regions to the Irkutsk region is projected at 27.4 thousand tons.

Based on the analysis, the problems of its functioning were identified. The problems are in reducing the production of pig products with the existing unsecured demand for these products, as well as poor development of the infrastructure of the regional market. It includes the interaction between all market participants from production to sale of finished pig products.

4. Discussion

To solve the identified problems, improving the infrastructure of the market under investigation is necessary. Thus, in order for the manufacturers to receive information about various kinds of support in a timely manner and to use it, they could easily purchase the necessary equipment, feed and biomaterial, they would have simplified methods for entering the market, both internal and external. At the same time, the buyers of final products in this system would have the opportunity to purchase high-quality products without a huge mark-up, working directly with the manufacturers. Creating a pig cluster will solve the identified problems.

The producers of primary pig production – agricultural organizations, peasant farms, and households – will represent the core of the pig breeding cluster. In the inner part of the cluster, the secondary pig producers will enter. From the cluster core, the manufactured products that meet all the necessary requirements will be recycled within the framework of the agreements concluded. Research centers and breeding organizations will provide advice on animal breeding, implement highly productive livestock to increase productivity. Financial organizations such as banks, insurance companies, leasing companies, investment and guarantee funds will provide financial support to pig producers and research

IOP Conf. Series: Earth and Environmental Science **395** (2019) 012098 doi:10.1088/1755-1315/395/1/012098

centers. Financial support will be provided through the provision of loans on concessional terms, insurance programs with an expanded package of risks and reduced tariffs, which are not included in the state support programs for minor remarks, as well as investing in the development of innovative production [3, 7].

One of the innovations for the entire cluster system of the region's pig production will be the creation of a network of transport and logistics hubs and centers, storage and distribution centers. It will cover all territorial transport hubs and will increase the availability of products with the possibility of delivery to other regions and countries. Sales services will help to promote the products, providing an opportunity for small and medium producers to enter the market.

The non-governmental organizations, such as the Union of Industrialists and Entrepreneurs of the Irkutsk Region, the Agro-Industrial Union of the Irkutsk Region, and the Public Union of Consumers will act as coordinators of the general activities of the cluster. The non-governmental organizations will coordinate together with the Ministry of Agriculture of the Irkutsk Region (represents the interests of the state). For the purposes of communication interaction between all cluster members and organization of the smooth operation of all cluster systems, the development of information and communication services is possible on the basis of common Internet platforms with the subsequent formation of a system with blockchain elements. Information and communication service will allow applying new marketing technologies to find customers and suppliers, quickly exchange the necessary information, technologies and products between cluster members with the possibility of reducing costs.

The outer contour of the cluster will also combine regional and municipal authorities with an internal system. These include profile ministries, regional and municipal administrations, departments and various services (tax service, customs, veterinary, sanitary and epidemiological, etc.). In the framework of cooperation, manufacturers will be able to receive the necessary services from public services in a simplified mode. This circumstance will reduce the time of movement of products to final consumers. In addition, the pig products produced will have all the necessary documents confirming its quality and safety for entering the foreign market [10].

The cluster will be fully formed after it includes educational and personnel structures, suppliers of production equipment, feed, veterinary drugs, packaging, biological material. Interaction with educational and personnel organizations will allow forming effective labor collectives in all organizations-members of the cluster [11].

5. Conclusion

Creating a pig cluster of the Irkutsk region is subject to one of the main tasks. The main task is the effective production of pig products to provide the residents of the region with quality products, as well as the development of the market for pig products in general in order to fulfill the tasks set by the state to increase exports and increase the income of organizations, the region and the country. These measures will improve the competitiveness of products in the food market of the region.

References

[1] Altukhov A I 2011 Strategy of agro-industrial complex development as the main condition of the national agri-food policy implementation *J. Economy of Region* **4** pp 35-44

[2] Anishchenko A N 2017 Current problems and perspectives for the development of pig farming *J. Problems of Territory's Development* **4**(90) pp 146-160

[3] Bashmachnikov V F, Drokin, V V, and Zhuravlev A S 2018 Involvement of rural housholds in solving the problems of import substitution *J. Economy of Region* **14**(2) pp 663-675

[4] Borodin K 2018 Economic Access to Food: Factors and Methods of Assessment *J. HSE Economic Journal* **4** pp 563-582

[5] *Federal State Statistics Service of the Russian Federation* Available at: http://cbsd.gks.ru (Accessed 20 12 2018)

[6] Kamilova Z, and Udin E 2014 Functioning of the market of meat and meat products of Russia in the conditions of the WTO *J. Economics of Agriculture of Russia* **3** pp 36-43

[7] Kundius V A 2011 Cluster Approach to realization of Innovation Development Strategy for the agro-

industrial complex of the Region *Economy of Region* **4** pp 117-133

[8] Maydanevich P N 2017 Priority Directions of state regulation of the organizational and economic mechanism of pig-breeding production *Science and Education: Economy, Enterprisem Law and Management* **6**(85) pp 14-22

[9] Mokhnachev S A, Shumkova Y G, and Shumkov N V 2015 Prospects of the pig industry in the region *Fundamental Research* **11** pp 595-599

[10] Mukhin N Y 2016 Import substitution processes in Russian's agro-industrial complex *Russian Forein Economic Journal* **6** pp 106-117

[11] Necheukhina N S, and Mustafina O V 2018 Competitiveness of various segments of the regional consumer market *Economy of Region* **3** pp 836-850

[12] Shevkunova E S 2014 Analysis of Food Products Consumption Level *Scientific Journal of KubSAU* **101**(07) pp 1-16

[13] Valiev Sh Z, and Isaeva N V 2018 Biotechnological clusters as a tool for innovative territorial Development of the republic of Bashkortostan *Bulletin USPTU. Science, education, economy* (Series economy), **1**(23) pp 28-36

Prerequisites for implementing the principles and tools of "green" economy in the Altai Republic

L Ischuk[1], T Kuttubaeva[1*] and O Shvakova[1]

[1] Gorno-Altai State University, 1 Lenkina str., Gorno-Altaisk 649000 Russia

E-mail: tosk-k@yandex.ru

Abstract. The study aims to identify the prerequisites and features of the implementation of the principles and tools of a "green" economy in the Altai Republic, one of the constituent entities of the Russian Federation. The paper suggests a number of approaches to the formation of the main stages in the implementation of the principles and tools of a "green" economy in a region. The authors also systematize the key prerequisites necessary for introducing the principles and tools of a "green" economy in the Altai Republic. The study presented in this paper will contribute to the development of priority "green" sectors in the economy of the Altai Republic.

1. Introduction

In contemporary conditions, the development of the economy and social sphere in many countries and their regions is characterized by such problems as environmental pollution; growing anthropogenic load on ecosystems; irrational resource consumption; inefficient recycling systems and consumption wastes; population poverty, etc. The need to solve these problems determines the search for new models of economic growth, the important features of which should be the effective use of natural resources; the preservation and increase of natural capital; reduction of environmental pollution; low carbon emissions; preventing loss of ecosystem services and increasing incomes and employment. As of today, one of these economic growth models is the "green" economy model. The purpose of this model is the socio-economic development of the region through the effective use of natural resources, ensuring the reproduction of resources and the preservation of ecological systems. However, the effectiveness of the implementation of the principles and tools of the "green" economy in individual regions depends on its successful adaptation to the specific conditions of the region's development and the consideration of differences in natural, human, and financial resources.

The aim of the study is to determine the prerequisites and features of the implementation of the principles and tools of "green" economy at the regional level. The object of the research is the process of introducing the model of "green" economy in the Altai Republic.

2. Materials and Methods

In the course of the study, the authors used such general scientific methods as system analysis, descriptive, comparative research methods. The paper is based on the works of domestic and foreign authors on the problems of building a model of "green" economy. More than that, we relied on materials of international, Russian, and regional conferences, as well as legislative and regulatory acts of the Russian Federation and the Altai Republic.

Content from this work may be used under the terms of the Creative Commons Attribution 3.0 licence. Any further distribution of this work must maintain attribution to the author(s) and the title of the work, journal citation and DOI.
Published under licence by IOP Publishing Ltd

3. Results

The study of the theoretical foundations of the problem of introducing the "green" economy model in individual regions clearly demonstrates that, at present, there are many definitions of this concept. But most authors are inclined to the point of view shared by the UN experts in the field of environmental protection. According to this view, a "green" economy is "an economy that provides long-term improvements in people's well-being and reduced inequality without risks to the environment, or an economy that promotes social progress, sustainable consumption and production" [1].

Therefore, the following principles can be attributed to the basic principles of a "green" economy:

- Ensuring equality, fairness and objectivity in relations between countries, regions, within one generation and between generations;
- Ensuring interconnection and consistency of the principles of "green" economy with the principles of sustainable development;
- Improving the quality of life of the population: reducing poverty, increasing the level of incomes of the population, ensuring equal access of the population to universal benefits;
- Ensuring sustainable and efficient use of resources, rational consumption and resource-efficient production;
- Providing employment, including through the creation of "green" jobs;
- Providing a comprehensive assessment of natural and social capital (internalization of externalities, the organization of "green" accounting and auditing, etc.);
- Searching for funds to achieve the goals of a "green" economy (attracting "green" investments, selling "green" bonds, developing a system of state financial support, etc.) [2].

In our opinion, the process of introducing the principles and tools of a "green" economy in the region should be carried out in stages (Figure 1).

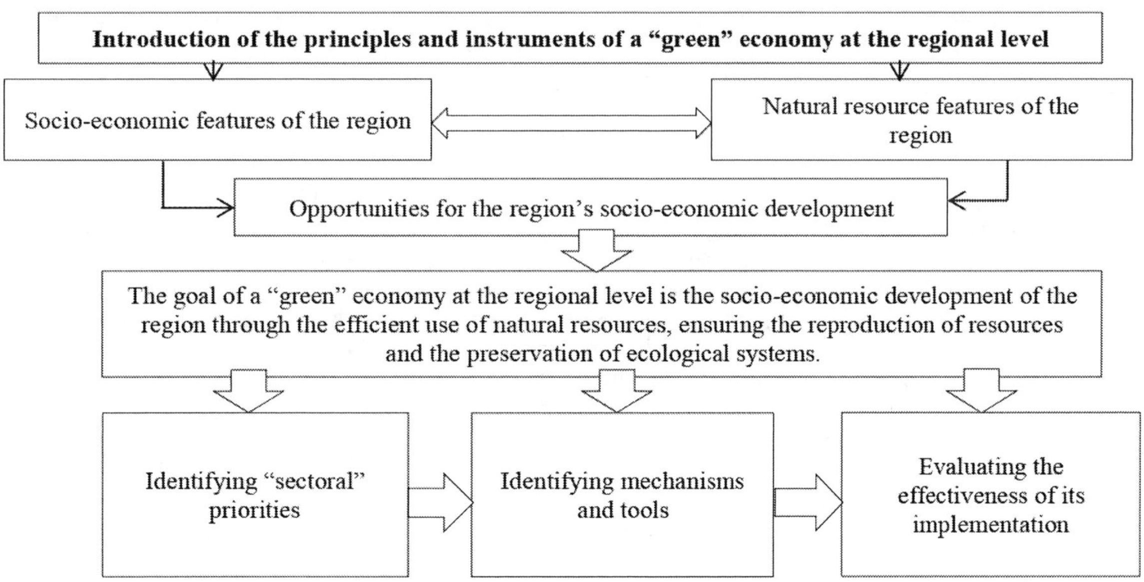

Figure 1. The process of introducing the "green" economy model at the regional level.

At the first stage of the implementing the principles and tools of a "green" economy at the regional level, the particularities of the region should be taken into account. This will allow to determine the balance between the possibilities of the socio-economic development of the region and its natural resource potential, which is the goal of a "green" economy.

At the same time, analyzing and defining the features of regional development, which are a prerequisite for the "green" economy model, an assessment and analysis of important indicators of the region in the following areas should be carried out:

- Availability, level, and condition of the region's natural capital (mineral resources, land resources, forest resources, biological and water resources);
- Assessing the socio-economic situation in the region (GRP per capita; investments in fixed capital per capita; a sectoral structure of GRP and investment in fixed capital; a degree of depreciation of fixed assets, an income level; unemployment rates; life expectancy at birth, etc.);
- Assessing the environmental situation in the region (polluting emissions; wastewater discharges; production wastes; emissions from motor vehicles; a share of specially protected natural territories, etc.);
- Evaluating the institutional environment of the region (a set of factors that influence the functioning of economic entities through the activities of political, economic, social and other institutions).

Based on the analysis and determination of the features of regional development, specific goals and objectives of the "green" economy model in the region should be formulated. At the same time, when shaping the goals and objectives of the "green" economy model, goals and objectives should be defined in three areas of regional development: environmental, economic, and social.

On the basis of identifying features of regional development, "sectoral" priorities of a "green" economy should be clarified and shaped at the regional level. In turn, for the development of priority "green" sectors of the economy, effective mechanisms and tools should be identified. When developing mechanisms and tools for a "green" economy in the region, a number of circumstances must be taken into account, including the peculiarities of the region's socio-economic development, the types of financial and economic instruments implemented in the region, the ability of governments to develop tools for a "green" economy and to ensure effective monitoring of the implementation of the goals of a "green" economy [3]. The main instruments of the "green" economy that can be applied in the region are presented in Table 1.

Table 1. Main instruments of the "green" economy.

"Green" economy mechanisms	"Green" economy tools
Regulatory mechanism	- Developing a regulatory framework governing the economic, technological, and environmental bases of the "green" economy model; - Developing a set of technical regulations and technological standards, enshrining the mandatory and other norms and standards for energy efficiency, the use of energy from renewable sources, in the field of ecology, etc.; - Developing a system of legal liability for violation of environmental legislation, etc.
Financial and economic mechanism	- Reforming the system of taxes and payments through the introduction of taxes or payments for environmental pollution ("green" taxes) or the use of certain types of resources; - Stimulating the development of "green" business and "green" technologies ("green" innovation) through government subsidies, government procurement and other forms of direct support; - Developing a package of measures to stimulate "green" investments in the economy (tax, property and other benefits, the introduction of simplified administrative procedures, information support and investment projects, etc.); - Using new sources of funding for activities in the field of "green" economy ("green" bonds, etc.); - Encouraging the creation of "green" jobs (assistance in training and retraining of the personnel, its evaluation, etc.); - Formation of a system of environmental certification and labeling (for organic

	food, energy efficiency, etc.).
Organizational and institutional mechanism	- Creating the organizational structures to ensure interaction of legislative, executive authorities, market, and social institutions for implementing the principles of "green" economy; - Establishing an information system that accumulates and transmits information on the implementation of the "green" economy principles; - Developing policies in environmental education and environmental awareness.

The effectiveness of applying the "green" economy principles and tools at the regional level (based on its goals) is determined by the degree of achievement of results, which are subject to the development of priority "green" sectors of the economy. More than that, it is characterized by the socio-economic and environmental consequences of its implementation, taking into account the effectiveness of economic, institutional, and legal mechanisms to ensure the transition to a "green economy".

4. Discussion

As noted earlier, when introducing the green economy model in a particular region, the specifics of regional development should be taken into account. Thus, when introducing the principles and tools of the "green" economy in the Altai Republic, the following features of regional development and prerequisites should be taken into account:

- The Altai Republic has a low level of economic development. Thus, in the overall ranking of the Russian Federation in 2017, the Altai Republic ranks 82nd among 85 constituent entities of the Russian Federation (83rd in 2016, 82nd in 2015);
- The Republic of Altai has a rich natural resource capital, including mineral resources, land, water and biological resources. At the same time, the use of the region's natural resource capital is characterized by a number of features: reserves of mineral resources are not sufficiently studied in terms of geological, economic, and environmental features of their development; a significant part of the region's agricultural land is subject to erosion processes; the main environmental problem of wastewater disposal is the low degree of sewerage in settlements, etc.;
- The natural resource potential and the predominance of the rural population led to the agrarian and tourist-recreational specialization of the region. At the same time, agriculture and tourism are identified as priority areas for the socio-economic development of the Altai Republic. This is enshrined in the Strategy for the Socio-Economic Development of the Region until 2035;
- One of the acute problems of the socio-economic development of the Altai Republic is the energy supply. Until 2015, the Altai Republic did not generate electricity on its territory, with the exception of ten small diesel power plants, a wind farm, and two small hydropower plants with a total capacity of 1.3 MW. They were intended for local power supply of objects in remote and remote settlements of the region, and not connected to the general power grid of the Altai Republic. In 2015-2017, a number of solar power plants were launched in the region [5]. Dependence on the electricity supply of the region increases the relevance of the development of local energy based on renewable energy sources;
- The ecological situation in the Altai Republic is recognized as quite favorable. However, we can identify a number of environmental problems: the increase in the level of air pollution in the republic by motor transport, which, along with boiler houses, is becoming one of the main sources of pollution; a specific factor of environmental impact of the Republic of Altai is the objects of tourist infrastructure, which are characterized by such types of environmental violations as: illegal cutting and damage to trees and shrubs, non-compliance with the requirements for the protection of water bodies from pollution and depletion, non-compliance with the requirements in the field of waste management and in the field of land legislation; growth in the volume of municipal waste in the absence of waste processing facilities in the region.

The features of each region will determine the "sectoral" structure of the "green" economy model. At the same time, the basic principle of the development of a "green" economy is to ensure a balanced interaction of the main priority "green" sectors. These industries define the essence of regional development on the basis of a "green" economy, as ensuring a favorable environment for economic growth in the region. Thus, the main areas of the Republic of Altai, contributing to the "greening" of the entire economy, include: "green" agriculture (organic agriculture), forestry, sustainable water management, fisheries, alternative energy, tourism development, waste management and their utilization.

5. Conclusion

One of the new models of economic growth, ensuring the socio-economic development of the region through the effective use of natural resources, ensuring the reproduction of resources and the preservation of ecological systems, is the model of "green" economy. At the same time, the introduction of the principles and tools of the "green" economy will be effective only if it adapts to the specific conditions of the regional development, taking into account the level of economic development, resource potential, management system, etc. At the same time, it is precisely the consideration of the specific features of regional development that will determine the types and applicability of "green" economy instruments.

6. Acknowledgments

This work was financially supported by the Russian Foundation for Basic Research, the project №18-410-040001р_а "Formation of mechanisms for the introduction of a model of "green" economy at the regional level".

References

[1] Zakharova T V 2015 Green economy and sustainable development of Russia: contradictions and prospects *Vestnik of Tomsk State University. Economy* **2**(30) Available at: https://cyberleninka.ru/article/n/zelenaya-ekonomika-i-ustoychivoe-razvitie-rossii-protivorechiya-i-perspektivy (Accessed 04 04 2019)

[2] Zomonova E M 2016 Concept and principles of a "green" economy *ANI: Economics and Management* 5 1(14) pp 13-17

[3] Pakhomova N V, Richter K K, and Malyshkov G B 2012 Structural transformations in the context of a "green" economy: challenges for the Russian state and business *Problems of the Modern Economy* **3**(43) Available at: http://www.m-economy.ru/art.php?nArtId=4155 (Accessed 04 04 2019)

[4] Bogoviz A V, Lobova S V, Alekseev A N, Koryagina I A, Aleksashina T V 2018 Digitization and internetization of the Russian economy: Achievements and failures *Advances in Intelligent Systems and Computing* **622** pp 609-616

[5] Matraeva L, Solodukha P, Erokhin S, Babenko M 2019 Improvement of Russian energy efficiency strategy within the framework of "green economy" concept (based on the analysis of experience of foreign countries) *Energy Policy* **125** pp 478-486

[6] Alekseev A N, Lobova S V, Bogoviz A V, and Ragulina, Y V 2019 A critical review of Russia's energy efficiency policies in the construction and housing sector *International Journal of Energy Economics and Policy* **9**(4) pp 166-172

[7] TASS 2012 The Republic of Altai will reach full energy supply with solar stations by 2020 Available at: https://tass.ru/sibir-news/6313394 (Accessed 04 04 2019)

Export of dairy products of the Altai Region: opportunities, prospects, and problems

V Belyaev[1], M Butakova[1*] and O Goryaninskaya[1]

[1] Altai State University, 61 Lenin Ave., Barnaul, 656049 Russia

E-mail: marina.butakova.59@mail.ru

Abstract. The article is devoted to the study of the export potential of dairy products of the Altai region. The factors that determine the possibility of expanding the presence of enterprises in the markets of the near and far abroad are identified and carefully analyzed in the paper.

1. Introduction

One of the key drivers of the regional socio-economic development is the presence of the products of their enterprises in the national and global markets. The economic development of the Russian Federation and its regions depends largely on the quality of trade relations, the intensity and distribution of commodity trade flows. Domestic researchers in the food market note that in the context of globalization, it is necessary and appropriate to focus on improving the competitiveness of agro-industrial products of deep processing and increasing its share in exports of domestic products [1-2]. The need for these measures is confirmed by the strategic objectives set for the agro-industrial complex of Russia [3].

2. Materials and Methods

The works of researchers in regional markets, including dairy markets, served as the methodological basis of the study [1], [2], [4], [5], [6]. To analyze the dynamics of the production of milk and dairy products in the Russian Federation and the Altai region, the authors used the balance method and methods of economic and statistical analysis. Their application made it possible to assess the role and place of the Altai region in the domestic and foreign markets, to analyze the problems of the industry in the Altai Region, to justify the feasibility and necessity of export-oriented specialization of the dairy industry in the Altai region. The materials of the Federal State Statistics Service of the Russian Federation, the territorial body of the Federal State Statistics Service for the Altai Region, the Altai Region Administration for the Food, Processing, Pharmaceutical Industry and Biotechnology served as background information [7], [8].

3. Results

A comparative analysis of the volume of milk production in the Russian Federation, the Siberian Federal District, and the Altai Territory allowed the authors to conclude that it is possible to sell dairy products of the Altai Region in the markets of the near and far abroad. The Altai region ranks fourth in this indicator in the country, and it is the first in the Siberian Federal District. In 2018, in the Altai Region, an amount of 1,178 thousand tons of milk was produced, which amounted to 3.84% of the all-Russian indicator (Fig. 1).

Content from this work may be used under the terms of the Creative Commons Attribution 3.0 licence. Any further distribution of this work must maintain attribution to the author(s) and the title of the work, journal citation and DOI.
Published under licence by IOP Publishing Ltd

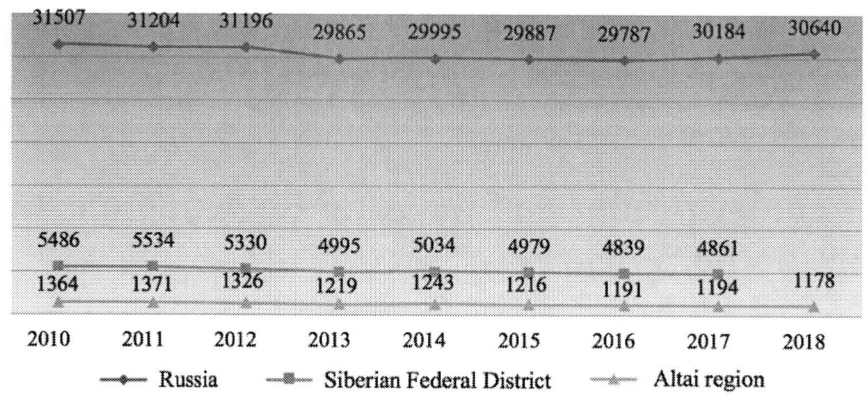

Figure 1. Dynamics of milk production in 2010-2018, thousand tons.

Table 1. The "production- demand" demand of cheese and butter in the Altai region.

Year	Population, thous. people	Production, thous. tons	Requirement, thous. tons	"Production-demand" balance, thous. tons	Self-sufficiency, %
			Cheese and cheese products		
2011	2407	68.6	16.8	51.8	407
2012	2399	67.5	16.8	50.7	402
2013	2391	61.1	16.7	44.4	365
2014	2385	71.3	16.7	54.6	427
2015	2377	83.1	16.6	66.5	499
2016	2366	83.4	16.6	66.8	504
2017	2350	86	16.5	69.6	523
2018	2333	91.2	16.3	74.9	558
			Butter		
2011	2407	12.2	4.8	7.4	253
2012	2399	11.2	4.8	6.4	233
2013	2391	11.8	4.8	7.0	247
2014	2385	16	4.8	11.2	335
2015	2377	19.1	4.8	14.3	402
2016	2366	20.1	4.7	15.4	425
2017	2350	20.3	4.7	15.6	432
2018	2333	18.4	4.7	13.7	394

The authors note that, the milk production per capita was 504.9 kg per year in the Altai region in 2018, which exceeds the average figure for Russia by 2.5 times and significantly exceeds the rational norms of milk consumption approved by the Ministry of Health of the Russian Federation (325 kg per year) [9]. The dairy industry of the region meets the needs of its population and other geographically close regions. This is confirmed by the author's calculations of the "production–demand" balance (Table 1), which is based on a comparison of the volume of dairy products production and the needs of the population of the Altai Region for them, taking into account rational consumption norms.

The analysis shows the overproduction of dairy products of deep processing in the region, which allows to conclude that there is a high export potential of dairy production. In the face of declining consumer demand, this fact determines the need to increase exports of dairy products, encourages manufacturers to largely focus on foreign markets, and consider the possibility of their presence as a necessary condition for the growth of dairy production in the region. In 2018, the share of exports of dairy products in the total volume was 13.9%, amounting to 3.85 million dollars in value terms (Fig. 2).

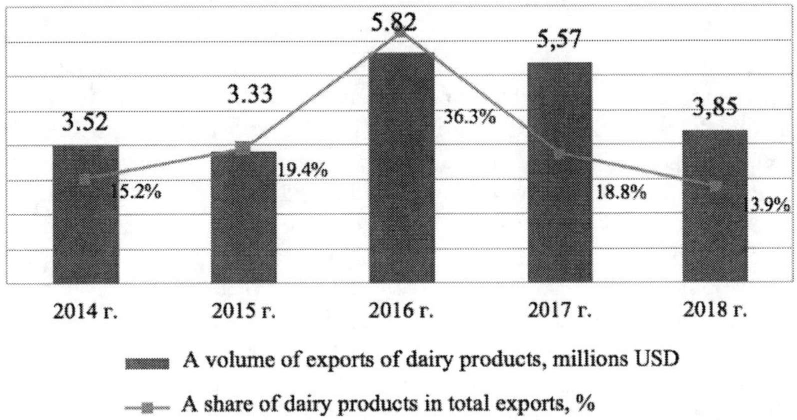

Figure 2. Exports dynamics of dairy products of the Altai region in 2014-2018, millions USD.

At the present time, the geography of export supplies of dairy products of the Altai Territory covers the CIS countries (Kazakhstan, Belarus, Kyrgyzstan, Tajikistan) and other countries (China, Mongolia, France). Preserving the presence of Kazakhstan and China in the most capacious markets, searching for new markets should be evaluated not only as a way to increase sales, but also as a means of reducing the risks of introducing restrictive measures in the partner countries. The structure of the export of dairy products of the Altai region (Fig. 3) corresponds to the logic of the formation of an effective foreign economic policy involving the export of products with higher value added (cheese, milk and condensed cream, ice cream, butter).

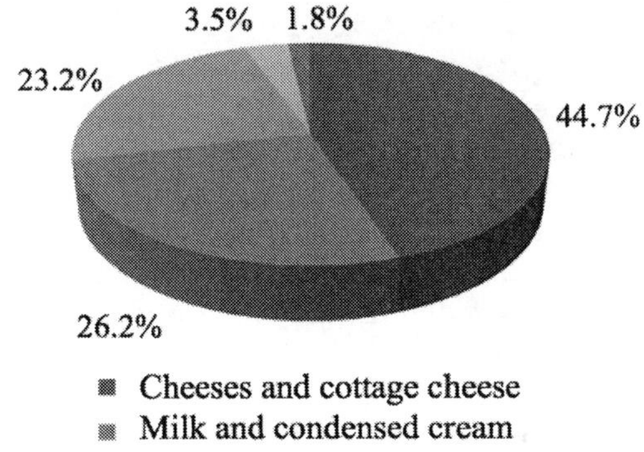

Figure 3. Export structure of dairy products of the Altai region in 2018, %.

The authors identified constraints that adversely affect the development of the industry and its export potential. First of all, this is the low efficiency of the raw material base, the evaluation indicator of which is milk yield per cow. During 2010-2017, it ranged from 76 to 86% of the level in the Russian Federation in the Altai region, and from 94 to 99% of the level in the Siberian Federal District. The lower value of the indicator is largely offset by the size of the livestock in the Altai region. Reducing the attractiveness of cattle breeding for producers actualizes the problem of shortage

IOP Conf. Series: Earth and Environmental Science **395** (2019) 012100 doi:10.1088/1755-1315/395/1/012100

of raw milk, speaks about the need to solve it by attracting investment, improving the breeding base and feeding rations. The authors consider the pronounced negative impact of seasonality on the volume and quality of raw milk, the cost of its production as the second deterrent. The climatic features of the Altai region force manufacturers to make appropriate adjustments to the product and sales policy, which is reflected in economic performance.

The authors believe that changes in processing and storage technologies have a positive effect on the ability to export domestic dairy products. These changes in technology have allowed to expand the boundaries of the market by increasing the shelf life of milk and dairy products. In addition, a conscious orientation to the preferences of customers in various market segments contributes to the competitiveness of domestic manufacturers. Expanding and deepening the range of dairy products is a reflection of the orientation on the preferences of customers. The popularity of a healthy lifestyle among different categories of the population is also becoming a factor contributing to the growth of the dairy market. The result of its influence is expressed in the growing need for natural, environmentally friendly products, products with biologically and physiologically active substances, with reduced calorie content, with therapeutic and prophylactic properties, and in products with a mixed raw material composition. Some enterprises expand their product line by including other types of milk (goat, mare) along with cow's milk as products with unique properties. Expansion and updating of the export range should be carried out by creating new export-oriented products that are competitive in the world market. Prerequisites for increasing the export potential of enterprises in the Altai region include an increase in the proportion of durable products in the assortment of dairy products, as well as a corresponding change in the production technology of products, the use of the latest developments in packaging and storage. In the context of globalization, it is economically feasible for producers to reduce the share of raw materials and low value-added goods in the export structure by developing the production of butter and cheese. It should be noted that the market for this product is less stable. In addition, the customs tariff legislation of some countries provides for an increase in the rate of customs duty as the degree of processing of goods increases.

4. Discussion

The recent studies of the market of milk and dairy products pay close attention to the features of the demand formation in this market [5], [6], the prospects for exporting the dairy products of the Russian producers [1], [2]. Analyzing the market for milk and dairy products, some experts note a different price elasticity of demand, which implies a low elasticity for drinking milk and whole milk products and a high elasticity for its processed products (cheese, butter). Also, these experts note that the demand for dairy products is strictly dependent on the level of income and the purchasing power of the population [5], [6]. In the scientific literature, considerable attention is paid to the development of regional dairy products markets and the assessment of export potential [1], [2], [4]. But the issues of formation and valuation of value added when exporting raw materials and finished products are not sufficiently covered. The study of the involvement of regional producers in global value chains in the supply of raw materials and intermediate products to the international market is of scientific interest. The problems of stimulating and supporting the export of products were the subject of discussion in the works of domestic and foreign authors.

Our analysis of the main achievements of scientific discussion in the theory of regional economics and spatial development on the problems of interregional interaction allowed to justify the author's position on the need and desirability of deepening the processes of interregional economic interaction. As a result of theoretical and practical research questions, the authors came to the conclusion that it is necessary to form a scientifically grounded approach to the formation of an effective model for managing the region's export potential in the context of economic sectors. The authors substantiated the export potential of the dairy industry of the Altai region in the international market based on the balance method and methods of economic and statistical analysis.

5. Conclusion

The article reflects the results of a study focuses on the possibilities of expanding the presence of enterprises, producers of milk and dairy products of the Altai Region in foreign markets. The authors systematized information about the specifics of the demand for dairy products, opportunities and problems of expanding the geographical boundaries of the market. Despite the difficulties of entering the foreign markets of dairy products, we note the continuing high capacity of this market with various possibilities of our own production. This preserves the potential attractiveness of the market for producers of dairy products in the Altai Region.

Increasing export volumes of dairy products of deep processing and the geographical expansion of foreign markets will give a multiplicative effect to the sectoral and regional development. The multiplicative effect is characterized by the growth of value added for producers of dairy products, increasing the efficiency of the economic system of the region and shaping the image of the region as a region specializing in the production and export of dairy products.

6. Acknowledgments

The article was prepared with the support of the RFBR grant, № 18-010-00593 A "A Study of Global Value Chains as a Factor in Improving the Efficiency of the Regional Economy".

References

[1] Krylatykh E N and Frolova E Yu 2017 Creating the export potential of the dairy complex of Russia *Ostrovskiye Chteniya* **1** pp 55-60

[2] Anfinogentova A A, Morekhanova M Yu, Blinova T V and Andryushchenko S A 2018 *Russia's national food security: strategic priorities and conditions for ensuring* (Saratov, Russia: Saratov source)

[3] Presidential Executive Office 2017 *Decree on the main directions of state policy on the development of competition: the national plan for the development of competition in the Russian Federation for 2018-2020* (dated December 21, 2017, No. 618) Available at: http:// www.consultant.ru (Accessed 06 04 2019)

[4] Altukhov A I 2017 Improving inter-regional food ties in the country: questions of methodology and practice *Russia's Agri-Food Policy* **2**(62) pp 2-11

[5] Suray N M, Levichev V E and Buzoverov S Yu 2016 Analysis of the impact of socio-economic factors on the production and consumption of dairy products *Vestnik of the Altai State Agrarian University* **5**(139) pp.186-192

[6] Yakovenko N A and Ivanenko I S 2018 Overcoming regional differentiation in the dairy market *Agrarian Scientific Journal* **9** pp 96-100

[7] Federal State Statistics Service 2018 *Regions of Russia: socio-economic indicators* Available at: http://www.gks.ru/wps/wcm/connect/rosstat_main/rosstat/ru/statistics/publications/catalog/doc_1138623506156 (Accessed 23 03 2019)

[8] *EMISS: unified inter-agency information and statistical information* Available at: https://fedstat.ru/ (Accessed 06 04 2019)

[9] Ministry of Health of the Russian Federation 2016 *Order of the Ministry of Health of the Russian Federation on approval of recommendations on rational norms of food consumption that meet modern requirements for healthy nutrition* (dated August 19, 2016, No. 614) Available at: https://www.garant.ru/products/ipo/prime/doc/71385784/ (Accessed 23 03 2019)

IOP Conf. Series: Earth and Environmental Science **395** (2019) 012101 doi:10.1088/1755-1315/395/1/012101

Modeling regional economic growth with the formation of innovative ecosystems of technological entrepreneurship as a factor of sustainable development

I Korchagina[1*] **and R Korchagin**[1]

[1] Kemerovo State University, 6 Krasnaya str., Kemerovo 650043 Russia

E-mail: korchagina-i@mail.ru

Abstract. The article analyzes the potential of growth theories in the context of determining sources for further development of the Russian regions, taking into account the "new normality" in the economy. As a result of a comparative analysis of endogenous and exogenous theories, their similarity has been proven. Modification of the Aguion-Howitt model for a region subjected to the development of an innovative technological entrepreneurship ecosystem has been developed. This model will allow to evaluate the impact of the innovation ecosystem as a factor of endogenous growth of the region.

1. Introduction

Sustainable development of any type of region, including cross-border ones, is based on economic growth of the productive use of labor and capital. In context of the "new normality" in modern economy, the purposeful formation of variables that increase the returns on labor and capital (human capital and the technological level of the economy) are of particular importance.

Currently, the lack of research that links innovation policies at the meso-level with the territory's economic dynamics is observed in regional science. Thus, the formation of innovative ecosystems for the development of technological entrepreneurship in theory and in practice is poorly evaluable from the standpoint of influence on regional growth. The purpose of this article is to develop a model of endogenous growth of the region using a technology entrepreneurship factor.

2. Materials and Methods

Recognized theories and models of endogenous and exogenous growth are widely presented and discussed in foreign scientific literature (Solow, Ramsey, Mankew - Romer - Weil, Aerrow - Romer, Uzava - Lucas, Aghion - Howitt, etc.). They serve as are the material for research.

We used the following research methods: comparative analysis, logical-structural analysis, mathematical modeling, differential calculus to identify differences between models, as well as model allocation with the highest heuristic potential.

3. Results

The Solow model is a classic exogenous model. This model explores the stationary solution of a differential equation, showing that if the actual investment coincides with the necessary one, economic growth depends on the rate of technical progress and population growth. In the absence of a rapid growth in the number of able-bodied citizens, usually uncharacteristic for developed economies, it is

Content from this work may be used under the terms of the Creative Commons Attribution 3.0 licence. Any further distribution of this work must maintain attribution to the author(s) and the title of the work, journal citation and DOI.
Published under licence by IOP Publishing Ltd

IOP Conf. Series: Earth and Environmental Science **395** (2019) 012101 doi:10.1088/1755-1315/395/1/012101

scientific and technical progress that would determine the dynamics of macroeconomic indicators (according to R. Solow, technological shifts explained 87.2% growth in the United States) [1], which, however, must be given exogenously. The Ramsey (Ramsey-Kass-Kupmans) and Mankew-Romer-Weil models became the development or modification of the Solow model.

However, exogenous models do not provide convincing explanations regarding differences in the influence of classical growth factors in countries and regions, and also they do not allow to quantify the impact of innovative technological effects. In fact, the introduction of new explanatory parameters changes their axiomatics and brings them closer to endogenous approaches. Another problem of exogenous models is the inability to constructively substantiate any pronounced economic policy [2]. Therefore, a greater heuristic potential of endogenous growth theories can be assumed.

P. Romer proposed one of the first modern models of endogenous growth (Arrow-Romer model). According to the model, the self-developing process of accumulation and use of knowledge occurs within the economy and individual firms [3]. The level of knowledge (technology development) is an independent argument of the function of output; therefore the growth of this value causes increasing returns on capital. This can explain the steady self-sustaining growth of countries with mature market economies. The spread of technology among firms and countries is also of independent importance [4].

Endogenous scientific and technological progress explains economic growth in the model of the creative destruction of P. Aghion and P. Howitt. It put economic growth in dependence on three parameters: (1) the flow of innovation (generation of innovations in the research and development sector, subject to the theory of probability), (2) the share of skilled labor (the model separates the number of skilled and unskilled workers), (3) the impact of innovation on the economy [5]. Thus, the growth in the model was explained only by the technical progress arising during the interfirm competition.

In 1998, P. Aghion and P. Howitt added to the number of growth factors also the accumulation of capital, not in its classical interpretation (as an autonomous source of growth), but as a driving force of technological changes. Now, the rate of flow of innovation has become dependent on the capital stock per employee in the field of research and development, and the innovation process has begun to be seen as capital-intensive [6]. As a result, the model began to largely overlap with the construction of Ramsey, but with an endogenous task of speed of technological development. In the 2000s, analyzing modern applications of the Schumpeterian growth paradigm, some new important effects were highlighted.

1. The growth rate depends positively on the rate of creation and liquidation of firms and jobs. This thesis points out contradictions between economic growth and living standards, which would not necessarily correlate in the medium term.
2. In many developing countries and countries with economies in transition, there is significant growth potential while reducing the boundaries and barriers of creative destruction, as well as institutional or educational reforms. The recommendation to many states to develop at an accelerated pace either higher or secondary education follows from here.
3. The market structure and size of firms are subject to deep institutional features of different countries. For example, the slow growth of India is largely due to the deliberate preservation of the small size of firms within the range of control of one family. With a weak social capital, low trust, poorly working laws, it is very risky to entrust managerial powers and assets to people outside their family. This constant circle of small firms uses unproductive capital and, moreover, has no incentive to innovate [7, 8].

We believe that it is the Aghion-Howitt model that can most adequately characterize the current problems of economic growth, both at the national and subnational levels.

4. Discussion

A comparative analysis of endogenous and exogenous growth models suggests their convergence. Theories of endogenous and exogenous growth do not differ according to the fact that production factors belong to residents or non-residents. Differentiation of growth patterns occurs on the basis of explaining a different range of factors from the model itself. Exogenous models consider only labor and capital, suggesting a certain unexplained remainder ("Solow's remainder"), which is caused by factors not included in the model, i.e. external to it. Scientific and technical progress in a broad sense is most often such a factor.

At some point, since the share of "Solow's remainder" turns out to be too large, the inclusion of human capital and technological development in the models as new internal variables becomes logical. This underlines the continuity of exogenous and endogenous growth patterns. They have no conceptual differences. It is not by chance that the Aghion-Howitt model with adding the "capital" parameter makes it possible to obtain the Ramsey model as a trivial special case. Thus, the axiomatic of different growth models differs only in a different explanation of the variables associated with scientific and technological progress (from the external environment or the internal logic of the model itself).

In context of implementing the model of endogenous growth, the formation of innovative ecosystems of technological entrepreneurship can take a significant place. As was shown above, since productive technological entrepreneurship can be carried out only in the presence of an innovation ecosystem, the creation of the latter is a significant factor of endogenous growth. Based on the structural and logical analysis, identifying the following areas of the relationship between innovation ecosystems and the endogenous model of territory's economic growth seems possible (see Table 1).

Thus, under the constraints of labor and capital, which are characteristic of Russia and its regions, the potential for economic growth is mainly associated with an increase in technological level, as well as human capital. The increase is also due to internal sources of self-development generated by the innovative ecosystem of technological entrepreneurship. In this case, a significantly higher return will be achieved from investments, if one manages to attract them in a significant amount. Consequently, the development and testing of models reflecting the influence of individual components of scientific and technological progress, including technological entrepreneurship, on economic growth and development of territories is necessary.

To simulate the contribution of technology entrepreneurship to regional growth, it seems most appropriate to modify the Aghion-Howitt model with differentiation of various components of the territory's technological development. In general, this model looks like this (1) [8]:

$$Y(t)=\left[\int_0^1 A(v,t)x^{\alpha}(v,t)dv\right]L^{1-\alpha}, \quad (1)$$

where x –capital stocks in the industry v, used at time t,
A – a technological level in the industry v, existing at time t,
L – labor amount,
α – a capital elasticity ratio,
$1-\alpha$ – an elasticity coefficient (according to labor).

Table 1. The impact of the innovation ecosystem of technology entrepreneurship on the territory's endogenous growth.

Specificity of endogenous growth type	Potential of the innovation ecosystem of technology entrepreneurship
Use of already existing (endogenously) specified classical factors of production (labor and capital)	Increasing the effect of using existing growth factors
	Increasing the territorial attractiveness with a developed innovation ecosystem both for investments and carriers of valuable human capital
	Retention of own factors of production (human

IOP Conf. Series: Earth and Environmental Science **395** (2019) 012101 doi:10.1088/1755-1315/395/1/012101

	capital, investments) at the expense of an attractive image of an innovatively active territory
The internal (endogenous) origin of new growth factors – human capital (skill level, competence), technology	Increasing the multiplier effect of technological level due to own research and development
	Human capital development in the innovation ecosystem
"Stream of innovations" (according to Aghion-Howitt), particularly its speed and power	Increasing the likelihood of innovation. Increasing the number of developments
Mediating a technological level of the economy by improving the institutional environment, including the speed at which firms are created and liquidated	A developed innovation ecosystem creates favorable conditions for creating technological and other new firms

Given the regional level of analysis, the exogenous variable *T*, which takes into account the effect of interregional technology transfer, should be included into the model (2):

$$Y(t)=\left[\int_0^1 A(v,t)x^a(v,t)dv\right]L^{1-a}+T \quad (2)$$

The growth rate *g* due to the increasing technological level of equipment due to innovations:

$$g(t) = \dot{A}(t) / A(t) = (\gamma - 1)\lambda n(t), \quad (3)$$

where $\dot{A}$ is the technological level after introducing innovation,
$(\gamma - 1)$ is the rate of increasing technological level, where $\gamma>1$,
λn – the probability of inventing (implementating) innovation due to one element of the whole innovation system *n*.

As can be seen from the above data, increasing a technological level depends on the probability of inventing λ for one element *n* (for example, making a discovery by one researcher), as well as due to the number of elements *n*. In other words, the probability of invention (innovation) in the industry *v* in the period of time *t* is $\lambda n\ (v,\ t)$. At the same time, the parameters *n* and λ are characteristics of the innovation ecosystem of technological entrepreneurship.

The value of *n* will be associated with the state of "hard" innovation infrastructure, i.e. the presence of researchers, innovation support institutions, and technological entrepreneurship, the release of funding. The value of λ depends on the "soft" infrastructure, which most determines the productivity of the regional innovation ecosystem.

Then the parameter λn of the Aghion-Howitt model is proposed to be written as follows (4):

$$\lambda n(v,\ t)=\frac{\acute{s}(t)}{s(t)}=\ (\gamma\ -\ 1)E\sum_{i=1}^{k}n_i\sum_{j=1}^{k}\lambda_j \quad (4)$$

where $\acute{s}$ – a number of startups implemented after time point *t*,
s – a number of implemented startups up to the moment of time *t* (before exiting the equilibrium state),
E – a coefficient reflecting the degree of interaction in the innovation ecosystem by the triple helix model,
n_i – a normalized (correlated with the maximum or minimum) value of the severity of the *i*- th element of the "hard" innovation infrastructure,
λ_j – a normalized value of the *j*- th element of the "soft" innovation ecosystem.

As a result, this model reflects the influence of the innovation ecosystem and its qualitative characteristics on the values of the coefficients of the Aghion-Howitt model, details, and clarifies the latter.

5. Conclusion

The proposed model allows a differentiated assessment of the impact of different elements of the innovation ecosystem on regional economic growth and start-up generation. In turn, this provides an opportunity to more reasonably develop and stimulate regional innovation ecosystems.

6. Acknowledgments

This article was written based on the results of research work financed by the Kemerovo State University funds on the topic "Formation of an Innovative Ecosystem of Technological Entrepreneurship as a Factor in Reducing the Mono-Dependence of a Region."

References

[1] Solow R M 1957 Technical change and the aggregate production function *Review of Economics and Statistics* **39**(3)pp 312–320
[2] Romer P M 1990 Endogenous technological change *Journal of Political Economy* **98**(5) pp 71-102
[3] Romer P M 1988 Increasing returns and long-run growth *Journal of Political Economy* **94**(5) pp 1002-1037
[4] Rivera-Batiz L A, and Romer P M 1991 Economic integration and endogenous growth *Quarterly Journal of Economics* **106**(2) pp 531-555
[5] Aghion P, and Howitt P 1992 A model of growth through creative destruction *Econometrica* **60**(2) pp 323-351
[6] Howitt P, and Aghion P 1998 Capital accumulation and innovation as complementary factors in long-run growth *Journal of Economic Growth* **3**(2) pp 111-130
[7] Aghion P, Akcigit U, and Howitt P 2015 The Schumpeterian growth paradigm *Annual Review of Economics* **7**(2) pp 557-575
[8] Howitt P, and Aghion P 2009 *The economics of growth* (Cambridge, MA: The MIT Press)
[9] Pozdnyakova U A, Bogoviz A V, Lobova S V, Ragulina J V, and Popova E V 2019 The model of well-balanced taxation for overcoming the shadow economy in modern Russia *Studies in Systems, Decision and Control* **182** pp 207-215

IOP Conf. Series: Earth and Environmental Science **395** (2019) 012102 doi:10.1088/1755-1315/395/1/012102

Benchmarking sustainable energy technologies in cross-border regions: issues of economic efficiency

O Chernikova[1*], Y Zlatitskaya[1] and Y Klimashina[1]

[1] Siberian State Industrial University, 42 Kirov str., Novokuznetsk 654006 Russia

E-mail: chernikovaop@yandex.ru

Abstract. The long-term trends of world energy development envisage a radical increase in the energy efficiency, an increase in its quality indicators, and a transition from fossil fuels to the use of sustainable energy technologies. The article presents an algorithm developed by a team of authors for planning, implementing, and evaluating economic efficiency of benchmarking sustainable energy technologies in cross-border regions. Its use will allow optimizing the energy intensity of enterprises, ensuring energy security, economic efficiency and environmental stability of adjacent territories, preserving the mineral resource base, and improving the quality of life of the population.

1. Introduction

At present, the principles of sustainable development, which envisage achieving the parity of economic, ecological, and social values of modern society, are integrated into the world economic practice in the planning and management of organizations, which is reflected in various international studies [1-4]. The basic idea of the concept of sustainable development of a cross-border region provides that business processes aimed at improving the efficiency of using natural resources of the territory and using new technologies should be accompanied by an increase in the level and quality of life, compliance with environmental requirements, environmental management in the interests of present and future generations, using a number of opportunities for cross-border cooperation [12].

At the same time, territorial planning should focus on using sustainable energy technologies as spatially linked, environmentally friendly, reducing dependence on fossil fuel resources and providing opportunities for mitigating greenhouse gas emissions [5].

Possessing similar natural resources and climatic conditions, cross-border regions can carry out technology benchmarking, jointly solve problems of power supply of territories, and implement projects of interest to both parties.

2. Materials and Methods

The empirical basis of this study was the study and analysis of Russian and international experience in the use of alternative energy types. A systematic approach has allowed a comprehensive and holistic study of this issue. The methodological base of the research was the scientific and educational literature given in the list of sources.

In setting the goal and defining the main tasks at the initial stage, the team of authors used an abstract-logical research method to systematically explore the possibilities of benchmarking sustainable energy technologies in cross-border regions.

Content from this work may be used under the terms of the Creative Commons Attribution 3.0 licence. Any further distribution of this work must maintain attribution to the author(s) and the title of the work, journal citation and DOI.
Published under licence by IOP Publishing Ltd

IOP Conf. Series: Earth and Environmental Science **395** (2019) 012102 doi:10.1088/1755-1315/395/1/012102

In the process of studying the advantages and disadvantages, as well as the possibilities of applying sustainable energy technologies, the method of comparative analysis and synthesis was used to identify patterns of development of its types and cause-effect relationships between them.

At the stage of developing an algorithm for planning, implementing, and evaluating the economic efficiency of benchmarking sustainable energy technologies in cross-border regions, the method of modeling accounting processes was used, as well as the methods of scientific knowledge – scientific abstraction, dialectical development, and statistical generalization.

In the course of our discussion, heuristic and logical methods were applied. The reliability and validity of the results are confirmed by using such methods as statistical and experimental ones.

3. Results

Results of the study is an algorithm for planning, implementing, and evaluating the economic efficiency of benchmarking sustainable energy technologies in cross-border regions include 5 steps.

Stage 1 "Planning results of production and economic activities of the enterprise for the upcoming period." Based on the implementation of which, the enterprise management sets the planned indicators of production volume and unit cost of production.

Stage 2 "Assessing the possibilities of using and planning indicators of the economic efficiency of benchmarking sustainable energy technologies in cross-border regions." This stage provides for: (a) determining the energy intensity of business processes of an enterprise, ensuring the achievement of planned indicators of production volume and production costs; (b) identifying factors that influence the energy intensity of processes (energy system characteristics, used technical means and equipment, etc.); (c) a study of benchmarking opportunities for using sustainable energy technologies in cross-border regions (collection and analysis of information on natural conditions and resources of transboundary regions, on the experience of using renewable (solar energy, water, wind, biomass, etc.) and other sustainable energy sources, such as biogas, when recycling urban waste; (d) calculating capital costs and evaluating effectiveness of an investment project for introducing sustainable energy technologies.

Stage 3 "Development of measures for introducing sustainable energy technologies" involves (a) planning a set of measures for introducing sustainable energy technologies at the responsibility centers; (b) assessing the resource intensity of carrying out activities in each responsibility center; (c) assessing the impact of the proposed measures on the company's financial result (cost savings in the areas of heat supply and electricity, profit, net present value, return on investment, payback period of capital investments) and of the region (subsidies and investments for developing the region's energy potential, savings in budget funds for heat and power supply), indicators of the organization's energy efficiency (energy savings, energy intensity of business processes and organization's overall activities, specific energy consumption per unit of product) and of the region.

Stage 4 "Implementing measures for introducing sustainable energy technologies" involves the organization of work on the introduction of sustainable energy technologies for the responsibility centers; improving professional skills of workers involved in the introduction and use of sustainable energy technologies; monitoring results of the measures being implemented.

Stage 5 "Analyzing results of introducing and using sustainable energy technologies, assessing the replication experience" includes studying the results of investment projects, implementing corrective measures, and working on investment projects involving the extension of business practices to other business processes or organizational structures.

4. Discussion

In modern conditions, the companies seek to ensure sustainable, competitive, and financially successful operation in the market, including through the use of management tools based on an analysis of Russian and international experience in the application of sustainable energy technologies [6, 7].

The hypothesis of the study was to identify the ability of the company's management on the basis of the developed algorithm for planning, implementing, and evaluating the economic efficiency of benchmarking sustainable energy technologies in cross-border regions to achieve the planned financial and economic indicators, to respond promptly and accurately to the deviations arising to ensure the enterprise's stable development, as well as to improve the quality and standard of living as a whole.

As a result of the study, we found that implementing the concept of reliable, efficient, socially acceptable, and environmentally safe energy supply with universal access to sustainable energy technologies without compromising the ability of future generations to meet their needs is designed to reduce the total energy consumption by 40%, ensuring the growing share of sustainable energy up to 30% and increasing energy efficiency of activities of organizations and society as a whole [8].

Incentives for using sustainable energy technologies include ensuring energy security; environmental protection; conservation of mineral fuel resources; coverage of new markets. The types of sustainable energy sources include solar energy, water, wind, biomass, geothermal energy, wave and sea energy, hydrogen. In addition, nuclear energy and waste of cities, while not being renewable, are at the same time sustainable as it would not run out in the foreseeable future.

In the research process it was found that the main directions of the energy policy of most countries of the world include the development and improvement of wind and photogeneration technologies, biomass processing, and searching for safe technologies to use hydrogen [9]. With new construction, the development of eco-city technologies is possible, with the aim of providing a closed cycle of energy supply [10]. Recycling through anaerobic decomposition with the production of biogas (which is based on methane) can solve the problems of recycling and replace fossil fuels in the energy balance of cities [10]. Also, the use of biomass (agricultural and forestry waste) can significantly increase the production of fuel for energy content (in terms of energy intensity, the total amount of 420-550 million tons of biomass waste in Russia is equivalent to 85-128 million tons of gasoline). At the same time, the main task is to develop efficient technologies for their processing [11].

The studied experience of applying sustainable energy technologies of the world is not systematized in terms of the impact of their application on the energy efficiency of regions and presents isolated studies on the analysis of the results of the implementation of certain alternative energy technologies.

This study proposes a comprehensive algorithm for planning, implementing, and evaluating the economic efficiency of benchmarking sustainable energy technologies in cross-border regions. The results of the study can be used to further improve the methodology for assessing the economic efficiency of using sustainable energy technologies in cross-border regions in order to minimize energy costs and environmental damage.

5. Conclusion

The algorithm proposed by the authors for planning, implementing, and evaluating the economic efficiency of benchmarking sustainable energy technologies in cross-border regions will allow for more efficient management, strengthening the responsibility of enterprise employees, on the one hand, and reducing the level of relevant costs, on the other. At the same time, benchmarking sustainable energy technologies of companies with best business practices is important not only for those enterprises interested in saving energy resources and increasing operational efficiency. Its results are beneficial to the state, which finances energy development through direct investment and through tariff regulation, indirectly. And the state is interested in increasing the efficiency of spending these funds.

References

[1] Akhtar Hussain, Syed Muhammad Arif, Muhammad Aslam 2017 Emerging renewable and sustainable energy technologies *Renewable and Sustainable Energy Reviews* **71** pp 12-28

[2] Rosemary N Wojuola, and Busisiwe P 2019 Alant sustainable development and energy education in Nigeria *Renewable Energy* **139** pp 1366-1374

[3] Pibin Guo, Juan Kong, Yanshan Guo, and Xiuli Liu 2019 Identifying the influencing factors of the sustainable energy transitions in China *Journal of Cleaner Production* **215** pp 757-766
[4] Chernikova O P, and Baranov P P 2018 Environmental reporting in coal mining *Gornyi Journal* **3** pp 82-85
[5] Kuznetsova D I, and Chernikova O P 2016 Opportunities for using solar energy in the Kemerovo Region In *Proceedings of the Conference on Energy, Informatics, Innovations-2016* (pp 130-134) (Smolensk, Russia)
[6] *International year of sustainable energy for all* 2012 Available at: https://www.un.org/ru/events/sustainableenergyforall/background.shtml (Accessed 29 03 2019)
[7] *UN Chronicle on Sustainable Energy* 2015 Available at: https://unchronicle.un.org/issue/sustainable-energy (Accessed 10 02 2019)
[8] Liuyang Zhan, Meiting Ju, and Jinpeng Liu 2011 Improvement of China energy label system to promote sustainable energy consumption *Energy Procedia* **5** pp 2308-2315 Available at: https://www.researchgate.net/publication/251712497_Improvement_of_China_Energy_Label_System_to_Promote_Sustainable_Energy_Consumption (Accessed 10 02 2019)
[9] Shevtsova S V, and Zholud D S 2010 Analysis of foreign experience in the use of alternative energy sources *Energy Saving. Energy. Energy Audit* **6**(76) pp 49-53
[10] Phillips L, and Smith P 2015 *Sustainable urban energy is the future* Available at: https://unchronicle.un.org/article/sustainable-urban-energy-future/
[11] Varfolomeev S D, and Shevaleevsky O I 2010 Biofuels and photoelectricity – Russian perspectives *The Chemical Journal* **10** pp 38-41
[12] Alekseev A N, Lobova S V, Bogoviz A V, and Ragulina, Y V 2019 A critical review of Russia's energy efficiency policies in the construction and housing sector *International Journal of Energy Economics and Policy* **9**(4) pp 166-172

Prospects for sustainable development of Russia: the case of eco-innovations

M Imekova[1*] and E Boltanova[1]

[1] National Research Tomsk State University, 32 Lenin Avenue, Tomsk 634050 Russia

E-mail: imekova_maria@mail.ru

Abstract. The article discusses current global trends of sustainable development. Our research clearly demonstrates that environmental innovation is one of the main tools for such development. However, this tool is not widely spread in Russia, including among cross-border regions. The article identifies the causes and possible solutions to the current problem.

1. Introduction

On September 25, 2015, the UN General Assembly (UN GA) adopted Resolution A/70/L.1 *Transforming Our World: The Sustainable Development Agenda for the period up to 2030* by unanimous vote [1] (*Agenda for the Period Until 2030*). This agenda included 17 global goals and 169 related tasks that are complex and indivisible, balancing all three components of sustainable development: economic, social, and environmental ones.

In the *Agenda for the Period Until 2030*, science, technology, and innovation are recognized as one of the main factors for achieving the goals of sustainable development. A similar conclusion was made in another Resolution of the UN GA A/RES/72/242 of December 22, 2017 *The Impact of Rapid Technological Progress on Achieving Sustainable Development Goals* [2]. According to the UN GA, "Science, technology, and innovation, including environmental innovation, can play a crucial role in the development and facilitation of efforts to solve global problems, in particular efforts to eradicate poverty, ensuring food security and nutrition and more efficient development of agriculture, expanding access to energy resources and improving energy efficiency, fighting diseases, improving the education system, protecting the environment, accelerating economic diversification and the transformation process and increasing productivity and competitiveness, and supporting sustainable development."

Given the importance of this issue, the UN GA proposes to focus on environmental innovation (eco-innovation) to achieve the goals of sustainable development. More than that, the motivation of many states to pay attention to the problem of developing and introducing eco-innovations is quite understandable. However, in contrast to such states, an understanding of the need for their accelerated development appeared relatively recently in Russia. Since it is impossible to improve the competitiveness of the country, to solve the tasks of economic, social, and environmental development without relying on them. Thus, in the rating of countries' competitiveness in 2017–2018, which is published by the World Economic Forum, indicates that 36 countries of the world are at the stage of innovative development of the economy, which are mainly the countries of Europe (23 countries), as well as the USA, Canada, Australia, and a number of countries Asia (Japan, Singapore, Republic of

Content from this work may be used under the terms of the Creative Commons Attribution 3.0 licence. Any further distribution of this work must maintain attribution to the author(s) and the title of the work, journal citation and DOI.
Published under licence by IOP Publishing Ltd

Korea, Bahrain, Hong Kong, Taiwan, Qatar, etc.). Another 20 countries, including Russia, are at the stage of transition to an innovative development path [3].

In recent years, to ensure the transition of the Russian Federation to such a development path, a number of strategic planning documents (federal concepts, strategies, programs, and projects), offering various measures to stimulate innovative activity of environmental users, have been developed and approved. Within the framework of this article, we are invited to analyze these documents in order to determine the prospects for sustainable development of Russia.

2. Materials and Methods

The study was conducted on the basis of the following materials: (a) UN GA resolutions; (b) domestic legal acts of the Russian Federation, including strategic planning documents (*Concept of the Long-Term Socio-Economic Development of the Russian Federation for the Period up to 2020*, approved by the Decree of the Government of the Russian Federation of November 17, 2008 No. 1662-p; *Strategy of Innovative Development of the Russian Federation for the Period up to 2020*, approved by the Order of the Government of the Russian Federation No. 2227-p dated December 8, 2011; *Basis of the State Policy in the Field of Environmental Development of the Russian Federation for the Period up to 2030*, approved by the Decree of the President of the Russian Federation on April 30, 2012; *Strategy of Environmental Safety of the Russian Federation for the Period up to 2025*, approved by the Decree of the President of the Russian Federation No. 176 of April 19, 2017, etc.); (c) regulatory legal acts (Tax Code of the Russian Federation, Federal Law on Environmental Protection, etc.); and (d) contemporary scientific literature.

Both general scientific and private scientific methods were used in the research process. The general scientific methods used in the study include a dialectical method (when studying the development of legislation of the Russian Federation in the field of eco-innovations), deduction and induction methods (when determining measures to stimulate innovative activity of environmental users). The private scientific methods used in the study include the formal legal method (when studying UN GA resolutions, domestic acts of the Russian Federation), the method of legal interpretation (in understanding the content and determining the practical feasibility of the norms contained in the internal acts of the Russian Federation), as well as the method of legal forecasting (in determining the prospects for sustainable development of Russia).

3. Results

Strategic planning documents provide for various measures to stimulate innovative activity of environmental users. However, the study clearly shows that only the following were implemented of all the proposed measures:

1. Tax policy measures;
2. A new system of environmental regulation and minimizing a negative impact on the environment, providing for the issuance of integrated environmental permits to enterprises (hereinafter referred to as IEP) was put in place. More than that, a number of standards and plans for a phased reduction of pollution to the levels corresponding to the best available technologies were adopted;
3. Ecological certification of eco-innovations. Up to date, in the Russian Federation, a number of national standards aimed at encouraging the introduction of eco-innovations by environmental users have been developed. However, it should be noted that in our country, these standards are not mandatory, are a voluntary mechanism for confirming the compliance of products and processes with environmental requirements. In this connection, voluntary environmental certification in Russia is not widely used in the activities of business entities, in contrast to the developed foreign countries;
4. A program approach in the field of environmental protection and environmental management. In recent years, in the Russian Federation, a programmatic approach in the field of environmental protection has acquired a special scope. After adopting the *Agenda for the Period*

IOP Conf. Series: Earth and Environmental Science **395** (2019) 012103 doi:10.1088/1755-1315/395/1/012103

until 2030, Russia took up its adaptation to the Russian legal system by embedding it in its national policies, strategies, and plans. In this case, the three main areas of action were identified: (1) Conducting an audit of existing strategies and plans at national, regional, and local levels and comparison with global goals of sustainable development and objectives to identify inconsistencies and opportunities for change; (2) determining their own national goals guided by global goals, but taking into account national conditions and having achievable objectives; (3) developing strategies and plans based on sustainable development goals.

The following documents are the results of the adaptation work:

1. Presidential Decree No. 642 of December 1, 2016 *On the Strategy for the Scientific and Technological Development of the Russian Federation*. At the moment, the Implementation Plan for the first stage of the Strategy for Russia's Scientific and Technological Development has been approved. The documents provides fundamentals for the development of the scientific and technological complex. These documents are also reflected in the State Program *Development of Science and Technology for 2013–2020*, including federal targeted programs and subprogrammes.
2. Presidential Decree No. 204 of 07 May 2018 *On the National Goals and Strategic Objectives of the Development of the Russian Federation for the Period up to 2024*. The main national development goals of Russia, as well as national projects (programs) that must be developed to achieve the goals set, are defined in this act as a result of adapting the *Agenda for the Period up to 2030*. It is noteworthy that in the aforementioned Presidential Decree, environmental problems are not designated as the national goal of the development of Russia, although the Government of the Russian Federation was instructed to develop a passport of the *Ecology* national project.

The main disadvantages of all the documents listed are the following:

1. Basically, they have a "declarative" character, i.e. they lack an effective implementation mechanism;
2. In these documents, the concept of sustainable development has not been fully implemented: environmental priorities are not combined with the country's economic objectives. In addition, there is no link between innovation and sustainable development of cross-border regions, cities and settlements, country's economic growth try, increased productivity in the economy, efficient use of resources, and innovative modernization of industry, etc. In other words, the documents weakly link the problems of ecology and economic development of Russia.

The considered issues lead to the conclusion that in our state, at the present stage, insufficiency of the existing measures to stimulate innovative activity of environmental users is obvious. The legislative and executive bodies should move away from the "declarative" approach of legally regulating relations in the field of eco-innovations and take additional measures aimed at implementing the ideas proposed in the strategic planning documents, including through the adoption of laws and other legal acts.

In addition, in the strategic planning documents of Russia, the concept of sustainable development requires a deeper rethinking. Three components form its basis (economic, social, and environmental), which should be developed in conjunction with each other. Economic measures should provide both economic and environmental benefits that would reduce environmental risks, including depletion of natural resources, increased man-made pressures, and environmental pollution. In other words, the transition of Russia to sustainable development is possible only with the interconnectedness and consistency of the economic and environmental measures being taken.

4. Discussion

In recent years, a lot of attention among the Russian scholars (E. S. Boltanova [4], O. P. Burmatova [5], Ya. D. Vishnyakov, S. P. Kiseleva [6], D. M. Khloptsov [7]) has been given to the problems of sustainable, ecologically oriented innovative path of development of Russia. At the same time, researchers consider these problems from various aspects: both in terms of the imperfection of Russian legislation, as well as in terms of the lack of effective economic instruments. At the same time, there is a definite unifying principle, or thought in all these works. According to the authors of these works, Russia undoubtedly has a high potential necessary for the transition to a sustainable, environmentally oriented innovative development path, and the key role in this transition belongs to the state. The responsibility for implementing a systematic policy of state regulation of the development of environmentally oriented innovation processes lies with the state.

So, E. S. Boltanova believes that Russia has created a strategic base for the implementation of eco-innovations in various sectors of the economy. However, since eco-innovations are not investment-attractive for businesses, the state plays a dominant role in this process. The state through stimulation and coercion in regulatory legal acts, regulatory technical acts can create necessry conditions for promoting energy-efficient economic activity [4].

Ya. D. Vishnyakov, and S. P. Kiseleva express a similar point of view. In addition, they propose, on the part of the state, to introduce additional measures of state regulation of environmentally oriented innovative development, which is aimed at increasing the interest of all business entities in introducing environmental innovations.

Among such measures are the following:

1. Developing a modern mechanism for the formation of a state order to develop environmentally oriented production processes and environmentally oriented innovative products (services);
2. Establishing and conducting short and medium-term state support for domestic producers;
3. Developing a national integrated information and communication infrastructure, ensuring the formation of a single information space of innovation activities in priority areas, etc. [6].

Thus, according to a number of Russian scholars, it is the state that is the main regulator, which can ensure the transition of the Russian Federation to an environmentally oriented innovative development path. To achieve this, the development and implementation of additional measures to stimulate innovative activity of economic entities is strongly required.

5. Conclusion

One of the modern global trends of sustainable development is the accelerated development of eco-innovations. In Russia, an understanding of the need for their development has appeared relatively recently. In this regard, in the past few years, a lot of strategic planning documents (federal concepts, strategies, programs and projects), offering various measures to stimulate innovative activity of environmental users, have been developed and approved in our country. However, the study showed that the main shortcomings of such documents are their "declarative" nature, as well as the insufficiency of measures provided for in them to stimulate innovative activity of environmental users. The legislative and executive authorities of the Russian Federation need to abandon this approach, develop, and introduce additional incentive measures.

Also, the concept of sustainable development requires deep rethinking in strategic planning documents. This concept has not received its full realization. The documents weakly link the problems of ecology and economic development, while economic measures should provide not only economic but also environmental benefits. Only under the condition of consistent implementation of the main ideas of the concept, in the short term, our country will be able to move to an eco-oriented way of innovative development.

6. Acknowledgments

The article was prepared with the support of the RFBR grant 18-011-00612.

References

[1] General Assembly 2015 *Transforming Our World: the 2030 Agenda for Sustainable Development* –

Resolution (September 25, 2015) Available at: https://undocs.org/ru/A/RES/70/1 (Accessed 05 04 2019)

[2] General Assembly 2017 *Impact of rapid technological change on the achievement of the Sustainable Development Goals – Resolution* (December 22, 2017) Available at: https://undocs.org/ru/A/RES/72/242 (Accessed 05 04 2019)

[3] *Global competitiveness index 2017–2018* 2018 Available at: http://reports.weforum.org/global-competitiveness-index-2017-2018/appendix-a-methodology-and-computation-of-the-global-competitiveness-index-2017-2018 (Accessed 05 04 2019)

[4] Boltanova E S 2018 Legal support of environmental innovation on the example of the construction industry *Environmental Law* **4** pp 39-45

[5] Burmatova O P 2012 Ecologization of production in the light of innovation development *Region: Economics and Ecology* **4** pp 257-277

[6] Vishnyakov Ya D, and Kiseleva S P 2014 Improving the economic mechanism for stimulating business entities to reduce a technogenic load on the environment in the context of innovative development *World of Science* **3** pp 1-12

[7] Khloptsov D M 2018 Economic analysis of legal regulation in the field of innovative development In N A Simchenko (Ed) *The economic basis for the development of science and technology in Russia* (pp. 235-241) (Simferopol: ARIAL)

IOP Conf. Series: Earth and Environmental Science **395** (2019) 012104 doi:10.1088/1755-1315/395/1/012104

The state of food industry as a factor in food security of transboundary territories

V Pomozova[1], T Kiseleva[1] and N Stepakova[2*]

[1] Kemerovo State University, 6 Krasnaya str., Kemerovo 650000 Russia
[2] Far Eastern Higher Combined-Arms Command School, 158 Lenin str., Blagoveshchensk 675000 Russia

E-mail: kitex74@mail.ru

Abstract. The article presents an analysis of the current state in the food industry in the Far Eastern Federal District. Also, it analyzes its impact on ensuring food security in the region.

1. Introduction

One of the primary problems at the present stage of development of society is the provision of high-quality food to the population. The food industry is one of the most important strategic sectors of the economy, on the state and development of which food security of the state depends. The availability of high-quality and safe food for all groups of the country's population in the volumes and assortment necessary to maintain a full-fledged lifestyle and performance is a factor determining the quality of life of the population and directly affecting the change in demographic situation [7].

In the framework of the Food Security Doctrine of the Russian Federation (adopted in 2010), the main measures to ensure food security of the Russian Federation are the organization of continuous monitoring and control of internal and external threats to food security, ensuring food independence through the stable development of domestic industrial and agricultural production, the introduction of innovative technologies and high-tech equipment at the food industry enterprises, quality assurance and food safety [1].

The purpose of this article is to analyze the state of the food industry in the Far Eastern Federal District in the framework of food security.

2. Materials and Methods

The theoretical basis of the study is regulatory documents. The information base is the reporting data of the Federal State Statistics Service, the Federal Customs Service, the Unified Interdepartmental Information and Statistical Systems. Research methods: analytical and statistical.

3. Results

The food industry of Russia is part of the manufacturing industry, and it is one of its key industries. By the volume of shipped goods of own production, the food industry is in the top three among the manufacturing industries, competing with such industries as metallurgical production and the production of coke and petroleum products [2]. Analyzing the state of the food industry from 2009 to 2018, it should be noted that in the pre-crisis period (2009-2014), the growth rate of the food industry in terms of the volume of products shipped did not exceed the growth rate of manufacturing industries (Figure 1). With

Content from this work may be used under the terms of the Creative Commons Attribution 3.0 licence. Any further distribution of this work must maintain attribution to the author(s) and the title of the work, journal citation and DOI.
Published under licence by IOP Publishing Ltd

IOP Conf. Series: Earth and Environmental Science **395** (2019) 012104 doi:10.1088/1755-1315/395/1/012104

the onset of the crisis and the subsequent imposition of sanctions on the export to Russia of various types of products, the growth rate of the food industry increases significantly: in 2014, the food industry shows an increase of 113.3%, which is 2.7% more than in the manufacturing industry; in 2016, growth rates have a value of 106.7%, this is 3.8% more than in the manufacturing sector in spite of a decline [2].

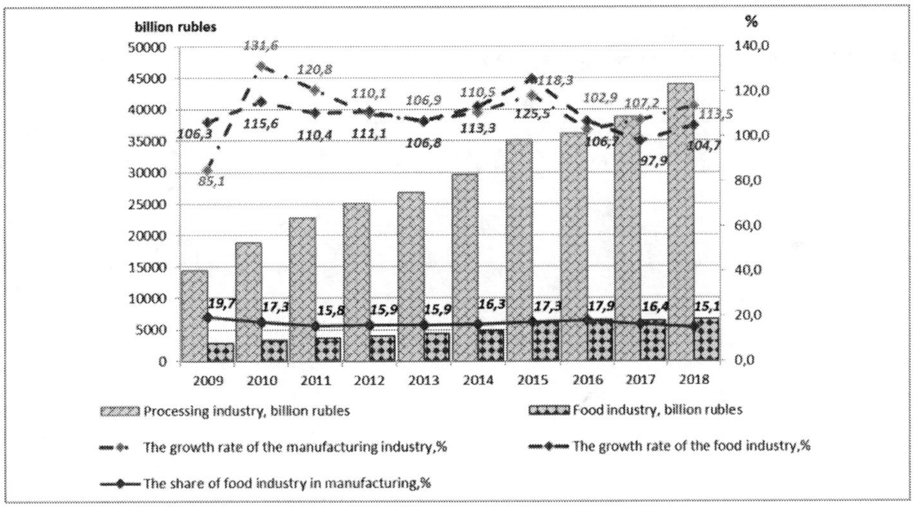

Figure 1. Dynamics of the volume of shipped goods of own production by branches of manufacturing industries and food industry [2-4].

The development of food production industries leads to a change in the volume of shipped products in the regions of Russia. In many regions, there is a tendency to maintain the share of food industry products shipped among the manufacturing industries (Table 1) [2–4].

Table 1. Regional indicators of the shipment of food products in the total shipments of the manufacturing industry, in%.

Federal District	2009	2010	2011	2012	2013	2014	2015	2016	2017	2018
Russian Federation	19.8	17.2	15.2	15.6	15.5	15.9	17.3	17.7	15.6	16.4
Central	26.5	23.8	20.9	21.6	20.7	20.2	22.9	22.2	20.4	21.8
Northwestern	18.5	15.1	11.4	12.6	12.6	13.2	15.4	16.0	15.1	16.4
Southern	30.8	26.8	24.2	23.6	23.3	23.9	25.3	26.0	25.8	27.8
North Caucasus	32.4	36.8	43.1	39.6	36.1	39.6	38.7	38.9	36.8	39.2
Volga	13.5	11.8	10.4	10.9	11.3	12.1	13.0	14.0	11.2	12.6
Ural	8.6	6.8	6.6	6.3	6.5	7.7	6.8	7.2	6.4	7.5
Siberian	17.8	15.1	14.2	14.5	14.8	14.3	14.2	14.7	14.3	16.1
Far Eastern	38.5	37.2	27.1	34.2	32.5	31.1	37.5	41.5	36.8	37.7
Crimean							53.1	45.5	44.7	45.3

In the Southern, North Caucasus, and Far Eastern Federal Districts, the percentage of food production shipments among the processors is at least 25%. From the perspectives of the state and development of food production, the Far Eastern District is of particular interest, which is due to a number of features: geographical location, climatic conditions, socio-economic development.

The Far Eastern Federal District is one of the largest districts of Russia. It occupies 40.6% of the country's territory, where 5.6% of the population of Russia lives. It is the most remote from the developed regions of the country. The climatic conditions of the Far East Federal District are characterized by sudden temperature changes, high humidity in the warm season, long winter weather with low temperatures. This, in turn, negatively affects both the physical and mental state of people. Urbanization of the population amounted to 75.43% in 2017. Urbanization contributes to the increasing concentration of anthropogenic stress, as a result of which the environment is polluted more intensively. The urban environment may not keep up with its rapid growth. For example, in many cities of the Far Eastern Federal District, the quality of water and air does not meet the necessary standards [5]. Such environmental conditions dictate increased requirements for providing the population of the region with healthy and safe food.

Table 2. The structure of the resource base for basic foodstuffs in the Far Eastern Federal District [2].

Resource type	2009	2010	2011	2012	2013	2014	2015	2016	2017
Meat and meat products, thousand tons	83621	81742	84510	82109	81188	86955	78604	77705	79842
per capita, kg / year	13.26	12.99	13.44	13.10	12.98	13.96	12.65	12.54	12.91
import share, including import, %	41.7	42.30	43.05	51.44	53.19	46.02	52.63	54.48	53.29
Milk and dairy products, thousand tons	44367	43722	42039	44876	48500	48719	53721	53596	54594
per capita, kg / year	70.56	69.48	66.89	71.62	77.58	78.24	86.49	86.52	88.30
import share, including import, %	16.21	16.38	17.42	16.88	16.24	15.97	14.07	14.44	15.16
Potatoes, thousand tons	13114	12838	12884	13221	10377	13076	11925	11212	11973
per capita, kg / year	203.0	204.0	205.0	211.0	166.0	210.0	192.0	181.0	193.7
import share, including import, %	7.22	7.07	9.02	9.09	12.18	11.45	10.51	16.13	14.16
Vegetables and melons, thousand tons	3889	4012	4300	4338	3913	4503	4012	4138	4136
per capita, kg / year	60.2	65.8	70.6	71.8	64.6	74.6	66.7	66.8	66.9
import share, including import, %	56.3	58.3	57.0	54.7	56.6	57.5	57.8	57.6	57.1
Eggs, mln.	1127	1189	1167	1185	1149	1189	1149	1191	1173
per capita, pieces	175	189	187	191	185	193	191	191	193
import share, including import, %	39.6	38.1	40.2	34.9	40.3	34.4	38.7	39.5	48.5

One of the main factors of food safety is the degree of dependence on the importation and importation of food into the region. The data in table 2 indicate that only milk and dairy products show positive dynamics. In the period from 2012 to 2016, the imports of this type of products decreased by 2.4%, and over the past 9 years there was a steady increase in the production of milk and milk products per capita. Between 2012 and 2017, this indicator increased by 16.68 kg / year. The critical situation is in such types of products as meat and meat products and vegetables and melons and gourds. The volume of production of meat and meat products per capita is not able to satisfy the physiological norms of consumption of these products even by 20% [6]. From 2012 to 2017, the share of imported and imported meat products exceeds 50%. The volume of vegetables and gourds grown on the territory of the Far-Eastern Federal District averages 50% of the recommended physiological norm of consumption, while the share of imported resources remains consistently high and amounts to 55-58% relative to its own production.

According to the state of the resource base of basic food products, not only food independence, but also physical accessibility of food products is determined. One of the criteria for assessing the physical availability of food is the actual energy value and composition of substances included in the daily diet of the population.

Table 3. The actual nutritional and energy value of the daily diet of the inhabitants of the Far Eastern Federal District, per person per day.

Indicator	2009	2010	2011	2012	2013	2014	2015	2016	2017
Squirrels, g	72.9	77.1	78.3	79.4	80.4	79.9	76.1	74.0	73.5
Fat, g	98.5	105.8	107.0	108.0	109.0	108.4	103.4	101.6	102.0
Carbohydrates, g	321.4	338.8	338.4	337.2	334.7	331.6	313	303.7	301.6
Energy value, kcal	2474.2	2627.4	2641.2	2650.9	2652.9	2633.3	2498.4	2435.8	2429.9
% to normal	82.5	87.6	88.0	88.3	88.4	87.7	83.3	81.2	80.9

Analyzing dynamics of the total energy value and the quantitative composition of substances in the consumed grocery set, which is presented in Table 3, it should be noted that there is a lack of nutritional balance. Inadequate carbohydrate intake is observed. The actual carbohydrate intake of DFO residents is no more than 75% of the recommended norm, while the fat intake exceeds the recommended rate by 10-15%. The most stable is the consumption of proteins, the dynamics of consumption is in the range of 90-100%. The total energy value of the food basket averages 85% of the recommended norm [7]. These indicators indicate that physical accessibility of food products is not fully implemented in the region.

4. Discussion

The analysis of quantitative indicators of food production in the Far Eastern Federal District allows us to draw the following conclusions:

- The volume of shipment of food products among the processing industries remains stable with minor fluctuations, which corresponds to the indicators for this criterion for Russia as a whole;
- From a consumer point of view, the DFO only partially fulfills its intended function of providing the population with high-quality food;
- Ensuring food security, as one of the functions of the food industry in the region, is not fully implemented, as evidenced by the volumes of imported and imported food products.

5. Conclusion

For successfully implementating objectives of the food industry in the Far East region and increasing the volume of its own food production, a complex system of measures is needed, consisting both in state support of the region, development of agricultural production as a raw material base of food production, and in the development of interregional ties in the food industry security.

References

[1] Presidential Executive Office 2010 *Decree "On Approval of the Doctrine of Food Security of the Russian Federation"* (January 30, 2010 No. 120) Available at: http://base.garant.ru/12172719/#ixzz5jTt7kAfa
[2] Federal State Statistics Service (2019) *Statistical database* Available at: http://www.gks.ru
[3] Rosstat 2016 *Industrial production in Russia in 2016* (Moscow, Russia: Rosstat)
[4] Rosstat 2018 *Russian statistical yearbook - 2018* Available at: http://www.gks.ru/free_doc/doc_2018/year/year18.pdf
[5] Shkrabtak N, Frolova N, Kiseleva T, Sergeeva I, and Pomozova V 2019 Impact of environmental conditions on the health of the Far Eastern region population. *Appl. Sci.* **9**, p 1354 doi:10.3390/app9071354
[6] Ministry of Health of the Russian Federation 2016 *Order "On approval of Recommendations on rational norms of food consumption that meet modern requirements for healthy nutrition"* (August 19, 2016 No. 614) Available at: http://www.garant.ru/products/ipo/prime/doc/71385784/
[7] MR 2.3.1.2432-08 2008 *Norms of physiological needs for energy and nutrients for various groups of the population of the Russian Federation (December 18, 2008)* Available at: http://docs.cntd.ru/document/1200076084

IOP Conf. Series: Earth and Environmental Science **395** (2019) 012105 doi:10.1088/1755-1315/395/1/012105

State and problems of exporting the products of the agro-industrial complex in the Altai region

Yu Bugai[1], **A Minenko**[1] **and M Khorunzhin**[1*]

[1] Altai State Agricultural University, Krasnoarmeysky 98, Barnaul 656049 Russia

E-mail: khorunzin@yandex.ru

Abstract. The article reflects the export orientation of the Altai Region and its relationship with the production of agro-industrial products. Export promotion instruments, including subsidies for compensation of a part of the costs of transporting agricultural products and foodstuffs, as well as a preferential tariff for the transportation of grain by rail are considered. The structure of exports of agricultural products are analyzed. Infrastructure and logistics problems of the grain market are also reviewed.

1. Introduction

The agro-industrial complex of the Altai Region is the main structural sub-system of the regional economy. Agriculture defines the basis of the potential of agriculture. Over 15% of gross regional product is accounted for by agricultural production only [2]. Export potential is directly related to the production of agro-industrial products and the development of agro-industrial complex. To expand access to foreign markets, new export promotion tools are needed by organizations at the regional level.

2. Materials and Methods

When conducting research, the following general scientific and special methods and techniques of economic research were used: monographic and economic-statistical methods, the method of institutional analysis.

3. Results

One of the key directions in strategic development of the agro-industrial complex is international cooperation and export. In order to support the export of agricultural products, raw materials and foodstuffs, the Ministry of Agriculture of Russia has developed a priority project "Export of Agricultural Products" (approved by the Presidium of the Council under the President of the Russian Federation on Strategic Development and Priority Projects, Protocol No. 11 of 30 November 2016). This project is included in the State Program for the Development of Agriculture and Regulation of Agricultural Products, Raw Materials and Food Markets for 2013–2020.

The goal of the priority project "Export of agricultural products" is to create an industry-wide system for supporting and promoting Russian agricultural products on foreign markets and ensuring its compliance with their requirements. The plan is to increase the volume of exports of agricultural products in terms of value by at least 12.5% by the end of 2018, by no less than 26.7% by the end of 2020.

Content from this work may be used under the terms of the Creative Commons Attribution 3.0 licence. Any further distribution of this work must maintain attribution to the author(s) and the title of the work, journal citation and DOI.
Published under licence by IOP Publishing Ltd

The Center for the Analysis of Exports of Agricultural Products has been created to provide comprehensive information and analytical support to domestic producers and exporters of agricultural products. In cooperation with the industry associations of manufacturers of the agro-industrial complex and the "Russian Export Center" JSC, the Center prepared and approved programs to promote and increase the export of certain types of beet sugar, grain processing, oil and fat, confectionery products of the industry, food products of the agro-industrial complex with high margins, as well as grain and poultry products.

At the federal level, new export promotion tools were launched in 2017 in order to expand access for Russian organizations to foreign markets:

- Subsidies are provided to a number of Russian organizations to reimburse part of the cost of transporting agricultural and food products by land, including rail. They contribute to reducing the cost of transporting agricultural and food products by rail involving road transportation, as well as increasing exports of agricultural products (Resolution of the Government of the Russian Federation of September 15, 2017 No. 1104);
- The preferential tariff is provided for the transportation of grain by rail. In this case, the grain is related to the codes of the Unified tariff-statistical nomenclature of cargo 011005, 014003, 015006 (wheat, barley, corn) (Decree of the Government of the Russian Federation of December 20, 2017 No. 1595).

The Altai region belongs to the export-oriented regions. Realization of its export potential is directly related to the production of agro-industrial products and the development of agro-industrial complex [3].

New export support mechanisms launched in 2017	
Subsidies from the federal budget provided to Russian organizations for reimbursement of part of all costs of transporting agricultural and food products by land, including rail transport (Resolution of the Government of the Russian Federation No. 1104 of September 15, 2017)	Subsidies are provided from the federal budget of "Russian Railways" OJSC; and losses in income resulting from the establishment of tariffs for the transportation of grain are reimbursed (Resolution of the Government of the Russian Federation No. 1595 of December 20, 2017)
The following products were subsidized (before the amendment of May 4, 2018): grain grinding products (flour, groats, etc.) of sunflower, rapeseed, flax seeds, potatoes, vegetables, meat and meat products, dairy products, etc.	The following products are subsidized: wheat, barley, and corn.
↓	↓
Altai enterprises exported agricultural products in the amount of 315.5 million rubles, the number of subsidies received was 25.2 million rubles (up to 50% of transportation costs are compensated)	Inclusion of the Altai region in the list of regions will allow compensating transportation fees for grain cargo (freight charge), i.e. 50% for grain transportation

Figure 1. New export support tools launched in 2017.

In order to ensure the successful promotion of Altai goods (non-commodity exports) and services to foreign markets, the Action Plan ("road map") entitled "Development of Export and Transit and Logistics Activities in the Altai region was approved by the Order of the Administration of the Altai Region No. 278-p on September 28, 2016.

The implementation of the "road map" will increase the number of exporting organizations, significantly increase the volume of non-oil exports, increase a share of highly processed goods in the total volume. Consequently, this would positively affect the state of competition and the business climate in the region [4].

Implementation of the "Cooperation Agreement between the Government of the Altai Region and the Key State Grain Market Operators" signed for 2017-2018 will help reduce barriers to export activities and provide competitive conditions for promoting Altai products in the foreign market, and most importantly in the markets of the Asia-Pacific countries region [6].

Food and processing industry enterprises of the region successfully implement export-oriented policies. The range of goods supplied for export is very wide: from grain processing products to complex food additives, including products of antler reindeer husbandry.

Significant volumes of export deliveries fall at the milling enterprises. Every year, the milling industry produces about 1.1 million tons of flour and about 0.5 million tons of cereals, processes almost all commodity grain grown in the region. Already today, the Altai grain processing products are represented in China, Mongolia, and Central Asian republics; experience in supplying it to Vietnam, South Korea, Indonesia and a number of other countries in the Asia-Pacific region is developing. The markets of near and far abroad for the supply of vegetable oils are actively mastered.

The Altai region was one of the first to have an agreement with Federal State Institution (or FGU) "Emerkom" Agency," which enabled our processors to participate in humanitarian operations of the Russian Federation.

An important step in the promotion of products to foreign markets of the region is to market the image of the region and its food. In order to present the most valuable resources of the Altai Region to residents of the whole country, work on promoting brand products is carried out, including through large-scale food festivals. Thus, in the Altai region, the following events are held annually: "Days of Altai cheese," "Meat Gourmet's Day," "Honey Spas in Altai," "Feast of Bread," "AltaiFest," and others. Every year, the number of participants in these holidays is growing, and the most popular festivals become the reasons for event tourism.

According to the Siberian Customs Administration, the export of food products and food raw materials from the region amounted to about $ 186 million in 2017 ($ 184 million in 2016). In the commodity structure of exports of the Altai Territory, the share of food products and agricultural raw materials was 19.6% (only 5.8% in the Russian Federation).

Among the main trading partners of the Altai Region, such countries act as the Republic of Kazakhstan, the People's Republic of China, the Republic of Belarus, the United States of America, Canada, Uzbekistan, Ukraine, Afghanistan, Germany, Mongolia, Tajikistan, Kyrgyzstan, Latvia, Lithuania, the Netherlands, Italy, and others.

The main commodity items in the structure of exports of agricultural products in terms of value were: vegetable oils (26.2%, including sunflower oil - 14.0%, rapeseed oil - 9.0%), oilseeds (17.3%, including rapeseed - 8.3%, flax seeds - 6.1%), products of the milling industry (15.7%, including cereals, processed in other ways (flakes, chopped, etc.) - 11.5%), wheat flour - 3.0%), beans dried, peeled (6.3%, including peas - 3.5%, lentils - 2.5%), cereals (3,8%, including wheat -1.16%, buckwheat - 2.0%).

In comparison to 2016, exports of the following types of agricultural and food products increased in 2017: meat and edible meat by-products in 3.6 times (in value terms) (up to 901.3 tons), milk and cream in 1.5 times (up to 3.6 thousand tons), dried leguminous vegetables in 1.8 times (up to 34.0 thousand tons), cereals in 2.3 times (up to 30.0 thousand tons (without rice), including wheat and meslin in 3 times (up to 11.1 thousand tons), barley in 7.7 times (up to 1.6 thousand tons), buckwheat in 2.3 times (up to 12.6 thousand tons), oilseeds in 2 times (up to 73.6 thousand tons, including rapeseed in 13 times (up to 37.5 thousand tons), sunflower in 3 times (up to 6.3 thousand tons), beet sugar in 1.7 times (up to 164 tons), bran in 2 times (up to 9.7 thousand tons) [1].

At the same time, a decrease in export deliveries of the following products was observed: butter by 60.8% (to 43.5 tons), cheese and curd by 26.2% (to 656 tons), wheat flour by 43.3% (to 13.4 thousand tons), vegetable oils - by 14.5% (to 70.3 thousand tons).

The Altai region was actively involved in the implementation of the Government Decree No. 1104 of September 15, 2017, which provides for subsidizing the costs of transporting agricultural and food products in the direction of China. The region's proposals for expanding border crossing stations and tanks for transporting products were taken into account when finalizing a federal decree, which allowed Altai enterprises to export products worth more than 315.5 million rubles and receive more than 25.0 million rubles in subsidies. This is more than half of the budget spent in this area in Russia.

By setting a preferential tariff for the transportation of grain, the process was launched at the end of last year within the framework of Government Resolution No. 1595 in order to stabilize the price situation on the grain market. The limits of grain export volumes were determined for 13 regions of Russia, including the regions of Siberia (Novosibirsk and Omsk regions). The Resolution provides for the maintenance of export of 3.2 million tons of grain, including 200 thousand tons from the regions of the Siberian Federal District. The regions of Siberia were most active in using this tool of support, having almost completely mastered the limit by the end of February. In this regard, the export limit of grain of the Novosibirsk and Omsk regions increased by 250 thousand tons (change from April 25, 2018).

The implementation of Resolution No. 1595 revealed the following serious infrastructural and logistical problems of the grain market: shortage of grain wagons, low rolling stock turnover, high tariffs for grain transportation and the cost of services for grain wagons, lack of infrastructure facilities for storage, processing and handling of grain. All this hinders the export deliveries of the regions, reduces the efficiency of the internal grain market and predetermines the need to develop and implement additional measures for the formation of an efficient logistics system for the supply of grain products and the development of market infrastructure.

The Altai region has repeatedly appealed to the Ministry of Agriculture of Russia with a request to include our region among the participants eligible for preferential transportation of goods and consider expanding the list of transported goods. According to the federal agency, the inclusion of the Altai Territory in the list of regions for the transport of grain cargo by rail without charge is impossible due to budget constraints.

In 2017, according to customs statistics, about 30.0 thousand tons of grain (excluding rice) was exported from the Altai Region, which was 3.7 times more than in 2016. The largest increase was recorded for wheat (4.4 times) and buckwheat (4.1 times).

In the structure of exports, these crops occupy 37% and 42%, respectively. Oats is in third place, accounting for 15%.

In 2017, the main export destinations in terms of supply volumes were: Lithuania (26.4%, buckwheat), Italy (23.4%, wheat), Mongolia (14.7%, oats), Latvia (14%, wheat) [5].

The positive dynamics of exports and the rise in prices on the world grain market will positively affect the situation in the domestic grain market.

4. Discussion

The implementation of the Cooperation Agreement between the Government of the Altai Region and the Key State Operators of the Grain Market signed for 2017-2018 will help reduce barriers to export activities and provide competitive conditions for the promotion of Altai products in the foreign market. An important step in the promotion of products to foreign markets of the region is a set of measures aimed at marketing the region's image and its food.

The establishment of a preferential tariff for the transportation of grain revealed serious infrastructure and logistics problems of the grain market. All this hinders the export deliveries of the regions, reduces efficiency of the internal grain market and predetermines the need to develop and implement additional measures for the formation of an efficient logistics system for the supply of grain products and the development of market infrastructure.

5. Conclusion

In the course of the study, it was revealed that the sectoral system of support and promotion of Russian agricultural products on foreign markets and ensuring its compliance with their requirements will increase the volume of exports of agricultural products. The Center for the Analysis of Exports of Agricultural Products has been established to provide comprehensive information and analytical

IOP Conf. Series: Earth and Environmental Science **395** (2019) 012105 doi:10.1088/1755-1315/395/1/012105

support to domestic producers and exporters of agricultural products. In cooperation together with the industry associations of manufacturers of the agro-industrial complex and the "Russian Export Center" JSC, the Center prepared and approved programs to promote and increase the export volumes of certain types of products with high margins. In 2017, in order to expand access for Russian organizations to the markets of foreign countries, new export promotion tools were launched at the federal level. The Action Plan ("road map") "Development of Export and Transit and Logistics Activities in the Altai Region" was approved in order to ensure the successful promotion of Altai goods (non-primary exports) and services to foreign markets. The implementation of Resolution No. 1595 revealed serious infrastructural and logistical problems of the grain market, such as a shortage of grain wagons, low rolling stock turnover, high tariffs for grain transportation and the cost of services for supplying grain wagons, lack of infrastructure facilities for storage, processing and handling of grain.

References

[1] Bogoviz A V, Bugai Yu A, and Minenko A V 2016 Import substitution as a target guideline for state regulation of the agro-industrial complex *AIC: Economics, Management* **12** pp 67-73

[2] Bugai Yu A 2017 Economic analysis of the effectiveness of existing instruments of state support for the agro-industrial complex of the region *Vestnik of Volgograd State University (Series 3: Economics. Ecology)* **19** 3(40) pp 174-183

[3] Bugai Yu A, and Khimochka V S 2018 Development trends of the agglomeration center of an agrarian-oriented region (on the example of Barnaul) *Regional Economy. South of Russia* **2** pp 40-48

[4] Gritsenko G, Vernigor N, Minenko A, and Apalkova O 2018 The role of state support infrastructure in the development of small rural entrepreneurship in the Altai Region *AIC: Economics, Management* **11** pp 58-67

[5] Petrikov A V 2017 Development of export of products of the Russian agro-industrial complex: problems and solutions *Nikonovskiye Chteniya* **22** pp 3-4

[6] Kaishev V G, and Seryogin S N 2018 Expansion of exports of agro-industrial complex products is a priority for the state's agri-food policy *Meat Technology* **3**(183) pp 54-57

[7] Khorunzhin M G, and Minenko A V 2018 *Support for technical and technological modernization of agricultural production in the Altai region* Available at: http://www.vectoreconomy.ru/images/publications/2018/12/regionaleconomy/Khorunzhin_Minenko2.pdf (Accessed 30 03 2019)

Problems and prospects of cheese production in Russia

I Ugarova[1], O P Pavlusenko[1*] and N I Davydenko[2]

[1] Altai State Technical University, 46 Lenina prosp., Barnaul 656038 Russia
[2] Kemerovo Institute of Food Science and Technology (University), 47 Boulevard Stroiteley, Kemerovo 650056 Russia

E-mail: oxsalex@yandex.ru

Abstract. The Russian cheese market is in constant dynamics. A range of products is available on the market, a wide price range makes cheese a product affordable for all segments of the population; and a steady growth in this industry can be predicted. In this regard, an analysis of the prospects for the production and sale of cheese on the Russian market is of scientific and practical interest. Analysis of statistical data of the Federal State Statistics Service, the Territorial Body of the Federal State Statistics Service for the Altai Region, as well as analytical materials of the Altai Region Administration for the development of entrepreneurship and market infrastructure are presented and critically analyzed. Trends in the production and sale of cheese products are formulated. Factors hindering their development identified; development vectors in terms of import substitution are defined. The following general scientific research methods were used: analysis and synthesis, comparison, statistical methods, modeling, expert assessment method, description, etc. The scientific novelty of the study consists in assessing the factors determining the development of cheese production in the Russian Federation, as a whole, and in the Altai Region, in particular, and in developing the institutional, industry, and local vectors for its development.

1. Introduction

In nutrition, cheese is recognized as a highly nutritious, a biologically complete and easily digestible product, because it contains the necessary human proteins, fats, carbohydrates, and their derivatives, mineral salts, trace elements and vitamins [1, 2]. According to the Agency International Dairy Federation, France (25.9 kg/year), Iceland (25.2 kg/year), Finland (24.7 kg/year) are among the three leaders in per capita cheese consumption.

In Russia, cheese is traditionally not so popular, this is due to national traditions, low cheese involvement as an ingredient in dishes of national cuisine. However, the active expansion of foreign, mainly European national cuisines into the restaurant industry, the expansion of the product range in the food retail industry, the growth of outbound tourism has formed a steady trend towards increasing demand for this product, made it popular and even “fashionable”. In the first years after food embargo, the range of cheese on the Russian market was significantly reduced. However, there is an increase in the production of cheese products by domestic producers over the past two years. Today, we can state with confidence that the fashion for cheese is unlikely to pass; it is firmly established in the diet of the modern Russian consumer. The Russian cheese market is in constant dynamics, the range of products on the market, the wide price range makes cheese a product accessible to all segments of the population; and we can predict the steady growth of the whole industry. In this regard, an analysis of the prospects for the production and sale of cheese on the Russian market is of scientific and practical interest.

Content from this work may be used under the terms of the Creative Commons Attribution 3.0 licence. Any further distribution of this work must maintain attribution to the author(s) and the title of the work, journal citation and DOI.
Published under licence by IOP Publishing Ltd

2. Materials and Methods

An analytical review of scientific literature on the research topic and analysis of statistical data of the Federal State Statistics Service, the Territorial Body of the Federal State Statistics Service for the Altai Region, as well as of analytical materials of the Altai Region Administration for the development of entrepreneurship and market infrastructure, was conducted. The boundary conditions of the study aimed at identifying trends in the production and sale of cheese products and factors hindering their development are analyzed; development vectors in terms of import substitution are defined. To solve the set tasks, the following general scientific research methods were used: analysis and synthesis, comparison, statistical methods, modeling, expert assessment method, description, etc.

The scientific novelty of the study consists in assessing the factors determining the development of cheese production in the Russian Federation and in the Altai Region, in particular, and in developing the institutional, industry and local vectors for its development.

3. Results

In 2016, the Ministry of Health of the Russian Federation, taking into account changes in the production and consumption of food products, developed new recommended food consumption standards to meet modern requirements for healthy nutrition [3]. The recommended annual consumption of cheese is 7 kg per person, which is 1 kg more than it was previously recommended by the Ministry of Health of the Russian Federation [4]. The real average per capita consumption of cheese in Russia is significantly lower, both new and old, and amounts to 5.4 kg in 2014 (90.0% of the norm recommended in 2014) and 5.7 kg, according to the International Dairy Federation (IDF) in 2016 (81.4% of the norm recommended in 2016) (Table 1). The reasons for this situation are the unformed culture of cheese consumption due to the acute shortage of this product during the Soviet era, the lack of national dishes with cheese.

Table 1. Dynamics of consumption of cheese in Russia (according to [5, 6]).

Indicator	2000	2005	2010	2011	2012	2013	2014	2015
Cheese sales, thousand tons	339	689	808	820	858	850	789	762
Population, thousand people	146597	143519	142849	142961	143202	143507	146091	146406
Per capita cheese consumption, kg / person / year	2.31	4.80	5.66	5.74	5.99	5.92	5.40	5.20

Before the introduction of restrictions on the import of food from a number of foreign countries [7], that is, until the autumn of 2014, imported products accounted for about 50% of the cheese consumed in Russia. According to TrendEconomy.ru [12], from 2002 to 2014, Russia occupied the leading position as an importing country on the world cheese and cottage cheese market (Table 2).

Table 2. Indicators characterizing Russian cheese imports.

Year	2002	2003	2004	2005	2006	2007	2008	2009	2010	2011	2012	2013	2014	2015
Place in the world ranking of importing countries [12]	11	10	10	9	10	8	8	8	5	5	4	4	5	10
The share of cheese imports in their commodity resources [13], %	…	…	…	46.5	40.9	42.6	41.3	41.2	47.4	46.1	47.8	48.0	37.3	23.3

Cheese was delivered to the Russian Federation from more than 30 countries of the world. Over the years, Belarus, Germany, Ukraine, Lithuania, Finland, the Netherlands, and other countries were the main importers of cheese and cottage cheese. About 70% of all imports accounted for the products of countries that fell under the food embargo in August 2014. In this regard, the volume of imports decreased significantly, and the structure of imports in the context of the counterparty countries changed. According to the Federal Customs Service, more than 192 thousand tons (192130319 kg) or 86.24% of the cheese was imported to Russia from Belarus. Swiss cheeses are an alternative to French, Italian, and German cheeses. Cheese is also supplied from South America (Argentina, Brazil, Uruguay, Chile), the Caucasus (Armenia, Azerbaijan, Georgia), and Asia (Kyrgyzstan, Kazakhstan), Africa (Tunisia, Morocco, Egypt), and other. The share of each of the listed importing countries in the total volume of cheese imports does not exceed 4%.

Foreign manufacturers, in order not to lose their positions in the Russian market, transfer or expand their production in the Russian territory (for example, the brands "Svjalea" from Lithuania, "Valio" from Finland) or in states that are not affected by the food embargo (for example, the brand "Rokiskio" from Lithuania, because it is produced in Belarus).

The introduction of countersanctions favorably affected the development of the production of cheese and cheese products in Russia. Domestic producers began to increase production volumes. In comparison to 2010, the production of cheese and cheese products in the Russian Federation increased by 38.38% in 2016. However, this growth does not meet the requirements [8].

Traditionally, the Altai Region is a leader in the production of cheese in the Siberian Federal District (about 70% of the production of cheese and cheese products in the Siberian Federal District) and the Russian Federation (about 15% of the production of cheese and cheese products in the Russian Federation). The development trends of cheese production in the Altai Region are similar to the all-Russian.

8 cheese-producing enterprises with the status of a legal entity registered in the Altai Region.

The production capacity for cheese production in Russia is increasing annually, including due to the growth of small enterprises in this area. Mini-cheese factories are opening, the production of cheese in farms is organized.

In connection with the introduction of countersanctions, at the end of 2016 and the beginning of 2017, despite the positive trends in the development of cheese production, the trend of reducing the growth rate of the industry was outlined in Russia. Since the production of cheese is "milk-rich" (10 kg of milk is necessary for the production of 1 kg of cheese), the determining factor is the shortage of raw materials provoked by the decline of agricultural production in the post-Soviet period. So, the volume of milk production in farms of all categories in Russia decreased from 31,847.3 thousand tons to 30,757.9 thousand tons for 2010-2016 [14], which amounted to -3.42%. There is a trend of decline in the production of raw milk (-6.23% and -1.58%, respectively) as well in the Siberian Federal District and in the Altai region. Therefore, experts note, milk processing production facilities in Russia are approximately loaded by 60% [15].

Compliance with the following microbiological indicators of milk is extremely important in the production of cheese: a level of bacterial contamination (in the reductase test is not lower than the first class), KMAFanM (not more than 5.0 • 105 CFU / cm^3), a number of spores of mesophilic anaerobic lactate digesting oil-acid microorganisms (for cheeses with a low second heating temperature, no more than 13,000, with a high second heating temperature, no more than 2,500 spores per 1 dm^3) [16]. However, these requirements are not being complied with in the current state of animal husbandry to the full, which reduces production efficiency and negatively affects the consumer characteristics of the finished product.

4. Discussion

A significant factor hindering the development of the cheese industry in Russia is the rise in average prices for raw milk sold by agricultural producers of all categories. Over an average of 7 years, prices for cow's milk rose by 83,34%. In the Siberian Federal District, an increase is about 63.14%, and it is around 83.84% in the Altai region. This directly affects the increase in the cost of production from milk, as well as the selling and retail prices for finished products, including cheese (the correlation coefficient between the prices of raw milk and prices for rennet cheese is 0.98 in the period 2011-2015). This makes it uncompetitive in price compared to imported cheese. Given the decline in real disposable income in Russia, consumer preferences of the population are focused on cheaper dairy products, which cheese a priori cannot be.

The production capacity of the Russian enterprise-producers of cheese is significantly worn out, and technical re-equipment is not available to everyone, since significant financial investments with a long payback period require for this. Industry experts note that at present, for the production of products, expensive equipment of foreign production is mainly used due to the lack of analogues of domestic. It is purchased for a currency at an unattractive rate that has been formed lately. This problem is aggravated by

IOP Conf. Series: Earth and Environmental Science **395** (2019) 012106 doi:10.1088/1755-1315/395/1/012106

the fact that when using imported technological equipment for primary production, the use of imported packaging equipment for packaging the finished product of a certain configuration and parameters is necessary.

Partially, the ingredients (bacterial starter cultures, concentrates, and milk-clotting enzyme preparations) used in cheese production are of imported origin, which also leads to an increase in the production costs.

Due to the shortage of raw materials and the desire to reduce the cost of production, producers use poor quality milk with the addition of stabilizers or vegetable substitutes. Experts attribute the growth in the physical production of cheese and cheese products in Russia to the increased growth rates of palm oil imports. At the same time, unscrupulous commodity producers do not indicate the use of fats with a non-dairy origin in the labeling of products, which is a violation and is recognized as counterfeit with appropriate checks.

In a hurry to occupy the vacated niche formed as a result of the introduction of the food embargo, some manufacturers use the technology of accelerated ripening of cheese. This leads to the fact that the resulting product does not meet consumer expectations.

5. Conclusion

Factors hindering the development of cheese production in Russia can be divided into the following: "design and development of recipes," "raw materials," "production technology," "packaging and labeling," "storage," "implementation," "transportation," and "staff" (Table 3).

Table 3. Factors constraining the development of cheese production in Russia.

Factors	Influence
Design and formulation development	Long innovation cycle in the development, production and marketing of new varieties of cheese; "Gap" of the innovation cycle, low commercialization of developments [9].
Raw materials	Lack of basic raw materials, both quantitative and from qualitative points of view; Seasonality; High level and increase in prices for basic raw materials; Dependence on imported ingredients (bacterial starters, concentrates and milk-clotting enzyme preparations).
Production technology	Limited own technologies and recipes; High depreciation of fixed assets; Dependence on foreign manufacturers of technological equipment; Limited investment in the production of cheese, their high cost and a long payback period.
Packaging and labeling	Dependence on foreign manufacturers of packaging equipment; Low competitiveness of domestic packaging materials; Failure to comply with the requirements of technical regulations [10] on labeling by unscrupulous manufacturers.
Storage	High capital intensity and energy intensity due to the long production cycle
Implementation	High share of imports; Dumping by foreign suppliers; A high level and rising prices of cheese; Low competitiveness in price and quality of domestic products; Expansion of large retail and difficult cooperation conditions for small manufacturing enterprises; Elasticity of demand for cheese; Decrease in the purchasing power of the population; Underdevelopment of cheese consumption culture.
Transportation	High transport rates.
Staff	The lack of highly qualified specialists in the production of cheese.

Despite these difficulties, many experts believe that the domestic cheese market will continue to grow. Further development of the cheese market in the medium and long term will depend on institutional, industrial, and local trends. The increase in production will be ensured by the development of domestic production, namely: strengthening the position of existing enterprises-manufacturers and the emergence of new enterprises. Thus, there is a scientific and production potential for the development of domestic cheese production. This potential needs to be raised and correctly used in the current economic conditions.

References

[1] Bjerklie D 2003 The milk and cheese diet *Time* **161**(16) p 80
[2] Ash A, and Wilbey A 2010 The nutritional significance of cheese in the UK diet *International Journal of Dairy Technology* **63**(3) pp 305-319
[3] Ministry of Health of Russia 2016 *The Order "On approval of recommendations on rational norms of food consumption that meet contemporary requirements for healthy nutrition"* (August 19, 2016 No. 614) *Rossiyskaya Gazeta* **7059**(191)
[4] Ministry of Health and Social Development of the Russian Federation 2010 *Order "On approval of recommendations to rational norms of food consumption that meet contemporary requirements for healthy nutrition"* (August 2, 2010 No. 593n) *Rossiyskaya Gazeta* **234**
[5] Rosstat 2017 *Russia in figures - 2017* (Moscow, Russia: Rosstat- M)
[6] Rybalova T I 2015 Production of goat cheeses in the world and Russia *Cheese and Butter Production* **6** pp 14-18
[7] Government of the Russian Federation 2014 *Resolution of the Government of the Russian Federation of 07.08.2014 N 778 "On measures to implement the decrees of the President of the Russian Federation of August 6, 2014 N 560, of June 24, 2015 N 320, of June 29, 2016 N 305, of June 30, 2017 . N 293 and July 12, 2018 N 420"* Available at: http://www.consultant.ru/ (Accessed 02 03 2019)
[8] Presidential Executive Office 2010 *Decree "On Approving the Doctrine of Food Security of the Russian Federation"* (January 30, 2010 No. 120) *Rossiyskaya Gazeta* **21**
[9] Osetskaya I A 2016 Possibilities for assessing the innovative potential of students in priority areas of study at the School of Natural Sciences FEFU *Scientific and Methodical Electronic Journal "Concept"* **1** pp 21-25
[10] Council of the Eurasian Economic Commission 2013 *Decision "On the Technical Regulations of the Customs Union "On the Safety of Milk and Dairy Products" (together with "TR CU 033/2013. Technical Regulations of the Customs Union")* (October 9, 2013 No. 67) Available at: http://www.consultant.ru/ (Accessed 02 03 2019)
[11] Hawkins J. 2011 *Creative economy: how to turn ideas into money* (Moscow, Russia: Classic-XXI)
[12] TrendEconomy.ru 2016 *World imports of cheeses and cottage cheese in 2016* Available at: http://data.trendeconomy.ru/dataviewer/trade/statistics/commodity_h2?kf=kfv_commodity_h2&time_period=2002,2003,2004,2005,2006,2007,2008,2009,2010,2011,2012,2013,2014,2015,2016&trade_flow=Import&commodity=0406 (Accessed 02 03 2019)
[13] EMISS n. d. *The share of imports of certain goods in their commodity resources* Available at: https://www.fedstat.ru/indicator/37191
[14] EMISS n. d. *Milk production in farms of all categories* Available at: https://www.fedstat.ru/indicator/40694 (Accessed 02 03 2019)
[15] EMISS n. d. *The level of use of the average annual production capacity from 2010 to 2016 (in accordance with the All-Russian classifier of products by type of economic activity)* Available at: https://www.fedstat.ru/indicator/40641 (Accessed 02 03 2019)
[16] STO VNIIMS 019-2014 "Raw cow milk. Technical conditions"

Indicators of the "green economy" as a tool for monitoring the regional economy

R Adarina[1*], Yu Gazukina[1] and K Yankovskaya[1]

[1] Gorno-Altaysk State University, 1 Lenkina str., Gorno-Altaysk 649000 Russia

E-mail: aarrtt16@mail.ru

Abstract. The paper draws attention to the need to measure a level of "green economy" for monitoring, which is part of the state regulation of the regional economy. The authors note the difficulties and problems in the formation of indicators, indices used for assessing a level of the "green economy" development. The paper proposes an approach for calculating the integral indicator, which can become an indicator of the regional economy, taking into account the contribution of economic, social, and environmental factors to achieving sustainable development.

1. Introduction

Evaluating the implementation of the green economy concept is a prerequisite for stimulating the sustainable environmental and economic development of regions. The economic potential of a region is "the totality of all resources within its borders, both those involved in the processes of social production and social development, and those that can be used to grow the economy and improve the quality of life of the population of a given territory" [1].

The development of territorial policies that woulf be aimed at achieving sustainable development should be based on reliable information on the availability of regional resources, existing disparities in their use and promising directions. In this context, the development of indicators of sustainable development in the regional programs will stimulate regional authorities to carry out economic reforms aimed at greening activities that will facilitate and accelerate the transition to a green economy [2].

The need to modernize the structure of the economy in order to ensure its sustainable development is considered as the most important task in the concept of a "green economy," which is a model of the economy aimed at improving human well-being and ensuring social justice while significantly reducing environmental risks and degradation [3].

2. Materials and Methods

When justifying lines of transition to a green economy in Russia, one should assess the current level of its formation, differentiation, and development of the regions, which dictates the need to consider regional peculiarities in conducting such an assessment.

Therefore, it is necessary to substantiate the list of indicators evaluating the transition results to a green economy. This is due to the fact that there is no established system of indicators assessing the green economy, either in Russian or in foreign practice.

Content from this work may be used under the terms of the Creative Commons Attribution 3.0 licence. Any further distribution of this work must maintain attribution to the author(s) and the title of the work, journal citation and DOI.
Published under licence by IOP Publishing Ltd

The development of sustainable development indicators is actively involved in such international structures as the UN, the World Bank, the OECD, etc. At present, almost all regions of the Russian Federation monitor economic, social and environmental activities, have positive experience in developing indicators of sustainable development, and observe the similarity of most assessment indicators.

As a rule, these indicators are calculated from the initial statistical data and allow to draw conclusions about the various changes occurring in the region.

The information base for the calculation of indicators of a "green" economy are the official data of the Federal State Statistics Service (Rosstat) and the territorial body of the Federal State Statistics Service for the Republic of Altai and the Altai Territory (Altaikraistat), as well as annual regional reports on the state and environment protection. Most of the indicators from the statistical compilations are reflected in absolute values, which does not always allow us to assess the real situation of sustainable development of the region.

When developing indicators, one should take into account that each region is individual, within the framework of a specific subject of the Russian Federation, it may be necessary to further develop some specific indicators taking into account geographical, economic, climatic and other features of the territory. It seems appropriate to split the indicators into three groups: indicators of economic development, indicators of social well-being, indicators of environmental development. Each group will have an equal number of indicators, and the weight of the groups will be proportional.

Proportionality is an essential tool for obtaining a balanced integral indicator that takes into account the level of economic and social development, as well as environmental well-being. Otherwise, when there is no balance in the set of indicators, the ratings derived from them give a distorted picture of the processes of implementing the green economy model in each specific region.

So, in the ecological rating of the All-Russian public organization "Green Patrol", the Altai Republic ranked from 2nd to 4th place during 2008-2018. [6] The eco-economic index, which was formed once in 2012, shows the leadership of the Altai Republic in the ranking based on it [7]. And in the ranking of cities developed by the consulting company Ernst & Young, in 2013 the city of Gorno-Altaisk, which is the administrative center of the Republic of Altai, ranked second, giving primacy to the city of Moscow. At the same time, it is known that the Republic of Altai, by socio-economic indicators, belongs to regions with a low level of economic development. Thus, in the ranking of the socio-economic situation of the constituent entities of the Russian Federation according to the results of 2017, the Altai Republic ranks 82nd among 85 constituent entities of the Russian Federation (83rd place in 2016, and 82nd place in 2015).

So, the first rule for calculating the integral indicator characterizing the movement of a region to a green economy model is to use a number of indicators that are indicators of social, environmental, and economic development. In fact, this consists of indicator blocks, each of which includes a number of indicators. The second rule is that the three blocks should have the same weight, which ensures the equivalence of all three areas of the regional economy. The equal value of the three blocks of indicators does not mean an equal number of indicators in the block. The composition of the indicators by which the region can be assessed and analyzed should reflect the level of environmental friendliness of the economy and the social benefits derived from the introduction of the green economy model. The third rule concerns the set of indicators themselves. In our opinion, indicators should be informative, reflect the most important factors and development trends of the green economy. The number of indicators should not be excessive, which makes it possible to identify the main factors and obstacles in development, among which are the use of dirty productions, the low quality of life of the population, and the weak economic potential.

3. Results

The basic rules that must be followed when assessing a "green economy" represents methodological approaches to developing indicators that allow monitoring, and on its basis one should understand the real processes of greening economic processes and make the necessary management decisions. Table 1

IOP Conf. Series: Earth and Environmental Science **395** (2019) 012107 doi:10.1088/1755-1315/395/1/012107

presents the indicators included in the integral indicator, as well as the weight values of one or another indicator and unit of measurement.

For calculating the integral indicator and comparability of data, it is necessary that the indicators are to be measured as a percentage, i.e. a fraction indicator. This solves the problem of data comparability, and it is the most common technique for integrating estimated parameters [4].

In this way, the analyzed indicators either represent a share indicator (N4, N5, N7, N8, N9), or they are found by dividing by the same average indicator (all other indicators).

The average is either the average for Russia or for the Siberian Federal District. This allows one to assess the performance of the region in comparison with other levels of the territorial organization of the economy, or with other regions in general. Comparing the indicator for the region with the average Russian or the average for the federal district, we use the reference approach.

The comparison with the reference (threshold) value is as follows:

$$n_{ij} = \frac{N_{ij}}{N_0} \tag{1}$$

where n_{ij} – a normalized indicator, obtained as a result of comparison with the reference (threshold) value, which is the average value; N_{ij} – an initial indicator characterizing a certain aspect of the socio-economic or environmental development of a region, for example, GRP per capita in the region; N_0 – a reference value of the indicator with which the initial indicator characterizing a certain aspect of the socio-economic or environmental development of the region is compared, for example, the average value of GRP per capita on average in Russia.

Moreover, if the indicator is a negative factor in the socio-economic development of the region (for example, unemployment), then the average value should be divided by the value of the indicator for the region.

$$n_{ij} = \frac{N_0}{N_{ij}} \tag{2}$$

Table 1. Indicators of the integral indicator of a "green economy" at the regional level.

Indicators	Units	Designation	Weighted indicator
Economic indicators			
1. GRP per capita in the region to the average GRP per capita	%	N1	0.10
2. Capital investment per capita to average investment per capita	%	N2	0.7
3. A capital-labor ratio in the region to the average capital-labor ratio	%	N3	0.07
4. A share of "green industries" in the region's GRP	%	N4	0.09
Total (block)			0.33
Environmental performance			
5. A proportion of recycled waste in the production and consumption waste	%	N5	0.07
6. Energy intensity of regional GRP	%	N6	0.07
7. A share of specially protected natural areas (SPNA) in the total area of the region	%	N7	0.06
8. A reforestation ratio for the five previous years of observation	%	N8	0.06
9. A share of electricity generated by alternative sources	%	N9	0.08
Total (block)			0.34
Social indicators			
10. A ratio of cash income per capita per month in the region to the average value	%	N10	0.09
11. A ratio of average unemployment to unemployment in the region	%	N11	0.09

12. Investments in education and health care per capita to the average value of investments in education and health care per capita	%	N12	0.08
13. Life expectancy in the region to the average life expectancy	%	N13	0.07
Total (block)			0.33
The sum of weights for all blocks			1.00

Table 2 presents the integral indicator of the developed "green economy" in the Altai Republic, calculated over a number of years.

Table 2. An integrated indicator of the "green economy" for the Republic of Altai.

Block indicator	**2013**	**2014**	**2015**	**2016**	**2017**	**Growth rates in 2017/2013**
Summary indicator	0,560	0,574	0,605	0,560	0,603	107,7%
Block of economic indicators	0,135	0,146	0,144	0,145	0,142	104,8%
Environmental performance block	0,149	0,152	0,185	0,141	0,179	119,9%
Social indicators block	0,275	0,276	0,276	0,275	0,282	102,5%

We see that the obtained values are less than one, which means the lag in the level of development of the "green economy" in comparison with the average Russian value. A large contribution to the final indicator made a social unit, and this was due to investment in education and life expectancy at birth. "Ecology" and "economy: make about the same contribution, with a slight lag in the economic bloc. Over the past five years, the value of the final indicator has increased, and the reason for this was the increased greening of the regional economy.

4. Discussion

Discussion questions of the conducted research are the method of reference comparison of individual indicators, the weight values of individual indicators, and the set of indicators themselves. On the one hand, the correlation of actual regional indicators with the average Russian value allows us to conduct a comparative analysis and designate a development vector. On the other hand, the development strategies of the region, thus defined, do not take into account the specifics of the territory. The weight values of the region in this study are set so that the weights of the indicators are approximately the same. At the same time, the authors understand that the proposed weights can be changed taking into account the significance of each indicator. Finally, the set of indicators themselves can also be changed. However, our position is that the number of indicators should not be too large; otherwise, the calculated indicator will not be suitable for making decisions regarding the regulation of the regional economy.

5. Conclusion

A "green economy" is such a model, which is formed with the active participation of the federal state and local government. The process of monitoring is one of the essential elements in building a "green economy". The monitoring tools should be based on indicators, including on integral ones. They take into account various aspects of sustainable development: economic, environmental and social spheres. The calculation of integral indicators is associated with various difficulties of a methodological, methodical, informational nature. Their solution is to create such monitoring tools that would reveal the strengths and weaknesses of the region, searching for mechanisms for ensuring a balanced development and making effective management decisions.

6. Acknowledgment

The study was carried out with the support of the RFBR, the project "Formation of mechanisms for the introduction of a green economy model at the regional level" (No. 18-410-040001).

References

[1] Varavin E V, and Kozlova M V 2018 Evaluation of the development of a green economy in the region. On the example of the Republic of Kazakhstan *Economy of the Region* **14**(4) pp 1282-1297

[2] Bobylev S, and Perelet R 2013 Sustainable development and green economy in Russia: current situation, problems and prospects (Russian-German Bureau of Environmental Information)

[3] Bina O 2013 The green economy and sustainable development: an uneasy balance? *Environment and Planning C: Government and Policy* **31** pp 1023-1047

[4] Yashalova N N 2014 Development of green economy indicators at the regional level *National Interests: Priorities and Security* **40**(277) pp 26-34

[5] Zakharova T V 2015 Green economy and sustainable development of Russia: contradictions and prospects *Bulletin of Tomsk State University* (Economy) **2**(30) pp 116-126

[6] *Ecological rating of the Russian regions* 2018 Available at: http://www.greenpatrol.ru (10 03 2019)

[7] Bobylev S N, Minakov V S, Solovyova S V, and Tretyakov V V 2012 *An ecological-economic index of regions of the Russian Federation: methods and indicators for calculation* Available at: https://wwf.ru/upload/iblock/dc8/index.pdf (Accessed 05 03 2019)

[8] Mityakov E S, and Kornilov D A 2011 On the choice of weights when finding integral indicators of economic dynamics *Works NSTU n.a. R.E. Alekseeva* pp 289-294

Formation and implementation of the strategy of food import substitution in ensuring sustainable development of the border region (on the materials of the Amur region)

N Vlasova[1*], and V Lazareva[1,2]

[1] Ural State University of Economics, 62 8th Marta str., Yekaterinburg 620144 Russia
[2] Amur State University, 21 Ignat'evskoe shosse, Blagoveshchensk 675000 Russia

E-mail: nat-vlasova@yandex.ru

Abstract. The article substantiates the importance and importance of the role of the implementation of the strategy of import substitution in ensuring the food security of the border region and its sustainable development. The hypothesis of the study is that the implementation of the import substitution strategy, which involves reducing food imports and organizing their production within the region, can be achieved through more efficient use of the potential of the border position. On the example of the Amur region, the article analyzes the labor supply for increasing agricultural production and calculates the need for labor resources for the future. It was justified the need for widespread use of foreign labor. The calculations can be used by regional authorities in the formation and implementation of import substitution strategies, as well as in determining the size of quotas for attracting foreign labor.

1. Introduction

One of the important factors of sustainable development and the condition for ensuring food security in the region is the formation and implementation of the import substitution strategy for food products.

The organization of food supply of the population is among the priority tasks recorded in the programs of socio-economic development of the Russian Federation and its regions in particular. This problem is most acute in the regions of the Far East, which, first of all, is associated with significant restrictions, due to the level of socio-economic development of the territory, infrastructure equipment, as well as natural and climatic conditions. For this reason, along with the development of the agro-industrial complex, much attention is paid to the organization of the delivery of food products from other regions of the country and from abroad.

In the context of the region's remoteness from the developed and densely populated centers of the country, especially a significant role in the provision of public procurement of food purchased outside the country. This situation has a negative impact on domestic producers of food and agricultural raw materials, as they do not withstand competition and are forced to close production. The acceptability of the current situation is significantly reduced in the context of international sanctions against Russia. Even taking into account the fact that the main supplier of food is China, which does not share the sanctions policy, increasing the level of self-sufficiency in the region by strengthening the competitiveness of local agricultural producers is becoming a priority regional and state task. To solve this problem, Russia imposed sanctions restricting food imports and creating conditions for domestic producers of agricultural products to increase their own food production.

Content from this work may be used under the terms of the Creative Commons Attribution 3.0 licence. Any further distribution of this work must maintain attribution to the author(s) and the title of the work, journal citation and DOI.
Published under licence by IOP Publishing Ltd

However, the weakening of competitive restrictions does not exhaust the whole range of problems for implementating the task of import substitution in the food market. The existing potential of agriculture, in particular, does not allow us to count on an accelerated increase in production. Among the factors that impede this is personnel shortage. The Amur region is traditionally one of the labor-deficient regions.

As part of this work, the authors set the task, based on an analysis of the socio-demographic situation in the Amur Region, to assess the prospects for ensuring the process of increasing agricultural production from internal sources of labor resources and predict the need for attracting foreign seasonal labor in the Russian-Chinese border area.

2. Materials and Methods

Ensuring food security, being a global problem, at the global, state, regional and local level, differs in the set goals and objectives used by the mechanisms for achieving it [3]. Since this is about creating conditions for providing the population with food based on their own production, food security is determined primarily by the development of the agro-industrial complex, i.e. agriculture and the emerging socio-demographic situation.

Based on population censuses for 11979, 1989, 2002, and 2010, as well as current statistical reporting, a study of the socio-demographic situation was conducted in the Amur Region. The study analyzed the following indicators: changes in population size, localization in the region (remoteness from the district center), as well as favorable conditions for farming (belonging to a certain natural-climatic zone). As a result of the study, a database was formed for all municipal districts of the Amur region. On the basis of the obtained base, a typology of municipal districts was built, combined with the territorial localization of abandoned agricultural land, which can be used for production [3].

3. Results

The climatic conditions and the structure of the land resources of the Amur Region are the most favorable for agriculture among other subjects of the Russian Federation within the Far Eastern Federal District. About 60 percent of all arable land in the Far Eastern Federal District is in the Amur Region. Agricultural land, much of which remains deduced from economic circulation, is one of the largest underutilized resources in the region. In 2017, the sown area of the Amur region was only 63.8% of the 1990 level. Moreover, in 4 districts out of 20 less than 30% of the arable wedge in 1990 was used, in 4 districts from 30 to 50% were used, and only in 5 districts it exceeded 75%" [4].

The analysis shows that it is possible to return to agricultural use in the Amur Region in the amount of 267 thousand hectares of the arable land which is not used currently. At present, one of the main restrictions on the increase in agricultural production is the shortage of personnel directly related to the demographic situation in the villages of the region. The situation is influenced by the following factors: natural, economic and socio-economic parameters, uneven distribution of social facilities, availability of jobs, etc. Without setting the task to exhaustively describe the socio-demographic situation in the region, we note only its most significant elements:

1. *Significant spatial differentiation of demographic processes*. In Blagoveshchensk (the administrative center of the region) and the adjacent areas over the past 30 years, there has been a steady increase in population. However, in other districts of the Amur Region, the number of residents has been steadily decreasing. In 2017, in the most distant districts, the population compared to 1989 was: in Shimanovsky – 71.1%, Magdagachinsky – 60.2%, Skovorodinsky – 50.1%, Arkharinsky – 45.6%.

2. *Dependence of a number of demographic indicators on migration processes*. In the municipal districts of the region, the level of demographic old age (an indicator of the proportion of the population aged 60 years and older) was more than 12%. Over time, this trend will only get worse. In some districts, the proportion of people over 60 may differ by 5.1 points. In the areas with high migration growth of the working population, the lowest rate of aging is observed.

3. *In small villages, the decline in the population is on average 1.5 times higher than in the centers of village councils.* In 2002, the small settlements concentrated in themselves 27% of the population of the Amur region. The population census in 2010 showed a 41% loss in the number of small settlements made up of the total number of losses in the region. From 2002 to 2017, in the southern and central natural-climatic zones of the region, where the vast majority of agricultural organizations are concentrated, a decrease in the population of more than 50% was observed mainly in small agricultural-type settlements, which were the former branches of state farms with a population of 7 to 563 people. In 2002, 9% of the population lived in settlements of the Amur region with a population of up to 500. In the total number of losses, their share was 20%. The lack of a developed social infrastructure is one of the most important reasons for the low attractiveness of small remote settlements. All social infrastructure is concentrated, as a rule, only in the district center. This fact, in particular, draws the attention of S. A. Kovalev [5].

The reserves of arable land are concentrated in the peripheral part of the Amur Region, and there is a particularly shortage of personnel there. As a result, in these municipalities, there is no possibility for locating and developing enterprises producing and processing rural products that are very labor intensive. All this is reflected in the level of food security of the Amur region as a whole and its hotel municipalities. The experience of resettlement in the region does not allow us to count on a sufficiently large influx of personnel to the agriculture of the region and to those settlements where there are reserves of arable land.

High labor intensity eliminates the possibility of large-scale production of fruits and vegetables without the use of seasonal labor. The analysis of the migration policy shows that there is a tendency to reduce the number of attracted workers from the People's Republic of China, which does not quite meet the needs of the Amur Region and acts as a deterrent to the development of agriculture.

To determine the required number of foreign labor in the Amur Region, the need for labor resources for agriculture in the region was calculated according to the planned indicators of agricultural production using formula 1 [6].

$$E = P/Lb \quad (1)$$

where E – industry employment for the planning period; P – a production volume for the planned period; Lb – a labor productivity in the industry.

Table 1. Forecasting the need for labor resources for agricultural production of the Amur region.

	A number of people employed in the industry in 2016, thousand people	Labor productivity in the industry in 2016, mln rubles / thousand people *	The volume of production for the planned period of 2020, million rubles			The industry's need for labor resources for the planning period until 2020, thousand people		
			Conservative forecast	Basic forecast	Target forecast	Conservative forecast	Basic forecast	Target forecast
Agriculture	29.4	1795.5	62417.2	64984.3	66727.5	34.76	36.19	37,16

4. Discussion

The analysis confirms the hypothesis that in the coming years, many border areas of the Russian Federation, including the Amur Region, will face an acute shortage of labor resources in the development of agriculture [7]. This should form new approaches in calculating the required amount of foreign labor recruited annually. It seems that the regions should make maximum use of the benefits of the border position and, with an increase in the demand for foreign labor, first of all, consider the possibilities of attracting it from neighboring countries. In the example of the Amur Region under consideration, it is reasonable to attract the necessary workers from the neighboring PRC.

Since the task is to eliminate the shortage of seasonal workers, the decision on quotas should be developed on the basis of the tasks and interests of the employers determined by the state. At the same time, attracting low-skilled labor resources is limited to certain types of work, including planting, harvesting, processing and packing the harvest (for example, vegetables, berries and flowers), caring for livestock (for example, sheep, pigs, or on-farm processing of poultry). Also, wages must be specified, not less than the minimum set at the national level, the relevant living conditions. All this should lead to the need to adjust migration policy.

5. Conclusion

The region has significant resources to increase agricultural production in the framework of the implementation of the import substitution strategy, which is based on land resources. The negative factors include the demographic situation in the rural areas of the region, the shortage of qualified personnel. The lack of internal sources of personnel for the development of agriculture in the region has no alternative to attracting the Chinese labor for seasonal works, especially when implementing large-scale projects for the production and processing of agricultural products. As noted by V. A. Kolosov, in the border areas of Northeastern China, a significant segment of the economy has been formed, focused on the Russian market. Many Chinese citizens who speak Russian can potentially become the "frontellers" (cross-border commuting migrants) in a number of Russian border regions [8]. However, this requires certain steps to form an appropriate regulatory framework at the federal level.

References

[1] Bazgă B 2015 Food security component of sustainable development–prospects and challenges in the next decade *Procedia Economics and Finance* **32** pp 1075-1082

[2] Clapp J 2017 Food self-sufficiency: Making sense of it, and when it makes sense *Food Policy* **66** pp 88-96

[3] Dyachenko V, and Lazareva V 2016 Import substitution in the agricultural production of the region: the demographic aspect (on the example of the Amur region) *National Interests: Priorities and Security* **9** pp 129-138

[4] Lazareva V 2016 Factors of development of the regional agricultural sector (on the example of the Amur region) *Nikon readings* **21** pp 209-211

[5] Kovalev S 1960 *Geographical study of rural settlement (tasks, methods, materials, special maps of settlement)*

[6] Komartseva V 2012 Impact of labor migration on the economic security of the host region *National Interests: Priorities and Security* pp 46-52

[7] Chernova V Yu 2017 Import substitution and reproduction potential of modernization: problems and prospects *Manager* **2**(66) pp 12-20

[8] Kolosov V 2016 Cross-Border regionalization and frontalier migration: european experience for Russia? *Regional Studies* pp 83-93

Factors of socio-economic development of rural areas

S Vorobyov[1*] and Yu Bugai[2]

[1]Altai State University, 61 Lenina prosp., Barnaul 656049 Russia
[2]Altai State Agricultural University, 98 Krasnoarmeysky prosp., Barnaul 656049 Russia

E-mail: servsp@mail.ru

Abstract. The socio-economic development of rural areas is determined by a complex of factors. The paper proposes the author's classification of factors, suggesting their division into external (regulated and unregulated by the main business entities and management structures) and internal ones. The characteristics of the main internal factors were given with respect to the sectoral specialization, infrastructure development, as well as considering certain features of the social sphere and the involvement of business entities in the processes of cooperation and economic integration.

1. Introduction

In recent years, the development of policies and mechanisms for the socio-economic development of rural areas has become one of the current areas of socio-economic research, which is determined by contradictory trends. On the one hand, deindustrialization is observed in many countries. On the other hand, many rural areas are the zones of unhappiness and poverty, and it is difficult to implement development mechanisms used in urban areas.

2. Materials and Methods

The theoretical and methodological basis of the research was the research of scientists in the field of sustainable development of rural areas and the functioning of social infrastructure. In the process of research, general scientific (scientific abstraction, inductive, deductive) and special approaches (comparison, monographic) were used.

3. Results

The socio-economic development of rural areas is rather difficult to measure, since it is influenced by a number of factors. Most authors distinguish internal and external factors. Internal factors depend on the activities of local authorities, business entities operating in the territory of the municipality. External factors do not directly depend on the activities of entrepreneurs and local management structures, but they have a direct and/or indirect impact on the development of the territory. At the same time, there is a contradiction in the assignment of certain factors to groups of influence.

Thus, I. V. Mishchenko proposes to subdivide all factors into external and internal. The author refers to *external* factors nationwide (stable agrarian economy (support of agrarian and industrial complex), foreign policy (state protectionism), a legal base, stimulating (restraining) development of competitive relations, a level of development of market institutions (tax, credit, budget, innovation policy), pricing policy (price disparity), social policies and *regional* conditions (climatic conditions,

Content from this work may be used under the terms of the Creative Commons Attribution 3.0 licence. Any further distribution of this work must maintain attribution to the author(s) and the title of the work, journal citation and DOI.
Published under licence by IOP Publishing Ltd

industry specialization, developed infrastructure of the region, financial sustainability (investment attractiveness). The *internal* factors include the volume and dynamics of production, financial and logistical support, use of natural resources, availability of natural and recreational resources, provision of social services, infrastructure condition, having an efficient management team, environmental sustainability, working age population, unemployment rate, cultural and moral level of the population [1].

O. Ya. Frolova proposed a similar classification. According to her classification, the production potential of enterprises of the agricultural sector, demographic condition, development of processes of cooperation of small and medium business forms, integration processes in major industries of the region, economic ties and relationships, innovation activity, investment attractiveness. In turn, the *external* factors include climatic conditions, social protection and support, fiscal, credit and tax policy of the state, a level of scientific and technological progress, state protectionism [2].

E. A. Kuklina follows a slightly different classification. Its classification distributes all the factors on natural resources: objective and subjective; innovation factors: technological, socio-economic, environmental, state-legal environment, the spiritual sphere, security, law and order [3].

The sustainable regional development can be carried out only with a balanced balance of economic, social, political, legal (effectiveness of civil society structures and political institutions, political culture, a system of relations between federal and regional authorities) and environmental (assimilation potential of the region, features of reproduction of natural resources, a state of the environment) of groups of factors. The economic component included the efficiency of reproduction processes in the region, the degree of infrastructure development (market and production), spatial differentiation of the territory, involvement and the processes of globalization and economic integration of economic entities. The socio-cultural factors are represented by the socio-economic genotype of a given society.

Based on the systematization of perspectives among the scientists and practitioners on the factors of socio-economic development of rural territories, we propose the following classification of factors: external (regulated and unregulated) and internal (Fig. 1).

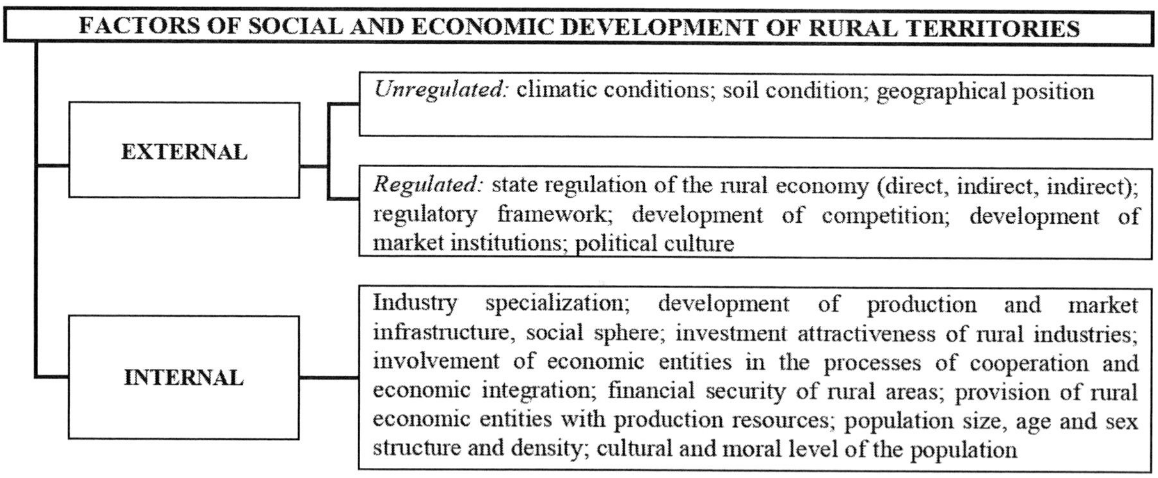

Figure 1. Factors of socio-economic development of rural areas.

State regulation of the sectors of the rural economy is one of the main among external regulated factors. At the same time, the following forms of support are distinguished: (a) *direct* – subsidies for production and material and technical resources, interest rate on loans, for compensation of a part of producers' costs, various debt restructuring schemes, etc.; (b) *indirect* – availability of benefits for a number of taxes, fees, payments; purchase of agricultural products and food for state needs; making purchasing and commodity interventions in the market of agricultural products, etc.).

IOP Conf. Series: Earth and Environmental Science **395** (2019) 012109 doi:10.1088/1755-1315/395/1/012109

Sectoral specialization is one of the main internal factors of socio-economic development of rural areas. The direction of activity of entrepreneurial structures determines the balance of the labor market, physical and financial capital, financial security (sufficiency) of budgets of all levels, incomes of the population (through wages and pensions) and, as one of the consequences, the development of the social sphere. The vast majority of Russian scientists in the field of regional, municipal, sectoral economy note the mono-branch development of the rural economy, since the main branch of its specialization is agricultural production. At the same time, they note the underdevelopment of alternative areas of employment for the rural population, which is reduced only to informal services in the intra-village market of services, embryonic rural tourism [5].

Tourism can be attributed to the real source of diversification of the rural economy, which is new to the rural economy of Russia. However, its development is hampered by a number of problems associated with the presence of a large number of norms and restrictions, such as a need to license a number of areas (hotel business, etc.), a lack of infrastructure and related services, a marketing system, a lack of knowledge of the rural population in the field of business, information technologies. The development of tourism is hindered by the low consumer demand of potential consumers, the lack of real savings of the population for investment, and the training of unskilled workers and consulting.

Therefore, the improvement of sectoral specialization based on a combination of industries, diversification of the rural economy is today one of the conditions for the further development of rural areas. This allows one to use the internal resources of the territories, create conditions for the development of the social sphere, production and market infrastructure, the development of integrated structures, processes of cooperation of small and medium businesses, and also, on a qualitative basis, increasing the availability of production resources and the investment attractiveness of the sectors of the rural economy, financial support not only for the federal, regional, but also for the local budgets.

Thus, the demographic process observed in rural areas predetermines the state of the social sphere and the development of the relevant infrastructure. Despite the large number of works devoted to the study of the organization and functioning of rural infrastructure, many of the problems and prospects for its development in a systemic mode remain little studied. The main reasons for the negative dynamics of infrastructure provision were: (a) the transfer of infrastructure to the municipal level and the financial failure of agricultural producers; (b) the lack of a multi-channel system for financing infrastructure development adequate to the market; (c) the lack of a unified integrated system of rural infrastructure provision that takes into account the particularities of rural development; (d) the imperfection of the system of scientific, methodological, and regulatory framework that solves the problems of the integrated development of rural infrastructure.

One of the most important problems of infrastructure development is the departmental disunity of the management of rural infrastructure development. The solution of tasks for the development of rural infrastructure is concentrated in a number of federal target programs, however, there is practically no proper coordination between the ministries and departments responsible for their implementation. The Ministry of Agriculture coordinates only the solution of individual tasks in the development of the social sphere, engineering infrastructure and infrastructure coordinating complex. The interdepartmental coordination of these programs is not carried out, their actions are not implemented within the framework of the general direction, since the Ministry of Agriculture does not have sufficient powers to coordinate activities. To solve the problems of complex infrastructure development, taking into account the foreign experience of EU countries, the USA, and Canada, it is advisable first of all to form an infrastructure coordinating complex in the Ministry of Agriculture system, consisting of formations of various directions.

4. Discussion

In 2002, the federal targeted program "Social Development of the Village until 2010" was adopted, which identified factors hindering the real transition of rural areas to sustainable development, namely the disunity in their management between ministries and departments; a limited access of rural residents to life-support and development resources; a lack of a system of scientifically based social standards; a weak development of civil society institutions, etc.

The lack of a scientifically-based system of social standards in such sectors as medical care, education, and social services for the population that would take into account the particularities of the settlement of rural residents affects the development of the social sphere in rural areas. The reasons, as a result of which the standardization of social services for the population still has a very weak influence on the introduction of the regulatory approach to the practice of social services, can be divided into three groups: political, theoretical, and methodological.

The reasons of a political nature lie, above all, in the uncertainty of the goals of the social policy of the government of the Russian Federation. The EU countries have legislation on minimum social standards, but it is absent in Russia. However, at the level of some regions, such standards have already been developed. For example, in the Belgorod region in 2007, the Government introduced the concept of a rural social cluster, which refers to a group of objects of social, engineering and transport infrastructure in a rural settlement, which together can fully meet the minimum needs of the population in various services, including education, culture, life, housing and communal services [6]. The list of objects included in the rural social cluster is determined by social standards based on the number of rural population. In accordance with social standards, rural settlements of the Belgorod region were divided into settlements, in which a social cluster is fully present and represented by individual elements.

The cultural and moral level of the population is also an important factor in the socio-economic development of rural areas. Today, there is a shortage of skilled, physically and morally healthy workforce capable of productive employment in agriculture. The problem of the quality of labor capital is the main obstacle to sustainable development.

5. Conclusion

Thus, among the factors that determine the socio-economic development of rural areas, we should highlight the factors of external and internal environments. At the same time, if implementing internal factors directly affects the level of socio-economic development of municipalities, the implementation of external factors only creates the necessary conditions for this.

References

[1] Mishenko I V 2010 The main factors of sustainable development of rural areas *Available at:* http://izvestia.asu.ru/2010/2-1/econ/TheNewsOfASU-2010-2-1-econ-12.pdf (Acessed on 13.03.2019)

[2] Frolova O Ya 2009 Aspects of sustainable development of rural areas *News of the Orenburg State Agrarian University* **1**(21) pp 202-205

[3] Kuklina E A 2008 *Methodology of sustainable development of natural resource regions* (Abstract. dis. ... Dr. Econ. of science) (St. Petersburg, Russia)

[4] Kovaleva G A 2006 Assessment of the potential of competitiveness as a tool for managing the socio-economic development of municipalities *Economy of the Region* **4** pp 159-165

[5] Voytyuk M M 2009 Comprehensive development of rural infrastructure *Regional Economics: Theory and Practice* **5** pp 60-63

[6] "Knowledge" Society of Russia 2008 Social arrangement of rural areas based on a cluster approach *Informational Bulletin of the Belgorod Regional Public Organization – "Knowledge" Society of Russia* **6** pp 1-4

IOP Conf. Series: Earth and Environmental Science **395** (2019) 012110 doi:10.1088/1755-1315/395/1/012110

Prospects for the socio-economic development of rural areas of the Altai territory

V Vorobyova[1*], S Vorobyov[1] A. Sorokina[2]

[1] Altai State University, 61 Lenina prosp., Barnaul 656049 Russia
[2] The Altai Branch of the Russian Presidential Academy of National Economy and Public Administration, 187 Partizanskaya str., Barnaul 656008 Russia

E-mail: vvvtoria@mail.ru

Abstract. The paper substantiates the need for differentiation of regional policy measures regarding the directions of development of rural areas of different specialization. On the example of rural areas of the Altai region, the paper discusses the key issues of agricultural specialization, the main development directions of agriculture, social, and market infrastructure.

1. Introduction

The development of directions and mechanisms for the policy of socio-economic development of rural areas, taking into account regional specifics, has been in recent years one of the current areas of socio-economic research. In the Altai Territory, the agricultural areas dominate among rural areas, but there is no uniformity even among them. A part of the territories have a developed food industry, and in part of the territories farms of the population dominate in the structure of agricultural producers. In the South-East, municipalities have a tourist focus. The diversity of rural areas by specialization predetermines the difficulties of pursuing a common regional policy for their development.

2. Materials and Methods

The research relies the research of scientists in the field of sustainable development of rural areas and the functioning of social infrastructure. In the process of research, general scientific (scientific abstraction, inductive, deductive, comparative analysis) and special approaches were used. Special methods includes comparison, monographic, economic and statistical methods. For the analysis of statistical data, we usedthe Microsoft Office software package. The sources of statistical information were Rosstat and its territorial bodies, data from the Ministry of Agriculture of the Altai Territory and the Ministry of Health of the Altai Territory.

3. Results

The Resolution of the Administration of the Altai Territory "On approval of the long-term target program "Sustainable development of rural territories of the Altai Territory for 2012-2020" (August 2, 2011 No. 420) approved the following blocks of activities:

- Stimulating the development of agricultural and non-agricultural activities, creating new jobs (the collection and processing of wild plants, fish farming, beekeeping, support for crafts, rural tourism and agrotourism, the development of family dairy and meat farms). The types of

Content from this work may be used under the terms of the Creative Commons Attribution 3.0 licence. Any further distribution of this work must maintain attribution to the author(s) and the title of the work, journal citation and DOI.
Published under licence by IOP Publishing Ltd

mechanisms for implementing these measures are (1) subsidies on the interest rate on loans at the expense of the regional budget (at the first stage 2/3 of the Central Bank refinancing rate, at the second stage of 5% rate), as well as (2) the grant support system;
- Acceleration of housing construction in rural areas due to the allocation of social benefits for the construction and purchase of housing, complex compact development of rural settlements (co-financing the development of development projects);
- Improving the quality of educational services, healthcare, culture, sports, including through the reconstruction and construction of new social and engineering infrastructure in the village;
- Organizing competitions among rural areas and communities (national, cultural, andeducational), sports competitions and competitions, competitions for the most comfortable village, at the expense of the regional budget.

At the same time, assessing the current situation in rural areas of the Altai Territory predetermines the need to complement these activities with the following:

- Promoting the development of peasant (farm) farms as small and medium-sized businesses, including through the re-registration of part of the high-value subsidiary farms of the population into farms. For this, it is necessary to teach citizens the entrepreneurial business, accounting, taxation. In the sector of the agrarian economy one should not expect a strong increase in the taxable base, however, the rural population will significantly increase incomes, food self-sufficiency, decrease unemployment, crime, and the influence of other factors hindering rural development.
- Increasing the marketability of production of the main products produced in personal subsidiary and peasant (farmer) farms (potatoes, milk, meat, livestock and poultry, marketable honey and other products) by increasing their production potential, introducing innovative processes to reduce the labor intensity of production, development cooperation;
- Creating conditions for the development of agricultural consumer cooperation, including credit;
- Conducting financial rehabilitation of peasant (farmer) farms and agricultural enterprises, creating conditions for the development of market-oriented and efficiently using production resources of business structures;
- Monitoring the bankruptcy of agricultural enterprises, peasant (farmer) farms, prevention of criminal bankruptcies of entrepreneurial village-forming structures, leading to deprivation of the rural community;
- Developing "social standards" that take into account the peculiarities of rural settlement, lifestyle and ensure an increased access to healthcare, culture, and sports services for the population of rural areas regardless of their place of residence.
- Expanding the range of services provided by information and consulting services, developing information centers based on schools, village administrations, libraries, developing local media, involving the population in the implementation of municipal rural development programs, in order to overcome the information vacuum of the rural population;
- Overcoming interdepartmental fragmentation in rural development issues by introducing the practice of concluding interdepartmental agreements in the field of health, transport, trade, culture, education, information and advisory services to the population;
- Monitoring the development of rural areas, the formation of a federal data bank in the context of municipalities through the improvement of state accounting and statistics software [1].

First of all, it is necessary to create conditions for increasing the productivity of workers of large-scale production, increasing the marketability of production in agriculture, creating new jobs through cooperation of small business forms, training in mass professions most demanded by production, and the fundamentals of small business of persons released from and intangible production.

To improve the management system of rural development, it is necessary to ensure coordination of the rural development issues at the federal and regional levels, increase the degree of reasonableness

IOP Conf. Series: Earth and Environmental Science **395** (2019) 012110 doi:10.1088/1755-1315/395/1/012110

of plans for socio-economic development of rural areas, find local "growth points," and develop development plans for the constituent entities of the Russian Federation formations included in them.

Legal support of rural development involves the introduction of a number of amendments to the existing Federal Laws "On the Fundamentals of Local Self-Government in the Russian Federation," "On Peasant (Farm) Farming," "On Personal Subsidiary Farming," "On Credit Cooperation," "On Agricultural Land Turnover appointments," and a number of others, as well as in the regulatory legal acts ensuring their implementation.

These activities should be carried out taking into account the following principles: expediency, equality, sustainability, right of choice, performance. Also, it is necessary to focus on the investment potential of rural areas, determined by the specialization of municipalities (agriculture, industry, services), implemented investment projects and other factors.

4. Discussion

Prospects for developing the most of the rural areas of the Altai Territory, specializing in agricultural production (Zavyalovsky, Aleysky, Zarinsky districts, etc.), are associated with the development of large-scale large-scale production and industrial processing of agricultural products. Monitoring and preventing the occurrence of unreasonably large enterprises will allow creating conditions for the development of market-oriented small business entities. In rural areas specializing in the production of products in small business forms (Zarinsky, Bystroistoksky, Kosikhinsky, Ust-Pristansky, Soloneshensky, Soltonsky, Shipunovsky, Novichikhinsky, Krasnoshchekovsky, Yeltsovsky, Kurychinsky, Baevsky, Burlinsky, Kruychinsky regions, regions of the region, Kruchikhinsky, Yeltsovsky, Kurinsky, Baevsky, Burlinsky, Kruychinsky regions, regions of the region, Kruchikhinsky, Yeltsovsky, Kurychinsky, Baevsky, Burlinsky, Kruychinsky regions, areas of the region, Kruchikhinsky district into the market turnover) [2].

Contrary to the generally accepted opinion of the close interaction of personal subsidiary farms among the population with large-scale rural business, in most designated areas of the Altai Territory, this connection has been broken not only with agricultural enterprises and peasant (farmer) farms due to the bankruptcy of the former and underdeveloped latter, but also with the processing area. The main market participants in these areas are intermediaries who concentrate the bulk of value added. To solve these problems, it is necessary to create a cooperation of local households, farms, agricultural organizations to ensure the most effective marketing of products. For this, the following forms of interaction between agricultural enterprises and personal subsidiary farms of local residents in the development of consumer cooperation are possible:

1. For financially unsustainable agricultural organizations, this is a reorganization by creating on its base a cooperative servicing personal subsidiary farms and small peasant (farmer) farms.
2. For financially unsustainable agricultural organizations, the provision of services on a reimbursable basis to small business forms (consulting, veterinary, etc.), as well as providing them with working capital (coal, seeds, feed, etc.);
3. For financially unstable but operating agricultural enterprises, this is the involvement of the products of small forms of business on partnership relations in their commodity circulation;
4. A combination of the second and third forms of participation of agricultural enterprises in the development of consumer cooperation.

It is also necessary to create conditions for the development of personal subsidiary farms of residents of municipalities of the first group of socio-economic development (Blagoveshchensky, Yeltsovsky, Loktevsky districts), in which the large-scale agricultural production is practically not developed. At the same time, in the Eltsovsky district, it is additionally recommended to develop beekeeping. In Blagoveshchensk and Loktevskiy districts, it is necessary to strengthen the monitoring of compliance with environmental legislation by business entities (OJSC Kuchuksulfat, CJSC KPF Neverovskaya DSF, etc.), taking into account the significant emissions of air pollutants.

For individual rural areas (Kalmansky, Biysky, Egoryevsky, Zarinsky, Kulundinsky, Zonalny, Pervomaysky, Novichikhinsky, Aleysky, Burlinsky, Krutikhinsky, Troitsky, Zalesovsky districts), it is recommended to strengthen the measures to reduce the real unemployment through a system of grants for the development of small business in the field of social services, small processing of agricultural processing, light industry. One should also consider the possibility of simplified registration of individual entrepreneurs with the possibility of paying only contributions to extra-budgetary funds, a tax on imputed income in the first years of operation. These measures will significantly reduce the cost of the state to pay unemployment benefits, reduce the deficit of the pension fund, medical insurance fund, increase residents' incomes.

For all rural areas it is necessary to create conditions for the development of a social service system. Recreation of the lost infrastructure in small and remote villages is economically impossible, and inappropriate. To improve the quality of life for the residents of small and remote settlements and to provide them with a complex of necessary services, we recommend using the experience of organizing mobile social assistance brigades. These mobile social services are designed to provide social services to all categories of citizens (pensioners, the disabled, families with children, the unemployed and working, who need social support) [3].

The central direction in the development of rural health care and the main tasks of the state social policy (to ensure constitutional guarantees for the protection of the health of the population of rural territories) include ensuring a universal access to free qualified primary, emergency ,and specialized medical care. The Ministry of Health of the Altai Territory in the development of departmental targeted programs, aimed at ensuring preventive and anti-epidemic measures to prevent the most dangerous social diseases, should take into account their different distribution throughout the region.

To improve trade services, it is necessary to increase the role of consumer cooperation and the private sector in providing the population with food with the organization of mobile forms of trade in small settlements, trade on pre-orders for home delivery, fairs with the participation of neighboring and urban trade institutions.

In the area of improving consumer services for the rural population, it is necessary to take measures to restore and develop a comprehensive system of consumer services, basic production facilities in regional centers, including household services for the population, performing repair of household appliances, electronic equipment, dry cleaning of clothes and washing clothes, repair and tailoring of footwear and clothing, repair and construction of housing for individual orders of the population. It is advisable to restore and improve the range and forms of mobile and stationary services by service companies, to increase the volume of services.

5. Conclusion

The establishment of priority directions for the development of social infrastructure does not mean that its other branches should not be developed. The establishment of priority directions for the development of social infrastructure in various rural areas of the Altai Territory means only to identify the most "bottlenecks," with the aim of eliminating them and in a similar way achieving social and economic development not only of the municipalities themselves, but of the whole region.

References

[1] Gritsenko G M 2013 Background of the development of agricultural consumer cooperatives in the Altai Territory *Bulletin of the Altai State Agrarian Universitya* **8** pp 153-157

[2] Kalugina Z I 2007 Deprivation of rural communities in the context of insolvency of agricultural enterprises *Region: Economics and Sociology* **1** pp 186-198

[3] Bogoviz A V, Vorobyov S P, and Vorobyova V V 2016 The efficiency of branches combination in in agrarian economy *AIC: Economy, Management* **8** pp 45-51

The development of the agro-industrial complex of transboundary regions within the territory of Greater Altai

E Permyakova[1*], I Svistula[1] and I Sycheva[1]

[1] I. I. Polzunov Altai State Technical University, 46 Lenina prosp., Barnaul 656038 Russia

E-mail: elena_gushina200@mail.ru

Abstract. In this article, the subject of the research is the prospects of integration relations in the border region of Russia based on the concept of polarized development of the territories. In particular, we focus on the Altai Krai with the adjacent territories of the Republic of Kazakhstan. Our analysis of the existing prerequisites and the current state of the integration processes, as well as the specifics of the development of the regions, shows the possibility of using the agriculture sector as a propulsion industry. More than that, an algorithm for creating a cross-border development axis was proposed.

1. Introduction

Under market conditions, the regional division of labor in border areas of neighboring countries acquires crucial importance, which determines the rational use and integrated development of the productive forces of individual economic regions, as well as promotes regional cooperation and the development of interregional economic cooperation with least cost. The formation of new economic realities caused by the increasing inter-regional competition, as well as the increasing trend of international economic integration, leads to the interdependence of regions to such an extent that the actions of one of them affect the interests of neighboring countries [5]. At the same time, border regions acquire particular importance, for which a specific system of state planning and management should be developed, taking into account their cross-border features and the potential of neighboring regions.

The territory within the Altai Mountains or the territory of the Great Altai, including the border regions of the six adjacent regions of Russia, Kazakhstan, China, and Mongolia, in the context of international cooperation, taking place in the framework of the interstate program "Our Common Home - Altai", can be defined as a complex of several large cross-border regions modern management tools that should be specific. In this study, as an example of the cross-border development of the border regions of the Greater Altai are considered the Altai Krai and the neighboring region of Kazakhstan – the East Kazakhstan region. For the Altai Krai, as a border region, located away from the main transit corridors, the development of foreign economic relations will compensate for its distance from the main world markets [6].

The Altai Krai closely cooperates with the Republic of Kazakhstan in the trade, economic, scientific, and cultural spheres. Traditionally, for a long time, Kazakhstan is the main foreign trade partner of the Altai Krai. Kazakhstan's share in the total turnover of the Altai Krai is about 30%. As can be seen in the graph (Figure 1), the trade turnover between the Altai Krai and the Republic of

Content from this work may be used under the terms of the Creative Commons Attribution 3.0 licence. Any further distribution of this work must maintain attribution to the author(s) and the title of the work, journal citation and DOI.
Published under licence by IOP Publishing Ltd

IOP Conf. Series: Earth and Environmental Science **395** (2019) 012111 doi:10.1088/1755-1315/395/1/012111

Kazakhstan has grown significantly in 2017. However, such dynamics are relative, especially if we analyze a longer period, which is much more objective when it comes to foreign trade, the volume of trade is now only approaching the level of the pre-crisis period of 2014. Now we see a picture of growth in the context of 2015-2016, when the drop in mutual trade was about 40%.

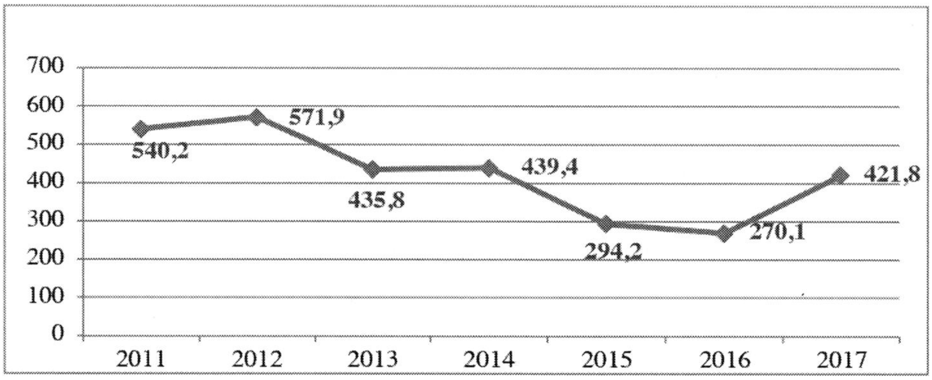

Figure 1. Trade turnover of the Altai Krai and the East Kazakhstan region (mln. USD) [8].

Of course, such a decline in commodity turnover in 2015 in monetary terms was also influenced by the growth of the dollar against the ruble at the end of 2014, without taking into account this factor, a decrease in commodity turnover in physical terms in 2015 would not be so big. In recent years, the currency rates have been more stable, business concerns related to the volatility of the national currencies of Russia and Kazakhstan have decreased, and the Central Bank's refinancing rates have decreased as compared to the level of 2015, which leads to a revival of the economic activity of business, promising relations of the Altai Krai and the Republic of Kazakhstan to a new qualitative level.

2. Materials and Methods

As was noted, the border regions play a special important role in the economy of their states as a liaison with world markets; therefore, it is necessary to take into account the specific conditions and possibilities for their development [7].

The factor analysis of the indicators of socio-economic development of the Altai Krai and the East Kazakhstan Region confirms the possibility of using an integrated approach in regional planning in the cross-border area under consideration. Most of the development factors of border regions largely characterize them as a similar cross-border territory, as part of the Great Altai with a similar landscape, climate, population, socio-economic problems of the regional economy becoming in market conditions, allows considering these border regions as an integrated economic system and developing proposals for the integrated use of resources for the development of the agro-industrial complex [2].

A cross-border region is interpreted as "adjacent border territories of states characterized by a certain natural, economic, socio-cultural, ethnic unity" [5]. To study the potential of socio-economic development of border regions and comprehensive planning of their regional economies, we propose to use the technique of "cross-border growth poles" and "cross-border development axes." At the same time, the cross-border propulsion industry is understood as "a set of enterprises of the same industry located in the local territories of the neighboring states, which are able to perceive innovations and have a positive impact on the development of economies of border regions" [5].

The algorithm for creating a cross-border development axis in the border region should include three stages (Figure 2).

An analysis of regional economies has shown that the point of growth or the "cross-border propulsion industry" can be the agriculture sector.

Training	• Analyzing the compliance of border regions and cities located on their territory with growth poles criteria; • Selecting the cities; • Assessing the industrial potential of cross-border growth poles to identify the propulsion industry.
Formation	• Choosing those enterprises which make up the industrial production, research, financial basis of the propulsion industry as part of the cross-border development axis; • Distributing the ties of specialization and cooperation, as well as the range of products.
Control	• Choosing the organizational-legal form of functioning of enterprises as a single corporation; • Financing the creation and operation of the project, defining a socio-economic effect.

Figure 2. An algorithm for creating a cross-border development axis [5].

3. Results

The Altai region has a unique natural and economic potential, its land area is 16.8 million hectares, among which 11 million hectares are of agricultural land, including 6.5 million hectares of arable land [8]. In the agricultural sector of the region, 5.6% of all available agricultural land in Russia is used, 5.5% is arable land, more than 1.5% of labor resources are used, 4.5% of the total cattle livestock are contained. Crop production is one of the main branches of the agro-industrial complex of the region.

The agro-industrial complex of the Republic of Kazakhstan has good prospects for further development: export positions of the oilseed and meat sectors are increasing. For the export of grain and flour, Kazakhstan is among the largest exporting countries in the world. In 2017, agriculture in the share of GRP of the East Kazakhstan region was 8%. For 14 out of 15 districts of the East Kazakhstan region, agriculture is the basic sector of the economy.

In the Altai region, with seven natural-economic zones, the concentration of production of dairy and meat raw materials was uneven. The livestock of cows and the production of raw milk are concentrated in five natural-economic zones: Kulundinskaya, Prialeiskaya, Biysko-Chumyshskaya, Priobskaya and Prialtayskaya. Among these zones, the Biysk-Chumysh zone has the smallest area of agricultural land. Due to this fact, the indicators of the density of the original livestock and output per 100 hectares of agricultural land are among the highest in the region.

The livestock of agricultural animals and the production of raw meat are concentrated in the Priobskoy and Kulundinskaya natural-economic zones. These disproportions are one of the main reasons for the loss of raw materials during transportation and storage, as well as high-cost production as a whole. Consequently, there are reserves for increasing the livestock of farm animals and the production of dairy and meat raw materials in each natural-economic zone of the region.

The region has all the necessary prerequisites for the further development of agricultural production and increasing its efficiency, meeting the population's need for basic foodstuffs through local production, which is one of the main tasks of the region's agricultural sector. The main volume of milk (more than 80%), harvested in the region, is processed by three large producers of whole-milk products: Barnaul Dairy Plant JSC (Barnaul), Kholod LLC (Zarinsk), Altai Burenka LLC (Zonalny District) [1]. Products of Altai milk processing enterprises are exported to the following republics: Kazakhstan, Mongolia, Uzbekistan, Kyrgyzstan, Tajikistan; the volume of shipments is insignificant.

The largest producers of meat products in the Altai Krai are: Prutskoy LLC (Pavlovsky district), Altayskiy Meat-Processing Plant LLC (Kosikhinsky district), Aleysk Meat-Processing Plant LLC (Aleysk), Rubtsovsk Meat-Processing Plant JSC (Rubtsovsk), Brukke LLC (German National District), Zarinsk Meat-Processing Plant LLC (Zarinsk), and others.

Dairy and meat products production capacity is less than 50%; therefore, there are significant reserves to increase the production of final products. The concentration of various capacities of the dairy and meat industry in the territory of the region, without taking into account the location of the raw material base, leads to the loss of dairy and meat raw materials, its unsustainable transportation and low utilization of the production potential.

In the Altai region, the arable land occupies an area of 6.5 million hectares, there are 5.5 million hectares in acreage. Therefore, a significant increase in the acreage is possible [4].

4. Discussion

For cross-border regions, as part of the Greater Altai, the development of the agro-industrial complex is a way to resolve the acute problems of budget revenues at all levels, household incomes, increasing the investment attractiveness of agriculture, increasing employment, developing and arranging rural life. Prospects for developing the agricultural production due to the presence of effective demand for the products of the industry [3].

Favorable prerequisites for the formation of the conditions necessary to accelerate the innovation process in the agro-industrial complex of the region can be created on the basis of support by the governing bodies of the regional agro-industrial complex, not individual organizations and enterprises, but their aggregates. Associations created taking into account the climatic, historical, and social factors are considered the most productive. This approach is reflected in the concept of particularly significant agrarian territories of the Russian Federation.

Analyzing the state of the cross-border region' agro-industrial complex, it can be concluded that the main reasons for the lack of competitiveness of products and the factors hindering the industrial development are:

- A low equipment with modern technological equipment does not allow to produce high-quality products that meet standards and reduce production costs;
- A highly expensive, energy-intensive, technically backward production, designed for processing large quantities of agricultural raw materials;
- Inaccessibility of long-term, concessional lending and high interest rates when lending production;
- Large distances for the transportation of dairy and meat raw materials are several times higher than the optimal performance.

For the development of a cross-border region, an important fact is the prospects for the complementarity of agricultural development. The East Kazakhstan region traditionally specializes in animal husbandry, the development of which in Kazakhstan is inherently associated with the expansion of the feed base. Therefore, the development of crop production in the Altai Krai can serve as a complement to the Kazakhstan feed base become a factor in the development of animal husbandry in the East Kazakhstan region. Moreover, the products of Russian agrarians can find a large market for Kazakhstani agricultural producers.

5. Conclusion

Speaking about prospects, in the first place, it is important to remember that the Altai Krai is a border region and the strategic planning of its development is impossible without taking this factor into account. The lack of an integrated approach in the strategic planning of cross-border areas in practice leads to an unacceptable disregard for the advantages of the location of border regions, their particular geopolitical significance for states and, as a consequence, the omission of a number of promising development directions that cannot be realized within just one border region. Therefore, the prospects for cooperation, if we talk about a new qualitative level, we see in the orientation towards the joint development of points of growth, priority sectors, the development of joint industrial facilities, for example, in the field of agriculture.

6. Acknowledgment

The article was completed in the framework of the regional competition for fundamental research projects (type of competition r_a) of grant No. 19-410-220010 "Research and Formation of a Model for the Effective Development of the Agro-industrial Complex of the Border Region on the Basis of Economic Integration with Adjacent Rerritories (on the Example of the Altai Krai)."

References

[1] Svistula I A, and Belaya N V 2017 Problems of the integration development of the agro-industrial complex of the mountain territories of the Altai Krai *Russian Economic Internet-Journal* **3** Available at: http://www.e-rej.ru/upload/iblock/01b/01b069350139dc3a25cab32e38f1bea5.pdf (Accessed 31 03 2019)

[2] Sycheva I N, Belaia N V, Permyakova E S 2017 Problems of manpower formation and use in a region with agro-industrial specialization *Espacios* **38** Available at: http://www.revistaespacios.com/a17v38n49/17384917.html (Accessed 31 03 2019)

[3] Sycheva I N, Kuzmina N N, Permyakova E S, and Svistula I A 2015 Labor theory of value in the methodology of researching economic systems *Journal of Applied Economic Sciences* **6**(36) pp 973-979

[4] Sycheva I N, and Svistula I A 2017 Cross-sectoral approach to improving integration processes in agro-industrial complexes *Espacios* **38** Available at: http://www.revistaespacios.com/a17v38n33/a17v38n33p06.pdf (Accessed 31 03 2019)

[5] Permyakova E S 2012 *Development of the border agro-industrial region in the context of economic integration with adjacent territories* (Cand. of Sci. dissertation) (Barnaul, Russia: Altai State University)

[6] Sycheva I N, Permyakova E S and Kuzmina N N 2016 *Cross-border development of the regional economy in the context of the Eurasian Economic Union: State, problems and prospects* (Barnaul, Russia: Altai State Technical University)

[7] Sycheva I N, Mihailushkin P M, Poltarykhin A L, and Permyakova E S 2013 *Cross-border agro-industrial region: background and prospects for economic integration* (Barnaul, Russia: Azbuka)

[8] Official statistics of the Territorial Institution of the Federal State Statistics Service for the Altai Territory n.d. Available at: http://akstat.gks.ru/ (Accessed 31 03 2019)

The impact of modernization on the formation of the poultry meat market in the Russian Federation

M Kh Barcho[1*]

[1] Kuban State Agrarian University named after I.T. Trubilin, 13 Kalinina str., Krasnodar 350044 Russia

E-mail: barcho_m@mail.ru

Absract. Food production and meeting the needs of the population in high-quality food products are among the most important practical tasks of state importance. A significant contribution to the development of domestic agriculture makes poultry. For a short period of time (2001-2018), there was an increase in poultry meat production by 5.5 times and eggs by 27% in Russia. The main factor in increasing production in the poultry industry was the implementation of innovative processes and integrated technical and technological modernization. In order to maintain the momentum and continue the progressive development of poultry production, it is necessary to ensure that its competitiveness is increased, taking into account the existing and forecasted future challenges. In order to determine the impact of modernization processes on the formation of the poultry meat market, the article analyzed the dynamics of the number of new, modernized and reconstructed poultry facilities and estimated the volume of increase in poultry production for slaughter at these enterprises. The proposed measures of state support in terms of modernization of the industry, the implementation of which, in the author's opinion, would contribute to the achievement of full provision of the domestic market for poultry products and the transformation of the sub-sector into an export-oriented business.

1. Introduction

The poultry farming is one of the most dynamically developing branches of the agro-industrial complex, as evidenced by high production growth rates. According to the Rosptitsesoyuz, the volume of poultry meat production was 4.98 million tons in the slaughter mass in farms of all categories in 2018, which is 3.05 times (or 3.59 million tons) more than the 2006 level in which implementation began Priority National Project "Development of the Agricultural & Industrial Complex." Later adopted, the State Program for the Development of Agriculture for 2008-2012 and the targeted program of the department for 2010-2012 and the subsequent state and departmental programs continued the course taken to modernize the industry [1].

The main factor in increasing production in the poultry industry was the implementation of innovative processes and complex technical and technological modernization, including the construction of new modern poultry farms, the updating and improvement of bird crosses, the transition to innovative technologies for their maintenance, as well as poultry meat processing.

In the modern conditions of globalization, the conditions of Russia's membership in the WTO and the deepening of its participation in the global integration processes, the risks of domestic producers of poultry products have increased. In this regard, issues of improving the competitiveness of products

Content from this work may be used under the terms of the Creative Commons Attribution 3.0 licence. Any further distribution of this work must maintain attribution to the author(s) and the title of the work, journal citation and DOI.
Published under licence by IOP Publishing Ltd

based on innovative approaches to the development and continuous process of further modernization of the industry, as well as the development of state support measures are updated.

2. Methods

The following research methods were used: observation, comparison, abstract-logical, retrospective, economic, and statistical. The empirical information was obtained from official statistics, the Ministry of Agriculture, Rosptitsesouz, and Rosstat.

3. Results

More than 90% of the production in the meat poultry industry accounts for industrial enterprises, about 70% in large integrated formations that ensure the modernization of the technological process throughout the food chain (production-processing-logistics-sales) [7]. The possibilities of such formations in the work on quality and branding of products, promotion in retail chains determined their competitiveness in the domestic and, in some cases, in the external market.

Over the past 5 years (2013-2017), 74 new poultry farms were commissioned in Russia, 65 farms were modernized, more than 1,151.6 thousand tons of poultry meat were additionally produced (Table 1). During this period, the share of products manufactured using innovative technologies increased to 18.4%, which made it possible to increase its competitiveness [2].

Table 1. Increase in poultry meat production for slaughter (in live weight) on the newly built, reconstructed, and modernized farms.

Indicators	Years										2017 to
	2008	2009	2010	2011	2012	2013	2014	2015	2016	2017	2008, %
Total number of objects	85	61	61	40	26	31	30	26	20	32	37.6
Including:											
introduced new objects	22	17	18	9	7	10	19	13	14	18	81.82
reconstructed and modernized objects	63	44	43	31	19	21	11	13	6	14	22.22
Total poultry production for slaughter (in live weight), thousand tons	220.1	125.6	181.9	114.1	61.9	154.2	194.6	232	251.5	319.3	145.07
Including:											
due to the commissioning of new objects, thousand tons	91.7	75.7	81.2	49.2	20.4	100.3	178.6	191	238.8	313.4	341.77
due to reconstruction and modernization of objects, thousand tons	128.3	49.9	100.7	64.9	41.5	53.9	16	41	12.7	5.9	4.60
The share of additional production at the constructed, reconstructed, and modernized facilities in the total volume of poultry production for slaughter (in live weight), %	7.3	3.6	4.7	2.6	1.3	3	3.5	3.9	4.1	4.8	-2.50 p.p.

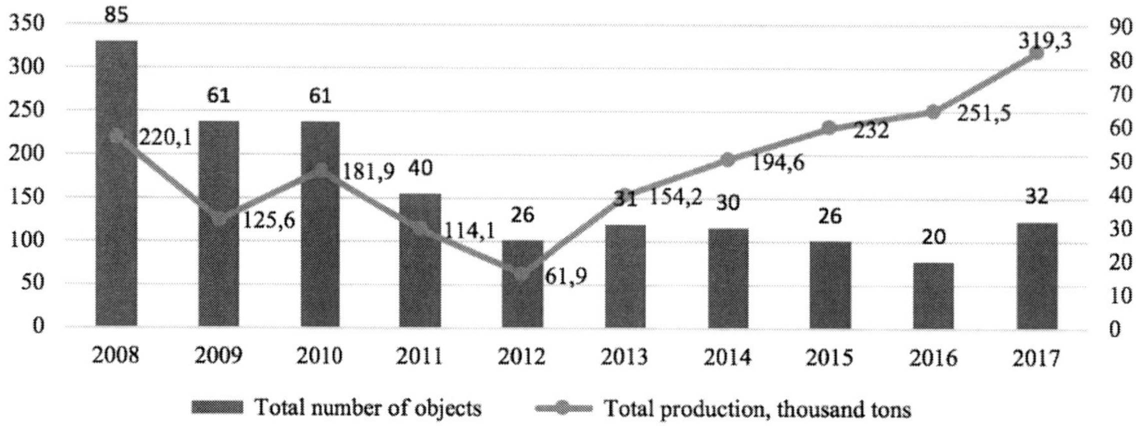

Figure 1. Dynamics of the number of new, modernized and reconstructed poultry facilities and the volume of poultry production for slaughter (in live weight).

Our analysis of the largest poultry organizations clearly shows that the leading position in the ranking is occupied by those organizations that modernize their production and introduce innovations. The successful implementation of state support programs in terms of modernization and restructuring of the industry led to the fact that poultry meat production continues to be the fastest growing in the meat market in Russia. Domestic poultry farmers are ranked fourth in the ranking of the largest poultry meat producing countries in the world, behind the primacy of the United States, China and Brazil.

The total capacity of the poultry meat market in the Russian Federation was 5.02 million tons in the slaughter mass in 2018, which was 16.2% higher than in 2013, for example (Fig. 3). In 2018, the poultry meat consumption per capita reached 34.2 kg, exceeding this figure for other types of meat [2]. There was a sharp decline in imports of poultry meat to 270 thousand tons in 2015, while about 473 thousand tons (- 57%) were brought to Russia in 2014.

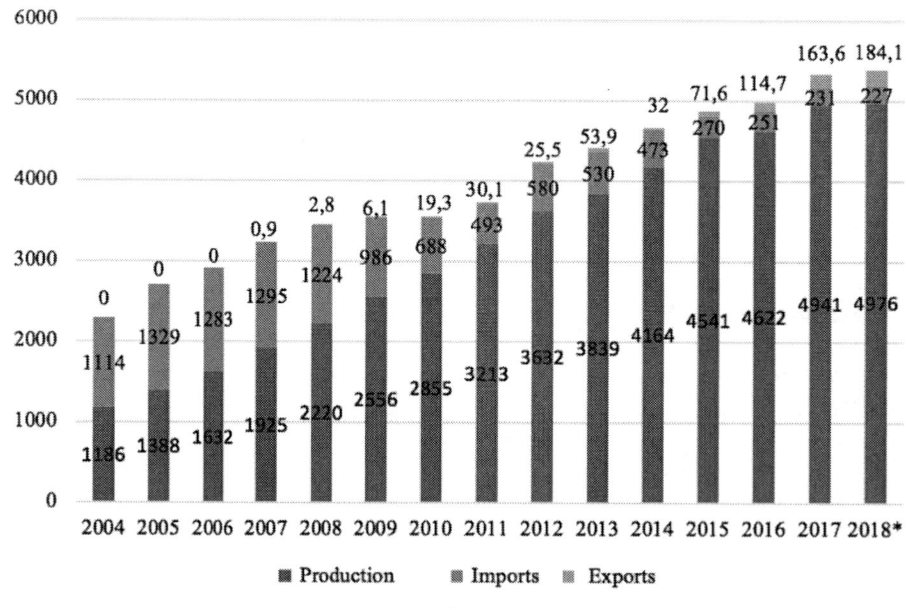

Figure 2. Formation of the poultry meat market in the Russian Federation, thousand tons.

The main reason for the decline in poultry meat imports was the devaluation of the ruble, which made imported products uncompetitive. The share of imports fell to 4.5% in 2018. However, the devaluation of the ruble did not stimulate the export of poultry meat (Fig. 3). Nevertheless, in comparison with 2017, export volumes increased slightly in 2018, reaching 184.1 thousand tons. It should be noted that at present, imports of poultry meat significantly exceed export volumes, although the industry has a significant potential for export growth.

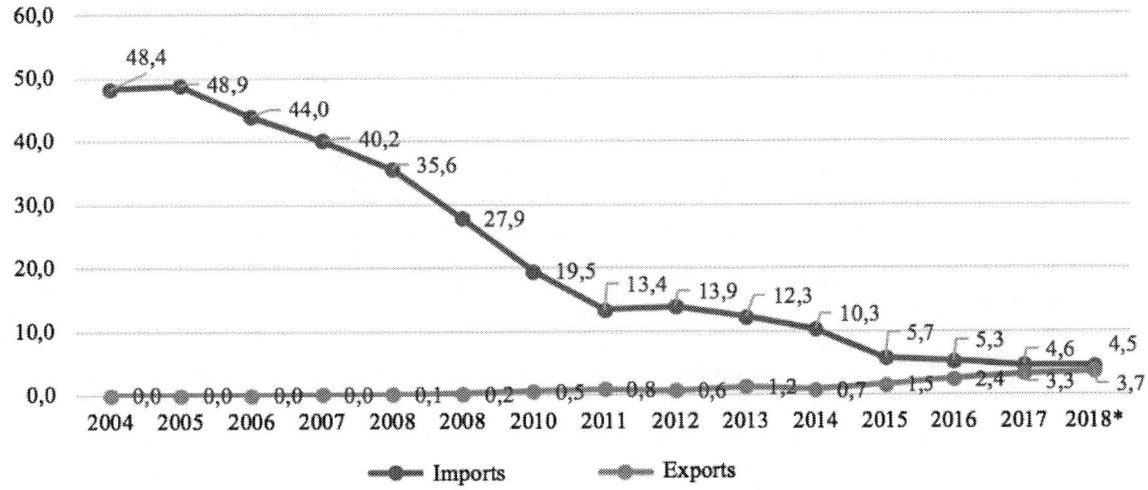

Figure 3. Dynamics of the share of exports and imports of poultry meat in the market, %.

The intensification of modernization processes affected the state of alternative types of poultry farming, such as the production meat of turkey, duck, and others, although their share in the total volume of the poultry meat market is about 6-8%. However, in recent years, the turkey farming sector has been growing rapidly, where the successful implementation of several large projects has reached a production level of 233.0 thousand tons in the slaughter mass in 2018, which is 68.7% more than in 2013. According to forecasts of BusinesStat, the production of turkey meat in Russia will grow by 11.6-18.1% per year in 2018-2022 [3]. Growth up to 471.4 thousand tons in slaughter weight is projected at 2022, which would exceed 2 times the value of 2017. The market will grow both due to technical and technological modernization and, accordingly, to increase production at existing enterprises, and due to the emergence of new enterprises.

Also, the first large project for the production of duck meat is actively developing ("Donstar" in the Rostov region), and other large enterprises in this industry appear, for example, "Chicken-Dak" in the Altai Region, and "Utinyye fermy" in the Chelyabinsk Region [4].

4. Discussion

As practice shows, the increase in poultry meat production is closely connected with the modernization of the industry along the entire technological chain, as a result of which poultry crosses are updated, new technologies of feeding and housing of poultry are introduced, and its productivity is increased.

The analysis showed that the distinctive feature of the present stage of development is the saturation of the poultry production market and the danger of an overproduction crisis. The consumption of food (poultry meat, egg) meets the standards, the level of the Food Safety Doctrine for these types of products reached back in 2011. Moreover, according to the Rosptitsesoyuz, the forecast for the development of the industry envisages bringing the volume of poultry meat production to 5.2 million tons, as well as reducing imports to 230 thousand tons in 2024.

It should be noted that the purchasing power of the population in poultry products is growing, but its growth rates are lower than the growth rates of poultry meat production, which does not allow for an increase in consumption.

5. Conclusions

In the future, the Russian poultry meat market will face serious intra-industry competition. In this regard, manufacturers are interested in the development of exports. The way out is as follows:

- Expansion (diversification) of exports of poultry products, namely poultry meat in the form of halal products, to the EEU countries (by-product are exported to China);
- Expanding the range of poultry products;
- Expanding the range of poultry products of deep processing.

Modernization of the agrarian sector of the economy requires a multiple increase in the share of expenditures for these purposes in the budgets of all levels. However, the actual size of state support is much lower than in developed countries and does not meet the task of modernizing the economy.

For state support of the poultry industry, the following is necessary:

- Prolong subsidization of interest rates on loans attracted for the construction, reconstruction, and modernization of poultry facilities;
- To carry out the selection of investment projects for the construction, reconstruction, and modernization of poultry facilities, in regions where the production of poultry meat per capita is below the average Russian level;
- Provide priority support to projects for the development of breeding reproducers, the development of non-traditional types of poultry meat – turkey breeding, goose breeding, duck breeding, (a 10% growth in the share is planned for 2020);
- To stimulate measures to ensure environmental safety, processing and disposal of poultry waste;
- Support for the development, creation, and reconstruction of breeding and genetic centers for poultry farming;
- Income support for agricultural producers in the poultry industry [2].

References

[1] Government of the Russian Federation 2012 *Decree "On the State Program for the Development of Agriculture and Regulation of Agricultural Products, Raw Materials and Food Markets for 2008-2012"* (July 14, 2007 No. 446) Available at: http://www.consultant.ru/document (Accessed 02 04 2019)

[2] Ministry of Agriculture of the Russian Federation 2017 *National report "On the progress and results of the implementation in 2017 of the state program for the development of agriculture and the regulation of the markets for agricultural products, raw materials and food for 2013–2020"*Available at: http://mcx.ru (Accessed 02 04 2019)

[3] Completed Market Reviews "Busines Stat" n.d. *Analysis of turkey meat market in Russia in 2014-2018, forecast for 2019-2023* Available at: https://businesstat.ru/catalog/id44760/ (Accessed 02 04 2019)

[4] Institute for Agricultural Market Studies 2019 *Results of the year 2018: meat industry* Available at: http://ikar.ru/lenta/676.html (Accessed 02 04 2019)

[5] *Official site of the Federal State Statistics Service of the Russian Federation* n.d. Available at: http://www.gks.ru (Accessed 02 04 2019)

[6] Institute for Agricultural Market Studies n.d. *Poultry market* Available at: www.ikar.ru/poultry/profile.html (Accessed 02 04 2019)

[7] Sokolov N A, and Kuzmitskaya A A 2016 Innovative and technological development of meat poultry farming in terms of import substitution *Vestnik of the Bryansk State Agricultural Academy* **5** pp 50-58

[8] Rosinformagrotech 2016 *Technological processes and equipment used in intensive breeding of poultry* (Moscow, Russia: "Rosinformagrotekh" Federal State Scientific Institution)

[9] Fisinin V I, Egorov I A, Buyarov V S, and Buyarov A V 2014 Innovative and technological development of the poultry industry *Vestnik of OrelSAU* **5**(50) pp 50-58

IOP Conf. Series: Earth and Environmental Science **395** (2019) 012113 doi:10.1088/1755-1315/395/1/012113

Organizational and economic aspects of technical and technological modernization of the poultry farming

M Kh Barcho[1*]

[1] Kuban State Agrarian University named after I.T. Trubilin, 13 Kalinina str., Krasnodar 350044 Russia

E-mail: barcho_m@mail.ru

Abstract. The article deals with the problems of development of domestic poultry farming on the basis of technical and technological modernization. The dynamics of the number of new, modernized, and reconstructed poultry facilities was analyzed and the volume of increase in poultry production for slaughter at these enterprises was assessed. Factors constraining the development of modernization processes in the sub-industry are considered. Government support measures for the modernization of the industry have been proposed. According to the author, the implementation of these measures will contribute to the full provision of the domestic market for poultry products and the transformation of the poultry industry into a high-tech sub-industry competitive in the domestic and foreign markets.

1. Introduction

Under the existing macroeconomic conditions of the agrarian economy and the openness of national food markets, poultry farming plays a significant role, as it is able to restore and increase production and increase the average per capita consumption of such products as eggs and poultry meat in a relatively short period of time.

Addressing the issues of raising the level of modernization in all areas (from equipment for production and processing of products to infrastructure facilities and resource support of production processes) is extremely important for maintaining the industry's leading position in the food market and ensuring the country's food security. Addressing the issues of improving the technical and technological equipment of production, the safety of poultry products, improving its quality and competitiveness are directly related to the implementation of renovation and modernization projects by producers in the sub-sector.

2. Materials and Methods

The study was conducted using such methods as monographic, economic and statistical, abstract-logical, and design-constructive. The analysis was based on data from the Federal State Statistics Service of the Russian Federation, data from the Russian Poultry Union, reporting data from agricultural organizations, Internet resources, and personal research and observations of the author.

The methodological basis of this study was the works of domestic scientists on the problems of modernization and development of agriculture of the Russian Federation, including poultry farming. Among them are the following works [1], [4], [6], [7], [8].

Content from this work may be used under the terms of the Creative Commons Attribution 3.0 licence. Any further distribution of this work must maintain attribution to the author(s) and the title of the work, journal citation and DOI.
Published under licence by IOP Publishing Ltd

3. Results

Poultry farming is one of the most dynamically developing branches of the agro-industrial complex, as evidenced by high production growth rates. According to the estimates made by the Rosptitsesoyuz, the production of poultry meat reached 4.98 million tons in the slaughter mass in farms of all categories in 2018, which was 3.05 times (or 3.59 million tons) more than the 2006 level in which it began implementation of the Priority National Project "Development of the Agricultural & Industrial Complex" (Figure 1).

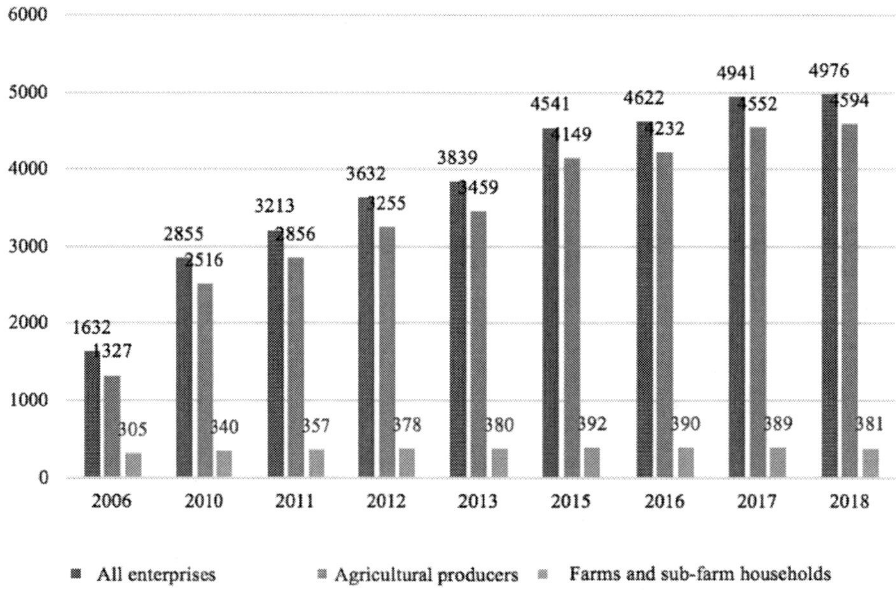

Figure 1. Dynamics of poultry meat production, thousand tons in a slaughter mass [This is calculated by the author based on materials of Rosptitsesoyuz (2018)]

It should be noted that the contemporary poultry farming is a production based on intensive poultry rearing, which is characterized by a large concentration of livestock in enterprises, and an in-depth on-farm and inter-farm specialization of production units, the rhythm and precision of the technological process of poultry housing and rearing with the help of a wide range of contemporary resource-saving technical equipment.

The data in Figure 1 shows that more than 90% of the production in the poultry meat industry is accounted for by industrial enterprises, about 70% are large integrated formations that ensure the modernization of the technological process throughout the food chain "production-processing-logistics-sales" [7]. The possibilities of such formations in the work on quality and branding of products, promotion in retail chains determined their competitiveness in the domestic and, in some cases, in the external market.

In addition, in world practice, due to a number of reasons, independent farmers can hardly compete with integrated production, which is tied to the production and value chains. And one of the reasons for this is the problem of obtaining the necessary biological materials (bird crosses) with high genetic potential, high-quality feed, available only to those producers who are participants in the production and value chain.

The growth of production in the poultry industry was largely made possible as a result of the successful implementation of government support programs for modernization and restructuring of the industry. Over the past 5 years (2013-2017), 74 new poultry farms were commissioned in Russia, 65 were modernized, more than 1,151.6 thousand tons of poultry meat were additionally produced in them (Table 1).

Table 1. Increase in the poultry meat production for slaughter (in live weight) on the newly built, reconstructed, and modernized farms in the Russian Federation.

Indicators	Years										2017 *vs*
	2008	2009	2010	2011	2012	2013	2014	2015	2016	2017	2008, %
The number of objects, including:											
new	22	17	18	9	7	10	19	13	14	18	81.82
reconstructed and modernized	63	44	43	30	19	21	11	13	6	14	22.22
Poultry production for slaughter (live weight) at the commissioned facilities, thousand tons	91.7	75.7	81.2	49.2	20.4	100.3	178.6	191	238.8	313.4	341.77
Poultry production for slaughter (in live weight) due to reconstruction and modernization of facilities, thousand tons	128.3	49.9	100.7	64.9	41.5	53.9	16	41	12.7	5.9	4.60
Poultry production for slaughter (in live weight), obtained through the introduction of new, reconstruction, and modernization of existing facilities, thousand tons	220.1	125.6	181.9	114.1	61.9	154.2	194.6	232	251.5	319.3	145.07
The share of additional poultry production for slaughter at the constructed, reconstructed, and modernized facilities in the total volume of poultry production for slaughter (in live weight), %	7.3	3.6	4.7	2.6	1.3	3	3.5	3.9	4.1	4.8	-2.50 p.p.

Source: Data calculated by the author according to the FSSS RF (2018).

During this period, the share of products manufactured using innovative technologies increased to 18.4%, which made it possible to increase its competitiveness [3].

Despite some success, the poultry enterprises are experiencing the need for further modernization and renewal of basic production assets. In some poultry farms, equipment wear is 50% or more, production technology is also obsolete.

On the example of poultry farms in the Krasnodar region, we calculated indicators of the efficiency of using fixed assets, analysis of which showed that capital productivity is low even per 100 rubles of the value of funds (Table 2).

The highest capital productivity in "Novomyshastovskaya Poultry Farm" LLC is 1.5 kopecks per 1 rub. funds (2016). At the same factory, the coefficient of renewal of fixed assets was 1.5 in 2016 against 0.003 in 2013.

Modernization has affected the growth of production performance, which affected the capital productivity. Capital-labor ratio in farms is high. In PF "Belorechenskaya" LLC, it is 2.11 million rubles / person. The profitability of the use of funds is the highest in "Rassvet" CJSC, it is 71.67%.

In our opinion, in conditions when the degree of depreciation of fixed production assets already exceeds 60%, the enterprises need higher rates of their renewal and higher "thresholds" of profitability, therefore.

IOP Conf. Series: Earth and Environmental Science **395** (2019) 012113 doi:10.1088/1755-1315/395/1/012113

Table 2. The effectiveness of the using the funds in poultry farms of the Krasnodar region.

Enterprice	2013				2016			
	The capital-labor ratio, rub. / person.	Capital productivity (per 100 rubles of the value of funds), rub.	Capital intensity (per 100 rubles of gross output), rub.	Profitability (unprofitability) of using funds (net profit per 100 rubles of the value of funds), rub.	The capital-labor ratio, rub. / person	Capital productivity (per 100 rubles of the value of funds), rub.	Capital intensity (100 rubles gross output), rub.	The profitability of using funds (net profit per 100 rubles of the value of funds), rub.
PF "Belorechenskaya" LLC	2761224	0.24	0.41	14.16	2113180	0.30	0.33	2.28
"Harvest XXI century" LLC	740261	0.00	0.58	31.10	1003740	0.02	0.46	61.71
"Druzhba Poultry Farm" JSC	543075	0.53	0.19	122.52	522476	0.65	0.15	27.27
"Agrocomplex" named N. I. Tkachev JSC	1261553	0.02	1.00	13.04	1363616	0.03	0.69	9.49
"Yugmelprodukt" LLC	714389	0.44	0.23	52.43	441674	0.75	0.13	33.47
PF "Yubileinoye" LLC	232303	0.36	0.27	-151.03	478616	0.17	0.52	6.24
PPF "Caucasus" CJSC	694782	0.13	0.43	5.56	1233065	0.28	0.28	5.27
"Novomyshastovskaya Poultry Farm" LLC	293813	1.11	0.09	-701.01	127219	1.53	0.06	58.97
OPH "Slava Kubani" LLC	3347686	0.00	0.22	101.18	914063	0.00	0.49	14.96
"PPZ "Labinsky" JSC	398083	0.60	0.17	92.96	487355	0.40	0.25	63.25
"Pervomayskaya IPS" LLC	331186	0.33	0.30	39.18	748587	0.35	0.28	56.76
"Rassvet" CJSC	666296	0.04	0.47	43.75	1349813	0.03	0.47	71.67
"Poultry Farm "Primorskaya" LLC	6235938	0.27	0.37	3.25	750357	0.46	0.21	-17.37
"PPF Timashevskaya" CJSC	2456995	0.15	0.67	13.82	2108478	0.20	0.49	12.38

Source: Data calculated by the author according to the FSSS RF (2018).

The main limiting factor of scientific and technological development based on high-tech technologies in animal husbandry in general and in poultry farming is the disparity of prices (Fig. 1), the statistical values of which have been steadily increasing for 25 years [8].

The most important factor in the growth of the cost of poultry products are feed. Now, the situation for poultry farmers is not as good as an increase in grain prices, which has a negative impact on the profitability of poultry enterprises, since the share of feed is about 70% in the cost structure of poultry meat production (Table 3).

So, in the period 2015-2018, the cost of feed increased by an average of 20-25%, and the cost of production grew by 20%. At the same time, an average selling price for the industry has changed only by 1.78% for poultry meat, and for eggs has decreased by 6.12%.

Table 3. Dynamics of prices for consumed resources and poultry products.

Indicator	2015	2016	2017	2018	2018 vs 2015, %
An average producer price for poultry meat per year, rub / kg **	96.55	98.57	93.37	98.27	101.78
An average producer egg price per year, rub / kg **	41.5	41.7	35.7	38.96	93.88
An average price to purchase the following resources for the year, rubles / kg:					
- *wheat*	9.56	9.73	8.04	9.25	96.76
- *corn*	9.15	10.01	8.27	9.49	103.72
- *barley*	8.27	9.04	7.7	8.82	106.65
- *premixes of domestic producers*	112.33	129.58	112.68	134.0	119.29
- *soybean meal*	32.2	36.63	33.7	40.0	124.22
- *sunflower cake*	12.91	17.41	13.87	17.00	131.68

Source: Data calculated by the author according to the Federal State Statistics Service of the Russian Federation (2018), prices do not include VAT and logistics costs.

The functioning of the sub-industry in such conditions led to a decrease in the profitability of enterprises (on average, the profitability decreased 2-2.5 times) and the lack of working capital. In the face of rising costs and limited opportunities to raise prices for the sale of products, manufacturers face financial problems and a number of them, including market leaders, will be forced to reduce production.

It is known that there are certain threshold values of the industry profitability during reproduction, below which even simple reproduction is impossible. The slowdown in the reconstruction of enterprises and the renewal of basic production assets become a consequence, and the industry is starting to deteriorate. However, the level of profitability depends not only on the level of costs for the production and sale of products, but also on its prices. It should be noted here that the purchasing power of the population in poultry products is growing, but its growth rates are lower than the growth rates of poultry meat production. This circumstance does not allow to increase consumption and to raise prices. In Russia, an average per capita consumption of poultry meat exceeds 30 kg per capita. In the near future, this indicator is to grow, because we do not observe an increase in real incomes of the population. Accordingly, domestic demand for products will not increase [6].

The current situation in the industry is expected to lead to a decrease in the level of profitability from the sale of poultry meat and eggs (Fig. 2).

Thus, the main challenge that poultry enterprises facing in the coming years is a gradual decrease in the profitability of production and sales of products.

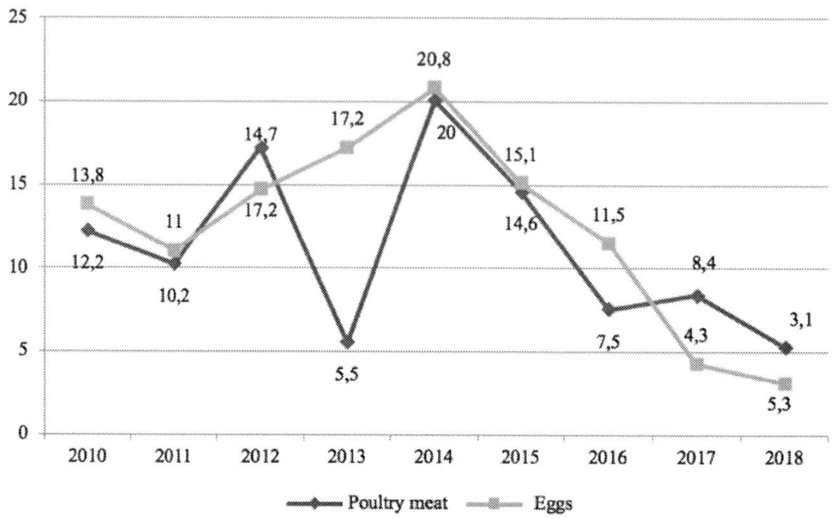

Figure 2. The level of profitability from the sale of poultry meat and eggs, % [Calculated by the author according to the FSSS RF (2018)].

In the face of declining profitability, enterprises are experiencing difficulties with the implementation of the modernization of production. In this regard, as a number of authors rightly point out, "... from January 1, 2015, the cessation of subsidies to new investment projects under the *State Program for the Development of Agriculture and Regulation of Agricultural Products, Raw Materials and Food Markets for 2013-2020* will significantly complicate the update and modernization of material and technical bases of poultry enterprises" [4].

The most serious problem hampering the development of the poultry sub-sector is the current state of breeding, which is characterized by a catastrophic shortage of poultry baselines. At the beginning of 2017, the share of foreign producers accounted for more than 95% of the total required volume of hatching eggs. The only Russian supplier of incubation eggs was the FGBU SGC "Smena". Over the past five years, according to the Institute for Agricultural Market Studies (IKAR), the volume of imports of hatching eggs more than doubled (from 300-330 million units / year to 700 million units in 2015) [2]. It is obvious that the need to expand and modernize the domestic breeding base has matured at a fairly high rate of increase in domestic broiler production.

4. Discussion

Continuing investment in the sub-industry with the use of reconstruction activities and modernization of the existing poultry farms with a simultaneous increase in technical and technological support allows us to predict a further increase in production volume, which would also contribute to the replacement of imported products with domestic production. At the same time, a distinctive feature of the current stage of development is the saturation of the poultry market and the danger of an overproduction crisis. In the future, the Russian poultry meat market will face serious intra-industry competition; therefore, producers are interested in export development. At the same time, the exacerbation of the situation in accessing export markets is possible, which is associated with stringent quality control requirements and veterinary documentation, and with epizootic situations that arise.

Risks and threats to the further development of the sub-industry are associated with the unsatisfactory financial and economic condition of enterprises in the industry, insufficient investment and the unacceptably low level of its state support, which is several times lower than in the West. Thus, in the European Union, an aggregate authorized support for agriculture from the national and general budgets is about $ 100 billion, 39.6 billion in Japan, 19.1 billion in the United States. Based on 1 hectare of arable land, the state support in the EU is about 250 euros, and it is 434 rubles in Russia (5-6 euros at the current rate) [1].

The high level of imports of not only tribal, but also material resources creates serious risks for the competitiveness of the domestic agriculture. Unfortunately, the share of imported veterinary drugs, vitamins, microelements, amino acids, probiotics, vaccines, as well as corn and soybean meal remains significant, which exceed 50% of the need.

5. Conclusion

In the future, the development of the poultry industry should ensure the achievement of the main goal. It consists in the formation of competitive, efficient and stably developing production that meets international standards. This necessitates the implementation of a complex of organizational and economic measures provided by significant investments from both the state and private business.

In 2019, the main priority issues for the development of the industry and for the future include the following:

- Extension of interest rate subsidies for loans attracted for the construction, reconstruction, and modernization of poultry facilities;
- An elimination of disparity of prices for the resources used and prices for finished products;
- Support for the development, creation, and reconstruction of breeding and genetic centers for poultry farming;
- An improving the infrastructure of commodity markets and pricing in the food market;
- Changes in foreign economic policy that would ensure the formation of quotas for the import of poultry meat on the basis of the annual balance of production and consumption;
- Stimulation of measures to ensure environmental safety, processing and disposal of poultry waste;
- Creating plants for the production of biologically active substances (vitamins, microelements, amino acids, probiotics, vaccines, diagnostics, etc.);
- Creating the Russian state reserve of feed grain;
- Developing the mechanisms for the export of agricultural products;
- An availability and cost of credit resources.

References

[1] Dolgushkin N K 2016 Technological modernization is the basis of the efficiency of the agro-industrial complex and the sustainable development of rural areas *Nivy of Russia* **8**(141) pp 58-62

[2] Institute for Agricultural Market Studies 2019 *Results of the year 2018: meat industry* Available at: http://ikar.ru/lenta/676.html (Accessed 02 03 2019)

[3] Ministry of Agriculture of the Russian Federation 2018 *National report "On the progress and results of the implementation in 2017 of the State Program for the Development of Agriculture and Regulation of Agricultural Products, Raw Materials and Food Markets for 2013-2020"* (April 28, 2018 No. 808-p) (Moscow, Russia)

[4] Nechaev V I, Bershitsky Yu I, Fetisov S D, and Slepneva T N 2014 The current state and trends in the development of the poultry industry in Russia *Izvestiya of the Timiryazev Agricultural Academy* **4** pp 102-111

[5] *Official site of the Russian Poultry Union* 2018 Available at: http://www.rps.ru/ (Accessed 02 03 2019)

[6] Smirnov A A 2018 Added weight: poultry production continues to grow *Rossiyskaya Gazeta* **53**(7516)

[7] Sokolov N A, and Kuzmitskaya A A 2016 Innovative-technological development of meat poultry farming in terms of import substitution *Vestnik of the Bryansk State Agricultural Academy* **5** pp 50-58

[8] Tarasov A N 2015 *Industry and technical and technological problems of import substitution: state and solutions* Available at: http://institutiones.com/agroindustrial/2847-otraslevye-i-tehniko-tehnologicheskie-problemy-importozamescheniya.html (Accessed 02 03 2019)

[9] Federal State Statistics Service 2018 *Russian statistical yearbook: agriculture* Available at: http://www.gks.ru/bgd/regl/b17_13/Main.htm (Accessed 02 03 2019)

Digitization of marketing activities of agricultural entrepreneurship in the conditions of the Kyrgyz Republic

M Akzholova[1*] and A Osmonova[1]

[1] International University of Kyrgyzstan, 17A/1, Tolstogo str., Bishkek, 720007 Kyrgyz Republic

E-mail: meimanby@bk.ru

Abstract: The problems associated with the current trend in the consumption of natural food products, the sale of manufactured products, increasing the competitiveness of domestic products, foreign product intervention, ensuring the institutionalization of Internet commerce by the state are discussed in the article. The analysis carried out features of the introduction and development of efficient, digital technologies in agriculture. Also, promising areas of digitalization of the agro-industrial complex are highlighted by the authors. It was recommended to entrepreneurs to create such sites, the effectiveness of which reached the audience at an appropriate level, the volume of orders was maximum, which ultimately would increase profits and strengthen market position.

1. Introduction

Based on the characteristics of agriculture, it is possible to distinguish those features of entrepreneurial activity that are not inherent in other industries, or are of particular importance for the development of entrepreneurial activity in agriculture. Today in Kyrgyzstan, the family farm is the main form of organization of agricultural production, which means that the farmer can only rely on himself.

In order to improve the efficiency of his business, he must possess knowledge in the field of the agro-industrial complex, know the equipment and be able to operate it properly, be knowledgeable with modern technologies of growing, processing and selling the manufactured products, know the accounting, know how to use computers, etc.

The victory in the fight against competitors for high quality products is strategically important for entrepreneurship, because, as you know, low quality products are not in high demand among consumers. In the aspect of international transformations, the identification of promising areas for improving the competitiveness of domestic products is one of the most important strategic issues of economic reform. Improving the quality and competitiveness of products is closely related to the improvement of the management system, standardization and certification of products. This problem cannot be solved without applying a systems approach to quality management at the enterprise level, as well as legal regulation at the macro level. The higher the quality, the better the needs of consumers are satisfied and the social and economic problems of the development of society are more effectively solved. Also, the interaction of information technology and entrepreneurship is the implementation of environmentally friendly products.

Therefore, the relevance of my research is related to the current trend in food consumption, the problems of sales of manufactured products and the increasing competitiveness of domestic products.

Content from this work may be used under the terms of the Creative Commons Attribution 3.0 licence. Any further distribution of this work must maintain attribution to the author(s) and the title of the work, journal citation and DOI.
Published under licence by IOP Publishing Ltd

IOP Conf. Series: Earth and Environmental Science **395** (2019) 012114 doi:10.1088/1755-1315/395/1/012114

In modern times, when air, water and land are polluted, despite all the efforts of mankind, human activity and the ecological situation continues to deteriorate, people are increasingly beginning to think about their health. In the conditions of an oversupply of food products in stores, city dwellers tend to get better and more natural products in order to maintain their health. It is important to note that sales of fruits and vegetables, meat and dairy products are the fastest growing segments of the organic food market. The growth in sales of organically produced products is evidence that buyers are ready for the added value of goods. Consumers are placing increasing demands on the quality of food. Natural products, lack of use of genetic engineering in the production and the possibility of harm to health is a priority for them.

For a number of reasons, the movement of farm products to the end customer is not effective through traditional distribution channels. As a distribution channel for products, markets are also not able to fully comply with the requirements of producers of natural products and customers. At such sites, the level of service required for the sale of premium organic goods is absent, and buyers do not receive guarantees of product quality due to the low level of reliability of sanitary examinations organized in the markets. The sale of farm products to processing enterprises is also not feasible due to low purchase prices and low profitability for farmers, as well as due to the high quality and low volume of production in farms..

2. Theory and Methods

Therefore, Internet entrepreneurship is an independent new and advanced form of entrepreneurship. The online store is the main element of the functioning of the "Internet Enterprise" system. The online store is a representation of entrepreneurship on the Internet, created on the basis of a web server. Its main goal is to ensure the sale of goods and services to other Internet users.

The Internet allows farmers to make decisions using an exhaustive amount of information, and this opportunity is an important advantage in the implementation of Internet processes. The sale of organic farm products directly from the farmer, through the manufacturer's online store, is being developed in Europe, the USA and China. Since it is the farmers who are involved in the processes of marketing and selling products without attracting distributors for this. It is important to note that prices for farm products are at a lower level than in supermarkets. This situation is achieved through constant monitoring of prices for organic products in supermarkets and savings in logistics.

Internet technologies will become the most important factor in the development of entrepreneurship. They allow you to create, store, process and provide effective ways to transfer information to the consumer. Creating the most favorable conditions for an entrepreneur to access information is the most important task of building the country's information infrastructure. Thus, the relevance is determined by the increasing value of information in the infrastructure of modern entrepreneurship. As information is needed for an entrepreneur about other producers, about potential consumers, about suppliers of raw materials, components and technologies, about prices, about the situation in the commodity and capital markets, about the situation in business life, about the general political and economic situation not only in their own country but all over the world.

Sales of products is the most important problem for farmers. This is due to foreign product intervention, extremely low technical potential of domestic production, insufficient state support, lack of information, insufficient marketing support, problems in the functioning of domestic wholesale markets and access to foreign markets. This not only deprives the farms of sufficient financial resources for development, but also the economic motives for increasing production volumes.

In this regard, online stores of agricultural products eliminate the above problems in the process of buying and selling. Online stores are ready to buy goods in small batches, not interested in long-term storage of the product. Online stores deliver goods to the buyer's table as soon as possible after they are purchased from a farmer, have all the relevant documentation. In online commerce, the ability to purchase goods from their location hundreds of kilometers away is an advantage. However, geographic limitations are characteristic of trade in the field of farm products. Delivery by freezing or preserving eliminates the advantageous qualities of an organic product that distinguishes it from a product of mass consumption. In this case, the online store is forced to cooperate only with local farms or farms from nearby areas. In addition, the Internet audience is growing every day, therefore, the number of buyers who are ready to purchase goods on the Internet is growing.

Therefore, we recommend entrepreneurs to create such sites, the effectiveness of which reached the audience at an appropriate level, the product range and order volumes were maximum and regularly engaged in the expansion of communication opportunities. As a rule, e-commerce will increase profits, strengthen market positions, and also allow your company to reach a new level.

Thus, you need to start with an analysis of the sites of your competitors in order to achieve effective development of entrepreneurship on the Internet. Competitors have a strong influence on prosperity, both in real and in online entrepreneurship. The main thing is to fill the site with unique, rare, valuable or just useful information.

The only way to ensure the institutionalization of Internet commerce on the part of the state is to advance the development of the national logistics infrastructure and improve the regulatory framework. That is why, today, such recognized leaders of Internet commerce as China, Europe and the United States give priority to this.

But it is impossible not to note the fact that rural residents with traditional conservative views are, as a rule, farmers in Kyrgyzstan. They are not always ready to agree with the placement of detailed information about the farm on the Internet, as they are wary of advanced technologies. At least two problems need to be solved in order to improve the quality of Internet entrepreneurship:

The first and most important problem is the question of delivery, that is, logistics. The higher the level of the Internet, the greater the coverage of online shopping.

The second problem is the low level of implementation of online payment. Opportunities already exist, but objective factors exist that inhibit their widespread adoption. This is the absence of such a culture among the majority of the population, that is, fear, weak spread of electronic money, low level of use of bank cards. These specific problems are complemented by low Internet penetration in the regions of the republic, power outages, unreliable mail, etc.

But the solution of the two main problems listed above will lead to the effect and growth of online trading. Since round-the-clock availability, the absence of queues, a wide range of products, the opportunity to familiarize yourself with the technical specifications and use the price matching system without getting up from the chair will lead to an increase in online trading.

If you analyze the possibility of using Internet technologies in the fight against competitors for high quality products, the Internet has tremendous marketing potential and allows you to access, in the shortest possible time, the audience you need. To do this, it is not necessary to spend money. You can start promoting Internet entrepreneurship by exchanging banners, free advertising on subject forums, message boards and creating official pages on social networks. As your business grows, you can start buying ads or order a professional promotion on the Internet. Initially, online trading in social networks is a very effective tool for promoting products and services. Ordinary people and big companies can use this service equally effectively.

Therefore, by monitoring the status of these indicators, the management and marketers of the online store have the opportunity to identify positive and negative characteristics and work on the errors made when creating a resource.

I hope that the recommendations I have listed will help you to improve and increase your sales.

Agriculture is an important component of the country's economy. In agriculture, a huge economic potential is concentrated and products that are vital for society are produced. The development of agriculture and the agro-industrial complex as a whole determines the state of the entire economic potential, the level of food security of the state and the socio-economic situation in society.

Today, the problem of ensuring the effective functioning of entrepreneurship in agriculture is poorly understood in the conditions of Kyrgyzstan. From this position it can be argued that the affected issue is relevant and timely. Therefore, the importance of entrepreneurship for the development of agriculture is great and at the same time, its underdevelopment determines the relevance of the study.

The purpose of this study is to study the characteristics, forms, problems and prospects of entrepreneurship in agriculture of the Kyrgyz Republic at the present stage. And in the long term, the goal is to provide the people of the country with plenty of high-quality and complete nutrition of their own production and turn Kyrgyzstan into one of the world leaders in organic agriculture, ensuring sustainable development of the economy.

To achieve the goal, the following tasks are set:

- Exploring the essence of the economic efficiency of entrepreneurship in agriculture;
- Studying the international experience of using Internet technologies in order to improve the efficiency of entrepreneurship;
- Identifying problems and prospects for the development of entrepreneurial activities in agriculture;
- Analyzing the use of information technology in agricultural entrepreneurship;
- Identifying the main factors affecting the development of entrepreneurship in agricultural production: legislation, taxation, resources, government support, market infrastructure and the formation of entrepreneurs.
- Developing a set of organizational and economic measures to ensure the implementation of operations for the purchase and sale of agricultural products;
- Organizing a training for entrepreneurs in the form of business seminars;
- Analyzing the competitiveness of environmentally friendly products in the international market;
- Identify the most effective ways to influence potential customers of the online store of agricultural products;
- Developing convenient and inexpensive ways to deliver goods, i.e. logistics.

3. Results

The practical significance of the work lies in the fact that the research results create a methodological basis for the organization of online trading of environmentally friendly products. This is done in order to increase the competitiveness of business entities engaged in business activities in agriculture.

By 2025, according to United Nations projections, food production will need to increase by 70%. Therefore, the agricultural sector is faced with the task of increasing labor productivity and competitiveness based on the application of the latest scientific achievements and best practices. Thus, in the second decade of the digital age, many areas of life and professions underwent transformation. Now, the image of the farmer, as a simple peasant, familiar to our imagination, has irrevocably sunk into the past. And the current agricultural worker is a modern manager who can manage economic activities through computer programs and work with electronic data.

Digitalization is the conversion of information into digital form. The digital economy is a global network of economic and social events. They are implemented through platforms such as the Internet, as well as mobile and sensor networks. In our time, the boundaries have already been erased, and all commercial companies can be called digital. After all, every person uses Internet access, mobile communications. Increasing labor productivity and reducing costs are two key indicators that will be achieved in the era of digitization of agriculture.

Experts predict that by 2020, 25% of the global economy will move to the introduction of digitalization technologies. They allow the state, business and society to function effectively.

Based on the above facts in the field of digitalization, the comment of foreign and domestic experts should be cited, as well as foreign experience of using Internet technologies in order to improve the efficiency of entrepreneurship.

According to the "Digital Kazakhstan" program, Amr Salem, global managing director for Smart Cities Internet of Everything Cisco, said that Kazakhstan has the right approach to digitalization as a strategy for developing the economy and improving people's lives. Two factors are important here - the creation of an infrastructure that connects all government agencies with each other, and a single platform for providing services to the population. The expert also stressed that the digitalization program is a great journey, and not a project with some kind of ending, it will live for many years and constantly improve [1].

IOP Conf. Series: Earth and Environmental Science **395** (2019) 012114 doi:10.1088/1755-1315/395/1/012114

The director of the Boston Consulting Group in the Middle East, Alexander Tyurpits, noted that the growth of the Internet economy in developing countries is 15-25% per year, which no other sector of the economy can show. He reported that 90% of all global data was created in just the last two years. And 99% of the world's data is already digitized. 35 billion devices worldwide generate data and share it every day. It is five times larger than the world's population and makes up only one percent of the devices that can be connected. According to him, the timely introduction of technology will increase productivity in various industries and give them a "second life", will create jobs. In addition, Alexander Tyurpits noted that the growing electronic commerce erases the borders of states. Each country follows its own unique path. And the country focuses on one of the technologies that changes our lives. For example, Germany is developing the industrial Internet of things, the USA is artificial intelligence, Sweden is the blockchain, China is 3D printing, and Saudi Arabia is big data. Kazakhstan is one of 15 countries in the world, which is launching a comprehensive program that includes all four components of success - the digital transformation of traditional industries, the development of human capital, the digitalization of state. services and digital infrastructure development.

He also noted the reasons for failures in this area. This is a limited participation of the private sector, focusing primarily on the ICT sector, focusing only on fast results and on "current" technologies. The expert also stresses that the fight against cybercrime is the other side of the process (digitalization). Last year, governments and corporations spent almost half a trillion dollars on countering cyber-attacks **[2]**.

Alexey Gordeev, Presidential Plenipotentiary Representative in the Central Federal District, said that digitization of agriculture is necessary for the further development of the industry in the world. "Digitization of agriculture is of great importance. This is one of the most promising areas for sustainable development of the agri-food sector ", he said during the 31st session of the Regional Conference of the Food and Agriculture Organization of the United Nations (FAO) for Europe. According to Gordeev, the feasibility of digitalization has matured due to the need to use natural resources productively, develop sustainable and inclusive food systems, reduce losses of agricultural raw materials and food, and pursue an open and sustainable agricultural policy in all countries [3].

At the St. Petersburg International Economic Forum 2018 "Internet technologies in the agricultural sector," Deputy Minister of Agriculture of Russia Ivan Lebedev said that digitalization is a fundamental trend in the development of agriculture, it allows you to increase agricultural production and ensure the profitability of the industry. Since according to preliminary estimates, the overall minimum economic effect from the introduction of IoT technology in agriculture for the period up to 2025 may be about 469 billion rubles due to the optimization of personnel costs; reduce crop losses (grain); reduce the loss of fuel and lubricants (POL) [4].

At the international conference "Digital Transformation of Agriculture," organized by the Internet Initiatives Development Fund (IIDF), the Ministry of Agriculture of Russia and the World Bank together with the Association of the Internet of Things, held on May 15, 2018 in Moscow, it was noted that over the past 5 years total global investment in digital technology for agriculture has reached $ 10.1 billion. Russia still occupies only 1.5% of the global volume of the Internet of Things, while in agriculture the share is even lower. Increased investment will increase the level of digitization of agriculture [5]. In addition, digital technologies allow farmers in different countries to be trained in best practices, as well as adhere to uniform production standards. For example, Nestle (Switzerland) has trained 10,000 farmers in West Africa with modern technology for farming and storing products. As a result, the company received raw materials of guaranteed quality, and farmers - access to the global market and product sales at high prices.

On May 29, 2014 in Astana, the agreement was signed on the establishment of the Eurasian Economic Union. From January 1, 2015, the EAEU began to function together with Russia, Belarus and Kazakhstan as its participants. From January 2, 2015, Armenia became a member of the EAEU, Kyrgyzstan became a full member of the Union on August 12, 2015. In addition, the concept of digital transformation of the Kyrgyz Republic for 2019-2023 was developed in Kyrgyzstan in 2018. Digitization focuses precisely on the activities of state bodies, contributing to the improvement of public administration, improving the quality of services to the population, optimal business management and the development of the economy as a whole. The head of Kyrgyzstan Sooronbay Zheenbekov also stressed that the implementation of the regional development strategy will continue. At the same time, in the coming year, emphasis will be placed on the introduction of new technologies in the economy and everyday life. To this end, the year 2019 was declared in Kyrgyzstan the "Year of the development of regions and the digitalization of the country". The President said that in order to support rural entrepreneurs, new enterprises were exempted from inspections of supervisory bodies for 3 years, a moratorium on unreasonable business inspections was introduced, laboratories for certification and standardization of products were built, significant tax exemptions were made to local enterprises engaged in processing agricultural products, agricultural finance improved. Kyrgyzstan plans to implement the program “Digitization of agriculture” for 2019-2023 [6].

In our country, often raw materials are the main attractive factor for foreign investors. Kyrgyzstan is rich in various types of raw materials bases in various directions, this attractive factor reinforces low prices for them. Two main problems are listed, such as the lack of new technologies for processing raw materials and the lack of understanding of how to maintain and develop the raw material base [7], [5].

4. Conclusion

In conclusion, I note that today is the time when intelligent digital solutions should help the country's agriculture to cope with the problems of increasing productivity and sustainable development. The agricultural sector is the most vulnerable sector of the economy, largely dependent on the vagaries of nature. And the impact of climate change on food security in the world will only increase. Digitization of the agro-industrial complex will reduce these risks, adapt to climate change, increase crop yields and animal productivity, and plan field work in a timely manner. And a farmer, in order to remain competitive, must be able to predict the supply of his products depending on demand and consumer preferences. In order to make the right, or, as they say now, “smart” management decision, the farmer needs to own digital technologies. So, it is necessary to change the system of professional training of highly qualified personnel, starting from school, college, university and review the policy in the field of education. Additional financial resources are needed by rural entrepreneurs to speed up digitization. The Kyrgyz-Russian Fund was established in Kyrgyzstan for the implementation of investment projects in such areas as the agro-industrial complex, clothing, textiles, manufacturing, mining and metallurgical industry, transport, housing, trade, development of entrepreneurship and infrastructure. However, I consider it expedient to create an independent fund of digitalization of the agro industrial complex. This will allow the country's agriculture to become attractive for investments. Digitization of agriculture is an increase in labor productivity and cost savings. The main goal is to increase the yield and product quality. The agro-industrial complex of Kyrgyzstan is a field of activity with increased risks and depends largely on climatic factors. Digitization of agriculture in Kyrgyzstan will help reduce the impact of climate and gradually move to precision agriculture. [8, p. 83]

With the successful implementation of the program “Digitization of Agriculture”, it will give new advantages in the following main areas: (a) availability of funding through process automation; (b) the issuance of accompanying and permitting documents will be automated, electronic grain receipts, product-to-field traceability are provided and conditions for the introduction of online trading are created; (c) the registration system of agricultural equipment of its pledge and traceability will be automated, and the time frame for obtaining public services will be reduced; (d) a geoportal is created using space monitoring of forest and water resources, wildlife, hunting and fisheries.

Most importantly, the introduction of Internet commerce will increase direct sales by at least 20% and eliminate unscrupulous intermediaries.

Agricultural producers and cooperatives will be able to quickly consolidate their products into bulk lots, ensure a systematic supply and effectively promote them in the domestic and foreign markets. Also, as part of the task of technological re-equipment, work on introducing elements of precision farming, creating SMART farms, and engaging entrepreneurs in the use of IT technologies will be carried out. The introduction of elements of precision farming will significantly improve the efficiency of production processes of farmers. These elements include electronic field maps, accurate weather data, sensors and sensors, space monitoring, etc. The introduction of SMART farms will allow monitoring livestock and herd management, provide autonomous greenhouse management, keep records and analyze costs in online mode.

Obviously, the digitalization of the agro-industrial complex will improve competitiveness and productivity, ensure the implementation of environmentally friendly products and food safety, as well as ensure investment attractiveness in the industry.

References

[1] Espolov T 2018 *Digitization of the agro industrial complex - the demand of the new time* Available at: http://kzvesti.kz/kv/thirdband/25528-cifrovizaciya-apk-trebovanie-novogo-vremeni.html (Accessed 10 03 2019)

[2] Expert from Electronic magazine IDK 2018 Ministry of Agriculture is developing a program "Digitization of agriculture" *Electronic magazine IDK* Available at: https://аграрная.рф/index.php?id=422 (Accessed 10 03 2019)

[3] Digitization of agriculture at the 31st session of the Regional Conference of the Food and Agriculture Organization of the United Nations (FAO) for Europe 31st Session, Voronezh, Russian Federation, May 16-18, 2018 Available at: http://www.fao.org/about/meetings/erc31/ru/ (Accessed 10 03 2019)

[4] Sidorov G 2017 *Internet of Things (IoT) in Russia. The technology of the future, available now* Available at: https://www.pwc.ru/ru/publications/iot/IoT-inRussia-research_rus.pdf (Accessed 10 03 2019)

[5] Ministry of Agriculture of the Russian Federation 2018 *The conference "Digital transformation of agriculture" took place in Moscow* Available at: http://mcx.ru/press-service/news/v-moskve-sostoyalas-tsifrovaya-transformatsiya-selskogo-khozyaystva/

[6] Public Foundation Civil Initiatives on Internet Policy 2018 *The concept of the National Digital Transformation Program "Digital Kyrgyzstan" – 2019-2023 (Bishkek, Kyrgyzstan)* Available at: https://clck.ru/HMF2U Accessed 10 03 2019

[7] Akzholova M 2018 Problems of improving the investment-business attractiveness in Kyrgyzstan *Colloquium-journal* **3**(14) Available at: https://elibrary.ru/item.asp?id=32720708 (Accessed 10 03 2019)

[8] SCIEURO 2018 *Proceedings from 7th the International Scientific-Practical Conference: "Education Transformation Issues"* (London, UK: SCIEURO Publishing)

[9] AITC 2014 *E-commerce Market in Russia 2012-2013* Available at: http://www.akit.ru/data-insight-e-commerce-russia/ (Accessed 10 03 2019)

[10] POLIT.RU 2018 *Digitization of agriculture* Available at: http://polit.ru/article/2018/02/21/sk_digital_farming/ (Accessed 10 03 2019)

[11] Salnikov S G 2018 Actual directions of digital transformation of the agro-industrial complex of Russia In *Digital economy: mantechnologies, institutions* (Moscow, Russia: Moscow State University) Available at: http://www.viapi.ru/ (Accessed 10 03 2019)

[12] Fedorenko V F, et al. 2017 *Improving the effectiveness of information support in the field of agriculture: scientific. analyte review* (Moscow, Russia: Federal State Scientific Institution Rosinformagrotekh)

IOP Conf. Series: Earth and Environmental Science **395** (2019) 012115 doi:10.1088/1755-1315/395/1/012115

Feeding capacity and meat productivity of sheep of different breeds in the Republic of Buryatia

S Biltuev[1], V Achituev[1] and B Zhamyanov[1*]

[1] Buryat State Agriculture Academy by V.R. Philippov, 8 Pushkin, Ulan-Ude 670034 Russia

E-mail: bair1717@mail.ru

Abstract. In this article, we studied the feeding and meat productivity of young coarse sheep (Buryat, Mongolian, Tuva, Edilbaev breeds) raised in 4 sheep farms of the Republic of Buryatia. In addition, the paper analyzes the economic efficiency of different breeds.

1. Introduction

Sheep breeding for the Republic of Buryatia is a traditional branch of animal husbandry, providing the population with ecologically clean mutton and also supplying wool and fur raw materials for the processing industry. The increase in the production of young lamb, high quality coarse wool and fur coat sheepskin is important for the economy of the republic.

Currently, the sheep population in the republic is mostly represented by the Buryat, Tuva, Mongolian, and Edilbaev coarse-wool sheep (300 thousand heads). In this regard, research on the study of productive and biological features of these breeds is a very topical issue.

According to the results of research, the best of the compared breeds for feeding and meat productivity were identified, which would improve the efficiency of production of environmentally friendly mutton in conditions of a year-round sheep in the Baikal region. More than that, it would help solve the problem of providing the population with food.

The results obtained during the research and production experiments and laboratory studies will be part of a larger scientific material on the identification of adaptive properties of domestic and imported breeds. They will be also used for further analysis and theoretical justification on the feasibility of using their gene pool in the Republic of Buryatia.

2. Research Methods

The purpose of the research is to study the feeding and meat productivity of young coarse wool sheep of different breeds bred in the Republic of Buryatia.

The research objectives include the following:

- To conduct a study on the growth and development of young stock in the lactation and after lactation periods;
- To study the slaughter and meat qualities at the age of 7 months, including the possibility of their use for the production of cheap and environmentally friendly mutton in the Republic of Buryatia;

Content from this work may be used under the terms of the Creative Commons Attribution 3.0 licence. Any further distribution of this work must maintain attribution to the author(s) and the title of the work, journal citation and DOI.
Published under licence by IOP Publishing Ltd

IOP Conf. Series: Earth and Environmental Science **395** (2019) 012115 doi:10.1088/1755-1315/395/1/012115

- To determine the economic efficiency of coarse-wool breeding in the conditions of the Republic of Buryatia.

In order to identify the feeding and meat productivity of the wethers in different breeds, we conducted an experiment in the farm “Dorzhiev I. F.” of the Selenginsky district. To do this, after pounding from different farms at the age of 4 months, a total of 4 groups of animals (30 animals in each0 were formed, depending on the average body weight in the herd. Group I included the Buryat coarse wool wethers. The Group II: Tuva short-fat-tail wethers. The Group III: Mongolian coarse wool wethers. Group IV: Edilbaev wethers. These animals were put on the field for 90 days (September-November).

The live weight was measured in the experimental animals aged 4 and 7 months; it was determined by individual weighing with an accuracy of 0.5 kg.

Meat productivity of the castrated coarse-wool wethers was studied by controlling slaughter at the age of 7 months according to the VIZh technique (1970, 1978). Cutting carcasses was carried out in accordance with GOST 7596-81. The organoleptic evaluation of the qualities of animals meat and broth being compared was carried out according to GOST 9959-2015 (“General conditions for conducting organoleptic evaluation”) on a 9-point scale.

The economic efficiency was determined by calculation based on the analysis of total costs and revenues according to the statistical reporting of an enterprise.

The digital material is processed biometrically according to the method developed by E. K. Merkurieva (1970) and N. A. Plokhinsky (1969), using the “Excel” computer program.

3. Research Results

The high growth rate of young coarse-wool sheep is one of the main conditions for increasing the profitability of lamb production. The growth rate of lambs was the highest during the lactation period, when mother’s milk was the main food for them [2].

The most intensive growth at the age from 4 to 7 months was observed in the young animals of the Tuvan short-fat-tail sheep. They had an advantage in the average daily gain over their peers of the Buryat and Mongolian coarse-wool sheep by 15.3 and 8.4 g, with Edilbaev – by 16.7 g or 5.8; 7.06 and 15.1% (Table 1).

In the after-lactation period, the largest decrease in the average daily weight gain (more than 2 times) was observed in the young Edilbaev breed, which was due to the reaction of their organism to the changing nature of their feeding. At the same time, over the entire feeding period, a relative weight gain was the smallest among the young Edilbaev breed (110.6%, or 8.3). In was 9.7 and 16.3% lower than that of the Mongolian, Buryat coarse-wool, and Tuva short-fat-tail peers. This circumstance can probably be explained by the late ripeness of this breed, which was formed in them under the influence of environmental conditions.

Table 1. Live weight, young sheep of different breeds, kg.

Indicators	Group			
	I	II	III	IV
Number of animals, heads	30	30	30	30
Live weight, kg:				
when feeding	31.3±0.94	28.4±0.45	27.3±0.39	32.66±0.34
when removed from feeding	37.66±0.28	36.14±0.64	32.45±0.20	36.13±0.25
Absolute increase, kg:	6.36	7.74	5.15	3.47
Average daily gain, g:	70.7	86.0	57.2	38.6

The slaughter quality and morphological composition of young sheep carcasses of various breeds of coarse-wool productivity. Studies by the Russian authors clearly show that the young animals for meat are most effective at 7-8 months of age [3], [4], [5]. A young lamb is tastier and more nutritious, it has no specific taste typical of adult sheep meat. It is less fat, and the existing fat is usually placed between the muscle fibers, which gives the lamb special juiciness and marbling.

The results of the control slaughter indicate a high meat condition of young sheep of the Buryat semi-coarse-wool, Tuva short-fat-tail, and Edilbaev meat-greasy breeds (Table 2).

Table 2. Slaughter quality wethers of different breeds at the age of 7 months.

Indicators	Group			
	I	II	III	IV
Pre-slaughter weight, kg	38.12±0.39	36.20±1.66	32.45±0.85	35.68±1.03
Carcass weight, kg	19.27±0.48	16.63±0.62	13.6±0.72	16.68±0.09
Mass of internal fat, kg	0.62±0.14	0.98±0.11	0.46±0.01	0.33±0.05
Slaughter weight, kg	19.89±0.62	17.61±0.77	14.06±0.64	17.02±0.11
Slaughter yield, %	52.16±1.09	48.74±1.79	43.3±1.24	47.76±1.18

Among the compared groups of animals, at the age of 7 months, the largest mass of carcass and slaughter yield were observed among the wethers sheep of the Buryat coarse-wool sheep, which by these indicators exceeded the Tuva short-fat-tail peers (by 1.92 kg and 3.42%), and Edilbaev (2.17 kg and 3.43%). The advantage of the Buryat coarse-wool wethers in these indicators was more significant over the Mongolian ones, with a significant difference of 5.67 kg and 8.86%.

In addition to the quantitative assessment of meat productivity, we need to know its quality characteristics. The meat of animals is determined not only by the ratio of live and carcass weight, but also by the proportion in the carcass of the most valuable cuts, by the correspondence in the carcass of edible and inedible parts [1]. The Buryat coarse-wool wethers surpassed other groups (II, III, and IV) by 5.54, 8.04, and 1.25 kg, according to the coefficient of meatiness – 1.1, 1.38, and 0.3 (Table 3).

Table 3. The morphological composition of wethers in different breeds.

Indicator	Group			
	I	II	III	IV
Chilled carcass weight, kg	19.27±0.48	16.63±0.62	13.6±0.72	16.68±0.09
Pulp yield, %	78.96	73.42	70.92	77.71
Bone yield, %	21.04	26.58	29.08	22.29
Meat ratio	3.9	2.8	2.52	3.6

The presented research data on the morphological composition of carcasses of wethers in different breeds showed relatively low indices of the morphological composition of carcasses in young Mongolian coarse-wool sheep.

The efficiency of breeding sheep of different breeds in the Republic of Buryatia is also studied by us. An indicator of the adaptive qualities of sheep of different breeds to the conditions of breeding is the level and quality of products obtained from them, the cost of production and the cost of implementation. The imported and domestic breeds in the Republic of Buryatia, which were compared by us, were differed among themselves in the safety of lambs, an increase in the young live weight, as well as the trimmed wool (Table 4).

When calculating the efficiency of aforementioned breeds, we used the following results: lambing of ewes, safety of lambs; increase in live weight of young stock from birth to 7 months of age; cost price and selling price of 1 kg of growing the trimmed wool; cost price and purchase price of 1 kg of wool in the original wool; profitability of production at 1 structural head at the beginning of the year.

In our studies, the largest absolute increase in live weight per structural head, taking into account the safety of the young animals, was obtained from the Buryat coarse-wool sheep (19.67 kg), and the smallest one was from Edilbaev meat-greasy (16.97 kg). According to this indicator, the Tuva short-fat-tailed sheep were inferior to the Buryat coarse-wool sheep by 0.98 kg, exceeding the Mongolian by 1.13 kg and the Edilbaev meat-grease by 1.72 kg. In terms of wool productivity, the best indicator of 2.1 kg was provided by the Edilbaev meat-greasy animals, surpassing the Tuva short-fat-tailed by 0.8 kg, and outperforming the Buryat and Mongolian coarse-wool by 0.9 and 0.5 kg. The cost of 1 kg increase in live weight and collection of dirty wool in sheep of comparable breeds were the same (73.5 and 33.7 rubles), because they did not differ in costs incurred for their year-round pasture maintenance. The cost of their products was also the same, amounting to 91.0 and 12.0 rubles.

Table 4. The efficiency of sheep from different breeds in the Republic of Buryatia, per 1 head.

Indicator	**Buryat coarse-wool**	**Tuva short-wool**	**Mongolian coarse-wool**	**Edilbaev**
Ewes seeded, heads	450	325	514	170
Ewes given birth, heads	450	316	510	137
Live lambs received, heads	441	309	503	137
Ewes in the herd structure, %	60.7	62.3	63.7	68.0
Saved lambs:				
heads	419	290	492	113
%	95	93.8	96.3	82.5
Increase in live weight of young stock, kg:				
from birth to 7 months	34.12	31.99	28.63	30.24
an inseminated ewe	32.41	30.01	27.57	24.95
a structural head	19.67	18.69	17.56	16.97
Cut wool on 1 head, kg	1.2	1.3	1.6	2.1
The cost of 1 kg, rubles:				
increase in live weight	73.5	73.5	73.5	73.5
cutting wool in physical mass	33.37	33.37	33.37	33.37
Total cost per head, rub.	1485.79	1417.09	1344.05	1317.37
The cost of 1 kg, rub:				
increase in live weight	91	91	91	91
wool	12.0	12.0	12.0	12.0
The cost of sales of 1 head, total, rubles including:	1804.37	1716.39	1617.16	1569.47
increase in live weight	1789.97	1700.79	1597.96	1544.27
wool	14.4	15.6	19.2	25.2
Profit from sales, rubles	318.59	299.3	273.11	252.1
increase in live weight	344.23	327.07	307.3	296.97
wool	-25.64	-27.77	-34.19	-44.87
Profitability level, %	23.17	23.08	22.86	19.1

The sales value of the total production of the Buryat coarse-wool sheep per head amounted to 1,804.37 rubles, at a cost of 1,485.79 rubles. Profits from selling the products amounted to 318.59 rubles, the level of profitability of production amounted to 23.17%, or 0.09, 0.31, and 4.07, which was more than the Tuvan short-fat-tailed, Mongolian coarse-wool, and Edilbaev meat-greasy sheep breeds.

4. Conclusion

The sheep form different coarse-wool breeds differ in the intensity of growth of lambs in the lactation and after-lactation periods, slaughter and meat qualities and in the effectiveness of breeding them in the Republic of Buryatia. The lowest level of profitability is 19.1%.

References

[1] Biltuev S I, Zhilyakova G M, and Achituev V A 2016 Efficiency of mutton production in the Baikal region *Sheep, Goats and Wool* **3** pp 38-39

[2] Lushnikov V P, Prosvirina E V, and Mikhailov I Yu 2007 Crossing - as a method of increasing the meat productivity of aboriginal breeds of sheep *Sheep, Goats and Wool* **2** pp 5-8

[3] Biltuev S I, Zhilyakova G M, and Achituev V A 2009 Meat productivity of fine-wool and semi-coarse-wool sheep raised in Buryatia *Vestnik of the Buryat State Agricultural Academy named after V. R. Filippov* **2** pp 86-89

[4] Biltuev S I, Zhilyakova G M, Achituev V A, and Zhamyanov B V 2018 Ways to improve the efficiency of sheep farming in the Republic of Buryatia *Vestnik of the Buryat State Agricultural Academy named after V. R. Filippov* **1** pp 109-116

[5] Zhilyakova G M, Biltuev S I, Achituev V A, and Zhamyanov B V 2017 Efficiency of sheep production in the Republic of Buryatia In *Collection of scientific papers of the All-Russian Research Institute for Sheep and Goat Breeding* **1**(10) pp 126-131

IOP Conf. Series: Earth and Environmental Science **395** (2019) 012116 doi:10.1088/1755-1315/395/1/012116

Culturological aspects of environmental policy in coal mining region

F E Kazakov[1], V I Markov[2*], and G N Minenko[2]

[1] Department of Philosophy, Institute of History, Public Administration and International Relations, Kemerovo State University, 6 Krasnaya str., Kemerovo 650000 Russia

[2] Department of Cultural Studies, Kemerovo State Institute of Culture, 17 Voroshilova str., Kemerovo 650029 Russia

E-mail: vikt-markov@yandex.ru

Abstract. The paper presents an analysis of the problem of environmental safety on the basis of the cultural vision of socio-ecological problems associated with the development of the coal industry in the Kuzbass region. It is emphasized that at the present time, there is a growing crisis in the relations between the man and nature, which is especially evident in such actively interacting with the natural environment regions as the Kuzbass. The article emphasize the statement that the coal industry of the Kuzbass plays such a significant role that its development appears to be socially and culturally important for the entire region. The culture of environmental safety in the coal mining industry affects life and health, in the literal sense of the word, every inhabitant of the region. Therefore, the formation of a high ecological culture should become a strategic priority of the regional policy covering all spheres of life and activity of the Kuzbass population. In this complicated situation, the choice of orientations and motivations in the formation of ecological consciousness, in ecological education acquires a huge role. However, in publications, only environmental education is most often emphasized, and sometimes norms of environmental ethics. The article proposes a broader approach based on the synthesis of rational impact, ethical and aesthetic motivation, as well as on involving young people in the practical volunteer movement in the sphere of ecology.

Keywords: culture, ecology, ecological crisis, globalization, coal industry, Kuzbass, environmental policy and education, people, nature.

1. Introduction

Currently, the environmental crisis is becoming global. The power of modern civilization is sometimes comparable with the power of natural processes. Increasing anthropogenic pressure on nature carries with it the danger of major catastrophic consequences. The problem of human interaction with nature is particularly relevant for such coal-mining regions as the Kuzbass, where the scale and intensity of anthropogenic impacts reaches critical values. And this must be put on the agenda the problem of environmental safety, affecting both the existence of nature and the person himself, and his basic components. The growing crisis in the relationship between man and nature is in direct connection with the transformation of culture into civilization, bearing the loss of moral guidelines, the reduction of humanistic values, supplanted by pragmatics and profit. Therefore, the task of finding a cultural paradigm in the relationship between man and nature seems to be the most relevant, where the

Content from this work may be used under the terms of the Creative Commons Attribution 3.0 licence. Any further distribution of this work must maintain attribution to the author(s) and the title of the work, journal citation and DOI.
Published under licence by IOP Publishing Ltd

Kuzbass can be a convincing example. But despite the growing acuteness of environmental problems, a little has changed in the attitude of man to nature in fact, in his mind, in the system of those values that are needed to preserve nature itself. The growth of consumer priorities continues, the rate of environmental degradation is increasing, and the rate of movement towards a global crisis is increasing. Unfortunately, the Kuznetsk coal basin fits into this trend.

2. Materials and Methods

Recently, the society has gone through the periods of formation of non-classical (an "ecosystem" by A. Tensley) and post-non-classical ecology (a theory of human invasion of ecosystems). In studying the problem of the relationship between man and nature, the transition took place from the traditional opposition "organism-environment" (classical ecology) to the opposition "natural - artificial" ("Social ecology" of R. Park and E. Bergessa, studying the "second", artificial human environment).

According to N. N. Moiseev, the entry into the epoch of the Noosphere should occur not in the form of a bifurcation jump (which can lead to unpredictable consequences), but in the form of co-evolution with the biosphere. As the dominant cultural paradigms, the concepts of a new humanism, such as A. Schweitzer's "ethics of reverence for life," should be chosen. In the study of the cultural aspects of ecological paradigms, their influence on the solution of ethological problems and the identification of the specifics of the specific changes of modern technologies, the following methods and approaches were used: the methods of systemic and comparative analysis; the structural-functional approach; the conceptual analysis, historical, and logical. From a methodological point of view, the application of a regional approach, linked to the specifics of the coal-mining region, and the extension of the principles of environmental education used are also important.

And this specificity, sufficiently developed in the studies of the Kuzbass scientists, indicates the pre-crisis state of the environment of the Kuzbass. Even in the capital city of Kemerovo, which has not been directly linked to coal mining for a long time, the content of harmful dust particles indicates a powerful anthropogenic stress on the atmosphere [1]. More broadly, these studies draw attention to the fact that in the annually increasing coal mining in the region, the share of the most dangerous open method is already 63.8%. And the outdated attitude of coal miners to the problems of environmental damage from their activities already causes in the world petitions to abandon the import of Kuzbass coal, which connects the problems of economic and environmental efficiency into a single node [2]. All this speaks of the urgent need to find new approaches to solving the long-standing environmental problems in the coal region [3].

3. Results and Discussion

A consumer attitude to nature cannot but be interrelated with a consumer attitude towards a man (as an important, special, but still part of nature). Hence, the consumer attitude towards nature turns into a consumer attitude of a man towards himself. When nature acts as a means, then, quite often, a person acts as a "means for a means." The attitude of a man to nature is extrapolated to the attitude of man to man, man to God, to the future, to himself. "The death of God" and "the death of a man" cannot be interrelated with each other and with the "death of nature." And the "death of nature" cannot but be in connection with the "death of the future." Between the "|dead man" of modern civilization and nature, life in general, the "dead" (deadening the living) relationships necessarily circulate.

To avoid crises, the systemic noospheric environmental thinking and the corresponding hierarchy of values-regulators of human activities need to be formed. These values set the level of ecological culture, which requires the formation of socio-cultural programs that ensure the harmonious development of the "nature-society" system on the way of their harmonious co-evolution [4] [5].

The contemporary researchers have expressed a number of important ideas on this topic. So, from the point of view of M. Heidegger, modern technologies have put at the service of people not only the whole nature, but also people themselves. According to K. Jaspers, a person becomes one of the types of raw materials to be processed and can no longer be freed from the power of the equipment created by him. As a result, both the nature and man are degraded, as they become simple functional elements and the material of the soulless, self-producing Megamachines. From the point of view of F. Rapp, the mechanisms of culture and human values are behind the technique, so the problem of technology can be solved only by improving social institutions, control, and education mechanisms. J. Habermas believes that within the framework of "strategic action" (associated with the categories of power and wealth), people are guided solely by success, by the results of their actions, putting economic pressure on both the natural environment and other people.

Creating a new "noosphere technology" and environmentally friendly technologies that can ensure a harmonious and safe co-evolutionary development of nature and society is the most urgent [6]. Thus, the ecological imperative of the "co-evolution of man and nature" (N. N. Moiseev) unfolds in unity with the moral imperative. Thus, ecology turns into *ecological ethics* in the ecosophical concepts of "deep ecology" (A. Naess, W. Fox, B. Devall, G. Sessions), for example, requiring the awareness of the falsity of a number of needs and destructiveness of many desires generated by the consumer society. Not only a change in attitude towards outer nature, but also a change in the attitude of man towards himself (toward his inner nature) appears here as the most important goal. In the framework of the "paradigm of the controlled evolution of the biosphere" (A.V. Yablokov) [7], not only the man's awareness of the catastrophic situation is associated with the environment, but also the development of specific measures and technologies to solve basic environmental contradictions is implied.

Under the conditions of an ecological crisis covering the entire planet, environmental policy moves into the category of strategic objectives. The cultural self-identification both in Russia as a whole and in each region in particular cannot but happen without their social and ecological self-identification. In this regard, the task of preserving natural systems, maintaining their integrity, improving the quality and standard of living of the population, ensuring environmental safety is a strategic task for the Kuzbass region [8; 9]. Otherwise, the increased anthropogenic and anthropogenic pressure on the environment can lead to its irreparable destruction, which would result in economic and social damage [10]. To prevent this from happening, a constant monitoring of technogenic pollution of natural resources and a number of systemic measures for their conservation and restoration (water purification, plantation, soil recultivation) are necessary [11].

The environmental sphere should become more susceptible to the achievements of scientific and technological progress; the environmental thinking (the formation of which must begin with preschool institutions, continue at school and university) should become "transparent," permeating all spheres of state and regional governance [12]. The ecologization of the entire system of upbringing and education, a widespread promotion of ecological culture and ideas of sustainable development is necessary. Moreover, the rich nature of the Kuzbass gives all grounds for this. The regulators of economic market are very important to step up in order to encourage entrepreneurial initiatives in the field of environmental protection [13].

Problems and achievements, including breakthroughs, can be identified in the problem of the environmental security of the Kuzbass [14]. Playing a leading role in coal mining, the Kemerovo region belongs to a group of regions with high environmental instability. The Kuzbass landscape image is experiencing obvious technological overloads, and the "ridges" of dumps throughout the region are the result of them. The fact indicates that the area of disturbed land is 10% of the entire Kuznetsk basin [15]. In this regard, first of all, we should talk about the quality of life of the population of the region, which directly depends on the state of the atmosphere, soil and water, flora and fauna, working conditions and the level of health. And here the picture is very disturbing: the Kuzbass region has one of the highest rates of cancer and tuberculosis in Russia; occupational morbidity indicators in the region significantly exceed the corresponding figures for Russia (1.47) [16, 84]; which can not affect the life expectancy. The active impact on the inhabitants of this region coming from the accumulated technogenic high-frequency radiation and chemicals leads to an increase in the number of people with pathological disabilities [17, 62].

At the same time, the Kuzbass has accumulated a great deal of experience in solving problems of environmental safety, including in the coal industry. So, the recultivation of the "coal ridges" is actively carried out. Though, it is mainly at the expense of the enterprises engaged in coal mining. In 2016, the enterprises allocated 126 million rubles, and they planned to allocate about 400 million rubles in 2017 [18]. Diagnostics of hazardous situations is being actively developed, which makes it possible to react in a timely manner to their occurrence, significantly increasing the chances of preventing local man-made disasters. As Yu. G. Markov rightly notes, the "relations with nature are not reduced to the field of rational knowledge. Much more fundamental factors are ethical principles and laws, including aesthetic requirements. If a person remained deaf to moral concepts, he would be deaf to nature" [19, p. 191].

4. Conclusion

Thus, we have revealed that a systemic view of the development of society should proceed from the interrelation of two fundamental activities: the creation of livelihoods and the development of man himself. Increasingly, the formation and development of a human is becoming the dominant party, causing the functioning and development of all other activities. The humanity is not a means of production, not only it subject, but it is the goal of production. If so, then the problem of the level and quality of human life, its environmental safety should be of paramount importance. The role and importance of the problem of environmental safety for the Kuznetsk coal basin is emphasized. The formation of high ecological culture should be a strategic priority of the regional policy covering all spheres of life and activities of the Kuzbass people. The environmental education of the young generation can rely not only on rational and ethical arguments, but also it should positive aesthetic stimuli of love and admiration for the beauty and harmony of nature, for which the nature of the Kemerovo region still represents a fertile material. Thus, it is the rational, ethical, and aesthetic synthesis that represents the most promising area of environmental education. This synthesis can become more effective if it would be implemented in the practical activities of the younger generation. In particular, if it is implemented on the basis of the development of the volunteer movement.

References

[1] Golokhvast K S, Manakov Yu A, Bykov A A, Chayka V V, Nikiforov P A, Rogulin R S, Romanova T Yu, Karabtsov A A, ... Semenikhin V A 2017 Some characteristics of dust particles in the atmosphere of the Kemerovo City according to the pollution data of snow cover. IOP Conference Series: Earth and Environmental Science **87** 042005

[2] Kopytov A I, Manakov Yu A, and Kupriyanov A N 2017 Coal mining and the issued of ecosystem preservation in the Kuzbass *Ugol* **3** pp 72-77

[3] Mikhailov V G, Golofastova N N, and Seredyuk I V 2016 Transformation of the ecological-economic system of the coal mining region. In *Proceedings of the 8th Russian-Chinese Symposium on Coal in the 21st Century - Mining, Processing, Safety* (Kemerovo, Russia) **92** pp 377-382

[4] Gritskevich T I, Kazakov E F, Konovalov A B 2017 The value of reforming social and economic models of interaction in society development *The European Proceedings of Social & Behavioural Sciences* **49** pp 402-428

[5] Zaruba N A, Egorova N, and Lean N 2016 Governance as a condition for the creation of the ntellectual Coal Industry *Proceedings of the 8th Russian-Chinese Symposium on Coal in the 21st Century - Mining, Processing, Safety* (Kemerovo, Russia) **92** pp 39-44

[6] Zolotukhin V M, Gogolin V A, Yazevich M Yu, Baumgarten M I and Dyagileva A V 2017 Environmental management: the ideology of natural resource rational use. Ecology and safety in the technosphere: current problems and solutions *IOP Conference Series: Earth and Environmental Science* **50** 012046

[7] Yablokov A V, Levchenko V F, Kerzhentsev A S 2015 Ocherki biosferologii *Philosophy & Cosmology* **14**

[8] Tyulenev M A, Lesin Yu V, and Vik S V 2016 Methodological bases of advanced geo-ecological problems resolving in neo-industrial clusters *Proceedings of the 8th Russian-Chinese Symposium on Coal in the 21st Century – Mining, Processing, Safety* (Kemerovo, Russia) **92** pp 333-336

[9] Kovalev V A, Khoreshok A A, and Litvin O I 2016 The role of innovative technologies in solving the Kuzbass coal industry's geo-ecological problems *Proceedings of the 8th Russian-Chinese Symposium on Coal in the 21st Century – Mining, Processing, Safety* (Kemerovo, Russia) **92**, pp 287-290

[10] Trifonov V, Loyko O, Nesteruk D, Zhironkin S, Strekovtsova E 2017 Managing a monotown as a priority social and economic development area *AIP Conference Proceedings* **1800** 050009 DOI: 10.1063/1.4973069

[11] Galanina T V, Baumgarten M I, Mikhailov V G, Koroleva T G, Mikhailov G S 2017 Environmental-socio-economic monitoring as a tool of region's environmental-economic system management *IOP Conference Series: Earth and Environmental Science* **50** 012030

[12] Kiseleva T V, Mikhailov V. G, Karasev 2016 Management of local economic and ecological system of coal processing company *IOP Conference Series: Earth and Environmental Science* **45** 012013

[13] Efremenkov A B, Khoreshok A A, Zhironkin S A, and Myaskov A V 2017 Coal mining machinery development as an ecological factor of progressive technologies implementation *IOP Conference Series: Earth and Environmental Science* **50**(1) 012009 DOI: 10.1088/1755-1315/50/1/012009

[14] Zhironkin S A, Kolotov K A, Genin A E, Agafonov F V, and Kovalevsky S A 2017 NBIC-convergence of machinery and basic technologies as the ecological factor of wellbeing *IOP Conference Series: Earth and Environmental Science* **50**(1) 012011 DOI: 10.1088/1755-1315/50/1/012011

[15] Smirnov A 2014 The Kuzbass coal: the flip side of the coin *Ecology and Law* Available at: http://bellona.ru/2014/01/22/kuzbasskij-ugol-oborotnaya-storona-m/ (Accessed 07 09 2017)

[16] Federal Service for Supervision of Consumer Rights Protection and Human Welfare 2017 *The State report "On the state of sanitary and epidemiological well-being of the population in the Russian Federation in 2016"* (Moscow, Russia: Federal Service for Supervision of Consumer Rights Protection and Human Welfare)

[17] Nipa S S 2017 The role of ecologism in the preservation of the natural reserve of noosphere development *Philosophy of Economy* **4**(112) pp 57-63

[18] The official website of the Kemerovo region 2017 *Tuleyev: the coal miners of Kuzbass will more than triple the costs of land reclamation* Available at: http://ecoyear.ru/2017/08/tuleev-ugolshhiki-kuzbassa-bolee-chem-vtroe-uvelichat-zatraty-na-rekultivatsiyu-zemel

[19] Markov Yu G 2011 *Ecology and ethics: landmarks of civilizational change* (Novosibirsk, Russia: SPSTL SB RAS)

IOP Conf. Series: Earth and Environmental Science **395** (2019) 012117 doi:10.1088/1755-1315/395/1/012117

The level and quality of food consumption as a factor in ensuring the social security of the population of the Amur region

N Koryakina[1*], N Stepakova[1] and V Pomozova[2]

[1] Far Eastern Higher All-Arms Command School, 158 Lenin str., Blagoveshchensk 675000 Russia
[2] Kemerovo State University, 6 Krasnaya str., Kemerovo 650000 Russia

E-mail: koryakina.nata1405@yandex.ru

Abstract. The article analyzes the level and quality of food consumption, identifies the main trends in the formation of the diet of the population of the Amur region. The factors affecting the consumption of basic foods are analyzed.

1. Introduction

In the economically developed countries of the world, one of the most important issues of state policy is the health of the population, which largely depends on a healthy diet. Food consumed by a person, its quantity, as well as its properties determine the physical development and growth of a person, starting from early childhood, creates a tendency to diseases, to reduce working ability, lays a lifetime resource and neuropsychic state [5, 6].

The Order of the Government of the Russian Federation approved the Basics of the state policy in the field of healthy nutrition of the population for the period up to 2020. On its basis, the work of the Federal Service for Supervision of Consumer Rights Protection and Social Security of Society is built. The problem of food security should be addressed not only at the national, but also at the regional level. Since there are significant differences between food production and the need for it in areas with different climatic and environmental conditions [1].

The purpose of the work is to study nutrition as a risk factor for the health of the population of the Amur Region. To study this problem, the following questions were analyzed: analysis of the balance of nutrition in terms of the consumption of basic products and the assessment of the economic affordability of the minimum food mix.

2. Materials and Methods

The study used data for the Amur region for the period 2011-2017. Were analyzed indicators of the volume of production and consumption of basic foodstuffs in the Amur region [3].

The study used methods: statistical, comparative, ranking.

Content from this work may be used under the terms of the Creative Commons Attribution 3.0 licence. Any further distribution of this work must maintain attribution to the author(s) and the title of the work, journal citation and DOI.
Published under licence by IOP Publishing Ltd

3. Results

To determine the level of food security of an individual in the world practice, standards of normal nutrition are used. Recently, this figure is equal to 2700 kcal per day. The categories of hungry people include the population consuming 1520 kcal, on the verge of starvation and malnutrition - 2150 kcal.

To determine the norms of proper nutrition, monitoring of the food security of an individual is carried out across countries and regions, taking into account the natural and social living conditions of the population.

When monitoring criteria are used:

- *Energy:* a daily caloric intake of a person. The critical limit is 0.5 of the physiological norm of the average person. There are three levels of vital activity associated with daily energy consumption: optimal 2500-3500 kcal, insufficient 1500-2500 kcal, critical <1500 kcal;
- *Component*, an amount of proteins, fats, carbohydrates, vitamins consumed by man per day;
- *Food safety*, a proportion of products produced with a content of contaminants is greater than the maximum allowable amount, which reduces the amount suitable for the consumption of food.

Standard nutrition standards existing in the Russian Federation are associated not only with the change in energy consumption per 1 kg of human body weight, but also with the standard of living of the population, with the peculiarities of climate, culture and traditions. Officially, this is spelled out in the Decree of the Government of the Russian Federation of January 28, 2013 No. 54 "on the approval of guidelines for determining the consumer basket for the main socio-demographic groups in the constituent entities of the Russian Federation." 10 natural-climatic zones have been established, each of which is distinguished by the specifics of the set of products, and hence the norms for ensuring the energy value of the food ration.

A set of food products is created that are necessary to preserve human health and ensure its vital activity, taking into account the possibility of satisfying the needs of the main socio-demographic groups of the population, in nutrients based on the chemical composition and energy value of food. But the minimum set of food products cannot be regarded as the recommended diet, because it is designed only to meet the needs of the subsistence minimum.

In the Russian Federation, to assess food security, the ratio of the actual consumption of certain types of food to scientifically based nutritional standards (energy, protein, fat, carbohydrates, vitamins) is considered [2]. Such an assessment shows a clear picture of the general malnutrition of the country's population, reveals a deficit of individual nutrients, which allows for the proper construction of the state agri-food policy and the food supply system.

For the Amur region in 2017 compared with 2016 there was a decrease in the consumption of a number of staple foods: vegetables and melons - by 1.7%; fish and fish products - by 2.2%, fruits and berries - by 3.4%, sugar and confectionery - by 5.9%. At the same time, consumption increased: bread and bread products - 1.5%, potatoes - by 3.1%, milk and dairy products - by 4.0%, vegetable oils and fats - by 4.5%, meat and meat products - by 5.8%; eggs - by 6.3%.

At the same time, the expenditures of the population on food in 2017 compared to 2016 decreased by 0.7% and amounted to an average of 5,531 rubles per month per person.

Despite a slight increase in the consumption of basic foodstuffs (meat and meat products, milk and dairy products), in comparison with physiological norms, the population of the Amur Region experiences a significant deficit in the consumption of meat, milk, fruits and vegetables. And the observed increase is not significant and is due to finished products imported from neighboring regions. The existing deficiency is covered by the excessive use of potatoes, bread and bread products, which indicates a poor quality of nutrition.

Table 1. Consumption of basic foodstuffs in the Amur region per capita per year, kg.

Product name	Physiological norms, kg	2011	2012	2013	2014	2015	2016	2017
Meat and meat products	74	61	63	66	65	64	63	67
Vegetable oil	12,8	15	17	14.1	14.2	14.1	14.8	15,5
Potatoes	100	144	144	144	146	147	144	148
Fruits and berries	80	56	59	59	60	60	61	62
Vegetables and gourds	139	127	130	117	127	132	131	129
Sugar	38	33	36	36	37	35	36	34
Bread and bakery products	110	122	127	133	135	135	135	137
Milk and dairy products	389	166	167	178	174	179	179	186
Fish and fish products	22	23	23	24	26	27	27	26
Eggs, pieces	260	309	310	310	316	276	271	287

The analysis showed that consumption of basic foodstuffs in the Amur Region is lower than the average in Russia. The Amur region ranks 48th in the rating list of regions of the Russian Federation, meat ranks 48th in milk, 68th in milk consumption, and 14th in potato consumption and 8th in bread and bread products (ranking in descending order).

The level of food consumption is closely related to the volume of production [3].

Table 2. The production of basic foodstuffs in the Amur region, in thousands of tons.

Product name	2011	2012	2013	2014	2015	2016	2017
Meat and meat products	14,2	14,9	14,7	14,9	13,8	18,5	18,8
Vegetable oil	4,7	6,5	4,3	2,8	1,7	0,8	1,2
Corn	20,1	19,1	9,0	17,9	22,7	26,6	26,7
Potatoes	325,3	352,3	158,2	324,3	367,4	270,4	336,5
Fruits and berries	3,6	3,2	2,6	3,1	3,1	2,7	3,1
Vegetables and gourds	135	165	102	166	176	158	188
Sugar	0	0	0	0	0	0	0
Milk and dairy products	154,8	148,1	147,6	126,0	128,9	128,1	124,0
Fish and fish products	57,3	57,5	57,6	57,7	57,8	57,9	57,2
Eggs, mln.	246,2	231,0	239,4	240,1	203,1	199,6	206,2

The absence of an increase in volumes, and often a reduction in agricultural production, requires additional measures in order to avoid a food threat to the region. Such low volumes of livestock production and crop harvesting have a negative impact on local food industry enterprises. Production capacity of food enterprises are used by less than 50%. For example, in 2017, the utilization rate of the average annual capacity of enterprises for the production of sausage products was 53%, and whole-milk products - 75.2%. Replenishment of the missing raw materials is carried out by importing livestock products from other regions of the country or imports. Dairy products are made from recovered a raw material, which negatively affects the taste and nutritional qualities of the final product.

The impossibility of self-sufficiency of food enterprises in the region with basic foodstuffs is explained by the harsh natural and climatic conditions of the region. The Amur region does not have the climatic conditions that are necessary for many crops, the growing season does not allow growing crops, and natural disasters (floods) dramatically reduce crop yields. Therefore, regional enterprises are largely dependent on the supply of imported food. When analyzing the supply of imported food can be divided into three main groups of food: fruits, vegetables and meat products. Thus, the lack of vegetables and fruits in the Amur Region is being replenished by importing products from the PRC. The development of cross-border trade allows us to provide the region's food market with fresh vegetables, fruits and citrus fruits. According to statistics, the import of this product is 40% [4]. In the future, if the volume of food production continues to decline, the region will fall into economic dependence on imports. The consumption of imported potatoes, vegetable oil, dairy products will exceed a critical figure. It should be noted that the quality of imported food does not always meet the hygienic standards for sanitary-chemical or microbiological indicators. On this basis, in order to ensure the food security of the region, it is necessary to conduct a thorough inspection of imported food, especially imported from China. An increase in the consumption of fruits and vegetables imported into the Amur region from the Republic of China may lead to an increase in the level of health risks due to their low quality and nitrate content.

To assess the food security of an individual in the regions of the Russian Federation, it is necessary to take into account the ratio of the actual consumption of certain types of food to scientifically based nutritional standards [4]. To review the quality and usefulness of nutrition of the population, the analysis of the average caloric level of the daily diet in the Amur region with a critical caloric value of the daily diet was conducted.

Table 3. The change in the ratio of the average daily calorie intake in the Amur region to the critical value of the daily intake of calorie.

The critical value of caloric intake, kcal		**2011**	**2012**	**2013**	**2014**	**2015**	**2016**	**2017**
The actual caloric content of the daily diet, kcal	2267	2474	2428	2481	2569	2512	2374	2610
Ratio		1,091	1,071	1,094	1,133	1,108	1,047	1,151

The table shows the annual increase in the ratio of the average daily caloric intake in the Amur Region to the critical value of the daily intake of calories, which is associated with an increase in the economic availability of food to the population of the region.

The cost of the food minimum included in the consumer basket in the period of 2017-2018 in the Amur Region is set at 4,515.2 - 4,551.42 rubles. The minimum wage in the region in 2017 was 7,800 rubles, while the average wage was declared in the amount of 3,5100 rubles. Such a significant difference in monthly income forces the residents of the region to form their own food basket from affordable foodstuffs. The population of the region with the average per capita cash income of up to 10,000 rubles is about 15%. Therefore, solving the problem of ensuring healthy and wholesome nutrition for the entire region's population requires a programmatic approach and is a regional problem.

4. Discussion

Thus, an analysis of the physical availability of food and nutritional adequacy in the Amur Region reveals significant deviations from the threshold values of the food safety indicators specified in the RF food security doctrine.

The tendency to an increase in the economic dependence of the Amur Region on food imports has been revealed.

In the diet of the population of the Amur region there is a shortage of milk and dairy products, fruits, vegetables, meat and meat products. In abundance of potatoes, bread and bakery products, eggs. Such features in the diet of the inhabitants of the region is the cause of a deficiency of vitamins, macro and microelements and an excess of fast carbohydrates. The current nutritional imbalance can be explained by the poverty of a significant part of the population. Solving this problem requires an integrated approach on which the country's food and social security depends.

5. Conclusion
Ensuring social security is one of the most important problems of our time. Social security is associated with the protection of the interests of the state and society, regions, social groups and the individual, with the system of life support and socialization of people. The problem of food security in the Russian Federation should be addressed not only at the national level, but also at the regional and municipal levels. Since there are differences between the need of the population for food and the possibility of food production. The ability of the regions to produce basic food products largely depends on the climatic conditions and, consequently, on the availability of agricultural raw materials and animal products.

References
[1] On the sanitary-epidemiological welfare of the population. The federal law. - put into effect with changes 2018-03-08 (Moscow, Russia): 2018.
[2] Federal State Statistics Service Available at: http://www.gks.ru
[3] Federal State Statistics Service (n.d.) *Consumption of basic food products in the Russian Federation* Available at: http://www.gks.ru/wps/wcm/ (Accessed 03 30 2019)
[4] *Recommendations on rational food consumption standards that meet modern requirements for healthy nutrition.* From 08/19/2016, number 614
[5] Lobova S V, Alekseev, A N, Bogoviz A V, and Ragulina J V 2019 Wireless future of the agrarian market as a basis of food security provision *Studies in Computational Intelligence* **826** pp 975-981
[6] Lobova S V, and Bogoviz A V 2018 The subject-object identification of personnel security threats *Espacios* **39**(24)

IOP Conf. Series: Earth and Environmental Science **395** (2019) 012118 doi:10.1088/1755-1315/395/1/012118

Introducing innovative farming methods in the adjacent territories of Russia and Kazakhstan

T Artamonova[1], S Bondarenko[2*] and K Kaliyeva[3]

[1] Altai State Agricultural University, 98 Krasnoarmeyskiy prosp., Barnaul 656049 Russia
[2] Altai State Agricultural University, 98 Krasnoarmeyskiy prosp., Barnaul 656049 Russia
[3] D. Serikbayev East Kazakhstan State Technical University, 19 Serikbayev str., Ust-Kamenogorsk 070003 Kazakhstan

E-mail: bonsvet@bk.ru

Abstract. Attitude to nature is the basic value of the peoples of Eurasia. This is especially clearly manifested in the example of land use. The development of virgin lands in the adjacent territories of Russia and Kazakhstan led to the solution of the food problem, but with social and environmental consequences. Violation of the traditional integrity of human and natural being lies at their core. The authors argue that it is possible to restore the proportionality of their coexistence, including through the introduction of innovative farming methods as the basis for sustainable development of agricultural territories.

1. Introduction

The unity of peoples in Eurasia rests on the commonality of their socio-cultural constants. Contemporary theorists of Eurasianism distinguish the attitude to nature as a basic value among other things. "Nature is the true constant of culture, it recognizes the existence of a spiritual principle and the possibility of a direct impact on a person" [1, p. 154]. Such an attitude to the natural beginning was reflected not only in the mythology, rituals and legends of antiquity, but also in the socio-economic structure of various ethnic groups of Eurasia. This was manifested particularly vividly by the use of land resources. It is well known that the sedentary peoples developed a cult of land, but the nomads of Eurasia also had to build harmonious interdependence in the "man-pasture" relationship in order to survive the harsh conditions of the steppe. For centuries, the nomadic economy has developed environmental standards of existence [2]. Respect for the land has always been determined not only by the need for physical, but also spiritual and moral survival of human. And today, the state of the land is an indicator of balanced environmental management and the basis of the socio-economic development of territories.

The balance of natural and human beginnings was disturbed with the beginning of the industrial development. The peak of this activity in the adjacent territories of Russia and Kazakhstan fell on the period of developing virgin lands. As known, the USSR government adopted a resolution in 1954 on the development of virgin and fallow lands in various regions of the country, in order to solve the acute grain problem in the country. A total of 2.9 million hectares of virgin lands were plowed in Altai Region, 25.5 million hectares in Kazakhstan. In an effort to fulfill their plans, the regional authorities

Content from this work may be used under the terms of the Creative Commons Attribution 3.0 licence. Any further distribution of this work must maintain attribution to the author(s) and the title of the work, journal citation and DOI.
Published under licence by IOP Publishing Ltd

IOP Conf. Series: Earth and Environmental Science **395** (2019) 012118 doi:10.1088/1755-1315/395/1/012118

plowed up the lands that were unsuitable for farming (sand, salt licks, soils with low humus, water protection areas). Having solved the food problem, at that time, the country's leadership could not foresee all possible social and environmental consequences, among which we can single out the destruction of the traditional integrity of the human and natural being. The restoration of proportionality of their coexistence today is possible, including through the introduction of innovative farming methods as the basis for the sustainable development of agricultural territories. The experience of developing virgin and fallow lands clearly showed the need to take into account the environmental factor when making economic decisions, forecasting long-term consequences for nature and society. Ignoring environmental aspects leads to economic and social crises.

2. Materials and Methods

When writing this article, we relied on the following methods: comparative analysis, as well as historical retrospective, descriptive, and analytical methods.

3. Results

Currently, in the agro-industrial complex of Siberia and Northern Kazakhstan, trends in land degradation and disturbances in the socio-ecological balance persist. A big problem for the regions is the increase in unused (waste) lands. In the Russian Federation, about 45 million hectares of arable land are considered abandoned, and it is 14 million hectares in the Republic of Kazakhstan [3, p. 26]. The continuing trend threatens to reduce agricultural production and depopulation of rural areas and consequently leads to problems with ensuring food security. Searching for new strategies for the restoration of land resources remains an urgent problem for the transboundary territories under consideration, which is due to the similarity of climatic conditions, historical genesis of these problems, and the experience of jointly solving them.

Already in the late 1950s, negative environmental consequences caused by the development of virgin land resources fully manifested themselves. For example, in the Altai Territory, only 1.2-1.4% of the 2.5-3.5% humus remained in the chestnut soils of Kulunda for about fifty years [4]. When plowing the land, the specificity of dry-steppe farming was completely ignored, the methods of land cultivation were mechanically transferred from the country's central regions. Terrible soil erosion in the 60s and dust storms that caused enormous environmental damage to nature were the result of this. In particular, as a result of wind erosion in the areas of the Kulunda steppe in 1963-1965, about 500 thousand hectares of arable land were taken out of processing, amounting to 658 thousand hectares of arable land in total [4]. In Northern Kazakhstan, 9 million hectares were taken out of economic circulation due to irrational land development by 1960 [5].

In the late 1950s, the regions began developing a number of protective systems. One of the first measures to protect the soil was the planting of forest shelter belts. As practice has shown, these measures have been effective only in conjunction with other protective measures. In 1960-70, the development of water and wind erosion was able to stop together with Kazakhstani scientists. Unfortunately, the problem could not be completely solved. As already indicated, at present, large tracts of land remain abandoned in both Russia and Kazakhstan. The effectiveness of their involvement in agricultural turnover remains a debatable moment.

Modern methods of scientific research, including the use of satellite imagery, provide more accurate scientific data. Thus, a group of scientists from Germany, Denmark, USA, Kazakhstan conducted a comprehensive assessment of changes in agricultural land use in the northern part of the Republic of Kazakhstan. As a result of large-scale research, scientists have concluded that only one-third of the 14 million hectares of abandoned land makes sense to engage in grain production. First of all, this is due to the low agro-climatic characteristics of such lands and the need to leave the rest of the land for cattle grazing, which is traditional for the region. Similar results were obtained by scientists on Russian abandoned lands. The main conclusion of the scientists was as follows: "It is necessary to concentrate on increasing the yield on the already cultivated lands and leave unproductive alone" [3].

IOP Conf. Series: Earth and Environmental Science **395** (2019) 012118 doi:10.1088/1755-1315/395/1/012118

Thus, an increase in soil fertility is a common problem in the contiguous territories of Russia and Kazakhstan under consideration. Currently, one of the most promising, but debatable methods of agrotechnology is organic farming. As practice shows, soil fertility increases with the introduction of organic farming methods.

Today, a group of scientists from Kazakhstan are actively studying organic farming methods, since Kazakhstan has a great potential for this. Since the area of arable land, on which mineral fertilizers are applied, is only 14%, and the area of arable land treated with herbicides is 31%. It is equally important that Kazakhstan imposed an official ban on the cultivation of genetically modified crops. The main coordinator of the Center for Traditional Knowledge Dissemination Talgat Absattar believes that there are several stages in the development of agriculture. In the 21st century, the model of organic farming replaces intensive agriculture. Organic farming is better, safer, and more sustainable [6]. Also, a return to organic farming allows one to restore the traditional attitude of the Eurasian people to the land as a basic value.

Responsibility to the land determines the specifics of agriculture; therefore, precision agriculture comes to replace the classical extensive agriculture. It includes a system of auto-driving machine and tractor units, differentiation of seeding rates and fertilizer doses, using plant protection products, yield mapping [7]. This direction has been successfully implemented in a number of farms in the Kuryinsky, Kosikhinsky, Kytmanovsky, Mamontovsky, and Rebrikhinsky districts of the Altai Region, as well as in the steppe regions of Kazakhstan, with the support of the Scientific Production Center named after A. I. Barayev [8]. The training class of precision agriculture was opened in the Altai State Agricultural University. It is equipped with modern equipment, which allows future specialists to master the subtleties of technology.

Consequences of the irrational use of land resources are also among the reasons for the outflow of rural population, which may threaten the depopulation of rural areas, leading to a decrease in the level of development of agricultural resources. This is a common problem of the cross-border territory of Russia. In 2001-2013, the entire Kazakhstan frontier (Omsk, Novosibirsk, and Altai Regions) lost population due to rural migration. Rural population decreased there by 14.5%, while the urban population decreased by 0.3%. [9, p. 22]. The Altai Region lost the most of its rural population. So, the share of the rural population of the Altai Territory decreased by 20% over the last 26 years. In numerical terms, it is from 2130 thousand people in 1989 to 995 thousand people in 2015 [10, p. 77]. Despite some different trends, in general, there was an increase in the rural population in Kazakhstan. But still, in some areas of North Kazakhstan, West Kazakhstan, Pavlodar and Akmola regions, the number of villagers decreased from 11 to 20 percent.

Also, changes in the social sphere are reflected in the value-ideological attitudes of the population of the adjacent territories of Russia and Kazakhstan. The transformation of sociocultural constants is reflected in the results of a sociological study of the life values of the Kazakh and Russian population of the East Kazakhstan region, The study was conducted by the employees of the Department of Ethnosocial Studies of the Institute of Philosophy and Law of the SB RAS (Yu. V. Popkov was the Head of Complex Research). So, in order of importance, such an indicator as "the value of living in rural areas" occupies one of the last places (22% in Kazakhstan, 17.1% in Russia). Thus, traditionally agrarian territories are no longer attractive for the majority of the population of the region. At the same time, "preservation of land, nature, resources" is among the top ten priorities and ranks 4th in importance both among the Kazakhs and Russians (Kazakhs – 74.5%, Russians – 77.8%) [11, p. 237-244]. This indicates that the value of nature remains an important dominant worldview, but it does not find its realization in the field of agriculture.

4. Discussion

Scientists at the Eurasian National University named after L. N. Gumilyov (Astana) and the Kazakh National Agrarian University (Almaty) of the Republic of Kazakhstan, when developing methods for integrated environmental assessment of landscape productivity, argue that it should be based on the fundamental laws of wildlife, including private assessments of soil and vegetation productivity. Based

IOP Conf. Series: Earth and Environmental Science **395** (2019) 012118 doi:10.1088/1755-1315/395/1/012118

on these principled positions, they determined the biological productivity of the natural system of Northern Kazakhstan [12, 13]. The transition to more economical resource-saving systems for autumn processing is a logical stage in the overall development of the technological process, which has become possible due to the emergence and development of new concepts in the field of tillage. Altai scientists came to the conclusion that with extensive agriculture, productivity depends on soil fertility and weather conditions by 50-60%, and it is with intensive agriculture by 30-35% [14].

The question of international cooperation in solving this problem was discussed at the International Congress of the Eurasian Federation of Soil Science Society "The Soil Resources and Environment Conservation", held in October 2018 in Almaty. The congress was organized by the Eurasian Federation of Soil Scientists Societies, the Kazakhstan Society of Soil Scientists and the Kazakh Research Institute of Soil Science and Agrochemistry named after U. U. Uspanov. Scientists from 27 countries from leading universities, research institutes, and the Eurasian Soil Partnership Assembly (ESPA) urged governments to give priority to solving the problems of soil degradation and desertification, to expand cooperation between the International Society of Soil Science and the Global Soil Partnership in the field of rational use of land and water resources. Scientists noted that at present, the scientific search for innovative solutions to fundamental, applied, methodological directions in soil science and environmental protection are priorities throughout the world [15]. Despite the controversial positions, it is obvious to scientists in Russia and Kazakhstan that contemporary farming systems must become highly adaptable, meet environmental requirements, creating conditions for the development of balanced land use and increasing soil fertility.

5. Conclusion

In conclusion, we can conclude that the development and introduction of innovative farming methods is important not only for solving agro-ecological and food problems, but also for preserving traditional values. For centuries, people living in the adjacent territories of Russia and Kazakhstan have developed traditions of commensurate coexistence. Today, within the framework of sustainable development, the most important task is to return to these principles on the basis of modern science-based technologies that contribute to an increase in yields on already cultivated lands.

The crisis of modern civilization has shown that only those peoples who preserve the vital forces of nature and their spiritual unity with it, will restore and improve traditions of sustainable management, have a chance to survive and develop organically in the future.

References

[1] Ivanov A V, Popkov Yu V, Tyugashev E A, and Shishin M Yu 2007 *Eurasianism: key ideas, values, political priorities* (Barnaul, Russia: Publishing House of the Altai State Agricultural University)

[2] Saparov A S, and Kozybayeva F E n.d. *The first ideas of Kazakh public figures, enlighteners, researchers, scholars and writers of the late 19th and early 20th centuries about the soils of the Kazakh steppes* Available at: http://soil.kz/kratkaya-informatsiya-o-dostizheniyah-ord/ (Accessed 08 04 2019)

[3] Kramer R, Prischepov A, and Mueller D *et al* 2015 Analyzing the long-term dynamics of land use change and the potential for expanding arable land due to abandoned land in the region of development of virgin lands of Kazakhstan *Earth From Space: The Most Effective Solutions* **5** pp 20-28

[4] Bondarenko S I 2014 *Historical aspects of the development of virgin and fallow lands: significance, experience, prospects (to the 60th anniversary of the virgin epic)* In a collection of articles IX International Scientific and Practical Conference *"Agrarian Science to Agriculture"* (vol 1) (pp 18-22) (Barnaul, Russia: Altai State Agricultural University)

[5] Alpyspaeva G A, and Sayakhimova Sh N 2017 Ecological problems during the development of virgin and fallow lands in Kazakhstan *Sociosphere* **17** pp 97-102

[6] AgroInfo 2016 *Principles of organic farming, or how to become an organic farmer?* Available at: http://agroinfo.kz/principy-organicheskogo-zemledeliya-ili-kak-stat-organicheskim-fermerom/ (Accessed 08 04 2019)

[7] Belyaev V I 2018 Contemporary technology and information technology in agriculture of the Altai Region *Vestnik of the Altai State Agricultural University* **8**(166) pp 158-162

[8] *Website* of *the Research and Production Center of grain farming named after A. I. Baraev* Available at: http://2.baraev.kz/ (Accessed 08 04 2019)
[9] Soboleva S V, Grigoriev Yu A, Smirnova N E, and Chudaeva O V 2014 Features in the formation of the population of the border territories of Siberia *All-Russian Economic Journal ECO* **11** pp 22-26
[10] Borisenko M A 2017 Demographic changes in rural areas of the Altai Region in the post-Soviet period *Vestnik of the Altai State Agricultural University* **8**(154) pp 75-79
[11] Popkov Yu V Ed 2010 *The Eurasian world: values, constants, self-organization* (Novosibirsk, Russia: Nonparel)
[12] Kundius V A 2017 *Economic preferences in the strategic development of the agro-industrial complex of the Altai Region* In *Agrarian Science to Agriculture* (vol 1) (pp 115-119) (Barnaul, Russia: Altai State Agricultural University)
[13] Adilbektegi G A, Mustafayev Zh S, and Kozykeeva A T 2018 *Environmental assessment of the productivity of the natural system of Northern Kazakhstan* In *Agrarian Science to Agriculture* (vol 2) (pp 9-11) (Barnaul, Russia: Altai State Agricultural University)
[14] Antonova O I 2018 *Problems of intensification of resource saving technologies* In *Agrarian Science to Agriculture* (vol 1) (pp 12-14) (Barnaul, Russia: Altai State Agricultural University)
[15] *Kazakh Research Institute of Soil Science and Agrochemistry named after U.U. Uspanov LLP International Congress* n.d. Available at: http://soil.kz/blog/2018/10/25/mezhdunarodnyiy-kongress/ (Accessed 08 04 2019)

The role of peasant farms in the crop industry of the Altai region

E V Rygalov[1*], A V Krotov[1] and A G Zinovev[1]

[1] Altai State University, 61 Lenina prosp., Barnaul 656049 Russia

E-mail: rugalov@mail.ru

Abstract. The article discusses the contribution of peasant (farmer) farms to the crop industry of the Altai Region. The paper estimates a share of the production of grain crops and sunflower, as well as analyzes the differentiation of these indicators within the region and the zonal change of the main indicators of farm activity.

1. Introduction

The peasant (farmer) farms (PFs) are a fairly common form of agricultural organizations in the Altai Region. PFs received its development in the early 90s of the twentieth century. Then, as part of the Land Reform, the land began to move from a purely state to non-state forms of ownership; and individuals, groups, and legal entities became land owners massively; the formation of new types of farms occurred on the basis of collective and state farms. One of these types of farms were PFs.

The legal foundations were enshrined in the RSFSR Law of November 22, 1990 No. 348-1 "On Peasant (Farm) Farming" [1], which is no longer valid. Now their activities are regulated by the 2003 Federal Law with amendments and additions. According to this document, a PF is "an association of citizens related by kinship and (or) property, having common property and jointly carrying out production and other economic activities (production, processing, storage, transportation and sale of agricultural products), based on their personal involvement" [2]. The advantage of this type of agricultural producers over larger forms of management is associated with preferential taxation, state support in the form of subsidies and grants.

According to the results of the All-Russian Agricultural Census, more than 2,000 PFs were registered in the region, and during the intercensal period (from 2006 to 2016), their number decreased by more than 2 times. At the same time, a total area of agricultural land in them increased 1.5 times. Thus, over 10 years, an average area of agricultural land of one PF increased in more than 3 times and amounted to 956 hectares (in agricultural organizations, this figure is 4,178 hectares) [3].

A PF is a rather effective and mobile form of agricultural production in terms of responding to changes in weather and market conditions [4], [5]. In our region, PFs make a significant contribution to the production of agricultural products, especially in the crop industry, since the main part of farms specializes in this sector.

2. Materials and Methods

The main materials of the study were the data of the 2016 All-Russian Agricultural Census (Census), as well as official data of the Office of the Federal Statistics Service for the Altai Region and the Republic of Altai. Mathematical and statistical, comparative geographical and analytical methods were used in

Content from this work may be used under the terms of the Creative Commons Attribution 3.0 licence. Any further distribution of this work must maintain attribution to the author(s) and the title of the work, journal citation and DOI.
Published under licence by IOP Publishing Ltd

IOP Conf. Series: Earth and Environmental Science **395** (2019) 012119 doi:10.1088/1755-1315/395/1/012119

the work. They allowed to get an idea of the development trends of PFs in the region, assess their real contribution to the production structure of the main types of agricultural products, reveal the internal differentiation of the Altai Region districts.

Analyzing the contribution of PFs to the general structure of agricultural land, the structure of arable land in the region, as well as to the production of main crop production (production of grain, fodder crops, and the main industrial crop, sunflower) was carried out. In addition, based on the average values for the period from 2008 to 2018, the share of PFs of the total production of grain crops, as well as sunflower, was calculated. Evaluation and analysis of the geographical differentiation of the main indicators of the natural-economic zones of the Altai Region was also conducted.

3. Results

Most K (F) X have a pronounced crop specialization (about 85% of farms are engaged in the cultivation of grain and leguminous, fodder, industrial crops, some farms are focused on the cultivation of vegetables and berries, gardening). Key indicators characterizing the contribution of K (F) X to the crop industry of the region are reflected in Table 1 (the distribution was made for the following main categories of crops: grains, as well as sunflower, as the main technical crop, this type of farms provides more than a third of total harvest). In the production of crop products in general (this also includes vegetable, fruit and berry crops), the share of K (F) X is about 25%.

Table 1. PFs in the structure of farms of all categories.

Indicators	The share of PFs,%
Area of agricultural land	32.8
Arable land	36.2
Grain production (gross yield)	38.1
Production of feed crops (gross yield)	28.3
Sunflower production (gross yield)	35.7

Intraregional differentiation in the production of main types of crops has some specifics. In the production of grain crops, a number of regions can be distinguished in which the share of K (F) X in the production of grain crops is more than 50% of the share of all types of farms: Klyuchevsky (77.6%), Uglovsky (72.6%), Volchihinsky (69.8%), Krutikhinsky (64.6%), Troitsky (63%), Kosikhinsky (61.7%), Eltsovsky (59.7%), Aleyskiy (58.7%), Krasnogorsky (57, 2%), Kulundinsky (53.8%), Ust-Pristansky (53.7%), Rebrikhinsky districts (51.2%). In some areas, the share of PFs the production of grain is insignificant (here the main grain producers are agricultural organizations), less than 10%: Zmeinogorsky (8.4%), Biysky (6.8%), Zalesovsky (6.4%), Zonal (1.7%), Soloneshensky (0.7%), Togulsky (0.2%). This distribution has individual characteristics and does not reveal the general relationship.

The total values of the PFs fraction in natural-economic zones reflect some features of the zonal distribution (Fig. 1). It can be noted that the West Kulundinskaya and East Kulundinskaya zones make the greatest contribution to the production of cereals, which is explained by more favorable climatic conditions for the cultivation of durum wheat. The Priobskaya zone is also highlighted by the indicators under consideration, this territory has a rather favorable combination of natural and climatic conditions for growing crops. In addition, the production here is a major consumer of products – the Barnaul urban agglomeration [5].

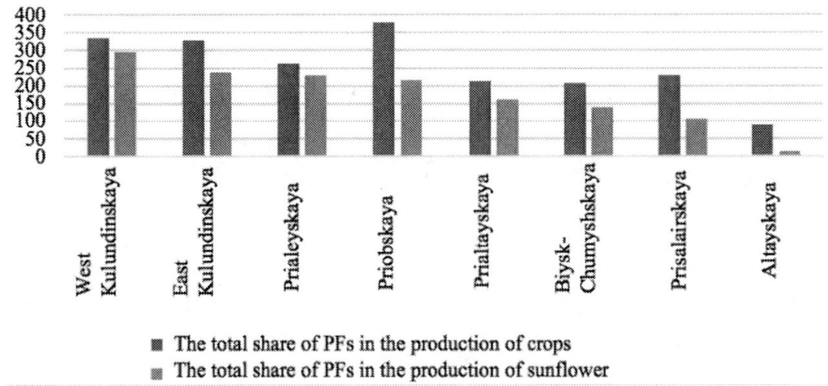

Figure 1. The contribution of PFs to the production of grain crops and sunflower in the natural-economic zones of the Altai region [6].

In the production of sunflower, the leaders are the PFs of the following districts: Klyuchevsky (76.5%), Uglovsky (75.6%), Ust-Pristansky (72.1%), Tyumen (63.9%), Shipunovsky (59.2 %), Rodinsky (56.0%), Bystroistoksky (53.4%), Aleysky (52.3%), Kosikhinsky (51.3%).

A zonal distribution reflects a direct relationship when moving from west to east (Fig. 1), this is due to favorable climatic conditions in the western part of the region. Sunflower requires a warm dry climate for normal growth; as for soil conditions, this crop has a high ability to adapt to the soil, and a rather late harvest period allows the crop to ripen until the first frost. Among the regions of the Siberian Federal District, the Altai region has the most favorable conditions for growing this crop, about 90% of sunflower crops are concentrated in the region [7]. In addition, sunflower is a highly demanded and highly profitable crop on the market, so even with low yields, farms can make a profit.

The change in such indicators characterizing the PFs of the region, as their number and size (an average area of the farm) also obeys a zonal dependence (when moving from west to east of the region) (Fig. 2). The rather high closeness of communication (0.9) of such indicators as the size of the farm and the gross yield of sunflower and grain indicates a fairly high efficiency in the use of sown areas.

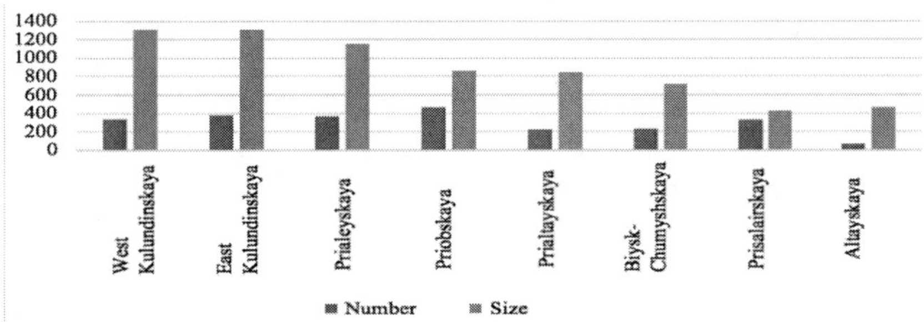

Figure 2. Changes in the number and average area of PFs in the natural-economic areas of the Altai region (compiled from [6]).

4. Discussion

The identified zonal change of the considered indicators may be associated with several aspects. On the one hand, the climatic conditions of the western part of the region favor the development of such sub-sectors of plant growing as the cultivation of grain crops and sunflower. On the other hand, the climatic factor has the following downside: the western regions of the region belong to the zone of risky farming due to the significant influence of climatic conditions on the yield of agricultural crops [8]. In combination with the difficult economic situation, many previously large farms cease to exist, numerous farms come to their land (either on loan or through the allocation of land owned by the farm). When moving to the east, natural-economic zones show a greater balance of economic specialization, which consists in the following: a wider range of crop production, an increase in the share of livestock (based on more succulent feed), as well as the emergence of consumer-oriented sub-sectors (production of garden crops, etc.). A wide range of sub-sectors and a combination of several industries require a more comprehensive approach, the presence of highly qualified specialists, and this is a higher level of management and larger forms of agricultural organizations (LLC, CJSC, OJSC).

5. Conclusion

The analysis performed allows us to draw some conclusions characterizing the contribution of PFs to the development of crop production in the Altai region. PFs make a significant contribution to the production of the main crop products of the Altai region; in some areas of the region, this type of farms makes a significant contribution to the gross yield of main agricultural crops (up to 75%). This type of economy is a fairly effective form in the conditions of contemporary market conditions. They are a more flexible form of management and easily adapt to the needs of the market and under the influence of climatic conditions. During periods of extreme climatic events (droughts), they show maximum work efficiency (have higher yields than other types of farms). In addition, the high closeness of relations between the performance indicators of farms (the average area of the farm and the gross yield of grain and sunflower) show a rather high efficiency of PFs.

Because of their organizational characteristics and personal responsibility and interest, PFs sometimes have a higher motivation for high productivity than larger forms of farms. Moreover, the majority of PFs have a fairly narrow specialization and the survival rate of the economy depends on the work efficiency.

Quite a clear zonal dependence (on the natural-economic zones) was found in the production of the main types of crops (grain, sunflower), as well as the number and size of farms, which is explained by certain peculiarities of the climatic nature.

References

[1] The Supreme Council 1990 *Law of the RSFSR "On Peasant (Farmer) Farming"* (November 22, 1990 No. 348-1) Available at: http://www.consultant.ru/document/cons_doc_LAW_11050/ (Accessed 27 06 2018)

[2] Russian Federation 2003 *Federal Law "On Peasant (Farmer) Farming"* (June 11, 2003 No. 74-FZ) Available at: http://www.consultant.ru/document/cons_doc_LAW_42662/ (Accessed 27 06 2018)

[3] Office of the Federal State Statistics Service for the Altai Region and the Altai Republic *The results of the 2016 All-Russian Agricultural Census by municipal areas and urban districts of the Altai Region* Available at: http://akstat.gks.ru/wps/wcm/connect/rosstat_ts/akstat/ru/census_and_researching/census/national_census_2016/score_2016/ (Accessed 02 04 2019)

[4] Kundius V A, and Streltsov A E 2012 Development of peasant (farmer) farms: current state, level and trends *Ideas and Ideals* **1**(11), pp 67-72

[5] Rygalov E V, and Rygalova N V 2018 Peasant (farmer) farms as the most effective form of agriculture in the Altai Region In *Current Trends in Spatial Development and Priorities of Social Geography* (vol 1) (pp. 389-395) (Barnaul, Russia: Publishing House of Altai State University)

[6] *Official website of the Office of the Federal State Statistics Service for the Altai Region and the Altai Republic* 2019 Available at: http://www.gks.ru/dbscripts/munst/munst01/DBInet.cgi#1 (Accessed 05 04 2019)

[7] Puzikov A N, and Suvorova Yu N 2017 New sunflower varieties in Western Siberia *Vestnik of Omsk State Agrarian University* **2**(26) pp 27-34

[8] Kharlamova N F, and Silantieva M M 2011 The dependence of the yield of grain crops in the Kulunda districts on climatic factors *Izvestiya of Altai State University* **3-2** pp 88-94

AUTHOR INDEX

Institute of Physics
Dirac House, Temple Back
Bristol BS1 6BE UK

ISSN: 1755-1307
ISBN 978-1-7138-0262-4

CURRAN ASSOCIATES INC.
proceedings.com